DIGITAL CONTROL SYSTEMS

HRW
Series in
Electrical and
Computer Engineering

M. E. Van Valkenburg, Series Editor

Benjamin C. Kuo DIGITAL CONTROL SYSTEMS

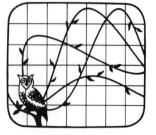

DIGITAL CONTROL SYSTEMS

BENJAMIN C. KUO
University of Illinois, Urbana

HOLT, RINEHART AND WINSTON, INC.
New York Chicago San Francisco Atlanta
Dallas Montreal Toronto London Sydney

Library of Congress Cataloging in Publication Data

Kuo, Benjamin C 1930-

Digital control systems.

(HRW series in electrical and computer engineering)

Includes bibliographical references and index.
1. Digital control systems. I. Title. II. Series.
TJ216.K812 1980 629.8'043 80-16455
ISBN 0-03-057568-0

Printed in the United States of America
0 1 2 3 1 4 4 9 8 7 6 5 4 3 2 1

CONTENTS

PREFACE

This book represents a fairly extensive revision of the 1977 edition of DIGITAL CONTROL SYSTEMS published by the SRL Publishing Company. Since 1963, the author has written several books on the subject of digital and sampled-data control systems. The first text, ANALYSIS AND SYNTHESIS OF SAMPLED-DATA CONTROL SYSTEMS, was published in 1963 by Prentice-Hall, Inc. This earlier book features the classical analysis and design of sampled-data control systems, and has a chapter on nonlinear systems. In general, the book treats a wide range of sampled-data control systems problems in a comprehensive manner. The second book, DISCRETE-DATA CONTROL SYSTEMS, was published in 1970 by Prentice-Hall, Inc. The book represents a completely new approach from the 1963 edition by concentrating on the state variable approach and modern control theory. DIGITAL CONTROL SYSTEMS is a complete revision and expansion of DISCRETE-DATA CONTROL SYSTEMS.

In the 1960's, at the height of the expansion of the aerospace industries, development of sampled-data control systems theory enjoyed the most prolific period. A large number of papers and books were published during this period on discrete-data and sampled-data control systems. In recent years although research and development activities in the aerospace sector have stabilized, the advances made in microcomputers and microprocessors and the process control industry have infused new life to the growth of digital control systems. However, the popularity of microprocessor control has put new emphasis on digital control systems theory. Because microprocessors are slow digital machines, and are usually equipped with a small wordlength, it is necessary to place importance on these constraints and the effects of time delays and amplitude quantization when designing a digital control system. Many practicing engineers apply microprocessors in control systems simply by implementing an analog control algorithm and using a very small sampling period, in order to keep the overall system stable. While this scheme may work in many non-critical situations, it would not be an acceptable solution for more critical design cases.

For those who are familiar with the first edition, DIGITAL CONTROL SYSTEMS, the following major revisions are made in the current edition:

1. The conventional analysis and design methods are expanded. This step was taken because a majority of the systems designed in the industry still rely on the use of the conventional design techniques. This happens to be true for both analog and digital control systems.
2. The design of digital control systems is greatly expanded. Chapter 10 presents a variety of the design techniques encompassing the conventional as well as the modern control theory.
3. A chapter on microprocessor control is introduced. It is difficult to draw a line on how much and what portion of the microprocessor applications should be included in a text such as this. Most electrical engineering curricula now have courses on the application of microprocessors. The author believes that the learning of the programming aspects should not belong to this book. However, it is impossible that every reader of this text is proficient in microcomputer programming. It is quite possible that the control system engineer is assigned with the task of designing the digital control system, and a programmer or software engineer would probably be more proficient in programming the digital controller. However, it is important that the control system designer understands how a microcomputer works, and all the programming ramifications, so that a realistic digital controller can be designed subject to all the physical constraints of the microcomputer. With this in mind, Chapter 14 introduces the basic components and the programming logic of a microprocessor without dwelling into specific details, but places emphasis on pointing out all the effects due to time delays and quantization due to a finite wordlength.
4. In order to conserve space and make room for all the expanded material, the chapters on statistical design have been eliminated. Since this new edition is now intended to be used at the senior or first-year graduate level, the omission of the material on statistical design may not be a serious handicap to the majority of the users.
5. The home problems have been expanded. Many practical analysis and design problems are introduced in the problem sections.

As a college text, the material presented should be more than adequate for a one-semester course. This will give the instructor some flexibility in selecting the subjects he would like to cover. All the material has been class-tested by the author teaching at the University of Illinois at Urbana-Champaign. The book is also prepared with the needs of the practicing engineers in mind; it should be suitable as a reference and for self-study purposes.

It is assumed that the reader has a knowledge on the basic principles of feedback control systems. Background on the basics of matrix algebra, Laplace transform is essential. A prior knowledge of the state-variable analysis and microprocessor programming would be helpful.

Chapter 1 gives a general introduction to the subject of digital control systems. Some typical examples of digital control systems are presented.

Chapter 2 covers signal conversion, signal processing, and the sampling operation. It gives the motivation to the mathematical and analytical treatment of digital data and signals in control systems. Chapter 3 gives a careful treatment of the z-transform theory. Modified z-transform, signal flow graph applications, and nonuniform and multirate digital control systems are included in the chapter. Chapter 4 gives a comprehensive treatment of state variable technique applied to digital systems. Time-invariant as well as time-varying systems are included. Chapter 5 covers the subject on stability. Various methods of testing the stability of linear digital control systems are given. The second method of Liapunov and its use in optimal system design is also covered. The subject on digital simulation is given in Chapter 6. Numerical integration and the z-form methods are discussed. Also included in this chapter is the subject on digital redesign, which is on how to approximate a continuous-data system by an equivalent digital system. Chapter 7 is on time-domain analysis of digital control systems. Emphasis is placed on the prediction of time response from pole-zero locations, root locus method, steady-state and transient analysis. Chapter 8 is on frequency-domain analysis of digital control systems. The conventional methods such as the Nyquist criterion, Bode plot, Nichols' chart, gain margin and phase margin are all extended to digital control systems. Chapter 9 is devoted to the subject on controllability, observability and time-optimal control. Chapter 10 covers a wide range of design techniques, from conventional to state-feedback and output-feedback controls. Chapters 11 through 13 are devoted to optimal control of digital control systems. The subjects include maximum principle, optimal linear regulator design, dynamic programming, sampling-period sensitivity, and observer design. The last chapter, Chapter 14, is on microprocessor control.

The book contains a large number of illustrative examples, and many are derived from the practical experiences of the author, although most problems are simplified in order to be suitable for textbook presentation. Many practical digital control systems problems are given in the problems section at the end of each chapter. A solution manual to the problems is available, and qualified users should make inquiry to the publisher.

The author is grateful to many of his graduate students whose assistance in the form of thesis work and classroom discussions has contributed in many ways to the preparation of this book. He wishes to acknowledge his daughter Tina for doing most of the illustrations in the book, and his other two daughters, Lori and Linda, for proofreading the typed manuscript. Special thanks go to Jane Carlton, his former secretary, for typing the complete manuscript in a professional and efficient manner. Finally, the author wishes to pay tribute to his wife, Margaret, for her contribution in many ways to this and many other projects throughout the years. Without her moral and physical assistance, these projects could never have been accomplished.

B. C. Kuo
Champaign, Illinois

DIGITAL CONTROL SYSTEMS

1

INTRODUCTION

1.1 INTRODUCTION

In recent years significant progress has been made in discrete-data and digital control systems. These systems have gained popularity and importance in all industries due in part to the advances made in digital computers, and more recently in microcomputers, as well as the advantages found in working with digital signals.

Discrete-data and digital control systems differ from the continuous-data or analog systems in that the signals in one or more parts of these systems are in the form of either a pulse train or a numerical code. The terms, sampled-data systems, discrete-data systems, discrete-data, discrete-time systems, and digital systems have been used loosely in the control literature. Strictly, sampled-data refers to signals that are pulse-amplitude modulated, that is, trains of pulses with signal information carried by the amplitudes, whereas digital data usually refer to signals which are generated by digital computers or digital transducers and thus are in some kind of coded form. However, it will be shown later that a practical system such as an industrial process control usually is of such complexity that it contains analog and sampled as well as digital data. Therefore, in this text we shall use the term discrete-data systems in a broad sense to describe all systems in which some form of digital or sampled signals takes place.

Figure 1-1 shows the basic elements of a typical closed-loop control system with sampled data. The sampler simply represents a device or operation at the output of which the signal is of the form of a periodic or aperiodic pulse train, with no information transmitted between two consecutive pulses. Figure 1-2 illustrates a common way of how a sampler operates. A continuous input signal e(t) is sampled by a sampler, and the output of the sampler is a sequence of pulses. In the present case the sampler is assumed to have a uniform sampling rate. The magnitudes of the pulses at the sampling instants represent the values of the input signal e(t) at the corresponding instants. In general, the sampling schemes exist in many variations; some of these are the periodic, cyclic-rate, multirate, random, and pulsewidth modulated samplings. The most common variations found in practical systems are the single-rate and multirate samplings. These various types of sampling and their effects will be discussed in the subsequent chapters of this book.

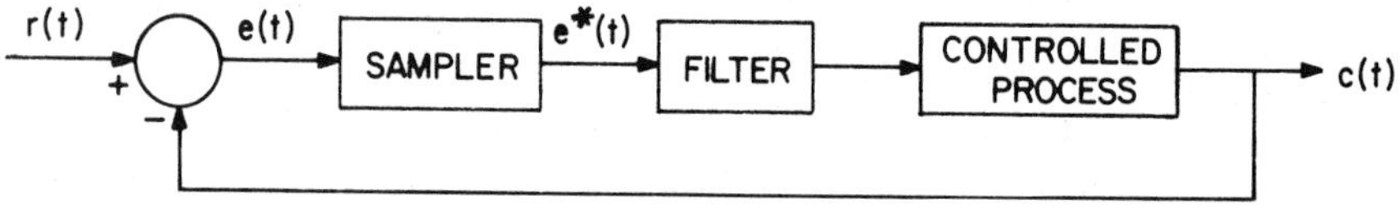

Fig. 1-1. Closed-loop sampled-data control system.

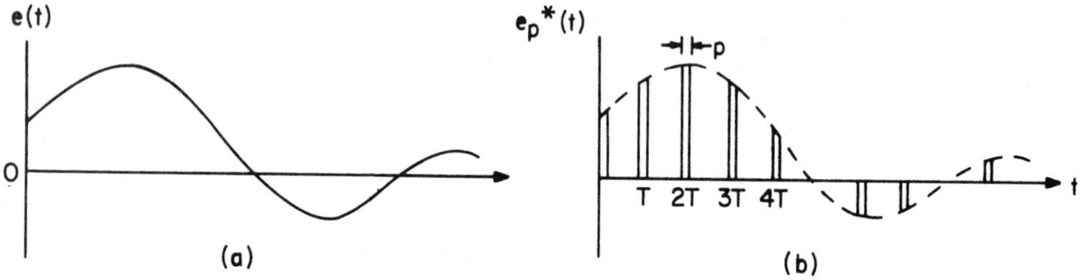

Fig. 1-2. (a) Continuous-data input to sampler.
(b) Discrete-data output of sampler.

The filter located between the sampler and the controlled process is used for the purpose of smoothing, since most controlled processes are naturally designed and constructed to receive analog signals.

A digital control system is defined as one in which the signal at one or more points of the system is expressed in a numerical code for digital-computer or digital-transducer processing in the system. The block diagram of a typical digital control system is shown in Fig. 1-3. The appearance of digitally-coded signals, such as binary-coded, in certain parts of the system requires the use of digital-to-analog (D/A) and analog-to-digital (A/D) converters. Although there are basic differences between the structures and components of a sampled-data and a digital control system, we shall show that from an analytical standpoint

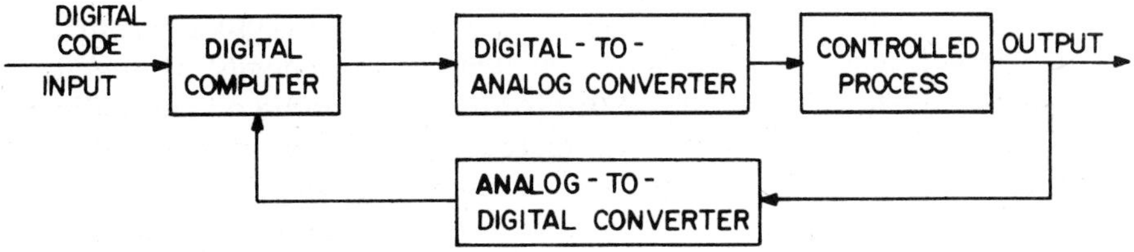

Fig. 1-3. A typical digital control system.

both types of systems are treated by the same analytical tools.

The use of sampled data in control systems can be traced back to at least seventy years ago. Some of these early applications of sampled data were for the purpose of improving the performance of the control system in one form or another. For instance, in the chopper-bar galvanometer described by Oldenbourg and Sartorious [1], (see Fig. 1-4) the sampling operation produces greater system sensitivity to a low-level input signal.

With reference to Fig. 1-4, a small signal is normally applied to the galvanometer coil. The chopper bar is lowered periodically, and the projected pointer of the galvanometer causes the load to be driven in proportion to the signal strength. The torque which is applied to the load shaft is thus determined by the chopper-bar drive rather than just by the torque developed in the galvanometer coil.

Another known example of an early application of the sampling concept in control systems is the constant-temperature oven devised by Gouy [2]. The system consists of the elements shown in Fig. 1-5. The purpose of this system

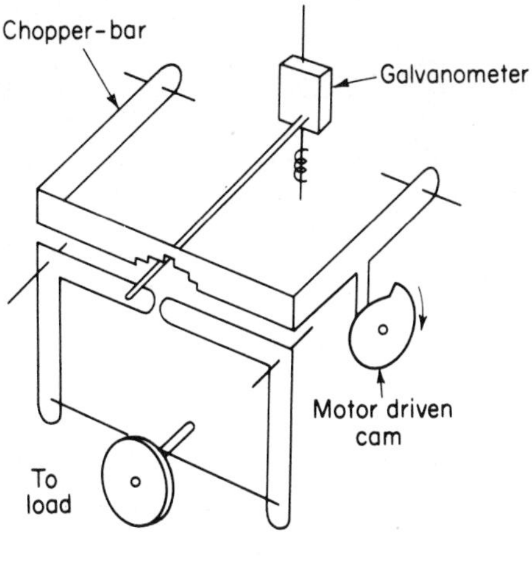

Fig. 1-4. Chopper-bar galvanometer.

is to maintain a constant temperature in the oven at all times. Whenever the electrical contact rod is immersed in the mercury, current flows in the relay coil causing the relay to open, and thus interrupt the heating current. Since the contact is periodically dipped into the mercury, the heating current consists of a sequence of pulses. Furthermore, since the time that the contact is immersed in the mercury depends upon the level of the mercury, and the level in turn depends upon the temperature of the oven, the pulse widths of the heating current are varied in proportion to the temperature of the oven, but the pulses all have equal amplitudes. A typical set of signals of the sampled-data temperature control system is shown in Fig. 1-6. In contrast to the conventional sampling scheme, such as the one described in Fig. 1-2, in which the amplitude of the signal is sampled and transmitted by the amplitude of a pulse, the sampled-data shown in Fig. 1-6 have a constant pulse amplitude, but with the signal information carried by the widths of the pulses. Therefore, we see that in practice there are many ways of sampling a signal, or perhaps more appropriately, of representing a signal in sampled form.

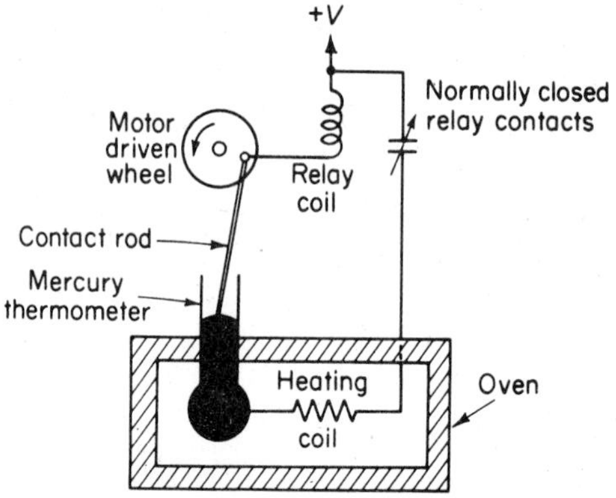

Fig. 1-5. A constant-temperature oven control system utilizing sampled-data.

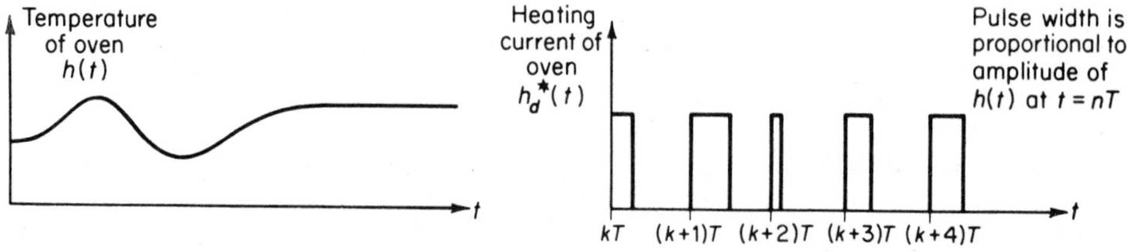

Fig. 1-6. Input and output of pulsewidth sampler.

1.2 ADVANTAGES OF SAMPLING IN CONTROL SYSTEMS

The two examples given in the preceding section on sampled-data systems merely illustrate the early applications of sampling in control systems. In order to fully understand the merits and advantages of using sampling in control systems, we should ask the natural question: *Why sampled-data?* In other words, what are the advantages and characteristics of sampling that have made sampled-data and digital control systems so important in modern control technology? To answer these questions we must first recognize that many physical systems have inherent sampling, or their behavior can be described by sampled-data or digital models. For instance, in a radar tracking system the signals transmitted and received by the system are in the form of pulse trains. The scanning operation of the radar performs the function of a sampler converting both the azimuth and the elevation information to sampled-data. There are numerous phenomena, social, economic or biological systems whose dynamics can be modelled by sampled-data system models.

Many modern control systems contain intentional sampling and digital processors. Some of the advantages of sampled-data and digital control are:

Improved sensitivity
Better reliability
No drift
Less effect due to noise and disturbance
More compact and lightweight
Less cost
More flexibility in programming

The chopper-bar galvanometer system described in the last section is an example on how sampling can improve sensitivity. In that case, a small signal is amplified for control purposes through the sampling operation.

One distinct advantage of digital controllers is that they are more versatile than analog controllers. The program which characterizes a digital controller can be modified to accommodate design changes, or adaptive performances, without any variations on the hardware. Digital components in the form of electronic parts, transducers and encoders, are often more reliable, more rugged in construction, and more compact in size than their analog equivalents. These and other glaring comparisons are rapidly converting the control system technology into a digital one.

1.3 EXAMPLES OF DISCRETE-DATA AND DIGITAL CONTROL SYSTEMS

In this section we shall illustrate several typical examples on discrete-data and digital control systems. The objective is simply to show some of the essential components of these systems, and the coverage is not intended to be exhaustive.

A Simplified Single-Axis Autopilot Control System

Figure 1-7 shows the block diagram of a simplified single-axis (pitch, yaw, or roll) analog autopilot control of an aircraft or missile. This is a typical analog or continuous-data system in which the signals can all be represented as functions of the continuous-time variable t. The objective of the control is that the attitude of the airframe follow the command signal. The rate loop is incorporated here for the improvement of system stability. Instead of using the analog controller as shown in Fig. 1-7, a digital controller with the necessary analog-to-digital (A/D) and digital-to-analog (D/A) converters can be used for the same objective as shown in Fig. 1-8. Since, other than the digital controller, all the remaining components of the system are still analog, the A/D and D/A converters are necessary for signal matching purposes.

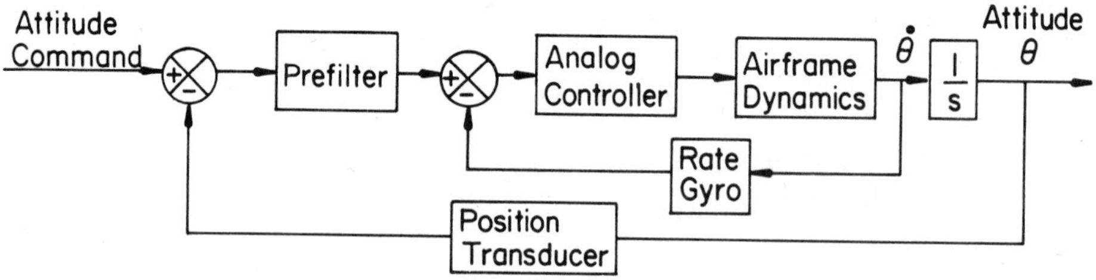

Fig. 1-7. A simplified single-axis autopilot control system with analog data.

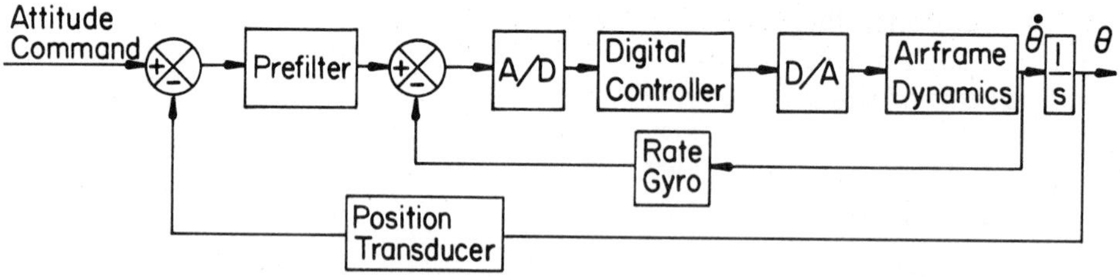

Fig. 1-8. A simplified single-axis autopilot control system with digital data.

Figure 1-9 shows the digital autopilot control system in which the position and rate information are obtained by digital transducers and the operations are represented on the block diagram by sample-and-hold devices. A sampler essentially samples an analog signal at some uniform sampling period, and the hold device simply holds the value of the signal until the next sample comes along. Figure 1-9 illustrates the situation in which the two samplers have different sampling periods, T_1 and T_2. In general, if the rate of variation of the signal in one loop is very much less than that of the other loop, the sampling period of the sampler used in the slower loop can be larger. The system shown in Fig. 1-9 which has samplers with different sampling periods is called a *multirate sampled-data system.*

One of the advantages of using sampling and multirate sampling is that certain expensive components of the system can be used on a time-sharing basis.

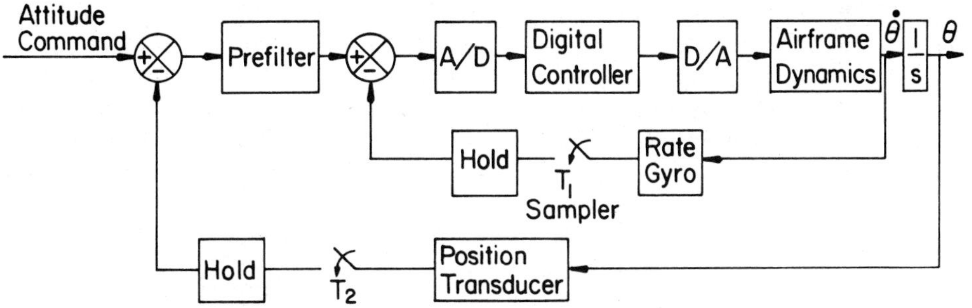

Fig. 1-9. A simplified single-axis digital autopilot control system with multirate sampling.

A Digital Computer Controlled Rolling Mill Regulating System

Many industrial processes are controlled and monitored by digital computers and digital transducers. Practically all the modern steel rolling mills are regulated and controlled by digital computers. Figure 1-10 illustrates the basic elements of such a system. Figure 1-11 shows a block diagram of the thickness control portion of the system.

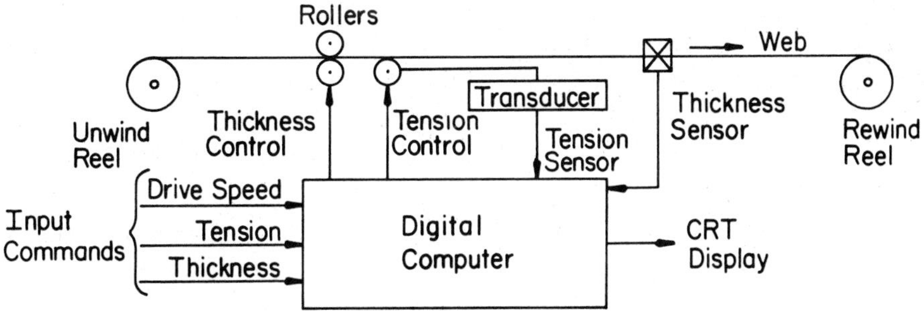

Fig. 1-10. A digital computer controlled rolling mill regulating system.

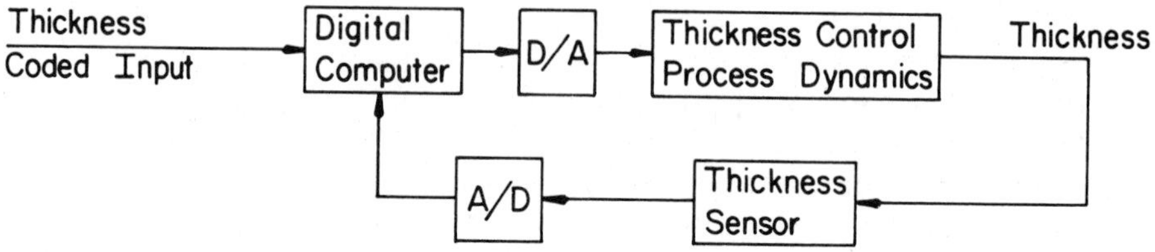

Fig. 1-11. Thickness control of a rolling mill regulating system.

A Digital Controller for a Turbine and Generator

Figure 1-12 shows the block diagram and the essential elements of a minicomputer system which is used for speed and voltage control as well as data acquisition of a turbine-generator unit. The D/A converters form the interface between the digital computer and the speed and voltage controls. The data-acquisition system performs the measurements of such variables as the generator speed, rotor angle, terminal voltage, field and armature currents, and real and reactive power. Some of these variables may be measured by digital transducers whose outputs are then digitally multiplexed and sent to the computer, as shown in Fig. 1-13. The quantities that are measured by analog transducers are first sent through an analog multiplexer which performs a time-division multiplexing operation between a number of different input signals. Each input-signal channel is sequentially connected to the output of the multiplexer

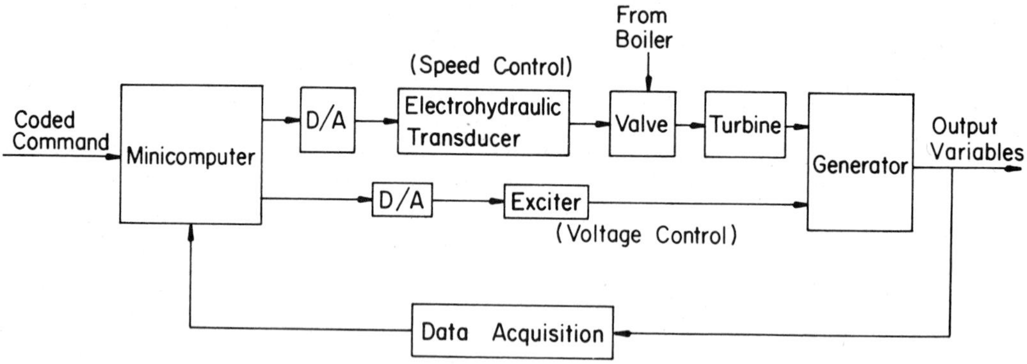

Fig. 1-12. Computer control of a turbine and generator unit.

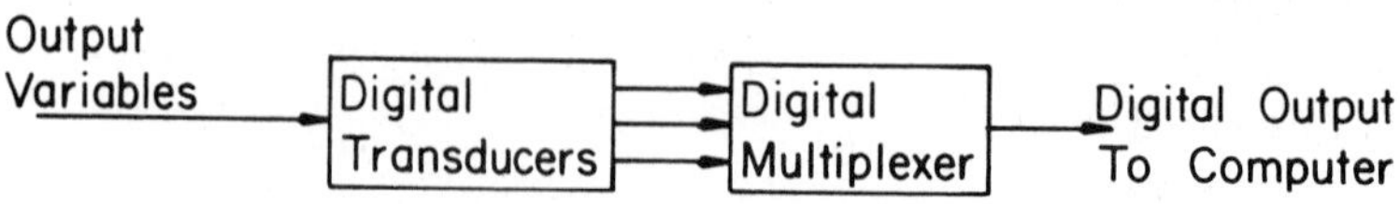

Fig. 1-13. A digital data acquisition system.

for a specific period of time. The system which follows the multiplexer is thus time shared between a number of signal channels. Figure 1-14 illustrates the data-acquisition system when the system variables are measured by analog transducers. The output of the analog multiplexer is connected to a sample-and-hold device which samples the output of the multiplexer at a fixed time interval and then holds the signal level at its output until the A/D converter completes the analog-to-digital conversion.

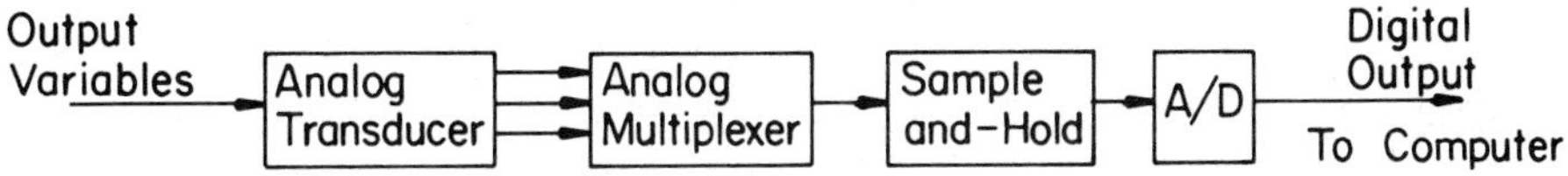

Fig. 1-14. A data-acquisition system with analog transducers.

A Step Motor Control System

Occasionally we may come across a system which contains all-digital elements so that the need for using A/D and D/A converters for signal matching may be avoided. Figure 1-15 shows such a system, which is used for the control of the read-write head of a computer memory disk. The prime mover used in the disk drive system is a step motor which is driven by pulse commands. The step motor moves one fixed displacement increment in response to each pulse input. Thus, the system may be considered to be an all-digital system.

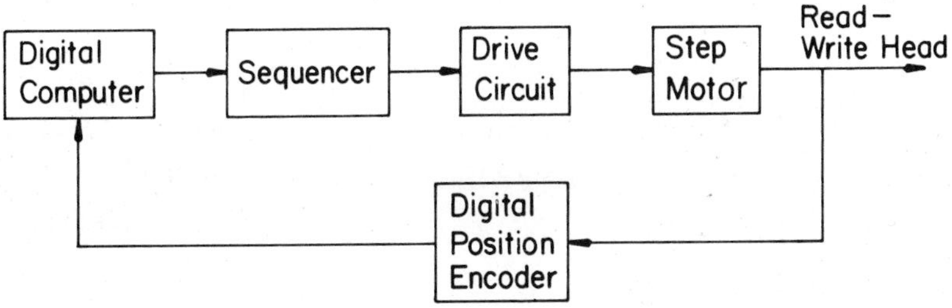

Fig. 1-15. A step motor control system for read-write head positioning on a disk drive.

A Discrete-Data Model of an Interest Payment Problem

Thus far we have illustrated digital systems which are physical in nature. There are numerous social and economic systems that can be modeled by a discrete-data system model. Specifically, we recognize that analog dynamic systems are described by differential equations with variables that are functions of the continuous variable t. For discrete-data systems, the system dynamics are described by difference equations. The variables for a discrete-data model are either functions of the discrete time variable kT (k = 0,1,2,...), where T is a constant, or simply functions of the discrete variable k.

Let us consider an interest payment problem which is defined as follows:

Consider that the amount of capital P_0 is borrowed initially. The interest rate on the unpaid balance is r percent per period. We assume that the principal and interest are to be paid back in N equal payments u over N periods.

Let P_k denote the amount owed after the kth period. Then the following difference equation can be written for the problem:

$$P_{k+1} = (1 + r)P_k - u \tag{1-1}$$

where P_{k+1} denotes the amount still owed at the end of the (k + 1)th period. The boundary conditions are P_0 given and $P_N = 0$. The difference equation in Eq. (1-1) can be solved for u by a recursive method or the z-transform method. The equation is also known as a first-order difference equation. A discrete system model is shown in Fig. 1-16 in the form of a block diagram.

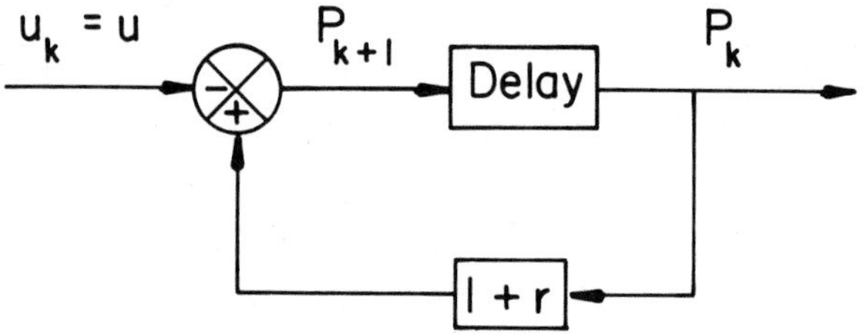

Fig. 1-16. A discrete-data system model for an interest-payment problem.

REFERENCES

1. Oldenbourg, R. C., and Sartorius, H., *The Dynamics of Automatic Control.* American Society of Mechanical Engineers, New York, 1948.
2. Gouy, G., "On a Constant Temperature Oven," *J. Physique*, 6, Series 3, 1897, pp. 479-483.

2

SIGNAL CONVERSION AND PROCESSING

2.1 INTRODUCTION

Digital computers are used by control systems engineers for two basic purposes. One aspect is that digital computers are used extensively for simulation and computation of control systems dynamics. Since most physical systems are of the high order and contain nonlinear elements, control engineers often rely on digital computer simulations to conduct the analysis and design of complex systems which do not conform to any of the established analytical tools. Computer simulations are also used to check the results obtained by analytical means.

Another important application of the digital computer in control systems is their use as controllers or processors. Since most of the controlled processes contain analog elements, the signals found in most digital control systems are both analog as well as digital. Therefore, the process of signal conversion is essential so that the digital and analog components can be interfaced in the same system. For instance, the output signals of an analog device must first undergo an *analog-to-digital* (A/D) conversion before they can be processed by a digital controller. Sometimes the analog-to-digital conversion process may be described as an ENCODING operation. Similarly, the coded signal from a digital controller or computer must first be decoded by a digital-to-analog (D/A) converter before it can be sent to an analog device for processing. It was illustrated in the last chapter that other standard operations such as TIME-SHARING require the use of such interfacing equipment as the MULTIPLEXER, SAMPLE-AND-HOLD, etc. Because these components are important for signal processing and conditioning in digital systems, they will be described briefly in this chapter. It should be borne in mind, however, that the main subject of this text is not on signal processing and is not hardware-oriented. The practical aspects and the details of the operations of A/D and D/A converters and digital signal processing devices are quite extensive and are beyond the main scope of this book. The reader may acquire more information on these devices by referring to the list of references given at the end of this chapter. The objective of the treatment of the subjects on A/D and D/A converters, multiplexers, and the sample-and-hold (S/H) is to establish the importance of signal processing in discrete-data and digital systems, and, most important, how these components are modelled and treated mathematically for analysis and design.

Before embarking on the discussions on D/A and A/D converters, multiplexers, and sample-and-hold, the basic definitions of these operations are first given as follows.

D/A Converter (D/A). A digital-to-analog converter (D/A) performs the function of DECODING on a digital-coded input. The output of the D/A converter is an analog signal, usually in the form of a current or a voltage. A D/A is needed as an interface between a digital channel or digital computer and an analog device. A D/A is also known as a DECODER.

A/D Converter (A/D). An analog-to-digital converter (A/D) is a device which converts an analog signal to a digital-coded signal. The device is necessary as an interface between an analog system or component whose output is to be processed by a digital processor or computer.

Sample-and-Hold (S/H). A sample-and-hold (S/H) is used for many purposes in discrete-data and digital systems. A S/H makes a fast acquisition of an analog signal and then holds this signal at a constant value until the next acquisition (sample) is made. We shall show later that S/H is an integral part of an A/D converter.

Multiplexer. When signals from several sources are to be processed by the same processor or communication channel, a multiplexer is used to couple the signals to the processor in some prescribed sequence. This way, the processor is time-shared by all the incoming signals. For example, if several digital-data sources are to be processed by a central computer unit, these input signals are usually coupled to the computer by a multiplexer through a common set of parallel lines.

2.2 DIGITAL SIGNALS AND CODING

Digital signals in digital computers and digital control systems usually are represented by digital words or codes. The information carried by the digital word is generally in the form of discrete bits (logic pulses of "0" or "1") coded in a serial or parallel format. The numerical value of the digital word then represents the magnitude of the information in the variable which this word represents.

Since it is simple to distinguish and implement just two states, "on" and "off", all modern digital computers are built on the basis of the binary number system. A digital signal can be stored in a digital computer as a binary number of 0's and 1's. Each of the binary digits (0 or 1) is referred to as a *bit*. The bit itself, however, is too small a number to be considered the basic unit of information. Typically, bits are strung together to form larger, more useful units; with 8 bits placed together to form a *byte*. Several of these 8-bit bytes may be placed together to form a *word*. In general, words may be of almost any bit length, from 4 bits to 128 bits or more. The distinction between a bit and a byte is somewhat like that between a letter of the alphabet and a word in the English language. While a letter may be said to be the smallest unit that can be written down, it usually has no meaning in itself until it is strung together with other letters to form a word. Figure 2-1 illustrates the relation between *word*, *byte* and *bit*. In this case, the length of the word is two bytes or 16 bits.

The capacity of the memory of a digital computer is often specified in terms of bytes. For example, the Intel 8080 microprocessor is so designed that up to 64K bytes can be attached to the machine. (64K bytes actually represents 2^{16} or 65,536 bytes to be exact.)

The accuracy of a digital computer in its ability to store and manipulate digital signals is indicated by its word length. For example, a computer with an 8-bit word can only store numbers with 8 bits of accuracy in its memory. Similarly, the registers and the accumulator all are 8 bits in length so that temporary storage and computation can be carried out only in 8 bits of accuracy unless double-precision algorithm is used.

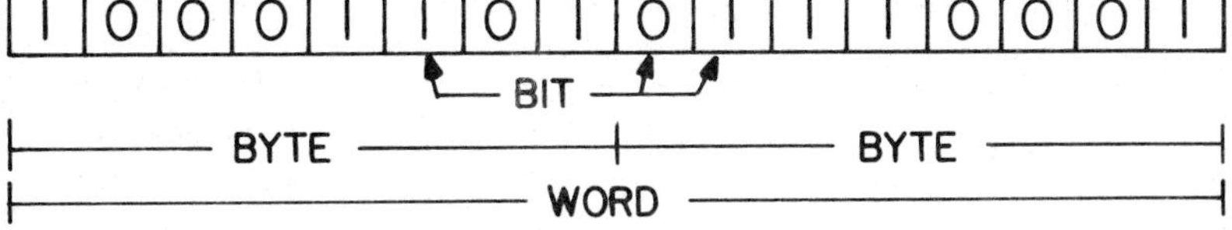

Fig. 2-1. Relations between word, byte and bit.

Digital signals in a computer can be represented as *fixed-point numbers* or *floating-point numbers*. These number systems are discussed as follows.

Fixed-Point Number Representation

If we use all 16 bits of the word shown in Fig. 2-1 to represent a number, where each bit can be a 0 or a 1, we have the fixed-point number representation. In general, an n-bit binary word representing a fixed-point integer number N is written as

$$N = a_{n-1}2^{n-1} + \ldots + a_2 2^2 + a_1 2^1 + a_0 2^0 \tag{2-1}$$

where the coefficients a_i, $i = 0,1,2,\ldots,n-1$ are either 0 or 1. The digits of the number in Eq. (2-1) are ordered from left to right with the most significant bit (MSB) being a_{n-1} on the left and the least significant bit (LSB), a_0, on the right.

As a simple example, let us consider a three-bit binary word,

$$N = a_2 2^2 + a_1 2^1 + a_0 2^0 \tag{2-2}$$

By assigning various combinations of 0 and 1 to the coefficients a_0, a_1, and a_2, the word N can represent eight (2^3) distinct integral states or numbers. The conversion relationship between binary and decimal integers for a three-bit word is shown in Table 2-1.

We can also use the fixed-point notation to represent nonintegers or fractions. Using a fictitious "binary point" in a data word, a part of the word can be used to represent integer, and the other part represents fraction. For example, the 8-bit word shown in Fig. 2-2 has its first five digits representing the integral part, and the last three digits representing the fractional part of the number. Note that the "binary point" is only symbolic so that it does not occupy

TABLE 2-1. CONVERSION RELATIONSHIP BETWEEN NATURAL BINARY AND DECIMAL CODES FOR INTEGER CODING

Decimal Integer	Binary Integer	Binary Integer Code		
		MSB (X 4)	(X 2)	LSB (X 1)
0	000	0	0	0
1	001	0	0	1
2	010	0	1	0
3	011	0	1	1
4	100	1	0	0
5	101	1	0	1
6	110	1	1	0
7	111	1	1	1

any bit.

With reference to the 8-bit word shown in Fig. 2-2, the number N is written

$$N = a_4 2^4 + a_3 2^3 + a_2 2^2 + a_1 2^1 + a_0 2^0 + a_{-1} 2^{-1} + a_{-2} 2^{-2} + a_{-3} 2^{-3} \quad (2\text{-}3)$$

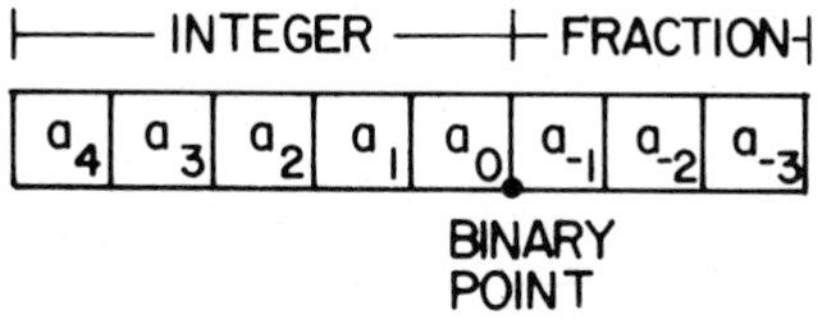

Fig. 2-2. Fixed-point representation of an 8-bit non-integer.

Thus, for instance, the binary number 01011.101 is equivalent to the decimal number

$$N = 0 \times 2^4 + 1 \times 2^3 + 0 \times 2^2 + 1 \times 2^1 + 1 \times 2^0 + 1 \times 2^{-1} + 0 \times 2^{-2} + 1 \times 2^{-3}$$

$$= 8 + 2 + 1 + \frac{1}{2} + \frac{1}{8} = 11.625 \quad (2\text{-}4)$$

In general, an n-bit *fraction* can be represented as

$$N = a_{-1} 2^{-1} + a_{-2} 2^{-2} + \ldots + a_{-n} 2^{-n} \quad (2\text{-}5)$$

where the coefficients a_i ($i = -1, -2, \ldots, -n$) assume the value of 0 or 1. The first coefficient a_{-1} represents the MSB and the LSB is a_{-n}. As a simple illustration, the eight distinct states or fractional numbers that are representable by a three-bit word are shown in Table 2-2.

Thus far we have only considered the binary representation of positive numbers. Negative numbers may be represented by assigning the first bit of the binary word as a *sign bit*; i.e., 0 for "+" and 1 for "−". An alternative method of representing negative numbers is to use the two's-complement algorithm (refer to any elementary text on digital computers).

If the first bit of a three-bit word is used as a sign bit, then the largest integer that the word can represent is $2^2 - 1 = 3$, and the smallest integer that can be represented is $-(2^2 - 1) = -3$. In general, the integer representable by an n-bit word with a sign bit lies between $2^{n-1} - 1$ and $-(2^{n-1} - 1)$, inclusive, zeros included. For example, for a three-bit word, with the first bit used for sign, the eight states or integers representable are tabulated in Table 2-3. Notice

TABLE 2-2. CONVERSION RELATIONSHIP BETWEEN NATURAL BINARY AND DECIMAL CODES FOR FRACTIONAL CODING

Decimal Fraction	Binary Fraction	Binary Fraction Code		
		MSB $(\times\frac{1}{2})$	$(\times\frac{1}{4})$	LSB $(\times\frac{1}{8})$
0	.000	0	0	0
1/8	.001	0	0	1
1/4	.010	0	1	0
3/8	.011	0	1	1
1/2	.100	1	0	0
5/8	.101	1	0	1
3/4	.110	1	1	0
7/8	.111	1	1	1

TABLE 2-3. CONVERSION RELATIONSHIP BETWEEN BINARY AND DECIMAL CODES FOR A 3-BIT INTEGER WITH A SIGN BIT

Decimal Integer	Binary Integer	Binary Integer Code		
		SIGN BIT	MSB (× 2)	LSB (× 1)
−3	111	1	1	1
−2	110	1	1	0
−1	101	1	0	1
−0	100	1	0	0
+0	000	0	0	0
+1	001	0	0	1
+2	010	0	1	0
+3	011	0	1	1

that in this case 0 (zero) is represented twice, +0 and −0. However, the three-bit word still defines eight distinct states, as in the case shown in Table 2-1.

In a similar fashion a sign bit can be used for the representation of a non-integer or fraction. The number that is representable by an n-bit word with a sign bit and m fractional bits (m bits after the "binary point") lies between $(2^{n-1} - 1)2^{-m}$ and $-(2^{n-1} - 1)2^{-m}$, inclusively.

The fixed-point representation of real numbers has a serious disadvantage due to the limited range the numbers can be represented by a given word length once the fictitious binary point has been assigned. When multiplying two large numbers, frequently the result will exceed the capacity of the word length of the fixed-point representation, and overflow will take place which causes inaccuracy in the computation.

Floating-Point Number Representation

A representation system of numbers which is easier to work with and has more range is the *floating-point* representation. This method is also known as the scientific notation where the first part of the data word is used to store a number called the *mantissa*, and the second part is the *exponent*. For instance, in the decimal system the number 5 can be written as 0.5×10^1, 50×10^{-1} or 0.05×10^2, etc. In digital computers and systems, binary floating-point numbers are usually represented as

$$N = M \times 2^E \tag{2-6}$$

where M is the mantissa and E is the exponent of the number N. Furthermore, M is usually scaled to be a fraction whose decimal value lies in the range of $0.5 \leqslant M < 1$.

Figure 2-3 shows a floating-point representation of an 8-bit word with a 5-bit mantissa and a 3-bit exponent. Since the mantissa and the exponent can both be positive or negative, the first bits of the mantissa and the exponent are the sign bits. (We can also use the first two bits of the entire word as the sign bits of the mantissa and the exponents, respectively.) For microcomputers that have small word lengths, two consecutive data words can be strung together to form a floating-point number, such as that shown in Fig. 2-4.

Since the mantissa is normalized to be a fraction between 1/2 and 1, *the first bit after the sign bit should always be a 1, and there is always a fictitious "binary point" just after the sign bit.*

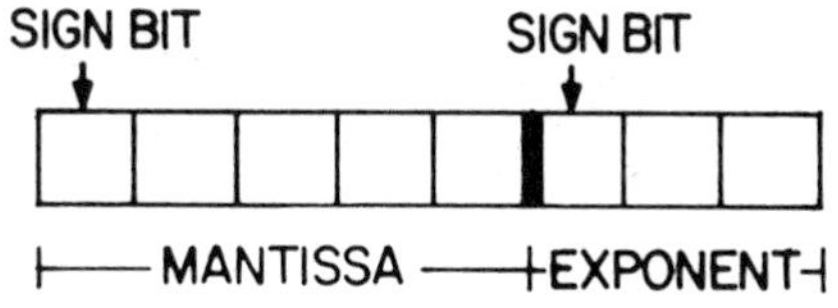

Fig. 2-3. Floating-point representations of an 8-bit number.

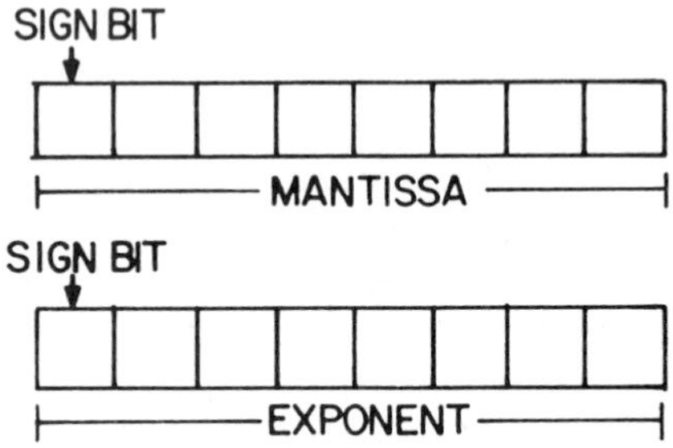

Fig. 2-4. Floating-point representation using two computer words.

The exponent E represents how many places the binary point should be shifted to the right (for $E > 0$) or to the left (for $E < 0$). As an illustrative example, the decimal number 6.5 is represented by the 8-bit floating-point binary word shown in Fig. 2-5. In this case, the mantissa consists of four bits for the fraction .1101, and the exponent is the two-bit binary integer 11 which represents the decimal number 3. Thus, the number represented is

$$
\begin{aligned}
N &= .1101 \times 2^{11} && \text{(binary)} \\
&= \left[\frac{1}{2} + \frac{1}{4} + \frac{1}{16}\right] 2^3 && \text{(decimal)} \\
&= 6.5 && \text{(decimal)}
\end{aligned}
$$

Also, we see that if we move the binary point to the right by 3 bits, where 3 is the exponent in decimal, we have the fixed-point binary number 110.1 which is 6.5 decimal.

For an n-bit word with an m-bit mantissa and an e-bit exponent, both including the sign bits, the *largest* number N_{max} that the word can represent is shown in Fig. 2-6(a). In this case all the non-sign bits are filled with ones, and N_{max} is expressed as

$$N_{max} = (1 - 2^{-m+1})2^{(2^{e-1}-1)} \tag{2-7}$$

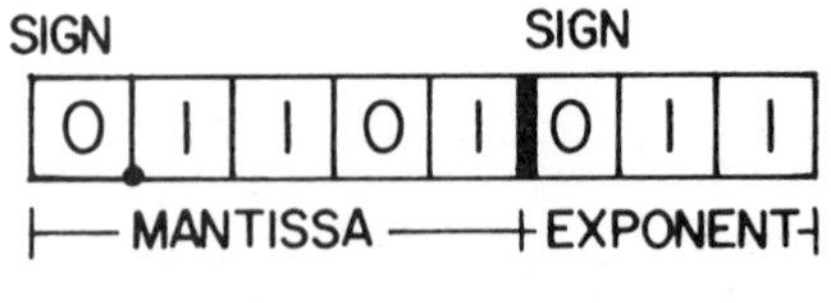

Fig. 2-5. Floating-point representation of the decimal number 6.5 by an 8-bit binary word.

For example, for the 8-bit floating-point word shown in Fig. 2-5, m = 5 and e = 3. The largest number that can be represented is

$$N_{max} = (1 - 2^{-4})2^{(2^2-1)} = (1 - 1/16)2^3 = 7.5 \tag{2-8}$$

For the same 8-bit word, if we allocate four bits for the mantissa and the same for the exponent, then

$$N_{max} = (1 - 2^{-3})2^{(2^3-1)} = (1 - 1/8)2^7 = 112 \tag{2-9}$$

Figure 2-6(b) shows the contents of the *smallest* positive number N_{min} that can be represented by an n-bit word with an m-bit mantissa and an e-bit exponent. In this case the first bit after the sign bit in the mantissa is a one, and the rest of the bits are all zeros. The bits of the exponent are all ones. Thus, N_{min} is written

$$N_{min} = 0.5 \times 2^{-(2^{e-1}-1)} \tag{2-10}$$

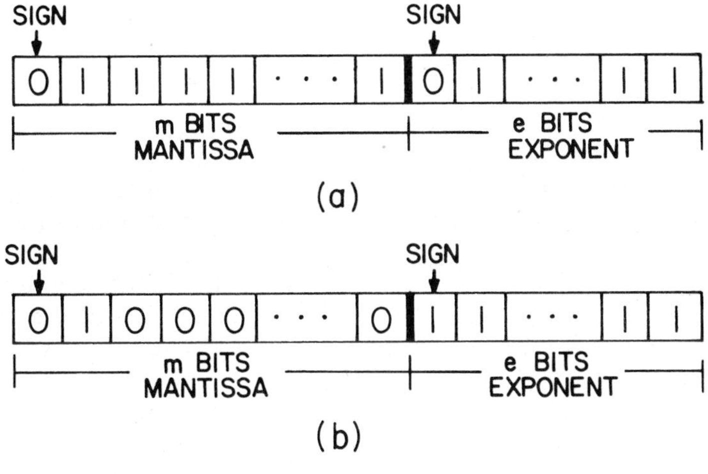

Fig. 2-6. Maximum and minimum magnitudes of floating-point numbers.

2.3 DATA CONVERSIONS AND QUANTIZATION

Since most digital control systems contain analog as well as digital components, it is necessary to convert analog signals into digital form by means of an analog-to-digital (A/D) converter, and digital signals into analog signals by digital-to-analog (D/A) converters.

In D/A and A/D conversions the MSB and the LSB and the weight of each bit in a digital-coded word are important in the understanding of the conversion process. The practical D/A and A/D converters that are based on the natural binary code make use of the fractional code. As shown in Table 2-2, for a three-bit binary fractional code the MSB has a weight of 1/2 of full scale (FS); the second bit has a weight of 1/4 FS; and the LSB has a weight of 1/8 FS. In general, for an n-bit binary fractional code the MSB still has a weight of 1/2 FS, but the LSB has a weight of 2^{-n} FS. *Regardless of whether integer or fractional coding is utilized, an n-bit binary word defines 2^n distinct states, thus the word provides a resolution of one part in 2^n.* For example, Fig. 2-7 illustrates the resolution of $2^{-3} = 1/8$ of a three-bit binary fractional code. The LSB in this case is 1/8 and full scale is 1. Resolution may be improved by using more bits. For instance, a four-bit binary fractional coded word still has a full-scale decimal equivalent of 1, but the LSB is 2^{-4} or 1/16, and the resolution is improved to one part in 16. The diagram shown in Fig. 2-7 would then have 16 distinct levels. It should be noted that the digital code used in general does not have to correspond to its analog signal, and vice versa. Therefore, the analog numbers represented on the vertical axis in Fig. 2-7 should be considered as a fractional part of full scale (FS). The same principle also works for the integer code. *It should be kept in mind that in order to improve the resolution by increasing the number of bits, the full-scale analog or digital signal value should be maintained the same.*

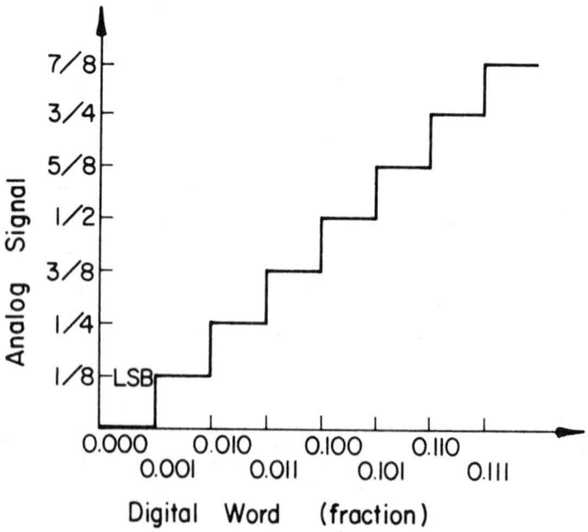

Fig. 2-7. Relation between digital binary fractional code and decimal number.

In converting an analog signal to a digital number, the operation is described as A/D conversion. If the number of bits in the digital word is finite, as it will be in a practical system, only a finite resolution can be attained by the A/D conversion. One of the major operations in the A/D conversion is *quantization*. Since the digital output can assume only a finite number of levels, it is necessary to *quantize* or *round off* the analog number into the nearest digital level. Figure 2-8 illustrates the relationship between the analog and digital binary integral code for a three-bit word for positive and negative numbers. The analog input is shown on the horizontal axis and the digital output levels are on the vertical axis.

As shown in Fig. 2-8, the analog signal has decision levels at the values of

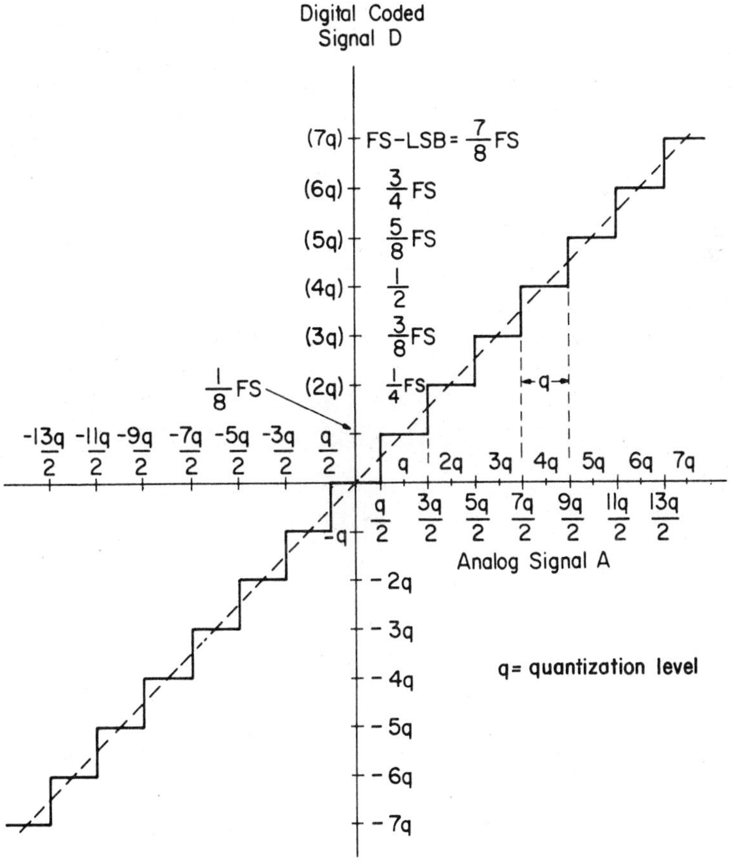

Fig. 2-8. Input-output relationship of an analog-to-digital three-bit quantizer.

0.5q, 1.5q, 2.5q, ..., 6.5q. In other words, the digital output number D is related to the analog number A through the relationships shown in Table 2-3.

It should be again noted that the analog-to-digital conversion shown in Fig. 2-8 is not generally a one-to-one relation. The parameter q which is equal to the least significant bit (LSB) is known as the *quantization level.* For this three-bit digital word, the LSB is 1/8 FS, as indicated in Table 2-4. The difference between the analog signal and the digital output is called the *quantization error.* The quantization error in general depends upon the number of quantization levels or the resolution of the quantizer. Figure 2-8 shows that the quantization error is zero when the analog signal value is equal to an integral multiple of q before reaching saturation. In general, the maximum quantization error is $\pm q/2$ before reaching saturation.

One important aspect with respect to the saturation level of the quantization process, as depicted in Fig. 2-8 and Table 2-4, is that the maximum digital output, which corresponds to the binary number 111, is 7/8 FS, not FS. This would not have any effect on the accuracy of the A/D conversion as long as the values of the analog inputs do not exceed 7/8 FS + q/2, since under this condition the maximum quantization error will be bounded by $\pm q/2$. For instance, if the full-scale reference voltage of a three-bit digital conversion process is set at 10 volts, referring to Fig. 2-8, FS = 10 volts;

$$q = \frac{1}{8}\ \text{FS} = 1.25\ \text{volts}$$

TABLE 2-4. RELATION BETWEEN ANALOG AND DIGITAL NUMBERS FOR A THREE-BIT ANALOG-TO-DIGITAL CONVERSION

ANALOG NUMBER A	BINARY-CODED DIGITAL NUMBER D			
	BINARY NUMBER			
	MSB (× 4q)	(× 2q)	LSB (× q)	
$\|A\| < 0.5q$	0	0	0	0
$0.5q \leq \|A\| < 1.5q$	0	0	1	1/8 FS = q = LSB
$1.5q \leq \|A\| < 2.5q$	0	1	0	1/4 FS = 2q
$2.5q \leq \|A\| < 3.5q$	0	1	1	3/8 FS = 3q
$3.5q \leq \|A\| < 4.5q$	1	0	0	1/2 FS = 4q
$4.5q \leq \|A\| < 5.5q$	1	0	1	5/8 FS = 5q
$5.5q \leq \|A\| < 6.5q$	1	1	0	3/4 FS = 6q
$6.5q \leq \|A\| < \infty$	1	1	1	7/8 FS = FS − LSB = 7q

Then, the maximum value of the analog signal that the quantizer can convert without exceeding the quantization error of ±0.625 volt is

$$\frac{7}{8}\,\text{FS} + \frac{q}{2} = 9.375 \text{ volts}$$

Although the natural binary code is the most commonly used in practice because it is the easiest one to understand and to be implemented by digital circuits, many other codes are also popular as well. These include the binary coded decimal (BCD) code, offset binary code, two's complement, the gray code, etc.

2.4 SAMPLE-AND-HOLD DEVICES

A sampler in a discrete-data or digital system is a device which converts an analog signal into a train of amplitude-modulated pulses or a digital signal. In Chapter 1 we have described some of the physical systems with inherent or intentional sampling. A hold device simply maintains or "freezes" the value of the pulse or digital signal for a prescribed time duration. In a majority of the practical digital operations, sample and hold are performed by a single unit, and the device is commercially known as sample-and-hold, or S/H.

Sample-and-hold devices are used extensively in digital systems. One of the main applications of sample-and-hold is to "freeze" fast-moving signals during all types of conversion operations. Another common usage of S/H is for the storage of multiplexer outputs while the signal is being converted. Peak detecting of a signal is another application of sample-and-hold. We shall show in the ensuing sections that sample-and-hold is often used in conjunction with A/D and D/A converters.

The simplest form of a S/H is illustrated by the circuit shown in Fig. 2-9. The opening and closing of the switch or sampler is controlled by a SAMPLE COMMAND. When the switch is closed the S/H output samples and TRACKS the input signal $e_s(t)$. When the switch is opened, the output is held at the voltage that the capacitor is charged to. Figure 2-10 illustrates typical input and output signals of the simple S/H shown in Fig. 2-9 when the source resistance is

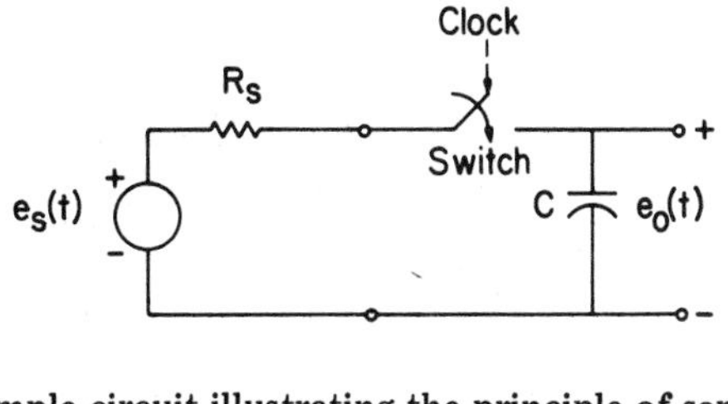

Fig. 2-9. A simple circuit illustrating the principle of sample-and-hold.

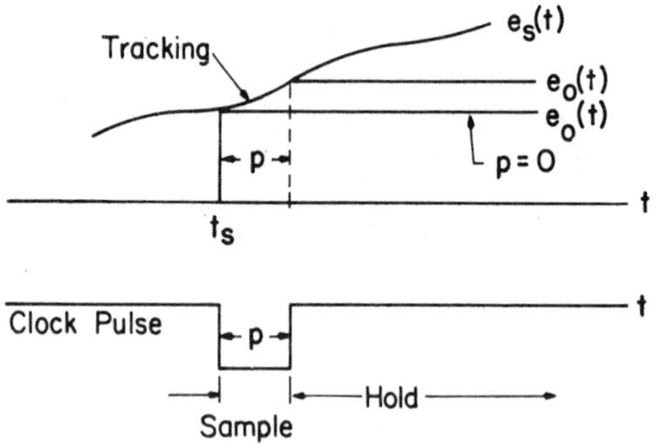

Fig. 2-10. Simplified sample-and-hold signals; $R_s = 0$.

zero. The time interval during which the sampler is closed is called the *sampling duration* p. In practice, the resistance R_s is not zero, and the capacitor will charge toward the sampled input signal with a time constant R_sC. Furthermore, the operation of the sampler is not instantaneous as it would take time for it to respond to the sample-and-hold commands.

In practice a sample-and-hold has many imperfections and errors, and the output of the device may deviate considerably from the ideal waveform illustrated in Fig. 2-10.

Figure 2-11 illustrates a typical input signal $e_s(t)$ and the corresponding S/H output of a practical S/H. The typical output signal of the S/H is characterized by several sources of time delays and imperfect holding during the hold mode. These characteristics are illustrated in Fig. 2-11 and are defined in the following.

Acquisition Time (T_a). When the SAMPLE COMMAND is given to the S/H the unit does not begin to track the input signal instantaneously. The acquisition time is measured from the instant SAMPLE COMMAND is given to the time when the output of the S/H enters and remains within a specified error band (say ±1%) around the input signal. Typical catalog information on acquisition time is given in terms of % FS for a certain size voltage step; e.g., 0.1% F.S. for 10 v step.

Aperture Time (T_p). When the HOLD command is given to the sample-and-hold while it is in the sample (track) mode, it will stay in the sample mode before reacting. The time between the issuance of the HOLD command and the time the sampler is opened is called the *aperture time*. This delay is usually caused by the switching circuit time delays within the S/H. For a given S/H, the aperture time is not constant, and the catalog information usually includes the worst-case figure. For example, the aperture time of a typical commercial S/H unit may be on the order of 10 nanoseconds.

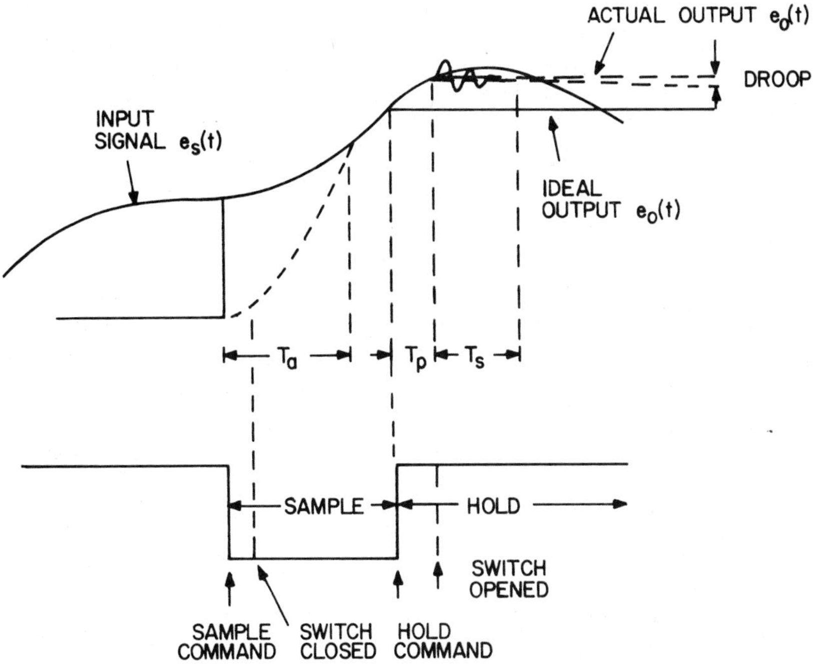

Fig. 2-11. Input and output signals of S/H with finite time delays.

Settling Time (T_s). In switching from the sample mode to the hold mode, transient caused by capacitance feedthrough from the digital logic circuitry through the electronic switch to the analog signal path can occur. The time required for the transient oscillation to settle to within a certain percent of FS is called the *settling time*. The typical settling time of a commercial S/H is on the order of several nanoseconds to several microseconds, depending on the final accuracy required.

Hold Mode Droop. During the HOLD mode the output voltage of the S/H may decrease slightly, due to the leakage currents with the FET switch and the buffer amplifier of the input circuit.

The droop in the output of the S/H can be greatly reduced by using a buffer amplifier with a very high input impedance at the output of the S/H. Similarly, we may use an input buffer amplifier so as to keep the input current of the S/H relatively constant. A S/H with input and output buffer amplifiers is illustrated by the block diagram of Fig. 2-12.

In digital systems the sample-and-hold operation is often controlled by a periodic clock. Figure 2-13 illustrates a uniform-rate sample-and-hold operation. Both the actual and the ideal S/H outputs are shown for the given input analog signal. The time duration between the sample commands is called the *sampling period* T.

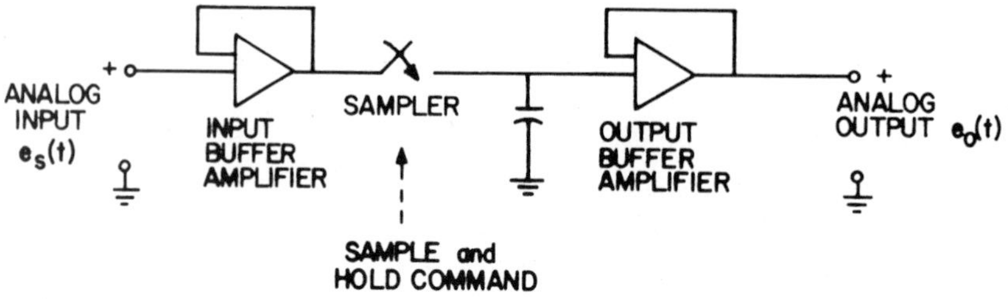

Fig. 2-12. Sample-and-hold with input and output buffer amplifiers.

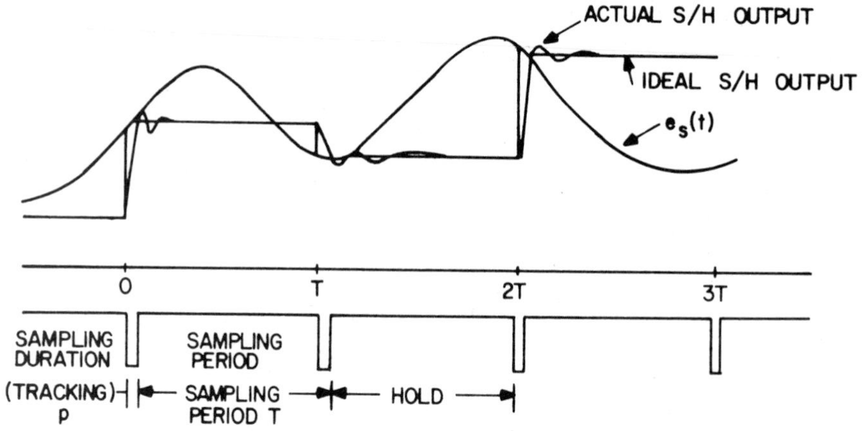

Fig. 2-13. Input and output signals of S/H with a uniform periodic sampling rate.

System Block Diagram Representation of S/H

Although sample-and-hold is available commercially as one unit, and in digital systems it is often represented by the functional block diagram element shown in Fig. 2-14, for analytical purposes it is advantageous to treat the sampling and the holding operations separately. Figure 2-15 illustrates an equivalent block diagram which isolates the sample-and-hold functions and the effects of all the delay times and transient oscillations. The sampler which can be regarded as a pulse-amplitude modulator has a pulse or sampling duration of p. The hold device simply holds the sampled signal during the holding periods. The pure time delay, T_d, approximates the acquisition time and the aperture time delays, whereas the filter is for the representation of the finite time constant and dynamics of the buffer amplifiers. In general, the transfer function of the filter can be expressed as

$$G_f(s) = \frac{\omega_n^2}{s^2 + 2\zeta\omega_n s + \omega_n^2}$$

The sampler shown in Fig. 2-15 is referred to as a *finite-pulsewidth sampler.* In practice, a majority of the sample-and-hold operations have a very small sampling duration p as compared with the sampling period T and the significant time constant of the input analog signal. The delay times of the S/H are also comparatively small so that these can also be neglected from the standpoint of the system dynamics. For example, the aperture time may be only 10 nanosec, and the acquisition time is 300 nanosec, of a given S/H. The settling time may be another 100 nanosec; i.e., the transient oscillations will decay to a prescribed accuracy within 100 nanosec. Thus, the total time delay is only 410 nanosec,

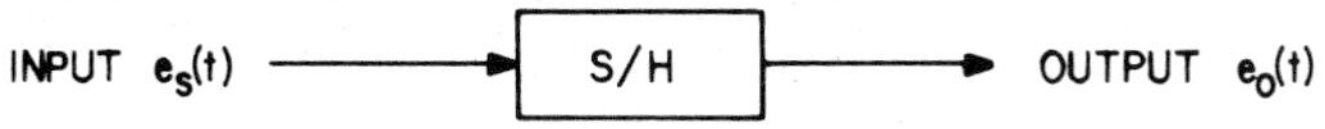

Fig. 2-14. A sample-and-hold.

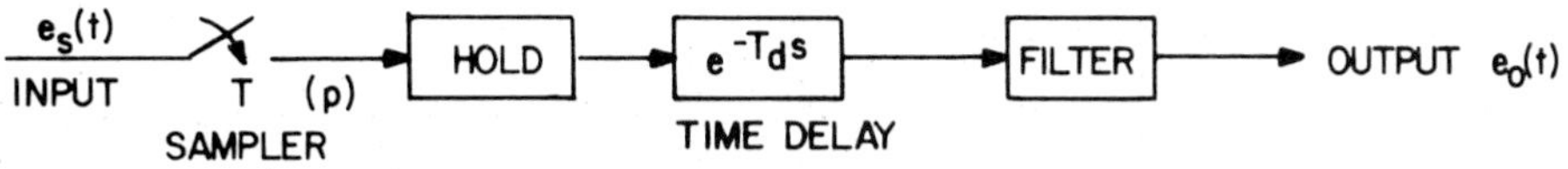

Fig. 2-15. A block diagram approximation of the sample-and-hold.

which may be neglected, since few control systems would respond to signals faster than this time frame.

Therefore, for all practical purposes, if $p \ll T$, and the time delay due to sample-and-hold is small, the S/H can be modelled by the block diagram shown in Fig. 2-16. In this case the sampler is called an *ideal sampler* since it is assumed to have a zero sampling duration; that is, $p = 0$. We shall show later that the ideal sampler leads to convenient mathematical modelling of the sample-and-hold operation. Figure 2-17 shows the input and output waveforms of an ideal sample-and-hold.

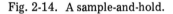

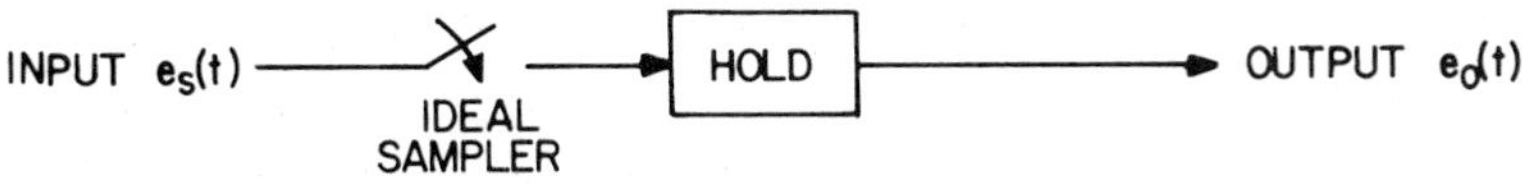

Fig. 2-16. An ideal sample-and-hold.

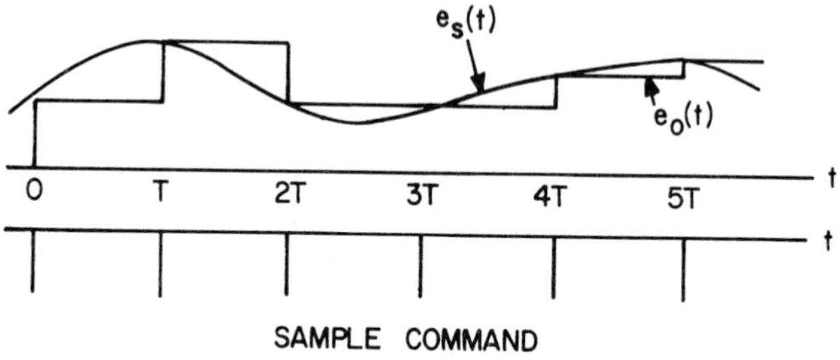

Fig. 2-17. Input and output signals of an ideal S/H.

2.5 DIGITAL-TO-ANALOG (D/A) CONVERSION

Digital-to-analog conversion, or simply decoding, consists of transforming the numerical information contained in a digitally coded word into an equivalent analog signal. The basic elements of a D/A are represented by the block diagram shown in Fig. 2-18. The function of the logic circuit is to control the switching of the precision reference voltage or current source to the proper input terminals of the resistor network as a function of the digital value of each digital input bit. Figure 2-19 illustrates a simple three-bit binary D/A. The values of the summing resistors of the operational amplifier are weighted in a binary fashion. Each of these resistors is connected through an electronic switch to the reference voltage or to ground. When a binary 1 appears at the control logic circuit of a switch it closes the switch and connects the resistor to the reference voltage. On the other hand, a binary 0 connects the resistor to ground. For the high-gain operational amplifier the input impedance is very low so that the voltage at the summing point 0 is practically zero. Then, if the switch of the MSB branch is connected to the reference voltage $-E_r$, and the other two switches are connected to ground, thus corresponding to a digital word of 100, the output voltage E_0 is

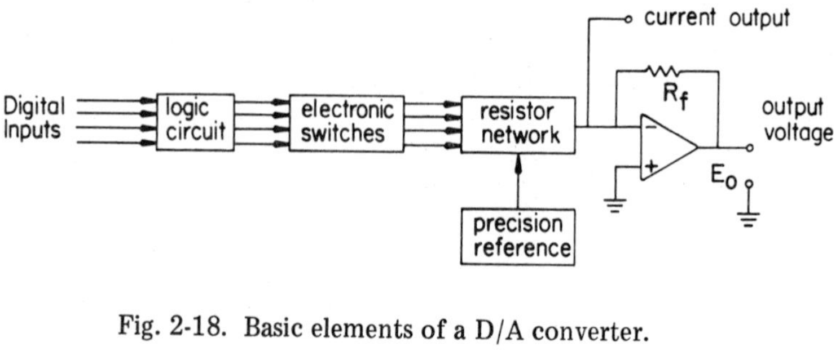

Fig. 2-18. Basic elements of a D/A converter.

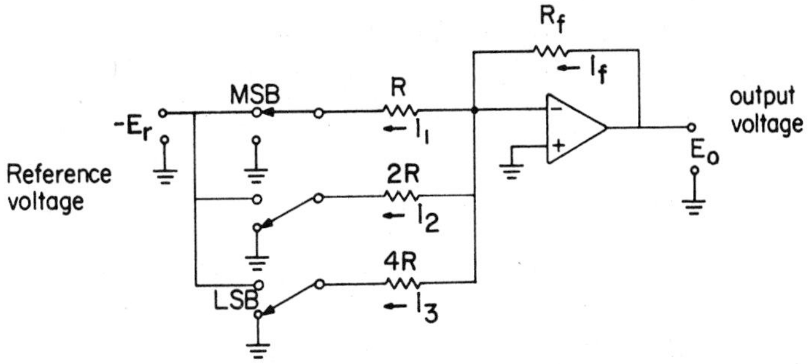

Fig. 2-19. A weighted-resistor three-bit D/A converter.

$$E_0 = R_f I_f \tag{2-11}$$

Since $I_f = I_1$ and $I_1 = E_r/R$, we have

$$E_0 = \frac{1}{R} R_f E_r \tag{2-12}$$

If the digital word 110 is to be converted, the switches in the MSB and the next significant branches are connected to the reference. The output voltage becomes

$$E_0 = \left[\frac{1}{R} + \frac{1}{2R}\right] R_f E_r = \frac{3}{2R} R_f E_r \tag{2-13}$$

The maximum three-bit word corresponds to the output voltage

$$E_0 = \left[\frac{1}{R} + \frac{1}{2R} + \frac{1}{4R}\right] R_f E_r$$

$$= \frac{7}{4R} R_f E_r \tag{2-14}$$

The LSB which corresponds to the digital word of 001 is $R_f E_r/4R$. Thus, full scale is $8R_f E_r/4R = 2R_f E_r/R$. Table 2-5 gives the output voltages of the D/A corresponding to all the three-bit binary words for a FS of 10 volts.

Extending to an n-bit binary word, the network in Fig. 2-19 should contain n parallel resistor branches. The resistor in the LSB branch has a value of $2^{n-1} R$. The output voltage is written as

$$E_0 = \left[\frac{a_0}{R} + \frac{a_1}{2R} + \ldots + \frac{a_{n-1}}{2^{n-1} R}\right] R_f E_r \tag{2-15}$$

where a_0, a_1, ..., a_{n-1} are either 1 or 0 depending upon the digital binary word

TABLE 2-5. OUTPUT VOLTAGES OF THE D/A OF FIG. 2-19 CORRESPONDING TO ALL DIGITAL INPUTS

Digital Word	Output Voltage E_0	Fraction of FS	Output Voltage for FS = 10 V
0 0 1	$\frac{R_f E_r}{4R}$	$\frac{FS}{8} = LSB$	1.25
0 1 0	$\frac{R_f E_r}{2R}$	$\frac{FS}{4}$	2.50
0 1 1	$\frac{3R_f E_r}{4R}$	$\frac{3}{8}$ FS	3.75
1 0 0	$\frac{R_f E_r}{R}$	$\frac{1}{2}$ FS	5.00
1 0 1	$\frac{5R_f E_r}{4R}$	$\frac{5}{8}$ FS	6.25
1 1 0	$\frac{6R_f E_r}{4R}$	$\frac{3}{4}$ FS	7.50
1 1 1	$\frac{7R_f E_r}{4R}$	$FS - LSB = \frac{7}{8}$ FS	8.75

which is to be converted.

The main disadvantage of the weighted-resistor type of D/A shown in Fig. 2-19 is that the accuracy and stability of the conversion depend on the accuracy and temperature stability of the resistors. Since the resistor values range from R to $2^{n-1}R$, for high resolution, the range between the largest and the smallest resistance values becomes quite large so that it is difficult to maintain accurate ratios between all the resistors.

A more practical D/A is the R – 2R ladder network. Figure 2-20 illustrates a D/A of this type for a three-bit digital word. Since all resistors have values of either R or 2R, they are easy to match and select. The network shown in Fig. 2-20 is also capable of giving a current output I_0 or a voltage output E_0.

The switch positions in Fig. 2-20 correspond to the digital word 100. The network can be simplified to the equivalent one as shown in Fig. 2-21. Since the input terminal to the operational amplifier is practically at ground potential, the currents in the resistor branches are easily calculated and are indicated in the circuit. The output current at the input to the operational amplifier is

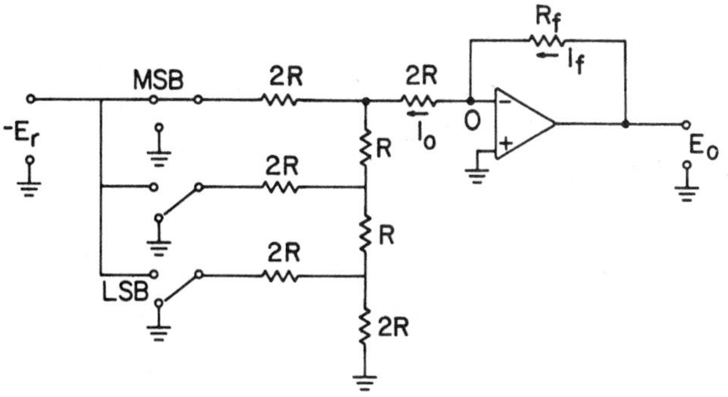

Fig. 2-20. An R-2R ladder D/A.

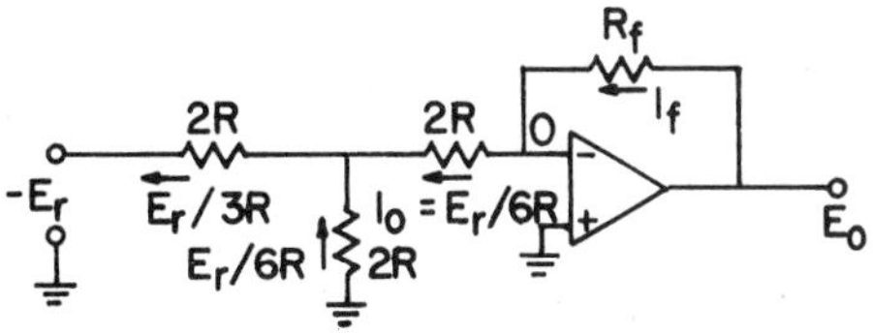

Fig. 2-21. Equivalent circuit of Fig. 2-20.

$$I_0 = \frac{E_r}{6R} \tag{2-16}$$

The output voltage is taken at the output of the operational amplifier, and is

$$E_0 = \frac{E_r R_f}{6R} \tag{2-17}$$

Now consider that the next significant-bit switch is closed with the resistor connected to the reference voltage, while all other terminals are grounded; this corresponds to the digital word of 010. The simplified equivalent circuit of the D/A is shown in Fig. 2-22. Again, because of the symmetry of the circuit, the currents in the branches are easily calculated and are indicated on the circuit of Fig. 2-22. In this case, the output current is

$$I_0 = \frac{E_r}{12R} \tag{2-18}$$

and the output voltage is

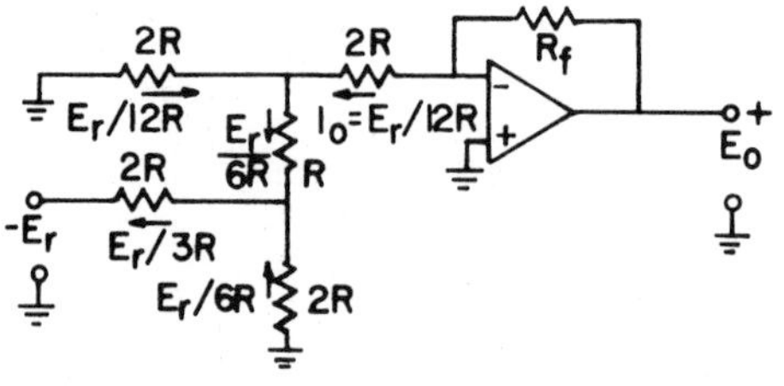

Fig. 2-22. Equivalent circuit of D/A for decoding the digital word 010.

$$E_0 = \frac{E_r R_f}{12R} \tag{2-19}$$

Using the same method we can show that for the digital word 001,

$$I_0 = \frac{E_r}{24R} \quad \text{(LSB)} \tag{2-20}$$

and

$$E_0 = \frac{E_r R_f}{24R} \quad \text{(LSB)} \tag{2-21}$$

In general, for the three-bit conversion,

$$I_0 = \frac{1}{3}\left[\frac{a_1}{2} + \frac{a_2}{2^2} + \frac{a_3}{2^3}\right]\frac{E_r}{R} \tag{2-22}$$

where a_1, a_2 and a_3 assume values of 1 or 0, depending upon the digital word to be converted.

For an n-bit digital word the R−2R D/A contains n parallel branches each with a resistor of value 2R. The output current is given by

$$I_0 = \frac{1}{3}\left[\frac{a_1}{2} + \frac{a_2}{2^2} + \ldots + \frac{a_n}{2^n}\right]\frac{E_r}{R} \tag{2-23}$$

Since a D/A essentially converts a digital signal into an analog signal of corresponding magnitude, from a functional standpoint, it may be regarded as a device which consists of a decoder and a S/H unit, as shown in Fig. 2-23. The decoder decodes the digital word into a number or an amplitude-modulated pulse. In reality, the sampler is redundant in the functional representation of Fig. 2-23. However, since the S/H is usually considered as one unit, the sampling operation is included even though it is not necessary. The transfer relation of the decoder is simply a constant gain, and in the ideal case this gain is unity.

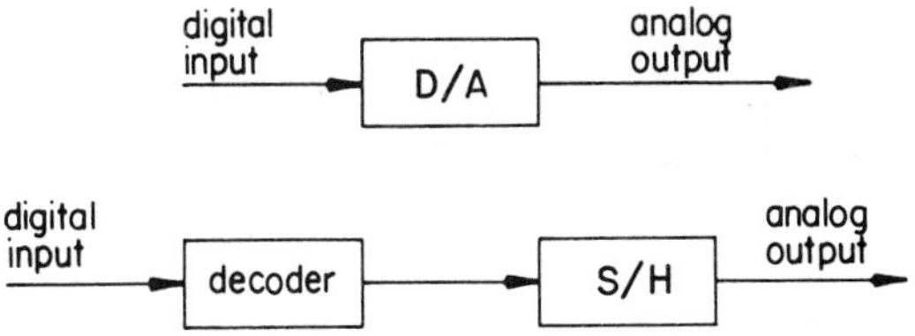

Fig. 2-23. Block diagram representation of a D/A.

It is interesting to note that the operational amplifier at the output of the D/A is capable of generating oscillations and spikes in the transient output, and in practice a S/H may be used to eliminate these spikes. Thus, the use of the S/H unit in Fig. 2-23 as a functional representation actually has a realistic justification.

2.6 ANALOG-TO-DIGITAL (A/D) CONVERSION

Analog-to-digital conversion, or simply encoding, consists of transforming the numerical information contained in an analog signal into a digital-coded word. The A/D conversion process is a more complex process than the D/A conversion discussed earlier, and requires more elaborate circuitry. In comparison to the D/A, the A/D converter (A/D) is generally more expensive and has slower response for the same conversion accuracy.

When a number is given as an input to an A/D, the converter performs the operations of *quantizing* and *encoding*. When a time-varying signal (voltage or current) is to be converted from analog to digital, the A/D usually performs the following sequential operations: **sample-and-hold, quantization** and **encoding.**

The sampling operation is needed to sample the analog signal at certain periodic intervals. Theoretically, the holding operation is not needed; however, the conversion time of an A/D is not zero, in order to reduce the effect of signal variation during conversion. The sampled signal is held until the conversion is completed. Figure 2-24 gives the block diagram representations of an A/D.

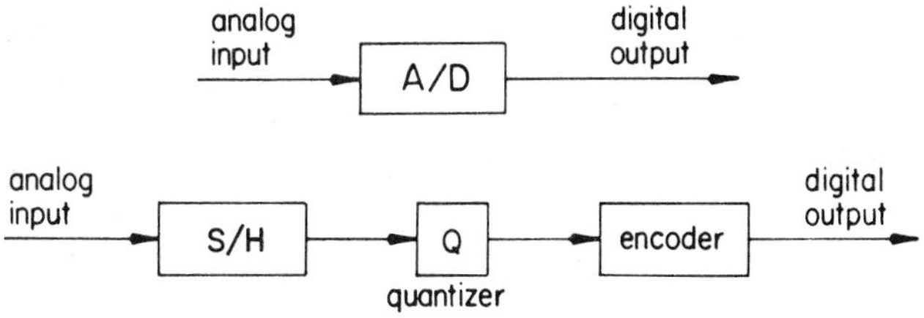

Fig. 2-24. Block diagram representation of an A/D.

The input to an A/D is usually in the form of a voltage or a current which is quantized during the conversion process. The ideal input-output relationship for a three-bit natural binary A/D is essentially identical to the quantization relation shown in Fig. 2-8. While all the input values comprise a continuum, the output is partitioned into 8 (2^3) discrete ranges. Thus, as Fig. 2-8 shows, there is an inherent ±1/2 LSB quantization uncertainty, in addition to the conversion errors which may exist in the process. If greater resolution is needed, the number of bits in the output signal should be increased. However, this will also increase the circuit complexity and possibly the total time of conversion.

Although there are a large number of A/D circuits available, only a few types are suitable for efficient and compact commercial units. The most commonly used A/D falls under the following types:

1. Successive-approximation
2. Integration (single and dual slope) type
3. Counter or servo type
4. Parallel type.

Each of the A/D types listed above has its own advantages and limitations. Each type is useful for a specific class of applications based on speed of conversion, cost, accuracy and size. Commercial A/D units are available from 6 bits to 16 bits. Typical conversion accuracy is 0.01% FS ± 1/2 LSB.

In order to illustrate the basic A/D conversion process, the successive-approximation type of A/D is briefly described in the following. Figure 2-25 shows the simplified block diagram of a successive-approximation type A/D. Basically, it consists of a comparator, a D/A, and some associated control logic. At the start of the conversion, all the bits of the output of the A/D are set to zero ("clearing"), and the MSB is then set to 1. The MSB, which represents one-half of full scale, is then converted by the D/A internally and compared with the analog input. If the input is greater than the converted MSB, the MSB = 1 is left on; otherwise, it is set to 0. The next significant bit is then turned on and the comparison is repeated. This process is continued until the LSB has been compared and set, then a status line indicates that comparison is completed and the digital output is available for transmission. A typical timing diagram for an A/D of the successive-approximation type is shown in Fig. 2-26.

The conversion time of an A/D acts as a time delay which is well-known to have adverse effects on the stability of closed-loop systems. In addition, the conversion time limits the frequency capability of the input signal. In general, the conversion time depends upon the resolution of the A/D and the conversion method used. Typical conversion times of commercial A/Ds range from 100 nsec to 200 μsec.

In the trivial case when the input analog signal is constant, the conversion time of the A/D is unimportant, since the input signal does not vary as it is

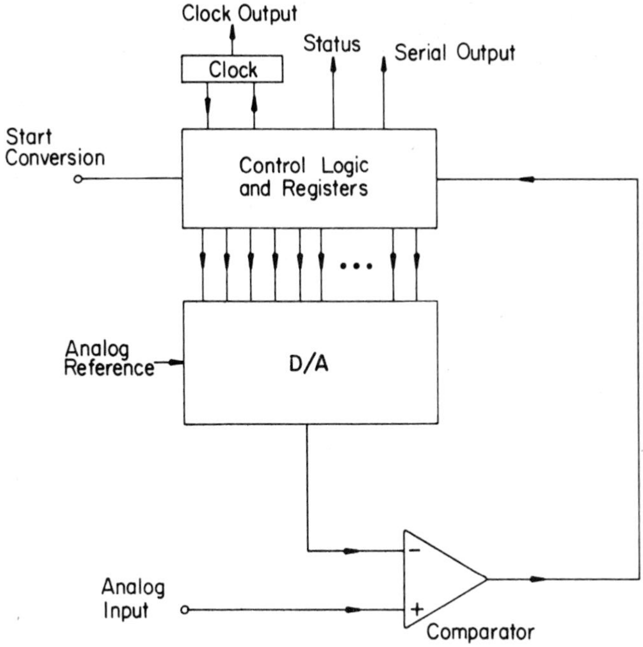

Fig. 2-25. Simplified block diagram of a typical successive-approximation A/D.

compared with the various bits of the A/D. However, in practice, the input signal usually varies with time; as mentioned earlier, the analog signal is first operated on by a S/H, which samples and holds the input until the conversion is completed.

To illustrate the implications of the conversion time, let us consider the analog signal shown in Fig. 2-27. The objective of using the A/D is to convert the value V_s into a digital number. However, if the time required to make a measurement or conversion is T_c^*, there is a corresponding uncertainty in the measured amplitude if the signal varies over the interval T_c. Let us assume that the full-scale value of the input signal shown in Fig. 2-27 is denoted by V_{FS}, and the A/D is an n-bit binary converter. This means that the signal is to be digitized to n bits resolution, or a resolution of 1 part in 2^n. The amplitude change of the signal over the conversion time T_c can be approximated by

$$\Delta V \cong \left. \frac{de_s(t)}{dt} \right|_{t=t_s} \times T_c \qquad (2\text{-}24)$$

*Strictly speaking, the conversion time T_c should include the time of conversion of the A/D, the aperture time, the acquisition time and the settling time of the S/H.

For a resolution of 2^{-n},

$$\frac{\Delta V}{V_{FS}} \leqslant 2^{-n} \tag{2-25}$$

Thus, using Eqs. (2-24) and (2-25), we have

$$\left.\frac{de_s(t)}{dt}\right|_{t=t_s} \leqslant 2^{-n} \frac{V_{FS}}{T_c} \tag{2-26}$$

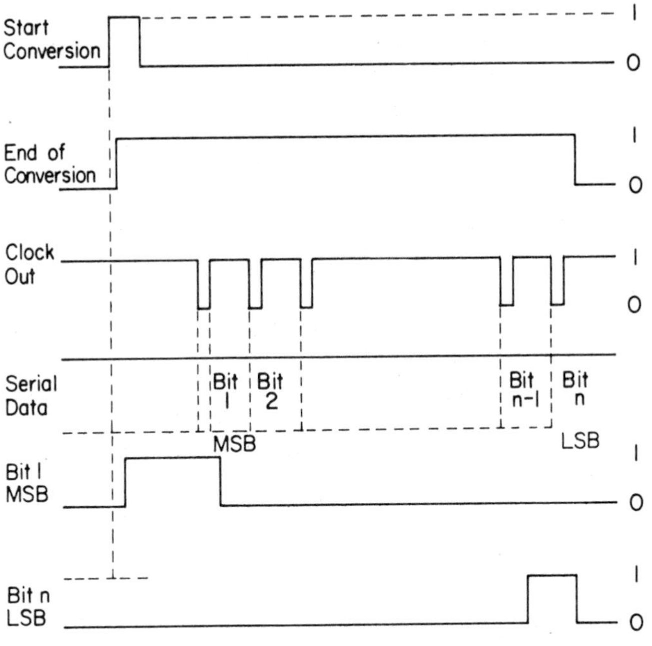

Fig. 2-26. A typical timing diagram of a successive-approximation A/D.

The right-hand side of Eq. (2-26) represents the upper limit on the rate of change of the input signal over any conversion period, so that the resolution of the conversion can be kept at 1 part in 2^n. The reason for this is that the digital output can thus represent any point of the input signal during the conversion period.

In practice we may measure an input signal by its highest frequency component of a sine wave. As an illustrative example, let us assume that the conversion time of a ten-bit binary A/D is 1 μsec. Representing $e_s(t)$ as

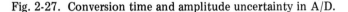

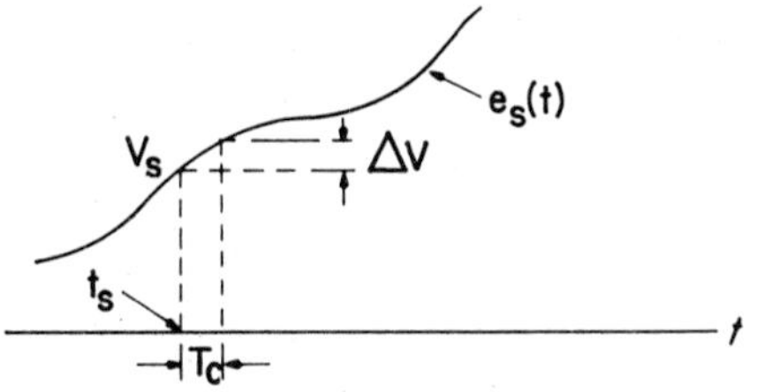

Fig. 2-27. Conversion time and amplitude uncertainty in A/D.

$$e_s(t) = \frac{V_{FS}}{2} \sin \omega t \tag{2-27}$$

we have from using Eq. (2-26),

$$V_{FS}\omega \cos\omega t\Big|_{t=0} \leqslant 2 \times 2^{-10} \frac{V_{FS}}{T_c} \tag{2-28}$$

In the last expression we have chosen the sampling instant to be at t = 0, since this gives the largest value for $\cos\omega t$ (or the steepest slope for $\sin\omega t$). Thus,

$$\omega \leqslant 2^{-9}/10^{-6} \cong 1953 \text{ rad/sec} \tag{2-29}$$

or the input signal should not vary at a rate in excess of 311 Hz. So, although the converter is capable of converting at a rate of 10^6 conversions per second, it cannot accurately encode signals whose frequency content is greater than 311 Hz. In general, the maximum frequency is given by

$$\omega_{max} = \frac{1}{2^{n-1} T_c} \tag{2-30}$$

where n is the number of bits of the conversion. Equation (2-30) shows that ω_{max} is inversely proportional to the conversion time T_c. As the number of bits (resolution) increases, the maximum allowable frequency decreases geometrically. Table 2-6 shows the frequency limitations for various combinations of word lengths and conversion times T_c.

Sampling Period Considerations

It is apparent from the preceding discussion that the conversion time T_c plays a dominant role in the effect of an A/D on system performance. While the effect of the conversion time cannot be significantly reduced except through the use of a faster converter, the restriction on the maximum input frequency can be relaxed by using a S/H at the input of the A/D. The function of the S/H is to sample the input signal and then hold the value of the signal

TABLE 2-6. FREQUENCY LIMITATION OF A/D DUE TO CONVERSION TIME T_c

Word Length n bits / Conversion Time T_c (sec)	4	6	8	10	12	14	16
	Maximum Frequency (rad/sec)						
10^{-3} (1 msec)	125	31.25	7.8125	1.953	0.488	0.122	0.0305
10^{-4} (100 μsec)	1250	312.5	78.125	19.53	4.88	1.22	0.305
10^{-5} (10 μsec)	12500	3125	781.25	195.3	48.8	12.2	3.05
10^{-6} (1 μsec)	125000	31250	7812.5	1953	488	122	30.5
10^{-7} (100 nsec)	1250000	312500	78125	19530	4880	1220	305

during the conversion period. In practice, the designer of a digital system is always confronted with the question: How fast should the sampling frequency be? Let us designate the sampling frequency as f_s in Hz or ω_s in rad/sec. The sampling frequency is related to the sampling period T (sec) through

$$f_s = 1/T \quad \text{(Hz)}$$

$$\omega_s = 2\pi f_s = 2\pi/T \quad \text{(rad/sec)}$$

In general, the sampling frequency of a S/H depends on many signal processing and system performance factors. In this section we can only address to the factors that are related to signal processing.

The minimum sampling period of a S/H is bounded by the conversion time of the A/D and the delay times of the S/H. For example, if the total conversion time of a 10-bit S/H and A/D combination is 1μsec, the minimum sampling period is also 1μsec. The corresponding maximum sampling frequency is 1 MHz. However, in practice, the minimum sampling period or the maximum sampling frequency is also governed by the characteristics of other system components as well as how fast the digital data can be processed. If multiplexing is involved or if the data are to be processed by a microprocessor, the S/H and A/D are not the only limiting factors on the sampling frequency. Since the microprocessor is a rather slow digital computer, data can be processed only at a certain maximum rate. Thus, in discrete-data and digital control systems the maximum sampling frequency is seldom limited by the S/H and the A/D.

In the other extreme, we can easily see that there is a lower limit on how slow the sampling frequency can be. It is apparent that the S/H should sample at a sufficiently fast rate so that the information contained in the input signal will not be lost through the sample-and-hold operation. We can use a sinusoidal signal to illustrate the point. Suppose we sample a 1 MHz sine wave only once every 1 μsec. If the sampling instants always occur while the sine wave goes through zero, the resulting sampled data would erroneously represent an input signal of zero volt. In fact, the sampled output would be zero if the sampling period is exactly 0.5 μsec with the sampling instants at the zero crossings of the sine wave. Therefore, to represent the sine wave characteristics accurately we have to sample at a frequency which is greater than twice the frequency of the input signal. In general, given an input signal whose highest frequency component is ω_h (rad/sec), in order to retain vital information through sampling, the minimum sampling frequency should be greater than $2\omega_h$. This criterion on minimum sampling frequency from the signal processing standpoint is presented in the next section as the SAMPLING THEOREM. For example, in the illustration given earlier on the 10-bit S/H and A/D, the total delay time is 1μsec. If we use a sampling frequency of 1 MHz (the theoretical maximum sampling rate), the input signal cannot contain frequency components in excess of 0.5 MHz.

In later chapters we will show that the stability of a closed-loop digital control system is closely related to the sampling period. In the majority cases low sampling periods have a detrimental effect on closed-loop stability. Therefore, the selection of a proper sampling period for a closed-loop digital control system is frequently made with stability implications.

Simplified Block Diagram Representations of A/D and D/A

It is interesting to compare the block diagram representations of the D/A (Fig. 2-23) and the A/D (Fig. 2-24). If the resolution of the A/D is very high, the nonlinear effect of the quantization can be neglected, and since the decoder and the encoder transfer relations can all be represented by constant gains, the two block diagrams essentially all reduce to a S/H. This is indeed the case in the analytical studies of digital systems. For instance, for analytical purposes, the digital autopilot system shown in Fig. 1-2 is represented by the block diagram of Fig. 2-28, where the A/D and the D/A are all replaced by S/H devices. In this case, the digital controller is represented by a discrete transfer function D(z) and the analog elements are represented by their respective analog transfer functions. The transfer relations of the S/H will be derived in Sec. 2-11.

2.7 MATHEMATICAL TREATMENT OF THE SAMPLING PROCESS

Since sample-and-hold is an important component in digital and sampled-data

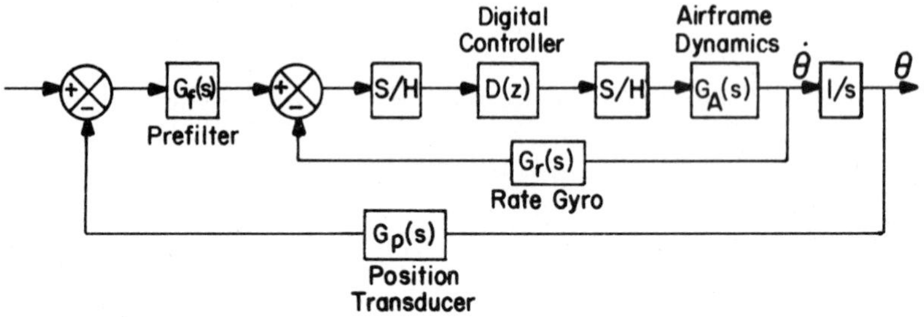

Fig. 2-28. An analytical block diagram of the digital autopilot in Fig. 1-8.

control systems, it is essential to establish realistic and yet mathematically simple models for analytical purposes. Although the hardware of a sample-and-hold device is usually available in one integrated unit, mathematically it is more convenient to model the sampling and the holding operations separately.

Sampling Process Modelled As a Pulse-Amplitude Modulator

In general, the operation of a sampler may be regarded as one which converts an analog or continuous-time signal into a pulse-modulated or digital signal. The most common type of modulation, as discussed in the sample-and-hold operation in the preceding sections, is the pulse-amplitude modulation (PAM). Figure 2-29 shows the block diagram representation of a periodic sampler with finite sampling duration. The pulse or sampling duration is p, and the sampling period is T. Of course, in practice the sampling duration p, or the time interval during which the sampling switch is closed (tracks) is finite.

For an input signal f(t) which is a function of the continuous-time variable t, the output of the sampler, designated as $f_p^*(t)$, is a train of finite-width pulses whose amplitudes are modulated by the input f(t). Figure 2-30 shows an equivalent block diagram representation of the sampler as a pulse-amplitude modulator. In this case, the input f(t) is considered to be multiplied by a carrier signal p(t) which is a train of periodic pulses, each with unit height. Figure 2-31 illustrates the typical waveforms of the input signal f(t), the carrier signal p(t), and the output $f_p^*(t)$.

In general, the sampling scheme of a sampler used in digital control sys-

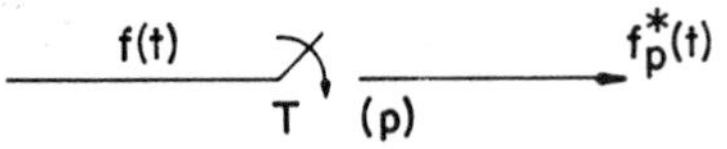

Fig. 2-29. A uniform-rate sampler with finite sampling duration.

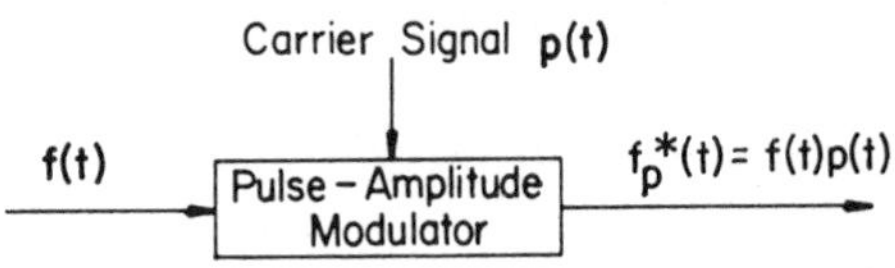

Fig. 2-30. Pulse-amplitude modulator as a sampler.

tems may take a great variety of forms. For instance, a sampler may have a nonuniform or cyclic-variable sampling rate. It is also common to find digital control systems which contain a multiple number of samplers with different sampling rates, as shown in the example of Fig. 1-3. This type of system is generally referred to as a MULTIRATE SAMPLING SYSTEM. The sampling rates of the multirate samplers may or may not be integrally related, and they may not be synchronized, that is, start at the same instant of time.

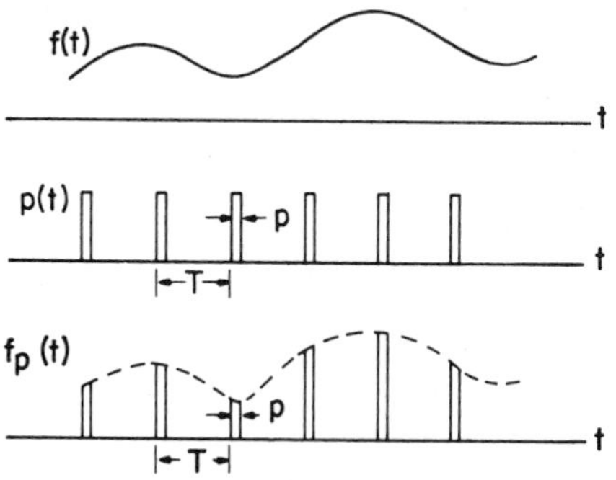

Fig. 2-31. Typical input and output waveforms of a uniform-rate sampler.

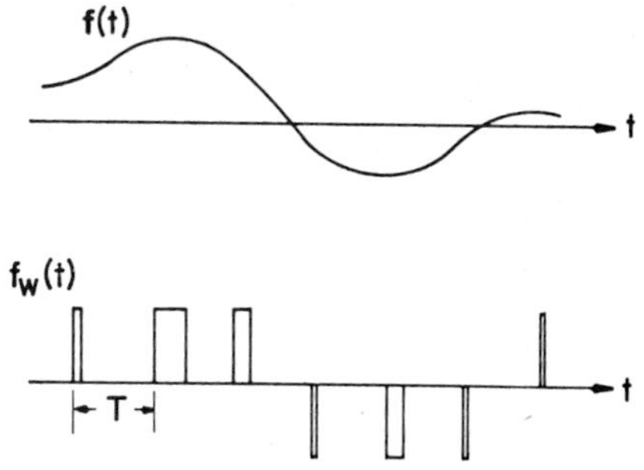

Fig. 2-32. Typical input and output signals of a pulsewidth modulator.

In certain inherent sampling systems the sampling operation may be entirely random; i.e., the time interval between successive samples may be thought of as to follow some random scheme.

The advantage of using the pulse-amplitude modulation representation as shown in Fig. 2-30 is apparent, since for the various sampling operations only the carrier signal p(t) needs to be changed.

There are sampling operations which may be described by a pulsewidth modulation (PWM) sampler. This type of sampler typically has outputs which are pulses with the pulsewidths varying as functions of the amplitudes of the inputs at the sampling instants. In this case, the magnitudes of the output pulses are constant. Typical input and output signals of a pulsewidth modulator are illustrated in Fig. 2-32. A still more elaborate scheme is one in which the amplitudes as well as the widths of the output pulses vary as some function of the input signal at the sampling instants. This type of sampler is known as a pulsewidth-pulse-amplitude modulator. In certain cases these more complex types of sampling may be used for improving the performance of digital control systems.

Only the mathematical treatment and modelling of the uniform-rate sampling operation are discussed in this chapter. However, once we have established the input-output transfer relation of the uniform-rate sampler, the method can be easily extended to certain types of nonuniform-rate samplers.

Input-Output Transfer Relations of Finite-Pulsewidth Samplers

The pulse-amplitude modulation concept introduced in the last section is useful for the mathematical analysis of the sampling operation. A sampler that can be described by a pulse-amplitude modulation process, such as that shown in Fig. 2-30 and Fig. 2-31, is known to be a *linear* device. The reason for this is because the sampler satisfies the principle of superposition. However, the analog-to-digital-conversion nature of the sampling operation precludes the possibility of arriving at a bonafide transfer function. Nevertheless, since the sampler considered here is linear, we can use Fourier transform and Laplace transform to arrive at the input-output transfer relation of the device.

The sampler considered here will also have a finite pulsewidth, as in the practical situation. As our development will show, the mathematics involved in considering finite pulsewidth in a digital control system can be quite complex. We shall show later that when sample-and-hold is considered, the sampler can be assumed to have a zero sampling duration, or zero pulsewidth, thus simplifying the mathematics considerably. However, for now we shall treat the general case so that a better understanding of all the ramifications of sampling can be achieved.

For the uniform-rate finite-pulsewidth sampler described in Figs. 2-29, 2-30 and 2-31, the output signal in response to an input f(t) can be regarded as the product of f(t) and the carrier p(t) which is a unit pulse train with period T. The unit pulse train p(t) contains pulses each with unit magnitude. Thus, p(t) can be expressed as

$$p(t) = \sum_{k=-\infty}^{\infty} [u_s(t - kT) - u_s(t - kT - p)] \qquad (p < T) \tag{2-31}$$

where $u_s(t)$ is the unit step function,

$$\begin{aligned} u_s(t) &= 0 \qquad t < 0 \\ u_s(t) &= 1 \qquad t > 0 \end{aligned} \tag{2-32}$$

In the present case we assume that the sampling operation begins at $t = -\infty$, and the leading edge of the pulse at $t = 0$ coincides with $t = 0$, as shown in Fig. 2-33. The output of the sampler is written as

$$f_p^*(t) = f(t)p(t) \tag{2-33}$$

Substituting Eq. (2-31) into Eq. (2-33), we get

$$f_p^*(t) = f(t) \sum_{k=-\infty}^{\infty} [u_s(t - kT) - u_s(t - kT - p)] \qquad (p < T) \tag{2-34}$$

Equation (2-34) gives a time-domain description of the input-output relation of the uniform-rate finite-pulsewidth sampler.

It is of interest to investigate the frequency-domain characteristics of the sampler output. Since, being a pulse train, $f_p^*(t)$ will generally contain much higher frequency components than f(t). Thus, the sampler may be regarded as a *harmonic generator*.

Since the unit pulse train p(t) is a periodic function with period T, it can be represented by a Fourier series,

$$p(t) = \sum_{n=-\infty}^{\infty} C_n e^{jn\omega_s t} \tag{2-35}$$

where ω_s is the sampling frequency in radians per second, and is related to the sampling period T by

$$\omega_s = 2\pi/T \tag{2-36}$$

C_n denotes the complex Fourier series coefficients, and is given by

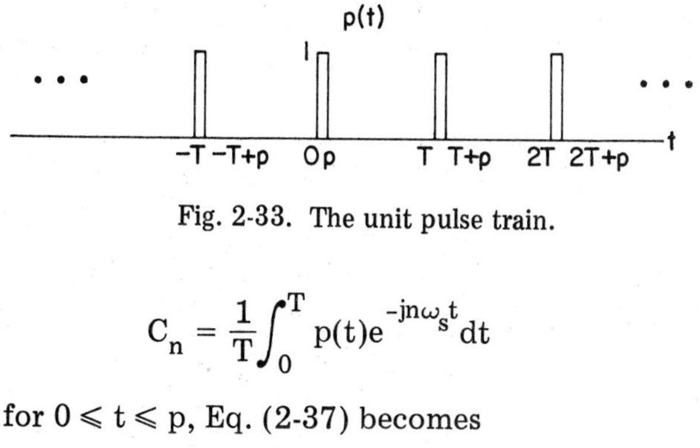

Fig. 2-33. The unit pulse train.

$$C_n = \frac{1}{T}\int_0^T p(t)e^{-jn\omega_s t}\,dt \tag{2-37}$$

Since p(t) = 1 for $0 \leqslant t \leqslant p$, Eq. (2-37) becomes

$$C_n = \frac{1}{T}\int_0^p e^{-jn\omega_s t}\,dt$$

$$= \frac{1 - e^{-jn\omega_s p}}{jn\omega_s T} \tag{2-38}$$

Using well-known trigonometric identities, C_n is written as

$$C_n = \frac{p}{T}\frac{\sin(n\omega_s p/2)}{(n\omega_s p/2)} e^{-jn\omega_s p/2} \tag{2-39}$$

Substitution of Eq. (2-39) into Eq. (2-35), we have

$$p(t) = \frac{p}{T} \sum_{n=-\infty}^{\infty} \frac{\sin(n\omega_s p/2)}{n\omega_s p/2} e^{-jn\omega_s p/2} e^{jn\omega_s t} \tag{2-40}$$

Substitution of p(t) from Eq. (2-40) into Eq. (2-33) gives

$$f_p^*(t) = \sum_{n=-\infty}^{\infty} C_n f(t) e^{jn\omega_s t} \tag{2-41}$$

where C_n is given by Eq. (2-39).

The Fourier transform of $f_p^*(t)$ is obtained as

$$F_p^*(j\omega) = \mathcal{F}[f_p^*(t)] = \int_{-\infty}^{\infty} f_p^*(t) e^{-j\omega t} dt \tag{2-42}$$

where the script letter $\mathcal{F}$ represents the Fourier transform operation. Using the complex shifting theorem of the Fourier transform which states that

$$\mathcal{F}[e^{jn\omega_s t} f(t)] = F(j\omega - jn\omega_s) \tag{2-43}$$

Eq. (2-42) is written as

$$F_p^*(j\omega) = \sum_{n=-\infty}^{\infty} C_n F(j\omega - jn\omega_s) \tag{2-44}$$

Equation (2-44) can also be written as

$$F_p^*(j\omega) = \sum_{n=-\infty}^{\infty} C_n F(j\omega + jn\omega_s) \tag{2-45}$$

since n extends from $-\infty$ to $+\infty$.

The significance of the sampling operation as viewed from the frequency domain is explained by the following development of Eq. (2-45).

First, taking the limit as $n \to 0$ of the Fourier coefficient in Eq. (2-39), we have

$$C_0 = \lim_{n \to 0} C_n = \frac{p}{T} \tag{2-46}$$

Taking only the $n = 0$ term in Eq. (2-45), we have

$$F_p^*(j\omega)\Big|_{n=0} = C_0 F(j\omega) = \frac{p}{T} F(j\omega) \tag{2-47}$$

The last equation shows the important fact that the frequency components contained in the original continuous-time input, f(t), are still present in the sampler output, $f_p^*(t)$, except that the amplitude is multiplied by the factor p/T.

For $n \neq 0$, C_n is a complex quantity, but the magnitude of C_n may be written as

$$|C_n| = \frac{p}{T}\left|\frac{\sin(n\omega_s p/2)}{n\omega_s p/2}\right| \tag{2-48}$$

The magnitude of $F_p^*(j\omega)$ is

$$|F_p^*(j\omega)| = \left|\sum_{n=-\infty}^{\infty} C_n F(j\omega + jn\omega_s)\right| \tag{2-49}$$

The frequency spectrum of the unit pulse train p(t) is simply the plot of the Fourier coefficient C_n as a function of ω, when n assumes various integral values between $-\infty$ and $+\infty$. The amplitude spectrum of C_n is shown in Fig. 2-34(a). We see that the amplitude spectrum of C_n is not a continuous function but rather a line spectrum at discrete intervals of ω, with $\omega = n\omega_s$, for $n = 0, \pm 1, \pm 2, \ldots$. The amplitude of the spectrum is described by the right-hand side of Eq. (2-49).

Equation (2-49) can also be written as

$$|F_p^*(j\omega)| \leqslant \sum_{n=-\infty}^{\infty} |C_n||F(j\omega + jn\omega_s)| \tag{2-50}$$

which may be used to illustrate the amplitude spectrum of $F_p^*(j\omega)$. If the amplitude spectrum of the continuous input f(t) is assumed to be as shown in Fig. 2-34(b), by use of Eq. (2-50), the spectrum of $F_p^*(j\omega)$ will be of the form shown in Fig. 2-34(c). It is interesting to note that $|F_p^*(j\omega)|$ contains not only the fundamental component of $F(j\omega)$ but also the harmonic or the complementary components, $F(j\omega + jn\omega_s)$, $n = \pm 1, \pm 2, \ldots$. The nth complementary component in the output spectrum is obtained by multiplying $|F(j\omega)|$ by its corresponding Fourier coefficient $|C_n|$ and shifting it by $n\omega_s$, $n = \pm 1, \pm 2, \pm 3, \ldots$. Therefore, the sampler may be visualized as a harmonic generator whose output contains the weighted fundamental components, plus all the weighted complementary components at all frequencies separated by the sampling frequency. The band around zero frequency still carries essentially all the information contained in the continuous input signal, but the same information also is repeated along the frequency axis; the amplitude of each component is weighted by the magnitude of its corresponding Fourier coefficient $|C_n|$. It should be pointed out that the frequency spectrum for $|F_p^*(j\omega)|$ shown in Fig. 2-34(c) is sketched with the assumption that the sampling frequency ω_s is greater than twice the highest

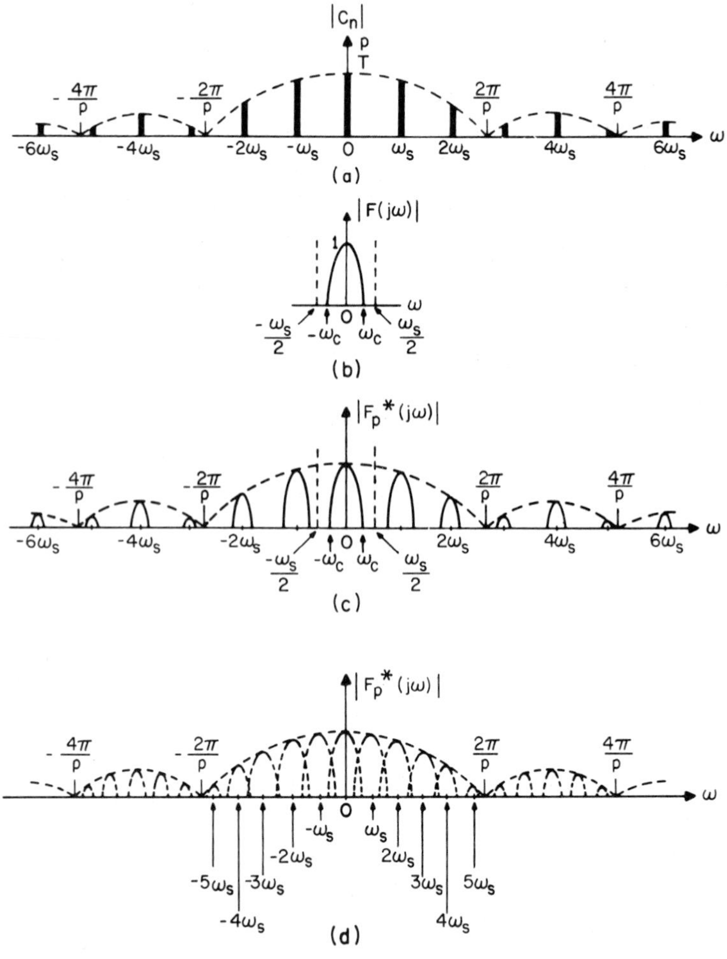

Fig. 2-34. Amplitude spectra of input and output signals of a finite-pulse-width sampler: (a) amplitude spectrum of unit pulse train p(t); (b) amplitude spectrum of continuous input signal f(t); (c) amplitude spectrum of sampler output ($\omega_s > 2\omega_c$); (d) amplitude spectrum of sampler output ($\omega_s < 2\omega_c$).

frequency contained in f(t); that is, $\omega_s > 2\omega_c$. If $\omega_s < 2\omega_c$, distortion in the frequency spectrum of $|F_p^*(j\omega)|$ will appear because of the overlapping of the harmonic components. Figure 2-34(d) shows that if $\omega_s < 2\omega_c$, the spectrum of $|F_p^*(j\omega)|$ around zero frequency bears little resemblance to that of the original signal. Therefore, theoretically the original signal can be recovered from the spectrum shown in Fig. 2-34(c) by means of an ideal low-pass filter, whereas the input signal cannot be recovered from the spectrum shown in Fig. 2-34(d), due to the overlapping of the complementary components.

The phenomenon of the overlapping of the high-frequency components with the fundamental component in the frequency spectrum of the sampled signal is sometimes called *folding*. The requirement that ω_s be at least twice as large as the highest frequency contained in the signal f(t) is formally known as the *sampling theorem*, which is stated more formally in Sec. 2.8. The frequency $\omega_s/2$ is sometimes known as the *folding frequency*. In reality, of course, most signals encountered in control systems do not have an abrupt cutoff in the response as shown in Fig. 2-34(b), so that some folding effect will exist even though the sampling frequency is made greater than the folding frequency. More on this subject and the physical implications of folding of frequency on the performance of digital control systems will be discussed later.

The Fourier transform of $f_p^*(t)$, $F_p^*(j\omega)$, given by Eq. (2-44) or Eq. (2-45) is useful in demonstrating the effects of sampling in the frequency domain. The expression shows that once the Fourier transform of the continuous-time signal f(t), $F(j\omega)$, is known, the Fourier transform of the sampled signal $f_p^*(t)$ is obtained by shifting $F(j\omega)$ by $jn\omega_s$ and weighing its amplitude by C_n. However, the expression in Eq. (2-45) for $F_p^*(j\omega)$ is not suitable for analytical purposes, since it is an infinite series.

An alternate description of the sampled signal $f_p^*(t)$ in the transform domain can be obtained by means of the complex-convolution method of the Laplace transform. With reference to Eq. (2-33), the Laplace transform of $f_p^*(t)$ is written

$$F_p^*(s) = \mathcal{L}[f_p^*(t)] = \mathcal{L}[f(t)p(t)] \tag{2-51}$$

where the script $\mathcal{L}$ denotes the Laplace transform operation.

Equation (2-51) is written as

$$F_p^*(s) = F(s) * P(s) \tag{2-52}$$

where the symbol "*" denotes complex convolution in Laplace transform, and F(s) and P(s) are the Laplace transforms of f(t) and p(t), respectively.

The Laplace transform of p(t) is determined first. We take the Laplace transform on both sides of Eq. (2-31), we get

$$P(s) = \sum_{k=0}^{\infty} \frac{1 - e^{-ps}}{s} e^{-kTs} \tag{2-53}$$

The summation on the right-hand side of Eq. (2-53) begins at k = 0, since the one-sided Laplace transform is defined for $0 \leq t < \infty$. The infinite series in Eq. (2-53) is written in closed form as

$$P(s) = \frac{1 - e^{-ps}}{s(1 - e^{-Ts})} \tag{2-54}$$

Substituting P(s) from the last equation into Eq. (2-52), we have

$$F_p^*(s) = F(s) * \frac{1 - e^{-ps}}{s(1 - e^{-Ts})} \tag{2-55}$$

It follows from the complex-convolution theorem of Laplace transform that

$$F_p^*(s) = \frac{1}{2\pi j}\int_{c-j\infty}^{c+j\infty} F(\xi)P(s - \xi)d\xi \tag{2-56}$$

Thus, using Eq. (2-54), Eq. (2-56) is written as

$$F_p^*(s) = \frac{1}{2\pi j}\int_{c-j\infty}^{c+j\infty} F(\xi)\frac{1 - e^{-p(s-\xi)}}{(s - \xi)[1 - e^{-T(s-\xi)}]}d\xi \tag{2-57}$$

where ξ is the variable of integration; $\sigma_1 < c < \sigma - \sigma_2$, $\sigma > \max(\sigma_1, \sigma_2, \sigma_1 + \sigma_2)$, and σ is the real part of s, and σ_1 and σ_2 are the abscissas of convergence of $F(\xi)$ and $P(\xi)$, respectively. The path of integration of the integral of Eq. (2-57) is along the straight line from $\xi = c - j\infty$ to $c + j\infty$ in the complex ξ-plane as shown in Fig. 2-35.

The convolution integral of Eq. (2-57) may be carried out by taking the contour integration along a path formed by the $\xi = c - j\infty$ to $c + j\infty$ line, and a semicircle of infinitely large radius going around either the right-half or the left-half ξ-plane. The contour integral is then evaluated by means of the residue theorem of complex variables. In other words, Eq. (2-57) may be written as

1.
$$F_p^*(s) = \frac{1}{2\pi j}\oint_{\Gamma_1} F(\xi)\frac{1 - e^{-p(s-\xi)}}{(s - \xi)[1 - e^{-T(s-\xi)}]}d\xi - \frac{1}{2\pi j}\oint F(\xi)\frac{1 - e^{-p(s-\xi)}}{(s - \xi)[1 - e^{-T(s-\xi)}]}d\xi \tag{2-58}$$

or

2.
$$F_p^*(s) = \frac{1}{2\pi j}\oint_{\Gamma_2} F(\xi)\frac{1 - e^{-p(s-\xi)}}{(s - \xi)[1 - e^{-T(s-\xi)}]}d\xi - \frac{1}{2\pi j}\oint F(\xi)\frac{1 - e^{-p(s-\xi)}}{(s - \xi)[1 - e^{-T(s-\xi)}]}d\xi \tag{2-59}$$

where Γ_1 is a closed path enclosing the entire finite left half of the ξ-plane and Γ_2 is a closed path enclosing the right half of the ξ-plane. The two paths are shown in Fig. 2-35.

In order to use the residue theorem, we have to investigate the poles and zeros of $F(\xi)$ and $P(s - \xi)$. Normally, the poles of $F(\xi)$ are considered to be in the left-half or on the imaginary axis of the ξ-plane and are finite in number, but $P(s - \xi)$ has simple poles at

$$\xi_n = s + \frac{2\pi nj}{T} = s + jn\omega_s \qquad -\infty < n \text{ (integer)} < \infty \tag{2-60}$$

where T is the sampling period and ω_s is the sampling frequency. Equation (2-60) shows that the poles of $P(s - \xi)$ are infinite in number and are located at frequency intervals of $n\omega_s$ ($n = 0, \pm 1, \pm 2, \ldots$) along the path of $\text{Re}(\xi) = \text{Re}(s)$ in the ξ-plane. Typical pole distributions of $F(\xi)$ and $P(s - \xi)$ are shown in Fig. 2-35. If the function $F(\xi)$ has at least one more pole than zero, that is,

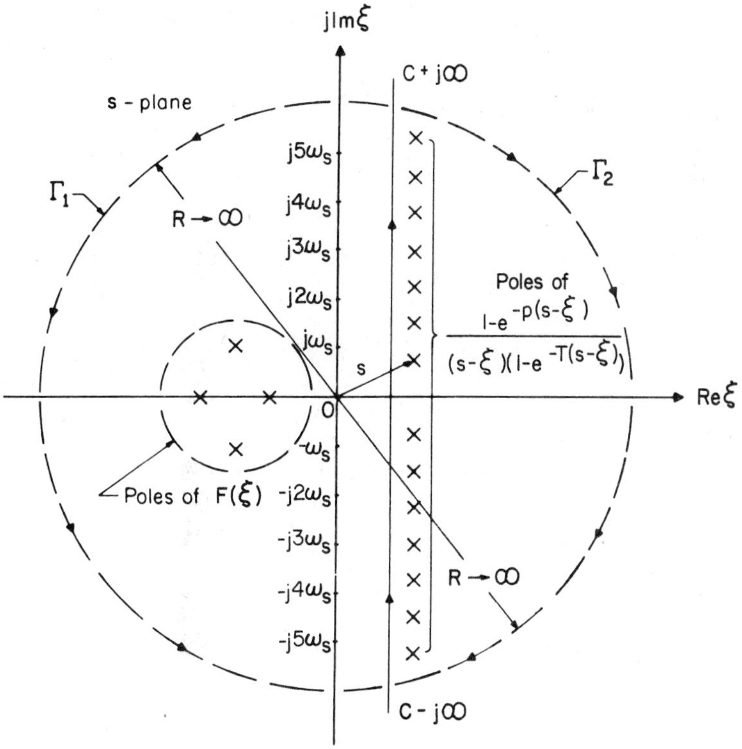

Fig. 2-35. Paths of contour integration in the right-half ξ-plane and the left-half ξ-plane.

$$\lim_{\xi \to \infty} F(\xi) = 0 \tag{2-61}$$

the second terms on the right side of Eqs. (2-58) and (2-59) are zero, since the integrals along the infinite-radius semicircles vanish. Equation (2-58) then becomes

$$F_p^*(s) = \frac{1}{2\pi j} \oint_{\Gamma_1} F(\xi) \frac{1 - e^{-p(s-\xi)}}{(s-\xi)[1 - e^{-T(s-\xi)}]} d\xi \tag{2-62}$$

From the residue theorem the contour integral of Eq. (2-62) is equal to the sum of the residues of the integrand evaluated at those poles enclosed by the closed path Γ_1. Therefore,

$$F_p^*(s) = \sum \text{Residues of } \left[F(\xi) \frac{1 - e^{-p(s-\xi)}}{(s-\xi)[1 - e^{-T(s-\xi)}]}\right] \text{at the poles of } F(\xi) \tag{2-63}$$

A more specific form can be arrived at for Eq. (2-63) if we assume that $F(\xi)$ is a rational function with k *simple* poles. Then Eq. (2-63) becomes

$$F_p^*(s) = \sum_{n=1}^{k} \frac{N(\xi_n)}{D'(\xi_n)} \frac{1 - e^{-p(s-\xi_n)}}{(s-\xi_n)\left[1 - e^{-T(s-\xi_n)}\right]} \tag{2-64}$$

where

$$F(\xi) = \frac{N(\xi)}{D(\xi)} \tag{2-65}$$

$$D'(\xi_n) = \left.\frac{dD(\xi)}{d\xi}\right|_{\xi=\xi_n} \tag{2-66}$$

and ξ_n denotes the nth pole of $F(\xi)$, $n = 1,2,\ldots,k$.

In general, let F(s) be a rational algebraic function having k distinct poles $s_1,s_2,\ldots,s_k$, with s_n of multiplicity m_n, $n = 1,2,\ldots,k$. $m_n \geqslant 1$. Then

$$F_p^*(s) = F(s) * P(s)$$

$$= \left.\sum_{n=1}^{k} \sum_{i=1}^{m_n} \frac{(-1)^{m_n-i} K_{ni}}{(m_n - i)!} \frac{\partial^{m_n-i}}{\partial s^{m_n-i}} P(s)\right|_{s=s-s_n} \tag{2-67}$$

where

$$K_{ni} = \left.\frac{1}{(i-1)!} \frac{\partial^{i-1}}{\partial s^{i-1}} \left[(s - s_n)^{m_n} F(s)\right]\right|_{s=s_n} \tag{2-68}$$

Now consider Eq. (2-59), which is the contour integral along the path Γ_2. If the condition in Eq. (2-61) is satisfied, Eq. (2-59) is written

$$F_p^*(s) = \frac{1}{2\pi j} \oint_{\Gamma_2} F(\xi) \frac{1 - e^{-p(s-\xi)}}{(s-\xi)[1 - e^{-T(s-\xi)}]} d\xi \qquad (2\text{-}69)$$

Applying the residue theorem, Eq. (2-69) is written

$$F_p^*(s) = -\sum \text{Residues of } F(\xi) \frac{1 - e^{-p(s-\xi)}}{(s-\xi)[1 - e^{-T(s-\xi)}]}$$

$$\text{at the poles of } \frac{1 - e^{-p(s-\xi)}}{(s-\xi)[1 - e^{-T(s-\xi)}]} \qquad (2\text{-}70)$$

where the minus sign comes from the fact that the contour integration along Γ_2 is taken in the clockwise direction. Since $P(s - \xi)$ has only simple poles that lie at periodic intervals along $\text{Re}(\xi) = \text{Re}(s)$ in the ξ-plane, Eq. (2-70) becomes

$$F_p^*(s) = -\sum_{n=-\infty}^{\infty} \frac{N(\xi_n)}{D'(\xi_n)} F(\xi_n) \qquad (2\text{-}71)$$

where $\xi_n = s + jn\omega_s$ denotes the poles of $P(s - \xi)$, $n = 0, \pm 1, \pm 2, \ldots$. Also, in this case,

$$N(\xi_n) = 1 - e^{-p(s-\xi)}\Big|_{\xi=\xi_n=s+jn\omega_s} = 1 - e^{jn\omega_s p} \qquad (2\text{-}72)$$

and

$$D'(\xi_n) = \frac{d}{d\xi}\left[(s-\xi)[1 - e^{-T(s-\xi)}]\right]\Big|_{\xi=s+jn\omega_s} = jn\omega_s T \qquad (2\text{-}73)$$

Therefore, Eq. (2-71) becomes

$$F_p^*(s) = \sum_{n=-\infty}^{\infty} \frac{1 - e^{jn\omega_s p}}{-jn\omega_s T} F(s + jn\omega_s) \qquad (2\text{-}74)$$

Notice that when s is replaced by $j\omega$, Eq. (2-74) is identical to Eq. (2-45) which is obtained by a different method.

As a summary to this section, we have derived several alternate expressions for the input-output transfer relations of the finite-pulsewidth sampler. These expressions are repeated as follows for quick reference.

Fourier transform:

$$F_p^*(j\omega) = \sum_{n=-\infty}^{\infty} \frac{p}{T} \frac{\sin(n\omega_s p/2)}{n\omega_s p/2} e^{-jn\omega_s p/2} F(j\omega + jn\omega_s) \qquad (2\text{-}45)$$

Laplace transform:

$$F_p^*(s) = \sum_{n=-\infty}^{\infty} \frac{p}{T} \frac{\sin(n\omega_s p/2)}{n\omega_s p/2} e^{-jn\omega_s p/2} F(s + jn\omega_s) \qquad (2\text{-}74)$$

Laplace transform [F(s) has k simple poles]:

$$F_p^*(s) = \sum_{n=1}^{k} \frac{N(\xi_n)}{D'(\xi_n)} \frac{1 - e^{-p(s-\xi_n)}}{(s - \xi_n)[1 - e^{-T(s-\xi_n)}]} \qquad (2\text{-}64)$$

$$F(\xi) = N(\xi)/D(\xi)$$

$$D'(\xi_n) = \left.\frac{dD(\xi)}{d\xi}\right|_{\xi=\xi_n}$$

$$\xi_n = \text{nth simple pole of } F(\xi),\ n = 1,2,\ldots,k$$

Laplace transform [F(s) has k poles with multiplicity $m_n \geqslant 1$]:

$$F_p^*(s) = \left.\sum_{n=1}^{k} \sum_{i=1}^{m_n} \frac{(-1)^{m_n-i} K_{ni}}{(m_n - i)!} \frac{\partial^{m_n-i}}{\partial s^{m_n-i}} P(s)\right|_{s=s-s_n} \qquad (2\text{-}67)$$

$$K_{ni} = \left.\frac{1}{(i-1)!} \frac{\partial^{i-1}}{\partial s^{i-1}} (s - s_n)^{m_n} F(s)\right|_{s=s_n} \qquad (2\text{-}68)$$

Flat-Top Approximation and The Ideal Sampler

In the analysis of the operation of a sampler, if the sampling duration p is very much smaller than the sampling period T and the smallest time constant of the input signal, f(t), the output of the finite-pulsewidth sampler, $f_p^*(t)$, can be approximated by a sequence of flat-topped pulses; that is,

$$\begin{aligned} f_p^*(t) &= f(kT) \qquad \text{for} \quad kT \leqslant t < kT + p \\ &= 0 \qquad\qquad kT + p \leqslant t < (k+1)T \end{aligned} \qquad (2\text{-}75)$$

where k = 0,1,2,3,... .

Then $f_p^*(t)$ is written

$$f_p^*(t) = \sum_{k=0}^{\infty} f(kT)[u_s(t - kT) - u_s(t - kT - p)] \qquad (2\text{-}76)$$

where $u_s(t)$ is the unit-step function.

Taking the Laplace transform on both sides of Eq. (2-76) gives

$$F_p^*(s) = \sum_{k=0}^{\infty} f(kT)\left[\frac{1 - e^{-ps}}{s}\right]e^{-kTs} \qquad (2\text{-}77)$$

If the sampling duration p is very small,

$$1 - e^{-ps} = 1 - \left[1 - ps + \frac{(ps)^2}{2!} - \ldots\right] \cong ps \qquad (2\text{-}78)$$

Then, Eq. (2-77) becomes

$$F_p^*(s) \cong p \sum_{k=0}^{\infty} f(kT)e^{-kTs} \qquad (2\text{-}79)$$

or, equivalently,

$$f_p^*(t) \cong p \sum_{k=0}^{\infty} f(kT)\delta(t - kT) \qquad (2\text{-}80)$$

where $\delta(t)$ is the unit-impulse function.

The right-hand side of Eq. (2-80) is recognized as a train of impulses with the strength of the impulse at $t = kT$ equal to $pf(kT)$. This means that the finite-pulsewidth sampler can be approximated by an "impulse modulator" whose diagram is that shown in Fig. 2-30 but with the carrier signal replaced by the unit-impulse train:

$$\delta_T(t) = \sum_{k=0}^{\infty} \delta(t - kT) \qquad (2\text{-}81)$$

Or, the finite-pulsewidth sampler can be approximated by an *ideal sampler* connected in series with an attenuator with attenuation p. The equivalent situation is illustrated in Fig. 2-36. Therefore, an ideal sampler is defined as a sampler which closes and opens instantaneously, for a zero time duration, every T seconds. It should be noted, however, that the attenuator is necessary only if the sampler is not followed by a hold device.

The output of the ideal sampler is written as

$$f^*(t) = \sum_{k=0}^{\infty} f(kT)\delta(t - kT) = f(t)\delta_T(t) \tag{2-82}$$

where f(t) is the input of the sampler, and sampling is assumed to begin at t = 0. Taking the Laplace transform on both sides of Eq. (2-82), we get

$$F^*(s) = \sum_{k=0}^{\infty} f(kT)e^{-kTs} \tag{2-83}$$

which is the Laplace transform of the output of the ideal sampler.

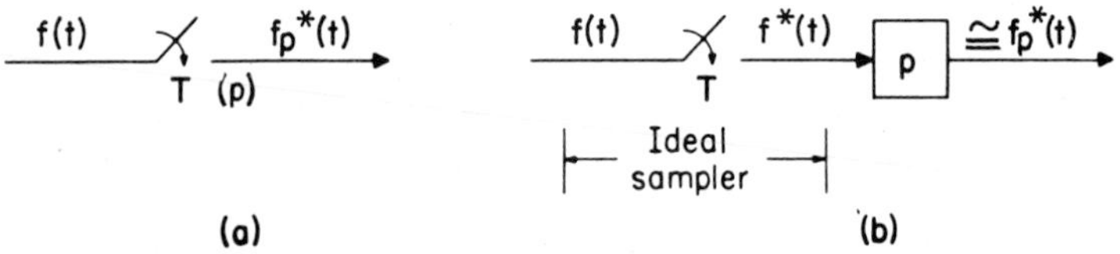

Fig. 2-36. (a) Finite pulsewidth sampler; (b) ideal sampler connected in series with an attenuator p. (a) and (b) are equivalent if p is much smaller than T and the smallest time constant of f(t).

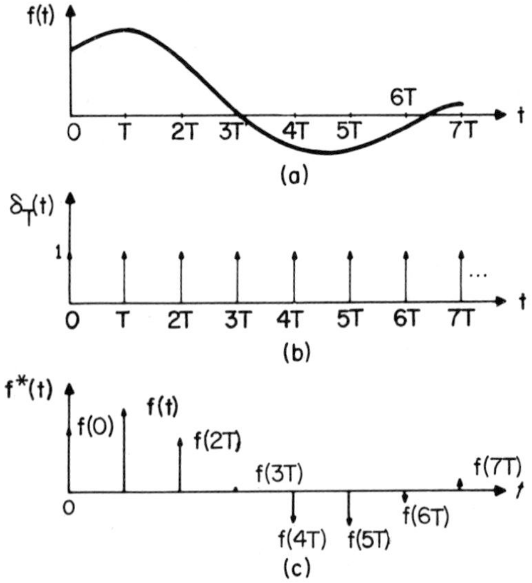

Fig. 2-37. (a) Input signal to ideal sampler; (b) unit impulse train; (c) output of ideal sampler. $f^*(t) = f(t)\delta_T(t)$.

Typical input and output signals of an ideal sampler are illustrated as shown in Fig. 2-37. The output of the ideal sampler is a train of impulses with the respective areas (strengths) of the impulses equal to the magnitudes of the input signal at the corresponding sampling instants. Since an impulse function is defined to have zero pulsewidth and infinite pulse amplitude, in Fig. 2-37 the impulses are represented by arrows with appropriate amplitudes representing the strengths of the impulses. We can also derive an alternate expression for $F^*(s)$ using Eq. (2-74). Since $\delta_T(t)$ and $p(t)$ are related by

$$\delta_T(t) = \lim_{p\to 0} \frac{1}{p}\, p(t) \tag{2-84}$$

$F^*(s)$ can be written as

$$F^*(s) = \lim_{p\to 0} \frac{1}{p} F_p^*(s) = \lim_{p\to 0} \frac{1}{p} \sum_{n=-\infty}^{\infty} \frac{1 - e^{jn\omega_s p}}{-jn\omega_s T} F(s + jn\omega_s)$$

$$= \frac{1}{T} \sum_{n=-\infty}^{\infty} F(s + jn\omega_s) \tag{2-85}$$

However, Eq. (2-74) is obtained from the contour integration around the path Γ_2 which is in the right-half ξ-plane. It is important to examine the convergence properties of the integral taken along the semicircle with infinite radius. Applying the limit as $p \to 0$ to $F_p^*(s)/p$ and using Eq. (2-69), we have

$$F^*(s) = \lim_{p\to 0} \frac{1}{2\pi jp} \oint_{\Gamma_2} F(\xi) \frac{1 - e^{-p(s-\xi)}}{(s - \xi)[1 - e^{-T(s-\xi)}]} d\xi$$

$$= \frac{1}{2\pi j} \oint_{\Gamma_2} F(\xi) \frac{1}{1 - e^{-T(s-\xi)}} d\xi \tag{2-86}$$

In this case since $1/1 - e^{-T(s-\xi)}$ has a simple pole at infinity in the ξ-plane, the part of the integral of Eq. (2-86) along the semicircle of infinite radius in the right half of the ξ-plane may not vanish. In fact, if the degree of the denominator of $F(\xi)$ in ξ is not higher than the degree of the numerator by at least two, the integral along the semicircle may have a finite value or may even not converge. Therefore, Eq. (2-85) is valid only if $F(s)$ has a pole-zero excess greater or equal to two. In other words, the input signal $f(t)$ must not have a jump discontinuity at $t = 0$.

A more general expression than Eq. (2-85) can be derived for $F^*(s)$ by defining $\delta_T(t)$ as an even function so that the unit impulse at the time origin $t = 0$ is considered to be a pulse with amplitude $1/p$, width extending from $t = -p/2$ to $t = p/2$, and in the limit $p \to 0$. Let the Fourier series of $\delta_T(t)$ for all t be designated by

$$\sum_{n=-\infty}^{\infty} C_n e^{jn\omega_s t}$$

where it can be easily shown that $C_n = 1/T$. Then the Fourier series for $\delta_T(t)$ for $t \geqslant 0$ is

$$\delta_T(t) = \frac{1}{2}\delta(t) + \sum_{n=-\infty}^{\infty} \frac{1}{T} e^{jn\omega_s t} \tag{2-87}$$

The $\delta(t)/2$ term on the right side of Eq. (2-87) is added because only half of the unit impulse was considered for $t > 0$. Now substituting Eq. (2-87) into Eq. (2-82) and taking the Laplace transform, we have

$$F^*(s) = \frac{f(0^+)}{2} + \frac{1}{T}\sum_{n=-\infty}^{\infty} F(s + jn\omega_s) \tag{2-88}$$

In a similar fashion we can show that for the case of the ideal sampler the expression for $F^*(s)$ which is analogous to Eq. (2-64) is (simple poles)

$$F^*(s) = \sum_{n=1}^{k} \frac{N(\xi_n)}{D'(\xi_n)} \frac{1}{1 - e^{-T(s-\xi_n)}} \tag{2-89}$$

where $N(\xi_n)$ and $D'(\xi_n)$ are defined in Eqs. (2-65) and (2-66), respectively.

The applications of Eqs. (2-83), (2-85), (2-88) and (2-89) will now be illustrated by an example.

Example 2-1.

Consider that a unit step function $f(t) = u_s(t)$ is sampled by an ideal sampler at intervals of T seconds. From Eq. (2-82), the output of the ideal sampler is

$$f^*(t) = \sum_{k=0}^{\infty} f(kT)\delta(t - kT) = \sum_{k=0}^{\infty} \delta(t - kT) \tag{2-90}$$

The Laplace transform of $f^*(t)$ is

$$F^*(s) = \sum_{k=0}^{\infty} e^{-kTs} = 1 + e^{-Ts} + e^{-2Ts} + \ldots$$

$$= \frac{1}{1 - e^{-Ts}} \quad \text{for} \quad |e^{-Ts}| < 1 \tag{2-91}$$

Now let us treat the same problem using Eq. (2-89). We have

$$F(s) = \frac{N(s)}{D(s)} = \frac{1}{s} \tag{2-92}$$

Thus, N(s) = 1 and D′(s) = 1, and Eq. (2-89) gives

$$F^*(s) = \frac{1}{1 - e^{-Ts}} \tag{2-93}$$

which is the same result as in Eq. (2-91).

We can also arrive at the same answer in Eq. (2-93) by using Eq. (2-88), but the mathematics involved is more complex. For this reason Eq. (2-88) is seldom used for the determination of a closed-form solution of F*(s).

A similar relation as in Eq. (2-67) can be derived for the multiple-order-pole case when the sampler is assumed to be ideal.

The mathematical transfer descriptions of the ideal sampler are summarized as follows:

Representation as an Infinite Impulse Train in the Time Domain:

Time domain:

$$f^*(t) = \sum_{k=0}^{\infty} f(kT)\delta(t - kT) \tag{2-82}$$

Laplace domain:

$$F^*(s) = \sum_{k=0}^{\infty} f(kT)e^{-kTs} \tag{2-83}$$

Representation as an Infinite Series in the Frequency Domain:

$$F^*(s) = \frac{f(0^+)}{2} + \frac{1}{T}\sum_{n=-\infty}^{\infty} F(s + jn\omega_s) \tag{2-88}$$

Laplace Transform [F(s) has k simple poles]:

$$F^*(s) = \sum_{n=1}^{k} \frac{N(\xi_n)}{D'(\xi_n)} \frac{1}{1 - e^{-T(s-\xi_n)}} \tag{2-89}$$

$$F(\xi) = N(\xi)/D(\xi)$$

$$D'(\xi_n) = \left.\frac{dD(\xi)}{d\xi}\right|_{\xi=\xi_n}$$

$$\xi_n = \text{nth simple pole of } F(\xi), \quad n = 1,2,\ldots,k$$

Laplace Transform [F(s) has k poles with multiplicity $m_n \geqslant 1$]:

$$F^*(s) = \sum_{n=1}^{k}\sum_{i=1}^{m_n} \frac{(-1)^{m_n-i}K_{ni}}{(m_n - i)!} \left.\frac{\partial^{m_n-i}}{\partial s^{m_n-i}} \Delta_T(s)\right|_{s=s-s_n} \tag{2-94}$$

$$K_{ni} = \frac{1}{(i-1)!} \frac{\partial^{i-1}}{\partial s^{i-1}} \left[(s - s_n)^{m_n} F(s) \right] \Bigg|_{s=s_n} \tag{2-95}$$

$$\Delta_T(s) = \frac{1}{1 - e^{-Ts}} \tag{2-96}$$

Example 2-2.

As an illustrative example on the application of Eq. (2-94) for the case of multiple-order poles, let us consider the sampling of a unit-ramp function by an ideal sampler. In this case,

$$f(t) = tu_s(t) \tag{2-97}$$

$$F(s) = \frac{1}{s^2} \tag{2-98}$$

Since F(s) has a double pole at $s = 0$, to evaluate $F^*(s)$, Eq. (2-94) should be used.

Since there is only one pole with a multiplicity of two, $k = 1$, and $m_n = m_1 = 2$. From Eq. (2-95),

$$K_{11} = s^2 F(s) \Big|_{s=0} = 1 \tag{2-99}$$

$$K_{12} = \frac{\partial}{\partial s} s^2 F(s) \Big|_{s=0} = 0 \tag{2-100}$$

Thus, from Eq. (2-94),

$$F^*(s) = (-1)K_{11} \left[\frac{\partial}{\partial s} \frac{1}{1 - e^{-Ts}} \right]$$

$$= \frac{Te^{-Ts}}{(1 - e^{-Ts})^2} \tag{2-101}$$

In general, Eqs. (2-83), (2-89) and (2-94) are useful for the determination of the Laplace transform of a signal that is sampled by an ideal sampler, whereas Eq. (2-88) is useful for frequency response plots and analysis.

Referring to Eqs. (2-85) and (2-88) we see that the ideal sampler is again a harmonic generator. The ideal sampler reproduces in its output the spectrum of the continuous input f(t) as well as the complementary components at integral multiples of the sampling frequency, and all components have the same amplitude of 1/T. Assuming that the amplitude spectrum of the continuous input signal is as shown in Fig. 2-38(a), the corresponding amplitude spectrum of the sampled signal $f^*(t)$ when $\omega_s > 2\omega_c$ is shown in Fig. 2-38(b), where ω_s is the sampling frequency and ω_c is the highest frequency contained in f(t). Again, if

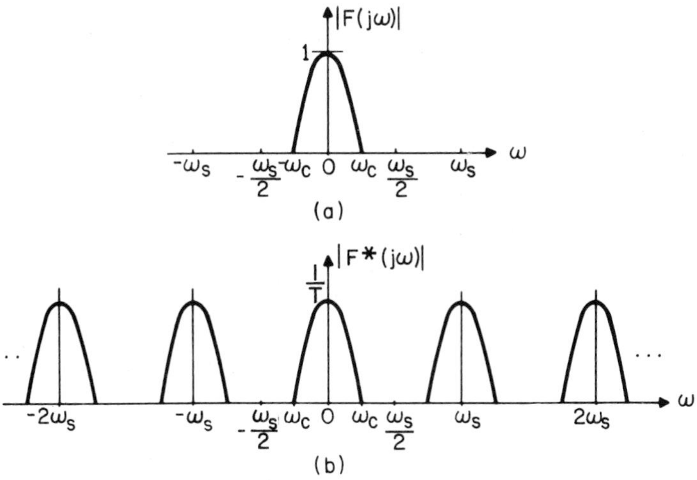

Fig. 2-38. Amplitude spectra of input and output signals of an ideal sampler: (a) amplitude spectrum of continuous input f(t); (b) amplitude spectrum of sampler output ($\omega_s > 2\omega_c$).

the sampling frequency is less than $2\omega_c$, distortion will occur in the output spectrum because of the overlapping of the sidebands in $|F^*(j\omega)|$.

2.8 THE SAMPLING THEOREM

The simple physical reasoning made in the preceding section concerning the minimum sampling frequency required for full recovery of the continuous signal actually answers a very basic yet important question in regard to the proper rate of sampling if sampling is applied intentionally. If sampling is used in a given system, we often ask the question: What are the limitations on the rate of sampling? Theoretically, there is no upper limit on the sampling frequency, although any physical sampler must have a finite maximum sampling rate. Theoretically, when the sampling frequency reaches infinity the continuous case is approached. We must treat the infinite sampling frequency carefully when ideal sampling is considered, since impulses are involved. As far as the lower limit of the sampling frequency is concerned, intuitively, we know that if the continuous signal changes rapidly with respect to time, sampling the signal at too low a rate may lose vital information on the signal between the sampling instants. Consequently, it may not be able to reconstruct the original signal from the information contained in the sampled data.

We can conclude from the amplitude spectra shown in Fig. 2-38 that the lowest sampling frequency for possible signal reconstruction is $2\omega_c$, where ω_c is the highest frequency contained in f(t). Formally, this is known as the *sampling theorem*. The theorem states that *if a signal contains no frequency higher than*

ω_c radians per second, it is completely characterized by the values of the signal measured at instants of time separated by $T = (1/2)(2\pi/\omega_c)$ second. In practice, however, stability of closed-loop systems and other practical considerations also dictate the choice of the sampling frequency and may make it necessary to sample at a rate much higher than this theoretical minimum. Furthermore, strictly speaking, a band-limited signal does not exist physically in communication or control systems. All physical signals found in the real world do contain components covering a wide frequency range. It is because the amplitudes of the higher frequency components are greatly diminished that a band-limited signal is assumed. Therefore, in practice, these factors plus the unrealizability of an ideal low-pass filter make it impossible to exactly reproduce a continuous signal from the sampled signal even if the sampling theorem is satisfied.

An interesting note on the sampling theorem may be inserted here. A signal can still be defined completely by sampling it at a rate less than $2\omega_c$ radians per second provided that the derivatives of the signal are known at the sampling instants as well as the amplitude information. L. J. Fogel [10] and others have proved that if a signal contains no frequency higher than ω_c radians per second, it is completely characterized by the values $f^{(n)}(kT)$, $f^{(n-1)}(kT)$, ..., $f^{(1)}(kT)$, and $f(kT)$, $(k = 0,1,2,\ldots)$ of the signal measured at instants of time separated by $T = (1/2)(n+1)(2\pi/\omega_c)$ second, where

$$f^{(n)}(kT) = \left.\frac{d^n f(t)}{dt^n}\right|_{t=kT} \tag{2-102}$$

This means that when the values of the first derivative of $f(t)$ at $t = kT$, $f^{(1)}(kT)$, for $k = 0,1,2,\ldots$, are known in addition to the values of $f(kT)$, the maximum allowable sampling time is $T = 2\pi/\omega_c$, which is twice the time required when $f(kT)$ alone is measured. The addition of each succeeding derivative allows the time between samples to become larger according to $T = (1/2)(n+1)(2\pi/\omega_c)$, where n is the order of the highest derivative when all lower ordered derivatives are observed for each sample.

2.9 SOME S-PLANE PROPERTIES OF F*(s)

Two important properties of the output of the ideal sampler can be observed from Eq. (2-85).

1. $F^*(s)$ is a periodic function with period $j\omega_s$.

The property of $F^*(s)$ as a periodic function is apparent in view of Eq. (2-85) and Fig. 2-38. Analytically, we can show this fact by substituting $s + jm\omega_s$ for s in Eq. (2-83), where m is an integer. We have

$$F^*(s + jm\omega_s) = \sum_{k=0}^{\infty} f(kT)e^{-kT(s+jm\omega_s)}$$

$$= \sum_{k=0}^{\infty} f(kT)e^{-kTs} \tag{2-103}$$

since $e^{-jkm\omega_s T} = 1$ for integral k and m. Therefore,

$$F^*(s + jm\omega_s) = F^*(s) \tag{2-104}$$

whenever m is an integer. In other words, given any point $s = s_1$ in the s-plane, the function $F^*(s)$ has the same value at all periodic points

$$s = s_1 + jm\omega_s$$

for m = any integer. This property is clearly illustrated by Fig. 2-39. Thus, the s-plane is divided into an infinite number of periodic strips as shown in Fig. 2-39. The strip between $\omega = -\omega_s/2$ and $\omega = \omega_s/2$ is called the *primary strip* and all others occurring at higher frequencies are designated as the complementary strips. The function $F^*(s)$ has the same value at all congruent points in the various periodic strips.

2. If the function $F(s)$ has a pole at $s = s_1$, then $F^*(s)$ has poles at $s = s_1 + jm\omega_s$ for m = integers from $-\infty$ to $+\infty$.

The fact that this statement is true can be observed easily from Eq. (2-85). A typical set of pole configurations of $F(s)$ and the corresponding $F^*(s)$ is shown in Fig. 2-40.

The periodic property of $F^*(s)$ as illustrated by Fig. 2-40 can be used to

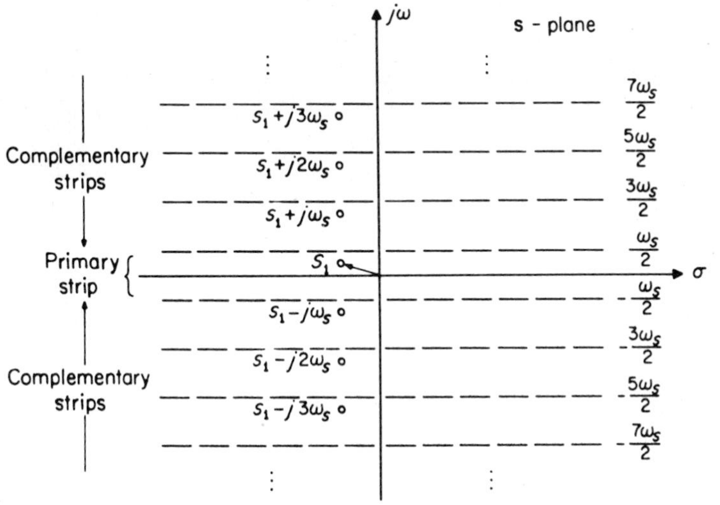

Fig. 2-39. Periodic strips in the s-plane.

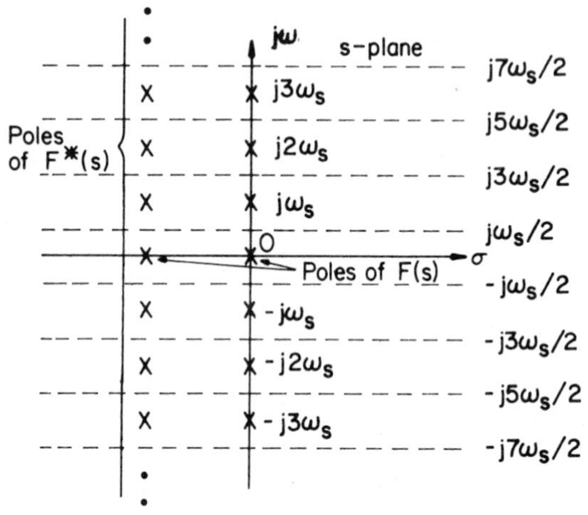

Fig. 2-40. Periodicity of the poles of F*(s).

explain the importance of satisfying the sampling theorem and the significance of folding frequency. Figure 2-40 clearly shows that the poles of F(s) are "folded" out about frequencies that are integral multiples of the folding frequency $\omega_s/2$, by the sampling operation, to form the poles of F*(s). As long as the poles of F(s) lie inside the primary strip of $-\omega_s/2 < \omega < \omega_s/2$, which corresponds to the situation that the sampling frequency is at least twice the highest frequency contained in F(s), these poles will be folded uniquely into the complementary strips by the sampling operation. Then, at least in principle, an ideal filter with a bandpass characteristic of $|\omega_c| < \omega_s/2$ would eliminate all the harmonic poles so that the end result would be the exact recovery of F(s). On the other hand, let us assume that the sampling theorem is not satisfied, so that the relative position of the poles of a given F(s) with respect to the folding frequency is as shown in Fig. 2-41(a). The sampled function F*(s) will now have poles that are "folded back" into the primary strip, as shown in Fig. 2-41(b). In control system design, the folding back of the poles to the low frequencies may cause design difficulties. Similarly, from a signal standpoint, if the sampling theorem is not satisfied, the folding back of the poles of F(s) in the primary strip would preclude the recovery of the original signal from F*(s) even by means of an ideal filter.

2.10 RECONSTRUCTION OF SAMPLED SIGNALS

In most discrete-data or digital control systems the high-frequency harmonic components in a signal f*(t) that resulted from the sampling operation must be removed before the signal is applied to the continuous-data portion of the sys-

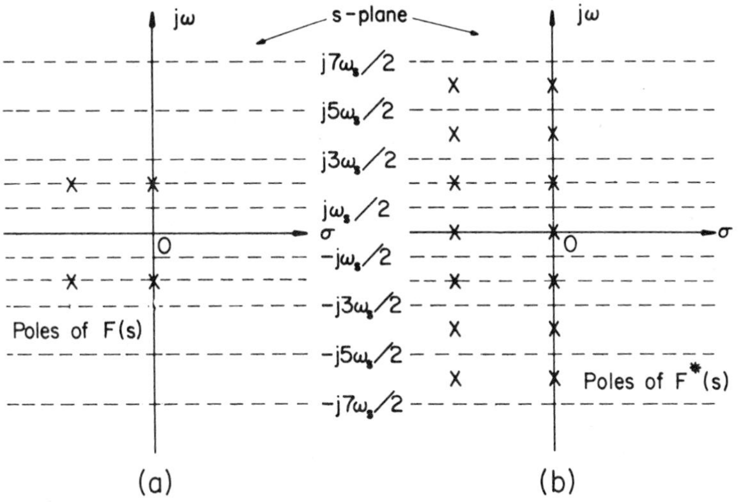

Fig. 2-41. Poles of F(s) and F*(s) to show the effects of folding.

tem. Most control systems have components which are designed to be actuated by continuous-data signals. Therefore, the smoothing of the pulse signals is necessary since otherwise these analog components may be subject to excessive wear. A data-reconstruction device, or, simply, a filter, is often used to interface between digital and analog components. The hold circuit used in conjunction with the sampling operation discussed in Sec. 2.4 is essentially the most common type of filtering device in discrete-data systems.

In addition, most commercial sample-and-holds come as one unit, and we have isolated the sampling operation for mathematical convenience, so now we must model the hold device to complete the entire sample-and-hold operation.

In order to study how data reconstruction may be accomplished, let us assume that an ideal sampler is sampling at a sampling frequency of ω_s which is at least twice as large as the highest frequency component contained in the continuous input being sampled. Figure 2-42 shows the amplitude spectrum of $F^*(s)$, from which it is apparent that in order to duplicate the continuous-data signal the sampled signal must be sent through an ideal low-pass filter with an amplitude characteristic as shown in Fig. 2-43. Unfortunately, an ideal filter characteristic is physically unrealizable, since it is well-known that its time response begins before an input is applied. However, even if we could realize the ideal filter response, we have mentioned previously that perfect reproduction of the continuous signal is still based on the assumption that f(t) is bandlimited. Therefore, for all practical considerations, it is impossible to recover a continuous signal perfectly once it is sampled. The best we can do in data

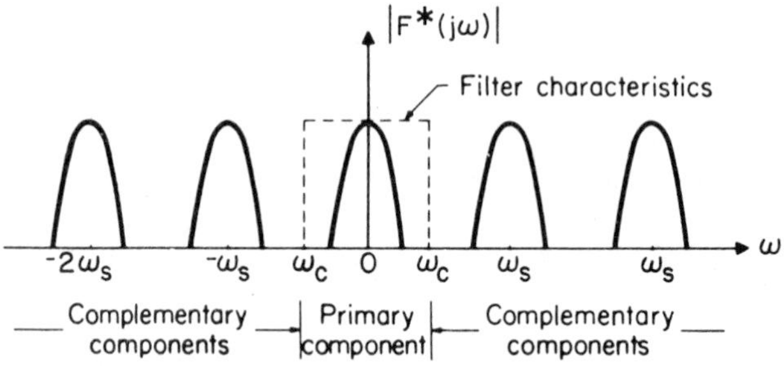

Fig. 2-42. Reconstruction of continuous data from digital data using an ideal low-pass filter.

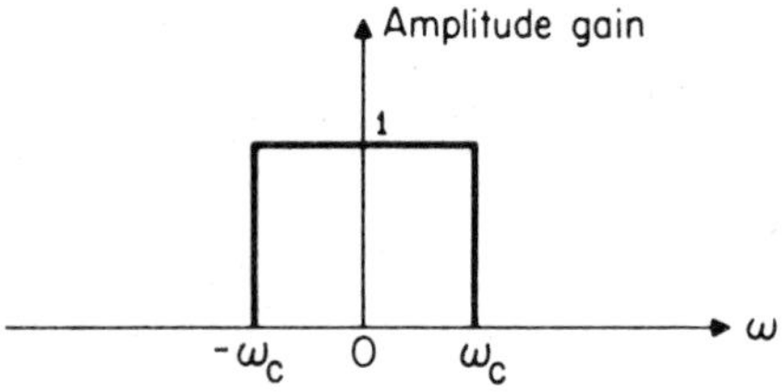

Fig. 2-43. Amplitude characteristic of an ideal filter.

reconstruction is to try to approximate the original time function as closely as possible. Furthermore, as will be shown later in this chapter, a better approximation of the original signal requires, in general, a long time delay that is undesirable in view of its adverse effect on system stability. Consequently, the design of a data-reconstruction device usually involves a compromise between the requirements of stability and the desirability of a close approximation of the continuous signal.

The problem confronting us is that we have a sequence of numbers, f(0), f(T), ..., f(kT), ..., or a train of impulses with the strength of the impulse occurring at $t = kT$ equal to f(kT), $k = 0,1,2,\ldots$, and the continuous-data signal f(t), $t \geqslant 0$, must be reconstructed from the information contained in the pulsed data. This data-reconstruction process may be regarded as an extrapolation process, since the continuous signal is to be constructed based on information available only at past sampling instants. For instance, the original signal f(t) between two consecutive sampling instants kT and $(k + 1)T$ is to be estimated based on the values of f(t) at all preceding sampling instants of kT, $(k - 1)T$, $(k - 2)T, \ldots, 0$; that is, f(kT), $f(k - 1)T$, $f(k - 2)T$, ..., f(0).

A well-known method of generating this desired approximation is based on the power-series expansion of f(t) in the interval between sampling instants kT and $(k + 1)T$; that is

$$f_k(t) = f(kT) + f'(kT)(t - kT) + \frac{f''(kT)}{2!}(t - kT)^2 + \ldots \tag{2-105}$$

where

$$f_k(t) = f(t) \qquad \text{for} \qquad kT \leqslant t < (k+1)T \tag{2-106}$$

$$f'(kT) = \left.\frac{df(t)}{dt}\right|_{t=kT} \tag{2-107}$$

and

$$f''(kT) = \left.\frac{d^2f(t)}{dt^2}\right|_{t=kT} \tag{2-108}$$

In order to evaluate the coefficients of the series given by Eq. (2-105), the derivatives of the function f(t) must be obtained at the sampling instants. Since the only available information concerning f(t) is its magnitudes at the sampling instants, the derivatives of f(t) must be estimated from the values of f(kT). A simple expression involving only two data pulses gives an estimate of the first derivative of f(t) at t = kT as

$$f'(kT) = \frac{1}{T}[f(kT) - f(k-1)T] \tag{2-109}$$

An approximated value of the second derivative of f(t) at t = kT is

$$f''(kT) = \frac{1}{T}\left[f'(kT) - f'[(k-1)T]\right] \tag{2-110}$$

Substitution of Eq. (2-109) into Eq. (2-110) yields

$$f''(kT) = \frac{1}{T^2}\left[f(kT) - 2f[(k-1)T] + f[(k-2)T]\right] \tag{2-111}$$

From these approximated values of f′(kT) and f″(kT), we see that the higher the order of the derivative to be approximated, the larger will be the number of delay pulses required. In fact, we can easily show that the number of delay pulses required to approximate the value of $f^{(n)}(kT)$ is n + 1. Thus, the extrapolating device described above consists essentially of a series of time delays, and the number of delays depends on the degree of accuracy of the estimate of the time function f(t). The adverse effect of time delay on the stability of feedback control systems is well-known. Therefore, an attempt to utilize the higher order derivative of f(t) for the purpose of more accurate extrapolation is often met with serious difficulties in maintaining system stability. Furthermore, a high-order extrapolation also requires complex circuitry and results in a high cost in construction. For these two reasons, quite frequently only the first term of Eq. (2-105) is used in practice. A device generating only f(kT) of Eq. (2-105), for the time interval $kT \leqslant t < (k+1)T$, is normally known as the *zero-order extrapolator*, since the polynomial used is of the zeroth order. It is also commonly referred to as the *zero-order hold* device, because it holds the value of the previous sample during a given sampling period, until the next sample arrives. The zero-order hold has exactly the same characteristics as the hold circuit discussed in Sec. 2.4. Similarly, a device that implements the first two terms of Eq. (2-105) is called a *first-order hold*, since the polynomial generated by this device is of the first order. Mathematical treatments of the zero-order hold are discussed in the following sections.

2.11 THE ZERO-ORDER HOLD

When only the first term of the power series in Eq. (2-105) is used to approximate a signal between two consecutive sampling instants, the polynomial extrapolator is called a *zero-order hold.* This is the type of hold that can be used to model the hold device of the sample-and-hold. Then, Eq. (2-105) is simply

$$f_k(t) = f(kT) \tag{2-112}$$

Equation (2-112) defines the impulse response of the zero-order hold to be as shown in Fig. 2-44. In principle, the operation of a sampler and zero-order hold device is illustrated by the simple circuit shown in Fig. 2-45. The capacitor is assumed to be charged up to the voltage f(kT) at the sampling instant $t = kT$ instantaneously. (The actual rate of charging up the capacitor is, of course, determined by the capacitance and the source impedance.) As the sampling switch is opened during the sampling period T, the capacitor holds the charge until the next pulse comes along from the sampler. The input impedance of the amplifier is assumed to be infinite so that there is no path for the capacitor to discharge. The zero-order hold circuit thus serves to stretch the input pulses into a series of rectangular pulses of width T. In practice, however, the ampli-

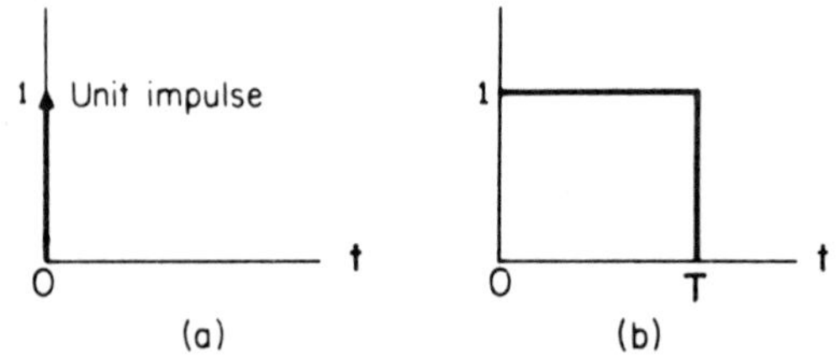

Fig. 2-44. (a) Unit impulse input to zero-order hold; (b) impulse response of zero-order hold.

fier does have a large finite input impedance. Therefore, the actual waveform of the output of the zero-order hold is not a train of exact rectangular pulses but rather, a series of exponentially decayed pulses having a large time constant. The input and output waveforms of an ideal zero-order hold device are shown in Fig. 2-46. Observe that the output of the zero-order hold is a step approximation to the continuous signal, and increasing the sampling rate of the sampler tends to improve the approximation of the continuous signal.

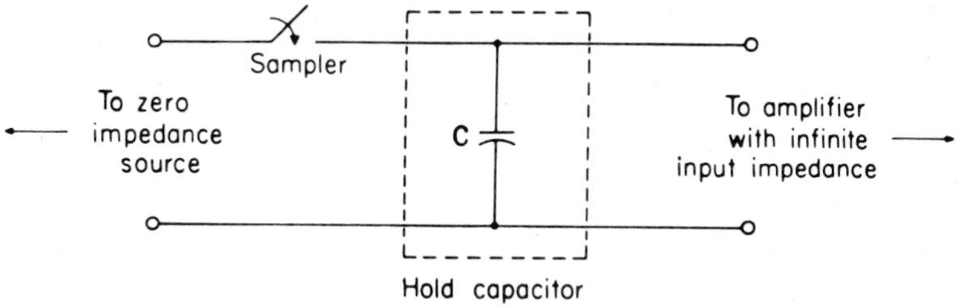

Fig. 2-45. Simplified diagram of a sampler and zero-order hold device.

From Fig. 2-44 the impulse response of the zero-order hold is written

$$g_{h0}(t) = u_s(t) - u_s(t - T) \tag{2-113}$$

where $u_s(t)$ is the unit-step function. The transfer function of the zero-order hold is then

$$G_{h0}(s) = \frac{1 - e^{-Ts}}{s} \tag{2-114}$$

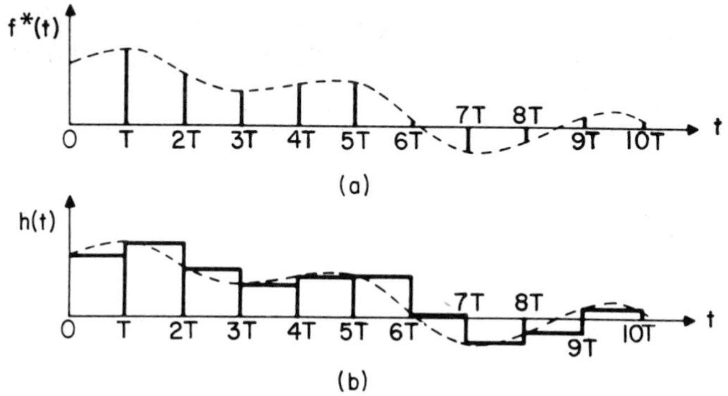

Fig. 2-46. Zero-order hold operation in the time domain: (a) input signal f(t) and sampled signal f*(t); (b) output waveform of zero-order hold.

Replacing s by $j\omega$ in the last equation, we have

$$G_{h0}(j\omega) = \frac{1 - e^{-j\omega T}}{j\omega} \tag{2-115}$$

Equation (2-115) can be written as

$$G_{h0}(j\omega) = \frac{2e^{-j\omega T/2}(e^{j\omega T/2} - e^{-j\omega T/2})}{j2\omega}$$

$$= \frac{2\sin(\omega T/2)e^{-j\omega T/2}}{\omega} \tag{2-116}$$

or

$$G_{h0}(j\omega) = T\frac{\sin(\omega T/2)}{(\omega T/2)} e^{-j\omega T/2} \tag{2-117}$$

Since T is the sampling period in seconds, and $T = 2\pi/\omega_s$, where ω_s is the sampling frequency in rad/sec, Eq. (2-117) becomes

$$G_{h0}(j\omega) = \frac{2\pi}{\omega_s}\frac{\sin\pi(\omega/\omega_s)}{(\omega\pi/\omega_s)} e^{-j\pi(\omega/\omega_s)} \tag{2-118}$$

The gain and phase characteristics of the zero-order hold as functions of ω are shown in Fig. 2-47. This figure shows that the zero-order hold behaves essentially as a low-pass filter. However, when compared with the characteristics of the ideal filter of Fig. 2-43, the amplitude response of the zero-order hold is zero at $\omega = \omega_s$, instead of cutting off sharply at $\omega_s/2$. At $\omega = \omega_s/2$, the magnitude of $G_{h0}(j\omega)$ is equal to 0.636.

Figure 2-46 clearly indicates that the accuracy of the zero-order hold as an extrapolating device depends greatly on the sampling frequency ω_s. The effect of the sampling frequency on the performance of the zero-order hold can also be shown from frequency response curves. The conclusion is that, in general, the filtering property of the zero-order hold is used almost exclusively in practice and in the ensuing chapters of this text, we shall refer to the sampler-zero-order-hold combination as the *sample-and-hold* or S/H.

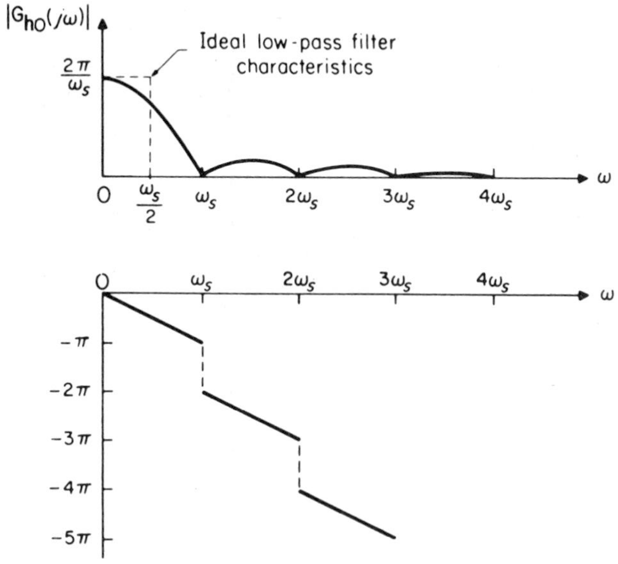

Fig. 2-47. Gain and phase characteristics of a zero-order hold device.

REFERENCES

Digital Signal Processing

1. Rabiner, L. R., and Rader, C. M., (ed.), *Digital Signal Processing.* IEEE Press, New York, 1972.
2. Helms, H. D., and Rabiner, L. R., (ed.), *Literature in Digital Signal Processing: Terminology and Permuted Title Index.* IEEE Press, New York, 1972.
3. Oppenheim, A. V., and Schafer, R. W., *Digital Signal Processing.* Prentice-Hall, Englewood Cliffs, N.J., 1975.

A/D and D/A Converters

4. Moeschele, D. F., Jr., *Analog-to-Digital/Digital-to-Analog Conversion Techniques.* John Wiley & Sons, New York, 1968.
5. Sheingold, D. M., (ed.), *Analog-Digital Conversion Handbook.* Analog Devices, Inc., Norwood, Massachusetts, 1972.
6. Schmid, M., *Electronic Analog/Digital Conversion.* Van Nostrand Rheinhold, New York, 1970.

Sampling and Data Reconstruction

7. Balakrishnan, A. V., "A Note on the Sampling Principle for Continuous Signals," *I.R.E. Trans. on Information Theory*, Vol. IT-3, June 1957, pp. 143-146.

8. Shannon, C. E., "Communication in the Presence of Noise," *Proc. I.R.E.*, Vol. 37, January 1949, pp. 10-21.

9. Shannon, C. E., Oliver, B. M., and Pierce, J. R., "The Philosophy of Pulse Code Modulation," *Proc. I.R.E.*, Vol. 36, November 1948, pp. 1324-1331.

10. Fogel, L. J., "A Note of the Sampling Theorem," *I.R.E. Trans. on Information Theory*, 1, March 1955, pp. 47-48.

11. Jagerman, and Fogel, L. J., "Some General Aspects of the Sampling Theorem," *I.R.E. Trans. on Information Theory*, 2, December 1956, pp. 139-146.

12. Linden, D. A., "A Discussion of Sampling Theorems," *Proc. I.R.E.*, Vol. 47, July 1959, pp. 1219-1226.

13. Oliver, R. M., "On the Functions Which are Represented by the Expansions of the Interpolation-Theory," *Proc. Royal Society* (Edinburgh), 35, 1914-1915, pp. 181-194.

14. Kuo, B. C., *Analysis and Synthesis of Sampled-Data Control Systems.* Prentice-Hall, Englewood Cliffs, N.J., 1963.

15. Barker, R. H., "The Reconstruction of Sampled-Data," *Proc. Conference on Data Processing and Automatic Computing Machines*, Salisbury, Australia, June 1957.

16. Porter, A., and Stoneman, F., "A New Approach to the Design of Pulse Monitored Servo Systems," *J.I.E.E.*, London, 97, Part II, 1950, pp. 597-610.

17. Ragazzini, J. R., and Zadeh, L. H., "The Analysis of Sampled-Data Systems," *Trans. A.I.E.E.*, 71, Part II, 1952, pp. 225-234.

18. Tsypkin, Y. Z., "Sampled-Data Systems With Extrapolating Devices," *Avtomat. Telemekh*, 19, No. 5, 1958, pp. 389-400.

19. Linden, D. A., and Abramson, N. M., "A Generalization of the Sampling Theorem," *Tech. Report No. 1551-2*, Solid-State Elec. Laboratory, Stanford University, August 1959.

20. Jury, E. I., "Sampling Schemes in Sampled-Data Control Systems," *I.R.E. Trans. on Automatic Control*, Vol. AC-6, February 1961, pp. 86-88.

21. Beutler, F. J., "Sampling Theorems and Bases in a Hilbert Space," *Information and Control*, Vol. 4, 1961, pp. 97-117.

PROBLEMS

2-1 Consider that a new car is purchased with a loan of P_0 dollars over a period of N months, at a monthly interest rate of r percent. The principal and interest are to be paid back in N equal payments. Determine the amount of each of the monthly payments u. The difference equation for this problem is given by Eq. (1-1). Solve the problem when $P_0 = \$5000$, N = 36 months, and r = 1 percent.

2-2 Consider that P_0 dollars are borrowed over a period of N months at a monthly interest rate on the unpaid balance of r percent. The principal and interest are to be paid back at the end of each period with each payment doubling the preceding one. Write a difference equation which expresses the amount owed at the end of the (k + 1)th period in terms of that of the previous period and the initial payment u_0. Find the solution for P_N in terms of P_0 and u_0. For $N = 5$, $r = 0.01$, $P_0 = 10{,}000$, find u_0, P_k, $k = 1, 2, \ldots, 4$, and the total amount paid over the entire period.

2-3 It is desired to determine a recursive expression (difference equation) for finding the rth root of a number N. We start by writing the Taylor expansion of a function f(x) about a point x_n:

$$f(x) = f(x_n) + (x - x_n)f'(x_n) + \frac{(x - x_n)^2}{2!} f''(x_n) + \ldots \tag{1}$$

If this series is truncated after two terms, we have

$$f(x) = f(x_n) + (x - x_n)f'(x_n) \tag{2}$$

Let x represent the next iterate x_{n+1} and let x also represent a solution of the equation $f(x) = 0$. Then Eq. (2) becomes

$$0 = f(x_n) + (x_{n+1} - x_n)f'(x_n) \tag{3}$$

or

$$x_{n+1} = x_n - \frac{f(x_n)}{f'(x_n)} \tag{4}$$

Equation (4) is a difference equation useful for solving $f(x) = 0$. Choose f(x) appropriately and determine the corresponding difference equation for finding the rth root of a number N. Do five iterations to calculate the cube root of 5 using $x_0 = 1$ as the initial guess.

2-4 Find the decimal equivalent of the following fixed-point binary numbers. The first bit is a sign bit.
(a) 11011011
(b) 0101.1111
(c) 001101011.011

2-5 Find the decimal equivalent of the following floating-point binary numbers.
(a) +.1011101 + 011
(b) +.1101 − 11

2-6 (a) Find the maximum conversion time required to digitize a 1-kilohertz sinusoidal signal $v(t) = V\sin\omega t$ to 10 bits resolution.
(b) Show that for a fixed resolution, the relation between conversion time and fre-

quency for the sinusoidal signal described in part (a) is a straight line in the log-log coordinates. Illustrate the cases for 10-bit and 16-bit resolutions for the conversion time scale of 10^{-9} sec to 10^{-3} sec as vertical axis, and frequency from 1 Hz to 100 kHz on the horizontal axis, on log-log coordinates.

2-7 Find the output voltages of an eight-bit DAC with 10 v full scale for the digital input words listed below. Also fill in the weighting factor for each bit in terms of fraction of full scale.

Binary Word	Fraction of FS	Output Voltage for FS = 10 v
00000001		
00000010		
00000100		
00001000		
00010000		
00100000		
10000000		
11111111		

2-8 Ten little monkeys,
Sitting in a tree,
Teasing the crocodile:
"You can't catch me!"
Up crept the crocodile,
Quiet as could be ... SNAP!
Nine little monkeys,
Sitting in a tree,
. . . .
. . . .

Upon encountering the above riddle, the inquiring engineer is at once consumed with the desire to find out: what system of equations best models the number of monkeys left on the tree as a function of the number of snaps? Eventually, one is interested in finding the solution of these equations.

Write a difference equation to represent the number of monkeys left as a result of the crocodile snaps. Let m(k) denote the number of monkeys left after k snaps, and m(0) be the initial number. (This problem was suggested by Steve Pollack.)

2-9 A machine gun mounted on a helicopter gunship may be modeled by the mass-spring-friction combination shown in Fig. 2P-9. The force acting on the mass is in the form of blows of impulses of strength F at time intervals of $t = kT$, $k = 0,1,2,\dots$. Assume that at $t = 0$ M is at rest in its equilibrium position of $x(0) = 0$. Write the equation of motion for the system and solve for the displacement x(t) of M for $t > 0$ by means of the Laplace transform method: K = linear spring constant = $M(u^2 + \omega_s^2)$, u = constant, and $\omega_s = 2\pi/T$, f = viscous friction coefficient = 2Mu.

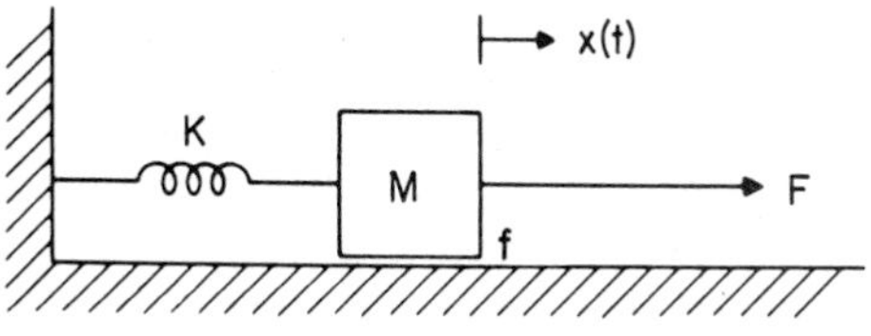

Fig. 2P-9.

2-10 The following signals are sampled by an ideal sampler with sampling period T. Determine the sampler output $f^*(t)$, and evaluate the pulse transform $F^*(s)$ by the Laplace transform method and the complex convolution method.
(a) $f(t) = te^{-at}$
(b) $f(t) = e^{-at}\sin t$ (a = positive constant).

2-11 Prove that if a signal f(t) contains no frequency components higher than ω_c rad/sec, it is completely characterized by the values $f^{(k)}(nT)$, $f^{(k-1)}(nT)$, ..., $f^{(1)}(nT)$, and $f(nT)$ $(n = 0,1,2,\ldots)$, of the signal measured at instants of time separated by $T = (1/2)(k+1)(2\pi/\omega_c)$ second, where

$$f^{(k)}(nT) = \left.\frac{d^k f(t)}{dt^k}\right|_{t=nT}$$

2-12 (a) Find $f^*(t)$ for $f(t) = e^{-a|t|}$, $(-\infty < t < \infty)$. Find the two-sided Laplace transform of $f^*(t)$, $F^*(s)$. Determine the region of convergence of $F^*(s)$.
(b) Find $f^*(t)$ of $f(t) = e^{-at} - e^{at}$, $(t \geqslant 0)$, and find $F^*(s)$.
Compare the two $F^*(s)$ functions obtained in (a) and (b). Discuss your findings.

2-13 A "delayed" sampler is considered to close for a short duration p at instants $t = \Delta$, $\Delta + T$, $\Delta + 2T$, $\Delta + 3T, \ldots$, $\Delta + nT$, ..., where $T > \Delta > 0$, and T is the sampling period in seconds. The input to the sampler, f(t), is continuous. The output of the sampler, denoted by $f_p^*(t)_\Delta$, is assumed to be a flat-topped pulsed train, since $\Delta \ll T$. Derive the pulse transform of $f_p^*(t)_\Delta$, $F^*(s,\Delta)$. [Note that this can also be referred to as the "delayed pulse-transform" of $f^*(t)$.] For small p, assume that $1 - e^{-ps} \cong ps$. Determine the "delayed pulse-transform" of $f(t) = e^{-at}$.

2-14 The "delayed pulse-transform" described in Problem 2-13 can be applied to finite-pulsewidth considerations in sampled-data systems. As shown in Fig. 2P-14(a), a pulse of width p can be considered as the resultant of N elementary pulses of width Δ, such that $N\Delta = p$, and Δ is very small. Therefore, a practical sampler can be represented by N samplers $S_0, S_\Delta, S_{2\Delta}, \ldots, S_{(N-1)\Delta}$, connected in parallel as shown in Fig. 2P-14(b). This implies that sampler S_0 samples at instants $0, T, 2T, \ldots, nT, \ldots$; sampler S_Δ is actuated at instants Δ, $\Delta + T$, $\Delta + 2T$, ..., $\Delta + nT$; etc. Derive the pulse transform of $f_p^*(t)$.

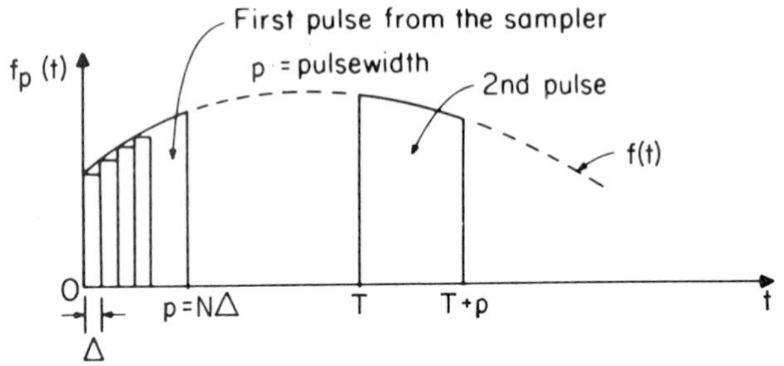

Fig. 2P-14(a).

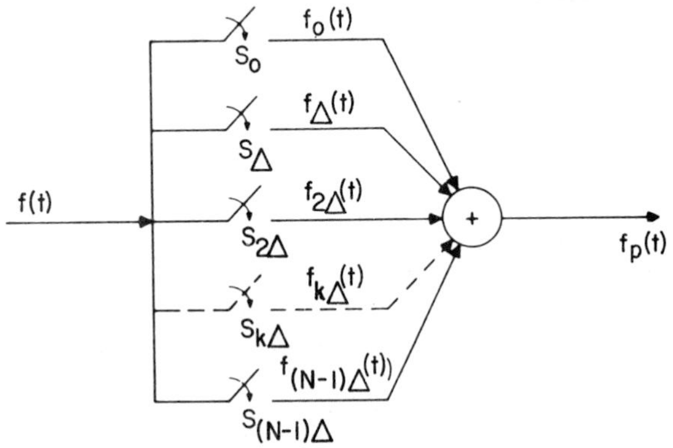

Fig. 2P-14(b).

2-15 It is pointed out in this chapter that the frequency characteristic of a zero-order hold device has a rapid attenuation for low-frequency signals, while that of a first-order hold has an overshoot at low frequency. It is suggested that a "fractional-order hold" can be devised by extrapolating the function in any given sampling interval with only a fraction of, rather than the full first difference. The process is illustrated in Fig. 2P-15. In the first sampling period from 0 to T second, the extrapolated function is a straight line whose slope is a ($a < 1$).

(a) Derive the transfer function of the fractional-order hold.

(b) Plot the frequency characteristics (amplitude and phase of the fractional-order hold for $a = 0.5, 0.3$, and 0.2). Discuss your results.

2-16 Two different schemes of triangular hold are shown in Figs. 2P-16(a) and 2P-16(b), respectively. Write the impulse responses and the transfer functions of the triangular holds. Which of these is physically realizable?

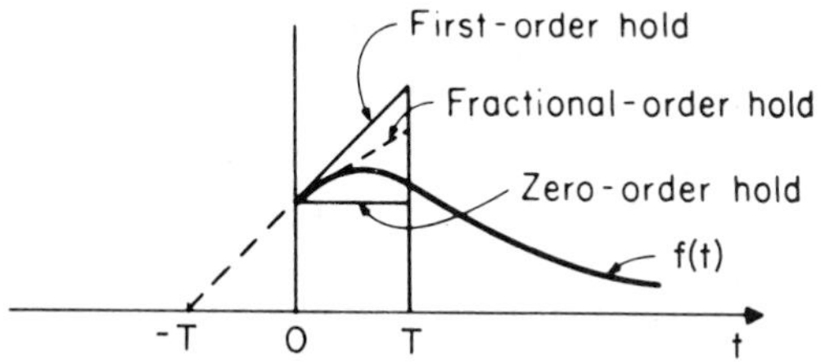

Fig. 2P-15.

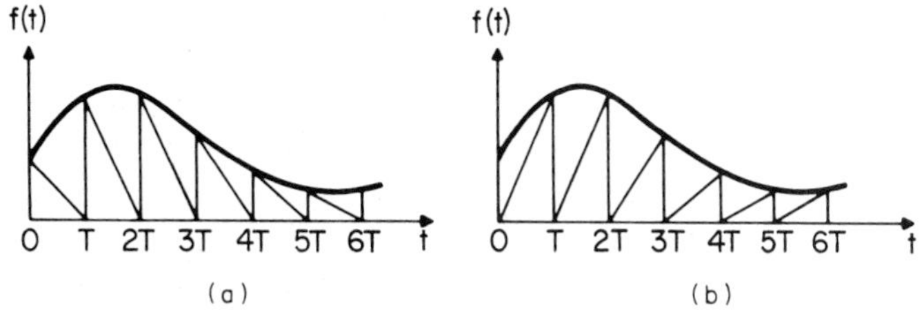

Fig. 2P-16.

2-17 Show that the polygonal approximation of a continuous signal shown in Fig. 2P-17 can be made by use of the two triangular holds of Problem 2-16. Derive the impulse response and the transfer function of the polygonal hold device.

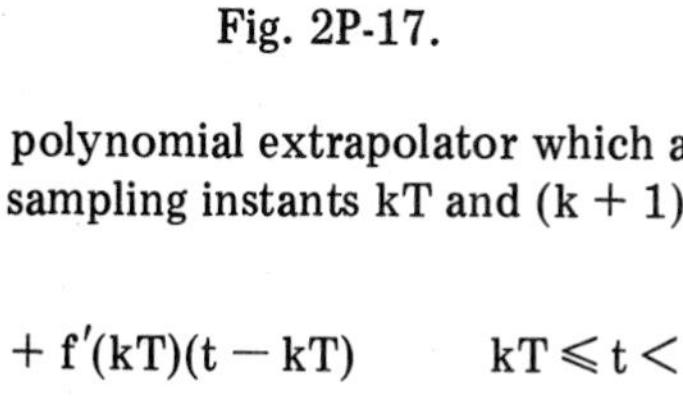

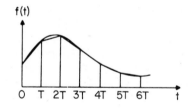

Fig. 2P-17.

2-18 The first-order hold is a polynomial extrapolator which approximates the function f(t) between two successive sampling instants kT and (k + 1)T by using the first two terms of Eq. (2-105). Thus,

$$f_k(t) = f(kT) + f'(kT)(t - kT) \qquad kT \leqslant t < (k+1)T$$

Approximate $f'(kT)$ by

$$f'(kT) \cong \frac{f(kT) - f[(k-1)T]}{T}$$

and show that the transfer function of the first-order hold is

$$G_{h1}(s) = \frac{Ts + 1}{Ts^2}(1 - e^{-Ts})^2$$

Sketch the output of the first-order hold when the input is a unit impulse.

2-19 This problem deals with the digital speed measurement of a control system with an incremental encoder. Consider an incremental encoder (such as an Inductosyn*) which generates two sinusoidal signals in quadrature as the encoder disc rotates. The outputs of the two channels are shown in Fig. 2P-19. Note that the two output signals generate four zero-crossings per cycle. These zero-crossings may be used for position indication or speed measurements in control systems. Consider that the encoder is mounted on the output shaft of a motor which directly drives a printwheel of an electronic typewriter or wordprocessor. The printwheel has 96 character positions, and the encoder disc has 480 cycles.

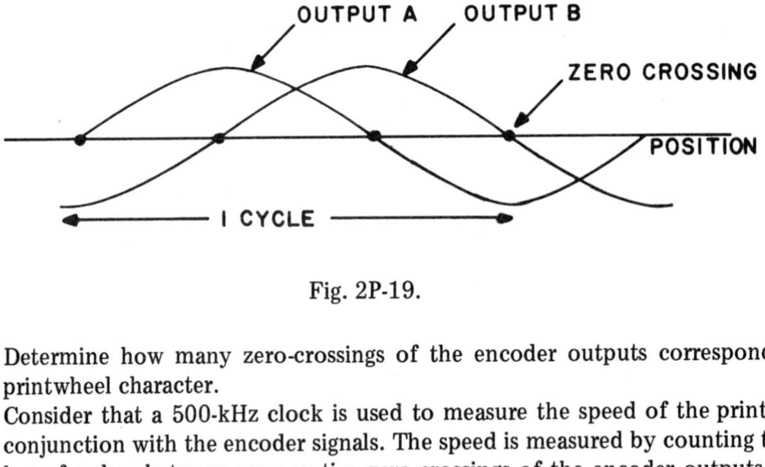

Fig. 2P-19.

(a) Determine how many zero-crossings of the encoder outputs correspond to one printwheel character.
(b) Consider that a 500-kHz clock is used to measure the speed of the printwheel in conjunction with the encoder signals. The speed is measured by counting the number of pulses between consecutive zero-crossings of the encoder outputs. What is the minimum velocity in RPM that can be measured by this setup without overflow if the counter has 12 bits?
(c) The maximum excursion of the printwheel is 180 degrees in either direction, or 48 characters. This must be accomplished in 30 milliseconds or less. Assuming that the printwheel represents a pure inertial load, and that the 48-character motion is accomplished with a constant acceleration profile for half the distance and constant deceleration for the remaining distance, find the maximum velocity in RPM the encoder setup must measure, in number of zero-crossings/sec.

*Registered trademark of Inductosyn Corporation, "Inductosyn Principles," Bulletin ER67301, Farrand Controls, Valhalla, New York; *Incremental Motion Control, Vol. I, DC Motors & Control Systems* (book), ed., B. C. Kuo and J. Tal, SRL Publishing Co., 1978.

(d) If the encoder-counter system is implemented in a control system for the control of the position of the printwheel, what is the range of the sampling periods in seconds that the system will subject to based on the minimum velocity constraint found in part (b), and the maximum velocity requirement found in part (c)? Each zero-crossing of the encoder corresponds to one sampling operation.

2-20 A digital speed measurement system makes use of an incremental shaft encoder and counts the number of pulses generated by the encoder over a given time interval. Let the resolution of the shaft encoder be P pulses/revolution. The encoder output is processed by a microprocessor which takes up to T_s second to count and register one pulse. The shaft speed is denoted by N RPM or ω rad/sec.

(a) Find the highest shaft speed that can be measured in RPM and in rad/sec in terms of the parameters P and T_s. Find the highest measurable shaft speed if $P = 1{,}000$ pulses/revolution and $T_s = 50$ microseconds.

(b) The shaft speed is computed by the microprocessor by counting the number of pulses during a sampling period of T sec. For an encoder with P pulses/revolution, find the resolution of this speed measuring system in RPM/pulse and in rad/sec/pulse in terms of P and T. Find the resolution with $P = 1{,}000$ pulses/rev and $T = 0.1$ sec.

(c) For analytical purposes it is desirable to devise a model for the speed measurement system described above. The sampled-data model of the speed measurement system is shown in Fig. 2P-20. The input variable is the shaft speed $\omega(t)$ in rad/sec. The output is sampled-data which represents the digital measurement of the shaft speed in pulses per sampling period T. Find the constant G_s in terms of P and T. Note that T_s does not enter the analysis in part (b) and part (c).

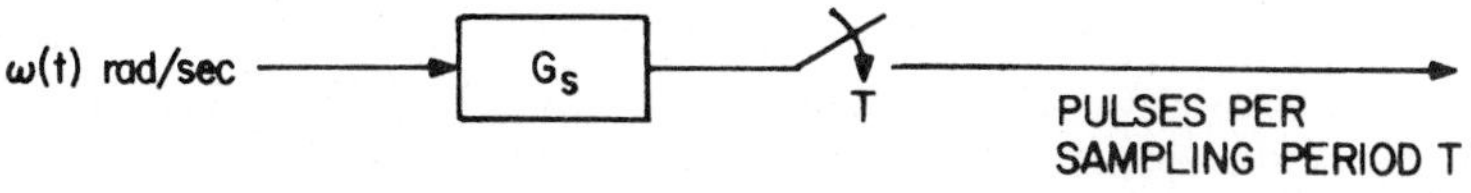

Fig. 2P-20.

3

THE Z-TRANSFORM

3.1 DEFINITION OF THE Z-TRANSFORM

One of the mathematical tools devised for the analysis and design of discrete-data systems is the z-transform. The role of the z-transform to digital systems is similar to that of the Laplace transform to continuous-data systems. In recent years, the state variable method has gained great significance in the studies of discrete-data systems due to its great versatility and unified approach to analysis and design problems. However, the importance of the z-transform method should not be underestimated, since the classical analysis and design of control systems will always have their place in the real world.

The motivation for using the z-transform for the study of discrete-data systems can be explained by referring to the Laplace transform of a sampled signal. Let the output of an ideal sampler be designated by $f^*(t)$ and is given by Eq. (2-82). The Laplace transform of $f^*(t)$ is given by Eq. (2-83),

$$\mathcal{L}[f^*(t)] = F^*(s) = \sum_{k=0}^{\infty} f(kT)e^{-kTs} \tag{3-1}$$

Since the expression of $F^*(s)$ normally contains the factor e^{-Ts}, unlike the majority of transfer functions of continuous-data systems, it is not a rational function of s. When factors involving e^{-Ts} appear in a transfer function, difficulties in taking the inverse Laplace transform may arise. Therefore, it is desirable to first transform the irrational function $F^*(s)$ into a rational function, say

F(z), through a transformation from the complex variable s to another complex variable z. An obvious choice for this transformation is

$$z = e^{Ts} \tag{3-2}$$

although $z = e^{-Ts}$ would serve the same purpose. Solving for s in Eq. (3-2), we get

$$s = \frac{1}{T} \ell n\, z \tag{3-3}$$

In these last two equations, T is the sampling period in seconds, and z is a complex variable whose real and imaginary parts are related to those of s through

$$\text{Re}\, z = e^{T\sigma} \cos \omega T \tag{3-4}$$

and

$$\text{Im}\, z = e^{T\sigma} \sin \omega T \tag{3-5}$$

with

$$s = \sigma + j\omega \tag{3-6}$$

The relation between s and z in Eq. (3-2) may be defined as the *z-transformation*. When Eq. (3-2) is substituted in Eq. (3-1), we have

$$F^*\left[s = \frac{1}{T} \ell n\, z\right] = F(z) = \sum_{k=0}^{\infty} f(kT) z^{-k} \tag{3-7}$$

which when written in closed form will be a rational function of z. Therefore, we can define F(z) as the *z-transform* of f(t), that is,

$$\begin{aligned} F(z) &= \text{z-transform of } f(t) \\ &= \mathcal{Z}[f(t)] \end{aligned} \tag{3-8}$$

where the symbol "$\mathcal{Z}$" denotes the "z-transform of." In view of the way the z-transform operation is defined in Eq. (3-7) and Eq. (3-1), we can also write

$$\begin{aligned} F(z) &= [\text{Laplace transform of } f^*(t)] \Big|_{s=\frac{1}{T}\ell n\, z} \\ &= [F^*(s)] \Big|_{s=\frac{1}{T}\ell n\, z} \end{aligned} \tag{3-9}$$

Since the z-transform of f(t) is obtained from the Laplace transform of f*(t) by performing the transformation $z = e^{Ts}$, we can say that, in general, any function f(t) that is Laplace transformable also has a z-transform.

In summary, the operation of taking the z-transform of a continuous-data function, f(t), involves the following three steps:

1. f(t) is sampled by an ideal sampler to give f*(t).
2. The Laplace transform of f*(t) is taken to give

$$F^*(s) = \mathcal{L}[f^*(t)]$$

$$= \sum_{k=0}^{\infty} f(kT)e^{-kTs}$$

3. Replace e^{Ts} by z in F*(s) to get F(z).

$$F(z) = \sum_{k=0}^{\infty} f(kT)z^{-k} \tag{3-10}$$

Equation (3-10) represents a useful expression for evaluating the z-transform of a function f(t). However, the disadvantage of the expression is that it is an infinite series, and additional effort is required to obtain the equivalent closed-form function.

An alternate expression for the z-transform of a function can be obtained by use of Eq. (2-89) once F(s) is given. Replacing e^{-Ts} by z^{-1} in Eq. (2-89), we have

$$F(z) = \sum_{n=1}^{k} \frac{N(\xi_n)}{D'(\xi_n)} \frac{1}{1 - e^{\xi_n T} z^{-1}} \tag{3-11}$$

where

$$F(s) = \frac{N(s)}{D(s)} \tag{3-12}$$

has only a finite number of simple poles. If F(s) has multiple-order poles, s_1, $s_2, \ldots, s_k$, with multiplicity, $m_1, m_2, \ldots, m_k$, respectively, the z-transform of F(s) is written by use of Eq. (2-94),

$$F(z) = \sum_{n=1}^{k} \sum_{i=1}^{m_n} \frac{(-1)^{m_n - i} K_{ni}}{(m_n - i)!} \left[\frac{d^{m_n - i}}{ds^{m_n - i}} \frac{1}{1 - e^{-Ts}} \right] \Bigg|_{s=s-s_n} \Bigg|_{z=e^{Ts}} \tag{3-13}$$

where

$$K_{ni} = \frac{1}{(i-1)!} \left[\frac{d^{i-1}}{ds^{i-1}} (s - s_n)^{m_n} F(s) \right] \Bigg|_{s=s_n} \tag{3-14}$$

3.2 EVALUATION OF Z-TRANSFORMS

Equations (3-10), (3-11), and (3-13) may be used for the evaluation of z-transforms. Equation (3-10) is used when f(t) or f(kT) is given. Strictly, the time function or time series can be of any function without restriction, although, to be able to express F(z) in a closed form, the infinite series of Eq. (3-10) must be convergent. Equations (3-11) and (3-13) give the z-transforms of functions which are originally in the form of F(s). Equation (3-11) is for a function F(s) which has only simple poles, whereas Eq. (3-13) is for F(s) which has at least one multiple-order pole.

The following examples illustrate how the z-transforms of several common functions are derived. For day-to-day engineering practice, a z-transform table, such as the one given in APPENDIX A, may be referred to.

Example 3-1.

Find the z-transform of the unit-step function $u_s(t)$. Following the steps outlined in the previous section,

(1) The unit-step function is sampled by an ideal sampler. The output of the ideal sampler is a train of unit impulses, described by

$$u_s^*(t) = \delta_T(t) = \sum_{k=0}^{\infty} \delta(t - kT) \tag{3-15}$$

(2) Taking the Laplace transform on both sides of Eq. (3-15), we get

$$U_s^*(s) = \Delta_T(s) = \sum_{k=0}^{\infty} e^{-kTs} \tag{3-16}$$

which converges for $|e^{-Ts}| < 1$.

To express $U_s^*(s)$ in a closed form, we multiply both sides of Eq. (3-16) by e^{-Ts} and then subtract the result from Eq. (3-16). The result is

$$U_s^*(s) = \Delta_T(s) = \frac{1}{1 - e^{-Ts}} \quad \text{for} \quad |e^{-Ts}| < 1 \tag{3-17}$$

(3) Replacing e^{Ts} by z in Eq. (3-17), we have

$$U_s(z) = \mathcal{Z}[u_s(t)] = \frac{1}{1 - z^{-1}} = \frac{z}{z - 1} \tag{3-18}$$

for $|z^{-1}| < 1$, or $|z| > 1$.

The same result can be obtained by use of Eq. (3-11). The Laplace transform of $u_s(t)$ is 1/s, which has a simple pole at s = 0. Thus, from Eq. (3-11), N(s) = 1, D(s) = s, D′(s) = dD(s)/ds = 1. The z-transform of the unit-step function is

$$F(z) = \left. \frac{1}{1 - e^{\xi T} z^{-1}} \right|_{\xi=0} = \frac{1}{1 - z^{-1}} \tag{3-19}$$

Example 3-2.

Find the z-transform of the exponential function $f(t) = e^{-at}$, where a is a real constant.

Without going through the detailed steps as in Example 3-1, we substitute f(t) into Eq. (3-10), and we have

$$F(z) = \sum_{k=0}^{\infty} f(kT) z^{-k} = \sum_{k=0}^{\infty} e^{-akT} z^{-k} \tag{3-20}$$

This infinite series converges for all values of z which satisfy

$$|e^{-aT} z^{-1}| < 1 \tag{3-21}$$

To obtain the closed-form expression of Eq. (3-20), we multiply both sides of the equation by $e^{-aT}z^{-1}$, and subtract the result from the same equation. After rearranging, the result is

$$F(z) = \frac{1}{1 - e^{-aT} z^{-1}} = \frac{z}{z - e^{-aT}} \tag{3-22}$$

for $|e^{-aT}z^{-1}| < 1$, or $|z^{-1}| < e^{aT}$.

We can demonstrate that Eq. (3-11) leads to the same result. The Laplace transform of e^{-at} is

$$F(s) = \frac{1}{s + a} \tag{3-23}$$

which has a simple pole at $s = -a$. In Eq. (3-11), $N(s) = 1$, $D(s) = s + a$, and $D'(s) = 1$. Thus, Eq. (3-11) gives

$$F(z) = \left. \frac{N(\xi_1)}{D'(\xi_1)} \frac{1}{1 - e^{\xi_1 T} z^{-1}} \right|_{\xi_1 = -a}$$

$$= \frac{1}{1 - e^{-aT} z^{-1}} = \frac{z}{z - e^{-aT}} \tag{3-24}$$

Example 3-3.

Find the z-transform of $f(t) = \sin \omega t$.

Equation (3-10) gives

$$F(z) = \sum_{k=0}^{\infty} \sin \omega kT z^{-k} \tag{3-25}$$

or

$$F(z) = \sum_{k=0}^{\infty} \frac{e^{j\omega kT} - e^{-j\omega kT}}{2j} z^{-k} \tag{3-26}$$

This infinite series is convergent for $|z^{-1}| < 1$, and can be written in closed form as

$$F(z) = \frac{1}{2j}\left[\frac{1}{1 - e^{j\omega T}z^{-1}} - \frac{1}{1 - e^{-j\omega T}z^{-1}}\right] \tag{3-27}$$

After simplification, the last equation is written

$$F(z) = \frac{z \sin \omega T}{z^2 - 2z \cos \omega T + 1} \tag{3-28}$$

Now using Eq. (3-11), with

$$F(s) = \mathcal{L}[\sin \omega t] = \frac{\omega}{s^2 + \omega^2} \tag{3-29}$$

we have

$$N(s) = \omega$$

$$D(s) = s^2 + \omega^2$$

$$D'(s) = 2s$$

The poles of F(s) are at $s = \xi_1 = j\omega$ and $s = \xi_2 = -j\omega$. Thus,

$$N(\xi_1) = N(\xi_2) = \omega$$

since N(s) is not a function of s.

$$D'(\xi_1) = 2j\omega$$

$$D'(\xi_2) = -2j\omega$$

Equation (3-11) gives

$$F(z) = \sum_{n=1}^{2} \frac{N(\xi_n)}{D'(\xi_n)} \frac{1}{1 - e^{-T\xi_n} z^{-1}}$$

$$= \frac{1}{2j}\left[\frac{1}{1 - e^{j\omega T}z^{-1}} - \frac{1}{1 - e^{-j\omega T}z^{-1}}\right] \tag{3-30}$$

which apparently leads to the same result as in Eq. (3-28).

Example 3-4.

Find the z-transform of the ramp function $f(t) = tu_s(t)$. Using Eq. (3-10), we have

$$F(z) = \sum_{k=0}^{\infty} kTz^{-k} = Tz^{-1} + 2Tz^{-2} + \ldots \tag{3-31}$$

To express F(z) in closed form, we first multiply both sides of Eq. (3-31) by z^{-1}, resulting in

$$z^{-1}F(z) = Tz^{-2} + 2Tz^{-3} + \ldots \tag{3-32}$$

Subtracting the last equation from Eq. (3-31), we have

$$(1 - z^{-1})F(z) = Tz^{-1} + Tz^{-2} + Tz^{-3} + \ldots \tag{3-33}$$

Now repeating the same process as described above, the closed form F(z) is written as

$$F(z) = \frac{Tz^{-1}}{(1 - z^{-1})^2} = \frac{Tz}{(z - 1)^2} \tag{3-34}$$

The Laplace transform of $f(t) = tu_s(t)$ is $F(s) = 1/s^2$, which has a double pole at $s = 0$. Thus, Eq. (3-13) should be used. In Eqs. (3-13) and (3-14), we identify that $k = 1$, $s_1 = 0$, and $m_1 = 2$. Equation (3-14) gives

$$K_{1i} = \frac{1}{(i-1)!}\left[\frac{d^{i-1}}{ds^{i-1}} s^2 \frac{1}{s^2}\right]\Bigg|_{s=0} \tag{3-35}$$

Thus, $K_{11} = 1$ and $K_{12} = 0$. Equation (3-13) gives

$$F(z) = \sum_{i=1}^{2} \frac{(-1)^{2-i}K_{1i}}{(2-i)!}\left[\frac{d^{2-i}}{ds^{2-i}} \frac{1}{1 - e^{-Ts}}\right]\Bigg|_{z=e^{Ts}} = \frac{Tz}{(z-1)^2} \tag{3-36}$$

which is the same result as in Eq. (3-34).

3.3 MAPPING BETWEEN THE S-PLANE AND THE Z-PLANE

The study of the mapping between the s-plane and the z-plane by the z-transformation $z = e^{Ts}$ is necessarily important. The design and analysis of continuous-data control systems often rely on the pole-zero configuration of the system transfer function in the s-plane. Similarly, the poles and zeros of the z-transform of the system transfer function determine the response of a discrete-data system at the sampling instants. In this section some of the important contours frequently used in the s-plane, such as the constant-damping-ratio line, the constant-frequency locus, etc., will be mapped into the z-plane by the transformation $z = e^{Ts}$.

The s-plane is first divided into an infinite number of periodic strips as shown in Fig. 2-39. The primary strip extends from $\omega = -\omega_s/2$ to $+\omega_s/2$, and the complementary strips extend from $-\omega_s/2$ to $-3\omega_s/2$, $-3\omega_s/2$ to $-5\omega_s/2$, ..., for negative frequencies, and from $\omega_s/2$ to $3\omega_s/2$, $3\omega_s/2$ to $5\omega_s/2$, ..., for positive frequencies. If we consider only the primary strip shown in Fig. 3-1(a), the path described by (1) — (2) — (3) — (4) — (5) — (1) in the left half of the s-plane is mapped into a unit circle centered at the origin in the z-plane by the transformation $z = e^{Ts}$, as shown in Fig. 3-1(b). Since

$$e^{(s+jn\omega_s)T} = e^{Ts}e^{j2\pi n} = e^{Ts} = z \tag{3-37}$$

for n = integer, all the other left-half s-plane complementary strips are also mapped into the same unit circle in the z-plane. All points in the left half of the s-plane are mapped into the region inside the unit circle in the z-plane. The

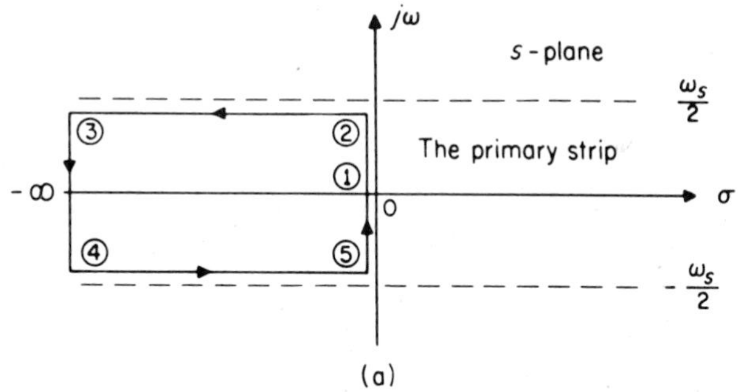

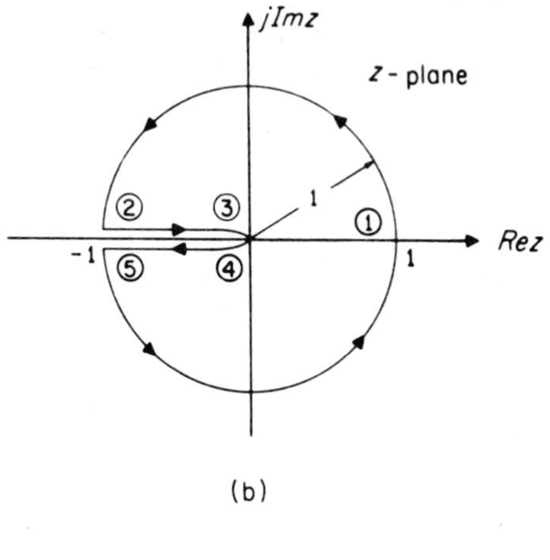

Fig. 3-1. Mapping of the primary strip in the left half of the s-plane into the z-plane by the z-transformation.

points in the right half of the s-plane are mapped into the region outside the unit circle in the z-plane.

The loci of the constant damping factor, constant frequency, and constant damping ratio in the z-plane, for digital systems, are discussed below.

1. The Constant Damping Loci

For a constant damping factor σ_1 in the s-plane, the corresponding z-plane locus is a circle of radius $z = e^{\sigma_1 T}$ centered at the origin in the z-plane as shown in Fig. 3-2.

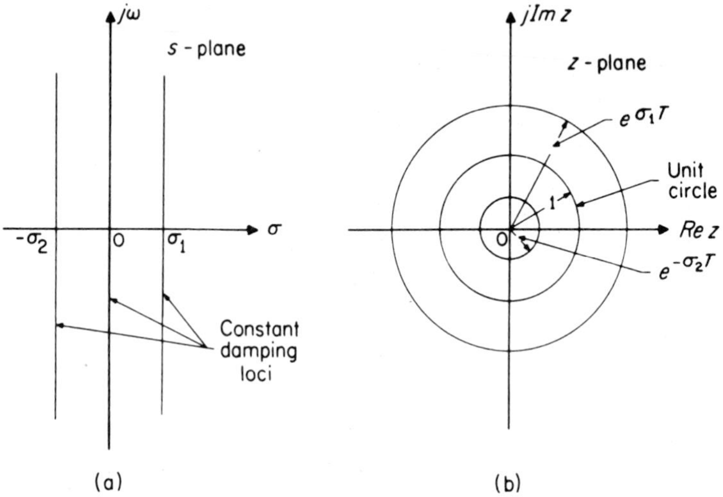

Fig. 3-2. Constant damping loci in the s-plane and the z-plane.

2. The Constant Frequency Loci

For any constant frequency $\omega = \omega_1$ in the s-plane, the corresponding z-plane locus is a straight line emanating from the origin at an angle of $\theta = \omega_1 T$ radian; the angle is measured from the positive real axis as shown in Fig. 3-3.

3. The Constant Damping Ratio Loci

For a constant damping ratio ζ, the s-plane loci are described by

$$s = -\omega \tan\beta + j\omega \tag{3-38}$$

The z-transform relation is

$$z = e^{Ts} = e^{-2\pi\omega \tan\beta/\omega_s} \angle 2\pi\omega/\omega_s \tag{3-39}$$

where

$$\beta = \sin^{-1}\zeta = \text{constant} \tag{3-40}$$

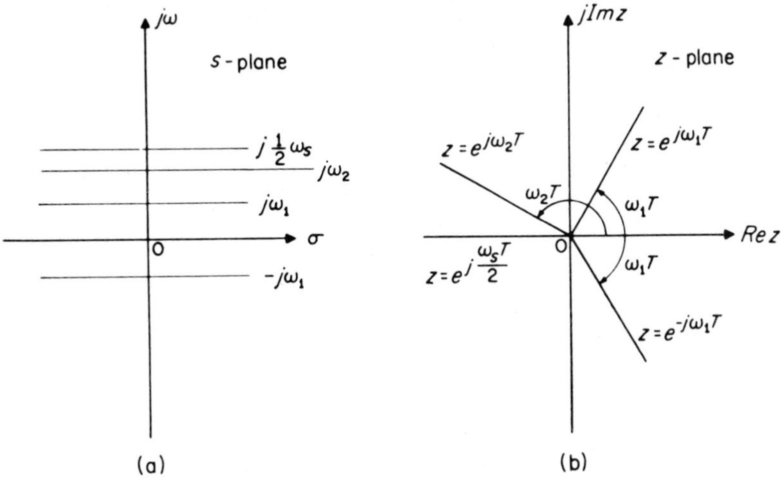

Fig. 3-3. Constant frequency loci in the s-plane and the z-plane.

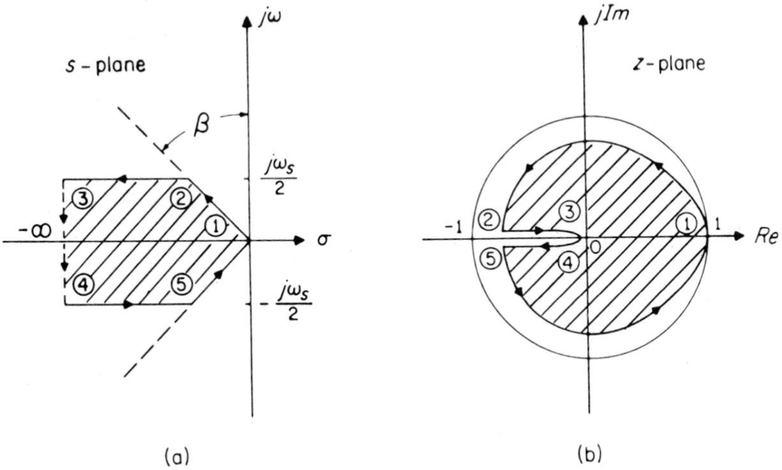

Fig. 3-4. Constant-damping-ratio loci in the s-plane and the z-plane.

For a given β, the constant-ζ path described by Eq. (3-39) is a logarithmic spiral in the z-plane, except for $\beta = 0°$ and $\beta = 90°$. The mapping of the constant damping ratio line from the s-plane to the z-plane is shown in Fig. 3-4. The region shown shaded in Fig. 3-4(a) corresponds to the shaded region of Fig. 3-4(b). In Fig. 3-5, the constant-ζ paths for $\beta = 30°$ are shown in both the s-plane and the z-plane. Each one-half revolution of the logarithmic spiral corresponds to the passage of the constant-ζ locus in the s-plane through a change of $\omega_s/2$ along the imaginary axis.

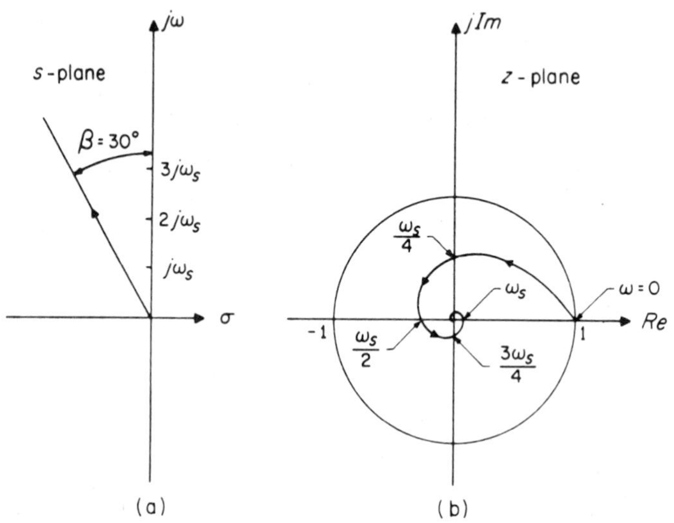

Fig. 3-5. Constant-damping-ratio loci for $\beta = 30°$ ($\zeta = 50\%$) in the s-plane and the z-plane.

3.4 THE INVERSE Z-TRANSFORM

It is well-known that the Laplace transform and its inverse transform are unique, so that if F(s) is the Laplace transform of f(t), then f(t) is the inverse Laplace transform of F(s). However, in the z-transformation, the inverse z transform is not unique. The z-transform of f(t) is denoted by F(z), but the inverse z-transform of F(z) is *not* necessarily equal to f(t). A correct result of the inverse z-transform of F(z) is f(kT) which is equal to f(t) only at the sampling instants $t = kT$. Figure 3-6 illustrates that while the z-transform of the unit-step function is $z/(z-1)$, which denotes a train of unit impulses, the inverse z-transform of $z/(z-1)$ can be of any time function which has a value

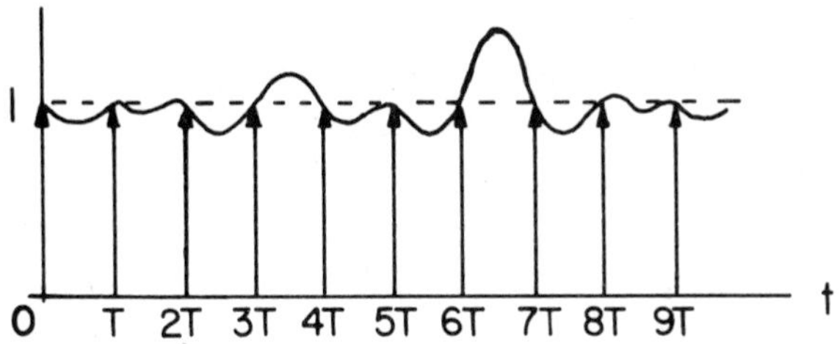

Fig. 3-6. Illustration of the nonuniqueness of the inverse z-transform.

of unity at $t = 0, T, 2T, \ldots$. The problem of nonuniqueness in the inverse z-transform is one of the limitations of the z-transformation, and this property should be remembered when applying the z-transform technique.

The inverse z-transform is denoted by

$$f(kT) = \mathcal{Z}^{-1}[F(z)] = \text{inverse z-transform of } F(z) \tag{3-41}$$

In general, the inverse z-transformation can be carried out with one of the following three methods.

1. Partial-Fraction-Expansion Method

This method is parallel to the partial-fraction-expansion method used in the Laplace transformation, with one minor modification. In the analysis of continuous-data systems, the inverse Laplace transform of a function F(s) is obtained by expanding F(s) as

$$F(s) = \frac{A}{s+a} + \frac{B}{s+b} + \frac{C}{s+c} + \ldots \tag{3-42}$$

where a, b, and c are the negative poles of F(s), and simple poles are assumed in this illustration; A, B, and C are the residues of F(s) at these poles. The inverse Laplace transform of F(s) is then

$$f(t) = Ae^{-at} + Be^{-bt} + Ce^{-ct} + \ldots \tag{3-43}$$

In the case of the z-transformation, we do not expand F(z) into a form similar to Eq. (3-42). The reason is that the inverse z-transform of a term such as $A/(z + a)$ is not found in the z-transform table, although it represents a delayed pulse train with exponentially decaying amplitude if a is positive. Furthermore, it is known that the inverse z-transform of $Az/(z - e^{-aT})$ is Ae^{-akT}. Therefore, it is more appropriate to expand the function F(z)/z by partial-fraction expansion. After the expansion, both sides of F(z)/z are multiplied by z to get F(z).

For functions which do not have any zeros at $z = 0$, it means that the corresponding time series has time delays. The partial-fraction expansion of $F(z)$ is performed in the usual manner, that is,

$$F(z) = \frac{A}{z+a} + \frac{B}{z+b} + \ldots \tag{3-44}$$

We then let

$$F_1(z) = zF(z) = \frac{Az}{z+a} + \frac{Bz}{z+b} + \ldots \tag{3-45}$$

Once the inverse z-transform of $F_1(z)$, $f_1(kT)$, is determined, the inverse z-transform of $F(z)$ is related to $f_1(kT)$ through

$$\begin{aligned} f(kT) = \mathcal{Z}^{-1}[F(z)] &= \mathcal{Z}^{-1}\left[z^{-1}F_1(z)\right] \\ &= f_1[(k-1)T] \end{aligned} \tag{3-46}$$

The identity in Eq. (3-46) is a direct result of Eq. (3-7) if $f(kT) = 0$ for all $k < 0$.

Example 3-5.

Given the z-transform

$$F(z) = \frac{(1-e^{-aT})z}{(z-1)(z-e^{-aT})} \tag{3-47}$$

where a is a positive constant, and T is the sampling period. Find the inverse z-transform of $F(z)$, $f(kT)$, by means of the partial-fraction-expansion method.

The partial-fraction expansion of $F(z)/z$ is written

$$\frac{F(z)}{z} = \frac{1}{z-1} - \frac{1}{z-e^{-aT}} \tag{3-48}$$

Therefore,

$$F(z) = \frac{z}{z-1} - \frac{z}{z-e^{-aT}} \tag{3-49}$$

From the z-transform table (APPENDIX A), the inverse z-transform of $F(z)$ is found to be a time function whose values at the sampling instants are described by

$$f(kT) = 1 - e^{-akT} \tag{3-50}$$

Therefore, the sampled time function is written

$$f^*(t) = \sum_{k=0}^{\infty} (1 - e^{-akT})\delta(t - kT) \tag{3-51}$$

Notice that the time function f(t) cannot be determined from the inverse z-transform, since no information on the behavior of the function is available between the sampling instants.

2. The Power-Series Method

The defining equation of the z-transform in Eq. (3-7) suggests that the inverse z-transform of F(z) can be determined simply by expanding the function into an infinite series in powers of z^{-1}. From Eq. (3-7),

$$F(z) = f(0) + f(T)z^{-1} + f(2T)z^{-2} + \ldots + f(kT)z^{-k} + \ldots \qquad (3\text{-}52)$$

Thus, the coefficients of the series expansion represent the values of f(t) at the sampling instants. The main difference between the partial-fraction-expansion method and the series-expansion method is that the former gives a closed-form solution to f(kT), whereas the latter usually gives a sequence of numbers. The two methods are, of course, equivalent, and we should be able to write a closed-form expression for the sequence of numbers.

Example 3-6.

Determine the inverse z-transform of the following function:

$$F(z) = \frac{(1 - e^{-aT})z}{z^2 - (1 + e^{-aT})z + e^{-aT}} \qquad (3\text{-}53)$$

which is the same function as in Eq. (3-47).

Dividing the denominator into the numerator of F(z) continuously yields

$$F(z) = (1 - e^{-aT})z^{-1} + (1 - e^{-2aT})z^{-2} + \ldots \qquad (3\text{-}54)$$

In this case it is easily observed that

$$f(kT) = 1 - e^{-akT} \qquad k = 0, 1, 2, \ldots, \qquad (3\text{-}55)$$

and thus

$$f^*(t) = \sum_{k=0}^{\infty} (1 - e^{-akT})\delta(t - kT) \qquad (3\text{-}56)$$

which is identical to the result in Eq. (3-51) obtained by the partial-fraction-expansion method.

3. The Inversion-Formula Method

It is of interest to compare the definitions of the Laplace transform and the z-transform. Given f(t) as a function of t, and that it is Laplace transformable, the Laplace transform and z-transform of f(t) are, respectively,

$$F(s) = \mathcal{L}[f(t)] = \int_0^\infty f(t)e^{-st}dt \tag{3-57}$$

$$F(z) = \mathcal{Z}[f(t)] = \sum_{k=0}^{\infty} f(kT)z^{-k} \tag{3-58}$$

The inverse Laplace transform of F(s) is

$$f(t) = \frac{1}{2\pi j}\int_{c-j\infty}^{c+j\infty} F(s)e^{ts}ds \tag{3-59}$$

where c denotes the abscissa of convergence and is chosen so that it is greater than the real parts of all the singularities of the integrand, $F(s)e^{ts}$. We shall show that the inverse z-transform is given by an analogous expression,

$$f(kT) = \frac{1}{2\pi j}\oint_{\Gamma} F(z)z^{k-1}dz \tag{3-60}$$

where Γ is a closed path (usually a circle) in the z-plane which encloses all the singularities of $F(z)z^{k-1}$. We now proceed with the derivation of Eq. (3-60).

Substituting t = kT in Eq. (3-59), we have

$$f(kT) = \frac{1}{2\pi j}\int_{c-j\infty}^{c+j\infty} F(s)e^{kTs}ds \tag{3-61}$$

The path of integration of the line integral of Eq. (3-61) is the straight line which extends from $-j\infty$ to $+j\infty$ along $\sigma = c$, as shown in Fig. 3-7(a). This path of integration is seen to pass through the periodic strips of the s-plane vertically, and, thus, the integral of Eq. (3-61) may be broken up into a sum of integrals, each performed within one of the periodic strips. Equation (3-61) can now be written as

$$f(kT) = \frac{1}{2\pi j}\sum_{i=-\infty}^{\infty}\int_{c+j\omega_s(i-\frac{1}{2})}^{c+j\omega_s(i+\frac{1}{2})} F(s)e^{kTs}ds \tag{3-62}$$

where $\omega_s = 2\pi/T$. Replacing s by $s + ji\omega_s$, where i is an integer, Eq. (3-62) becomes

$$f(kT) = \frac{1}{2\pi j}\sum_{i=-\infty}^{\infty}\int_{c-j\omega_s/2}^{c+j\omega_s/2} F(s + ji\omega_s)e^{kT(s+ji\omega_s)}d(s + ji\omega_s) \tag{3-63}$$

or

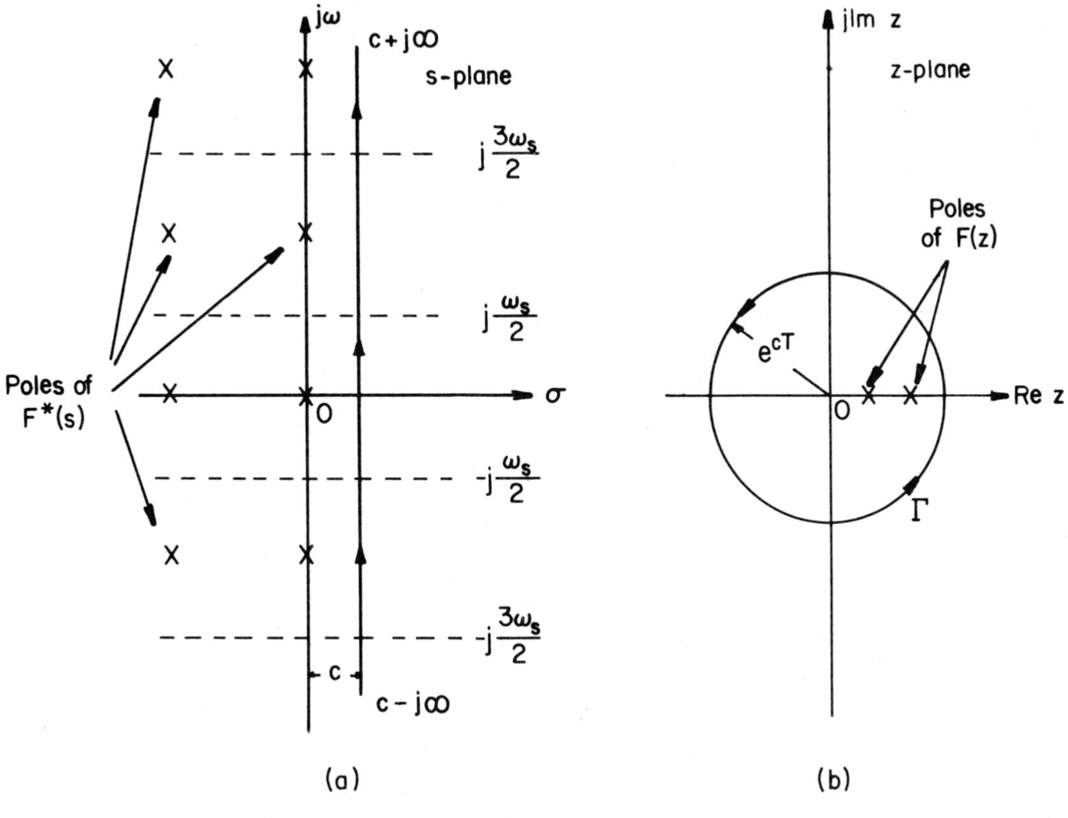

Fig. 3-7. Paths of integration of the inversion formula in the s-plane and the z-plane.

$$f(kT) = \frac{1}{2\pi j} \sum_{i=-\infty}^{\infty} \int_{c-j\omega_s/2}^{c+j\omega_s/2} F(s + ji\omega_s)e^{kTs}ds \tag{3-64}$$

Interchanging the summation and integration operations, Eq. (3-64) becomes

$$f(kT) = \frac{1}{2\pi j}\int_{c-j\omega_s/2}^{c+j\omega_s/2} \sum_{i=-\infty}^{\infty} F(s + ji\omega_s)e^{kTs}ds \tag{3-65}$$

Since from Eq. (2-85),

$$F^*(s) = \frac{1}{T} \sum_{k=-\infty}^{\infty} F(s + jk\omega_s) \tag{3-66}$$

Eq. (3-65) becomes

$$f(kT) = \frac{T}{2\pi j}\int_{c-j\omega_s/2}^{c+j\omega_s/2} F^*(s)e^{kTs}ds \tag{3-67}$$

Now substituting $z = e^{Ts}$ in Eq. (3-67), and since

$$F(s)\Big|_{s=\frac{1}{T}\ell n\, z} = F(z) \tag{3-68}$$

$$e^{kTs} = z^k \tag{3-69}$$

and

$$ds = d\left[\frac{1}{T}\ell n\, z\right] = \frac{1}{T}z^{-1}dz \tag{3-70}$$

Eq. (3-67) is written

$$f(kT) = \frac{1}{2\pi j}\oint_\Gamma F(z)z^{k-1}dz \tag{3-71}$$

The path of integration from $s = c - j(\omega_s/2)$ to $s = c + j(\omega_s/2)$ is mapped onto the circle $|z| = e^{cT}$ in the z-plane, as shown in Fig. 3-7(b). Since $F^*(s)$ does not have singularities on or to the right of the path of integration, $s = c + j\omega$, $\omega \in (-\infty, \infty)$, in the s-plane, all the singularities of $F(z)z^{k-1}$ must lie inside the circle Γ, $|z| = e^{cT}$, in the z-plane.

Example 3-7.

Determine the inverse z-transform of F(z) in Eq. (3-47) by means of the inversion formula of Eq. (3-60).

Substituting Eq. (3-47) into Eq. (3-60), we have

$$f(kT) = \frac{1}{2\pi j}\oint_\Gamma \frac{z(1-e^{-aT})}{(z-1)(z-e^{-aT})}z^{k-1}dz \tag{3-72}$$

where Γ is a circle which is large enough to enclose the poles of F(z) at $z = 1$ and $z = e^{-aT}$. The contour integral of Eq. (3-72) can be evaluated using the residue theorem of complex variable theory; that is,

$$f(kT) = \sum \text{residue of } F(z)z^{k-1} \text{ at the poles of } F(z) \tag{3-73}$$

Therefore, Eq. (3-72) becomes

$$f(kT) = \sum \text{residues of } \frac{z(1-e^{-aT})}{(z-1)(z-e^{-aT})}z^{k-1} \text{ at } z = 1 \text{ and } z = e^{-aT}$$

$$= 1 - e^{-akT} \tag{3-74}$$

This result agrees with that obtained in Eqs. (3-50) and (3-55) by the two preceding methods.

3.5 THEOREMS OF THE Z-TRANSFORM

The derivation and applications of the z-transformation can often be simplified when theorems of the z-transform are referred to. The proofs and illustrations of some of the basic theorems in z-transforms are given below.

1. Addition and Subtraction

If $f_1(t)$ and $f_2(t)$ have z-transforms

$$F_1(z) = \mathcal{Z}[f_1(t)] = \sum_{k=0}^{\infty} f_1(kT)z^{-k} \tag{3-75}$$

and

$$F_2(z) = \mathcal{Z}[f_2(t)] = \sum_{k=0}^{\infty} f_2(kT)z^{-k} \tag{3-76}$$

respectively, then

$$\mathcal{Z}[f_1(t) \pm f_2(t)] = F_1(z) \pm F_2(z) \tag{3-77}$$

Proof: By definition

$$\mathcal{Z}[f_1(t) \pm f_2(t)] = \sum_{k=0}^{\infty} [f_1(kT) \pm f_2(kT)]z^{-k}$$

$$= \sum_{k=0}^{\infty} f_1(kT)z^{-k} \pm \sum_{k=0}^{\infty} f_2(kT)z^{-k}$$

$$= F_1(z) \pm F_2(z) \tag{3-78}$$

2. Multiplication by a Constant

If $F(z)$ is the z-transform of $f(t)$, then

$$\mathcal{Z}[af(t)] = a\mathcal{Z}[f(t)] = aF(z) \tag{3-79}$$

where a is a constant.

Proof: By definition

$$\mathcal{Z}[af(t)] = \sum_{k=0}^{\infty} af(kT)z^{-k} = a\sum_{k=0}^{\infty} f(kT)z^{-k} = aF(z) \tag{3-80}$$

3. Real Translation (Shifting Theorem)

If f(t) has the z-transform F(z), then

$$\mathcal{Z}[f(t - nT)] = z^{-n}F(z) \tag{3-81}$$

and

$$\mathcal{Z}[f(t + nT)] = z^{n}\left[F(z) - \sum_{k=0}^{n-1} f(kT)z^{-k}\right] \tag{3-82}$$

where n is a positive integer.

Proof: By definition

$$\mathcal{Z}[f(t - nT)] = \sum_{k=0}^{\infty} f(kT - nT)z^{-k} \tag{3-83}$$

which can be written as

$$\mathcal{Z}[f(t - nT)] = z^{-n}\sum_{k=0}^{\infty} f(kT - nT)z^{-(k-n)} \tag{3-84}$$

Since f(t) is assumed to be zero for $t < 0$, Eq. (3-84) becomes

$$\mathcal{Z}[f(t - nT)] = z^{-n}\sum_{k=n}^{\infty} f(kT - nT)z^{-(k-n)} = z^{-n}F(z) \tag{3-85}$$

To prove Eq. (3-82), we write

$$\begin{aligned}\mathcal{Z}[f(t + nT)] &= \sum_{k=0}^{\infty} f(kT + nT)z^{-k}\\ &= z^{n}\sum_{k=0}^{\infty} f(kT + nT)z^{-(k+n)}\\ &= z^{n}\left[F(z) - \sum_{k=0}^{n-1} f(kT)z^{-k}\right]\end{aligned} \tag{3-86}$$

Example 3-8.

Find the z-transform of a unit-step function which is delayed by one sampling period T. Using the shifting theorem stated in Eq. (3-81), we have

$$\mathcal{Z}[u_s(t - T)] = z^{-1}\mathcal{Z}[u_s(t)] = z^{-1}\frac{z}{z-1} = \frac{1}{z-1} \tag{3-87}$$

4. Complex Translation

If f(t) has the z-transform F(z), then

$$\mathcal{Z}[e^{\mp at}f(t)] = [F(s \pm a)]\Big|_{z=e^{Ts}} = F(ze^{\pm aT}) \tag{3-88}$$

where a is a constant.

Proof: By definition

$$\mathcal{Z}[e^{\mp at}f(t)] = \sum_{k=0}^{\infty} f(kT)e^{\mp akT}z^{-k} \tag{3-89}$$

Let $z_1 = ze^{\pm aT}$, then Eq. (3-89) is written

$$\mathcal{Z}[e^{\mp at}f(t)] = \sum_{k=0}^{\infty} f(kT)z_1^{-k} = F(z_1) \tag{3-90}$$

Therefore,

$$\mathcal{Z}[e^{\mp at}f(t)] = F(ze^{\pm aT}) \tag{3-91}$$

Example 3-9.

Find the z-transform of $f(t) = e^{-at}\sin\omega t$ with the help of the complex translation theorem.

From the z-transform table in APPENDIX A, the z-transform of $e^{-at}\sin\omega t$ is found to be

$$\mathcal{Z}[e^{-at}\sin\omega t] = \frac{ze^{-aT}\sin\omega T}{z^2 - 2ze^{-aT}\cos\omega T + e^{-2aT}} \tag{3-92}$$

and the z-transform of $\sin\omega t$ is

$$\mathcal{Z}[\sin\omega t] = \frac{z\sin\omega T}{z^2 - 2z\cos\omega T + 1} \tag{3-93}$$

Apparently, the result in Eq. (3-92) may be obtained by substituting ze^{aT} for z in Eq. (3-93).

5. Initial-Value Theorem

If the function f(t) has the z-transform F(z), and if the limit

$$\lim_{z\to\infty} F(z)$$

exists, then

$$\lim_{k \to 0} f(kT) = \lim_{z \to \infty} F(z) \tag{3-94}$$

The theorem simply states that the behavior of the discrete signal f*(t) in the neighborhood of $t = 0$ is determined by the behavior of F(z) at $z = \infty$.

Proof: By definition, F(z) is written

$$F(z) = \sum_{k=0}^{\infty} f(kT)z^{-k} = f(0) + f(T)z^{-1} + f(2T)z^{-2} + \ldots \tag{3-95}$$

Taking the limit on both sides of the last equation as z approaches infinity, we get

$$\lim_{z \to \infty} F(z) = f(0) = \lim_{k \to 0} f(kT) \tag{3-96}$$

6. Final-Value Theorem

If the function f(t) has the z-transform F(z) and if the function $(1 - z^{-1})F(z)$ does not have poles on or outside the unit circle $|z| = 1$ in the z-plane, then

$$\lim_{k \to \infty} f(kT) = \lim_{z \to 1} (1 - z^{-1})F(z) \tag{3-97}$$

Proof: Let us consider two finite sequences,

$$\sum_{k=0}^{n} f(kT)z^{-k} = f(0) + f(T)z^{-1} + \ldots + f(nT)z^{-n} \tag{3-98}$$

and

$$\sum_{k=0}^{n} f[(k-1)T]\,z^{-k} = f(0)z^{-1} + f(T)z^{-2} + \ldots + f[(n-1)T]\,z^{-n} \tag{3-99}$$

We assume that $f(t) = 0$ for all $t < 0$ so that the term $f(-T)$ that would have appeared in Eq. (3-99) is zero. Now comparing Eq. (3-98) with Eq. (3-99), we see that the latter can also be written as

$$\sum_{k=0}^{n} f[(k-1)T]\,z^{-k} = z^{-1}\sum_{k=0}^{n-1} f(kT)z^{-k} \tag{3-100}$$

aking the limit as z approaches 1 of the difference between Eqs. (3-98) and 3-100) yields

$$\lim_{z \to 1}\left[\sum_{k=0}^{n} f(kT)z^{-k} - z^{-1}\sum_{k=0}^{n-1} f(kT)z^{-k}\right]$$

$$= \sum_{k=0}^{n} f(kT) - \sum_{k=0}^{n-1} f(kT) = f(nT) \qquad (3\text{-}101)$$

In the last equation if we take the limit as n approaches infinity, we have

$$\lim_{n\to\infty} f(nT) = \lim_{n\to\infty}\lim_{z\to 1}\left[\sum_{k=0}^{n} f(kT)z^{-k} - z^{-1}\sum_{k=0}^{n-1} f(kT)z^{-k}\right] \qquad (3\text{-}102)$$

Interchanging the limits on the right-hand side of the last equation, and since

$$\lim_{n\to\infty} \sum_{k=0}^{n} f(kT)z^{-k} = \lim_{n\to\infty} \sum_{k=0}^{n-1} f(kT)z^{-k} = F(z) \qquad (3\text{-}103)$$

we have

$$\lim_{n\to\infty} f(nT) = \lim_{z\to 1} (1 - z^{-1})F(z) \qquad (3\text{-}104)$$

which is the expression for the final-value theorem.

Example 3-10.

Given the z-transform

$$F(z) = \frac{0.792z^2}{(z-1)(z^2 - 0.416z + 0.208)} \qquad (3\text{-}105)$$

determine the final value of f(kT) by use of the final-value theorem.

Since

$$(1 - z^{-1})F(z) = \frac{0.792z}{z^2 - 0.416z + 0.208} \qquad (3\text{-}106)$$

does not have poles on or outside the unit circle $|z| = 1$, the final-value theorem may be applied. Thus, from Eq. (3-97)

$$\lim_{k\to\infty} f(kT) = \lim_{z\to 1} \frac{0.792z}{z^2 - 0.416z + 0.208} = 1 \qquad (3\text{-}107)$$

This result can be checked readily by expanding F(z) into a power series in z^{-1}.

$$F(z) = 0.792z^{-1} + 1.12z^{-2} + 1.091z^{-3} + 1.01z^{-4} + 0.983z^{-5}$$
$$+ 0.989z^{-6} + 0.99z^{-7} + \ldots \qquad (3\text{-}108)$$

In this case when sufficient terms are carried out in the expansion we can see that the coefficients of the series converge rapidly to the final steady-state value of unity.

7. Partial-Differentiation Theorem

Let the z-transform of the function f(t,a) be denoted by F(z,a), where a is an independent variable or a constant. The z-transform of the partial derivative of f(t,a) with respect to a is given by

$$\mathcal{Z}\left[\frac{\partial}{\partial a}[f(t,a)]\right] = \frac{\partial}{\partial a} F(z,a) \tag{3-109}$$

Proof: By definition,

$$\mathcal{Z}\left[\frac{\partial}{\partial a}[f(t,a)]\right] = \sum_{k=0}^{\infty} \frac{\partial}{\partial a} f(kT,a)z^{-k}$$

$$= \frac{\partial}{\partial a} \sum_{k=0}^{\infty} f(kT,a)z^{-k} = \frac{\partial}{\partial a} F(z,a) \tag{3-110}$$

The following example illustrates that the z-transforms of certain types of functions may be obtained easily if the partial-differentiation theorem is applied.

Example 3-11.

Determine the z-transform of $f(t) = te^{-at}$ by use of the partial-differentiation theorem. The z-transform of f(t) is

$$\mathcal{Z}[f(t)] = \mathcal{Z}[te^{-at}] = \mathcal{Z}\left[-\frac{\partial}{\partial a} e^{-at}\right] \tag{3-111}$$

From Eq. (3-109),

$$\mathcal{Z}\left[-\frac{\partial}{\partial a} e^{-at}\right] = \frac{-\partial}{\partial a}\mathcal{Z}[e^{-at}] = \frac{-\partial}{\partial a}\left[\frac{z}{z - e^{-aT}}\right] = \frac{Tze^{-aT}}{(z - e^{-aT})^2} \tag{3-112}$$

8. Real Convolution

If the functions $f_1(t)$ and $f_2(t)$ have the z-transforms $F_1(z)$ and $F_2(z)$, respectively, and $f_1(t) = f_2(t) = 0$ for $t < 0$, then

$$F_1(z)F_2(z) = \mathcal{Z}\left[\sum_{n=0}^{k} f_1(nT)f_2(kT - nT)\right] \tag{3-113}$$

Proof: The right-hand side of Eq. (3-113) is written as

$$\mathcal{Z}\left[\sum_{n=0}^{k} f_1(nT)f_2(kT - nT)\right] = \sum_{k=0}^{\infty} \sum_{n=0}^{k} f_1(nT)f_2(kT - nT)z^{-k}$$

$$= \sum_{k=0}^{\infty} \sum_{n=0}^{\infty} f_1(nT)f_2(kT - nT)z^{-k} \qquad (3\text{-}114)$$

Letting $m = k - n$, and interchanging the order of summation, we have

$$\mathcal{Z}\left[\sum_{n=0}^{k} f_1(nT)f_2(kT - nT)\right] = \sum_{n=0}^{\infty} f_1(nT)z^{-n} \sum_{m=-n}^{\infty} f_2(mT)z^{-m} \qquad (3\text{-}115)$$

Since $f_2(t) = 0$ for $t < 0$, the last equation becomes

$$\mathcal{Z}\left[\sum_{n=0}^{k} f_1(nT)f_2(kT - nT)\right] = \sum_{n=0}^{\infty} f_1(nT)z^{-n} \sum_{m=0}^{\infty} f_2(mT)z^{-m}$$

$$= F_1(z)F_2(z) \qquad (3\text{-}116)$$

The reader should recognize that the real-convolution theorem of the z-transformation is analogous to that of the Laplace transformation. An important fact to remember is that the inverse transform (Laplace or z) of the product of two functions is *not* equal to the product of the two functions in the real domain which are the inverse transforms of the two functions in the complex domain respectively; that is

$$\mathcal{Z}^{-1}[F_1(z)F_2(z)] \neq f_1(kT)f_2(kT) \qquad (3\text{-}117)$$

9. Complex Convolution

If the z-transforms of $f_1(t)$ and $f_2(t)$ are $F_1(z)$ and $F_2(z)$, respectively, then the z-transform of the product of the two functions is

$$\mathcal{Z}[f_1(t)f_2(t)] = \frac{1}{2\pi j} \oint_{\Gamma} \frac{F_1(\xi)F_2(z\xi^{-1})}{\xi} d\xi \qquad (3\text{-}118)$$

where Γ is a circle which lies in the region (annulus) described by

$$\sigma_1 < |\xi| < \frac{|z|}{\sigma_2}$$

and

$$|z| > \max(\sigma_1, \sigma_2, \sigma_1\sigma_2)$$

where

$$\sigma_1 = \text{Radius of convergence of } F_1(\xi)$$

$$\sigma_2 = \text{Radius of convergence of } F_2(\xi)$$

Proof: By definition the z-transform of $f_1(t)f_2(t)$ is written

$$\mathcal{Z}[f_1(t)f_2(t)] = \sum_{k=0}^{\infty} f_1(kT)f_2(kT)z^{-k} \tag{3-119}$$

In order that this z-transform series converges absolutely, the magnitude of z must be greater than the largest value of σ_1, σ_2, and $\sigma_1\sigma_2$; that is, $|z| > \max(\sigma_1, \sigma_2, \sigma_1\sigma_2)$.

We may write $f_1(kT)$ as an inverse z-transform relationship,

$$f_1(kT) = \frac{1}{2\pi j}\oint_\Gamma F_1(\xi)\xi^{k-1}d\xi \tag{3-120}$$

where Γ denotes a circle which encloses all the singularities of $F_1(\xi)\xi^{k-1}$; therefore, $|\xi| > \sigma_1$. Substituting Eq. (3-120) in Eq. (3-119), we obtain

$$\mathcal{Z}[f_1(t)f_2(t)] = \frac{1}{2\pi j}\oint_\Gamma \frac{F_1(\xi)}{\xi}d\xi \sum_{k=0}^{\infty} f_2(kT)(\xi^{-1}z)^{-k} \tag{3-121}$$

Since

$$F_2(\xi^{-1}z) = \sum_{k=0}^{\infty} f_2(kT)(\xi^{-1}z)^{-k} \tag{3-122}$$

and is absolutely convergent for $|\xi^{-1}z| > \sigma_2$, or $|\xi| < |z|/\sigma_2$, Eq. (3-121) becomes

$$\mathcal{Z}[f_1(t)f_2(t)] = \frac{1}{2\pi j}\oint_\Gamma \frac{F_1(\xi)}{\xi} F_2(z\xi^{-1})d\xi \tag{3-123}$$

with

$$\sigma_1 < |\xi| < \frac{|z|}{\sigma_2} \tag{3-124}$$

Example 3-12.

Determine the z-transform of $f(t) = te^{-at}$ by use of the complex-convolution theorem. Let $f_1(t) = t$ and $f_2(t) = e^{-at}$. Then

$$F_1(z) = \frac{Tz}{(z-1)^2} \qquad |z| > 1 = \sigma_1 \tag{3-125}$$

$$F_2(z) = \frac{z}{z - e^{-aT}} \qquad |z| > e^{-aT} = \sigma_2 \tag{3-126}$$

Substituting Eqs. (3-125) and (3-126) into Eq. (3-118) yields

$$\mathcal{Z}[f_1(t)f_2(t)] = \frac{1}{2\pi j}\oint_\Gamma \frac{T\xi}{\xi(\xi-1)^2}\,\frac{z\xi^{-1}}{z\xi^{-1}-e^{-aT}}\,d\xi \tag{3-127}$$

where Γ is a circular path which lies in the annular ring,

$$1 < |\xi| < \frac{|z|}{e^{-aT}} = |z|e^{aT} \tag{3-128}$$

and $|z| > 1$.

Therefore, the integration path of Eq. (3-127) encloses only the poles of the integrand at $\xi = 1$. Applying the residues theorem to Eq. (3-127), we get

$$\mathcal{Z}[f_1(t)f_2(t)] = \text{residue of } \frac{Tz\xi^{-1}}{(\xi-1)^2(z\xi^{-1}-e^{-aT})} \quad \text{at} \quad \xi = 1$$

$$= \frac{\partial}{\partial\xi}\left[\frac{Tz\xi^{-1}}{(z\xi^{-1}-e^{-aT})}\right]\Bigg|_{\xi=1} = \frac{Tze^{-aT}}{(z-e^{-aT})^2} \tag{3-129}$$

which agrees with the result obtained earlier in Example 3-11.

3.6 LIMITATIONS OF THE Z-TRANSFORM METHOD

In the preceding sections we have seen that the z-transformation is a very convenient tool for the representation of linear digital systems. However, the z-transform method has limitations; and in certain cases, care must be taken in the application and the interpretation of the results of the z-transforms.

The following considerations should be kept in mind when applying the z-transformation:

1. We must realize that the derivation of the z-transformation is based on the assumption that the sampled signal is approximated by a train of impulses whose areas are equal to the magnitude of the input signal of the sampler at the sampling instants. This assumption is considered valid only if the sampling duration of the sampler is small when compared with the significant time constant of the system.
2. The z-transform of the output of a linear system, $C(z)$, specifies only the values of the time function $c(t)$ at the sampling instants; $C(z)$ does not contain any information concerning the value of $c(t)$, between sampling instants. Therefore, given any $C(z)$, the inverse z-transform, $c(kT)$, describes $c(t)$ only at the sampling instants $t = kT$.
3. In analyzing a linear system by the z-transform method, the transfer function of the continuous-data system, $G(s)$, must have at least one

more pole than zeros, or equivalently, the impulse response of G(s) must not have any jump discontinuity at $t = 0$. Otherwise, the system response obtained by the z-transform method is misleading and sometimes even incorrect.

A complete description of any system response almost always necessitates a knowledge of the waveform between sampling instants. Several methods using the z-transformation have been developed for the evaluation of response of digital systems between sampling instants. Among these methods, the submultiple method and the modified z-transform method are the most common. These subjects are presented in Sec. 3.9.

3.7 THE PULSE TRANSFER FUNCTION

Thus far our examination of discrete-data systems has been confined to the discussion of the properties and mathematical representation of sampled signals. The situation must be investigated when a sampled signal is applied to the input terminals of a linear system.

For the linear open-loop system with continuous-data input r(t) as shown in Fig. 3-8(a), the transfer function

$$G(s) = \frac{C(s)}{R(s)} \tag{3-130}$$

describes the input-output relationship of the system. If the same system is now subjected to a sampled signal as shown in Fig. 3-8(b), the Laplace transform of the output of the system is written as

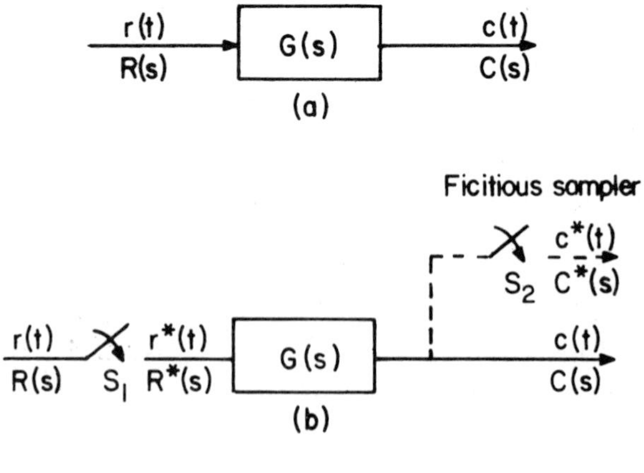

Fig. 3-8. (a) Linear system with continuous-data input. (b) Linear system with sampled-data input.

$$C(s) = R^*(s)G(s) \tag{3-131}$$

where $R^*(s)$ is the Laplace transform of the sampled-data input $r^*(t)$. Our objective here is to find a way of expressing the digital system in terms of the z-transforms $C(z)$, $R(z)$, and $G(z)$. A simple way of accomplishing this is to form $C^*(s)$ using Eq. (2-85); we obtain

$$C^*(s) = \frac{1}{T}\sum_{n=-\infty}^{\infty} C(s + jn\omega_s) = \frac{1}{T}\sum_{n=-\infty}^{\infty} R^*(s + jn\omega_s)G(s + jn\omega_s) \tag{3-132}$$

Using the identity of Eq. (2-104),

$$R^*(s + jn\omega_s) = R^*(s) \tag{3-133}$$

Equation (3-132) becomes

$$C^*(s) = R^*(s)\frac{1}{T}\sum_{n=-\infty}^{\infty} G(s + jn\omega_s) \tag{3-134}$$

Now defining

$$G^*(s) = \frac{1}{T}\sum_{n=-\infty}^{\infty} G(s + jn\omega_s) \tag{3-135}$$

we have from Eq. (3-134)

$$C^*(s) = R^*(s)G^*(s) \tag{3-136}$$

Upon transforming into z by $z = e^{Ts}$, Eq. (3-136) gives

$$C(z) = G(z)R(z) \tag{3-137}$$

which is the desired transfer function relation for a linear system with sampled-data input. Notice that the output of the system may still be a continuous-data signal, but in Eq. (3-137) $C(z)$ defines the output only at the sampling instants.

An alternate way of arriving at Eq. (3-137) would be to use the impulse response method. Suppose that at $t = 0$ a unit impulse is applied as input to the system of Fig. 3-8(b). The output of the system is simply the impulse response $g(t)$. If a fictitious sampler S_2 which is synchronized to S_1 and with the same sampling period, is placed at the system output, the output of S_2 is described by

$$c^*(t) = g^*(t) = \sum_{k=0}^{\infty} g(kT)\delta(t - kT) \tag{3-138}$$

where $g(kT)$, for $k = 0,1,2,3,\ldots$, is defined as the *weighting sequence* or the *impulse sequence* of the system.

When the discrete-data signal $r^*(t)$ is applied as input to the linear system, the output of the system is written as

$$c(t) = r(0)g(t) + r(T)g(t - T) + r(2T)g(t - 2T) + \dots \tag{3-139}$$

At $t = kT$, where k is a positive integer, Eq. (3-139) gives

$$c(kT) = r(0)g(kT) + r(T)g[(k-1)T] + \dots + r(kT)g(0)$$

$$= \sum_{n=0}^{k} r(nT)g(kT - nT) \tag{3-140}$$

Taking the z-transform on both sides of Eq. (3-140) and using the real-convolution theorem of Eq. (3-113), we have

$$C(z) = R(z)G(z) \tag{3-141}$$

where

$$G(z) = \sum_{k=0}^{\infty} g(kT)z^{-k} \tag{3-142}$$

is defined as the *z-transfer function* of the linear system. Therefore, the z-transfer function G(z) relates the z-transform of the input, R(z), to the z-transform of the output C(z) in the same way as the continuous-data transfer function G(s) does for R(s) and C(s). However, if the linear system contains all continuous-data elements, the output signal is a function of the continuous t. The z-transform expression of Eq. (3-141) gives information on the continuous c(t) only at the discrete time instants $t = kT$. This is one of the limitations of the z-transform method. In certain cases the loss of information between sampling instants may not be vital, e.g., if the value of c(t) does not vary appreciably between sampling instants. On the other hand, if c(t) contains large-amplitude oscillations between the sampling instants, the z-transform method would often give misleading results. Still another comment is that in view of Eqs. (3-135) and (3-142), the z-transfer function G(z) is derived from the impulse response g(t) in exactly the same manner as R(z) is derived from the signal r(t).

Discrete-Data Systems with Cascaded Elements

When a discrete-data system contains cascaded elements, care must be taken in deriving the transfer function for the overall system. Figure 3-9 illustrates a discrete-data system with cascaded elements G_1 and G_2. The two elements are separated by a second sampler S_2 which is identical and synchronized to the sampler S_1. The z-transfer function of the overall system is derived as follows: The transfer relationships of the two systems G_1 and G_2 are

$$D(z) = G_1(z)R(z) \tag{3-143}$$

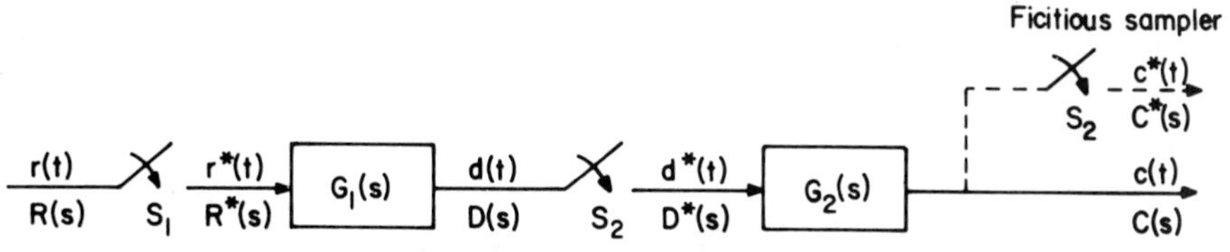

Fig. 3-9. Digital system with sampler-separated elements.

and

$$C(z) = G_2(z)D(z) \tag{3-144}$$

Substitution of Eq. (3-143) in Eq. (3-144) yields

$$C(z) = G_1(z)G_2(z)R(z) \tag{3-145}$$

Therefore, the z-transfer function of two linear elements separated by a sampler is equal to the product of the z-transfer functions of the two elements.

When a digital system contains two cascade elements not separated by a sampler, as shown in Fig. 3-10, the z-transfer function of the overall system should be written as

$$\mathcal{Z}[G_1(s)G_2(s)] = G_1G_2(z) = G_2G_1(z) \qquad \text{(3-146)*}$$

Notice that in general

$$G_1G_2(z) \neq G_1(z)G_2(z) \tag{3-147}$$

The input-output transforms of the system are related by

$$C(z) = G_1G_2(z)R(z) \tag{3-148}$$

The transfer relationships of digital control systems with feedback and a multiple number of synchronized samplers can be obtained by algebraic manipulation or by the signal-flow-graph method. In general, the algebraic method is

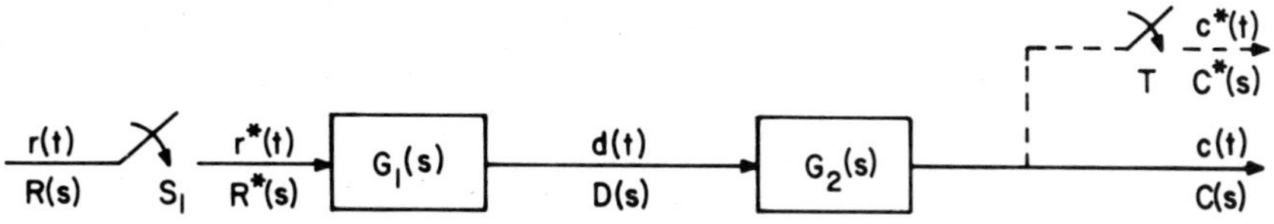

Fig. 3-10. Digital system with cascaded elements.

*For more than two functions the z-transform relation is expressed as $\mathcal{Z}[G_1(s)G_2(s)G_3(s)] = G_1G_2G_3(z)$. This type of notation is used in the rest of this book without further definition.

tedious and unreliable, especially in complex situations. The signal-flow-graph method is presented in Sec. 3.10.

3.8 PULSE TRANSFER FUNCTION OF THE ZERO-ORDER HOLD AND THE RELATION BETWEEN G(s) AND G(z)

It was established in Chapter 2 that the transfer function of a zero-order hold is given by [Eq. (2-114)],

$$G_{h0}(s) = \frac{1 - e^{-Ts}}{s} \tag{3-149}$$

The z-transform of $G_{h0}(s)$ is

$$G_{h0}(z) = \mathcal{Z}\left[\frac{1 - e^{-Ts}}{s}\right] = (1 - z^{-1})\mathcal{Z}\left[\frac{1}{s}\right] = 1 \tag{3-150}$$

This result is expected since the zero-order hold simply holds the sampled signal for one sampling duration, and taking the z-transform of the zero-order hold would revert to the original sampled signal again. However, in a great majority of the practical systems, the zero-order hold is followed by a continuous-data system as shown in Fig. 3-11. The z-transform of the output of the system is written as

$$C(z) = G_1(z)R(z) \tag{3-151}$$

where

$$G_1(z) = \mathcal{Z}[G_{h0}(s)G(s)] \tag{3-152}$$

Substituting Eq. (3-149) into Eq. (3-152), we have

$$G_1(z) = \mathcal{Z}\left[\frac{1 - e^{-Ts}}{s} G(s)\right]$$

$$= (1 - z^{-1})\mathcal{Z}[G(s)/s] \tag{3-153}$$

In this case the z-transform of the numerator of the zero-order hold is factored out, due to the shifting theorem in Eq. (3-81). However, the z-transform of G(s)/s must be taken as one operation.

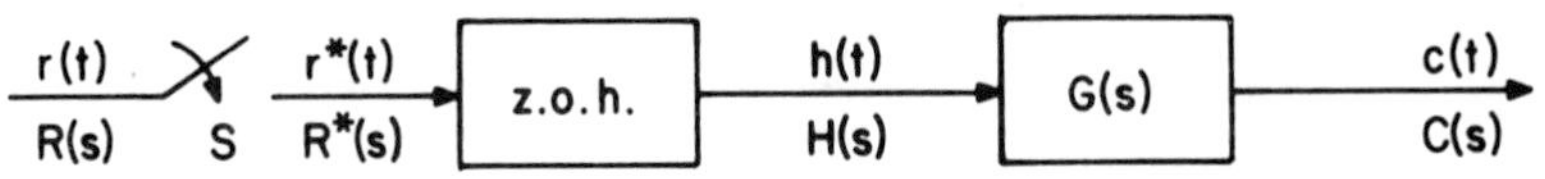

Fig. 3-11. A system with sample-and-hold.

A notion was made in Chapter 2 that in theory when the sampling frequency reaches infinity, a sampled-data system reverts to the corresponding continuous-data system. However, this *does not* imply that given

$$\mathcal{Z}[G(s)] = G(z) \tag{3-154}$$

then

$$\lim_{T\to 0} G(z) = G(s) \tag{3-155}$$

Since the theory of z-transform is based on the impulse-amplitude modulation of a continuous-data signal with a sampling period of T second, setting T to zero does not make much physical sense. In other words, if the signal r(t) is sampled by an ideal sampler to give r*(t), then setting the sampling period T to zero does not revert r*(t) to r(t). This explains why Eq. (3-155) is not true in general. However, if we send the sampled signal r*(t) through a zero-order hold with the output denoted by h(t), then

$$\lim_{T\to 0} h(t) = r(t) \tag{3-156}$$

or

$$\lim_{T\to 0} H(s) = R(s) \tag{3-157}$$

Thus, the significance of the last two equations is that if a continuous-data signal r(t) is sent through a sample-and-hold, the output of the latter can be reverted to r(t) by setting the sampling period T to zero.

Example 3-13.

As a simple illustrative example, consider that the input r(t) in the system of Fig. 3-11 is e^{-at}. Then $R(s) = 1/(s + a)$. The pulsed transform of r(t) is

$$R^*(s) = \frac{e^{Ts}}{e^{Ts} - e^{-aT}} \tag{3-158}$$

Thus, the output of the zero-order hold in the Laplace domain is

$$H(s) = G_{h0}(s)R^*(s) = \frac{1 - e^{-Ts}}{s} \frac{e^{Ts}}{e^{Ts} - e^{-aT}} \tag{3-159}$$

We can show that as the sampling period T approaches zero,

$$\lim_{T\to 0} H(s) = \frac{1}{s + a} = R(s) \tag{3-160}$$

Another important and useful property of the z-transform is

$$\lim_{T\to 0} \mathcal{Z}[G_{h0}(s)G(s)] = G(s) \tag{3-161}$$

This is proved simply by substituting Eq. (3-149) into Eq. (3-161), where we have

$$\lim_{T\to 0} \mathcal{Z}[G_{h0}(s)G(s)] = \lim_{T\to 0} \mathcal{Z}\left[\frac{1-e^{-Ts}}{s} G(s)\right] \tag{3-162}$$

Expanding e^{-Ts} into a power series and taking only the first two terms, the last equation is simplified to

$$\lim_{T\to 0} \mathcal{Z}[G_{h0}(s)G(s)] = \lim_{T\to 0} TG(z) \tag{3-163}$$

Now recalling the identity

$$G(z) = G^*(s)\Big|_{z=e^{Ts}} = \frac{1}{T}\sum_{n=-\infty}^{\infty} G(s + j2n\pi/T)\Big|_{z=e^{Ts}} \tag{3-164}$$

we can write Eq. (3-163) as

$$\lim_{T\to 0} \mathcal{Z}[G_{h0}(s)G(s)] = \lim_{T\to 0} \sum_{n=-\infty}^{\infty} G(s + j2n\pi/T)\Big|_{z=e^{Ts}} = G(s) \tag{3-165}$$

Example 3-14.

Consider that the transfer function of the sampled-data system shown in Fig. 3-11 is

$$G(s) = \frac{K}{s(s+a)} \tag{3-166}$$

where K and a are constants. The z-transform of the combination of the zero-order hold and G(s) is

$$G_1(z) = \mathcal{Z}[G_{h0}(s)G(s)] = (1 - z^{-1})\mathcal{Z}\left[\frac{K}{s^2(s+a)}\right] \tag{3-167}$$

Evaluating the z-transform in the last equation and simplifying, we have

$$G_1(z) = \frac{KT}{a(z-1)} - \frac{K(1-e^{-aT})}{a^2(z-e^{-aT})} \tag{3-168}$$

To show that the limit of $G_1(z)$ as $T \to 0$ is G(s), we set $z = e^{Ts}$ and approximate e^{Ts} by $1 + Ts$, the first two terms of the power-series expansion. Similarly, e^{-aT} is approximated by $1 - aT$. The limit is written as

$$\lim_{T\to 0} G_1(z) = \lim_{T\to 0} \frac{KT}{a(Ts+1-1)} - \lim_{T\to 0} \frac{K(1-1+aT)}{a^2(1+sT-1+aT)}$$

$$= \frac{K}{as} - \frac{K}{a(s+a)} = \frac{K}{s(s+a)} = G(s) \qquad (3\text{-}169)$$

3.9 RESPONSES BETWEEN SAMPLING INSTANTS

It was pointed out in Sec. 3.6 that the accuracy of the z-transform method is limited only to systems with signals that are well-behaved between the sampling instants. In other words, the z-transform method is effective only for systems in which the signals can be adequately represented by their values at the sampling instants.

When the z-transform analysis is inadequate, one must gain knowledge on the system responses between sampling instants. The *submultiple-sampling method* and the *modified z-transform method*, which are useful for the recovery of signal information between sampling instants, are discussed in the following. These methods are also useful as analytic tools for the studies of digital systems with nonuniform or multirate samplers.

1. The Submultiple-Sampling Method

Figure 3-12(a) shows the block diagram of a sampled-data system with a fictitious sampler placed at the output for z-transform analysis. The two samplers are denoted S_1 to indicate that they have the basic sampling period of T second. The input-output z-transform relation is given by Eq. (3-141) with the z-transfer function defined in Eq. (3-142). In order to recover the information between the sampling instants, we introduce two additional fictitious samplers S_N and insert them in the system as shown in Fig. 3-12(b). The sampling period of S_N is T/N, where N is a positive integer greater than one. These fictitious

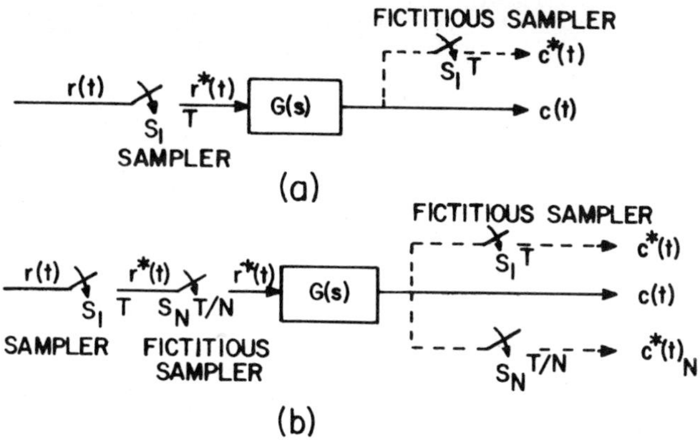

Fig. 3-12. (a) Sampled-data system. (b) Sampled-data system with fictitious samplers for the evaluation of response between sampling instants.

samplers are not physical components; they are used simply to facilitate the calculation of the system response between the sampling instants. Since the input signal to the fictitious sampler S_N at the input of the system is a train of impulses that are spaced T seconds apart, and S_N samples at a rate N times as fast as S_1, the insertion of these fictitious samplers does not in any way affect the performance of the original system.

The output of the system, c(t), may be written as a sum of impulse responses,

$$c(t) = \sum_{m=0}^{\infty} r(mT)g(t - mT) \tag{3-170}$$

where g(t) is the impulse response of the system. At any sampling instant t = kT/N of the sampler S_N, the value of c(t) is

$$c(kT/N) = \sum_{m=0}^{\infty} r(mT)g(kT/N - mT) \tag{3-171}$$

The output of the fictitious sampler S_N is written as

$$c^*(t)_N = \sum_{k=0}^{\infty} c(kT/N)\delta(t - kT/N) \tag{3-172}$$

Taking the z-transform on both sides of Eq. (3-172) yields

$$C(z)_N = \mathfrak{z}[c^*(t)_N] = \sum_{k=0}^{\infty} c(kT/N)z^{-k/N} \tag{3-173}$$

Comparison of Eq. (3-173) with the defining equation of the ordinary z-transform reveals that $C(z)_N$ can be obtained directly from C(z) by replacing z by $z^{1/N}$ and T by T/N; that is

$$C(z)_N = C(z)\Big|_{\substack{z=z^{1/N}\\ T=T/N}} \tag{3-174}$$

Substituting Eq. (3-171) into Eq. (3-173), we have

$$C(z)_N = \sum_{k=0}^{\infty}\sum_{m=0}^{\infty} r(mT)g(kT/N - mT)z^{-k/N} \tag{3-175}$$

Now let v/N = (k/N − m), where v is an integer, Eq. (3-175) becomes

$$C(z)_N = \sum_{v=0}^{\infty} g(vT/N)z^{-v/N} \sum_{m=0}^{\infty} r(mT)z^{-m} \tag{3-176}$$

Defining

$$G(z)_N = \sum_{v=0}^{\infty} g(vT/N)z^{-v/N} = G(z)\Big|_{\substack{z=z^{1/N}\\T=T/N}} \tag{3-177}$$

Eq. (3-176) is written

$$C(z)_N = G(z)_N R(z) \tag{3-178}$$

These derivations lead to the conclusion that the response between the regular sampling instants can be evaluated by changing z and T in the ordinary z-transform of the system transform function. It is important to point out that if the system has a zero-order hold, then the term, $1 - z^{-1}$, due to the hold *should not be affected* by the $T = T/N$ and $z = z^{1/N}$ operation.*

The number of points on c(t) at which the values are desired determines the value of N. In general, if m is the number of additional sampled values desired, $N = m + 1$.

Example 3-15.

Consider that the sampled-data system of Fig. 3-12(a) has the transfer function

$$G(s) = \frac{1}{s+1} \tag{3-179}$$

The input r(t) is a unit-step function. The sampling period is 1 second. It is desired to find the response of the system at time instants of $t = kT/3$, $k = 0,1,2,\ldots$.

From Eq. (3-178), the z-transform of the system at the submultiple-sampling instants is written as

$$C(z)_3 = G(z)_3 R(z) \tag{3-180}$$

where

$$G(z)_3 = G(z)\Big|_{\substack{z=z^{1/3}\\T=T/3}} = \frac{z}{z-e^{-T}}\Bigg|_{\substack{z=z^{1/3}\\T=T/3}} \tag{3-181}$$

Thus,

$$G(z)_3 = \frac{z^{1/3}}{z^{1/3} - e^{-1/3}} = \frac{z^{1/3}}{z^{1/3} - 0.717} \tag{3-182}$$

The z-transform of the unit-step input is $R(z) = z/(z-1)$. The z-transform of the submultiple-sampled output is

$$C(z)_3 = \frac{z^{1/3}}{z^{1/3} - 0.717}\,\frac{z}{z-1} \tag{3-183}$$

*If G(s) contains a zero-order hold, i.e., $G(s) = G_{ho}(s)G_1(s)$, then
$G(z)_N = (1 - z^{-1})\left[\mathcal{Z}[G_1(s)/s]\right]_{z=z^{1/N}, T=T/N}$.

However, one difficulty remains in that the last expression has fractional powers as well as integral powers of z. To overcome this difficulty, we introduce a new variable z_3, such that

$$z_3 = z^{1/3} \tag{3-184}$$

Equation (3-183) becomes

$$C(z)_3 = \frac{z_3}{z_3 - 0.717}\,\frac{z_3^3}{z_3^3 - 1} = \frac{z_3^4}{(z_3 - 0.717)(z_3^3 - 1)} \tag{3-185}$$

Expanding $C(z)_3$ into a power series in z_3^{-1}, we have

$$C(z)_3 = 1 + 0.717z_3^{-1} + 0.513z_3^{-2} + 1.368z_3^{-3} + 0.98z_3^{-4} + 0.703z_3^{-5}$$

$$+ 1.504z_3^{-6} + 1.08z_3^{-7} + 0.773z_3^{-8} + 1.55z_3^{-9} + \ldots \tag{3-186}$$

The coefficients of the power-series expansion of $C(z)_3$ are the values of $c^*(t)_3$ at $t = kT/3$, $k = 0,1,2,\ldots$. The response $c^*(t)_3$ is shown in Fig. 3-13. In this case, the value of the submultiple sampling method is clearly demonstrated, since the ordinary z-transform obviously would produce a misleading result.

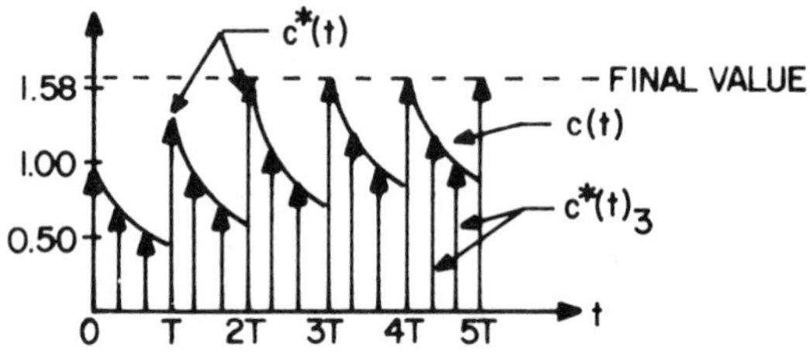

Fig. 3-13. Output responses of the system in Example 3-15.

The subject of the submultiple sampling method will be treated again when we deal with the problems of multirate sampled-data systems.

2. The Modified Z-Transform Method

Another method of evaluating the response between sampling instants of a sampled-data system is the modified z-transform method. The method represents a modification of the ordinary z-transform by using fictitious time delays in the sampled-data system.

Consider that it is desired to evaluate the response between the sampling instants of the system shown in Fig. 3-14(a). First we insert a fictitious time delay unit at the output of the system, and then the delayed output is sampled by a fictitious sampler, as shown in Fig. 3-14(b). The fictitious time delay is

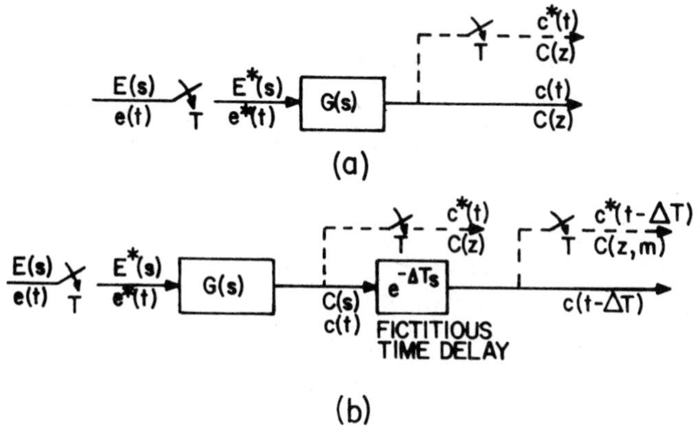

Fig. 3-14. (a) Open-loop sampled-data control system. (b) Sampled-data system with fictitious time delay.

ΔT, where Δ lies between zero and one ($0 \leqslant \Delta < 1$). The fictitious samplers have the same sampling rate and are synchronized with the sampler at the input of the system. As shown in Fig. 3-14(b), since the output c(t) is still available, the use of the fictitious sampler does not affect the operation of the system from the analytical standpoint.

The sampled output of the fictitious time delay is written

$$c^*(t - \Delta T) = \sum_{k=0}^{\infty} c(kT - \Delta T)\delta(t - kT) \tag{3-187}$$

The z-transform of $c^*(t - \Delta T)$ can be expressed as

$$\mathcal{Z}[c^*(t - \Delta T)] = C(z,\Delta) = \mathcal{L}[c(t - \Delta T)] * \mathcal{L}\left[\sum_{k=0}^{\infty} \delta(t - kT)\right]\Bigg|_{z=e^{Ts}} \tag{3-188}$$

or

$$C(z,\Delta) = \left[C(s)e^{-\Delta Ts} * \frac{1}{1 - e^{-Ts}}\right]\Bigg|_{z=e^{Ts}} \tag{3-189}$$

The convolution in the last equation is written as

$$C(z,\Delta) = \frac{1}{2\pi j}\left[\int_{c-j\infty}^{c+j\infty} C(\xi)e^{-\Delta T\xi}\, \frac{1}{1 - e^{-T(s-\xi)}}\, d\xi\right]\Bigg|_{z=e^{Ts}} \tag{3-190}$$

The line integral in the last expression can be evaluated along the line from $c - j\infty$ to $c + j\infty$ and the semicircle of infinite radius in either the left half or the right half of the complex ξ-plane. These contours of integration are similar to those of Fig. 2-35. The contour-integration method will yield a correct result for the line integral of Eq. (3-190) only if the integral along either one of the two semicircles is zero. However, even when the integrals along both semicircles are finite, the results from the two paths will still have different forms.

Let us first consider the contour integration along the infinite semicircle in the left-half ξ-plane. Since the term $e^{-\Delta T\xi}$ has a pole at $\xi = -\infty$, there is a pole on the infinite semicircle in the left-half plane. In order to overcome this difficulty, we introduce a parameter m, such that

$$m = 1 - \Delta \tag{3-191}$$

Since $0 \leqslant \Delta < 1$, the value of m also lies between zero and one. Then, if we define

$$C(z,m) = C(z,\Delta)\Big|_{\Delta=1-m} \tag{3-192}$$

Eq. (3-190) becomes

$$C(z,m) = \left[\frac{1}{2\pi j}\oint C(\xi)e^{-T\xi}e^{mT\xi}\,\frac{1}{1-e^{-T(s-\xi)}}\,d\xi\right]\Bigg|_{z=e^{Ts}} \tag{3-193}$$

Now the integral along the infinite semicircle in the left-half ξ-plane is zero, and the contour integral of Eq. (3-193) is equivalent to the line integral of Eq. (3-190).

Using the residue theorem of complex-variable theory, Eq. (3-193) is written

$$C(z,m) = z^{-1}\sum \text{Residues of } C(\xi)\,\frac{e^{mT\xi}}{1-e^{T\xi}z^{-1}}$$

$$\text{at the poles of } C(\xi) \tag{3-194}$$

Equation (3-194) gives one definition of the modified z-transform of the signal c(t). More important, the above discussion establishes the motivation for using the parameter m rather than the original delay factor Δ.

By taking the contour integral along the infinite semicircular path in the right-half ξ-plane, the line integral of Eq. (3-190) is written as

$$C(z,\Delta) = \frac{1}{2\pi j}\oint C(\xi)e^{-\Delta T\xi}\,\frac{1}{1-e^{-T(s-\xi)}}\,d\xi\Bigg|_{z=e^{Ts}} \tag{3-195}$$

In this case, the integral on the infinite semicircle alone is zero if $\lim_{s\to\infty} C(s) = 0$. Applying the residue theorem to Eq. (3-195), we have

$$C(z,\Delta) = -\sum \text{Residues of } C(\xi)e^{-\Delta T\xi} \left. \frac{1}{1 - e^{-T(s-\xi)}} \right|_{z=e^{Ts}}$$

$$\text{at the poles of } 1/[1 - e^{-T(s-\xi)}] \tag{3-196}$$

Since the poles of $1/[1 - e^{-T(s-\xi)}]$ are simple and are at $\xi = s \pm jn\omega_s$, $n = 0,1,\ldots$, Eq. (3-196) leads to the result

$$C(z,\Delta) = \frac{1}{T} \sum_{n=-\infty}^{\infty} C(s + jn\omega_s)e^{-\Delta T(s+jn\omega_s)} \Big|_{z=e^{Ts}} \tag{3-197}$$

Although in this case it is not necessary to introduce the parameter $m = 1 - \Delta$, for the purpose of being uniform in the modified z-transform definition we write Eq. (3-197) as

$$C(z,m) = \frac{1}{T} \sum_{n=-\infty}^{\infty} C(s + jn\omega_s)e^{-(1-m)T(s+jn\omega_s)} \Big|_{z=e^{Ts}} \tag{3-198}$$

Still another expression for the modified z-transform of c(t) is obtained by applying the definition of z-transform directly to Eq. (3-187). The result is

$$C(z,m) = \mathscr{Z}[c^*(t - \Delta T)] \Big|_{\Delta=1-m} = \sum_{k=0}^{\infty} c(kT + mT - T)z^{-k} \tag{3-199}$$

Using the shifting theorem of the z-transform, Eq. (3-81), the last equation is simplified to

$$C(z,m) = z^{-1} \sum_{k=0}^{\infty} c[(k + m)T]\, z^{-k} \tag{3-200}$$

We have determined three alternate expressions for the modified z-transform of a signal, Eqs. (3-194), (3-198) and (3-200). These equations are subject to different conditions of validity, and are useful for various purposes. Equation (3-194) is valid for any c(t) that has a Laplace transform C(s). The equation can be used for the purpose of evaluating the modified z-transform expression. Equation (3-198) is valid only if $c(0) = 0$, and thus it is not valid for time functions that have a jump discontinuity at $t = 0$. The expression in Eq. (3-200) is the most general in that it is defined for any c(t).

In summary, the modified z-transform of a function c(t) is denoted by

$$\text{modified z-transform of } c(t) = \mathscr{Z}_m[c(t)]$$

$$= C(z,m) \tag{3-201}$$

However, in reality, C(z,m) is also defined as the z-transform of the delayed function $c(t - \Delta T)$ or $c(t - T + mT)$, where $0 \leq \Delta < 1$ and $0 \leq m < 1$, that is,

$$C(z,m) = \mathcal{Z}[c(t - \Delta T)] = \mathcal{Z}[c(t - T + mT)] \tag{3-202}$$

The response of the signal between the sampling instants is determined from C(z,m) by varying the value of m (or Δ) between zero and one. It is clear that when $\Delta = 0$, the modified z-transform reverts to the z-transform. However, when $\Delta = 0$, $m = 1$, because of the mathematical conditions imposed on the derivations of the modified z-transform, the following relation is generally *not* true

$$C(z,m)\Big|_{m=1} = C(z) \tag{3-203}$$

Setting m = 1 in Eq. (3-200), we get

$$C(z,m)\Bigg|_{m=1} = z^{-1} \sum_{k=0}^{\infty} c[(k+1)T]z^{-k} = C(z) - c(0) \tag{3-204}$$

Thus, Eq. (3-203) is generally true only if $c(0) = 0$. On the other hand, setting $m = 1$ in Eq. (3-198) leads to

$$C(z,m)\Bigg|_{m=1} = \frac{1}{T} \sum_{n=-\infty}^{\infty} C(s + jn\omega_s)\Bigg|_{z=e^{Ts}} = C(z) \tag{3-205}$$

This is because Eq. (3-198) is valid only if $c(0) = 0$.

The evaluation of the modified z-transform of a signal is illustrated by the following example. A table of modified z-transforms is given in APPENDIX A, along with the table of z-transforms.

Example 3-16.

The time function $c(t) = e^{-at}$ $(t \geq 0)$, where a is a constant, is considered. First, using Eq. (3-200), the modified z-transform of c(t) is

$$C(z,m) = z^{-1} \sum_{k=0}^{\infty} e^{-a(k+m)T} z^{-k} \tag{3-206}$$

This infinite series can be put in a closed form:

$$C(z,m) = \frac{e^{-maT}}{z - e^{-aT}} \tag{3-207}$$

In this case, since $c(0) \neq 0$, $C(z,1) \neq C(z)$, as expected.
Now substituting the Laplace transform of c(t) into Eq. (3-194), we have

$$C(z,m) = z^{-1}\left[\text{Residue of } \frac{1}{\xi + a}\,\frac{e^{mT\xi}}{1 - e^{T\xi}z^{-1}}\right] \quad \text{at} \quad \xi = -a$$

$$= z^{-1}\left.\frac{e^{mT\xi}}{1 - e^{T\xi}z^{-1}}\right|_{\xi=-a} = \frac{e^{-maT}}{z - e^{-aT}} \tag{3-208}$$

The power-series expansion of $C(z,m)$ is

$$C(z,m) = e^{-maT}z^{-1} + e^{-(m+1)aT}z^{-2} + \ldots + e^{-(m+k)aT}z^{-(k+1)} + \ldots \tag{3-209}$$

The coefficient of the z^{-k} term in the infinite series represents the value of $c(t)$ between the sampling instants of $t = (k-1)T$ and $t = kT$ where $k = 1,2,\ldots,$ and $0 < m \leqslant 1$.

The relation of the modified z-transform in Eq. (3-198) can be used for the determination of the input-output transfer function of the system shown in Fig. 3-14. Since the Laplace transform of the output $c(t)$ is expressed as

$$C(s) = G(s)E^*(s) \tag{3-210}$$

substituting the last equation into Eq. (3-198) yields

$$C(z,m) = \frac{1}{T}\sum_{n=-\infty}^{\infty} G(s + jn\omega_s)E^*(s + jn\omega_s)e^{-(1-m)T(s+jn\omega_s)}\Big|_{z=e^{Ts}}$$

$$= E(z)\frac{1}{T}\sum_{n=-\infty}^{\infty} G(s + jn\omega_s)e^{-(1-m)T(s+jn\omega_s)}\Big|_{z=e^{Ts}}$$

$$= E(z)G(z,m) \tag{3-211}$$

where $G(z,m)$ denotes the modified z-transform of $G(s)$. The above development indicates that the modified z-transform of a transfer function already in the pulsed form is the z-transform of the function. The following relation is used to demonstrate this fact:

$$\mathscr{Z}_m[E^*(s)] = E(z) \tag{3-212}$$

Furthermore,

$$\mathscr{Z}_m[C(s)] = C(z,m) = \mathscr{Z}_m[G(s)E^*(s)] = G(z,m)E(z) \tag{3-213}$$

Example 3-17

Consider that the system shown in Fig. 3-14 has a transfer function

$$G(s) = \frac{1}{s + a} \tag{3-214}$$

where a is a constant. The input to the system is a unit-step function; i.e., $e(t) = u_s(t)$. The output of the system is to be determined by means of the modified z-transform.

The modified z-transform of the output is given by Eq. (3-213). The modified z-transform of G(s) is already determined in Eq. (3-207).

Substitution of Eq. (3-207) and the z-transform of the unit-step function into Eq. (3-213) yields

$$C(z,m) = G(z,m)E(z) = \frac{e^{-maT}}{z - e^{-aT}} \frac{z}{z-1} \tag{3-215}$$

When C(z,m) is expanded into a power series in z^{-1}, it can be shown that the coefficient of z^{-k} is $e^{-maT}(1 - e^{-kaT})/(1 - e^{-aT})$, which gives the output response for the time duration $(k-1)T < t \leqslant kT$, $k \geqslant 1$, when m is varied between 0 and 1. The output response for $t \geqslant 0$ should have the same form as in Fig. 3-13.

For a closed-loop sampled data system the response in between the sampling instants can be determined in much the same way as in an open-loop system. Figure 3-15 illustrates a closed-loop sampled-data system with a fictitious time delay positioned at the output. The following equations are written directly from the block diagram:

$$C(z,m) = G(z,m)E(z) \tag{3-216}$$

$$E(z) = R(z) - GH(z)E(z) \tag{3-217}$$

where GH(z) denotes the z-transform of G(s)H(s). Combining Eqs. (3-216) and (3-217) gives

$$C(z,m) = \frac{G(z,m)}{1 + GH(z)} R(z) \tag{3-218}$$

as the modified z-transform of the output.

Although the modified z-transform was originally devised for the purpose of describing signals between sampling instants, we shall show later in Sec. 3.11 that the method can also be used for the analysis of multirate sampled-data systems.

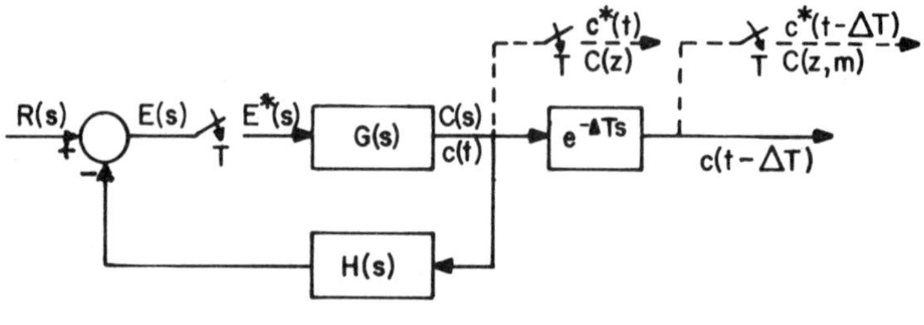

Fig. 3-15. A closed-loop sampled-data control system with fictitious time delay.

3.10 SIGNAL-FLOW-GRAPH METHOD APPLIED TO DIGITAL SYSTEMS

In the preceding sections closed-loop sampled-data systems are analyzed by the z-transform and the modified z-transform methods. For the single-loop feedback systems the input-output transfer relations are determined by algebraic means. However, for sampled-data control systems with multiple loops and samplers, the block diagram algebra may become too involved, and a more systematic approach is desired.

It is well-known that transfer functions of linear continuous-data systems can be determined from the signal flow graphs using Mason's gain formula [15, 16]. We shall now investigate the possibility of extending the signal-flow-graph technique to sampled-data and digital systems. However, most digital control systems contain digital as well as analog signals so that although a signal flow graph can be drawn as an equivalent to the block diagram, the Mason's gain formula cannot be applied directly. Two methods of extending the signal-flow-graph technique to the evaluation of input-output relationships of digital systems will be discussed. The first method relies on the forming of a "sampled signal flow graph" in which all the node variables are discrete quantities. Once all the node variables represent either continuous data or digital data uniformly, Mason's gain formula can again be applied. The second method, which is due to Sedler and Bekey [17], allows the derivation of the input-output transfer relations of a digital system directly from the system's signal flow graph. Since the signals in the system are usually mixed, a gain formula different from the original one by Mason has to be introduced.

The Sampled-Signal-Flow-Graph Method

The method involves the following steps:

1. With the system's block diagram as the starting point, construct an equivalent signal flow graph for the system. This is also known as the signal flow graph of the system, since it is entirely equivalent to the block diagram representation. As an illustration, Fig. 3-16 shows the block diagram of a system and its equivalent signal flow graph.
2. The second step concerns the construction of the "sampled signal flow graph" from the signal flow graph of the system. Several intermediate steps which lead to the drawing of the sampled signal flow graph are involved. With reference to Fig. 3-16(b), the following set of equations is written for all the noninput nodes of the signal flow graph:

$$E(s) = R(s) - G(s)H(s)E^*(s) \tag{3-219}$$

$$C(s) = G(s)E^*(s) \tag{3-220}$$

Notice that in this step Mason's gain formula is applied to all the noninput nodes with the inputs of the system as well as the outputs of the samplers considered as input variables. The samplers may be considered as being deleted from the signal flow graph once their output variables are defined.

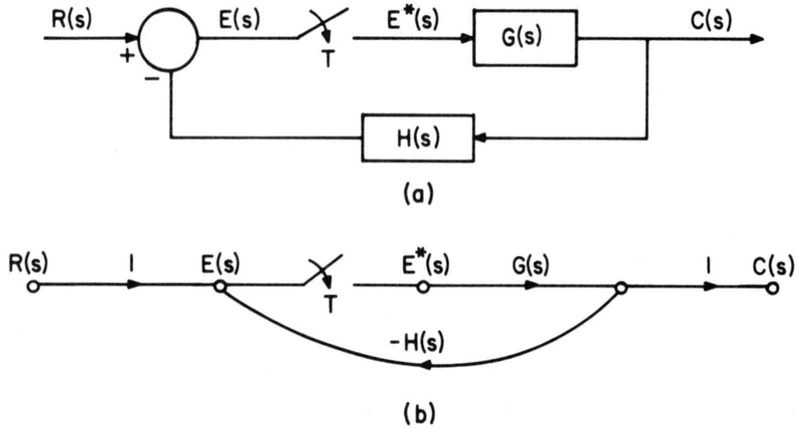

Fig. 3-16. (a) Block diagram of a closed-loop digital system; (b) equivalent signal flow graph of the system.

Taking the pulse transform on both sides of Eq. (3-219) and (3-220), we have

$$E^*(s) = R^*(s) - GH^*(s)E^*(s) \tag{3-221}$$

$$C^*(s) = G^*(s)E^*(s) \tag{3-222}$$

where it has been shown earlier that

$$[G(s)H(s)]^* = HG^*(s) = GH^*(s)$$

$$= \frac{1}{T} \sum_{n=-\infty}^{\infty} G(s + jn\omega_s)H(s + jn\omega_s) \tag{3-223}$$

and

$$[E^*(s)]^* = \frac{1}{T} \sum_{n=-\infty}^{\infty} E^*(s + jn\omega_s) = E^*(s) \tag{3-224}$$

Now since Eqs. (3-221) and (3-222) contain only sampled variables, the signal flow graph drawn using these equations is called the *sampled signal flow graph* of the system in Fig. 3-16. The sampled signal flow graph is shown in Fig. 3-17.

3. Once the sampled signal flow graph is drawn, the transfer function relation between any pair of input and output nodes* on this flow graph can be determined by use of Mason's gain formula. For instance, applying Mason's gain formula to the sampled signal flow graph of Fig. 3-17, with C*(s) and E*(s) regarded as output nodes, we get

$$C^*(s) = \frac{G^*(s)}{1 + GH^*(s)} R^*(s) \tag{3-225}$$

and

$$E^*(s) = \frac{1}{1 + GH^*(s)} R^*(s) \tag{3-226}$$

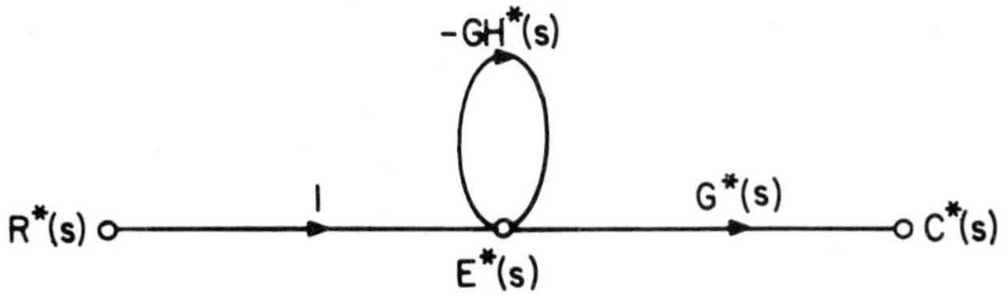

Fig. 3-17. Sampled signal flow graph of the system shown in Fig. 3-16.

4. The transfer relationships between inputs and continuous-data outputs can be determined from the *composite signal flow graph* [19]. The composite signal flow graph of a digital system is obtained by combining the equivalent and the sampled signal flow graphs. More specifically, the composite signal flow graph is formed by connecting the output nodes of samplers on the equivalent signal flow graph with unity-gain branches from the same nodes on the sampled signal flow graph. As an illustrative example, the composite signal flow graph of the system in Fig. 3-16 is drawn as shown in Fig. 3-18.

Application of Mason's gain formula to the composite signal flow graph yields the input-output transfer relations for all the digital and the continuous-data outputs. Therefore, from Fig. 3-18, we get

$$C(s) = \frac{G(s)}{1 + GH^*(s)} R^*(s) \tag{3-227}$$

and

*According to the definitions of the signal flow graph, an input node is one which has only outgoing branches, and an output node is one which has only incoming branches.

$$E(s) = R(s) - \frac{G(s)H(s)}{1 + GH^*(s)} R^*(s) \tag{3-228}$$

The procedure described above can be applied to linear multiloop, multisampler systems without difficulty, provided that the samplers are synchronized and of the same sampling frequency. The modified z-transform of C(s) or E(s) can be obtained directly from Eq. (3-227) or Eq. (3-228), respectively.

When one is more proficient with the method, equations such as (3-221) and (3-222) can be written directly from the equivalent signal flow graph.

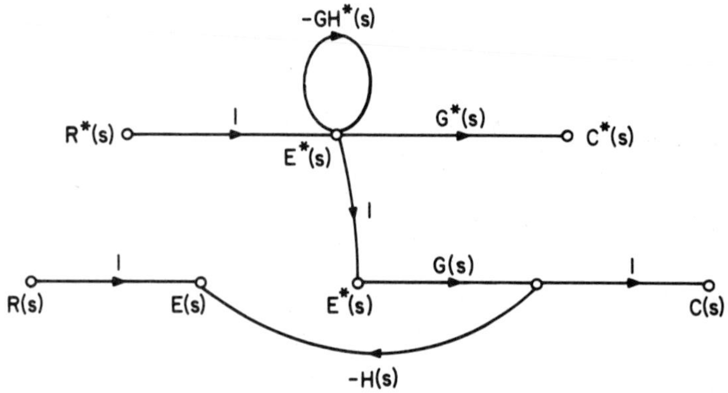

Fig. 3-18. Composite signal flow graph of the digital system in Fig. 3-16.

Example 3-18.

In this example the discrete and continuous-data outputs of the digital system shown in Fig. 3-19 will be determined by the sampled-signal-flow-graph method. The following steps are performed:

1. The signal flow graph of the system is drawn in Fig. 3-20.
2. The following equations are written directly from the signal flow graph of Fig. 3-20, using Mason's gain formula, and with E(s), $Y_1(s)$, and C(s) as output nodes, R(s) and E*(s) as input nodes.

$$E = \frac{R}{1 + G_2} - \frac{D^*G_1G_2}{1 + G_2} E^* \tag{3-229}$$

$$Y_1 = \frac{R}{1 + G_2} + \frac{D^*G_1}{1 + G_2} E^* \tag{3-230}$$

$$C = \frac{RG_2}{1 + G_2} + \frac{D^*G_1G_2}{1 + G_2} E^* \tag{3-231}$$

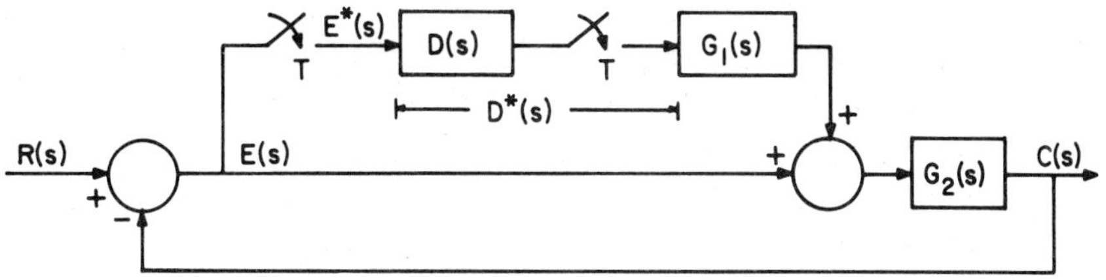

Fig. 3-19. Block diagram of digital control system.

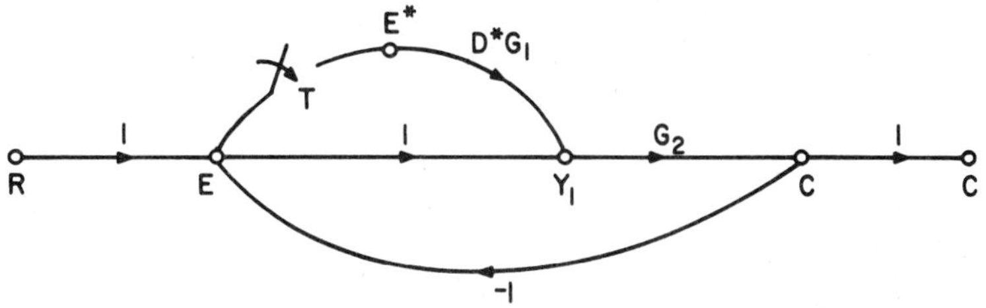

Fig. 3-20. Equivalent signal flow graph of the digital system in Fig. 3-19.

For simplicity, all symbols of functions of s have been dropped, that is, E*(s) is denoted by E*, R(s) by R, etc.

Taking the pulse transform on both sides of Eqs. (3-229), (3-230), and (3-231), we get

$$E^* = \left[\frac{R}{1+G_2}\right]^* - D^*\left[\frac{G_1G_2}{1+G_2}\right]^* E^* \tag{3-232}$$

$$Y_1^* = \left[\frac{R}{1+G_2}\right]^* + D^*\left[\frac{G_1}{1+G_2}\right]^* E^* \tag{3-233}$$

$$C^* = \left[\frac{RG_2}{1+G_2}\right]^* + D^*\left[\frac{G_1G_2}{1+G_2}\right]^* E^* \tag{3-234}$$

3. Using these last six equations, the composite signal flow graph of the system is drawn as shown in Fig. 3-21.
4. Now applying Mason's gain formula to the composite signal flow graph of Fig. 3-21, the discrete and continuous-data outputs of the system are determined.

$$C^* = \left[\frac{RG_2}{1+G_2}\right]^* + \frac{D^*\left[\frac{G_1G_2}{1+G_2}\right]^*}{1+D^*\left[\frac{G_1G_2}{1+G_2}\right]^*}\left[\frac{R}{1+G_2}\right]^*$$

$$= \frac{\left[\frac{RG_2}{1+G_2}\right]^* + \left[\left(\frac{RG_2}{1+G_2}\right)^* + \left(\frac{R}{1+G_2}\right)^*\right] D^* \left[\frac{G_1G_2}{1+G_2}\right]^*}{1 + D^*\left[\frac{G_1G_2}{1+G_2}\right]^*} \tag{3-235}$$

Since

$$\left[\frac{RG_2}{1+G_2}\right]^* + \left[\frac{R}{1+G_2}\right]^* = R^* \qquad (3\text{-}236)^*$$

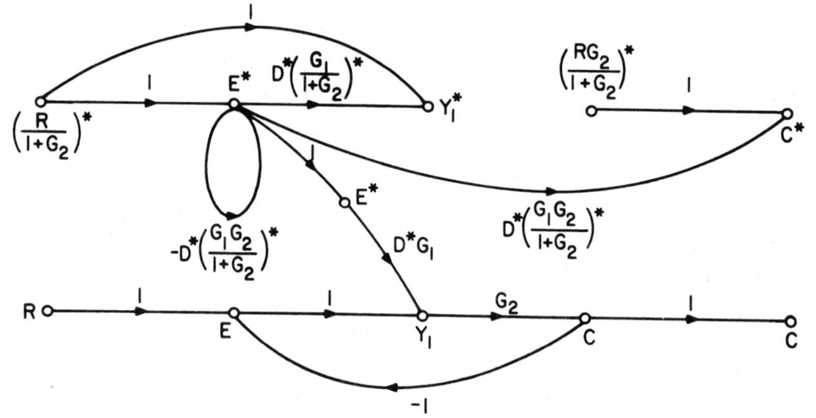

Fig. 3-21. Composite signal flow graph of the digital system in Fig. 3-19.

Eq. (3-235) is simplified to

$$C^* = \frac{\left[\frac{RG_2}{1+G_2}\right]^* + R^*D^*\left[\frac{G_1G_2}{1+G_2}\right]^*}{1 + D^*\left[\frac{G_1G_2}{1+G_2}\right]^*} \tag{3-237}$$

In z-transform notation, Eq. (3-237) becomes

$$C(z) = \frac{\frac{RG_2}{1+G_2}(z) + R(z)D(z)\frac{G_1G_2}{1+G_2}(z)}{1 + D(z)\frac{G_1G_2}{1+G_2}(z)} \tag{3-238}$$

*To prove this identity we simply "unstar" both sides of the equation and combine the two terms on the left-hand side of the equation.

where

$$\frac{RG_2}{1+G_2}(z) = \mathcal{Z}\left[\frac{R(s)G_2(s)}{1+G_2(s)}\right] \tag{3-239}$$

and $(G_1G_2/1 + G_2)(z)$ is defined similarly.

Using the continuous output C(s) as the output node on the composite signal flow graph, we have

$$C = \frac{RG_2\left[1 + D^*\left(\frac{G_1G_2}{1+G_2}\right)^*\right] + \left[\frac{R}{1+G_2}\right]^* G_1G_2D^*}{1 + G_2 + D^*\left[\frac{G_1G_2}{1+G_2}\right]^* + G_2D^*\left[\frac{G_1G_2}{1+G_2}\right]^*}$$

$$= \frac{RG_2}{1+G_2} + \frac{\left[\frac{G_1G_2}{1+G_2}\right]D^*\left[\frac{R}{1+G_2}\right]^*}{1 + D^*\left[\frac{G_1G_2}{1+G_2}\right]^*} \tag{3-240}$$

where for simplicity the function of s notation, (s), is dropped in the last equation.

The Direct Signal Flow Graph Method [17]

The method proposed by Sedlar and Bekey is a direct method which allows the evaluation of input-output relations from the signal flow graph of a digital control system without intermediate steps. This often makes the task of obtaining input-output relations for digital systems less tedious.

Since digital systems generally contain continuous and discrete signals, it is necessary to introduce a new node symbol which will be associated with the operation of sampling. Therefore, the following two types of nodes are defined:

White node. A white node ○ is used to represent continuous variables.
Black node. A black node ● is used to represent discrete-data or sampled-data variables.

With the introduction of the black node all samplers and digital operations in a digital system can be denoted by black nodes. Furthermore, all the node variables of this new signal flow graph may be represented by unstarred quantities. It should be interpreted that the variable represented by a black node is equal to the sampled form of the sum of all signals entering the node.

As a simple illustration, the signal flow graph shown in Fig. 3-16(b) is drawn as shown in Fig. 3-22 with the sampling operation replaced by a black node. Therefore, the variable at the black node is given by

$$Y_1(s) = E^*(s) \tag{3-241}$$

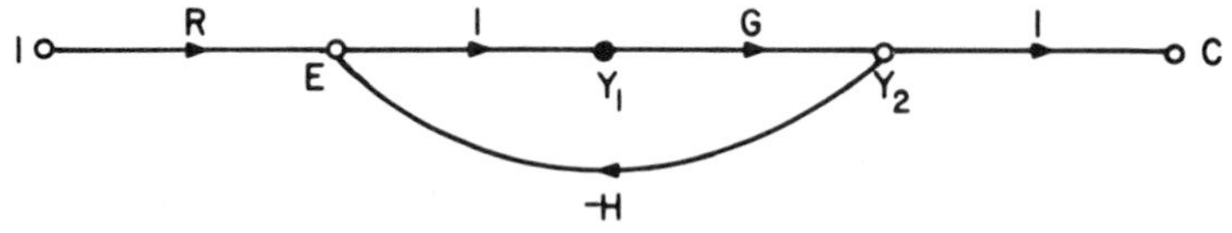

Fig. 3-22. Signal flow graph of Fig. 3-16(b) with sampling operation denoted by a black node.

As a matter of convenience, *the input variable is normalized to unity by introducing a branch with gain equal to the input* (see Fig. 3-22).

In addition to the definitions such as nodes and branches which have been established for conventional signal flow graphs, the following terms are necessary for the signal flow graphs used here.

Segment. A segment, denoted by $\sigma(y_m, y_n)$, is a path connecting between two nodes y_m and y_n, where

$$y_m \text{ is either an input node or a black node}$$

and

$$y_n \text{ is either an output node or a black node,}$$

and all other nodes of σ are white. Not one node of a segment should be traversed more than once.

Type-1 paths and loops. Type-1 paths and loops contain only white nodes, and are denoted by $u^{(1)}$ and $v^{(1)}$, respectively.

Type-2 paths and loops. Type-2 refers to paths and loops which contain at least one black node and are denoted by $u^{(2)}$ and $v^{(2)}$, respectively.

Type-1 elementary path. A type-1 path is elementary if it does not traverse any node more than once.

Type-1 elementary loop. A type-1 loop is elementary if it does not traverse any node (other than the coincident input and output nodes of the loop) more than once.

Type-2 elementary path. A type-2 elementary path is one which is composed of distinct segments such that no black node is met more than once.

Type-2 elementary loop. A type-2 elementary loop is one which is composed of distinct segments such that no black node is met more than once.

Forward path. A path connected between an input node and an output node. A forward path can be type-1 or type-2.

Type-1 connected (in touch). A type-1 path or loop is connected (in touch) with another type-1 path or loop or a segment if and only if they contain a node in common.

Type-2 connected. A type-2 path or loop is connected (in touch) with another type-2 loop or path if and only if they contain a black node in common.
Path gain. The path gain of a path u is the product of the individual gains of all the branches of the path. For a path with black nodes, sampled form of the gains preceding the black nodes should be taken. We shall use $P_i^{(1)}$ to denote the gain of the ith type-1 elementary path, and $P_i^{(2)}$ the gain of the ith type-2 elementary path.
Segment gain. The segment gain of a segment σ_i, S_i, is the product of the individual gains of all the branches of the segment.
Loop gain. The loop gain of the loop v is the product of the individual gains of all the branches of the loop. We define $L_i^{(1)}$ = gain of the ith type-1 elementary loop, and $L_i^{(2)}$ = gain of the ith type-2 elementary loop.

As a simple illustration of all the terms defined above, let us consider the signal flow graph shown in Fig. 3-22. The segments and the corresponding segment gains of the signal flow graph are

$$\sigma_1 = \sigma(1, Y_1) = (1, E, Y_1), \qquad S_1 = R^*$$

$$\sigma_2 = \sigma(Y_1, C) = (Y_1, C), \qquad S_2 = G$$

There are no type-1 elementary paths between the input node 1 and the output node C.

There is one type-2 elementary path between node 1 and node C.

$$u_1^{(2)} = (1, E, Y_1, C)$$

and the path gain is

$$P_1^{(2)} = R^*G$$

There are no type-1 elementary loops in the signal flow graph.

There is one type-2 elementary loop in the signal flow graph,

$$v_1^{(2)} = (Y_1, Y_2, E, Y_1)$$

and the loop gain is

$$L_1^{(2)} = -(GH)^*$$

Now let us consider a somewhat more elaborate example with reference to the signal flow graph shown in Fig. 3-23.

The following elementary paths exist between the input node 1 and the output node C:

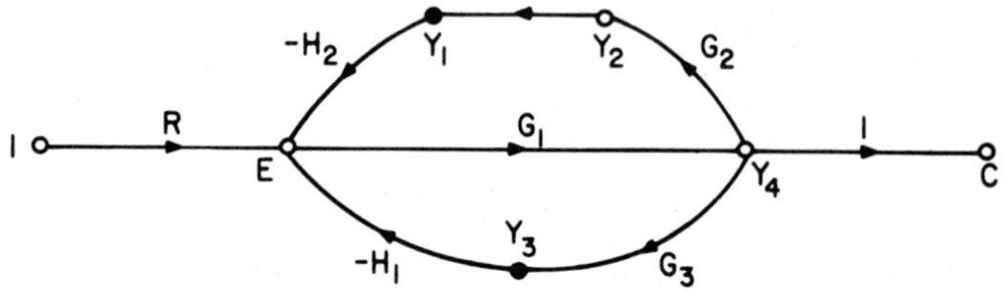

Fig. 3-23. Signal flow graph of a digital system.

Type-1: $u_1^{(1)} = (1, E, Y_4, C)$, $\qquad P_1^{(1)} = RG_1$

Type-2: $u_1^{(2)} = (1, E, Y_4, Y_2, Y_1, E, Y_4, C)$

$$P_1^{(2)} = -(RG_1G_2)^*H_2G_1$$

$$u_2^{(2)} = (1, E, Y_4, Y_3, E, Y_4, C)$$

$$P_2^{(2)} = -(RG_1G_3)^*H_1G_1$$

$$u_3^{(2)} = (1, E, Y_4, Y_3, E, Y_4, Y_2, Y_1, E, Y_4, C)$$

$$P_3^{(2)} = (RG_1G_3)^*(H_1G_1G_2)^*H_2G_1$$

$$u_4^{(2)} = (1, E, Y_4, Y_2, Y_1, E, Y_4, Y_3, E, Y_4, C)$$

$$P_4^{(2)} = (RG_1G_2)^*(H_2G_1G_3)^*H_1G_1$$

Notice that in selecting the type-2 elementary path, white nodes may be traversed more than once so long as each time a white node is passed it belongs to a different segment. For instance, for the path $u_2^{(2)}$, the first time the node Y_4 is traversed, it belongs to the segment $(1, E, Y_4, Y_3)$; the second time Y_4 is traversed, however, it belongs to the segment (Y_3, E, Y_4, C). Similarly, the type-2 elementary path $u_3^{(2)}$ contains the following distinct segments:

$$\sigma_1(1, Y_3) = (1, E, Y_4, Y_3)$$

$$\sigma_2(Y_3, Y_1) = (Y_3, E, Y_4, Y_2, Y_1)$$

$$\sigma_3(Y_1, C) = (Y_1, E, Y_4, C)$$

Therefore, although the nodes E and Y_4 are passed three times in path $u_3^{(2)}$, each time they belong to a different segment.

There are no type-1 elementary loops in the signal flow graph.

The following type-2 loops are observed:

$$v_1^{(2)} = (Y_3, E, Y_4, Y_3), \qquad L_1^{(2)} = -(G_1 G_3 H_1)^*$$

$$v_2^{(2)} = (Y_1, E, Y_4, Y_1), \qquad L_2^{(2)} = -(G_1 G_2 H_2)^*$$

$$v_3^{(2)} = (Y_3, E, Y_4, Y_2, Y_1, E, Y_4, Y_3), \qquad L_3^{(2)} = (G_1 G_2 H_1)^*(G_1 G_3 H_2)^*$$

The gain formula for the determination of input-output relationships on the new signal flow graph is given by

$$C = \frac{\Delta_j^{(1)}}{\Delta^{(1)}} \otimes \frac{\sum_{i=1}^{N} P_i \Delta_i^{(2)}}{\Delta^{(2)}} \tag{3-242}$$

where

C = Output variable of the signal flow graph (the values of input are normalized to unity).

P_i = Path gain of the ith elementary forward path (type-1 or type-2). The total number of elementary forward paths (type-1 and type-2) is N.

$\Delta^{(1)}$ = First determinant of the signal flow graph

$$= 1 - \sum L_i^{(1)} + \sum L_i^{(1)} L_j^{(1)} - \sum L_i^{(1)} L_j^{(1)} L_k^{(1)} + \ldots \tag{3-243}$$

where

$\sum L_i^{(1)}$ = Sum of loop gains of all type-1 elementary loops. (3-244)

$\sum L_i^{(1)} L_j^{(1)}$ = Sum of the products of loop gains of all non-connected (nontouching) type-1 elementary loops taken two at a time. (3-245)

$\sum L_i^{(1)} L_j^{(1)} L_k^{(1)}$ = Sum of the products of loop gains of all non-connected (nontouching) type-1 elementary loops taken three at a time. (3-246)

$\Delta^{(2)}$ = Second determinant of the signal flow graph

$$= 1 - \sum L_i^{(2)} + \sum L_i^{(2)} L_j^{(2)} - \sum L_i^{(2)} L_j^{(2)} L_k^{(2)} + \ldots \qquad (3\text{-}247)$$

The summation terms in Eq. (3-247) are interpreted the same way as above except that type-2 elementary loops are involved.

$\Delta_i^{(2)}$ = The second determinant, $\Delta^{(2)}$, of the part of the signal flow graph which is not *type-2 connected* (sharing at least one black node) with the ith forward path. Notice that if the ith forward path is type-1, $\Delta_i^{(2)} = \Delta^{(2)}$.

The symbol $\otimes$ represents the operation of multiplying $\Delta_j^{(1)}/\Delta^{(1)}$ by the gain of the jth segment for all the segments found in

$$\sum_{i=1}^{N} \frac{P_i \Delta_i^{(2)}}{\Delta^{(2)}}$$

If any of the segment gain appears in sampled form, the multiplication must be performed on the corresponding continuous quantities, and the product sampled.

$\Delta_j^{(1)}$ = The first determinant, $\Delta^{(1)}$ of that part of the signal flow graph which is not connected with the jth segment σ_j.

An illustrative example at this point should help explain the steps described thus far for the direct signal flow graph method.

Example 3-19.

Consider the digital system whose signal flow graph is drawn as shown in Fig. 3-22. The following information concerning the signal flow graph has been obtained earlier:

Type-1 Forward Elementary Paths: none

Type-2 Forward Elementary Paths: $u_1^{(2)} = (1, E, Y_1, C)$

Path Gain: $P_1^{(2)} = R^*G$

Type-1 Elementary Loops: none

Therefore, $\Delta^{(1)} = 1$

Type-2 Elementary Loops: $v_1^{(2)} = (Y_1, C, E, Y_1)$

Loop Gain: $L_1^{(2)} = -(GH)^*$

Therefore, $\Delta^{(2)} = 1 - L_1^{(2)} = 1 + (GH)^*$

Now using the gain formula in Eq. (3-242), we have

$$C = \frac{\Delta_j^{(1)}}{\Delta^{(1)}} \otimes \frac{P_1^{(2)}\Delta_1^{(2)}}{\Delta^{(2)}} \tag{3-248}$$

Since $u_1^{(2)}$ is type-2 connected to $v_1^{(2)}$, $\Delta_1^{(2)} = 1$; Eq. (3-248) becomes

$$C = \frac{\Delta_j^{(1)}}{\Delta^{(1)}} \otimes \frac{R^*G}{1+(GH)^*} \tag{3-249}$$

In the expression $R^*G/[1+(GH)^*]$ of Eq. (3-249), we recognize that R^* and G^* are gains of individual segments, and the $\otimes$ operation must be performed on these quantities. However, since $\Delta^{(1)} = 1$ and $\Delta_j^{(1)} = 1$ for all values of j, we have

$$C = \frac{R^*G}{1+(GH)^*} \tag{3-250}$$

which agrees with the result obtained earlier in Eq. (3-227). The z-transform of the output is obtained by taking the z-transformation on both sides of Eq. (3-250).

Example 3-20.

Let us determine the input-output relation of the digital system shown in Fig. 3-19 by means of the direct signal flow graph method. The signal flow graph of the system with black nodes representing the digital operations is shown in Fig. 3-24.

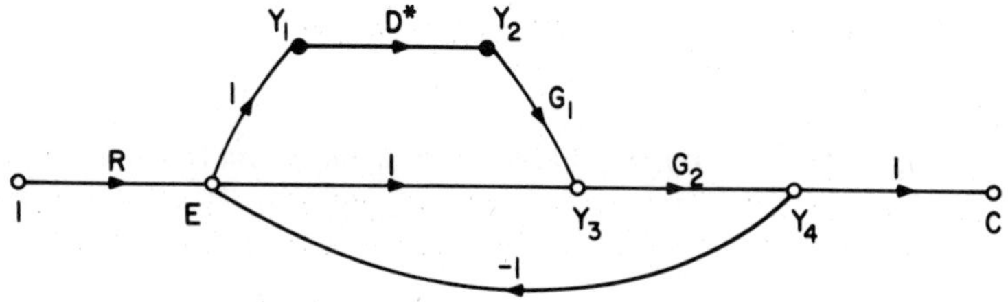

Fig. 3-24. Signal flow graph of the digital system shown in Fig. 3-19.

The following segments, forward paths, and loops are identified from the signal flow graph:

Segments:	$\sigma_1 = (1, E, Y_3, Y_4, C)$	$S_1 = RG_2$
	$\sigma_2 = (1, E, Y_1)$	$S_2 = R^*$
	$\sigma_3 = (Y_1, Y_2)$	$S_3 = D^*$
	$\sigma_4 = (Y_2, Y_3, Y_4, C)$	$S_4 = G_1G_2$

$$\sigma_5 = (Y_2, Y_3, Y_4, E, Y_1) \qquad S_5 = -(G_1G_2)^*$$

Elementary Forward Paths Between Nodes 1 and C:

Type-1: $u_1^{(1)} = (1, E, Y_3, Y_4, C)$ $\quad P_1^{(1)} = RG_2$

Type-2: $u_1^{(2)} = (1, E, Y_1, Y_2, Y_3, Y_4, C)$ $\quad P_1^{(2)} = R^*D^*G_1G_2$

This is the only type-2 elementary path between nodes 1 and C on the signal flow graph. The path, $(1, E, Y_1, Y_2, Y_3, Y_4, E, Y_3, Y_4, C)$, is not a type-2 elementary path although no black nodes are traversed more than once. The path violates the condition that it must contain distinct segments.

Type-1 Elementary Loop: $v_1^{(1)} = (E, Y_3, Y_4, E)$ $\quad L_1^{(1)} = -G_2$

Type-2 Elementary Loop: $v_1^{(2)} = (Y_1, Y_2, Y_3, Y_4, E, Y_1)$ $\quad L_1^{(2)} = -D^*(G_1G_2)^*$

First Determinant: $\Delta^{(1)} = 1 - L_1^{(1)} = 1 + G_2$ (3-251)

Second Determinant: $\Delta^{(2)} = 1 - L_1^{(2)} = 1 + D^*(G_1G_2)^*$ (3-252)

For $u_1^{(1)}$,

$$\Delta_1^{(2)} = 1 + D^*(G_1G_2)^*$$

and for $u_1^{(2)}$,

$$\Delta_2^{(2)} = 1$$

From Eq. (3-242), the input-output transfer relation of the system is written

$$C = \frac{\Delta_j^{(1)}}{\Delta^{(1)}} \otimes \frac{P_1^{(1)}\Delta_1^{(2)} + P_1^{(2)}\Delta_2^{(2)}}{\Delta^{(2)}}$$

$$= \frac{\Delta_j^{(1)}}{\Delta^{(1)}} \otimes \frac{RG_2[1 + D^*(G_1G_2)^*] + R^*D^*G_1G_2}{1 + D^*(G_1G_2)^*} \tag{3-253}$$

The second term on the right-hand side of Eq. (3-253) is identified to be

$$\frac{RG_2[1 + D^*(G_1G_2)^*] + R^*D^*G_1G_2}{1 + D^*(G_1G_2)^*} = \frac{S_1(1 - S_3S_5) + S_2S_3S_4}{1 - S_3S_5} \tag{3-254}$$

where S_j denotes the gain of the jth segment. The determinant $\Delta_j^{(1)}$ for $j = 1, 2, 3, 4, 5$ is found by determining the $\Delta^{(1)}$ of the part of the signal flow graph which is not connected to the jth segment. Therefore,

$$\sigma_1:\ \Delta_1^{(1)} = 1$$

$$\sigma_2:\ \Delta_2^{(1)} = 1$$

$$\sigma_3:\ \Delta_3^{(1)} = \Delta^{(1)} = 1 + G_2$$

$$\sigma_4:\ \Delta_4^{(1)} = 1$$

$$\sigma_5:\ \Delta_5^{(1)} = 1$$

Now multiplying the segment gains by the corresponding $\Delta_j^{(1)}/\Delta^{(1)}$ in Eq. (3-254) (if the segment gain appears in sampled form multiplication is performed on the corresponding continuous quantities, and then the sampled operation is applied to the product), the output of the system is written

$$C = \frac{RG_2}{1+G_2} + \frac{(R/1+G_2)^*D^*(G_1G_2/1+G_2)}{1+D^*(G_1G_2/1+G_2)^*} \tag{3-255}$$

which agrees with the result obtained earlier in Eq. (3-240).

Example 3-21.

Consider the digital control system shown in Fig. 3-25.

The following segments, forward paths, and loops are identified from the signal flow graph of Fig. 3-25(b).

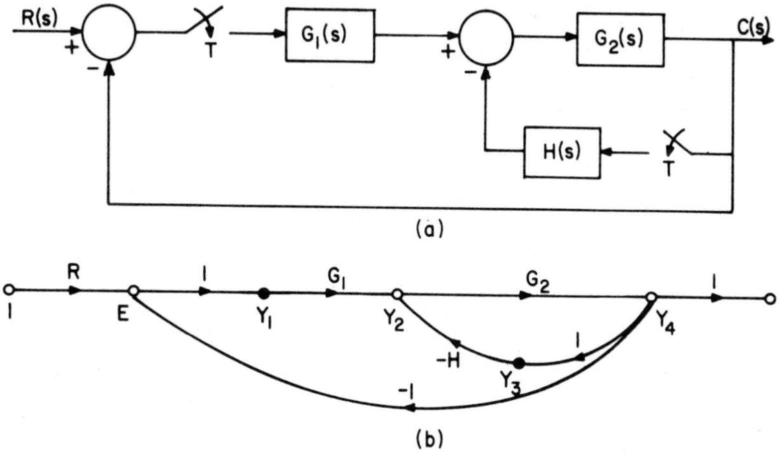

Fig. 3-25. (a) Block diagram of a digital control system; (b) signal flow graph of the system in (a).

Segments: $\sigma_1 = (1, E, Y_1)$ $\quad S_1 = R^*$

$\sigma_2 = (Y_1, Y_2, Y_4, C)$ $\quad S_2 = G_1G_2$

$\sigma_3 = (Y_3, Y_2, Y_4, C)$ $\quad S_3 = -G_2H$

$\sigma_4 = (Y_1, Y_2, Y_4, Y_3)$ $\quad S_4 = (G_1G_2)^*$

$\sigma_5 = (Y_3, Y_2, Y_4, E, Y_1)$ $\quad S_5 = (G_2H)^*$

Elementary Forward Paths Between Nodes 1 and C:

Type-1: none

$P_i^{(1)} = 0$ for all i

Type-2: $u_1^{(2)} = (1, E, Y_1, Y_2, Y_4, C)$ $\quad P_1^{(2)} = R^*G_1G_2$

$u_2^{(2)} = (1, E, Y_1, Y_2, Y_4, Y_3, Y_2, Y_4, C)$ $\quad P_2^{(2)} = R^*(G_1G_2)^*(-G_2H)$

Elementary Loops:

Type-1: none

$v_i^{(1)} = 0$ for all i

Type-2: $v_1^{(2)} = (E, Y_1, Y_2, Y_4, E)$ $\quad L_1^{(2)} = -(G_1G_2)^*$

$v_2^{(2)} = (Y_2, Y_4, Y_3, Y_2)$ $\quad L_2^{(2)} = -(G_2H)^*$

$v_3^{(2)} = (Y_1, Y_2, Y_4, Y_3, Y_2, Y_4, E, Y_1)$ $\quad L_3^{(2)} = (G_1G_2)^*(G_2H)^*$

First Determinant: $\Delta^{(1)} = 1$ since there are no type-1 loops

Second Determinant: $\Delta^{(2)} = 1 - L_1^{(2)} + L_2^{(2)} + L_3^{(2)} + L_1^{(2)}L_2^{(2)}$

The last term on the right side of the last equation denotes the product of the loop gains of $v_1^{(2)}$ and $v_2^{(2)}$ which are type-2 nontouching. Since $\Delta^{(1)} = 1$, $\Delta_j^{(1)} = 1$ for all values of j.

The input-output relation for the system is now obtained by using the gain formula of Eq. (3-242).

$$C = \frac{\Delta_j^{(1)}}{\Delta^{(1)}} \otimes \frac{P_1^{(2)}\Delta_1^{(2)} + P_2^{(2)}\Delta_2^{(2)}}{\Delta^{(2)}}$$

$$= \frac{R^*G_1G_2[1 + (G_2H)^*] - R^*(G_1G_2)^*G_2H}{1 + (G_1G_2)^* + (G_2H)^*} \tag{3-256}$$

The z-transform of the output is obtained by taking the z-transformation on both sides of Eq. (3-256).

$$C(z) = \frac{G_1G_2(z)[1 + G_2H(z)] - G_1G_2(z)G_2H(z)}{1 + G_1G_2(z) + G_2H(z)} R(z) \qquad (3\text{-}257)$$

From the illustrative examples worked out in this section, it appears that the direct signal flow graph method offers a simpler way of determining the input-output relations for digital systems than the sampled signal flow graph method. A further comparison of the two methods may show that the sampled signal flow graph method is perhaps easier to understand, whereas the direct signal flow graph method requires an understanding of the topological interpretation and definitions of the flow graph. However, we may find use in both methods for the purpose of checking the result obtained by each. The direct method is formulated in such a way that the analysis can be carried out on a digital computer.

3.11 MULTIRATE DIGITAL SYSTEMS

The digital control systems considered in the preceding sections all have a uniform sampling rate. When the data are digital, the rate of the input and the output data of digital processors and controllers throughout the system are assumed to be identical and uniform. For sampled-data systems, the samplers throughout the system all have the same uniform sampling rate.

In practice, the bandwidths of various portions of a digital system may be widely apart. Therefore, it is more appropriate to sample a slowly varying signal at a low sampling rate, while a signal with high-frequency content should be sampled at a high rate. When a digital system has different data rates at various locations, the system is referred to as a *multirate digital system.* Figure 3-26 illustrates the block diagram of a typical digital missile autopilot control system. Due to the difference between the dynamics in the position and the rate feedback loops, the samplers in the system have two sampling periods T_1 and T_2. It should be noted that the two data holds in the system of Fig. 3-26 have different characteristics, since they hold the sampled data for periods of T_1 and T_2, respectively.

The analysis and design of multirate digital systems are usually quite complex. In this chapter, the analysis of this type of system is centered around the z-transform method.

Since the multirate sampling adds new dimensions and complexity to digital systems, it is essential to first establish the methods of analysis with open-loop systems. Figure 3-27 illustrates three basic types of open-loop digital systems that require separate analytical approaches. The system shown in Fig.

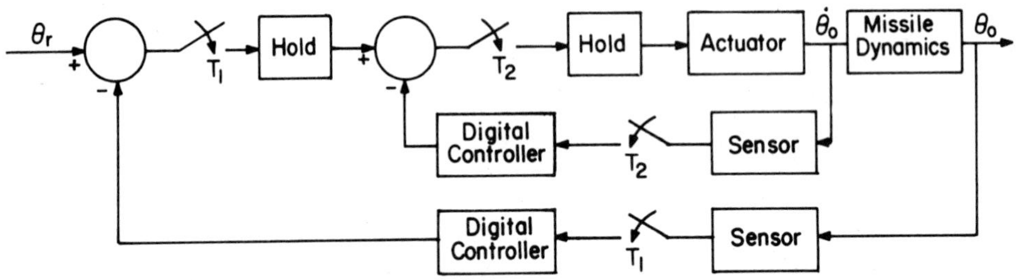

Fig. 3-26. A digital missile autopilot control system with multirate samplers.

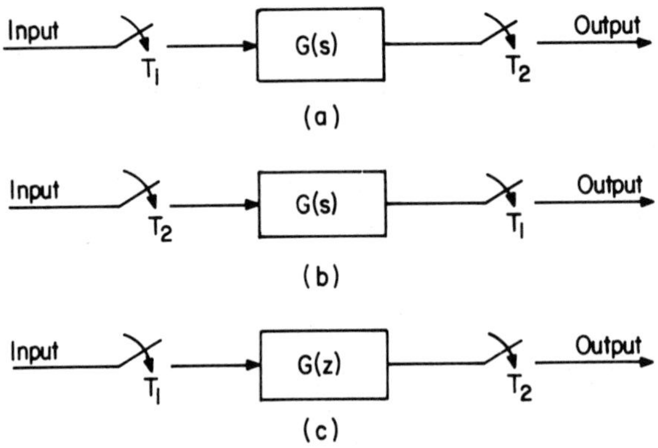

Fig. 3-27. Open-loop multirate digital systems. (a) $T_1 > T_2$, (b) $T_1 < T_2$, (c) all-digital system.

3-27(a) has a slow-rate sampler at the input and a fast-rate sampler at the output. This type of system is referred to as the *slow-fast multirate sampled system*. Figure 3-27(b) shows a system in which the sampler at the input has a faster rate than that of the output sampler, and the system is referred to as a *fast-slow multirate sampled system*. In general, the sampling periods T_1 and T_2 may not be related by an integral factor. However, for analytical purposes, it is convenient to assume that T_1 and T_2 are integrally related; that is, with reference to Fig. 3-27(a) and (b), $T_1 = NT_2$, where N is a positive integer > 1. Quite often, a multirate digital system may be a digital computer or controller whose input and output data rates are different. In this case, no sampling operation actually takes place. However, for analytical convenience we use the block diagram representation of Fig. 3-27(c) to indicate that the input and output data have different rates.

Slow-Fast Multirate Sampled Systems

Consider the multirate sampled system shown in Fig. 3-27(a), with $T_1 = NT_2$ where N is a positive integer greater than one. Several methods are available for the analytical study of this type of system. Each of these methods is preferred for a given situation, and the methods are described as follows.

1. Fictitious-Sampler Method

For this analysis, a fast-rate fictitious sampler is inserted at the input, as shown in Fig. 3-28. In this case, since the fictitious sampler samples at a faster rate than that of the input sampler, the net effect is zero. The situation is very similar to the submultiple sampling method discussed in Sec. 3.9. Since the system shown in Fig. 3-28 is now identical to the system configuration created in Fig. 3-12(b), we can use Eq. (3-178) to describe the input-output transform relation:

$$C(z)_N = G(z)_N R(z) \tag{3-258}$$

where

$$G(z)_N = \sum_{k=0}^{\infty} g(kT/N)z^{-k/N} \tag{3-259}$$

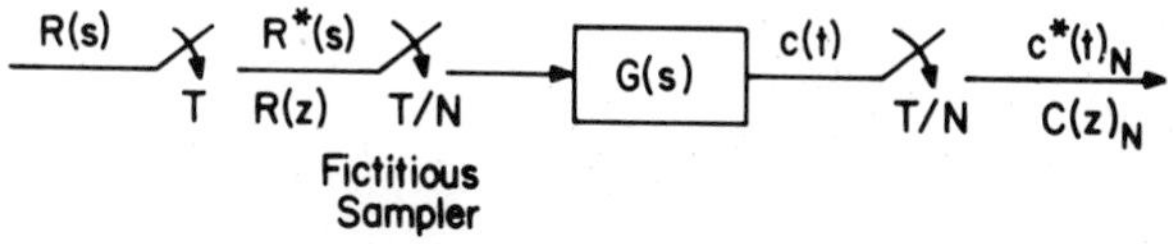

Fig. 3-28. Slow-fast multirate sampled system with a fictitious sampler.

and $C(z)_N$ is defined in the same way. Furthermore, $G(z)_N$ can be found from $G(z)$ through the following relation:

$$G(z)_N = G(z)\Big|_{\substack{z=z^{1/N} \\ T=T/N}} \tag{3-260}^*$$

2. Sampler-Decomposition Method

A versatile method of analyzing the slow-fast multirate system of Fig. 3-27(a) is to decompose the fast-rate sampler into N parallel-connected slow-rate samplers with time-delay and time-advance units, as shown in Fig. 3-29. Now, since the samplers are all at the same rate, the ordinary z-transform notation may be applied. The z-transform of the output is written as

*Refer to the footnote on page 113.

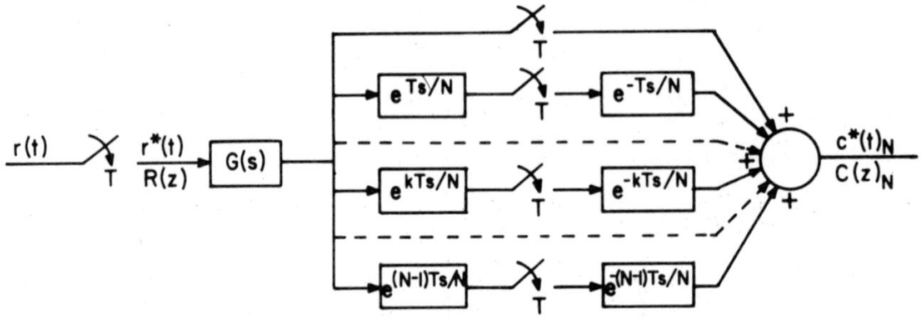

Fig. 3-29. Slow-fast multirate sampled system with the fast-rate sampler decomposed into N slow-rate samplers with time-advance and time-delay units.

$$C(z)_N = \sum_{k=0}^{N-1} z^{-k/N} \mathcal{Z}[e^{kTs/N} G(s)]\, R(z) \tag{3-261}$$

The last equation can be expressed in the form of Eq. (3-258) since by use of the definition of z-transform it can be shown that

$$G(z)_N = \sum_{k=0}^{N-1} z^{-k/N} \mathcal{Z}[e^{kTs/N} G(s)] \tag{3-262}$$

The z-transform of $e^{kTs/N}G(s)$ can be determined from the modified z-transform of G(s). By definition,

$$\mathcal{Z}[e^{kTs/N} G(s)] = \sum_{n=0}^{\infty} g\left[\left(n + \frac{k}{N}\right)T\right] z^{-n} \tag{3-263}$$

Since k/N is always less than unity, comparing the right-hand side of Eq. (3-263) with the defining equation of the modified z-transform in Eq. (3-200), we have

$$\mathcal{Z}[e^{kTs/N} G(s)] = [zG(z,m)]\Big|_{m=k/N}$$

$$= zG(z,k/N) \tag{3-264}$$

Substitution of Eq. (3-264) into Eq. (3-262) yields

$$G(z)_N = \sum_{k=0}^{N-1} z^{1-k/N} G(z,k/N) \tag{3-265}$$

Therefore, the output transform is written

$$C(z)_N = R(z) \sum_{k=0}^{N-1} z^{1-k/N} G(z,k/N) \tag{3-266}$$

Example 3-22.

Consider that the transfer function of the slow-fast multirate system in Fig. 3-27(a) is

$$G(s) = \frac{1}{s(s+1)} \tag{3-267}$$

The input is a unit step function, $T_1 = 3T_2$, and $T_1 = 1$ sec. The modified z-transform of G(s) for the sampling period of 1 sec is

$$G(z,m) = \frac{1}{z-1} - \frac{e^{-m}}{z - e^{-1}} \tag{3-268}$$

Using Eq. (3-266), the z-transform of the output with the fast-rate sampling is

$$C(z)_3 = \sum_{k=0}^{2} z^{1-k/3}\left[\frac{1}{z-1} - \frac{e^{-k/3}}{z-0.368}\right]\frac{z}{z-1} \tag{3-269}$$

Since Eq. (3-269) contains fractional powers of z in $z^{1/3}$ and $z^{2/3}$, we can again introduce the variable z_3 such that $z = z_3^3$. Then, $z^{1/3} = z_3$, and $z^{2/3} = z_3^2$. Expansion of $C(z)_3$ into a power series in z_3^{-1} will yield the response of c(t) at the fast-rate sampling instants where $t = kT/3$ and $k = 0,1,2,\ldots$.

3. Infinite-Series Representation

The input-output z transfer relation in Eq. (3-258) can be written as a pulse transfer relation in the Laplace domain; that is,

$$C^*(s)_N = G^*(s)_N R^*(s) \tag{3-270}$$

Using the infinite-series representation of G*(s) in Eq. (2-85), $G^*(s)_N$ is written as

$$G^*(s)_N = \frac{N}{T}\sum_{n=-\infty}^{\infty} G\left[s + j\,\frac{2\pi nN}{T}\right] \tag{3-271}$$

Substitution of the last equation in Eq. (3-270) yields an infinite-series representation of $C^*(s)_N$ in terms of the system transfer function G(s) and the input. Similar to Eq. (3-271), $C^*(s)_N$ can be written as

$$C^*(s)_N = \frac{N}{T}\sum_{n=-\infty}^{\infty} C\left[s + j\,\frac{2\pi nN}{T}\right] \tag{3-272}$$

The infinite-series representation is useful for the following development.

4. *Fast-Rate Sampled Signal in Terms of Slow-Rate Samples*

It is sometimes necessary to represent the fast-rate sampled output of the system in Fig. 3-28 in terms of the slow-rate samples. This is equivalent to placing a slow-rate sampler after the fast-rate sampler, as shown in Fig. 3-30. Physically, this is equivalent to replacing the fast-rate sampler by a slow-rate one, since the slow-rate sampler "picks off" only the output samples at intervals of T. We shall show that this is also true mathematically.

First, we write,

$$C^*(s) = \frac{1}{T} \sum_{n=-\infty}^{\infty} \left[C\ s + j\frac{2\pi n}{T}\right] \tag{3-273}$$

Letting $n = \ell N + m$, where $-\infty \leqslant \ell \leqslant \infty$, and $m = 0,1,2,\ldots,N-1$, the last equation becomes

$$C^*(s) = \frac{1}{T} \sum_{\ell N+m=-\infty}^{\infty} C\left[s + j\frac{2\pi \ell N}{T} + j\frac{2\pi m}{T}\right] \tag{3-274}$$

which is then written as

$$C^*(s) = \frac{1}{T} \sum_{m=0}^{N-1} \frac{T}{N} \cdot \frac{N}{T} \sum_{\ell=-\infty}^{\infty} C\left[s + j\frac{2\pi \ell N}{T} + j\frac{2\pi m}{T}\right] \tag{3-275}$$

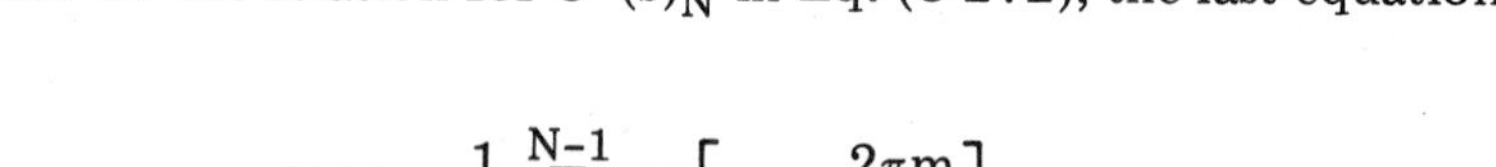

Fig. 3-30. Slow-fast multirate sampled system with an additional slow sampler at the output.

In view of the relation for $C^*(s)_N$ in Eq. (3-272), the last equation is written as

$$C^*(s) = \frac{1}{N} \sum_{m=0}^{N-1} C^*\left[s + j\frac{2\pi m}{T}\right]_N \tag{3-276}$$

Now using the relation in Eq. (3-270), Eq. (3-276) is written

$$C^*(s) = \frac{1}{N} \sum_{m=0}^{N-1} G^*\left[s + j\frac{2\pi m}{T}\right]_N R^*\left[s + j\frac{2\pi m}{T}\right] \tag{3-277}$$

Since m is an integer, it has been established in Chapter 2, Eq. (2-104), that $R^*(s + j2\pi m/T) = R^*(s)$, and using a relation similar to Eq. (3-276) for $G^*(s)$,

Eq. (3-277) is simplified to

$$C^*(s) = G^*(s)R^*(s) \tag{3-278}$$

or

$$C(z) = G(z)R(z) \tag{3-279}$$

Using the complex-translation theorem of the z-transform, Eq. (3-88), the z-domain equivalent of Eq. (3-276) is obtained as

$$C(z) = \frac{1}{N}\sum_{m=0}^{N-1} C\left[z_N e^{j2\pi m/N}\right]_N \tag{3-280}$$

Fast-Slow Multirate Sampled Systems

As mentioned earlier, a separate treatment is required if the fast-rate sampler is at the input and the slow-rate sampler is at the output of the system. Figure 3-31 shows the block diagram of such a system.

The alternate methods of analysis of a fast-slow multirate sampled system are discussed as follows.

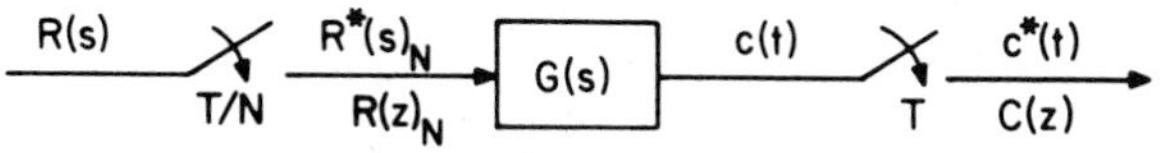

Fig. 3-31. Fast-slow multirate sampled system.

1. Fictitious-Sampler Method

One method of analyzing the fast-slow multirate sampled system shown in Fig. 3-31 is to insert a fictitious fast-rate sampler with sampling period T/N in front of the slow-rate sampler. The result is shown in Fig. 3-32. Since the output of the system is not affected by the fictitious fast-rate sampler, the two systems in Fig. 3-31 and Fig. 3-32 are the same from the analytical standpoint.

From Fig. 3-32, the z-transform of the output of the fictitious sampler is

$$C(z)_N = G(z)_N R(z)_N \tag{3-281}$$

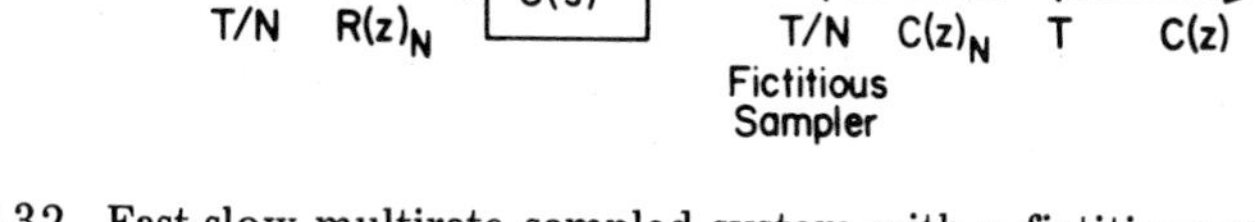

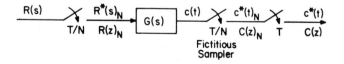

Fig. 3-32. Fast-slow multirate sampled system with a fictitious sampler.

where $G(z)_N$ is defined in Eqs. (3-259) and (3-260), and $R(z)_N$ is similarly defined. The z-transform of the output, $C(z)$, can be determined from $C(z)_N$ in a number of ways. Physically, the process of expressing $C(z)$ in terms of $C(z)_N$ corresponds to extracting the information on $c^*(t)$ from $c^*(t)_N$, or obtaining $c(kT)$ from the set $c(kT/N)$, $k = 0,1,2,\ldots$.

The z-transform of the output of the fictitious sampler in Fig. 3-32 is written

$$C(z)_N = \sum_{k=0}^{\infty} c(kT/N)z^{-k/N} \tag{3-282}$$

The inverse z-transform of $C(z)_N$ is

$$c(kT/N) = \frac{1}{2\pi j}\oint_{\Gamma} C(z_N)z_N^{k-1}dz_N \tag{3-283}$$

where $z_N = z^{1/N}$. The z-transform of the output of the system in Fig. 3-31 is

$$C(z) = \sum_{n=0}^{\infty} c(nT)z^{-n} \tag{3-284}$$

Letting $n = k/N$ and substituting Eq. (3-283) into Eq. (3-284), we have

$$\begin{aligned} C(z) &= \sum_{n=0}^{\infty} \frac{1}{2\pi j}\oint_{\Gamma} C(z_N)z_N^{nN-1}dz_N z^{-n} \\ &= \frac{1}{2\pi j}\oint_{\Gamma} C(z_N)\sum_{n=0}^{\infty} z_N^{nN}z^{-n}\frac{dz_N}{z_N} \end{aligned} \tag{3-285}$$

Writing the infinite series in closed form, the last equation is simplified to

$$C(z) = \frac{1}{2\pi j}\oint_{\Gamma} C(z_N)\frac{1}{1 - z_N^N z^{-1}}\frac{dz_N}{z_N} \tag{3-286}$$

The contour integral of Eq. (3-286) is carried out along the path Γ in the z_N-plane. Assuming that the contour Γ encloses the singularities of $C(z_N)z_N^{-1}$, from the residue theorem, Eq. (3-286) is written as

$$C(z) = \sum \text{Residues of } C(z_N)\frac{z_N^{-1}}{1 - z_N^N z^{-1}} \text{ at the poles of } C(z_N)z_N^{-1} \tag{3-287}$$

Example 3-23.

For the fast-slow multirate sampled system of Fig. 3-31, let

$$R(s) = \frac{1}{s} \quad \text{and} \quad G(s) = \frac{1}{s+1}$$

Then,

$$R(z)_N = \frac{z_N}{z_N - 1} \tag{3-288}$$

$$G(z)_N = \frac{z_N}{z_N - e^{-T/N}} \tag{3-289}$$

Equation (3-281) gives

$$C(z)_N = G(z)_N R(z)_N = \frac{z_N^2}{(z_N - 1)(z_N - e^{-T/N})} \tag{3-290}$$

The z-transform of the output of the system, C(z), is determined by using Eq. (3-287). We have

$$C(z) = \sum \text{Residues of } \frac{z_N^2}{(z_N - 1)(z_N - e^{-T/N})} \frac{z_N^{-1}}{(1 - z_N^N z^{-1})} \text{ at } z_N = 1,\ z_N = e^{-T/N}$$

$$= \frac{1}{1 - e^{-T/N}} \left[\frac{z}{z-1} - \frac{ze^{-T/N}}{z - e^{-T}} \right] \tag{3-291}$$

2. Series Representation

An alternate method of analyzing the system of Fig. 3-32 is to make use of the infinite-series expression in Eq. (3-273) which then leads to Eq. (3-276). Since

$$C^*(s)_N = G^*(s)_N R^*(s)_N \tag{3-292}$$

Eq. (3-276) becomes

$$C^*(s) = \frac{1}{N} \sum_{m=0}^{N-1} G^*\left[s + j\frac{2\pi m}{T}\right]_N R^*\left[s + j\frac{2\pi m}{T}\right]_N \tag{3-293}$$

The z-transform equivalent of the last equation is

$$C(z) = \frac{1}{N} \sum_{m=0}^{N-1} G\left[z_N e^{j2\pi m/N}\right]_N R\left[z_N e^{j2\pi m/N}\right]_N \tag{3-294}$$

3. Sampler-Decomposition Method

The fast-rate sampler is decomposed into N slow-rate samplers with time-advance and time-delay units, as shown in Fig. 3-33.

The input-output relation of the system shown in Fig. 3-33 is written in the z-domain:

$$C(z) = \sum_{k=0}^{N-1} \mathcal{Z}[R(s)e^{kTs/N}]\,\mathcal{Z}[G(s)e^{-kTs/N}] \tag{3-295}$$

The first term on the right side of Eq. (3-295) can be evaluated by the modified z-transform method described by Eq. (3-264); that is,

$$\mathcal{Z}[R(s)e^{kTs/N}] = zR(z,k/N) \tag{3-296}$$

The second term in Eq. (3-295) is also related to the modified z-transform. From the definition of the modified z-transform,

$$G(z,m)\Big|_{m=1-\Delta} = G(z,\Delta) \tag{3-297}$$

Thus, we have

$$\mathcal{Z}[G(s)e^{-kTs/N}] = G(z,m)\Big|_{m=1-k/N} \tag{3-298}$$

where $k/N < 1$. This means that the second term in Eq. (3-295) is determined by setting $m = 1 - k/N$ in the modified z-transform of G(s).

Equation (3-295) is now written as

$$C(z) = R(z)G(z) + \sum_{k=1}^{N-1} zR\left[z, \frac{k}{N}\right]G\left[z, 1-\frac{k}{N}\right] \tag{3-299}$$

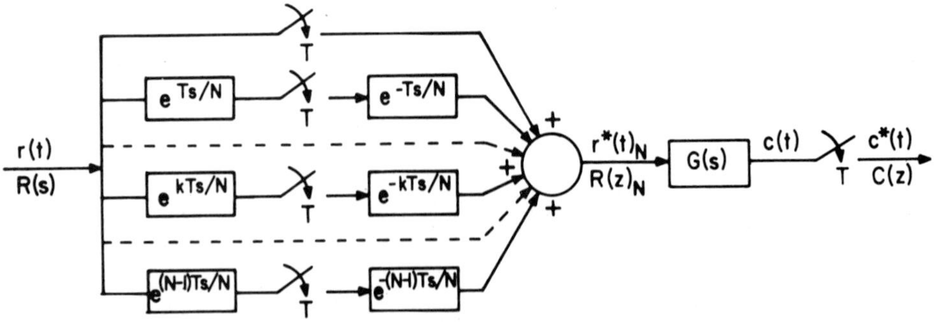

Fig. 3-33. Fast-slow multirate sampled system with the fast-rate sampler decomposed into N slow-rate samplers with time-advance and time-delay units.

Example 3-24.

Consider the multirate digital system described in Example 3-23; $R(s) = 1/s$ and $G(s) = 1/(s+1)$. The modified z-transforms of R(s) and G(s) are, respectively,

$$R(z,m) = \frac{1}{z-1} \tag{3-300}$$

$$G(z,m) = \frac{e^{-mT}}{z-e^{-T}} \tag{3-301}$$

Substitution of these transform expressions into Eq. (3-299) gives

$$C(z) = \frac{z^2}{(z-1)(z-e^{-T})} + \sum_{k=1}^{N-1} \frac{ze^{-(1-k/N)T}}{(z-1)(z-e^{-T})} \tag{3-302}$$

which can be shown to be equivalent to Eq. (3-291).

Multirate Systems With All-Digital Elements

It is common to have multirate digital systems in which some components are entirely digital so that the inputs and outputs of these digital components are digitally coded, and no physical samplings are involved. We shall show that the mathematical relations derived in the preceding sections for the multirate sampled systems are still applicable for this case.

Figure 3-34 shows the block diagram representation of an all-digital multirate system with a slow-rate input and a fast-rate output. In this case the input of the system consists of digital data that are represented by a pulse train with a period of T. The output pulse train has a period of T/N. The input-output z-transform relation is written

$$C(z)_N = G(z)_N R(z) \tag{3-303}$$

where $C(z)_N$ and $G(z)_N$ are defined only in the form of

$$G(z)_N = \sum_{k=0}^{\infty} g(kT/N)z^{-k/N} \tag{3-304}$$

The slow-rate output is written as

$$C(z) = G(z)R(z) \tag{3-305}$$

or C(z) may be determined from $C(z)_N$ using Eq. (3-280).

Figure 3-35 shows the block diagram representation of a fast-slow all-digital multirate system. In this case, we can first write

$$C(z)_N = G(z)_N R(z)_N \tag{3-306}$$

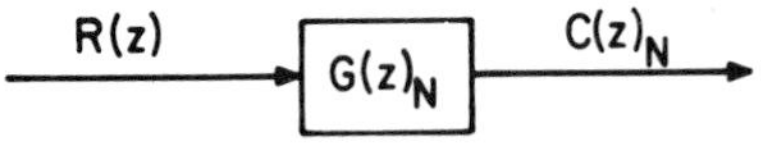

Fig. 3-34. A slow-fast all-digital multirate system.

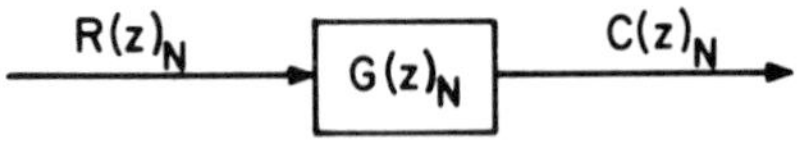

Fig. 3-35. A fast-slow all-digital multirate system.

and then the slow-rate output is given by

$$C(z) = \frac{1}{N} \sum_{m=0}^{N-1} G\left[z_N e^{j2\pi m/N}\right]_N R\left[z_N e^{j2\pi m/N}\right]_N \qquad (3\text{-}307)$$

Closed-Loop Multirate Digital Systems

In this section we apply some of the methods of analysis of multirate digital systems introduced in the preceding section to closed-loop system configurations.

Let us consider the closed-loop multirate digital system shown in Fig. 3-36(a). The fast-rate sampler is decomposed into synchronized slow-rate samplers with time-advance and time-delay units, as shown in Fig. 3-36(b). Notice that the zero-order hold of the fast-rate sampler has a different transfer function than that of the slow-rate sampler.

Let

$$G_{h0}(s) = \frac{1 - e^{-Ts}}{s} \qquad (3\text{-}308)$$

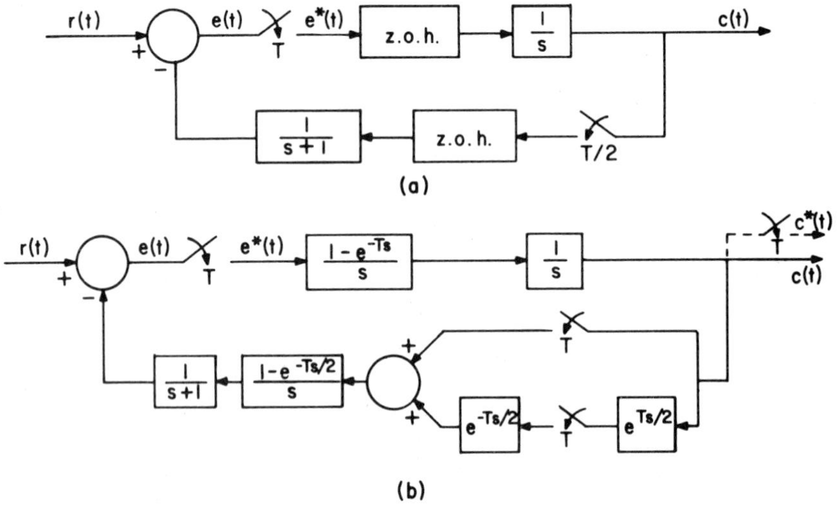

Fig. 3-36. Closed-loop multirate digital system.

and

$$G_{h1}(s) = \frac{1 - e^{-Ts/2}}{s} \tag{3-309}$$

The z-transfer function relation of the system is written from Fig. 3-36(b).

$$\frac{C(z)}{R(z)} = \frac{\mathcal{Z}[G_{h0}(s)G(s)]}{1 + \mathcal{Z}[G_{h1}(s)H(s)]\mathcal{Z}[G_{h0}(s)G(s)] + \mathcal{Z}[e^{Ts/2}G_{h0}(s)G(s)]\mathcal{Z}[e^{-Ts/2}G_{h1}(s)H(s)]} \tag{3-310}$$

where

$$\mathcal{Z}[G_{h0}(s)G(s)] = (1 - z^{-1})\mathcal{Z}\left[\frac{1}{s^2}\right] = \frac{T}{z-1} \tag{3-311}$$

$$\mathcal{Z}[G_{h1}(s)H(s)] = \mathcal{Z}\left[\frac{1 - e^{-Ts/2}}{s}\frac{1}{s+1}\right]$$

$$= \mathcal{Z}\left[\frac{1}{s(s+1)}\right] - \mathcal{Z}\left[\frac{e^{-Ts/2}}{s(s+1)}\right] \tag{3-312}$$

The last term in the last equation is evaluated by use of Eq. (3-298). Thus,

$$\mathcal{Z}\left[\frac{e^{-Ts/2}}{s(s+1)}\right] = \mathcal{Z}_m\left[\frac{1}{s(s+1)}\right]_{m=1/2} = \frac{1}{z-1} - \frac{e^{-T/2}}{z - e^{-T}} \tag{3-313}$$

After substituting Eq. (3-313) into Eq. (3-312) and simplifying, we have

$$\mathcal{Z}[G_{h1}(s)H(s)] = \frac{e^{-T/2} - e^{-T}}{z - e^{-T}} \tag{3-314}$$

Similarly,

$$\mathcal{Z}\left[e^{Ts/2}G_{h0}(s)G(s)\right] = \mathcal{Z}\left[e^{Ts/2}\frac{1 - e^{-Ts}}{s}\frac{1}{s}\right]$$

$$= (1 - z^{-1})\mathcal{Z}\left[\frac{e^{Ts/2}}{s^2}\right] \tag{3-315}$$

and using Eq. (3-264),

$$\mathcal{Z}\left[\frac{e^{Ts/2}}{s^2}\right] = z\mathcal{Z}_m\left[\frac{1}{s^2}\right]_{m=1/2}$$

$$= \frac{zT/2}{z-1} + \frac{Tz}{(z-1)^2} \tag{3-316}$$

$$\mathcal{Z}\left[e^{-Ts/2}G_{h1}(s)H(s)\right] = \mathcal{Z}\left[e^{-Ts/2}\frac{1-e^{-Ts/2}}{s}\frac{1}{s+1}\right]$$

$$= \mathcal{Z}\left[\frac{e^{-Ts/2}}{s(s+1)}\right] - \mathcal{Z}\left[\frac{e^{-Ts}}{s(s+1)}\right]$$

$$= \frac{1-e^{-T/2}}{z-e^{-T}} \tag{3-317}$$

Now substituting all these components into Eq. (3-310) and simplifying, the transfer function of the system is

$$\frac{C(z)}{R(z)} = \frac{T(z-e^{-T})}{z^2 - \left[1 + e^{-T} - \frac{T}{2} + \frac{T}{2}e^{-T/2}\right]z + e^{-T} - Te^{-T} + \frac{T}{2}e^{-T/2} + \frac{T}{2}} \tag{3-318}$$

As another illustrative example we consider the system shown in Fig. 3-37 which has sampled and all-digital elements.

The following equations are written from Fig. 3-37:

$$E(z)_N = -D_1(z)_N G_1(z)_N E(z)_N + D_2(z)_N [R(z) - C(z)] \tag{3-319}$$

$$C(z)_N = G_2(z)_N E(z)_N \tag{3-320}$$

Solving for $E(z)_N$ in Eq. (3-319) and substituting the result in Eq. (3-320), we get

$$C(z)_N = \frac{D_2(z)_N G_2(z)_N}{1 + D_1(z)_N G_1(z)_N}[R(z) - C(z)] \tag{3-321}$$

Let

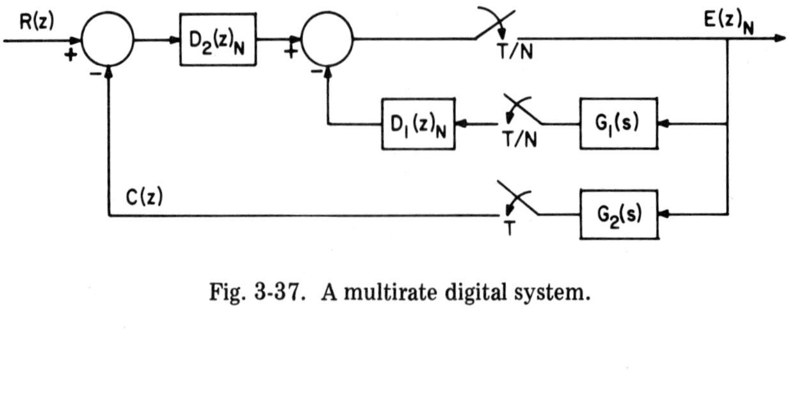

Fig. 3-37. A multirate digital system.

$$H(z)_N = \frac{D_2(z)_N G_2(z)_N}{1 + D_1(z)_N G_1(z)_N} \tag{3-322}$$

Eq. (3-321) becomes

$$C(z)_N = H(z)_N [R(z) - C(z)] \tag{3-323}$$

Then, it follows that

$$C(z) = H(z)[R(z) - C(z)]$$

which gives

$$\frac{C(z)}{R(z)} = \frac{H(z)}{1 + H(z)} \tag{3-324}$$

where H(z) can be found from Eq. (3-280); that is

$$H(z) = \frac{1}{N} \sum_{m=0}^{N-1} H\left[z_N e^{j2\pi m/N}\right]_N \tag{3-325}$$

or

$$H(z) = H(z)_N \Big|_{\substack{z^{1/N}=z \\ T/N=T}} \tag{3-326}$$

REFERENCES

Sampling and Z-Transforms

1. Lago, G. V., "Additions to Sampled-Data Theory," *Proc. National Electronics Conf.*, Vol. 10, 1954, pp. 758-766.
2. Helm, H. A., "The Z-Transformation," *Bell System Technical Journal*, Vol. 38, January 1959, pp. 177-196.
3. Ragazzini, J. R., and Zadeh, L. H., "The Analysis of Sampled-Data Systems," *Trans. AIEE*, Vol. 71, Part 2, 1952, pp. 225-234.
4. Kuo, B. C., *Analysis and Synthesis of Sampled-Data Control Systems.* Prentice-Hall, Englewood Cliffs, N.J., 1963.
5. Jury, E. I., "A General z-Transform Formula for Sampled-Data Systems," *IEEE Trans. on Automatic Control*, Vol. AC-12, 5, October 1967, pp. 606-608.
6. Jury, E. I., *Theory and Application of the z-Transform Method.* John Wiley & Sons, New York, 1964.
7. Kliger, I., and Lipinski, W. C., "The z-Transform of a Product of Two Functions," *IEEE Trans. on Automatic Control*, Vol. AC-9, October 1964, pp. 582-583.

8. Will, P. M., "Variable Frequency Sampling," *IRE Trans. on Automatic Control*, Vol. AC-7, October 1962, pp. 126.

9. Dorf, R. C., Farren, M. C., and Phillips, C. A., "Adaptive Sampling Frequency For Sampled-Data Control Systems," *IRE Trans. on Automatic Control*, Vol. AC-7, January 1962, pp. 38-47.

10. Cavin, R. K., III, Chenoweth, D. L., and Phillips, C. L., "The z-Transform of an Impulse Function," *IEEE Trans. on Automatic Control*, Vol. AC-12, February 1967, pp. 113.

11. Phillips, C. L., "A Note on Sampled-Data Control Systems," *IEEE Trans. on Automatic Control*, Vol. AC-10, October 1965, pp. 489-490.

12. Lago, G. V., "Additions to z-Transformation Theory for Sampled-Data System," *Trans. AIEE*, 74, Part 2, January 1955, pp. 403-407.

13. Hufnagel, R. E., "Analysis of Cyclic-Rate Sampled-data Feedback-Control Systems," *Trans. AIEE (Applications and Industry)*, Vol. 77, November 1958, pp. 421-425.

14. Jury, E. I., and Mullin, F. J., "The Analysis of Sampled-data Control Systems With a Periodically Time-varying Sampling Rate," *IRE Trans. on Automatic Control*, Vol. AC-4, May 1959, pp. 15-21.

Signal Flow Graphs

15. Mason, S. J., "Feedback Theory — Some Properties of Flow Graphs," *Proc. IRE*, 41, September 1953, pp. 1144-1156.

16. Mason, S. J., "Feedback Theory — Further Properties of Signal Flow Graphs," *Proc. IRE*, 44, July 1956, pp. 960-966.

17. Sedlar, M., and Bekey, G. A., "Signal Flow Graphs of Sampled-Data Systems: A New Formulation," *IEEE Trans. on Automatic Control*, Vol. AC-12, 2, April 1967, pp. 154-161.

18. Ash, R., Kim, W. E., and Kranc, G. M., "A General Flow Graph Technique for the Solution of Multiloop Sampled Systems," *Trans. of ASME Journal of Basic Engineering*, June 1960, pp. 360-370.

19. Kuo, B. C., "Composite Flow Graph Technique for the Solution of Multiloop, Multisampler, Sampled Systems," *IRE Trans. on Automatic Control*, Vol. AC-6, 1961, pp. 343-344.

20. Salzer, J. M., "Signal Flow Techniques for Digital Compensation," *Proc. Computers in Control Systems Conference*, October 1957.

21. Salzer, J. M., "Signal Flow Reductions in Sampled-Data Systems," *IRE WESCON Convention Record*, Part 4, 1957, pp. 166-170.

Pulse Transfer Function

22. Barker, R. H., "The Pulse Transfer Function and Its Application to Servo Systems," *Proc. IEE*, London, 99, Part 4, December 1952, pp. 302-317.

23. Lendaris, G. G., and Jury, E. I., "Input-Output Relationships for Multiple Sampled-Loop Systems," *Trans. AIEE*, 79, Part 2, January 1960, pp. 375-385.

24. Tou, J. T., "Simplified Technique for the Determination of Output Transforms of Multiloop, Multisampler, Variable-Rate Discrete-Data Systems," *Proc. of the IRE*, 49, March 1961, pp. 646-647.

25. Dejka, W. J., "The Generation of Discrete Functions with a Digital Computer," *IRE Trans. on Automatic Control*, Vol. AC-7, July 1962, pp. 56-57.

Modified Z-Transform

26. Jury, E. I., "Additions to the Modified z-Transform Method," *IRE WESCON Convention Record*, Part 4, 1957, pp. 136-156.

27. Jury, E. I., and Farmanfarma, "Tables of z-Transforms and Modified z-Transforms of Various Sampled-Data Systems Configurations," Univ. of California (Berkeley), Electronics Research Lab., Report 136A, Ser. 60, 1955.

28. Mesa, W., and Phillips, C. L., "A Theorem on the Modified z-Transform," *IEEE Trans. on Automatic Control*, Vol. AC-10, October 1965, pp. 489.

Multirate Sampled Systems

29. DuPlessis, R. M., "Two Digital Computer Programs With Multirate Sampled-Data System Analysis," *IRE Trans. on Automatic Control*, Vol. AC-6, February 1961, pp. 85-86.

30. Jury, E. I., "A Note on Multirate Sampled-Data Systems," *IEEE Trans. on Automatic Control*, Vol. AC-12, June 1967, pp. 319-320.

31. Boykin, W. H., and Frazier, B. D., "Multirate Sampled-Data Systems Analysis via Vector Operators," *IEEE Trans. on Automatic Control*, Vol. AC-20, August 1975, pp. 548-551.

32. Boykin, W. H,, and Frazier, B. D., "Analysis of Multiloop, Multirate Sampled-data Systems," *AIAA Journal*, Vol. 13, May 1975.

33. Flowers, D. C., and Hammond, J. L., Jr., "Simplification of the Characteristic Equation of Multirate Sampled Data Control Systems," *IEEE Trans. on Automatic Control*, Vol. AC-17, April 1972, pp. 249-250.

PROBLEMS

3-1 Determine the z-transforms of the following functions:

(a) $F(s) = \dfrac{1}{s(s^2 + 2)}$

(b) $F(s) = \dfrac{5}{s^2(s + 1)}$

(c) $F(s) = \dfrac{10}{s(s^2 + s + 1)}$

3-2 Given that the z-transform of g(t) for T = 1 second is

$$G(z) = \frac{1 - 3z^{-1} + 3z^{-2}}{(1 - 0.5z^{-1})(1 - 0.8z^{-1})}$$

Find the values of g(kT) for the first ten sampling periods.

3-3 Given $f_1(t) = t^2u_s(t)$ and $f_2(t) = e^{-2t}u_s(t)$. Find the z-transform of $f_1(t)f_2(t)$ by means of the partial differentiation theorem of z-transformation.

3-4 Given $f_1(t) = t^2u(t)$ and $f_2(t) = e^{-2t}u(t)$. Find the z-transform of $f_1(t)f_2(t)$ by means of the complex convolution theorem of z-transformation.

3-5 Determine the z-transform of the following sequences:

(a) $f(k) = 0$ for $k = 0$ and even integers

$f(k) = 1$ for $k =$ odd integers

(b) $f(k) = 1$ for $k = 0$ and even integers

$f(k) = -1$ for $k =$ odd integers

(c) $f(k) = e^{-k}(1 - \sin 2k)$

(d) $f(kT) = kTe^{-5kT}$

3-6 Given the following z-transforms F(z), find f(kT).

(a) $F(z) = \dfrac{z}{z^2 + 1}$

(b) $F(z) = \dfrac{10z}{z^2 - 1}$

(c) $F(z) = \dfrac{1}{z(z - 0.2)}$

3-7 Find the inverse z-transform of

$$F(z) = \frac{z(z + 1)}{(z - 1)(z^2 - z + 1)}$$

by means of the following methods:
(a) the real inversion formula
(b) power series expansion
(c) partial fraction expansion.

3-8 (a) Given the transfer function of a digital control system as

$$\frac{C(z)}{R(z)} = \frac{\sum_{k=0}^{M} b_k z^k}{\sum_{k=0}^{N} a_k z^k}$$

where M and N are positive integers, and $N \geq M$.
First, multiply the numerator and the denominator of the right-hand side of the last equation by z^{-N}/a_N, and then cross multiply both sides of the equation. Express C(z) in terms of the rest of the terms. Now show that the value of the output c(nT) for n = 0,1,2,... can be expressed as

$$c(nT) = \sum_{k=0}^{M} \frac{b_k}{a_N} r[(n-N+k)T] - \sum_{k=0}^{N-1} \frac{a_k}{a_N} c[(n-N+k)T]$$

where $r(jT) = 0$ and $c(jT) = 0$ for $j < 0$.
Note that the expression for c(nT) given above represents an alternate method of finding the inverse z-transform of C(z) given the values of the input sequence r(kT), k = 0,1,2,.... The characteristic of this method is that the z-transform of the input, R(z), need not be known or even exist, although the definition of R(z) must be used in the above proof. (The material used for this problem was contributed by Dr. S. M. Seltzer).

(b) Given the transfer function

$$\frac{C(z)}{R(z)} = \frac{z+1}{z^2 - z + 1}$$

with the input sequence $r(kT) = 1$ for $k \geq 0$. Find c(nT) for n = 0,1,2,..., using the expression given in part (a).

3-9 If $F_1(s)$ is band limited, that is, $|F_1(j\omega)| = 0$ for $|\omega| \geq \omega_1$, and the sampling frequency ω_s is greater than or equal to $2\omega_1$, show that

$$\mathcal{Z}[F_1(s)F_2(s)] = TF_1(z)\bar{F}_2(z)$$

where

$$F_2(z) = \mathcal{Z}[F_2(s)]$$

$$\bar{F}_2(z) = \mathcal{Z}[\bar{F}_2(s)]$$

and

$$\bar{F}_2(s) = F_2(s) \qquad \text{for } \omega < \omega_1$$

$$= 0 \qquad \text{for } \omega \geq \omega_1$$

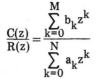

If $F_1(s)$ and $F_2(s)$ are both band limited with $|F_1(j\omega)| = 0$ for $|\omega| \geqslant \omega_1$ and $|F_2(j\omega)| = 0$ for $|\omega| \geqslant \omega_2$, and the sampling frequency $\omega_s \geqslant 2\max(\omega_1, \omega_2)$, show that

$$\mathscr{F}[F_1(s)F_2(s)] = TF_1(z)F_2(z)$$

3-10 The weighting sequence of a linear discrete-data system is

$$g(kT) = 5e^{-(k-1)T} \qquad \text{for } k \geqslant 1$$
$$= 0 \qquad \text{for } k = 0$$

Let the input to the system be $r(kT) = kT$ for $k \geqslant 0$. The sampling period is 0.5 second.
(a) Find the z-transform of the system output.
(b) Find the output sequence $c(kT)$ in closed form.

3-11 Find the output at the sampling instants for the system shown in Fig. 3P-11. The input is a unit-step function, and the sampling period is one second.

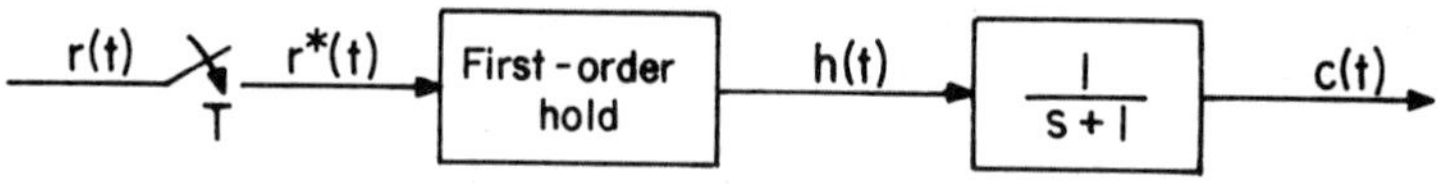

Fig. 3P-11.

3-12 Consider the sampled-data system with a nonintegral time delay as shown in Fig. 3P-12. Show that the z-transform of the output is

$$C_m(z) = z^{-1} \sum_{k=0}^{\infty} c[(k+m)T]z^{-k}$$

Assume that $c(t) = 0$ for $t < 0$.

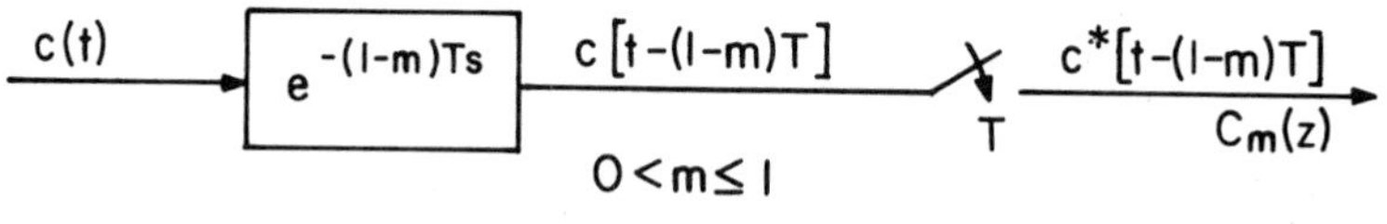

Fig. 3P-12.

3-13 The block diagram of a discrete-data control system is shown in Fig. 3P-13. Given that

$$G(s) = \frac{1}{s(s+0.2)}$$

$T = 0.2$ second, $r(t)$ = unit-step function, determine the following:
(a) the z-transform of the output, $C(z)$
(b) the output response at the sampling instants
(c) the final value of $c(kT)$.

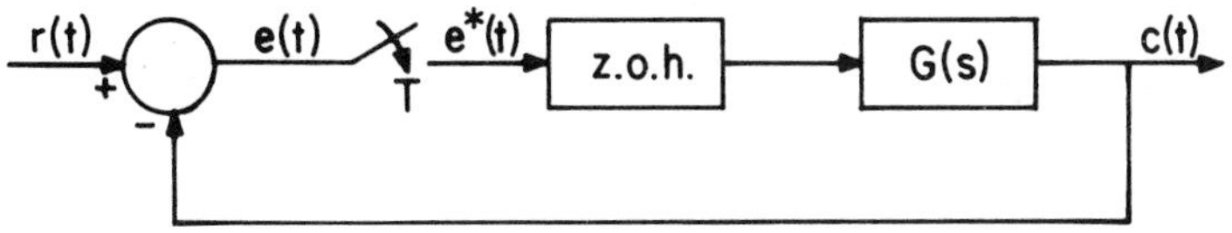

Fig. 3P-13.

3-14 The block diagram of a digital control system of the large space telescope is shown in Fig. 3P-14. The control incorporates proportional plus integral and rate feedback. Find the open-loop transfer function C(z)/E(s) and the closed-loop transfer function C(z)/R(z). A continuous-data counterpart of the digital system is obtained by deleting the sample-and-hold units. Find the open-loop and the closed-loop transfer functions of the continuous-data system by taking the limit of C(z)/E(z) and C(z)/R(z) as T approaches zero.

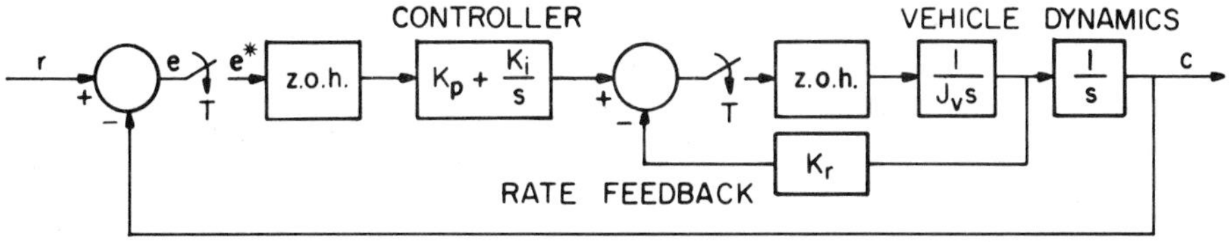

Fig. 3P-14.

3-15 The open-loop transfer function of the digital control system shown in Fig. 3P-15 is

$$G(z) = \frac{27.8(s+1)}{s(s^2+6s+48.6)}$$

The sampling period is 1 second.

(a) Determine the modified z-transform of the output, C(z,m), for a unit step input.
(b) Find C(z) by setting m = 1 in C(z,m).
(c) Find c(kT,m) by means of the inversion formula. Sketch c(kT,m) with several values of m ($0 < m < 1$) so that an accurate estimate of c(t) can be obtained.
(d) Find c(kT) from C(z) and compare with c(kT,m) for $m \neq 1$.

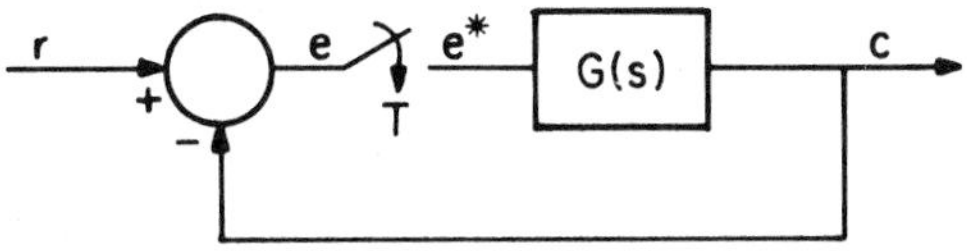

Fig. 3P-15.

3-16 The figure in Fig. 3P-16 shows a discrete-data system with a pulse-width sampler. If the input to the pulse-width sampler is r(t), then the sampled output $r_d(t)$ is written

$$r_d(t) = E \sum_{n=0}^{\infty} [\operatorname{sgn} r(nT)][u_s(t-nT) - u_s(t-nT-p_n]$$

where p_n, the variable pulse duration, is

$$p_n = a|r(nT)| \qquad \text{for } a|r(nT)| \leqslant T$$

$$p_n = T \qquad \text{for } |r(nT)| > T$$

a and E are constants, and T is the sampling period. R. E. Andeen suggested in his paper that the pulse-width sampler may be replaced by an equivalent arrangement with a pulse-amplitude sampler and a special hold device as shown in Fig. 3P-16, if the principle of equivalent area is considered. In other words, it can be shown that two signals $r_d(t)$ and $r'_d(t)$ are equivalent for suitably small T if

$$\int_{(n-1)T}^{nT} r_d(t)\,dt = \int_{(n-1)T}^{nT} r'_d(t)\,dt \tag{1}$$

where $r'_d(t)$ is now the output of the equivalent sample-and-hold device. The pulse-width of $r'_d(t)$ is chosen to be aR_{max} where R_{max} is the largest value of r(t) which is measurable.

(a) Is the pulse-width sampler a linear or a nonlinear device? Give reasons for your answer.

(b) Find the transfer function H(s) of the hold device so that Eq. (1) is satisfied.

(c) Find C(z), if G(s) = b/(s + b) and r(t) = u(t).

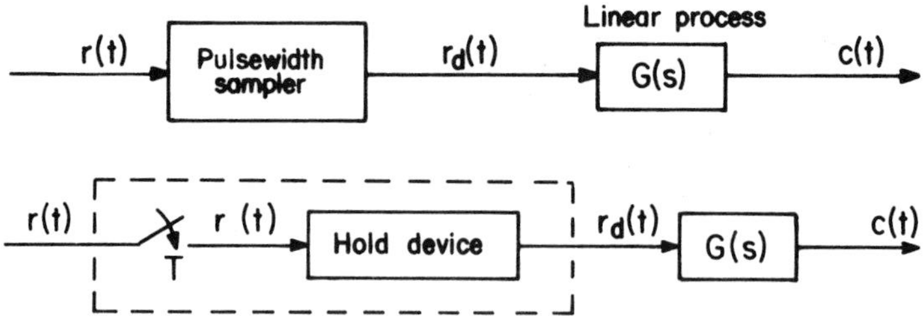

Fig. 3P-16.

3-17 For the digital systems shown in Fig. 3P-17, determine C(s) and C(z) using the sampled signal flow graph method. Check the answers with the direct signal flow graph method.

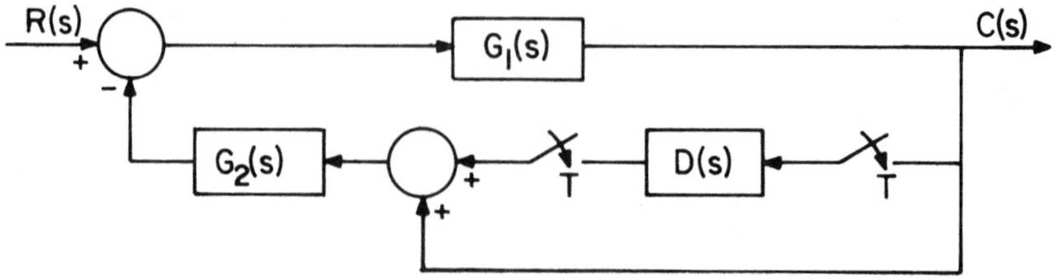

Fig. 3P-17(a).

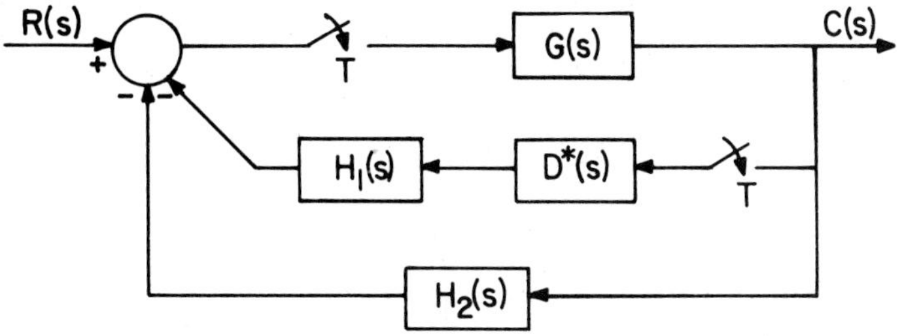

Fig. 3P-17(b).

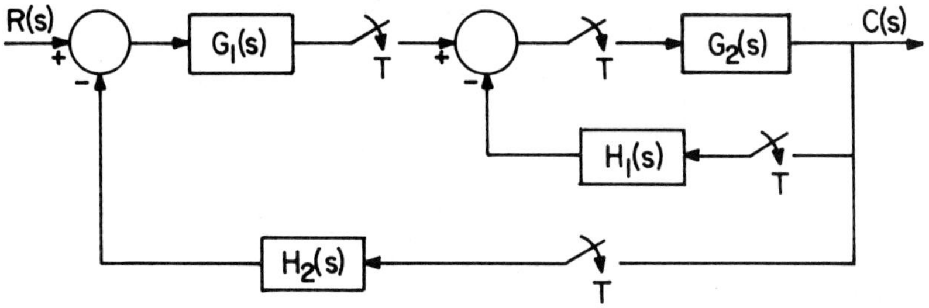

Fig. 3P-17(c).

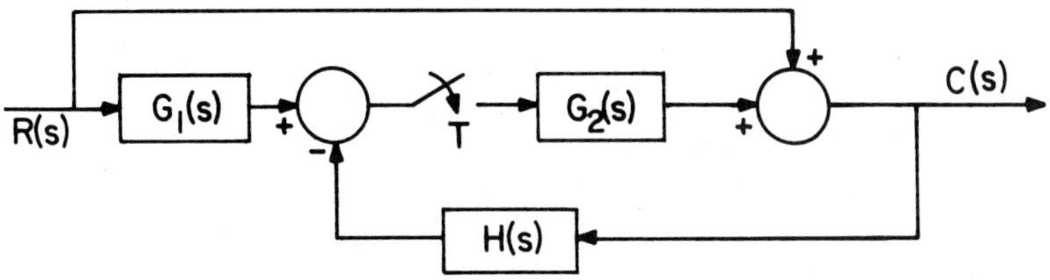

Fig. 3P-17(d).

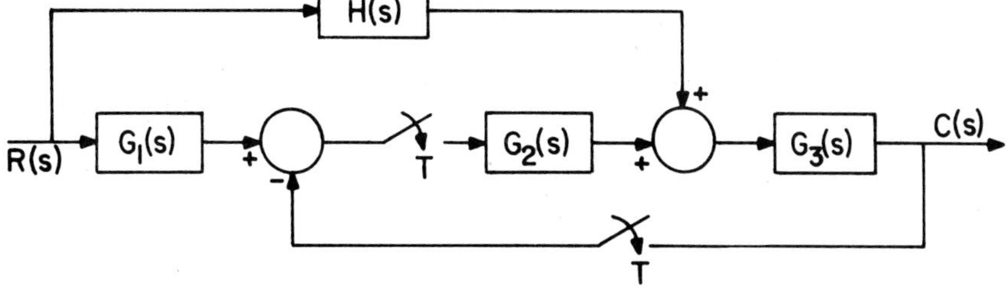

Fig. 3P-17(e).

3-18 The block diagram of a multirate sampled-data control system is shown in Fig. 3P-18. Find the closed-loop transfer function of the system, $C(z)/R(z)$. The sampling period T is one second, and N is an unspecified integer $\geqslant 1$. The transfer functions are:

$$D(s) = \frac{1}{s+1} \qquad G(s) = \frac{K}{s}$$

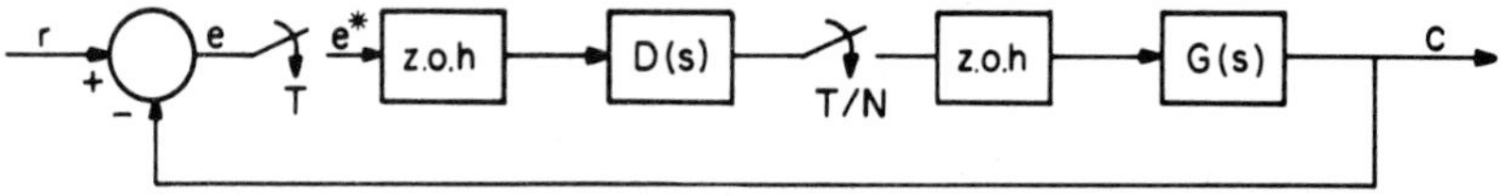

Fig. 3P-18.

3-19 A bilinear transformation is sometimes applied in the analysis of digital systems to transform from the z-plane to a w-plane so that the unit circle $|z| = 1$ is mapped onto the imaginary axis in the w-plane. Equations (3-286) and (3-287) give the z-transform of the slow-sampler output $C(z)$ in terms of the output of the fast sampler, $C(z_N)$. Let the bilinear transformation be $z = (w+1)/(w-1)$, such that

$$z_N = z^{1/N} = \frac{w_N + 1}{w_N - 1}$$

$$C(w) = C(z)\Big|_{z=(w+1)/(w-1)}$$

and

$$C(w_N) = C(z_N)\Big|_{z_N=(w_N+1)/(w_N-1)}$$

Show that $C(w)$ and $C(w_N)$ are related through

$$C(w) = - \quad \text{Residues of} \frac{2C(w_N)}{1 - \left[\frac{w_N+1}{w_N-1}\right]^N \frac{w-1}{w+1}} \frac{1}{w_N^2 - 1}$$

at the poles of $C(w_N)/(w_N + 1)$. Given $C(z_N) = z_N/(z_N - e^{-T/N})$, find $C(w)$ using the transformation and method described above.

3-20 The block diagram of a digital control system with a nonuniform sampling rate is shown in Fig. 3P-20. The sampler operates at time instants of $t = 0, T/N, T, (1+1/N)T, 2T, (2+1/N)T, 3T, \ldots$, where N is a positive number greater than one. Show that the z-transform of the output can be written as

$$C(z) = \frac{1}{\Delta(z)}\Big[\Big\{G_0(z)[1 + G_1(z)] - \mathcal{Z}[e^{Ts/N}G_0(s)]\mathcal{Z}[e^{-Ts/N}G_1(s)]\Big\}R(z)$$

$$+ \mathcal{Z}[e^{Ts/N}R(s)]\mathcal{Z}[1 + G_0(z)]\mathcal{Z}[e^{-Ts/N}G_1(s)] - \mathcal{Z}[e^{Ts/N}G_1(s)]G_0(z)\Big]$$

where

$$\Delta(z) = 1 + G_0(z) + G_1(z) - \mathcal{Z}[e^{Ts/N}G_0(s)]\mathcal{Z}[e^{-Ts/N}G_1(s)] + G_0(z)G_1(z)$$

$$G_0(z) = \mathcal{Z}\left[\frac{1 - e^{-Ts/N}}{s}\right]G(s)$$

$$G_1(z) = \mathcal{Z}\left[\frac{1 - e^{-Ts(1-1/N)}}{s}\right]G(s)$$

Find C(z) for $G(s) = K/(s + 1)$, $r(t)$ = unit step, and $T = 1$ sec. The values of K and N are not specified.

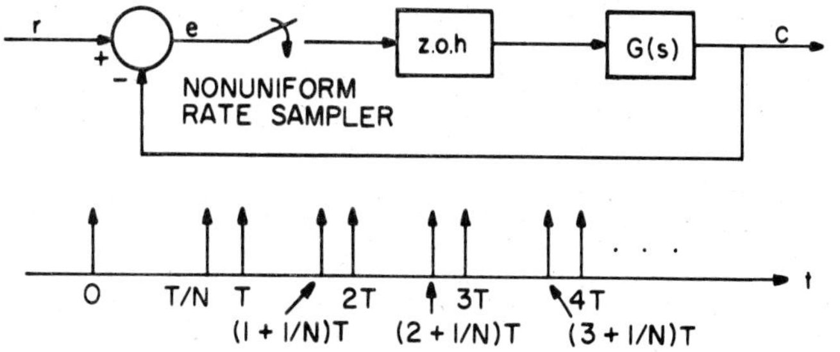

Fig. 3P-20.

3-21 A special type of multirate sampling scheme assumes that the input and output of a system are pulse trains with pulses occurring at $t = kT/N$, $k = 0,1,2,\ldots$, except that the input pulses change values only at $t = kT$, $k = 0,1,2,\ldots$. Figure 3P-21 shows the block diagram of the system with S_1 representing the special multirate sampler at the input. The sampler at the output is a regular multirate sampler with sampling period T/N. Show that the sampler S_1 can be decomposed into a combination of a sampler with period T and N parallel delays. Draw a block diagram for the equivalent system. Derive the z-transform of the output $c^*(t)_N$, $C(z_N)$, in terms of the transfer function of the system and the input transform R(z).

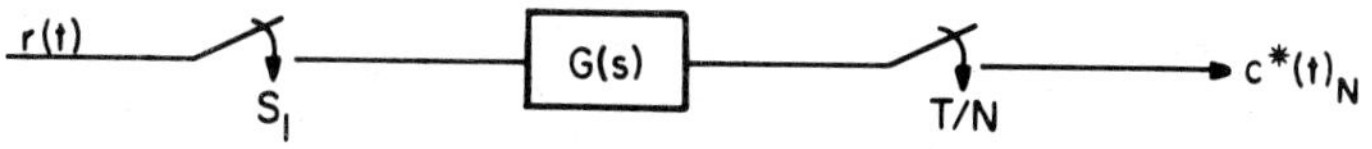

Fig. 3P-21.

3-22 Figure 3P-22 shows the block diagram of an all-digital data system, with the input and output data rates as described by the sampling operations of Problem 3-21. Draw a block diagram with samplers and delays as an equivalent analytical model for the system. Derive the output transform, $C(z_N)$, in terms of the system transform $G(z_N)$ and the input transform R(z).

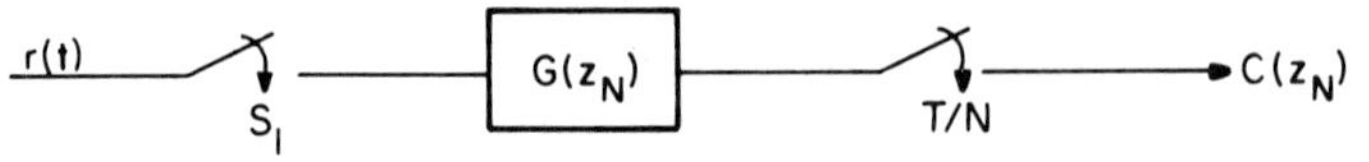

Fig. 3P-22.

3-23 The block diagram of a multirate digital control system is shown in Fig. 3P-23. Derive the transform relations $E(z)/R(z)$ and $C(z_N)/R(z)$ and show that the denominators of these two functions are different. Comment on this phenomenon. Add fictitious samplers in the system wherever it is necessary.

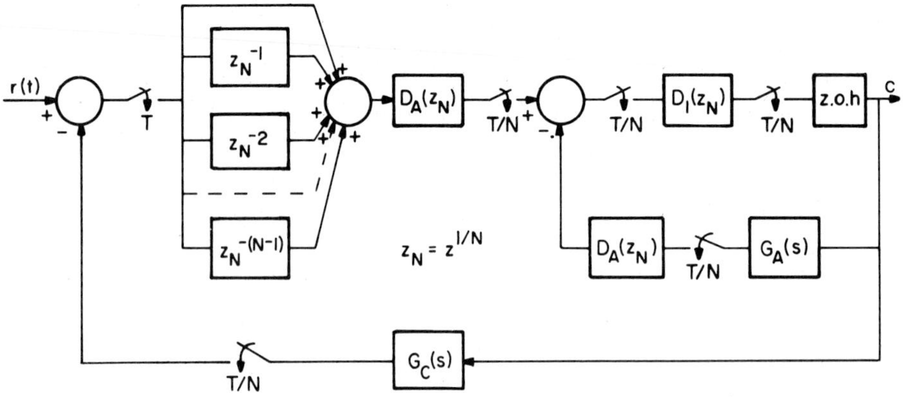

Fig. 3P-23.

3-24 The block diagram of a digital control system with nonuniform-rate sampling is shown in Fig. 3P-24. The sampler S_v samples at time instants of $t = 0, T/N, T, (1 + 1/N)T, 2T, (2 + 1/N)T, 3T, \ldots$, where N is a positive number greater than one. For $T = 1$ sec,

$$D(s) = \frac{1}{s+1} \qquad G(s) = \frac{K}{s}$$

find the closed-loop transfer function $C(z)/R(z)$ with K and N as parameters.

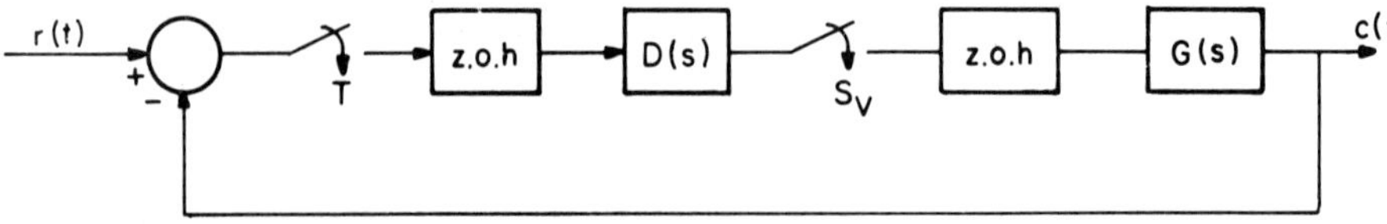

Fig. 3P-24.

3-25 Show that Eq. (3-302) is equal to Eq. (3-291).

4

THE STATE VARIABLE TECHNIQUE

4.1 INTRODUCTION

In general, the analysis and design of linear systems may be carried out by one of two major approaches. One approach relies on the use of Laplace and z-transforms, transfer functions, block diagrams or signal flow graphs. The other method, which is synonymous with modern control theory, is the state variable technique. The fact is that a great majority of modern control system design techniques are based on the state variable formulation and modeling of the system.

In a broad sense the state variable representation has the following advantages, at least in digital control systems studies, over the conventional transfer function method.

1. The state variable formulation is natural and convenient for computer solutions.
2. The state variable approach allows a unified representation of digital systems with various types of sampling schemes.
3. The state variable method allows a unified representation of single variable and multivariable systems.
4. The state variable method can be applied to certain types of nonlinear and time-varying systems.

In the state variable formulation a continuous-data system is represented by a set of first-order differential equations, called *state equations.* For a digital system with all discrete-data components, the state equations are first-order difference equations. As mentioned earlier, a digital system may contain analog as well as digital components, and the state equations of the system generally will consist of both first-order differential and difference equations. However, we do not intend to paint a picture in which there is always a clear-cut advantage in using the state variable formulation for control systems analysis and design. The conventional transfer function still has the advantage of being compact, and a large number of practical control systems design problems are still being carried out using well-established design techniques that are based on the transfer function formulation.

Since digital systems often contain analog elements, and difference equations are sometimes used to approximate the dynamics of analog systems, we shall begin by discussing the state equations and their solutions of continuous-data systems.

4.2 STATE EQUATIONS AND STATE TRANSITION EQUATIONS OF CONTINUOUS-DATA SYSTEMS

Consider that a continuous-data system with p inputs and q outputs as shown in Fig. 4-1 is characterized by the following set of n first-order differential equations, called state equations:

$$\frac{dx_i(t)}{dt} = f_i[x_1(t), x_2(t), \ldots, x_n(t), u_1(t), u_2(t), \ldots, u_p(t), t] \qquad (4\text{-}1)$$

$$(i = 1, 2, \ldots, n)$$

where $x_1(t)$, $x_2(t)$, ..., $x_n(t)$ are the state variables, $u_1(t)$, $u_2(t)$, ..., $u_p(t)$ are the input variables, and f_i denotes the ith functional relationship. In general, f_i can be linear or nonlinear.

The q outputs of the system are related to the state variables and the inputs through the *output equations* which are of the form,

$$c_k(t) = g_k[x_1(t), x_2(t), \ldots, x_n(t), u_1(t), u_2(t), \ldots, u_p(t), t] \qquad (4\text{-}2)$$

$$(k = 1, 2, \ldots, q)$$

Similar remarks can be made for g_k as for f_i.

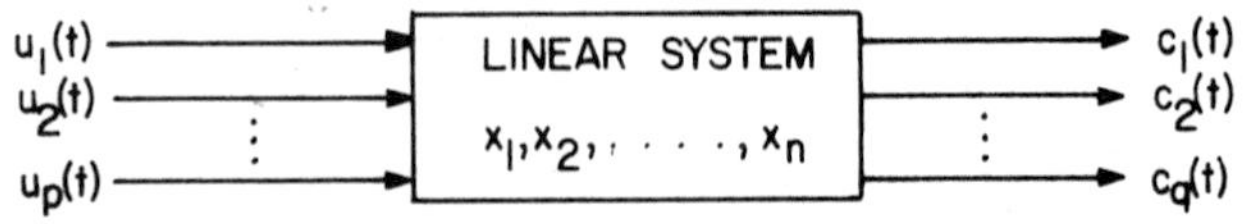

Fig. 4-1. A linear system with p inputs, q outputs, and n state variables.

The state equations and output equations as a set are called the *dynamic equations* of the system.

It is customary to write the dynamic equations in vector-matrix form:

State equation
$$\frac{d\mathbf{x}(t)}{dt} = \mathbf{f}[\mathbf{x}(t), \mathbf{u}(t), t] \tag{4-3}$$

Output equation
$$\mathbf{c}(t) = \mathbf{g}[\mathbf{x}(t), \mathbf{u}(t), t] \tag{4-4}$$

where $\mathbf{x}(t)$ is an $n \times 1$ column matrix, and is called the *state vector;* that is

$$\mathbf{x}(t) = \begin{bmatrix} x_1(t) \\ x_2(t) \\ \cdot \\ \cdot \\ \cdot \\ x_n(t) \end{bmatrix} \tag{4-5}$$

The input vector, $\mathbf{u}(t)$, is a $p \times 1$ column matrix, and

$$\mathbf{u}(t) = \begin{bmatrix} u_1(t) \\ u_2(t) \\ \cdot \\ \cdot \\ \cdot \\ u_p(t) \end{bmatrix} \tag{4-6}$$

The output vector, $\mathbf{c}(t)$, is defined as

$$\mathbf{c}(t) = \begin{bmatrix} c_1(t) \\ c_2(t) \\ \cdot \\ \cdot \\ \cdot \\ c_q(t) \end{bmatrix} \tag{4-7}$$

which is a $q \times 1$ column matrix.

If the system is linear but has time-varying elements, the dynamic equations of Eqs. (4-3) and (4-4) are written as

$$\frac{d\mathbf{x}(t)}{dt} = \mathbf{A}(t)\mathbf{x}(t) + \mathbf{B}(t)\mathbf{u}(t) \tag{4-8}$$

$$\mathbf{c}(t) = \mathbf{D}(t)\mathbf{x}(t) + \mathbf{E}(t)\mathbf{u}(t) \tag{4-9}$$

where $\mathbf{A}(t)$ is an $n \times n$ square matrix, $\mathbf{B}(t)$ is $n \times p$, $\mathbf{D}(t)$ is $q \times n$, and $\mathbf{E}(t)$ is $q \times p$. All the elements of these coefficient matrices are considered to be continuous functions of t.

If the system is linear and time-invariant, Eqs. (4-8) and (4-9) are of the form:

$$\frac{d\mathbf{x}(t)}{dt} = \mathbf{Ax}(t) + \mathbf{Bu}(t) \tag{4-10}$$

and

$$\mathbf{c}(t) = \mathbf{Dx}(t) + \mathbf{Eu}(t) \tag{4-11}$$

respectively. The matrices **A**, **B**, **D**, and **E** now all contain constant elements.

The State Transition Matrix of Time-Varying Systems

The *state transition matrix*, $\phi(t,t_0)$, is defined as an $n \times n$ matrix which satisfies the homogeneous state equation

$$\frac{d\mathbf{x}(t)}{dt} = \mathbf{A}(t)\mathbf{x}(t) \tag{4-12}$$

for any real t_0; that is,

$$\frac{d\phi(t,t_0)}{dt} = \mathbf{A}(t)\phi(t,t_0) \tag{4-13}$$

with initial condition $\phi(t_0,t_0) = \mathbf{I}$, where **I** is an $n \times n$ identity matrix.

The solution of the homogeneous state equation of Eq. (4-12) is then

$$\mathbf{x}(t) = \phi(t,t_0)\mathbf{x}(t_0) \tag{4-14}$$

for any t and t_0. This may be proved by taking the time derivative on both sides of Eq. (4-14) and then using Eq. (4-13).

For the time-varying state equation of Eq. (4-12), the state transition matrix is given by

$$\phi(t,t_0) = \exp\left[\int_{t_0}^{t} \mathbf{A}(\tau)d\tau\right] = \mathbf{I} + \int_{t_0}^{t} \mathbf{A}(\tau)d\tau + \frac{1}{2!}\int_{t_0}^{t} \mathbf{A}(\tau)d\tau\int_{t_0}^{t} \mathbf{A}(\lambda)d\lambda + \dots \tag{4-15}$$

if the matrix $\mathbf{A}(t)$ and $\int_{t_0}^{t} \mathbf{A}(\tau)d\tau$ commute; that is,

$$\mathbf{A}(t)\int_{t_0}^{t} \mathbf{A}(\tau)d\tau = \int_{t_0}^{t} \mathbf{A}(\tau)d\tau\,\mathbf{A}(t) \tag{4-16}$$

This is proven by taking the time derivative on both sides of Eq. (4-15),

$$\frac{d}{dt}\exp\left[\int_{t_0}^{t} \mathbf{A}(\tau)d\tau\right] = \frac{d}{dt}\left[\mathbf{I} + \int_{t_0}^{t} \mathbf{A}(\tau)d\tau + \frac{1}{2!}\int_{t_0}^{t} \mathbf{A}(\tau)d\tau\int_{t_0}^{t} \mathbf{A}(\lambda)d\lambda + \dots\right]$$

$$= A(t) + \frac{1}{2!} A(t)\int_{t_0}^{t} A(\tau)d\tau + \frac{1}{2!}\int_{t_0}^{t} A(\tau)d\tau A(t) + \ldots \quad (4\text{-}17)$$

The right-hand side of Eq. (4-13) is written

$$A(t)\phi(t,t_0) = A(t) + A(t)\int_{t_0}^{t} A(\tau)d\tau + \ldots \quad (4\text{-}18)$$

Comparing Eq. (4-18) with Eq. (4-17), we see that the two equations are equal only if Eq. (4-16) is satisfied by A(t).

Properties of the State Transition Matrix (Time Varying)

Several important properties of the state transition matrix $\phi(t,t_0)$ are stated below:

1. $\phi(t_1,t_2)\phi(t_2,t_3) = \phi(t_1,t_3)$ for any t_1, t_2, t_3. (4-19)
   This property stems from

$$\mathbf{x}(t_1) = \phi(t_1,t_2)\mathbf{x}(t_2) \quad (4\text{-}20)$$

$$\mathbf{x}(t_2) = \phi(t_2,t_3)\mathbf{x}(t_3) \quad (4\text{-}21)$$

$$\mathbf{x}(t_1) = \phi(t_1,t_3)\mathbf{x}(t_3) \quad (4\text{-}22)$$

   Substitution of Eq. (4-21) into Eq. (4-20) gives

$$\mathbf{x}(t_1) = \phi(t_1,t_2)\phi(t_2,t_3)\mathbf{x}(t_3) \quad (4\text{-}23)$$

   Comparison of Eq. (4-22) with Eq. (4-23) asserts the identity in Eq. (4-19).
2. $\phi(t_0,t_0) = \mathbf{I}$ (identity matrix). (4-24)
   This property follows directly from Eq. (4-15).
3. $\phi(t,t_0)$ is nonsingular; that is, the determinant of $\phi(t,t_0)$, $|\phi(t,t_0)|$, is nonzero.
   This is due to the Abel-Jacobi-Liouville Theorem [12] which states that

$$|\phi(t,t_0)| = \exp\left[\int_{t_0}^{t} \mathrm{tr}\, A(\tau)d\tau\right] \quad (4\text{-}25)$$

   where tr $A(\tau)$ denotes the trace of $A(\tau)$ and represents the sum of the elements on the main diagonal of $A(\tau)$. It is apparent that for $|\phi(t,t_0)|$ to be nonzero, $\int_{t_0}^{t} \mathrm{tr}\, A(\tau)d\tau$ cannot be equal to $-\infty$ for any t and t_0.
4. If $\phi(t_2,t_1)$ is nonsingular,

$$\phi(t_1, t_2) = \phi^{-1}(t_2, t_1) \text{ for any } t_1 \text{ and } t_2 \tag{4-26}$$

Setting $t_1 = t_3$ in Eq. (4-19) and using the property of Eq. (4-24), we have

$$\phi(t_1, t_2)\phi(t_2, t_1) = \mathbf{I} \tag{4-27}$$

Thus, Eq. (4-26) is obtained by postmultiplying both sides of the last equation by $\phi^{-1}(t_2, t_1)$.

The State Transition Matrix (Time Invariant)

For linear time-invariant systems, the homogeneous state equation is written as

$$\frac{d\mathbf{x}(t)}{dt} = \mathbf{A}\mathbf{x}(t) \tag{4-28}$$

In this case, from Eq. (4-15), the state transition matrix is given by

$$\phi(t - t_0) = \mathbf{I} + \mathbf{A}(t - t_0) + \frac{1}{2!}\mathbf{A}^2(t - t_0)^2 + \ldots$$

$$= e^{\mathbf{A}(t-t_0)} \tag{4-29}$$

For $t_0 = 0$,

$$\phi(t) = e^{\mathbf{A}t} = \sum_{k=0}^{\infty} \frac{\mathbf{A}^k t^k}{k!} \tag{4-30}$$

Since the Laplace transform is now applicable to the time-invariant system, we take the Laplace transform on both sides of Eq. (4-28) and get

$$s\mathbf{X}(s) - \mathbf{x}(0^+) = \mathbf{A}\mathbf{X}(s) \tag{4-31}$$

where $\mathbf{x}(0^+)$ is the initial state vector evaluated at $t = 0^+$.

Solving for $\mathbf{X}(s)$ in Eq. (4-31) yields

$$\mathbf{X}(s) = (s\mathbf{I} - \mathbf{A})^{-1}\mathbf{x}(0^+) \tag{4-32}$$

Taking the inverse Laplace transform on both sides of the last equation gives

$$\mathbf{x}(t) = \mathcal{L}^{-1}[(s\mathbf{I} - \mathbf{A})^{-1}]\mathbf{x}(0^+) \tag{4-33}$$

Thus, the state transition matrix is also given by

$$\phi(t) = \mathcal{L}^{-1}[(s\mathbf{I} - \mathbf{A})^{-1}] \tag{4-34}$$

Properties of the State Transition Matrix (Time-Invariant)

The properties of the state transition matrix for the time-varying case can all be reduced to simple special cases for time-invariant systems. These are listed as follows:

1. $\phi(t_1 - t_2)\phi(t_2 - t_3) = \phi(t_1 - t_3)$ for any t_1, t_2, t_3 (4-35)
2. $\phi(0) = I$ (4-36)
3. $\phi(t)$ is nonsingular for finite $\mathbf{A}$
4. $\phi^{-1}(t) = \phi(-t)$ (4-37)

Solution of the Nonhomogeneous State Equation

The nonhomogeneous time-varying state equation

$$\frac{d\mathbf{x}(t)}{dt} = \mathbf{A}(t)\mathbf{x}(t) + \mathbf{B}(t)\mathbf{u}(t) \tag{4-38}$$

is solved by noting that if the equation is first transformed into the form

$$\frac{d\mathbf{y}(t)}{dt} = \mathbf{f}(t) \tag{4-39}$$

the solution would be straightforward.

We let

$$\mathbf{x}(t) = \phi(t,t_0)\mathbf{y}(t) \tag{4-40}$$

where $\phi(t,t_0)$ is the state transition matrix. Then,

$$\frac{d\mathbf{x}(t)}{dt} = \frac{d\phi(t,t_0)}{dt}\mathbf{y}(t) + \phi(t,t_0)\frac{d\mathbf{y}(t)}{dt} \tag{4-41}$$

Equating Eq. (4-38) to Eq. (4-41) and using Eqs. (4-13) and (4-40), we have

$$\phi(t,t_0)\frac{d\mathbf{y}(t)}{dt} = \mathbf{B}(t)\mathbf{u}(t) \tag{4-42}$$

Thus,

$$\frac{d\mathbf{y}(t)}{dt} = \phi(t_0,t)\mathbf{B}(t)\mathbf{u}(t) \tag{4-43}$$

The solution of Eq. (4-43) is obtained by integrating both sides; we get

$$\mathbf{y}(t) = \mathbf{y}(t_0) + \int_{t_0}^{t} \phi(t_0,\tau)\mathbf{B}(\tau)\mathbf{u}(\tau)d\tau \tag{4-44}$$

Now using Eq. (4-40), the solution of Eq. (4-38) is found to be

$$\mathbf{x}(t) = \phi(t,t_0)\mathbf{x}(t_0) + \int_{t_0}^{t} \phi(t,\tau)\mathbf{B}(\tau)\mathbf{u}(\tau)d\tau \tag{4-45}$$

for any t and t_0.

For the time-invariant system, the solution of Eq. (4-10) is easily extended from Eq. (4-45).

$$\mathbf{x}(t) = \phi(t - t_0)\mathbf{x}(t_0) + \int_{t_0}^{t} \phi(t-\tau)\mathbf{B}(\tau)\mathbf{u}(\tau)d\tau \tag{4-46}$$

for any t and t_0.

These solutions of the state equation are also referred to as the *state transition equations.*

4.3 STATE EQUATIONS OF DIGITAL SYSTEMS WITH SAMPLE-AND-HOLD

An open-loop discrete-data system is obtained when sample-and-hold devices are added to the inputs of the system in Fig. 4-1. The resulting system is shown in Fig. 4-2.

Since the signals $u_i(t)$, $i = 1,2,\ldots,p$, are now outputs of sample-and-hold devices, they are described by

$$u_i(t) = u_i(kT) = e_i(kT) \qquad kT \leqslant t < (k+1)T \tag{4-47}$$

where $k = 0,1,2,\ldots$, and $i = 1,2,\ldots,p$.

Let the dynamics of the linear system be represented by the time-varying state equations of Eq. (4-8). The state transition equation is given by Eq. (4-45) for all t and t_0. Since the inputs are constant during the sampling period T, in

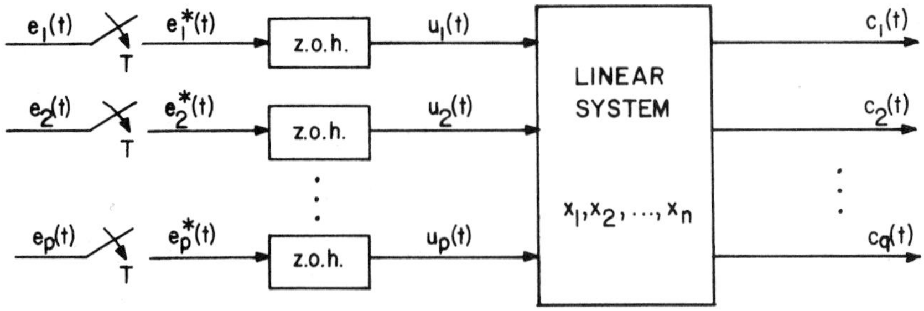

Fig. 4-2. A multivariable digital system with sample-and-hold.

Eq. (4-45) the input vector $\mathbf{u}(\tau)$ can be placed outside the integral sign. Thus,

$$\mathbf{x}(t) = \phi(t,t_0)\mathbf{x}(t_0) + \int_{t_0}^{t} \phi(t,\tau)\mathbf{B}(\tau)d\tau\mathbf{u}(kT) \tag{4-48}$$

where it is implied that $t_0 = kT$ and $kT \leqslant t \leqslant (k+1)T$.

Equation (4-48) defines the states during the sampling interval $kT \leqslant t \leqslant (k+1)T$. The equation can be further modified to describe the transition of states of the digital system at the sampling instants only. Letting $t_0 = kT$ and $t = (k+1)T$, Eq. (4-48) becomes

$$\mathbf{x}[(k+1)T] = \phi[(k+1)T,kT]\mathbf{x}(kT) + \theta[(k+1)T,kT]\mathbf{u}(kT) \tag{4-49}$$

for $kT \leqslant t \leqslant (k+1)T$, where

$$\theta[(k+1)T,kT] = \int_{kT}^{(k+1)T} \phi[(k+1)T,\tau]\mathbf{B}(\tau)d\tau \tag{4-50}$$

It should be noted that although $u_i(kT)$ is held constant only for the time interval from kT to $(k+1)T^-$, the solution in Eq. (4-48) is valid for the entire interval including $t = (k+1)T$, since $\mathbf{x}(t)$ is a continuous function of t.

Equation (4-49) is regarded as the *discrete state equation* of the digital system shown in Fig. 4-2. These state equations, however, describe the dynamics of the system only at the sampling instants. In other words, by setting $t_0 = kT$ and $t = (k+1)T$ in Eq. (4-48), all information on the system in between the sampling instants is lost.

In a similar manner, the output equation in Eq. (4-9) is discretized by setting $t = kT$. Then

$$\mathbf{c}(kT) = \mathbf{D}(kT)\mathbf{x}(kT) + \mathbf{E}(kT)\mathbf{u}(kT) \tag{4-51}$$

Equations (4-49) and (4-51) together form the dynamic equations of the digital system.

The dynamic equations in Eqs. (4-49) and (4-51) can be written in a simpler form by normalizing the sampling period to be $T = 1$. Then, the dynamic equations become

$$\mathbf{x}(k+1) = \phi(k+1,k)\mathbf{x}(k) + \theta(k+1,k)\mathbf{u}(k) \tag{4-52}$$

where

$$\theta(k+1,k) = \int_{k}^{k+1} \phi(k+1,\tau)\mathbf{B}(\tau)d\tau \tag{4-53}$$

$$\mathbf{c}(k) = \mathbf{D}(k)\mathbf{x}(k) + \mathbf{E}(k)\mathbf{u}(k) \tag{4-54}$$

Still another way of representing Eqs. (4-49) and (4-51) is to let $t = t_{k+1}$ and $t_0 = t_k$; then

$$\mathbf{x}(t_{k+1}) = \phi(t_{k+1}, t_k)\mathbf{x}(t_k) + \theta(t_{k+1}, t_k)\mathbf{u}(t_k) \tag{4-55}$$

where

$$\theta(t_{k+1}, t_k) = \int_{t_k}^{t_{k+1}} \phi(t_{k+1}, \tau)\mathbf{B}(\tau)d\tau \tag{4-56}$$

and

$$\mathbf{c}(t_k) = \mathbf{D}(t_k)\mathbf{x}(t_k) + \mathbf{E}(t_k)\mathbf{u}(t_k) \tag{4-57}$$

4.4 STATE EQUATIONS OF DIGITAL SYSTEMS WITH ALL-DIGITAL ELEMENTS

When a digital system is composed of all-digital elements and the inputs and the outputs of the system are all digital, the system may be described by the following discrete dynamic equations:

$$\mathbf{x}(k + 1) = \mathbf{A}(k)\mathbf{x}(k) + \mathbf{B}(k)\mathbf{u}(k) \tag{4-58}$$

$$\mathbf{c}(k) = \mathbf{D}(k)\mathbf{x}(k) + \mathbf{E}(k)\mathbf{u}(k) \tag{4-59}$$

where $\mathbf{A}(k)$, $\mathbf{B}(k)$, $\mathbf{D}(k)$, and $\mathbf{E}(k)$ are coefficient matrices with time-varying elements. The values of these elements can vary only at the discrete instants $k = 0,1,2,\ldots$. In reality, the dynamic equations of Eqs. (4-58) and (4-59) may represent a discrete-data system with k denoting "stages" or "sequence of events". Therefore, discrete time does not always have to be the independent variable in the dynamic equations.

4.5 STATE TRANSITION EQUATIONS OF DIGITAL SYSTEMS

Time-Varying Systems

It is easy to see that since the discrete state equations in Eqs. (4-49), (4-52), (4-55), and (4-58) are essentially of the same form, their solutions should also be of the same form. However, before embarking on the solutions of the discrete state equations, we should first point out some of the similarities and differences between the continuous-time and the discrete state equations. It should be noted that the solution of the continuous-time state equation of Eq. (4-38), which is given by Eq. (4-45), is valid for any t and t_0 if $\phi(t,t_0)$ is nonsingular. In other words, the solution is good for $t \geq t_0$ as well as $t \leq t_0$. This means that the transition of state for the continuous-time process can take place in forward time or reverse time. We can show that the discrete state equations in Eqs. (4-49), (4-52), and (4-55) also are bidirectional in k if $\phi(k+1,k)$ is nonsingular. These state equations are originally defined in the forward direction since they are obtained by setting $t_0 = kT$ and $t = (k+1)T$ in the continuous-time state transition equations.

There are at least two ways in which we can perform backward in time in the discrete state equations. Referring to the notation used in Eq. (4-52), if the state transition matrix $\phi(k+1,k)$ is nonsingular, we can write the state equation as

$$\mathbf{x}(k) = \phi^{-1}(k+1,k)\mathbf{x}(k+1) - \phi^{-1}(k+1,k)\theta(k+1,k)\mathbf{u}(k) \tag{4-60}$$

Using the properties of $\phi(k+1,k)$ stated in Eqs. (4-19) and (4-26), we have

$$\mathbf{x}(k) = \phi(k,k+1)\mathbf{x}(k+1) + \theta(k,k+1)\mathbf{u}(k) \tag{4-61}$$

where

$$\theta(k,k+1) = \int_{k+1}^{k} \phi(k,\tau)\mathbf{B}(\tau)d\tau \tag{4-62}$$

Equation (4-61) may be regarded as a state equation which describes the transition of state from $k+1$ to k, and $\mathbf{u}(k)$ denotes the input vector which has constant elements during this interval.

Another approach is to set $t_0 = k+1$ and $t = k$ in Eq. (4-45) with $\mathbf{u}(\tau) = \mathbf{u}(k)$, and the same equation in Eq. (4-61) results. In general, if the transition in state is to be bidirectional in k, $\phi(k+1,k)$ must be nonsingular.

It is interesting to note that since the state equations of Eqs. (4-49), (4-52), and (4-55) are obtained by sampling the state transition equation of a continuous-time system with sample-and-hold, the state transition matrix $\phi(k+1,k)$ is always nonsingular if the matrix $\mathbf{A}(t)$ in the original differential equation is continuous and finite.

The discrete state equations in Eq. (4-58) pose a different problem, since in principle there are no physical restrictions on the elements of $\mathbf{A}(k)$ and $\mathbf{B}(k)$. Therefore, unless $\mathbf{A}(k)$ is nonsingular for all k, the state equation in Eq. (4-58) can only be solved in the forward direction. In fact, if $\mathbf{A}(k)$ is nonsingular for $k \leqslant j$, then

$$\mathbf{x}(k) = \mathbf{A}^{-1}(k)\mathbf{x}(k+1) - \mathbf{A}^{-1}(k)\mathbf{B}(k)\mathbf{u}(k) \tag{4-63}$$

for $k = 0,1,2,\ldots,j$.

We shall now show that the discrete state equation can be solved by means of a recursive procedure. Let us choose the state equation in Eq. (4-58) because it will lead to a simpler notation, but keep in mind that the solution is equally valid for the other forms of discrete state equations if we replace $\mathbf{A}(k)$ by

$\phi(k+1,k)$, $\mathbf{B}(k)$ by $\theta(k+1,k)$, etc. Equation (4-58) is rewritten,

$$\mathbf{x}(k+1) = \mathbf{A}(k)\mathbf{x}(k) + \mathbf{B}(k)\mathbf{u}(k) \tag{4-64}$$

The following equations are written recursively:

$$k = 0 \qquad \mathbf{x}(1) = \mathbf{A}(0)\mathbf{x}(0) + \mathbf{B}(0)\mathbf{u}(0) \tag{4-65}$$

$$\begin{aligned} k = 1 \qquad \mathbf{x}(2) &= \mathbf{A}(1)\mathbf{x}(1) + \mathbf{B}(1)\mathbf{u}(1) \\ &= \mathbf{A}(1)\mathbf{A}(0)\mathbf{x}(0) + \mathbf{A}(1)\mathbf{B}(0)\mathbf{u}(0) + \mathbf{B}(1)\mathbf{u}(1) \end{aligned} \tag{4-66}$$

$$\vdots$$

$$\begin{aligned} k = N-1 \quad \mathbf{x}(N) &= \mathbf{A}(N-1)\mathbf{x}(N-1) + \mathbf{B}(N-1)\mathbf{u}(N-1) \\ &= \mathbf{A}(N-1)\mathbf{A}(N-2)\ldots\mathbf{A}(1)\mathbf{A}(0)\mathbf{x}(0) \\ &\quad + \mathbf{A}(N-1)\mathbf{A}(N-2)\ldots\mathbf{A}(1)\mathbf{B}(0)\mathbf{u}(0) \\ &\quad + \mathbf{A}(N-1)\mathbf{A}(N-2)\ldots\mathbf{A}(2)\mathbf{B}(1)\mathbf{u}(1) \\ &\quad + \ldots + \mathbf{A}(N-1)\mathbf{B}(N-2)\mathbf{u}(N-2) + \mathbf{B}(N-1)\mathbf{u}(N-1) \end{aligned} \tag{4-67}$$

Let

$$\phi(N,i+1) = \mathbf{A}(N-1)\mathbf{A}(N-2)\ldots\mathbf{A}(i+1) \tag{4-68}$$

for

$$i = -1,0,1,2,\ldots,N-2, \qquad N > i+1$$

and

$$\phi(N,i+1) = \mathbf{I} \tag{4-69}$$

for $N = i + 1$. Using these notations, Eq. (4-67) may be written as

$$\mathbf{x}(N) = \phi(N,0)\mathbf{x}(0) + \sum_{i=0}^{N-1} \phi(N,i+1)\mathbf{B}(i)\mathbf{u}(i) \tag{4-70}$$

which is the solution of Eq. (4-64) for $\mathbf{x}(N)$, $N \geqslant 0$, given the initial state $\mathbf{x}(0)$ and the input $\mathbf{u}(i)$, $i = 0,1,\ldots,N-1$.

The stages in Eq. (4-70) can be shifted forward by any positive integer M, so that

$$\mathbf{x}(N+M) = \phi(N+M,M)\mathbf{x}(M) + \sum_{i=M}^{N+M-1} \phi(N+M,i+1)\mathbf{B}(i)\mathbf{u}(i) \tag{4-71}$$

Now, if we let k = N + M, we have

$$\mathbf{x}(k) = \phi(k,M)\mathbf{x}(M) + \sum_{i=M}^{k-1} \phi(k,i+1)\mathbf{B}(i)\mathbf{u}(i) \tag{4-72}$$

where

$$\begin{aligned} \phi(k,M) &= \mathbf{A}(k-1)\mathbf{A}(k-2)\ldots\mathbf{A}(M) \qquad & k \geqslant M+1 \\ &= \mathbf{I} & k = M \end{aligned} \tag{4-73}$$

In general, we may write the state equation as

$$\mathbf{x}(k_{j+1}) = \mathbf{A}(k_j)\mathbf{x}(k_j) + \mathbf{B}(k_j)\mathbf{u}(k_j) \tag{4-74}$$

$$j = 0,1,2,\ldots,N-1$$

with the state transition equation as

$$\mathbf{x}(k_N) = \phi(k_N,k_0)\mathbf{x}(k_0) + \sum_{i=0}^{N-1} \phi(k_N,k_{i+1})\mathbf{B}(k_i)\mathbf{u}(k_i) \tag{4-75}$$

where

$$\begin{aligned} \phi(k_N,k_0) &= \mathbf{A}(k_{N-1})\mathbf{A}(k_{N-2})\ldots\mathbf{A}(k_1)\mathbf{A}(k_0) \qquad & k_N > k_0 \\ &= \mathbf{I} & k_N = k_0 \end{aligned} \tag{4-76}$$

In this case, the intervals between k_j and k_{j+1} need not be constant, and k_j can represent time or stages.

The n × n matrix $\phi(k_j,k_0)$ is defined as the state transition matrix of $\mathbf{A}(k_j)$ and satisfies the homogeneous state equation

$$\mathbf{x}(k_{j+1}) = \mathbf{A}(k_j)\mathbf{x}(k_j) \tag{4-77}$$

for $j \geqslant 0$. Thus, the following relation holds:

$$\phi(k_{N+1},k_0) = \mathbf{A}(k_N)\phi(k_N,k_0) \qquad k_N \geqslant k_0 \tag{4-78}$$

Properties of the State Transition Matrix $\phi(k_j,k_0)$

Similar to the continuous-time case, the state transition matrix $\phi(k_j,k_0)$ has the following properties which are useful for the analysis of digital systems:

1. $\phi(i,j)\phi(j,m) = \phi(i,m)$ for all stages i, j, m, (4-79)

   if $\mathbf{A}(k)$ is nonsingular for all k that lies within min(i,j,m) and max(i,j,m). If $\mathbf{A}(k)$ is singular for $k \geqslant p$, then Eq. (4-79) is true only for max(i,j,m) $< p$.

The steps in the proof of this property are very similar to the steps shown in Eqs. (4-20) through (4-23). We need the existence of $\mathbf{A}^{-1}(k)$ so that the state transition process may be carried out in both directions, since i, j, and m are arbitrary. If $\mathbf{A}(k)$ is singular for $k \geqslant p$, then, we can write

$$\phi(i,j)\phi(j,m) = \mathbf{A}(i-1)\mathbf{A}(i-2) \ldots \mathbf{A}(j)\mathbf{A}(j-1)\mathbf{A}(j-2) \ldots \mathbf{A}(m)$$
$$= \phi(i,m) \tag{4-80}$$

for $p > i \geqslant j \geqslant m$.

2. $\phi(k,k) = \mathbf{I}$ (4-81)

This property follows directly from the definition of $\phi(k,M)$ given in Eq. (4-73).

3. $\phi(i,j) = \phi^{-1}(j,i)$ for all i, j, (4-82)

if $\mathbf{A}(k)$ is nonsingular for

$$k = j-1, j-2, \ldots, i \qquad j > i$$
$$k = i-1, i-2, \ldots, j \qquad i > j$$

The proof of this property is left as an exercise (see Problem 4-16).

4.6 STATE TRANSITION EQUATIONS OF DIGITAL TIME-INVARIANT SYSTEMS

When a linear digital system is time-invariant, its dynamic equation can be written in one of the following forms:

$$\mathbf{x}[(k+1)T] = \phi(T)\mathbf{x}(kT) + \theta(T)\mathbf{u}(kT) \tag{4-83}$$

$$\mathbf{c}(kT) = \mathbf{D}\mathbf{x}(kT) + \mathbf{E}\mathbf{u}(kT) \tag{4-84}$$

where $\phi(T)$ is the state transition matrix.

$$\phi(T) = e^{\mathbf{A}T} = \mathbf{I} + \mathbf{A}T + \frac{\mathbf{A}^2T^2}{2!} + \ldots \tag{4-85}$$

$$\theta(T) = \int_0^T \phi(T-\tau)\mathbf{B}(\tau)d\tau \tag{4-86}$$

or

$$\mathbf{x}(k+1) = \phi(1)\mathbf{x}(k) + \theta(1)\mathbf{u}(k) \tag{4-87}$$

$$\mathbf{c}(k) = \mathbf{D}\mathbf{x}(k) + \mathbf{E}\mathbf{u}(k) \tag{4-88}$$

or

$$\mathbf{x}(k+1) = \mathbf{A}\mathbf{x}(k) + \mathbf{B}\mathbf{u}(k) \tag{4-89}$$

$$\mathbf{c}(k) = \mathbf{D}\mathbf{x}(k) + \mathbf{E}\mathbf{u}(k) \tag{4-90}$$

As verified earlier, $\phi(T)$ and $\phi(1)$ are always nonsingular if the elements of **A** are finite. However, in general, there is no control on the coefficient matrices of the all-digital state equation of Eq. (4-89), so that **A** may be a singular matrix.

Just as in the time-varying system, the time-invariant state equation can be solved by using a recursive procedure. For Eq. (4-83), the solution is

$$\mathbf{x}(NT) = \phi(NT)\mathbf{x}(0) + \sum_{k=0}^{N-1} \phi[(N-k-1)T]\theta(T)\mathbf{u}(kT) \tag{4-91}$$

where

$$\phi(NT) = \underbrace{\phi(T)\phi(T)\ldots\phi(T)}_{N} = \phi^{N}(T) \tag{4-92}$$

It should be noted that $\phi(NT)$ is only a notation used to represent the identity in Eq. (4-92). In general, $\phi(NT)$ is *not* equal to $\phi(T)$ with T replaced by NT, although in simple cases it may seem to be true.

For Eq. (4-89), the state transition equation is

$$\mathbf{x}(N) = \mathbf{A}^{N}\mathbf{x}(0) + \sum_{k=0}^{N-1} \mathbf{A}^{N-k-1}\mathbf{B}\mathbf{u}(k) \tag{4-93}$$

These state transition equations can also be shifted for nonzero initial time or stage. Then,

$$\mathbf{x}[(N+M)T] = \phi(NT)\mathbf{x}(MT) + \sum_{k=0}^{N-1} \phi[(N-k-1)T]\theta(T)\mathbf{u}[(M+k)T] \tag{4-94}$$

or

$$\mathbf{x}(N+M) = \mathbf{A}^{N}\mathbf{x}(M) + \sum_{k=0}^{N-1} \mathbf{A}^{N-k-1}\mathbf{B}\mathbf{u}(M+k) \tag{4-95}$$

4.7 DIGITAL SIMULATION AND APPROXIMATION

Discrete state equations may also be obtained by approximating an analog system by a digital model. Let us consider the following dynamic equations which are for an analog system.

$$\dot{\mathbf{x}}(t) = \mathbf{A}(t)\mathbf{x}(t) + \mathbf{B}(t)\mathbf{u}(t) \tag{4-96}$$

$$\mathbf{c}(t) = \mathbf{D}(t)\mathbf{x}(t) + \mathbf{E}(t)\mathbf{u}(t) \tag{4-97}$$

We may perform a digital approximation of the system at $t = t_k$. Let

$$t_{k+1} = t_k + \Delta t_k \tag{4-98}$$

The derivative of $\mathbf{x}(t)$ at $t = t_k$ can be approximated by the following relationship:

$$\dot{\mathbf{x}}(t_k) \cong \frac{1}{\Delta t_k} [\mathbf{x}(t_{k+1}) - \mathbf{x}(t_k)] \tag{4-99}$$

Then, Eq. (4-96) is approximated by

$$\frac{1}{\Delta t_k} [\mathbf{x}(t_{k+1}) - \mathbf{x}(t_k)] = \mathbf{A}(t_k)\mathbf{x}(t_k) + \mathbf{B}(t_k)\mathbf{u}(t_k) \tag{4-100}$$

Similarly, Eq. (4-97) becomes

$$\mathbf{c}(t_k) = \mathbf{D}(t_k)\mathbf{x}(t_k) + \mathbf{E}(t_k)\mathbf{u}(t_k) \tag{4-101}$$

Equation (4-100) is now written in the form of a discrete state equation

$$\mathbf{x}(t_{k+1}) = [\mathbf{I} + \Delta t_k \mathbf{A}(t_k)]\mathbf{x}(t_k) + \Delta t_k \mathbf{B}(t_k)\mathbf{u}(t_k) \tag{4-102}$$

4.8 SOLUTION OF THE TIME-INVARIANT DISCRETE STATE EQUATION BY THE Z-TRANSFORMATION

The z-transformation can be applied to solve the linear time-invariant discrete state equations. The discussions given in this section will also lead to another method of solving the discrete state transition matrix $\phi(kT)$.

Consider the discrete state equation

$$\mathbf{x}[(k+1)T] = \phi(T)\mathbf{x}(kT) + \theta(T)\mathbf{u}(kT) \tag{4-103}$$

Taking the z-transform on both sides of the equation, we have

$$z\mathbf{X}(z) - z\mathbf{x}(0) = \phi(T)\mathbf{X}(z) + \theta(T)\mathbf{U}(z) \tag{4-104}$$

where $\mathbf{X}(z)$ is defined as

$$\mathbf{X}(z) = \sum_{k=0}^{\infty} \mathbf{x}(kT)z^{-k} \tag{4-105}$$

and the same applies to $\mathbf{U}(z)$. Solving for $\mathbf{X}(z)$ from Eq. (4-104) yields

$$\mathbf{X}(z) = [z\mathbf{I} - \phi(T)]^{-1} z\mathbf{x}(0) + [z\mathbf{I} - \phi(T)]^{-1}\theta(T)\mathbf{U}(z) \tag{4-106}$$

The inverse z-transform of the last equation is

$$\mathbf{x}(kT) = \mathcal{Z}^{-1}\left[[z\mathbf{I} - \phi(T)]^{-1} z\right]\mathbf{x}(0) + \mathcal{Z}^{-1}\left[[z\mathbf{I} - \phi(T)]^{-1}\theta(T)\mathbf{U}(z)\right] \tag{4-107}$$

We shall show that the inverse z-transform of $[z\mathbf{I} - \phi(T)]^{-1} z$ is the discrete state transition matrix $\phi(kT)$.

The z-transform of $\phi(kT)$ is defined in the usual manner as

$$\Phi(z) = \sum_{k=0}^{\infty} \phi(kT) z^{-k} \tag{4-108}$$

Premultiplying both sides of the last equation by $\phi(T)z^{-1}$ and subtracting the result from Eq. (4-108), we have

$$[\mathbf{I} - \phi(T)z^{-1}]\Phi(z) = \mathbf{I}$$

Thus,

$$\Phi(z) = [\mathbf{I} - \phi(T)z^{-1}]^{-1} = [z\mathbf{I} - \phi(T)]^{-1} z \tag{4-109}$$

Taking the inverse z-transform on both sides of the last equation, we have

$$\phi(kT) = \mathcal{Z}^{-1}\left[[z\mathbf{I} - \phi(T)]^{-1} z\right] \tag{4-110}$$

Thus, Eq. (4-110) represents the z-transform method of determining the state transition matrix of a discrete state equation.

The last term of Eq. (4-107) is evaluated by use of the real convolution theorem of Eq. (3-113) and Eq. (4-110). Therefore, it can be shown that

$$\mathcal{Z}^{-1}\left[[z\mathbf{I} - \phi(T)]^{-1}\theta(T)\mathbf{U}(z)\right] = \sum_{i=0}^{k-1} \phi[(k-i-1)T]\theta(T)\mathbf{u}(iT) \tag{4-111}$$

The entire state transition equation is

$$\mathbf{x}(kT) = \phi(kT)\mathbf{x}(0) + \sum_{i=0}^{k-1} \phi[(k-i-1)T]\theta(T)\mathbf{u}(iT) \tag{4-112}$$

which is of the same form as Eq. (4-91). In a similar way state equations of the form of Eqs. (4-87) and (4-89) can be solved by means of the z-transform method just described.

Example 4-1.

In this example we shall illustrate the analysis of an open-loop digital system by the state variable method presented above. The block diagram of the system under consideration is shown in Fig. 4-3. The dynamic equations that describe the linear process are

$$\begin{bmatrix} \dfrac{dx_1(t)}{dt} \\ \dfrac{dx_2(t)}{dt} \end{bmatrix} = \begin{bmatrix} 0 & 1 \\ -2 & -3 \end{bmatrix} \begin{bmatrix} x_1(t) \\ x_2(t) \end{bmatrix} + \begin{bmatrix} 0 \\ 1 \end{bmatrix} u(t) \qquad (4\text{-}113)$$

$$c(t) = x_1(t) \qquad (4\text{-}114)$$

where $x_1(t)$ and $x_2(t)$ are the state variables, $c(t)$ is the scalar output, and $u(t)$ is the scalar input. Also, since $u(t)$ is the output of the zero-order hold,

$$u(t) = u(kT) = r(kT)$$

for $kT \leqslant t < (k+1)T$.

Comparing Eq. (4-113) with the standard state equation form of Eq. (4-10), we have

$$\mathbf{A} = \begin{bmatrix} 0 & 1 \\ -2 & -3 \end{bmatrix} \qquad (4\text{-}115)$$

and

$$\mathbf{B} = \begin{bmatrix} 0 \\ 1 \end{bmatrix} \qquad (4\text{-}116)$$

The following matrix is formed:

$$(s\mathbf{I} - \mathbf{A}) = \begin{bmatrix} s & -1 \\ 2 & s+3 \end{bmatrix} \qquad (4\text{-}117)$$

Therefore,

$$(s\mathbf{I} - \mathbf{A})^{-1} = \frac{1}{(s^2 + 3s + 2)} \begin{bmatrix} s+3 & 1 \\ -2 & s \end{bmatrix} \qquad (4\text{-}118)$$

The state transition matrix of $\mathbf{A}$ is obtained by taking the inverse Laplace transform of $(s\mathbf{I} - \mathbf{A})^{-1}$. Therefore, from Eq. (4-34),

$$\phi(t) = \mathcal{L}^{-1}[(s\mathbf{I} - \mathbf{A})^{-1}] = \begin{bmatrix} 2e^{-t} - e^{-2t} & e^{-t} - e^{-2t} \\ -2e^{-t} + 2e^{-2t} & -e^{-t} + 2e^{-2t} \end{bmatrix} \qquad (4\text{-}119)$$

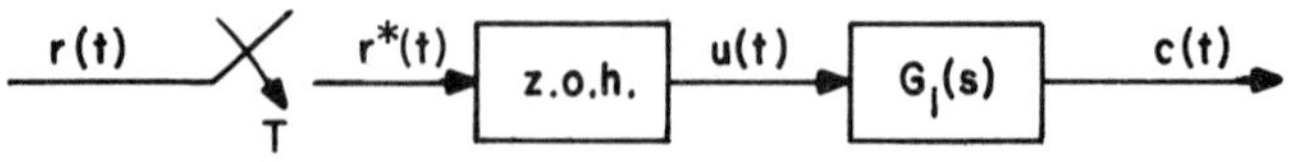

Fig. 4-3. An open-loop digital system.

Substitution of Eqs. (4-116) and (4-119) into Eq. (4-86) yields

$$\theta(T) = \int_0^T \phi(T-\tau)B d\tau$$

$$= \int_0^T \begin{bmatrix} e^{-(T-\tau)} - e^{-2(T-\tau)} \\ -e^{-(T-\tau)} + 2e^{-2(T-\tau)} \end{bmatrix} d\tau = \begin{bmatrix} \frac{1}{2} - e^{-T} + \frac{1}{2}e^{-2T} \\ e^{-T} - e^{-2T} \end{bmatrix} \tag{4-120}$$

Now substituting Eq. (4-119) with $t = T$, and Eq. (4-120) into Eq. (4-83), the discrete state equation of the system is written

$$\begin{bmatrix} x_1[(k+1)T] \\ x_2[(k+1)T] \end{bmatrix} = \begin{bmatrix} 2e^{-T} - e^{-2T} & e^{-T} - e^{-2T} \\ -2e^{-T} + 2e^{-2T} & -e^{-T} + 2e^{-2T} \end{bmatrix} \begin{bmatrix} x_1(kT) \\ x_2(kT) \end{bmatrix} + \begin{bmatrix} \frac{1}{2} - e^{-T} + \frac{1}{2}e^{-2T} \\ e^{-T} - e^{-2T} \end{bmatrix} u(kT) \tag{4-121}$$

Let the sampling period of the system be one second; Eq. (4-121) is simplified to

$$\begin{bmatrix} x_1(k+1) \\ x_2(k+1) \end{bmatrix} = \begin{bmatrix} 0.6 & 0.233 \\ -0.466 & -0.097 \end{bmatrix} \begin{bmatrix} x_1(k) \\ x_2(k) \end{bmatrix} + \begin{bmatrix} 0.2 \\ 0.233 \end{bmatrix} u(k) \tag{4-122}$$

Equation (4-121) can be identified with Eq. (4-83) or Eq. (4-89). Therefore, using Eq. (4-91), the solution of Eq. (4-122) is written as

$$\begin{bmatrix} x_1(N) \\ x_2(N) \end{bmatrix} = \begin{bmatrix} 2e^{-N} - e^{-2N} & e^{-N} - e^{-2N} \\ -2e^{-N} + 2e^{-2N} & -e^{-N} + 2e^{-2N} \end{bmatrix} \begin{bmatrix} x_1(0) \\ x_2(0) \end{bmatrix}$$

$$+ \sum_{k=0}^{N-1} \begin{bmatrix} 0.633e^{-(N-k-1)} - 0.433e^{-2(N-k-1)} \\ -0.633e^{-(N-k-1)} + 0.866e^{-2(N-k-1)} \end{bmatrix} u(k) \tag{4-123}$$

where N is any positive integer.

4.9 RELATION BETWEEN STATE EQUATION AND TRANSFER FUNCTION

It is of interest to investigate the relationship between the transfer function and the state variable methods.

Consider that a digital system with multiple inputs and outputs is described by the transfer relation:

$$\mathbf{C}(z) = \mathbf{G}(z)\mathbf{U}(z) \tag{4-124}$$

where

$$\mathbf{C}(z) = \begin{bmatrix} C_1(z) \\ C_2(z) \\ \cdot \\ \cdot \\ \cdot \\ C_q(z) \end{bmatrix} \tag{4-125}$$

is the q × 1 output transform vector;

$$\mathbf{U}(z) = \begin{bmatrix} U_1(z) \\ U_2(z) \\ \cdot \\ \cdot \\ \cdot \\ U_p(z) \end{bmatrix} \tag{4-126}$$

is the p × 1 input transform vector;

$$\mathbf{G}(z) = \begin{bmatrix} G_{11}(z) & G_{12}(z) & \dots & G_{1p}(z) \\ G_{21}(z) & G_{22}(z) & \dots & G_{2p}(z) \\ \cdot & & & \\ \cdot & & & \\ \cdot & & & \\ G_{q1}(z) & G_{q2}(z) & \dots & G_{qp}(z) \end{bmatrix} \tag{4-127}$$

is the q × p z-transfer function matrix.

The elements of $\mathbf{G}(z)$ may be of the form,

$$G_{ij}(z) = \frac{a_{n+1} + a_n z^{-1} + \dots + a_1 z^{-n}}{b_{m+1} + b_m z^{-1} + \dots + b_1 z^{-m}} \tag{4-128}$$

Now a digital system is described by the dynamic equations

$$\mathbf{x}[(k + 1)T] = \mathbf{Ax}(kT) + \mathbf{Bu}(kT) \tag{4-129}$$

$$\mathbf{c}(kT) = \mathbf{Dx}(kT) + \mathbf{Eu}(kT) \tag{4-130}$$

Taking the z-transform on both sides of Eq. (4-129) and solving for $\mathbf{X}(z)$, we get,

$$\mathbf{X}(z) = (z\mathbf{I} - \mathbf{A})^{-1} z\mathbf{x}(0) + (z\mathbf{I} - \mathbf{A})^{-1}\mathbf{BU}(z) \tag{4-131}$$

Substitution of Eq. (4-131) into the z-transform of Eq. (4-130) yields

$$\mathbf{C}(z) = \mathbf{D}(z\mathbf{I} - \mathbf{A})^{-1} z\mathbf{x}(0) + \mathbf{D}(z\mathbf{I} - \mathbf{A})^{-1}\mathbf{BU}(z) + \mathbf{EU}(z) \tag{4-132}$$

For the transfer function, we assume that the initial state $\mathbf{x}(0)$ is zero; therefore, Eq. (4-132) becomes

$$\mathbf{C}(z) = [\mathbf{D}(z\mathbf{I} - \mathbf{A})^{-1}\mathbf{B} + \mathbf{E}]\mathbf{U}(z) \tag{4-133}$$

Comparing Eq. (4-133) with Eq. (4-124) we see that the z-transfer function matrix of the system may be written as

$$\mathbf{G}(z) = \mathbf{D}(z\mathbf{I} - \mathbf{A})^{-1}\mathbf{B} + \mathbf{E} \tag{4-134}$$

If a discrete-data system has sample-and-hold operations and is described by the dynamic equations in Eqs. (4-10) and (4-11), we only have to replace $\mathbf{A}$ and $\mathbf{B}$ in Eq. (4-134) by $\phi(T)$ and $\theta(T)$, respectively. Of course, this is performed with the assumption that a z-transfer function can be written for the system with sample-and-hold. It was illustrated in Sec. 3-10 that sometimes transfer functions in the form of Eq. (4-124) cannot be defined for a system with sample-and-hold, and only input-output relations exist.

The inverse transform of the transfer function matrix is called the *impulse response matrix*. Taking the inverse z-transform on both sides of Eq. (4-134), we get

$$\mathbf{g}(kT) = \mathbf{D}\phi[(k-1)T]\mathbf{B} + \mathbf{E}\delta(0) \tag{4-135)*}$$

where $\delta(0)$ is a delta function at $t = 0$.

Since $\phi[(k-1)T] = \mathbf{0}$ for $k < 1$, $\mathbf{g}(kT)$ in Eq. (4-135) can be written as

$$\mathbf{g}(kT) = \mathbf{E} \qquad k = 0 \tag{4-136}$$

$$\mathbf{g}(kT) = \mathbf{D}\phi[(k-1)T]\mathbf{B} \qquad k \geqslant 1 \tag{4-137}$$

Example 4-2.

As an illustrative example, let us derive the transfer function of the open-loop system shown in Fig. 4-3 with the data of Example 4-1.

First, we shall use the state variable method described in the preceding paragraphs. From Example 4-1, we have

$$\phi(1) = \begin{bmatrix} 0.6 & 0.233 \\ -0.466 & -0.097 \end{bmatrix} \tag{4-138}$$

$$\theta(1) = \begin{bmatrix} 0.2 \\ 0.233 \end{bmatrix} \tag{4-139}$$

$$\mathbf{D} = [1 \quad 0] \qquad \text{and} \qquad \mathbf{E} = 0 \tag{4-140}$$

Substituting Eqs. (4-138), (4-139), and (4-140) into Eq. (4-134), we have the transfer function of the system

*The inverse z-transform of a constant is the delta function $\delta(t)$; see APPENDIX A.

$$G(z) = [1 \quad 0] \begin{bmatrix} z - 0.6 & -0.233 \\ 0.466 & z + 0.097 \end{bmatrix}^{-1} \begin{bmatrix} 0.2 \\ 0.233 \end{bmatrix}$$

$$= \frac{0.2z + 0.074}{(z - 0.135)(z - 0.368)} \tag{4-141}$$

To use the z-transform method we must first determine the transfer function of the linear process $G_1(s)$ of Fig. 4-3. From the dynamic equations which describe the process, Eqs. (4-113) and (4-114), the differential equation for the relation between u(t) and c(t) is written

$$\frac{d^2c(t)}{dt^2} + 3\frac{dc(t)}{dt} + 2c(t) = u(t) \tag{4-142}$$

where u(t) is the output of the zero-order hold. Therefore, the transfer function $G_1(s)$ is

$$G_1(s) = \frac{C(s)}{U(s)} = \frac{1}{s^2 + 3s + 2} \tag{4-143}$$

With reference to Fig. 4-3, the z-transfer function of the overall system is

$$G(z) = \mathcal{Z}\left[\frac{1 - e^{-Ts}}{s} G_1(s)\right] = \frac{C(z)}{R(z)} \tag{4-144}$$

or

$$G(z) = (1 - z^{-1})\mathcal{Z}\left[\frac{1}{s(s^2 + 3s + 2)}\right] \tag{4-145}$$

Evaluating the z-transform of Eq. (4-145), we have

$$G(z) = \frac{0.2z + 0.074}{(z - 0.135)(z - 0.368)} \tag{4-146}$$

which agrees with the result of Eq. (4-141).

The impulse sequence of the system can be obtained by taking the inverse z-transform of G(z). A partial fraction expansion of G(z) gives

$$G(z) = \frac{0.633}{z - 0.368} - \frac{0.433}{z - 0.135}$$

Therefore,

$$g(k) = 0.633e^{-(k-1)} - 0.433e^{-2(k-1)} \tag{4-147}$$

for $k > 0$. For $k = 0$, $g(0) = 0$. An alternate method of determining g(k) calls for the use of Eqs. (4-136) and (4-137). Therefore,

$$g(0) = E = 0$$

and

$$g(kT) = [1 \quad 0] \begin{bmatrix} 2e^{-(k-1)} - e^{-2(k-1)} & e^{-(k-1)} - e^{-2(k-1)} \\ -2e^{-(k-1)} + 2e^{-2(k-1)} & -e^{-(k-1)} + 2e^{-2(k-1)} \end{bmatrix} \begin{bmatrix} 0.2 \\ 0.233 \end{bmatrix}$$

$$= 0.633e^{-(k-1)} - 0.433e^{-2(k-1)} \qquad (4\text{-}148)$$

for $k > 0$.

4.10 CHARACTERISTIC EQUATION, EIGENVALUES AND EIGENVECTORS

Characteristic Equation

The characteristic equation of a linear time-invariant system can be defined with respect to the system's difference equation, transfer function, or the matrix **A**. These definitions are given as follows.

Difference Equation

Consider that a linear single-variable digital system is described by the nth-order difference equation

$$c(k+n) + a_n c(k+n-1) + a_{n-1} c(k+n-2) + \ldots + a_2 c(k+1) + a_1 c(k)$$
$$= b_{n+1} r(k+n) + b_n r(k+n-1) + \ldots + b_2 r(k+1) + b_1 r(k) \qquad (4\text{-}149)$$

The characteristic equation of the system is defined by the nth-order polynomial using the coefficients of the homogeneous part of the difference equation. Thus, the characteristic equation is written

$$\lambda^n + a_n \lambda^{n-1} + a_{n-1} \lambda^{n-2} + \ldots + a_2 \lambda + a_1 = 0 \qquad (4\text{-}150)$$

Transfer Function

The transfer function of the system described by Eq. (4-149) is obtained by taking the z-transform on both sides of the difference equation, setting the initial conditions to zero, and then forming the ratio $C(z)/R(z)$. Thus, the transfer function of the system described by Eq. (4-149) is

$$G(z) = \frac{C(z)}{R(z)} = \frac{b_{n+1} z^n + b_n z^{n-1} + \ldots + b_2 z + b_1}{z^n + a_n z^{n-1} + \ldots + a_2 z + a_1} \qquad (4\text{-}151)$$

The characteristic equation is defined as the equation obtained by equating the denominator of the transfer function to zero; that is,

$$z^n + a_n z^{n-1} + \ldots + a_2 z + a_1 = 0 \qquad (4\text{-}152)$$

The A Matrix

The characteristic equation of a linear time-invariant system can also be defined with reference to the state equation formulation. Consider that the digital system is described by the state equation

$$\mathbf{x}(k+1) = \mathbf{A}\mathbf{x}(k) + \mathbf{B}\mathbf{u}(k) \tag{4-153}$$

where $\mathbf{x}(k)$ and $\mathbf{A}$ are all n dimensional.

The characteristic equation of the system, in fact, often referred to as *the characteristic equation of A*, is defined as the determinant of $z\mathbf{I} - \mathbf{A}$ equated to zero; i.e.,

$$|z\mathbf{I} - \mathbf{A}| = 0 \tag{4-154}$$

Notice that the coefficient matrix $\mathbf{B}$ is not related to the characteristic equation at all, since the homogeneous part of Eq. (4-153) does not include $\mathbf{B}\mathbf{u}(k)$.

Equation (4-154) can also be derived from the transfer function matrix expression in Eq. (4-134).

Equation (4-134) can be written as

$$\mathbf{G}(z) = \mathbf{D}\frac{[\Delta_{ij}]'}{|z\mathbf{I} - \mathbf{A}|}\mathbf{B} + \mathbf{E}$$

$$= \frac{\mathbf{D}[\Delta_{ij}]'\mathbf{B} + |z\mathbf{I} - \mathbf{A}|\mathbf{E}}{|z\mathbf{I} - \mathbf{A}|} \tag{4-155}$$

where Δ_{ij} represents the cofactor of the ijth element of the matrix $z\mathbf{I} - \mathbf{A}$, and $[\Delta_{ij}]'$ is the matrix transpose of Δ_{ij}. Again, setting the denominator to zero gives the same result as in Eq. (4-154).

Eigenvalues

The roots of the characteristic equation are defined as the eigenvalues of the matrix A.

Therefore, the roots of Eq. (4-150) or of Eq. (4-152) are the same as the eigenvalues of $\mathbf{A}$.

For the system described in Example 4-2 it is simple to show that the characteristic equation of the system is

$$z^2 - 0.503z + 0.0497 = 0 \tag{4-156}$$

and the roots are $z_1 = 0.135$ and $z_2 = 0.368$.

Some of the important properties of eigenvalues are given as follows:

1. If the coefficients of the characteristic equation are scalar quantities, the eigenvalues are either real or in complex conjugate pairs.
2. The trace of **A** which is defined as the sum of the elements on the main diagonal of **A** is given by

$$\text{Trace of } \mathbf{A} = \text{Tr}(\mathbf{A}) = \lambda_1 + \lambda_2 + \dots + \lambda_n \tag{4-157}$$

   where λ_i, i = 1,2,...,n are the eigenvalues of **A**.
3. The determinant of **A** is related to the eigenvalues through

$$|\mathbf{A}| = \lambda_1\lambda_2\lambda_3\dots\lambda_n \tag{4-158}$$

4. If **A** is nonsingular with eigenvalues λ_i, i = 1,2,...,n, then $1/\lambda_i$, i = 1, 2,...,n, are the eigenvalues of $\mathbf{A}^{-1}$.
5. If λ_i is an eigenvalue of **A**, then it is an eigenvalue of $\mathbf{A}'$.
6. If **A** is a real symmetric matrix, then its eigenvalues are all real.
7. For square matrices **A** and **B**,

$$|\lambda\mathbf{I} - \mathbf{AB}| = |\lambda\mathbf{I} - \mathbf{BA}| \tag{4-159}$$

   Then the eigenvalues of **AB** are the same as that of **BA**.

Eigenvectors

The n × 1 vector $\mathbf{p}_i$ which satisfies the matrix equation

$$(\lambda_i\mathbf{I} - \mathbf{A})\mathbf{p}_i = \mathbf{0} \tag{4-160}$$

where λ_i, i = 1,2,...,n, denotes the ith eigenvalue of **A**, is called the *eigenvector of A associated with the eigenvalue* λ_i.

For distinct eigenvalues the eigenvectors can be solved directly from Eq. (4-160).

Example 4-3.

The eigenvectors of $\phi(1)$ in Example 4-2 are the vectors $\mathbf{p}_1$ and $\mathbf{p}_2$ that satisfy

$$[\lambda_i\mathbf{I} - \phi(1)]\mathbf{p}_i = 0 \qquad i = 1,2 \tag{4-161}$$

Or,

$$\begin{bmatrix} \lambda_i - 0.6 & -0.233 \\ 0.466 & \lambda_i + 0.097 \end{bmatrix}\mathbf{p}_i = 0 \tag{4-162}$$

where $\lambda_1 = 0.135$ and $\lambda_2 = 0.368$.

Letting

$$\mathbf{p}_1 = \begin{bmatrix} p_{11} \\ p_{12} \end{bmatrix} \qquad \mathbf{p}_2 = \begin{bmatrix} p_{21} \\ p_{22} \end{bmatrix} \tag{4-163}$$

from Eq. (4-162) we arrive at first, for λ_1, the following two scalar equations:

$$-0.466p_{11} - 0.233p_{12} = 0 \tag{4-164}$$

$$0.466p_{11} + 0.233p_{12} = 0 \tag{4-165}$$

Since these two equations are linearly dependent, we may assign an arbitrary value to either p_{11} or p_{12}. Let $p_{11} = 1$, then p_{12} is found to be equal to -2.

Similarly, for $\lambda_2 = 0.368$, Eq. (4-162) leads to

$$-0.233p_{21} - 0.233p_{22} = 0 \tag{4-166}$$

$$0.466p_{21} + 0.466p_{22} = 0 \tag{4-167}$$

Again, for these dependent equations if we let $p_{21} = 1$, we get $p_{22} = -1$. Thus, the eigenvectors are

$$\mathbf{p}_1 = \begin{bmatrix} 1 \\ -2 \end{bmatrix} \quad \text{for} \quad \lambda_1 = 0.135 \tag{4-168}$$

$$\mathbf{p}_2 = \begin{bmatrix} 1 \\ -1 \end{bmatrix} \quad \text{for} \quad \lambda_2 = 0.368 \tag{4-169}$$

For distinct eigenvalues, the eigenvectors of $\mathbf{A}$ can also be determined by using any nonzero columns of the matrix, $\text{Adj}(\lambda_i\mathbf{I} - \mathbf{A})$, $i = 1,2,\ldots,n$.

Example 4-4.

Using the same example given in Example 4-3, we have

$$[\lambda_i\mathbf{I} - \phi(1)] = \begin{bmatrix} \lambda_i - 0.6 & -0.233 \\ 0.466 & \lambda_i + 0.097 \end{bmatrix} \tag{4-170}$$

The adjoint of the last matrix is written

$$\text{Adj}[\lambda_i\mathbf{I} - \phi(1)] = \begin{bmatrix} \lambda_i + 0.097 & 0.233 \\ -0.466 & \lambda_i - 0.6 \end{bmatrix} \tag{4-171}$$

Then,

$$\text{Adj}[\lambda_1\mathbf{I} - \phi(1)] = \begin{bmatrix} 0.233 & 0.233 \\ -0.466 & -0.466 \end{bmatrix} \tag{4-172}$$

$$\text{Adj}[\lambda_2\mathbf{I} - \phi(1)] = \begin{bmatrix} 0.466 & 0.233 \\ -0.466 & -0.233 \end{bmatrix} \tag{4-173}$$

From Eq. (4-172) we set $\mathbf{p}_1$ equal to one column of the adjoint matrix, and after dividing by a common factor,

$$\mathbf{p}_1 = \begin{bmatrix} 1 \\ -2 \end{bmatrix} \quad \text{for} \quad \lambda_1 = 0.135 \tag{4-174}$$

Similarly, from Eq. (4-173),

$$\mathbf{p}_2 = \begin{bmatrix} 1 \\ -1 \end{bmatrix} \quad \text{for} \quad \lambda_2 = 0.368 \tag{4-175}$$

Some of the important properties related to eigenvectors are given as follows.

1. The rank of $(\lambda_i\mathbf{I} - \mathbf{A})$, where λ_i, $i = 1,2,\ldots,n$, denotes distinct eigenvalues of $\mathbf{A}$, is $n - 1$. This property has been illustrated by the example worked out above.
2. The eigenvector $\mathbf{p}_i$ is given by any nonzero column of the matrix $\text{Adj}(\lambda_i\mathbf{I} - \mathbf{A})$, where λ_i denotes the ith distinct eigenvalue of $\mathbf{A}$.
3. If the matrix $\mathbf{A}$ has n distinct eigenvalues, then the set of n eigenvectors, $\mathbf{p}_i$, $i = 1,2,\ldots,n$, are linearly independent.
4. If $\mathbf{p}_i$ is an eigenvector of $\mathbf{A}$, then $k\mathbf{p}_i$ is also, where k is a scalar quantity.

When one or more eigenvalues of $\mathbf{A}$ is a multiple-order root of the characteristic equation, a full set of n linearly-independent eigenvectors may or may not exist. The number of linearly-independent eigenvectors associated with a given eigenvalue λ_i of multiplicity m_i is equal to the *degeneracy* d_i of $\lambda_i\mathbf{I} - \mathbf{A}$. The degeneracy d_i of $\lambda_i\mathbf{I} - \mathbf{A}$ is defined as

$$d_i = n - r_i \tag{4-176}$$

where n is the dimension of $\mathbf{A}$ and r_i is the rank of $\lambda_i\mathbf{I} - \mathbf{A}$. *There are always d_i linearly-independent eigenvectors associated with λ_i.* Furthermore,

$$1 \leqslant d_i \leqslant m_i \tag{4-177}$$

The eigenvectors of a matrix $\mathbf{A}$ with multiple-order eigenvalues are determined according to the methods described as follows.

Full Degeneracy ($d_i = m_i$)

For the eigenvalue λ_i which has m_i multiplicity, the fully degenerated case has a full set of m_i eigenvectors associated with λ_i. These eigenvectors are found from the nonzero columns of

$$\frac{1}{(m_i - 1)!}\left[\frac{d^{m_i-1}}{d\lambda^{m_i-1}}[\mathrm{Adj}(\lambda\mathbf{I} - \mathbf{A})]\right]\Bigg|_{\lambda=\lambda_i} \tag{4-178}$$

Example 4-5.

Consider the matrix

$$\mathbf{A} = \begin{bmatrix} 3 & 0 & 0 \\ 2 & 4 & 1 \\ 2 & 1 & 4 \end{bmatrix} \tag{4-179}$$

The characteristic equation of **A** is

$$|\lambda\mathbf{I} - \mathbf{A}| = (\lambda - 3)^2(\lambda - 5) = 0 \tag{4-180}$$

The eigenvalues of **A** are $\lambda_1 = \lambda_2 = 3$ and $\lambda_3 = 5$. Thus, the eigenvalue $\lambda_1 = 3$ has a multiplicity of two, and $\lambda_3 = 5$ is simple. To check the degeneracy of $\lambda\mathbf{I} - \mathbf{A}$ for $\lambda_1 = 3$, we form

$$\lambda_1\mathbf{I} - \mathbf{A} = \begin{bmatrix} \lambda - 3 & 0 & 0 \\ -2 & \lambda - 4 & -1 \\ -2 & -1 & \lambda - 4 \end{bmatrix}\Bigg|_{\lambda=3} = \begin{bmatrix} 0 & 0 & 0 \\ -2 & -1 & -1 \\ -2 & -1 & -1 \end{bmatrix} \tag{4-181}$$

which has a rank of one. Thus, the degeneracy of $\lambda_1\mathbf{I} - \mathbf{A}$ is

$$d_1 = n - r_1 = 3 - 1 = 2 \tag{4-182}$$

Since λ_1 is of multiplicity two, we say that $\lambda_1\mathbf{I} - \mathbf{A}$ is of full degeneracy.

Now using Eq. (4-178), we get

$$\left[\frac{d}{d\lambda}\mathrm{Adj}(\lambda\mathbf{I} - \mathbf{A})\right]\Bigg|_{\lambda=\lambda_1} = \frac{d}{d\lambda}\begin{bmatrix} (\lambda-3)(\lambda-5) & 0 & 0 \\ 2(\lambda - 3) & (\lambda-3)(\lambda-4) & \lambda - 3 \\ 2(\lambda - 3) & \lambda - 3 & (\lambda-3)(\lambda-4) \end{bmatrix}\Bigg|_{\lambda=\lambda_1}$$

$$= \begin{bmatrix} 2\lambda - 8 & 0 & 0 \\ 2 & 2\lambda - 7 & 1 \\ 2 & 1 & 2\lambda - 7 \end{bmatrix}\Bigg|_{\lambda=\lambda_1}$$

$$= \begin{bmatrix} -2 & 0 & 0 \\ 2 & -1 & 1 \\ 2 & 1 & -1 \end{bmatrix} \tag{4-183}$$

Therefore, the two independent columns of the last matrix are selected as the eigenvectors,

$$\mathbf{p}_1 = \begin{bmatrix} -1 \\ 1 \\ 1 \end{bmatrix} \qquad \mathbf{p}_2 = \begin{bmatrix} 0 \\ -1 \\ 1 \end{bmatrix}$$

For the eigenvalue $\lambda_3 = 5$, the eigenvector is found in the usual manner by setting $(\lambda_3\mathbf{I} - \mathbf{A})\mathbf{p}_3 = 0$ and solving for $\mathbf{p}_3$, or by using any nonzero column of $\text{Adj}(\lambda_3\mathbf{I} - \mathbf{A})$. The result is

$$\mathbf{p}_3 = \begin{bmatrix} 0 \\ 1 \\ 1 \end{bmatrix}$$

Simple Degeneracy ($d_i = 1$)

When the degeneracy of $\lambda_i\mathbf{I} - \mathbf{A}$ is equal to one (simple degeneracy), there is only *one* eigenvector associated with λ_i regardless of the multiplicity of λ_i. The eigenvector associated with λ_i can be determined using the same method as for the distinct eigenvalue case. However, for the m_ith-order eigenvalue, there are $m_i - 1$ additional vectors called the *generalized eigenvectors.* These $m_i - 1$ generalized eigenvectors $\mathbf{p}_{i2}, \mathbf{p}_{i3}, \ldots, \mathbf{p}_{im_{i-1}}$ are found from the following $m_i - 1$ equations:

$$\begin{aligned} (\lambda_i\mathbf{I} - \mathbf{A})\mathbf{p}_{i2} &= -\mathbf{p}_{i1} \\ (\lambda_i\mathbf{I} - \mathbf{A})\mathbf{p}_{i3} &= -\mathbf{p}_{i2} \\ &\vdots \\ (\lambda_i\mathbf{I} - \mathbf{A})\mathbf{p}_{im_i} &= -\mathbf{p}_{im_{i-1}} \end{aligned} \tag{4-184}$$

where $\mathbf{p}_{i1}$ is the eigenvector of λ_i that is determined by solving

$$(\lambda_i\mathbf{I} - \mathbf{A})\mathbf{p}_{i1} = \mathbf{0}$$

Example 4-6.

Consider the matrix

$$\mathbf{A} = \begin{bmatrix} 1 & 2 \\ -2 & -3 \end{bmatrix} \tag{4-185}$$

The characteristic equation of **A** is

$$|\lambda \mathbf{I} - \mathbf{A}| = \lambda^2 + 2\lambda + 1 = 0 \tag{4-186}$$

The eigenvalues of **A** are $\lambda_1 = \lambda_2 = -1$. Thus, the eigenvalue $\lambda_1 = -1$ has a multiplicity of two. To check the degeneracy of the matrix $\lambda_1 \mathbf{I} - \mathbf{A}$, we form

$$\lambda_1 \mathbf{I} - \mathbf{A} = \begin{bmatrix} \lambda_1 - 1 & -2 \\ 2 & \lambda_1 + 3 \end{bmatrix} = \begin{bmatrix} -2 & -2 \\ 2 & 2 \end{bmatrix} \tag{4-187}$$

which has a rank of one. Thus, the degeneracy of $\lambda_1 \mathbf{I} - \mathbf{A}$ is one. This means that we can find only one independent eigenvector for the eigenvalue $\lambda_1 = -1$, from

$$\operatorname{Adj}(\lambda_1 \mathbf{I} - \mathbf{A}) = \begin{bmatrix} 2 & 2 \\ -2 & -2 \end{bmatrix} \tag{4-188}$$

Thus, the eigenvector of $\lambda_1 = -1$ is

$$\mathbf{p}_1 = \begin{bmatrix} 1 \\ -1 \end{bmatrix}$$

To find the generalized eigenvector, we set

$$(\lambda_1 \mathbf{I} - \mathbf{A})\mathbf{p}_2 = -\mathbf{p}_1 = \begin{bmatrix} -1 \\ 1 \end{bmatrix} \tag{4-189}$$

or

$$\begin{bmatrix} -2 & -2 \\ 2 & 2 \end{bmatrix} \mathbf{p}_2 = \begin{bmatrix} -1 \\ 1 \end{bmatrix} \tag{4-190}$$

Solving the last equation, we get a solution for $\mathbf{p}_2$ as

$$\mathbf{p}_2 = \begin{bmatrix} 0 \\ \frac{1}{2} \end{bmatrix}$$

As a summary of the discussions on the state variable analysis given in the preceding sections, a tabulation and comparison of the results are given in Table 4-1.

4.11 DIAGONALIZATION OF THE A MATRIX

It would have been a simple matter to solve the state equations of a linear time-invariant digital system if these equations had been decoupled from each other, that is, if **A** were a diagonal matrix. For instance, if the state equations of an nth-order digital system are of the form

TABLE 4-1. TABULATION OF RESULTS OF STATE VARIABLE ANALYSIS OF LINEAR SYSTEMS

	Continuous-Data System	Digital System with Sample-and-Hold	Digital System
State Variables	$\mathbf{x}(t)$	$\mathbf{x}(kT)$	$\mathbf{x}(k)$
State Equations (Time-Invariant)	$\dot{\mathbf{x}}(t) = \mathbf{A}\mathbf{x}(t) + \mathbf{B}\mathbf{u}(t)$	$\mathbf{x}[(k+1)T] = \phi(T)\mathbf{x}(kT) + \theta(T)\mathbf{u}(kT)$	$\mathbf{x}(k+1) = \mathbf{A}\mathbf{x}(k) + \mathbf{B}\mathbf{u}(k)$
State Equations (Time-Varying)	$\dot{\mathbf{x}}(t) = \mathbf{A}(t)\mathbf{x}(t) + \mathbf{B}(t)\mathbf{u}(t)$	$\mathbf{x}[(k+1)T] = \phi[(k+1)T, kT]\mathbf{x}(kT) + \theta[(k+1)T, kT]\mathbf{u}(kT)$	$\mathbf{x}(k+1) = \mathbf{A}(k)\mathbf{x}(k) + \mathbf{B}(k)\mathbf{u}(k)$
State Transition Matrix (Time-Invariant)	$\phi(t) = e^{\mathbf{A}t}$	$\phi(kT) = [\phi(T)]^k$	$\phi(k) = \mathbf{A}^k$
State Transition Matrix (Time-Varying)	$\phi(t, t_0) = \exp \int_{t_0}^{t} \mathbf{A}(\tau)d\tau$	$\phi[(k+1)T, kT] = \exp \int_{kT}^{(k+1)T} \mathbf{A}(\tau)d\tau$	$\phi(N,k) = \mathbf{A}(N-1)\mathbf{A}(N-2) \ldots \cdot \mathbf{A}(k+1)\mathbf{A}(k)$
Transforms of State Transition Matrix	$\Phi(s) = (s\mathbf{I} - \mathbf{A})^{-1}$	$\Phi(z) = [z\mathbf{I} - \phi(T)]^{-1} z$	$\Phi(z) = (z\mathbf{I} - \mathbf{A})^{-1} z$
Impulse Response Matrix	$\mathbf{g}(t) = \mathbf{D}\phi(t)\mathbf{B} + \mathbf{E}\delta(t)$	$\mathbf{g}(kT) = \mathbf{D}\phi[(k-1)T]\mathbf{B} \quad k \geqslant 1$ $= \mathbf{E} \quad k = 0$	$\mathbf{g}(k) = \mathbf{D}\phi(k-1)\mathbf{B} \quad k \geqslant 1$ $= \mathbf{E} \quad k = 0$
Transfer Matrix	$\mathbf{G}(s) = \mathbf{D}(s\mathbf{I} - \mathbf{A})^{-1}\mathbf{B} + \mathbf{E}$	$\mathbf{G}(z) = \mathbf{D}[z\mathbf{I} - \phi(T)]^{-1}\mathbf{B} + \mathbf{E}$	$\mathbf{G}(z) = \mathbf{D}(z\mathbf{I} - \mathbf{A})^{-1}\mathbf{B} + \mathbf{E}$
State Transition Equation (Time-Invariant)	$\mathbf{x}(t) = \phi(t - t_0)\mathbf{x}(t_0) + \int_{t_0}^{t} \phi(t - \tau)\mathbf{B}\mathbf{u}(\tau)d\tau$	$\mathbf{x}(NT) = \phi(NT)\mathbf{x}(0) + \sum_{k=0}^{N-1} \phi[(N-k-1)T]\theta(T)\mathbf{u}(kT)$	$\mathbf{x}(N) = \mathbf{A}^N\mathbf{x}(0) + \sum_{k=0}^{N-1} \mathbf{A}^{N-k-1}\mathbf{B}\mathbf{u}(k)$
State Transition Equation (Time-Varying)	$\mathbf{x}(t) = \phi(t, t_0)\mathbf{x}(t_0) + \int_{t_0}^{t} \phi(t,\tau)\mathbf{B}(\tau)\mathbf{u}(\tau)d\tau$	$\mathbf{x}(NT) = \phi(NT, 0)\mathbf{x}(0) + \sum_{k=0}^{N-1} \phi[NT, (N-k)T] \cdot \theta[(N-k)T, (N-k-1)T]\mathbf{u}[(N-k-1)T]$	$\mathbf{x}(N) = \phi(N, 0)\mathbf{x}(0) + \sum_{k=0}^{N-1} \phi(N, k+1)\mathbf{B}(k)\mathbf{u}(k)$
Characteristic Equation	$\lvert s\mathbf{I} - \mathbf{A} \rvert = 0$	$\lvert z\mathbf{I} - \phi(T) \rvert = 0$	$\lvert z\mathbf{I} - \mathbf{A} \rvert = 0$

$$x_i(k+1) = \lambda_i x_i(k) + \sum_{j=1}^{r} \gamma_j u_j(k)$$

$i = 1,2,\ldots,n$, the solutions of these state equations given $x_i(0)$ and $u_j(k)$ for $k \geqslant 0$ are simply

$$x_i(k) = \lambda_i^k x_i(0) + \sum_{j=1}^{r} \sum_{m=0}^{k} \gamma_j \lambda_i^{m-1} u_j(k-m) \tag{4-191}$$

Therefore, the state transition matrix $\phi(k)$ is also a diagonal matrix with elements λ_i^k, $i = 1,2,\ldots,n$, on the main diagonal.

In general, if $\mathbf{A}$ has distinct eigenvalues, it can be diagonalized by a similarity transformation. Let us consider the discrete state equation

$$\mathbf{x}(k+1) = \mathbf{A}\mathbf{x}(k) + \mathbf{B}\mathbf{u}(k) \tag{4-192}$$

where $\mathbf{x}(k)$ is an n-vector, $\mathbf{u}(k)$ an r-vector, and $\mathbf{A}$ has distinct eigenvalues λ_1, $\lambda_2,\ldots,\lambda_n$. Let $\mathbf{P}$ be a nonsingular matrix that transforms the state vector $\mathbf{x}(k)$ into $\mathbf{y}(k)$, i.e.,

$$\mathbf{x}(k) = \mathbf{P}\mathbf{y}(k) \tag{4-193}$$

and

$$\mathbf{y}(k) = \mathbf{P}^{-1}\mathbf{x}(k) \tag{4-194}$$

The desired state equation is

$$\mathbf{y}(k+1) = \mathbf{\Lambda}\mathbf{y}(k) + \mathbf{\Gamma}\mathbf{u}(k) \tag{4-195}$$

where

$$\mathbf{\Lambda} = \begin{bmatrix} \lambda_1 & 0 & 0 & \ldots & 0 \\ 0 & \lambda_2 & 0 & \ldots & 0 \\ 0 & 0 & \lambda_3 & \ldots & 0 \\ \cdots & \cdots & \cdots & \cdots & \cdots \\ 0 & 0 & 0 & \ldots & \lambda_n \end{bmatrix} \quad (n \times n) \tag{4-196}$$

The decoupled state equations in Eq. (4-195) are known as the *canonical forms.* In order to find the matrix $\mathbf{P}$, we substitute Eq. (4-193) into Eq. (4-192), and using the identity of Eq. (4-194), we have

$$\mathbf{\Lambda} = \mathbf{P}^{-1}\mathbf{A}\mathbf{P} \tag{4-197}$$

and

$$\Gamma = \mathbf{P}^{-1}\mathbf{B} \qquad (n \times r) \tag{4-198}$$

There are several methods of determining the matrix **P** given the matrix **A** and its eigenvalues. However, we shall show that the columns of **P** are always the eigenvectors of **A**. Let $\mathbf{p}_i$ represent the eigenvector of **A** that corresponds to λ_i. Then,

$$\mathbf{P} = [\mathbf{p}_1 \quad \mathbf{p}_2 \quad \cdots \quad \mathbf{p}_n] \tag{4-199}$$

The proof of this relation is carried by use of the definition of the eigenvector, Eq. (4-160), which is written as

$$\lambda_i \mathbf{p}_i = \mathbf{A}\mathbf{p}_i \qquad i = 1,2,\ldots,n \tag{4-200}$$

Forming the n × n matrix,

$$\begin{aligned} [\lambda_1 \mathbf{p}_1 \quad \lambda_2 \mathbf{p}_2 \quad \cdots \quad \lambda_n \mathbf{p}_n] &= [\mathbf{A}\mathbf{p}_1 \quad \mathbf{A}\mathbf{p}_2 \quad \cdots \quad \mathbf{A}\mathbf{p}_n] \\ &= \mathbf{AP} \end{aligned} \tag{4-201}$$

This equation is also written as

$$[\mathbf{p}_1 \quad \mathbf{p}_2 \quad \cdots \quad \mathbf{p}_n]\Lambda = \mathbf{P}\Lambda = \mathbf{AP} \tag{4-202}$$

which leads to Eq. (4-197).

It can be shown that if **A** is of the phase-variable canonical form, the matrix **P** is the Vandermonde matrix

$$\mathbf{P} = \begin{bmatrix} 1 & 1 & \cdots & 1 \\ \lambda_1 & \lambda_2 & \cdots & \lambda_n \\ \lambda_1^2 & \lambda_2^2 & \cdots & \lambda_n^2 \\ \cdots & \cdots & \cdots & \cdots \\ \lambda_1^{n-1} & \lambda_2^{n-1} & \cdots & \lambda_n^{n-1} \end{bmatrix} \tag{4-203}$$

4.12 JORDAN CANONICAL FORM

When the **A** matrix has multiple-order eigenvalues, it cannot be diagonalized unless the matrix is symmetric. However, this does not mean that the system cannot be transformed into a form in which the solution of the state equations is written by inspection. When **A** cannot be diagonalized, there exists a similarity transformation $\Lambda = \mathbf{P}^{-1}\mathbf{AP}$ such that Λ is the Jordan canonical form, which is almost a diagonal matrix. As an illustrative example of the Jordan canonical form, consider that **A** has eigenvalues λ_1, λ_2, λ_3, λ_3, λ_3, the last three being identical. The Jordan canonical form matrix Λ is given by

$$\Lambda = \left[\begin{array}{c|c|ccc} \lambda_1 & 0 & 0 & 0 & 0 \\ \hline 0 & \lambda_2 & 0 & 0 & 0 \\ \hline 0 & 0 & \lambda_3 & 1 & 0 \\ 0 & 0 & 0 & \lambda_3 & 1 \\ 0 & 0 & 0 & 0 & \lambda_3 \end{array}\right] \tag{4-204}$$

As another example, for an eigenvalue with a multiplicity of four,

$$\Lambda = \begin{bmatrix} \lambda_1 & 1 & 0 & 0 \\ 0 & \lambda_1 & 1 & 0 \\ 0 & 0 & \lambda_1 & 1 \\ 0 & 0 & 0 & \lambda_1 \end{bmatrix} \tag{4-205}$$

A Jordan canonical matrix generally has the following properties:

1. The diagonal elements of the matrix Λ are the eigenvalues of $\mathbf{A}$.
2. All the elements below the principal diagonal are zeros.
3. Some of the elements immediately above the principal diagonal are ones.
4. The submatrices that are formed by each eigenvalue in Λ as shown by the dotted sections in Eq. (4-204) are called *Jordan blocks.* The matrix in Eq. (4-205) is one Jordan block by itself.

There is a good reason for using the Jordan canonical form even though it is not a diagonal matrix. Consider the matrix Λ in Eq. (4-205) for the state equation

$$\mathbf{y}(k+1) = \Lambda\mathbf{y}(k) \qquad (4 \times 1) \tag{4-206}$$

It is possible to find the state transition matrix $e^{\Lambda t}$ in a systematic manner with almost the same ease as in the case of a diagonal matrix. The last state equation in Eq. (4-206) is entirely decoupled from the other equations. Therefore, the solution to $y_4(k)$ is

$$y_4(k) = \lambda_1^k y_4(0) \tag{4-207}$$

The third state equation reads

$$y_3(k+1) = \lambda_1 y_3(k) + y_4(k) \tag{4-208}$$

Since $y_4(k)$ is already given in Eq. (4-207), Eq. (4-208) is easily solved to yield

$$y_3(k) = \lambda_1^k y_3(0) + k\lambda_1^{k-1} y_4(0) \tag{4-209}$$

Similarly, the second state equation is

$$y_2(k+1) = \lambda_1 y_2(k) + y_3(k) \tag{4-210}$$

Again, substituting $y_3(k)$ from Eq. (4-209) into Eq. (4-210) and solving, we have

$$y_2(k) = \lambda_1^k y_2(0) + k\lambda_1^{k-1} y_3(0) + \frac{k(k-1)}{2!}\lambda_1^{k-2} y_4(0) \tag{4-211}$$

Continuing the same process, the solution of $y_1(k)$ is

$$y_1(k) = \lambda_1^k y_1(0) + k\lambda_1^{k-1} y_2(0) + \frac{k(k-1)}{2!}\lambda_1^{k-2} y_3(0) + \frac{k(k-1)(k-2)}{3!}\lambda_1^{k-3} y_4(0) \tag{4-212}$$

In matrix form, the state transition equation is written as

$$\mathbf{y}(k) = \phi(k)\mathbf{y}(0) \tag{4-213}$$

where

$$\phi(k) = \lambda_1^k \begin{bmatrix} 1 & k\lambda_1^{-1} & \frac{k(k-1)}{2!}\lambda_1^{-2} & \frac{k(k-1)(k-2)}{3!}\lambda_1^{-3} \\ 0 & 1 & k\lambda_1^{-1} & \frac{k(k-1)}{2!}\lambda_1^{-2} \\ 0 & 0 & 1 & k\lambda_1^{-1} \\ 0 & 0 & 0 & 1 \end{bmatrix} \tag{4-214}$$

In general, if λ_1 is of mth multiplicity, the state transition matrix is of the form:

$$\phi(k) = \lambda_1^k \begin{bmatrix} 1 & k\lambda_1^{-1} & \frac{k(k-1)}{2!}\lambda_1^{-2} & \cdots & \frac{k(k-1)\ldots(k-n+2)}{(n-1)!}\lambda_1^{-n+1} \\ 0 & 1 & k\lambda_1^{-1} & & \vdots \\ 0 & 0 & 1 & & \vdots \\ \vdots & & & & \frac{k(k-1)}{2!}\lambda_1^{-2} \\ \vdots & & & & k\lambda_1^{-1} \\ 0 & \cdots & \cdots & \cdots & 1 \end{bmatrix} \tag{4-215}$$

Now we shall determine the matrix **P** that will transform **A** with multiple-order eigenvalues into a Jordan canonical form. The matrix **P** is again formed using the eigenvectors of **A**, as in Eq. (4-199). The eigenvectors associated with the distinct eigenvalues of **A** are determined in the usual manner. The eigenvectors associated with an mth-order Jordan form are found by referring to the Jordan block being of the form

$$\Lambda = \begin{bmatrix} \lambda_j & 1 & 0 & \dots & 0 \\ 0 & \lambda_j & 1 & \dots & 0 \\ \dots & \dots & \dots & \dots & \dots \\ 0 & 0 & 0 & \dots & 1 \\ 0 & 0 & 0 & \dots & \lambda_j \end{bmatrix} \qquad (m \times m) \tag{4-216}$$

Then, using $\Lambda = \mathbf{P}^{-1}\mathbf{AP}$, the following relation must hold:

$$[\mathbf{p}_1 \quad \mathbf{p}_2 \quad \cdots \quad \mathbf{p}_m]\Lambda = \mathbf{A}[\mathbf{p}_1 \quad \mathbf{p}_2 \quad \cdots \quad \mathbf{p}_m] \tag{4-217}$$

which is expanded to

$$\begin{aligned} \lambda_j \mathbf{p}_1 &= \mathbf{A}\mathbf{p}_1 \\ \mathbf{p}_1 + \lambda_j \mathbf{p}_2 &= \mathbf{A}\mathbf{p}_2 \\ \mathbf{p}_2 + \lambda_j \mathbf{p}_3 &= \mathbf{A}\mathbf{p}_3 \\ &\vdots \\ \mathbf{p}_{m-1} + \lambda_j \mathbf{p}_m &= \mathbf{A}\mathbf{p}_m \end{aligned} \tag{4-218}$$

With rearranging, these equations become

$$\begin{aligned} (\lambda_j \mathbf{I} - \mathbf{A})\mathbf{p}_1 &= \mathbf{0} \\ (\lambda_j \mathbf{I} - \mathbf{A})\mathbf{p}_2 &= -\mathbf{p}_1 \\ (\lambda_j \mathbf{I} - \mathbf{A})\mathbf{p}_3 &= -\mathbf{p}_2 \\ &\vdots \\ (\lambda_j \mathbf{I} - \mathbf{A})\mathbf{p}_m &= -\mathbf{p}_{m-1} \end{aligned} \tag{4-219}$$

The generalized eigenvectors, $\mathbf{p}_1, \mathbf{p}_2, \ldots, \mathbf{p}_m$ are determined from these equations.

4.13 METHODS OF COMPUTING THE STATE TRANSITION MATRIX

In the early sections of this chapter the state transition matrix of a digital control system and its ramifications were presented. It is worthwhile to emphasize the difference between the formulations of the state transition matrix under two different conditions.

For the sampled-data system

$$\dot{\mathbf{x}}(t) = \mathbf{A}\mathbf{x}(t) + \mathbf{B}\mathbf{u}(t) \tag{4-220}$$

where

$$\mathbf{u}(t) = \mathbf{u}(kT) \qquad kT \leqslant t < (k+1)T \tag{4-221}$$

the state transition matrix of the discretized system

$$\mathbf{x}[(k+1)T] = \phi(T)\mathbf{x}(kT) + \theta(T)\mathbf{u}(kT) \tag{4-222}$$

is

$$\phi(NT) = \phi(T) \cdot \phi(T) \ldots \phi(T) \tag{4-223}$$

where

$$\phi(T) = e^{\mathbf{A}T} = \phi(t)\Big|_{t=T} \tag{4-224}$$

In the last equation, $\phi(t)$ is also known as the state transition matrix of **A**. It was pointed out earlier that, in general, $\phi(NT)$ is *not* equal to $\phi(t)$ with t replaced by NT.

Therefore, for the sampled-data system described by Eqs. (4-220) and (4-221), to find the state transition matrix $\phi(NT)$, we must first find $\phi(t)$, replace t by T, and then use Eq. (4-223). The problem essentially is that of finding the state transition matrix of **A**, $\phi(t)$.

For the digital control system

$$\mathbf{x}(k+1) = \mathbf{A}\mathbf{x}(k) + \mathbf{B}\mathbf{u}(k) \tag{4-225}$$

the state transition matrix is defined as

$$\phi(N) = \mathbf{A}^N = \underbrace{\mathbf{A} \cdot \mathbf{A} \cdot \mathbf{A} \cdot \mathbf{A} \ldots \mathbf{A}}_{N} \tag{4-226}$$

The problem of finding $\phi(N)$ involves only that of multiplying **A** by itself N times.

We shall first present two alternate methods of computing $\phi(NT)$ given $\phi(T)$ in the sampled-data case, or $\phi(NT)$ given **A** in the digital case.

The Cayley-Hamilton Theorem Method

Given the matrix $\phi(T)$ or **A**, as the case may be, Eq. (4-223) or Eq. (4-226) can be computed using the Cayley-Hamilton Theorem.

The theorem states that every square matrix must satisfy its own characteristic equation, or, in general, $\mathbf{A}^N$ can be written as

$$\mathbf{A}^N = \sum_{i=0}^{N-1} a_{Ni}\mathbf{A}^i \qquad \text{for any positive integer N} \qquad (4\text{-}227)$$

and n is the dimension of **A**, a_{Ni} denotes the coefficients of the characteristic equation of **A**. Similarly, if $\phi(T)$ is given,

$$\phi(NT) = [\phi(T)]^N = \sum_{i=0}^{N-1} a_{Ni}[\phi(T)]^i \qquad (4\text{-}228)$$

where a_{Ni} are the coefficients of the characteristic equation of $\phi(T)$.

Example 4-7.

To illustrate the use of the Cayley-Hamilton Theorem, consider that the **A** matrix in the system of Eq. (4-225) is given as

$$\mathbf{A} = \begin{bmatrix} 3 & 2 \\ 2 & 3 \end{bmatrix} \qquad (4\text{-}229)$$

The characteristic equation of A is

$$|\lambda\mathbf{I} - \mathbf{A}| = \lambda^2 - 6\lambda + 5 = 0 \qquad (4\text{-}230)$$

Applying the Cayley-Hamilton Theorem, we have the matrix equation,

$$\mathbf{A}^2 - 6\mathbf{A} + 5\mathbf{I} = 0 \qquad (4\text{-}231)$$

from which we have

$$\mathbf{A}^2 = 6\mathbf{A} - 5\mathbf{I} \qquad (4\text{-}232)$$

Thus $\mathbf{A}^2$ is expressed in terms of A. The crux of the Cayley-Hamilton Theorem is that $\mathbf{A}^N$ can be expressed as an algebraic sum of $\mathbf{A}^{N-1}, \mathbf{A}^{N-2}, \ldots$, A, and by repeatedly applying the theorem, $\mathbf{A}^N$ can eventually be expressed in terms of A.

Thus,

$$\begin{aligned} \mathbf{A}^3 = \mathbf{A}\cdot\mathbf{A}^2 &= \mathbf{A}(6\mathbf{A} - 5\mathbf{I}) \\ &= 6\mathbf{A}^2 - 5\mathbf{A} \\ &= 6(6\mathbf{A} - 5\mathbf{I}) - 5\mathbf{A} \\ &= 31\mathbf{A} - 30\mathbf{I} \end{aligned} \qquad (4\text{-}233)$$

Similarly, we can show that

$$A^4 = 156A - 155I \tag{4-234}$$

Continuing the same process we can compute A^N for any large N in a recursive manner.

The z-Transform Method

For the sampled-data system of Eq. (4-222) the state transition matrix of $\phi(T)$ is expressed in terms of z-transform in Eq. (4-110); that is,

$$\phi(NT) = \mathcal{Z}^{-1}\left[[z\mathbf{I} - \phi(T)]^{-1}z\right] \tag{4-235}$$

For the digital control system of Eq. (4-225),

$$\phi(N) = \mathcal{Z}^{-1}[(z\mathbf{I} - \mathbf{A})^{-1}z] \tag{4-236}$$

In these last two equations the evaluation of the state transition matrix involves the matrix inverse and then the inverse z-transform. For second-order systems these analytical steps can be carried out with ease. However, for higher-order systems the amount of work involved can be quite tedious.

The task of finding $(z\mathbf{I} - \mathbf{A})^{-1}z$ given $\mathbf{A}$ can be simplified by the method described as follows.

Let

$$(z\mathbf{I} - \mathbf{A})^{-1}z = \mathbf{F} \tag{4-237}$$

Then

$$z\mathbf{F} = z\mathbf{I} + \mathbf{AF} \tag{4-238}$$

Premultiplying both sides of the last equation by $z\mathbf{I} + \mathbf{A}$, we have after simplifying,

$$z^2\mathbf{F} = \mathbf{A}^2\mathbf{F} + z\mathbf{A} + z^2\mathbf{I} \tag{4-239}$$

Again multiplying both sides of the last equation by $z\mathbf{I} + \mathbf{A}$, we get after simplifying,

$$z^3\mathbf{F} = \mathbf{A}^3\mathbf{F} + z\mathbf{A}^2 + z^2\mathbf{A} + z^3\mathbf{I} \tag{4-240}$$

Continuing the process, the following equations are written:

$$z^4\mathbf{F} = \mathbf{A}^4\mathbf{F} + z\mathbf{A}^3 + z^2\mathbf{A}^2 + z^3\mathbf{A} + z^4\mathbf{I} \tag{4-241}$$

$$\vdots$$

$$z^n\mathbf{F} = \mathbf{A}^n\mathbf{F} + z\mathbf{A}^{n-1} + z^2\mathbf{A}^{n-2} + \ldots + z^{n-1}\mathbf{A} + z^n\mathbf{I} \tag{4-242}$$

Thus, the following equations are formed:

$$
\begin{aligned}
a_1 \mathbf{F} &= a_1 \mathbf{F} \\
a_2 z\mathbf{F} &= a_2 \mathbf{AF} + a_2 z\mathbf{I} \\
a_3 z^2 \mathbf{F} &= a_3 \mathbf{A}^2 \mathbf{F} + a_3 z\mathbf{A} + a_3 z^2 \mathbf{I} \\
a_4 z^3 \mathbf{F} &= a_4 \mathbf{A}^3 \mathbf{F} + a_4 z\mathbf{A}^2 + a_4 z^2 \mathbf{A} + a_4 z^3 \mathbf{I} \\
&\vdots \\
a_n z^{n-1} \mathbf{F} &= a_n \mathbf{A}^{n-1} \mathbf{F} + a_n z\mathbf{A}^{n-2} + a_n z^2 \mathbf{A}^{n-3} + \ldots + a_n z^{n-2} \mathbf{A} + a_n z^{n-1} \mathbf{I} \\
z^n \mathbf{F} &= \mathbf{A}^n \mathbf{F} + z\mathbf{A}^{n-1} + z^2 \mathbf{A}^{n-2} + \ldots + z^{n-1} \mathbf{A} + z^n \mathbf{I}
\end{aligned}
\tag{4-243}
$$

where $a_1, a_2, \ldots, a_n$ are the coefficients of the characteristic equation of **A**,

$$\sum_{i=0}^{n} a_{i+1} z^i = 0 \qquad (a_{n+1} = 1) \tag{4-244}$$

The equations in Eq. (4-243) are summed on both sides to give

$$\sum_{i=0}^{n} a_{i+1} z^i \mathbf{F} = \sum_{i=0}^{n} a_{i+1} \mathbf{A}^i \mathbf{F} + \sum_{i=1}^{n} a_{i+1} \mathbf{A}^{i-1} z + \sum_{i=2}^{n} a_{i+1} \mathbf{A}^{i-2} z^2$$

$$+ \ldots + \sum_{i=n-1}^{n} a_{i+1} \mathbf{A}^{i-n+1} z^{n-1} + z^n \mathbf{I} \tag{4-245}$$

where $a_{n+1} = 1$.

Because of the Cayley-Hamilton theorem, the first term on the right-hand side of the last equation is a null matrix. Thus, Eq. (4-245) leads to

$$\mathbf{F} = \frac{\displaystyle\sum_{j=1}^{n} z^j \sum_{i=j}^{n} a_{i+1} \mathbf{A}^{i-j}}{\displaystyle\sum_{i=0}^{n} a_{i+1} z^i} \tag{4-246}$$

or

$$\mathbf{F} = \frac{\displaystyle\sum_{j=1}^{n} z^j \sum_{i=j}^{n} a_{i+1} \mathbf{A}^{i-j}}{|z\mathbf{I} - \mathbf{A}|} \tag{4-247}$$

The numerator of the last equation is also known as the matrix, $[\mathrm{Adj}(z\mathbf{I} - \mathbf{A})]z$.

Once **F** is evaluated from Eq. (4-247), $\phi(N)$ is given by

$$\phi(N) = \mathcal{Z}^{-1}[\mathbf{F}] \tag{4-248}$$

For the case in Eq. (4-235), we simply replace **A** by $\phi(T)$ in Eq. (4-247).

Example 4-8.

Consider that the digital control system described by Eq. (4-225) has the A matrix,

$$\mathbf{A} = \begin{bmatrix} 0 & 1 & 0 \\ 0 & 0 & 1 \\ -6 & -11 & -6 \end{bmatrix} \tag{4-249}$$

The characteristic equation of **A** is

$$\begin{aligned} |\lambda\mathbf{I} - \mathbf{A}| &= \lambda^3 + 6\lambda^2 + 11\lambda + 6 \\ &= (\lambda - 1)(\lambda - 2)(\lambda - 3) = 0 \end{aligned} \tag{4-250}$$

The coefficients of the characteristic equation are: $a_4 = 1$, $a_3 = 6$, $a_2 = 11$ and $a_1 = 6$.
Using Eq. (4-247), **F** is written as

$$\mathbf{F} = (z\mathbf{I} - \mathbf{A})^{-1}z = \frac{\sum_{j=1}^{3} z^j \sum_{i=j}^{3} a_{i+1}\mathbf{A}^{i-j}}{|z\mathbf{I} - \mathbf{A}|}$$

$$= \frac{a_4 z^3\mathbf{I} + (a_3\mathbf{I} + a_2\mathbf{A})z^2 + (a_2\mathbf{I} + a_3\mathbf{A} + a_4\mathbf{A}^2)z}{|z\mathbf{I} - \mathbf{A}|} \tag{4-251}$$

Substitution of the coefficients a_4, a_3, a_2, a_1 and **A** into the last equation, we have

$$\mathbf{F} = \frac{z^3\mathbf{I} + \begin{bmatrix} 6 & 1 & 0 \\ 0 & 6 & 1 \\ -6 & -11 & 0 \end{bmatrix} z^2 + \begin{bmatrix} 11 & 6 & 1 \\ -6 & 0 & 0 \\ 0 & -6 & 0 \end{bmatrix} z}{(z-1)(z-2)(z-3)} \tag{4-252}$$

Performing the partial-fraction expansion, Eq. (4-252) gives

$$\mathbf{F} = \frac{z}{2(z-1)}\begin{bmatrix} 18 & 7 & 1 \\ -6 & 7 & 1 \\ -6 & -17 & 1 \end{bmatrix} - \frac{z}{z-2}\begin{bmatrix} 27 & 8 & 1 \\ -6 & 16 & 2 \\ -12 & -28 & 4 \end{bmatrix} + \frac{z}{2(z-3)}\begin{bmatrix} 38 & 9 & 1 \\ -6 & 27 & 3 \\ -18 & -39 & 9 \end{bmatrix} \tag{4-253}$$

Now taking the inverse z-transform on both sides of the last equation, we get

$$\phi(k) = \frac{1}{2}\begin{bmatrix} 18 & 7 & 1 \\ -6 & 7 & 1 \\ -6 & -17 & 1 \end{bmatrix} - \begin{bmatrix} 27 & 8 & 1 \\ -6 & 16 & 2 \\ -12 & -28 & 4 \end{bmatrix} e^{-0.693k} + \frac{1}{2}\begin{bmatrix} 38 & 9 & 1 \\ -6 & 27 & 3 \\ -18 & -39 & 9 \end{bmatrix} e^{-1.1k} \tag{4-254}$$

We can check this result by verifying that $\phi(0) = \mathbf{I}$, the identity matrix.

Computing The State Transition Matrix $\phi(T)$

The discussions conducted in this section thus far have been concentrated on the computation of the state transition matrix $\phi(NT)$ or $\phi(N)$ of the sampled-data or digital control system. When a sampled-data system is encountered, the starting point is Eq. (4-220) in which the matrices **A** and **B** are given. In order to find $\phi(NT)$ using Eq. (4-223), we must first evaluate $\phi(T)$ which is given by Eq. (4-224).

Earlier in this chapter the state transition matrix of **A**, $\phi(t)$, is given by Eqs. (4-30) and (4-34). Equation (4-30) is the power-series representation, whereas Eq. (4-34) gives the Laplace transform solution. We shall investigate various methods of computing $\phi(t)$, given the matrix **A**.

The Laplace Transform Method

Repeating Eq. (4-34),

$$\phi(t) = \mathcal{L}^{-1}[(s\mathbf{I} - \mathbf{A})^{-1}] \tag{4-255}$$

The matrix inverse of $s\mathbf{I} - \mathbf{A}$ can be conducted using essentially the same procedure as described in Eqs. (4-237) through (4-246). The result can be written as

$$(s\mathbf{I} - \mathbf{A})^{-1} = \frac{\sum_{j=0}^{n-1} s^j \sum_{i=j+1}^{n} a_i \mathbf{A}^{i-j-1}}{|s\mathbf{I} - \mathbf{A}|} \tag{4-256}$$

where n is the dimension of **A**, and

$$|s\mathbf{I} - \mathbf{A}| = a_{n+1}s^n + a_n s^{n-1} + \ldots + a_2 s + a_1 \tag{4-257}$$

Direct Power Series Expansion Method

The power series representation of $\phi(T)$ is

$$\phi(T) = e^{\mathbf{A}T} = \mathbf{I} + \mathbf{A}T + \frac{\mathbf{A}^2T^2}{2!} + \ldots \tag{4-258}$$

This expression can be programmed recursively. For instance, the kth term of the series is $A^kT^k/k!$, and the (k + 1)st term is $A^{k+1}T^{k+1}/(k+1)!$. Thus, we can write

$$(k+1)\text{th term} = \frac{AT}{k+1} \times k\text{th term} \tag{4-259}$$

k = 0,1,2,... . Normally, when computing the series, a check is made on the convergence, and the recursive iteration can be stopped after N terms.

Truncated Power-Series Expansion Method
(Cayley-Hamilton Theorem)

The power series of $\phi(T)$ can be truncated after N terms, where N is some positive integer. Then,

$$\phi(T) \cong \phi_N(T) = \sum_{k=0}^{N} \frac{A^k T^k}{k!} \tag{4-260}$$

Applying the Cayley-Hamilton theorem to the last equation, we can write

$$\phi_N(T) = \sum_{j=0}^{n-1} a_j \mathbf{A}^j \tag{4-261}$$

where n is the dimension of **A**, and a_j are dependent on the coefficients of the characteristic equation of **A** and can be computed recursively. In general, N and n are not related, so that N can be greater than n.

The Eigenvalue Method
(Sylvester's Expansion Theorem, A has distinct eigenvalues)

For systems with distinct eigenvalues, Sylvester's expansion theorem states that if

$$f(\mathbf{A}) = \sum_{k=1}^{\infty} c_k \mathbf{A}^k \tag{4-262}$$

then

$$f(\mathbf{A}) = \sum_{i=1}^{n} f(\lambda_i)\mathbf{F}(\lambda_i) \tag{4-263}$$

where λ_i, i = 1,2,...,n are the eigenvalues (all distinct) of **A**.

$$\mathbf{F}(\lambda_i) = \sum_{\substack{j=1 \\ i \neq j}}^{n} \frac{\mathbf{A} - \lambda_j \mathbf{I}}{\lambda_i - \lambda_j} \qquad i = 1,2,\ldots,n \tag{4-264}$$

For the state transition matrix problem, we have

$$f(\mathbf{A}) = \phi(T) = \sum_{k=0}^{\infty} \frac{\mathbf{A}^k T^k}{k!} = e^{\mathbf{A}T} \tag{4-265}$$

Thus,

$$f(\lambda_i) = e^{\lambda_i T} \tag{4-266}$$

and from Eq. (4-263),

$$\phi(T) = \sum_{i=1}^{n} e^{\lambda_i T} \mathbf{F}(\lambda_i) \tag{4-267}$$

where $\mathbf{F}(\lambda_i)$ is given by Eq. (4-264).

Example 4-9.

To illustrate the method, we use the same **A** matrix given in Eq. (4-229). The eigenvalues of **A** are $\lambda_1 = 5$ and $\lambda_2 = 1$, which are distinct. Then, from Eq. (4-266),

$$f(\lambda_1) = e^{5T} \tag{4-268}$$

$$f(\lambda_2) = e^{T} \tag{4-269}$$

Equation (4-264) gives

$$\mathbf{F}(\lambda_1) = \frac{\mathbf{A} - \lambda_2 \mathbf{I}}{\lambda_1 - \lambda_2} = \begin{bmatrix} 0.5 & 0.5 \\ 0.5 & 0.5 \end{bmatrix} \tag{4-270}$$

$$\mathbf{F}(\lambda_2) = \frac{\mathbf{A} - \lambda_1 \mathbf{I}}{\lambda_2 - \lambda_1} = \begin{bmatrix} 0.5 & -0.5 \\ -0.5 & 0.5 \end{bmatrix} \tag{4-271}$$

Thus,

$$\phi(T) = f(\lambda_1)\mathbf{F}(\lambda_1) + f(\lambda_2)\mathbf{F}(\lambda_2)$$

$$= 0.5 \begin{bmatrix} e^{5T} + e^{T} & e^{5T} - e^{T} \\ e^{5T} - e^{T} & e^{5T} + e^{T} \end{bmatrix} \tag{4-272}$$

4.14 DIGITAL ADJOINT SYSTEMS

Just as in continuous-data control systems, it is useful to consider the adjoint systems in linear digital systems. In the design of optimal control systems using the variations approach, quite often the necessary conditions lead to the requirement of satisfying the adjoint system equations.

For a linear continuous-data system that is described by

$$\dot{\mathbf{x}}(t) = \mathbf{A}(t)\mathbf{x}(t) + \mathbf{B}(t)\mathbf{u}(t) \tag{4-273}$$

the adjoint system is defined as

$$\dot{\mathbf{y}}(t) = -\mathbf{A}'(t)\mathbf{y}(t) + \mathbf{B}(t)\mathbf{u}(t)$$

which is arrived at by requiring that the homogeneous system satisfies the condition that the inner product of $\mathbf{x}(t)$ and $\mathbf{y}(t)$ be constant; that is $\mathbf{x}'(t)\mathbf{y}(t) =$ constant. It can be shown that if $\phi(t, t_0)$ is the state transition matrix of $\mathbf{A}(t)$, then the state transition matrix of $-\mathbf{A}'(t)$ is $\phi'(t_0, t)$.

We can define digital adjoint systems using the relations given above as a starting base. The discrete state equation as derived from Eq. (4-273) is [as in Eq. (4-75)],

$$\mathbf{x}(k_N) = \phi(k_N, k_0)\mathbf{x}(k_0) + \sum_{j=0}^{N-1} \phi(k_N, k_{j+1})\mathbf{B}(k_j)\mathbf{u}(k_j) \tag{4-274}$$

The state equation of the digital adjoint system is defined as

$$\mathbf{y}(k_N) = \phi^*(k_N, k_0)\mathbf{y}(k_0) + \sum_{j=0}^{N-1} \phi^*(k_N, k_{j+1})\mathbf{B}(k_j)\mathbf{u}(k_j) \tag{4-275}$$

where

$$\phi^*(k_N, k_0) = \phi'(k_0, k_N) \tag{4-276}$$

For strictly digital systems, given the system

$$\mathbf{x}(k_{i+1}) = \mathbf{A}(k_i)\mathbf{x}(k_i) + \mathbf{B}(k_i)\mathbf{u}(k_i) \tag{4-277}$$

we can represent the adjoint system as

$$\mathbf{y}(k_{i+1}) = \mathbf{A}^*(k_i)\mathbf{y}(k_i) + \mathbf{B}(k_i)\mathbf{u}(k_i) \tag{4-278}$$

Again, defining the adjoint system as one whose homogeneous system satisfies the condition that $\mathbf{x}'(k_i)\mathbf{y}(k_i) =$ constant, we can find the necessary relation between $\mathbf{A}(k_i)$ and $\mathbf{A}^*(k_i)$. First we must assume that $\mathbf{A}(k_i)$ and $\mathbf{A}^*(k_i)$ are nonsingular. Then, forming the inner product of $\mathbf{x}(k_i)$ and $\mathbf{y}(k_i)$, we have

$$\mathbf{x}'(k_i)\mathbf{y}(k_i) = \mathbf{x}'(k_{i+1})[\mathbf{A}'(k_i)]^{-1}[\mathbf{A}^*(k_i)]^{-1}\mathbf{y}(k_{i+1}) \tag{4-279}$$

For the above inner product to be constant, we must have

$$\mathbf{x}'(k_i)\mathbf{y}(k_i) = \mathbf{x}'(k_{i+1})\mathbf{y}(k_{i+1}) \tag{4-280}$$

Thus,

$$\mathbf{A}^*(k_i) = [\mathbf{A}'(k_i)]^{-1} = [\mathbf{A}^{-1}(k_i)]' \tag{4-281}$$

Furthermore, the state transition matrix of $\mathbf{A}(k_i)$ is [Eq. (4-76)],

$$\phi(k_i,k_0) = \mathbf{A}(k_{i-1})\mathbf{A}(k_{i-2}) \ldots \mathbf{A}(k_0) \qquad k_i > k_0$$
$$= \mathbf{I} \qquad k_i = k_0 \qquad (4\text{-}282)$$

The state transition matrix of $\mathbf{A}^*(k_i)$ is

$$\phi^*(k_i,k_0) = \mathbf{A}^*(k_{i-1})\mathbf{A}^*(k_{i-2}) \ldots \mathbf{A}^*(k_0) \qquad k_i > k_0$$
$$= \mathbf{I} \qquad k_i = k_0 \qquad (4\text{-}283)$$

Using Eq. (4-281), the following relation results:

$$\phi^*(k_i,k_0) = \phi'(k_0,k_i) \qquad (4\text{-}284)$$

which is valid for all k_i and k_0, and is the same as Eq. (4-276).

4.15 RELATIONSHIP BETWEEN STATE EQUATIONS AND HIGH-ORDER DIFFERENCE EQUATIONS

In preceding sections we discussed the state equations and their solutions for linear digital systems. In general, although it is usually possible to write the state equations directly from a system, in practice, the digital system may already be described by a high-order difference equation or a transfer function. Therefore, it is useful to investigate how state equations can be written directly from the difference equation or the transfer function.

The procedure of arriving at the state equations from the transfer function is defined as *decomposition* and is discussed in Sec. 4.18. It will be demonstrated that it is generally simpler to first transform the high-order difference equation into a transfer function and then apply a decomposition scheme to get the state equations. However, in this section we shall establish some basic relations in the formulation of the state equations of a high-order system.

Let us consider that a single-variable linear digital system is described by the following nth-order difference equation:

$$c(k+n) + a_n c(k+n-1) + a_{n-1} c(k+n-2) + \ldots + a_2 c(k+1) + a_1 c(k) = r(k) \qquad (4\text{-}285)$$

where $c(k)$ is the output variable and $r(k)$ is the input. The coefficient $a_1, a_2, \ldots, a_n$ can actually be time varying.

The problem is to represent Eq. (4-285) by n state equations and an output equation. The first step involves the definition of the state variables as functions of the output variable $c(k)$. Although the state variables of a given system are not unique, the most convenient way to define the state variables in the case of Eq. (4-285) is

$$x_1(k) = c(k)$$
$$x_2(k) = c(k+1) = x_1(k+1)$$
$$\vdots$$
$$x_n(k) = c(k+n-1) \tag{4-286}$$

After substitution of the relations in Eq. (4-286) into Eq. (4-285) and rearrangement, the state equations are

$$x_1(k+1) = x_2(k)$$
$$x_2(k+1) = x_3(k)$$
$$\vdots$$
$$x_n(k+1) = -a_1x_1(k) - a_2x_2(k) - \ldots - a_{n-1}x_{n-1}(k) - a_nx_n(k) + r(k) \tag{4-287}$$

The output equation is simply

$$c(k) = x_1(k) \tag{4-288}$$

For time-invariant systems, Eq. (4-277) is written as

$$\mathbf{x}(k+1) = \mathbf{Ax}(k) + \mathbf{B}r(k) \tag{4-289}$$

where $\mathbf{x}(k)$ is the $n \times 1$ state vector and $r(k)$ is the scalar input. The coefficient matrices are

$$\mathbf{A} = \begin{bmatrix} 0 & 1 & 0 & 0 & 0 & \ldots & 0 \\ 0 & 0 & 1 & 0 & 0 & \ldots & 0 \\ 0 & 0 & 0 & 1 & 0 & \ldots & 0 \\ \cdots & \cdots & \cdots & \cdots & \cdots & \cdots & \cdots \\ 0 & 0 & 0 & 0 & 0 & \ldots & 1 \\ -a_1 & -a_2 & -a_3 & -a_4 & -a_5 & \ldots & -a_n \end{bmatrix} \quad (n \times n) \tag{4-290}$$

$$\mathbf{B} = \begin{bmatrix} 0 \\ 0 \\ \vdots \\ 0 \\ 1 \end{bmatrix} \quad (n \times 1) \tag{4-291}$$

The output equation in vector-matrix form is

$$c(k) = \mathbf{D}\mathbf{x}(k) \tag{4-292}$$

where

$$\mathbf{D} = [1 \quad 0 \quad 0 \; \dots \; 0] \qquad (n \times 1) \tag{4-293}$$

The state equation of Eq. (4-289) with the coefficient matrices **A** and **B** in the basic forms of Eqs. (4-290) and (4-291) is called the *phase-variable canonical form.*

In the sections to follow we shall show that a system that is represented in the phase-variable canonical form has certain unique and useful characteristics which are helpful in the analysis and design of linear digital systems.

It should be noted that the high-order difference equation in Eq. (4-285) is not general since there are no high-order terms on the side of the input. In general, when terms such as r(k + 1), r(k + 2), ... are on the right-hand side of Eq. (4-285), the procedure of assigning state variables would not be as straightforward as in Eq. (4-286). In such a situation, it is more convenient to first transform the high-order difference equation into a transfer function and then conduct a decomposition to get the dynamic equations.

4.16 TRANSFORMATION TO PHASE-VARIABLE CANONICAL FORM

It was indicated in the last section that for a linear single-variable system, if the coefficient matrices **A** and **B** assume the forms of Eqs. (4-290) and (4-291), respectively, the system is referred to as the phase-variable canonical form. As will be shown later, a system in the phase-variable canonical form has characteristics which facilitate certain analysis and design procedures. The following theorem shows that a system in the phase-variable canonical form can always have its eigenvalues assigned arbitrarily by state feedback.

Theorem 4-1. Let the state equations of a linear digital system be represented by

$$\mathbf{x}(k+1) = \mathbf{A}\mathbf{x}(k) + \mathbf{B}u(k) \tag{4-294}$$

where

$$\mathbf{A} = \begin{bmatrix} 0 & 1 & 0 & 0 & 0 & \dots & 0 \\ 0 & 0 & 1 & 0 & 0 & \dots & 0 \\ 0 & 0 & 0 & 1 & 0 & \dots & 0 \\ \dots & \dots & \dots & \dots & \dots & \dots & \dots \\ 0 & 0 & 0 & 0 & 0 & \dots & 1 \\ -a_1 & -a_2 & -a_3 & -a_4 & -a_5 & \dots & -a_n \end{bmatrix} \qquad (n \times n) \tag{4-295}$$

$$\mathbf{B} = \begin{bmatrix} 0 \\ 0 \\ \vdots \\ 0 \\ 1 \end{bmatrix} \qquad (n \times 1) \tag{4-296}$$

The eigenvalues of the system can be arbitrarily assigned by the state feedback

$$u(t) = -\mathbf{G}\mathbf{x}(t) \tag{4-297}$$

where

$$\mathbf{G} = [g_1 \quad g_2 \quad \cdots \quad g_n] \tag{4-298}$$

and $g_1, g_2, \ldots, g_n$ are real constants.

A block diagram portraying the state feedback is shown in Fig. 4-4.

Proof: Substituting the input u(t) from Eq. (4-297) into Eq. (4-294), we have the state equations of the closed-loop system,

$$\mathbf{x}(k+1) = (\mathbf{A} - \mathbf{BG})\mathbf{x}(k) \tag{4-299}$$

The characteristic equation of the closed-loop system is

$$|\lambda\mathbf{I} - \mathbf{A} + \mathbf{BG}| = 0 \tag{4-300}$$

Substitution of Eqs. (4-295), (4-296), and (4-298) into Eq. (4-300) yields

$$\begin{vmatrix} \lambda & -1 & 0 & 0 & 0 & \cdots & 0 \\ 0 & \lambda & -1 & 0 & 0 & \cdots & 0 \\ 0 & 0 & \lambda & -1 & 0 & \cdots & 0 \\ \cdots & \cdots & \cdots & \cdots & \cdots & \cdots & 0 \\ 0 & 0 & 0 & 0 & 0 & \cdots & -1 \\ a_1 + g_1 & a_2 + g_2 & a_3 + g_3 & a_4 + g_4 & a_5 + g_5 & \cdots & \lambda + a_n + g_n \end{vmatrix} = \mathbf{0} \tag{4-301}$$

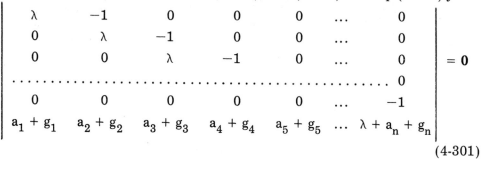

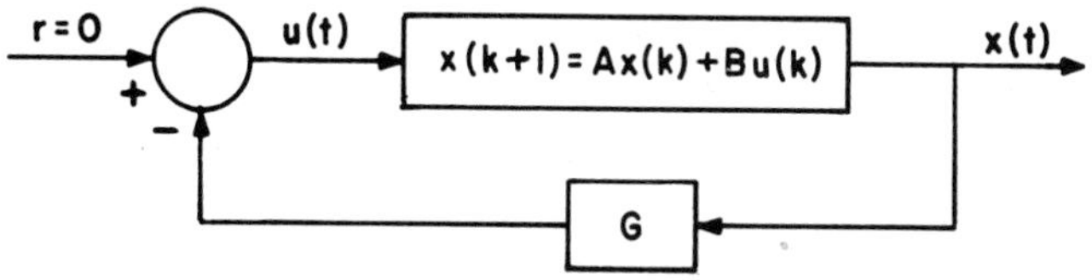

Fig. 4-4. A linear digital system with state feedback.

This characteristic equation is evaluated to be

$$\lambda^n + (a_n + g_n)\lambda^{n-1} + (a_{n-1} + g_{n-1})\lambda^{n-2} + \ldots + (a_2 + g_2)\lambda + (a_1 + g_1) = 0 \tag{4-302}$$

Since each of the feedback gains is associated with only one of the coefficients of the last equation, it is apparent that if the eigenvalues of the characteristic equation of the closed-loop system are arbitrarily assigned, which corresponds to the coefficients of the equation being arbitrary, then equating these coefficients to the corresponding ones in Eq. (4-302) would yield n linearly independent equations for solving the feedback gains, $g_1, g_2, \ldots, g_n$.

When a system is not in the phase-variable canonical form, there exists a similar transformation which transforms the **A** and **B** matrices into the proper forms of Eqs. (4-295) and (4-296), respectively. The following theorem describes a method of achieving this.

Theorem 4-2. Let the state equations of a linear digital system be represented by

$$\mathbf{x}(k + 1) = \mathbf{A}\mathbf{x}(k) + \mathbf{B}u(k) \tag{4-303}$$

where $\mathbf{x}(k)$ is an n-vector, u(t) is a scalar input, **A** is any $n \times n$ coefficient matrix, and **B** is any $n \times 1$ matrix, such that

$$\mathbf{S} = [\mathbf{B} \quad \mathbf{AB} \quad \mathbf{A}^2\mathbf{B} \ \ldots \ \mathbf{A}^{n-1}\mathbf{B}] \qquad (n \times n) \tag{4-304}$$

is nonsingular. Then there exists a nonsingular transformation

$$\mathbf{y}(k) = \mathbf{M}\mathbf{x}(k) \tag{4-305}$$

or

$$\mathbf{x}(k) = \mathbf{M}^{-1}\mathbf{y}(k) \tag{4-306}$$

which transforms Eq. (4-303) to the phase-variable canonical form

$$\mathbf{y}(k + 1) = \mathbf{A}_1\mathbf{y}(k) + \mathbf{B}_1u(k) \tag{4-307}$$

where

$$\mathbf{A}_1 = \begin{bmatrix} 0 & 1 & 0 & 0 & \ldots & 0 \\ 0 & 0 & 1 & 0 & \ldots & 0 \\ 0 & 0 & 0 & 1 & \ldots & 0 \\ \cdots & \cdots & \cdots & \cdots & \cdots & \cdots \\ 0 & 0 & 0 & 0 & \ldots & 1 \\ -a_1 & -a_2 & -a_3 & -a_4 & \ldots & -a_n \end{bmatrix} \tag{4-308}$$

$$\mathbf{B}_1 = \begin{bmatrix} 0 \\ 0 \\ \vdots \\ 0 \\ 1 \end{bmatrix} \tag{4-309}$$

The matrix **M** is given by

$$\mathbf{M} = \begin{bmatrix} \mathbf{M}_1 \\ \mathbf{M}_1\mathbf{A} \\ \vdots \\ \mathbf{M}_1\mathbf{A}^{n-1} \end{bmatrix} \qquad (n \times n) \tag{4-310}$$

where

$$\mathbf{M}_1 = [0 \quad 0 \ \ldots \ 1][\mathbf{B} \quad \mathbf{AB} \quad \mathbf{A}^2\mathbf{B} \ \ldots \ \mathbf{A}^{n-1}\mathbf{B}]^{-1} \tag{4-311}$$

Proof: Let the matrix **M** be written as

$$\mathbf{M} = \begin{bmatrix} m_{11} & m_{12} & \cdots & m_{1n} \\ m_{21} & m_{22} & \cdots & m_{2n} \\ \cdots & \cdots & \cdots & \cdots \\ m_{n1} & m_{n2} & \cdots & m_{nn} \end{bmatrix} = \begin{bmatrix} \mathbf{M}_1 \\ \mathbf{M}_2 \\ \vdots \\ \mathbf{M}_n \end{bmatrix} \qquad (n \times n) \tag{4-312}$$

where

$$\mathbf{M}_i = [m_{i1} \quad m_{i2} \ \cdots \ m_{in}] \qquad i = 1,2,\ldots,n \tag{4-313}$$

Equating the first row on both sides of Eq. (4-305) gives

$$y_1(k) = \mathbf{M}_1\mathbf{x}(k) \tag{4-314}$$

Advancing both sides of the last equation in time or stage by one yields

$$y_1(k+1) = \mathbf{M}_1\mathbf{x}(k+1) \tag{4-315}$$

Substitution of the state equation of Eq. (4-303) in the last equation and in view of Eq. (4-308), we have

$$\begin{aligned} y_1(k+1) &= \mathbf{M}_1\mathbf{A}\mathbf{x}(k) + \mathbf{M}_1\mathbf{B}u(k) \\ &= y_2(k) \end{aligned} \tag{4-316}$$

Since Eq. (4-305) stipulates that $\mathbf{y}(k)$ is a function of $\mathbf{x}(k)$ only, in Eq. (4-316) $\mathbf{M}_1\mathbf{B} = \mathbf{0}$. Thus, Eq. (4-316) becomes

$$y_1(k+1) = y_2(k) = \mathbf{M}_1\mathbf{A}\mathbf{x}(k) \tag{4-317}$$

Advancing the time or stage by one again in the last equation gives

$$\begin{aligned} y_1(k+2) = y_2(k+1) = y_3(k) &= \mathbf{M}_1\mathbf{A}\mathbf{x}(k+1) \\ &= \mathbf{M}_1\mathbf{A}^2\mathbf{x}(k) \end{aligned} \tag{4-318}$$

where it has been established that $\mathbf{M}_1\mathbf{A}\mathbf{B} = 0$.

Repeating the above procedure a total of $n-1$ times leads to

$$y_{n-1}(k+1) = y_n(k) = \mathbf{M}_1\mathbf{A}^{n-1}\mathbf{x}(k) \tag{4-319}$$

with $\mathbf{M}_1\mathbf{A}^{n-2}\mathbf{B} = 0$. Therefore, collecting the above results, we have

$$\mathbf{y}(k) = \begin{bmatrix} y_1(k) \\ y_2(k) \\ \vdots \\ y_n(k) \end{bmatrix} = \mathbf{M}\mathbf{x}(k) = \begin{bmatrix} \mathbf{M}_1 \\ \mathbf{M}_1\mathbf{A} \\ \vdots \\ \mathbf{M}_1\mathbf{A}^{n-1} \end{bmatrix} \mathbf{x}(k) \tag{4-320}$$

Thus, for any $\mathbf{x}(k)$,

$$\mathbf{M} = \begin{bmatrix} \mathbf{M}_1 \\ \mathbf{M}_1\mathbf{A} \\ \vdots \\ \mathbf{M}_1\mathbf{A}^{n-1} \end{bmatrix} \tag{4-321}$$

and in addition, $\mathbf{M}_1$ should satisfy the condition

$$\mathbf{M}_1\mathbf{B} = \mathbf{M}_1\mathbf{A}\mathbf{B} = \dots = \mathbf{M}_1\mathbf{A}^{n-2}\mathbf{B} = 0 \tag{4-322}$$

From Eq. (4-305) we write

$$\begin{aligned} \mathbf{y}(k+1) = \mathbf{M}\mathbf{x}(k+1) &= \mathbf{M}\mathbf{A}\mathbf{x}(k) + \mathbf{M}\mathbf{B}u(k) \\ &= \mathbf{M}\mathbf{A}\mathbf{M}^{-1}\mathbf{y}(k) + \mathbf{M}\mathbf{B}u(k) \end{aligned} \tag{4-323}$$

Comparing Eq. (4-323) with Eq. (4-307), we have

$$\mathbf{A}_1 = \mathbf{M}\mathbf{A}\mathbf{M}^{-1} \tag{4-324}$$

and

$$\mathbf{B}_1 = \mathbf{MB} \tag{4-325}$$

Then, from Eq. (4-321)

$$\mathbf{B}_1 = \begin{bmatrix} \mathbf{M}_1\mathbf{B} \\ \mathbf{M}_1\mathbf{AB} \\ \cdot \\ \cdot \\ \cdot \\ \mathbf{M}_1\mathbf{A}^{n-1}\mathbf{B} \end{bmatrix} = \begin{bmatrix} 0 \\ 0 \\ \cdot \\ \cdot \\ \cdot \\ 1 \end{bmatrix} \tag{4-326}$$

Since $\mathbf{M}_1$ is a $1 \times n$ matrix, Eq. (4-326) is written

$$\mathbf{M}_1 [\mathbf{B} \quad \mathbf{AB} \quad \mathbf{A}^2\mathbf{B} \ \ldots \ \mathbf{A}^{n-1}\mathbf{B}] = [0 \quad 0 \ \ldots \ 1] \tag{4-327}$$

Thus, $\mathbf{M}_1$ is written as

$$\begin{aligned} \mathbf{M}_1 &= [0 \quad 0 \ \ldots \ 1] [\mathbf{B} \quad \mathbf{AB} \quad \mathbf{A}^2\mathbf{B} \ \ldots \ \mathbf{A}^{n-1}\mathbf{B}]^{-1} \\ &= [0 \quad 0 \ \ldots \ 1] \mathbf{S}^{-1} \end{aligned} \tag{4-328}$$

if the matrix $\mathbf{S}$ is nonsingular. It will be shown later that the condition of $\mathbf{S}$ being nonsingular is the condition of complete state controllability. Once $\mathbf{M}_1$ is determined from Eq. (4-328), the matrix $\mathbf{M}$ is given in Eq. (4-321).

Theorem 4-3. The two systems represented by the state equations of Eqs. (4-303) and (4-307), which are related through the similar transformation of Eqs. (4-324) and (4-325), have the same eigenvalues.

Proof: The eigenvalues of the system in Eq. (4-303) are roots of the characteristic equation

$$|\lambda \mathbf{I} - \mathbf{A}| = 0 \tag{4-329}$$

For the system of Eq. (4-307), the characteristic equation is

$$|\lambda \mathbf{I} - \mathbf{A}_1| = |\lambda \mathbf{I} - \mathbf{MAM}^{-1}| = 0 \tag{4-330}$$

Equation (4-330) is written as

$$|\lambda \mathbf{MM}^{-1} - \mathbf{MAM}^{-1}| = 0 \tag{4-331}$$

or

$$|\mathbf{M}(\lambda \mathbf{I} - \mathbf{A})\mathbf{M}^{-1}| = 0 \tag{4-332}$$

Since the determinant of a product of matrices is equal to the product of the determinants, we have

$$|\mathbf{M}(\lambda \mathbf{I} - \mathbf{A})\mathbf{M}^{-1}| = |\mathbf{M}||\lambda \mathbf{I} - \mathbf{A}||\mathbf{M}^{-1}| = |\lambda \mathbf{I} - \mathbf{A}| \tag{4-333}$$

This proves that the characteristic equations of the two systems are identical, and thus their eigenvalues must be identical.

The significance of Theorem 4-3 is that, given any digital system with a single input and its state equation given by Eq. (4-303), if the matrix **S** of Eq. (4-304) is nonsingular, the eigenvalues of the system can be arbitrarily assigned by use of the state feedback $u(k) = -\mathbf{Gx}(k)$. Furthermore, we can transform the system into a phase-variable canonical form so that the constant feedback gains of $u(k) = -\mathbf{G}_1\mathbf{y}(k)$ of the transformed system can be easily determined once the eigenvalues are assigned. The feedback gains of the original system are finally determined from the relation of

$$\mathbf{G} = \mathbf{G}_1\mathbf{M} \tag{4-334}$$

For the general multiple-input case, the necessary and sufficient conditions for eigenvalue assignment of a system with state feedback is still that the matrix **S** ($n \times nr$) must be of rank n. The proof is mathematically more involved, and it is not given here.

The following example illustrates the eigenvalue-assignment design using the phase-variable canonical form.

Example 4-10.

Consider that the state equations of a second-order digital system are represented by

$$\mathbf{x}(k+1) = \mathbf{Ax}(k) + \mathbf{Bu}(k) \tag{4-335}$$

where

$$\mathbf{A} = \begin{bmatrix} 1 & -1 \\ 0 & 1 \end{bmatrix} \qquad \mathbf{B} = \begin{bmatrix} 1 \\ 1 \end{bmatrix} \tag{4-336}$$

Find the feedback gain matrix **G** with the state feedback $u(k) = -\mathbf{Gx}(k)$ such that the eigenvalues of the closed-loop system are at $\lambda_1 = 0.4$ and $\lambda_2 = 0.6$. The eigenvalues of **A** are $\lambda = 1, 1$, so that the state feedback moves these eigenvalues inside the unit circle in the z-plane, thus stabilizing the system.

Although it is possible to solve for the elements of **G** by equating the coefficients of the desired characteristic equation,

$$(\lambda - 0.4)(\lambda - 0.6) = \lambda^2 - \lambda + 0.24 = 0 \tag{4-337}$$

to those of

$$|\lambda \mathbf{I} - \mathbf{A} + \mathbf{BG}| = 0 \tag{4-338}$$

let us first transform the system into the phase-variable canonical form. From Eq. (4-304),

$$S = [B \quad AB] = \begin{bmatrix} 1 & 0 \\ 1 & 1 \end{bmatrix} \tag{4-339}$$

which is nonsingular. Then,

$$M_1 = [0 \quad 1]S^{-1} = [-1 \quad 1] \tag{4-340}$$

$$M = \begin{bmatrix} M_1 \\ M_1 A \end{bmatrix} = \begin{bmatrix} -1 & 1 \\ -1 & 2 \end{bmatrix} \tag{4-341}$$

For the system in canonical form,

$$y(k+1) = A_1 y(k) + B_1 u(k) \tag{4-342}$$

where

$$A_1 = MAM^{-1} = \begin{bmatrix} 0 & 1 \\ -1 & 2 \end{bmatrix} \tag{4-343}$$

$$B_1 = MB = \begin{bmatrix} 0 \\ 1 \end{bmatrix} \tag{4-344}$$

The characteristic equation of the transformed closed-loop system is

$$|\lambda I - A_1 + B_1 G_1| = 0 \tag{4-345}$$

where

$$G_1 = [g_1^* \quad g_2^*] \tag{4-346}$$

is the feedback matrix. Expanding Eq. (4-345) gives

$$\lambda^2 + (g_2^* - 2)\lambda + (g_1^* + 1) = 0 \tag{4-347}$$

Equating the corresponding coefficients of Eqs. (4-337) and (4-347), we have

$$G_1 = [g_1^* \quad g_2^*] = [-0.76 \quad 1] \tag{4-348}$$

The feedback matrix of the original system is determined from

$$G = G_1 M = [-0.76 \quad 1]\begin{bmatrix} -1 & 1 \\ -1 & 2 \end{bmatrix} = [-0.24 \quad 1.24] \tag{4-349}$$

It is a simple matter to show that the eigenvalues of $\mathbf{A} - \mathbf{BG}$ are at 0.4 and 0.6. Thus, the state feedback control that gives the desired eigenvalues is $u(k) = [0.24 \quad -1.24]\mathbf{x}(k)$.

4.17 THE STATE DIAGRAM

The signal-flow-graph method presented in Chapter 3 applies only to algebraic equations. Therefore, the conventional signal flow graph can be used only for the derivation of the input-output relations of a linear system in the transform domain. In this section we shall apply the method of the state transition signal flow graph to the representation of difference state equations. For simplicity, the *state transition signal flow graph* is called the *state diagram* in this text.

For a continuous-data system, the state diagram resembles the block diagram of the analog computer program. Therefore, once the state diagram is drawn, the problem can be solved either by an analog computer or by pencil and paper. For a digital system, the state diagram describes the operations of a hybrid or digital computer, so that the problem can again be solved either by machine or analytical methods.

State Diagram of Continuous-Data Systems

The fundamental linear operations that can be performed on an analog computer are *multiplication by a constant*, *addition*, *polarity inversion*, and *integration*. We shall now show that these computer operations are closely related to the elements of a state diagram.

1. *Multiplication by a constant.* Multiplication of a machine variable by a constant on an analog computer is done by potentiometers and amplifiers. Consider the operation

$$x_2(t) = ax_1(t) \tag{4-350}$$

where a is a constant. If a lies between zero and unity, a potentiometer is used for the purpose. If a is a negative integer less than or equal to -1, an operational amplifier is used to realize Eq. (4-350). The analog computer block diagram symbols of the potentiometer and the operational amplifier are shown in Figs. 4-5(a) and 4-5(b), respectively. Since Eq. (4-350) is an algebraic equation it can be represented by a signal flow graph as shown in Fig. 4-5(c). This signal flow graph can be regarded as a state diagram or an element of a state diagram if the variables $x_1(t)$ and $x_2(t)$ are state variables or linear combinations of state variables. Since the Laplace transform of Eq. (4-350) is

$$X_2(s) = aX_2(s) \tag{4-351}$$

the state diagram symbol of Fig. 4-5(c) also portrays the relation between

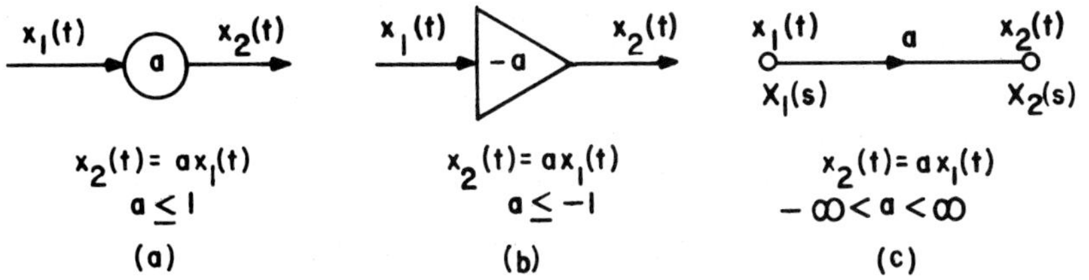

Fig. 4-5. (a) Block diagram symbol of a potentiometer; (b) block diagram symbol of an operational amplifier; (c) state diagram of the operation of multiplying a variable by a constant.

the transformed variables.

2. *Algebraic sum of two or more machine variables.* The algebraic sum of two or more machine variables of an analog computer is obtained by means of an operational amplifier as shown in Fig. 4-6(a). The illustrated case portrays the algebraic equation

$$x_4(t) = a_1 x_1(t) + a_2 x_2(t) + a_3 x_3(t) \tag{4-352}$$

where all the coefficients are assumed to be less than or equal to 0. The state diagram representation of Eq. (4-352) as well as its Laplace transform is shown in Fig. 4-6(b).

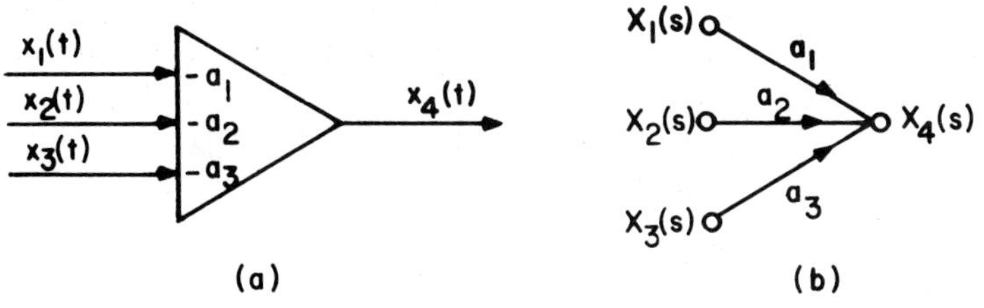

Fig. 4-6. (a) Block diagram symbol of the operational amplifier used as a summing device; (b) state diagram representation of the summing operation.

3. *Integration.* The integration of a machine variable on an analog computer is achieved by means of a computer element called the *integrator.* If $x_1(t)$ is the output of the integrator with the initial condition $x_1(t_0)$ given at $t = t_0$, and $x_2(t)$ is the input, the integrator performs the following operation:

$$x_1(t) = \int_{t_0}^{t} a x_2(\tau) d\tau + x_1(t_0) \tag{4-353}$$

where a is a constant $\leqslant 0$. The block diagram symbol of the integrator is shown in Fig. 4-7(a). In general, the integrator can be used simultaneously as a summing and amplifying device. To determine the state diagram symbol of the integrating operation we take the Laplace transform on both sides of Eq. (4-353), and we have

$$X_1(s) = a\frac{X_2(s)}{s} + a\int_{t_0}^{0} x_2(\tau)d\tau + \frac{x_1(t_0)}{s} \tag{4-354}$$

Since the past history of the integrator prior to t_0 is represented by $x_1(t_0)$, and the state transition is considered to begin at $t = t_0$, $x_2(\tau) = 0$ for $0 < \tau < t_0$. Thus, Eq. (4-354) becomes

$$X_1(s) = \frac{aX_2(s)}{s} + \frac{x_1(t_0)}{s} \qquad (t \geqslant t_0) \tag{4-355}$$

It should be pointed out here that the transformed equation of Eq. (4-355) is defined only for the period $t \geqslant t_0$. Therefore, the inverse Laplace transform of $X_1(s)$ should read

$$x_1(t) = \mathcal{L}^{-1}[X_1(s)] = a\int_{t_0}^{t} x_2(\tau)d\tau + x_1(t_0) \tag{4-356}$$

A state diagram representation of the integration operation is obtained from the signal-flow-graph description of Eq. (4-355). Two parallel versions of the state diagram symbols of an integrator are shown in Figs. 4-7(b) and 4-7(c).

Besides leading to computer solutions, a state diagram can also provide the following information on a linear dynamic system:

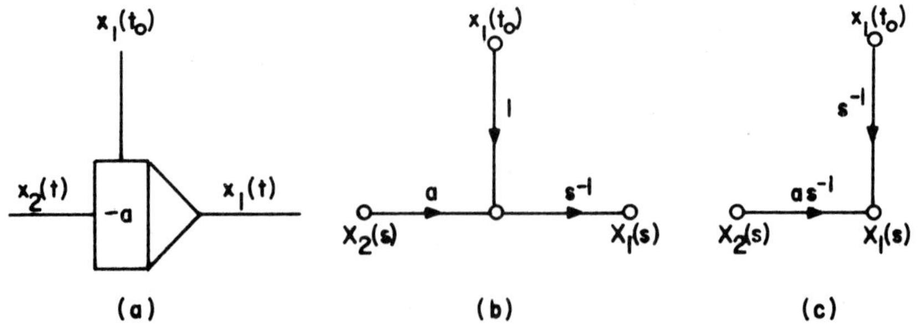

Fig. 4-7. (a) Block diagram representation of an integrator; (b), (c) state diagram symbols of an integrator.

1. Dynamic equations; that is, state equations and output equations
2. State transition equation
3. Transfer function
4. State variables.

We shall use the following numerical example to illustrate the use of the state diagram for analyzing linear continuous-data systems.

Example 4-11.

Consider that a linear dynamic system is described by the differential equation

$$\frac{d^2c(t)}{dt^2} + 3\frac{dc(t)}{dt} + 2c(t) = r(t) \qquad (4\text{-}357)$$

where c(t) is the output and the input r(t) is a unit-step function. The initial conditions of the system are represented by $c(t_0)$ and $dc(t_0)/dt$, all evaluated at $t = t_0$.

A state diagram of the system is drawn by first equating the highest-order derivative term to the rest of the terms in Eq. (4-357), as shown in Fig. 4-8. The state variables of the system, $x_1(t)$ and $x_2(t)$, are defined as output variables of the integrators. The state transition equations of the system in transformed form are written directly from the state diagram using Mason's gain formula.

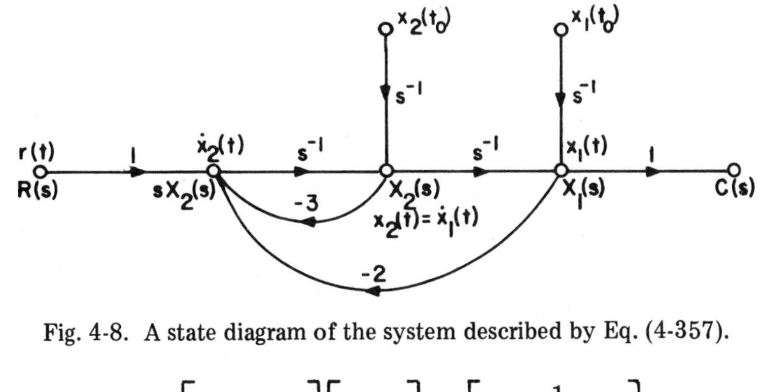

Fig. 4-8. A state diagram of the system described by Eq. (4-357).

$$\begin{bmatrix} X_1(s) \\ X_2(s) \end{bmatrix} = \frac{1}{(s+1)(s+2)}\begin{bmatrix} s+3 & 1 \\ -2 & s \end{bmatrix}\begin{bmatrix} x_1(t_0) \\ x_2(t_0) \end{bmatrix} + \begin{bmatrix} \dfrac{1}{(s+1)(s+2)} \\ \dfrac{s}{(s+1)(s+2)} \end{bmatrix} R(s) \qquad (4\text{-}358)$$

For a unit-step input, $R(s) = 1/s$, the inverse Laplace transform of Eq. (4-358) is

$$\begin{bmatrix} x_1(t) \\ x_2(t) \end{bmatrix} = \begin{bmatrix} 2e^{-(t-t_0)} - e^{-2(t-t_0)} & e^{-(t-t_0)} - e^{-2(t-t_0)} \\ -2e^{-(t-t_0)} + 2e^{-2(t-t_0)} & -e^{-(t-t_0)} + 2e^{-2(t-t_0)} \end{bmatrix}\begin{bmatrix} x_1(t_0) \\ x_2(t_0) \end{bmatrix}$$

$$+\begin{bmatrix} \frac{1}{2} - e^{-(t-t_0)} + \frac{1}{2} e^{-2(t-t_0)} \\ e^{-(t-t_0)} - e^{-2(t-t_0)} \end{bmatrix} \qquad t \geqslant t_0 \tag{4-359}$$

which is the state transition equation of the system.

Notice that in deriving the last equation the time origin is taken to be at $t = t_0$, so that the following inverse Laplace transform relation has been used:

$$\mathcal{L}^{-1}\left[\frac{1}{s+a}\right] = e^{-a(t-t_0)} \qquad t \geqslant t_0 \tag{4-360}$$

The reader may derive the state transition equation using the analytic expression in Eq. (4-46) for comparison.

The state equations of the system are obtained from the state diagram by applying Mason's gain formula to the nodes $\dot{x}_1(t)$ and $\dot{x}_2(t)$ with $r(t)$, $x_1(t)$, and $x_2(t)$ as inputs. The branches with the gain s^{-1} are deleted from the state diagram in this operation. Therefore, the dynamic equations of the system are written

$$\begin{bmatrix} \frac{dx_1(t)}{dt} \\ \frac{dx_2(t)}{dt} \end{bmatrix} = \begin{bmatrix} 0 & 1 \\ -2 & -3 \end{bmatrix} \begin{bmatrix} x_1(t) \\ x_2(t) \end{bmatrix} + \begin{bmatrix} 0 \\ 1 \end{bmatrix} r(t) \tag{4-361}$$

$$c(t) = [1 \quad 0] \begin{bmatrix} x_1(t) \\ x_2(t) \end{bmatrix} \tag{4-362}$$

The transfer function of the system is ordinarily obtained by taking the Laplace transform of Eq. (4-357). However, applying Mason's gain formula to the state diagram of Fig. 4-8 between R(s) and C(s) and setting the initial states to zero, we get

$$\frac{C(s)}{R(s)} = \frac{1}{s^2 + 3s + 2} \tag{4-363}$$

State Diagram of Digital Systems

When a digital system is described by difference equations or discrete state equations, a state diagram may be drawn to represent the relations between the discrete state variables. In contrast to the similarity between an analog computer diagram and a continuous-data state diagram, the digital state diagram portrays operations on a digital computer.

Some of the linear operations of a digital computer are: multiplication by a constant; addition of several variables; and time delay or storage. The mathematical descriptions of these basic digital operations and their corresponding z-transform expressions are given below:

1. *Multiplication by a constant*

$$x_2(kT) = ax_1(kT) \tag{4-364}$$

$$X_2(z) = aX_1(z) \tag{4-365}$$

2. *Summing*

$$x_3(kT) = x_1(kT) + x_2(kT) \tag{4-366}$$

$$X_3(z) = X_1(z) + X_2(z) \tag{4-367}$$

3. *Time delay or storage*

$$x_2(kT) = x_1[(k+1)T] \tag{4-368}$$

$$X_2(z) = zX_1(z) - zx_1(0) \tag{4-369}$$

or

$$X_1(z) = z^{-1}X_2(z) + x_1(0)$$

The state diagram representations and the corresponding digital computer diagrams of these operations are shown in Fig. 4-9. The following example serves to illustrate the construction and applications of the state diagram of a digital system.

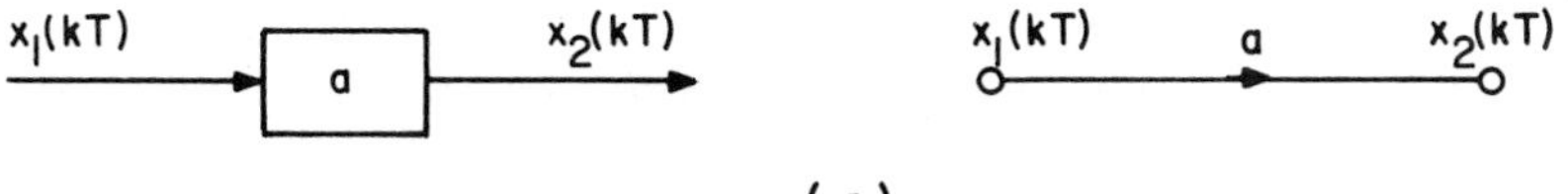

(a)

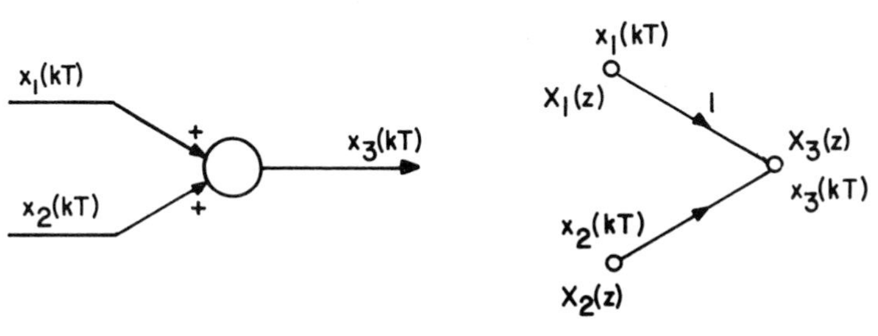

(b)

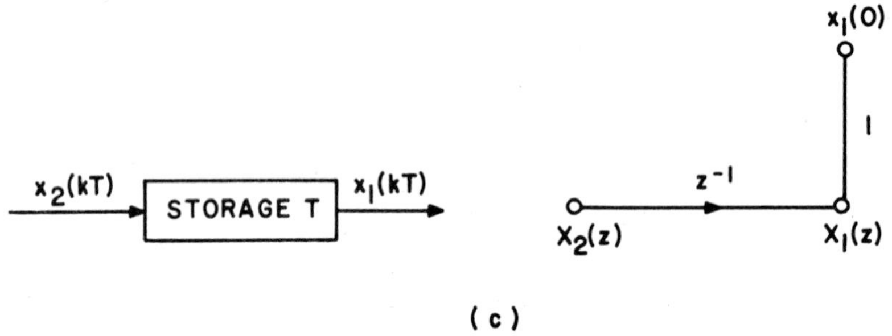

Fig. 4-9. Basic elements of a digital state diagram and the corresponding digital computer operations.

Example 4-12.

Consider that a digital system is described by the difference equation

$$c(k+2) + 2c(k+1) + 3c(k) = r(k) \tag{4-370}$$

The state diagram of the system is drawn as shown in Fig. 4-10 by first equating the highest-order term to the rest of the terms in Eq. (4-370).

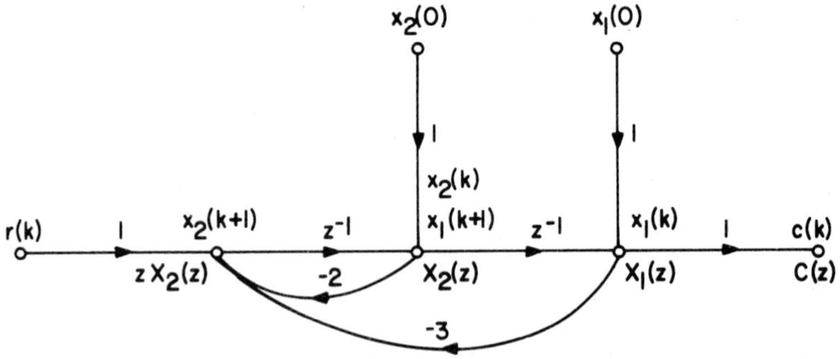

Fig. 4-10. A state diagram of the digital system described by Eq. (4-370).

The state variables of the system are designated as the output variables of the time delay units of the state diagram. Now neglecting the initial states and the branches with the gain z^{-1} on the state diagram, the state equations of the system are written

$$\begin{bmatrix} x_1(k+1) \\ x_2(k+1) \end{bmatrix} = \begin{bmatrix} 0 & 1 \\ -3 & -2 \end{bmatrix} \begin{bmatrix} x_1(k) \\ x_2(k) \end{bmatrix} + \begin{bmatrix} 0 \\ 1 \end{bmatrix} r(k) \tag{4-371}$$

and the output equation is

$$c(k) = [1 \quad 0]\begin{bmatrix} x_1(k) \\ x_2(k) \end{bmatrix} \tag{4-372}$$

The state transition equation of the system, which is the solution of the state equation in Eq. (4-371), is obtained from the state diagram using Mason's gain formula. Using $X_1(z)$ and $X_2(z)$ as output nodes, and $R(z)$, $x_1(0)$ and $x_2(0)$ as input nodes, we have

$$\begin{bmatrix} X_1(z) \\ X_2(z) \end{bmatrix} = \frac{1}{\Delta}\begin{bmatrix} 1 + 2z^{-1} & z^{-1} \\ -3z^{-1} & 1 \end{bmatrix}\begin{bmatrix} x_1(0) \\ x_2(0) \end{bmatrix} + \frac{1}{\Delta}\begin{bmatrix} z^{-2} \\ z^{-1} \end{bmatrix} R(z) \tag{4-373}$$

where

$$\Delta = 1 + 2z^{-1} + 3z^{-2} \tag{4-374}$$

Equation (4-373) is the state transition equation in the z-transform domain. The general form of the equation is given by Eq. (4-131), that is,

$$X(z) = (zI - A)^{-1}zx(0) + (zI - A)^{-1}BR(z) \tag{4-375}$$

One advantage of using the state diagram is that Eq. (4-373) is obtained simply by use of Mason's gain formula. This saves the effort of performing the matrix inverse of $(zI - A)$ which is required if Eq. (4-375) is used.

The state transition equation in the time domain is obtained from Eq. (4-373) by taking the inverse z-transform.

The transfer function between the output and the input of the system is determined from the state diagram,

$$\frac{C(z)}{R(z)} = \frac{X_1(z)}{R(z)} = \frac{z^{-2}}{1 + 2z^{-1} + 3z^{-2}} = \frac{1}{z^2 + 2z + 3} \tag{4-376}$$

4.18 DECOMPOSITION OF DIGITAL SYSTEMS

In general, the transfer function of a digital controller or system, $D(z)$, may be realized by pulsed-data RC networks, a general-purpose digital computer, or a special-purpose digital computer. Because of the advantages in computing speed, storage capacity, and flexibility, the use of digital computers in control systems has become increasingly important. Furthermore, due to the discrete nature of the signals received and processed by a digital controller, the realization of a discrete-data transfer function by a digital computer is quite natural.

The a priori requirement for the transfer function $D(z)$ before attempting implementation is that it must be physically realizable. The condition of physical realizability of any linear system implies that no output signal of the system can appear before an input signal is applied. Now consider that the transfer function D(z) of a digital controller is expressed as a ratio of two polynomials in z

$$D(z) = \frac{a_{m+1} z^m + a_m z^{m-1} + \ldots + a_1}{b_{n+1} z^n + b_n z^{n-1} + \ldots + b_1} \tag{4-377}$$

It is known that by expanding D(z) into a power series in z^{-1}, the coefficients of the series represent the values of the weighting sequence of the digital system. The coefficient of the z^{-k} term corresponds to the value of the weighting sequence at $t = kT$. Clearly, for the digital system to be physically realizable, the power series expansion of D(z) must not contain any positive power in z. Any positive power in z in the series simply indicates "prediction," or that the output precedes the input. Therefore, *for the D(z) in Eq. (4-377) to be a physically realizable transfer function, the highest power of the denominator must be equal to or greater than that of the numerator, or simply* $n \geqslant m$.

Quite often D(z) has an equal number of poles and zeros and is written

$$D(z) = \frac{a_{n+1} + a_n z^{-1} + \ldots + a_1 z^{-n}}{b_{m+1} + b_m z^{-1} + \ldots + b_1 z^{-m}} \tag{4-378}$$

where n and m are positive integers. In this case, the denominator of D(z) must not contain a common factor of z^{-1}, or, in Eq. (4-378), $b_{m+1} \neq 0$.

The implementation of a discrete-data transfer function by a digital computer can generally be effected in three different ways: direct decomposition, cascade decomposition, and parallel decomposition. These three methods of decomposition are illustrated in terms of state diagrams in the following.

1. Direct Decomposition

Consider that the transfer function of a digital controller is

$$D(z) = \frac{C(z)}{R(z)} = \frac{a_{n+1} + a_n z^{-1} + a_{n-1} z^{-2} + \ldots + a_1 z^{-n}}{b_{m+1} + b_m z^{-1} + b_{m-1} z^{-2} + \ldots + b_1 z^{-m}} \tag{4-379}$$

where $b_{m+1} \neq 0$, m and n are positive integers, and C(z) and R(z) are the z-transforms of the output and input of the controller, respectively. We shall find a state diagram for the system since it also represents the digital computer realization of the system.

Let us multiply the numerator and the denominator of the right-hand side of Eq. (4-379) by a variable X(z); we have

$$\frac{C(z)}{R(z)} = \frac{(a_{n+1} + a_n z^{-1} + a_{n-1} z^{-2} + \ldots + a_1 z^{-n})X(z)}{(b_{m+1} + b_m z^{-1} + b_{m-1} z^{-2} + \ldots + b_1 z^{-m})X(z)} \tag{4-380}$$

Now equating the numerators on both sides of the last equation gives

$$C(z) = (a_{n+1} + a_n z^{-1} + a_{n-1} z^{-2} + \ldots + a_1 z^{-n})X(z) \tag{4-381}$$

The same operation on the denominator brings

$$R(z) = (b_{m+1} + b_m z^{-1} + b_{m-1} z^{-2} + \ldots + b_1 z^{-m})X(z) \tag{4-382}$$

In order to construct a state diagram, Eq. (4-382) must first be written in a cause-and-effect relation. Therefore, solving for X(z) in Eq. (4-382) gives

$$X(z) = \frac{1}{b_{m+1}} R(z) - \frac{b_m}{b_{m+1}} z^{-1} X(z) - \frac{b_{m-1}}{b_{m+1}} z^{-2} X(z) - \ldots - \frac{b_1}{b_{m+1}} z^{-m} X(z) \tag{4-383}$$

The state diagram portraying Eqs. (4-382) and (4-383) is now drawn as shown in Fig. 4-11 for m = n = 3. For simplicity, the initial states are excluded from the state diagram. A digital computer program can easily be derived from the state diagram of Fig. 4-11, where the branches with the gain z^{-1} are realized by a time delay or storage of T seconds.

The state diagram of Fig. 4-11 can, of course, be used for analytical purposes. By defining the variables at the output nodes of all the time-delay units as state variables, the dynamic equations and the state transition equation can

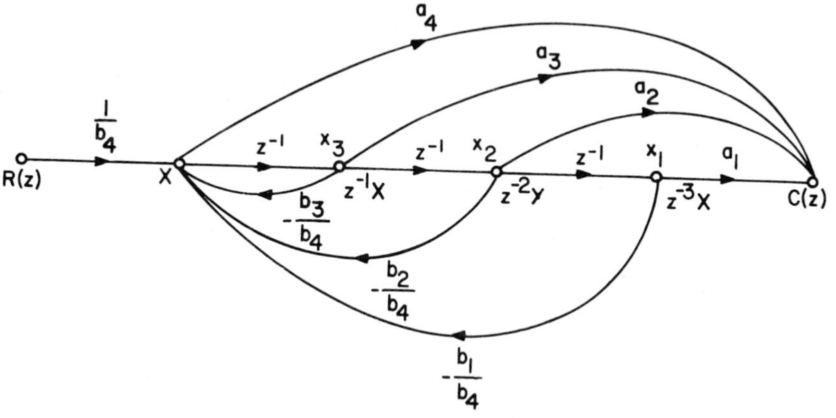

Fig. 4-11. State diagram representation of the transfer function of Eq. (4-379) with m = n = 3 by direct decomposition.

all be determined directly from the state diagram by applying the Mason's gain formula.

Since the discrete state equations are a set of first-order difference equations, the time-delay units with the branch gain of z^{-1} on the state diagram should be neglected when applying the gain formula.

Following the procedure described above and applying Mason's gain formula to the state diagram of Fig. 4-11, the state equations are written as

$$x_1(k+1) = x_2(k)$$

$$x_2(k+1) = x_3(k)$$

$$x_3(k+1) = -\frac{b_1}{b_4}x_1(k) - \frac{b_2}{b_4}x_2(k) - \frac{b_3}{b_4}x_3(k) + \frac{1}{b_4}r(k) \tag{4-384}$$

From these results we can conclude that direct decomposition will always yield a state-variable model for the system in the phase-variable canonical form.

2. Cascade Decomposition

If a transfer function D(z) is given in factored form, it can be written as the product of a number of first-order transfer functions, each realizable by a simple digital program or state diagram. The entire D(z) is then represented by the cascade connections of the digital programs or state diagrams of the first-order transfer functions.

Consider that the transfer function of a digital system is written as

$$D(z) = \frac{C(z)}{R(z)} = \frac{K(z+c_1)(z+c_2)\dots(z+c_m)}{(z+d_1)(z+d_2)\dots(z+d_n)} \tag{4-385}$$

where $n \geqslant m$; $-c_i$, $i = 1,2,\dots,m$, and $-d_j$, $j = 1,2,\dots,n$, are the zeros and poles of D(z), respectively. These poles and zeros, in general, can be real or complex numbers, although the computation of complex numbers in a digital program may be inconvenient. Therefore, the factored form of D(z) given in Eq. (4-385) is best suited for real poles and zeros.

Writing D(z) as the product of the constant K and n first-order transfer functions, we get

$$D(z) = KD_1(z)D_2(z)\dots D_n(z) \tag{4-386}$$

where

$$D_k(z) = \frac{z+c_k}{z+d_k} = \frac{1+c_kz^{-1}}{1+d_kz^{-1}} \qquad k = 1,2,\dots,m \tag{4-387}$$

and

$$D_k(z) = \frac{1}{z + d_k} = \frac{z^{-1}}{1 + d_k z^{-1}} \qquad k = m + 1, m + 2, \ldots, n \qquad (4\text{-}388)$$

The state diagram representation of Eq. (4-387) is shown in Fig. 4-12(a) and that of Eq. (4-388) is shown in Fig. 4-12(b). The overall digital program for D(z) is obtained by connecting an element with gain $K = a_0/b_0$ in series with the first-order digital programs of $D_k(z)$ shown in Fig. 4-12.

Once the state diagram is completed, the state equation and the output equation can be written directly from the state diagram by use of Mason's gain formula.

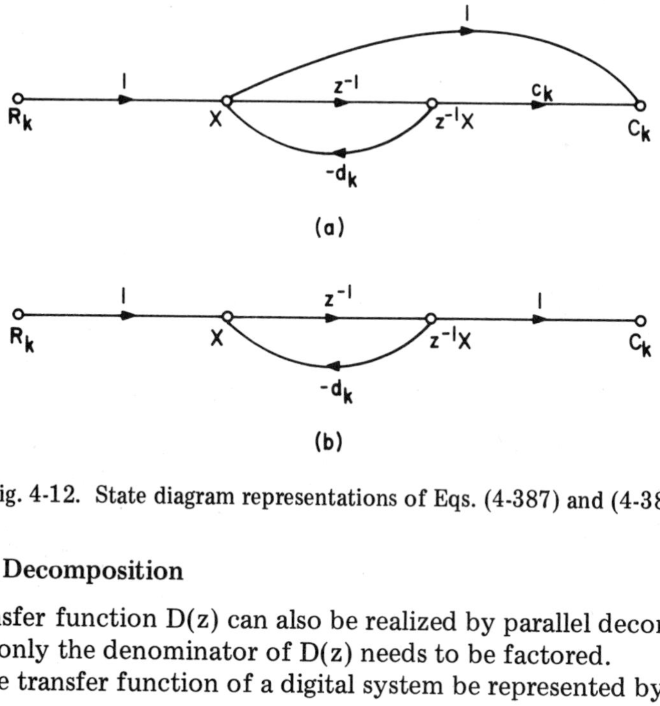

Fig. 4-12. State diagram representations of Eqs. (4-387) and (4-388).

3. Parallel Decomposition

A transfer function D(z) can also be realized by parallel decomposition, in which case only the denominator of D(z) needs to be factored.

Let the transfer function of a digital system be represented by

$$D(z) = \frac{C(z)}{R(z)} = K \frac{z^m + a_m z^{m-1} + \ldots + a_2 z + a_1}{z^n + b_n z^{n-1} + \ldots + b_2 z + b_1} \qquad (4\text{-}389)$$

where $n > m$. Assuming that there are no pole-zero cancellations in D(z), and that among the n eigenvalues i are distinct and the rest are of multiplicity $n - i$, then, by partial fraction expansion, D(z) is written as

$$D(z) = \sum_{k=1}^{i} \frac{K_k}{z + d_k} + \sum_{k=i+1}^{n} \frac{K_k}{(z + d_k)^{k-i}} \tag{4-390}$$

|← distinct eigenvalues →|← multiple-order eigenvalues →|

where d_k, $k = 1,2,\ldots,i$ are the distinct eigenvalues, and $d_{i+1} = d_{i+2} = \ldots = d_n$ denotes the eigenvalue of multiplicity $n - i$.

For state diagram representation, we rewrite Eq. (4-390) as

$$D(z) = \sum_{k=1}^{i} \frac{K_k z^{-1}}{1 + d_k z^{-1}} + \sum_{k=i+1}^{n} \frac{K_k z^{i-k}}{(1 + d_k z^{-1})^{k-i}} \tag{4-391}$$

The transfer function D(z) is now decomposed to a state diagram which consists of basic building blocks of the form of Fig. 4-12(b), all connected in parallel. It is important to point out that the parallel decomposition will lead to a set of state equations which is in the canonical form for distinct eigenvalues, or Jordan canonical form in general. The following example illustrates the important features of the parallel decomposition.

Example 4-13.

Consider the transfer function

$$D(z) = \frac{C(z)}{R(z)} = \frac{10(z^2 + z + 1)}{z^2(z - 0.5)(z - 0.8)} \tag{4-392}$$

which has eigenvalues at $z = 0, 0, 0.5$, and 0.8. We want to decompose this transfer function by parallel decomposition, and then determine the dynamic equations of the system.

Performing partial fraction expansion on D(z), we have

$$D(z) = \frac{-233.33}{z - 0.5} + \frac{127.08}{z - 0.8} + \frac{25}{z^2} + \frac{106.25}{z}$$

The transfer function of D(z) is realized by the parallel connection of first-order components shown in Fig. 4-13.

It should be noted that, since D(z) is of the fourth order, there should be only four time-delay units in the state diagram. Figure 4-13 also represents the minimum-order realization of the transfer function. The state equations of the system are now written with the established technique. We have

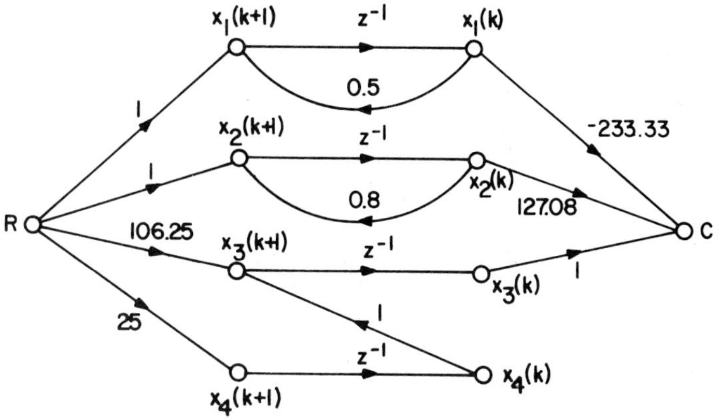

Fig. 4-13. State diagram of Eq. (4-392) by parallel decomposition.

$$\begin{bmatrix} x_1(k+1) \\ x_2(k+1) \\ x_3(k+1) \\ x_4(k+1) \end{bmatrix} = \begin{bmatrix} 0.5 & 0 & 0 & 0 \\ 0 & 0.8 & 0 & 0 \\ 0 & 0 & 0 & 1 \\ 0 & 0 & 0 & 0 \end{bmatrix} \begin{bmatrix} x_1(k) \\ x_2(k) \\ x_3(k) \\ x_4(k) \end{bmatrix} + \begin{bmatrix} 1 \\ 1 \\ 106.25 \\ 25 \end{bmatrix} r(kT) \tag{4-393}$$

which is recognized as Jordan canonical form. The output equation is written as

$$c(k) = [-233.33 \quad 127.08 \quad 1 \quad 0]\mathbf{x}(k) \tag{4-394}$$

4.19 STATE DIAGRAMS OF SAMPLED-DATA CONTROL SYSTEMS

A sampled-data control system usually contains digital as well as analog elements. The two types of elements are coupled together by sample-and-hold devices. As an illustration the block diagram of a sampled-data control system is shown in Fig. 4-14. The system is composed of a digital controller and zero-order hold, and a continuous-data process which is to be controlled. We shall show how state diagrams and a state variable analysis of digital systems of this type can be obtained. However, before the state diagram for the entire system in Fig. 4-14 can be drawn, we must establish the state diagram representation of the zero-order hold.

State Diagram of the Zero-Order Hold

Let the input and the output of the zero-order hold be denoted by $e^*(t)$ and $h(t)$, respectively. Then, for the time interval $kT \leqslant t < (k+1)T$,

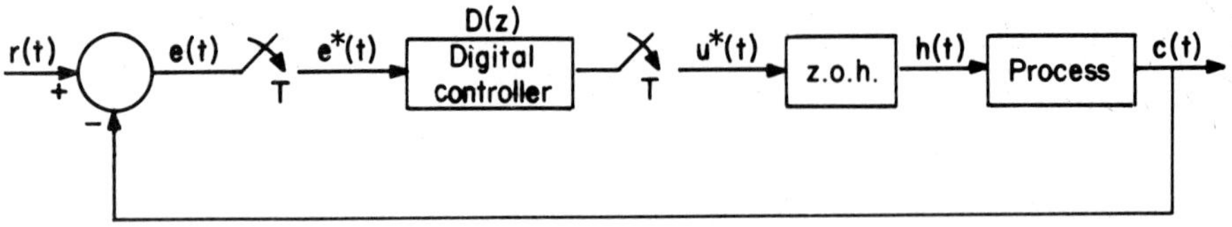

Fig. 4-14. A sampled-data control system.

$$h(t) = e(kT) \tag{4-395}$$

Taking the Laplace transform on both sides of the last equation, we get

$$H(s) = \frac{e(kT)}{s} \tag{4-396}$$

for $kT \leqslant t < (k + 1)T$. Therefore, the state diagram of the zero-order hold consists of a single branch connected between the nodes e(kT) and H(s) as shown in Fig. 4-15. The gain of the branch is s^{-1}.

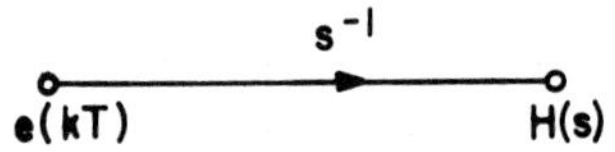

Fig. 4-15. State diagram of the zero-order hold for $kT \leqslant t < (k + 1)T$.

Example 4-14.

Consider the sampled-data control system shown in Fig. 4-16. We want to construct a state diagram and write the state equations for the system.

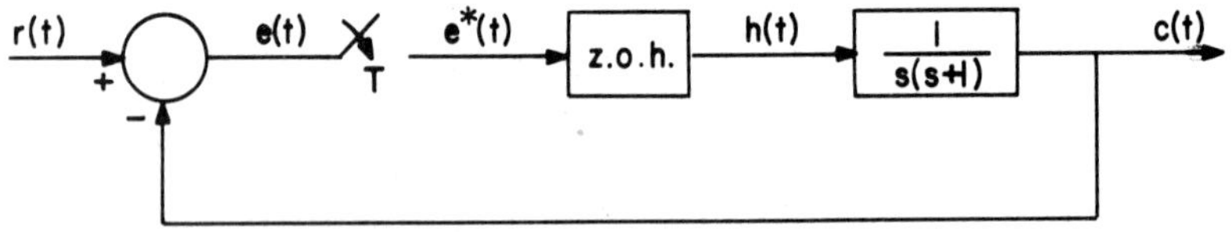

Fig. 4-16. A sampled-data control system.

Using the method of direct decomposition, the transfer function of the controlled process

$$G(s) = \frac{1}{s(s + 1)} \tag{4-397}$$

is portrayed by the state diagram of Fig. 4-17. The state diagram of the overall system is constructed by connecting the state diagram of Fig. 4-17 with that of the zero-order hold and using the relationship

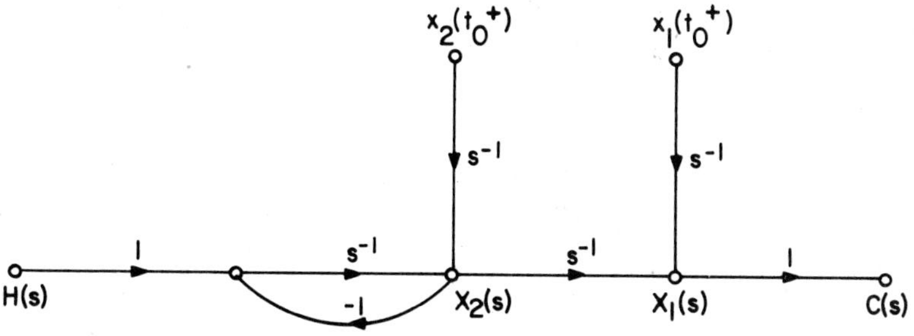

Fig. 4-17. A state diagram of G(s) = 1/s(s + 1).

$$\begin{aligned} e(kT) &= r(kT) - c(kT) \\ &= r(kT) - x_1(kT) \end{aligned} \tag{4-398}$$

Also, letting $t_0 = kT$, and

$$h(kT^+) = h(t) = e(kT) \qquad kT \leqslant t < (k+1)T \tag{4-399}$$

the complete state diagram of the system is shown in Fig. 4-18.

The state transition equations in transformed vector-matrix form are written directly by inspection from the state diagram using Mason's gain formula.

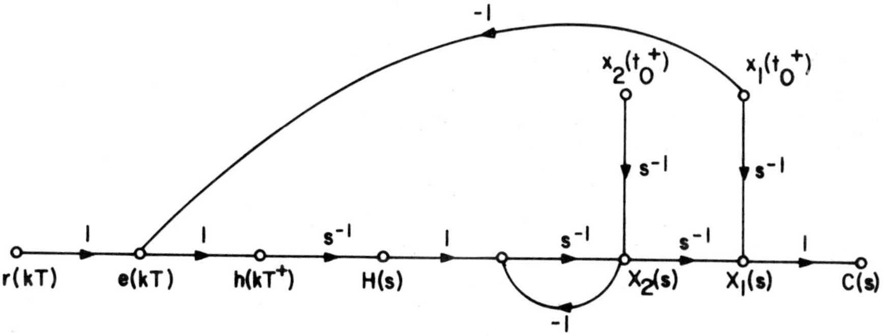

Fig. 4-18. A state diagram of the sampled-data control system shown in Fig. 4-16.

$$\begin{bmatrix} X_1(s) \\ X_2(s) \end{bmatrix} = \begin{bmatrix} \dfrac{1}{s} - \dfrac{1}{s^2(s+1)} & \dfrac{1}{s(s+1)} \\ -\dfrac{1}{s(s+1)} & \dfrac{1}{s+1} \end{bmatrix} \begin{bmatrix} x_1(kT) \\ x_2(kT) \end{bmatrix} + \begin{bmatrix} \dfrac{1}{s^2(s+1)} \\ \dfrac{1}{s(s+1)} \end{bmatrix} r(kT) \tag{4-400}$$

for $kT \leqslant t \leqslant (k+1)T$.

Taking the inverse Laplace transform on both sides of the last equation, we get

$$\begin{bmatrix} x_1(t) \\ x_2(t) \end{bmatrix} = \begin{bmatrix} 2-(t-kT)-e^{-(t-kT)} & 1-e^{-(t-kT)} \\ -1+e^{-(t-kT)} & e^{-(t-kT)} \end{bmatrix} \begin{bmatrix} x_1(kT) \\ x_2(kT) \end{bmatrix} + \begin{bmatrix} (t-kT)-1+e^{-(t-kT)} \\ 1-e^{-(t-kT)} \end{bmatrix} r(kT) \tag{4-401}$$

for $kT \leqslant t \leqslant (k+1)T$.

If only the responses at the sampling instants are of interest, we let $t = (k+1)T$; then Eq. (4-401) becomes

$$\begin{bmatrix} x_1(k+1)T \\ x_2(k+1)T \end{bmatrix} = \begin{bmatrix} 2-T-e^{-T} & 1-e^{-T} \\ -1+e^{-T} & e^{-T} \end{bmatrix} \begin{bmatrix} x_1(kT) \\ x_2(kT) \end{bmatrix} + \begin{bmatrix} T-1+e^{-T} \\ 1-e^{-T} \end{bmatrix} r(kT) \tag{4-402}$$

Notice that Eq. (4-402) is of the form of Eq. (4-83):

$$\mathbf{x}[(k+1)T] = \phi(T)\mathbf{x}(kT) + \theta(T)\mathbf{r}(kT) \tag{4-403}$$

For a sampling period of one second, $T = 1$ sec, and a unit-step input, $r(t) = u_s(t)$, we have

$$\phi(1) = \begin{bmatrix} 0.632 & 0.632 \\ -0.632 & 0.368 \end{bmatrix} \tag{4-404}$$

and

$$\theta(1) = \begin{bmatrix} 0.368 \\ 0.632 \end{bmatrix} \tag{4-405}$$

Then,

$$[z\mathbf{I} - \phi(1)]^{-1}z = \begin{bmatrix} z-0.632 & -0.632 \\ 0.632 & z-0.368 \end{bmatrix}^{-1} z$$

$$= \frac{z}{z^2 - z + 0.632} \begin{bmatrix} z-0.368 & 0.632 \\ -0.632 & z-0.632 \end{bmatrix} \tag{4-406}$$

and

$$\phi(k) = \mathcal{Z}^{-1}\left[[z\mathbf{I} - \phi(1)]^{-1}z\right]$$

$$= \begin{bmatrix} e^{-0.23k}(-0.378\sin 0.88k + \cos 0.88k) & e^{-0.23k}\sin 0.88k \\ -e^{-0.23k}\sin 0.88k & e^{-0.23k}(-0.786\sin 0.88k + \cos 0.88k) \end{bmatrix} \tag{4-407}$$

Also,

$$\phi[(N-k-1)]\theta(1) = \begin{bmatrix} e^{-0.23(N-k-1)}[0.493\sin 0.88(N-k-1) + 0.368\cos 0.88(N-k-1)] \\ e^{-0.23(N-k-1)}[-0.865\sin 0.88(N-k-1) + 0.632\cos 0.88(N-k-1)] \end{bmatrix} \tag{4-408}$$

Therefore, the discrete state transition equation of the system is

$$\mathbf{x}(N) = \begin{bmatrix} e^{-0.23N}(-0.378\sin 0.88N + \cos 0.88N) & e^{-0.23N}\sin 0.88N \\ -e^{-0.23N}\sin 0.88N & e^{-0.23N}(-0.786\sin 0.88N + \cos 0.88N) \end{bmatrix}$$

$$\times \begin{bmatrix} x_1(0) \\ x_2(0) \end{bmatrix} + \sum_{k=0}^{N-1} \begin{bmatrix} e^{-0.23(N-k-1)}[0.493\sin 0.88(N-k-1) + 0.368\cos 0.88(N-k-1)] \\ e^{-0.23(N-k-1)}[-0.865\sin 0.88(N-k-1) + 0.632\cos 0.88(N-k-1)] \end{bmatrix} \tag{4-409}$$

for $N = 1,2,3,\ldots$.

Example 4-15.

In this example we shall conduct a state analysis of a sampled-data control system which has a digital controller. Let us consider the block diagram shown in Fig. 4-14. The digital controller, which actually may be a digital computer, is described by

$$D(z) = \frac{a_2 + a_1 z^{-1}}{1 + b_1 z^{-1}} \tag{4-410}$$

and G(s) is as given in Eq. (4-397).

We want to draw a state diagram and to obtain the state transition equations for the system.

Applying the direct decomposition scheme to D(z), we have

$$D(z) = \frac{U(z)}{E(z)} = \frac{(a_2 + a_1 z^{-1})X(z)}{(1 + b_1 z^{-1})X(z)} \tag{4-411}$$

Let

$$U(z) = (a_2 + a_1 z^{-1})X(z) \tag{4-412}$$

and

$$E(z) = (1 + b_1 z^{-1})X(z) \tag{4-413}$$

From Eq. (4-413), we have

$$X(z) = E(z) - b_1 z^{-1} X(z) \tag{4-414}$$

The state diagram for the digital controller is constructed as shown in Fig. 4-19 using Eqs. (4-412) and (4-414).

From Fig. 4-19, the dynamic equations which characterize D(z) are written

$$x_3(k+1)T = e(kT) - b_1 x_3(kT) \qquad \text{(State Equation)} \tag{4-415}$$

$$u(kT) = a_2 e(kT) + (a_1 - a_2 b_1) x_3(kT) \qquad \text{(Output Equation)} \tag{4-416}$$

The state diagram for G(s) and the zero-order hold were established earlier in Example 4-14. The state diagram for the entire system is now drawn as shown in Fig. 4-20.

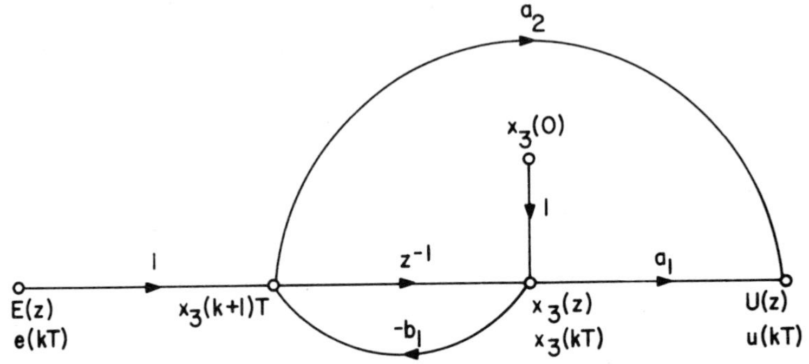

Fig. 4-19. State diagram for $D(z) = (a_2 + a_1 z^{-1})/(1 + b_1 z^{-1})$.

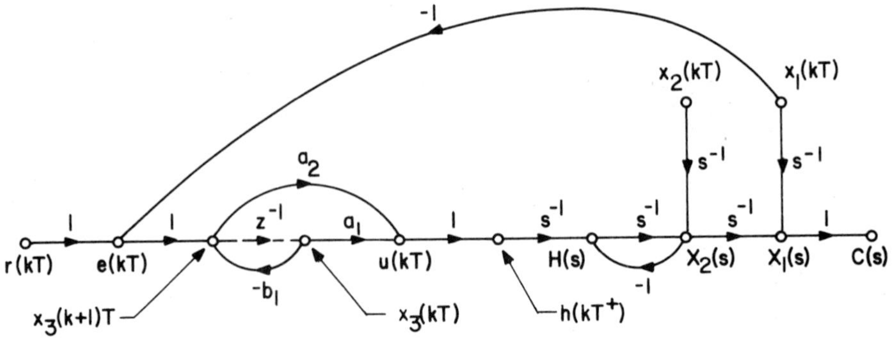

Fig. 4-20. State diagram for the sampled-data control system shown in Fig. 4-14.

Applying Mason's gain formula to Fig. 4-18 with $X_1(s)$ and $X_2(s)$ as output nodes, we get

$$X_1(s) = \frac{1}{s} - \frac{a_2}{s^2(s+1)} x_1(kT) + \frac{1}{s(s+1)} x_2(kT) + \frac{a_1 - a_2 b_1}{s^2(s+1)} x_3(kT) + \frac{a_2}{s^2(s+1)} r(kT) \tag{4-417}$$

$$X_2(s) = \frac{-a_2}{s(s+1)} x_1(kT) + \frac{1}{s+1} x_2(kT) + \frac{a_1 - a_2 b_1}{s(s+1)} x_3(kT) + \frac{a_2}{s(s+1)} r(kT) \tag{4-418}$$

Notice that $x_1(kT)$, $x_2(kT)$, $x_3(kT)$, and $r(kT)$ are regarded as inputs in this case. Also applying Mason's formula to $x_3(k+1)T$, we have

$$x_3(k+1)T = -x_1(kT) - b_1 x_3(kT) + r(kT) \tag{4-419}$$

Equation (4-419), together with Eqs. (4-417) and (4-418) when the latter equations are inverse-transformed and with $t = (k+1)T$, form the discrete state equations of the system.

$$\begin{bmatrix} x_1(k+1)T \\ x_2(k+1)T \\ x_3(k+1)T \end{bmatrix} = \begin{bmatrix} 1 - a_2(T - 1 + e^{-T}) & 1 - e^{-T} & (a_1 - a_2 b_1)(T - 1 + e^{-T}) \\ -a_2(1 - e^{-T}) & e^{-T} & (a_1 - a_2 b_1)(1 - e^{-T}) \\ -1 & 0 & -b_1 \end{bmatrix} \begin{bmatrix} x_1(kT) \\ x_2(kT) \\ x_3(kT) \end{bmatrix}$$
$$+ \begin{bmatrix} a_2(T - 1 + e^{-T}) \\ a_2(1 - e^{-T}) \\ 1 \end{bmatrix} r(kT) \tag{4-420}$$

The output equation is $c(kT) = x_1(kT)$.

Equation (4-420) can now be solved by means of the same method as described in the last example.

4.20 RESPONSE OF SAMPLED-DATA SYSTEMS BETWEEN SAMPLING INSTANTS USING THE STATE APPROACH

The state-variable method may be applied to the evaluation of systems response between sampling instants. The method represents an alternative to the modified z-transform method.

From Eq. (4-46), the state vector $\mathbf{x}(t)$ for any $t > t_0$ is represented by

$$\mathbf{x}(t) = \phi(t - t_0)\mathbf{x}(t_0) + \int_{t_0}^{t} \phi(t - \tau)\mathbf{B}\mathbf{u}(\tau)d\tau \tag{4-421}$$

when $\mathbf{x}(t_0)$ and $\mathbf{u}(\tau)$ for $t \geqslant t_0$ are specified. When $\mathbf{u}(t)$ is constant for $t_0 \leqslant \tau < t$, Eq. (4-421) is written as

$$\mathbf{x}(t) = \phi(t - t_0)\mathbf{x}(t_0) + \theta(t - t_0)\mathbf{u}(t_0) \tag{4-422}$$

where

$$\theta(t - t_0) = \int_{t_0}^{t} \phi(t - \tau)\,\mathbf{B}d\tau$$

and

$$\mathbf{u}(t_0) = \mathbf{u}(\tau) \qquad t_0 \leqslant \tau < t$$

Now if the response in between the sampling instants is desired, we let

$$t = kT + \Delta T = (k + \Delta)T \tag{4-423}$$

where $k = 0,1,2,\ldots$, and $0 \leqslant \Delta \leqslant 1$. Then, Eq. (4-422) gives

$$\mathbf{x}(kT + \Delta T) = \phi(\Delta T)\mathbf{x}(kT) + \theta(\Delta T)\mathbf{u}(kT) \tag{4-424}$$

where the initial time has been taken as $t_0 = kT$.

Thus, by varying the value of Δ between 0 and 1, essentially all information on $\mathbf{x}(t)$ for all t can be obtained.

Example 4-16.

For the system described in Example 4-14, the state transition equation in Eq. (4-402) gives values of the state variables at the sampling instants only. However, letting $t = (k + \Delta)T$, Eq. (4-402) becomes

$$\begin{bmatrix} x_1(kT + \Delta T) \\ x_2(kT + \Delta T) \end{bmatrix} = \begin{bmatrix} 2 - \Delta T - e^{-\Delta T} & 1 - e^{-\Delta T} \\ -1 + e^{-\Delta T} & e^{-\Delta T} \end{bmatrix} \begin{bmatrix} x_1(kT) \\ x_2(kT) \end{bmatrix} + \begin{bmatrix} \Delta T - 1 + e^{-\Delta T} \\ 1 - e^{-\Delta T} \end{bmatrix} r(kT) \tag{4-425}$$

To complete the iteration procedure, it is necessary to let $t = (k + 1)T$ and $t_0 = (k + \Delta)T$, and Eq. (4-402) becomes

$$\mathbf{x}[(k + 1)T] = \phi(T - \Delta T)\mathbf{x}[(k + \Delta)T] + \theta(T - \Delta T)\mathbf{r}[(k + \Delta)T] \tag{4-426}$$

The state transition equation for the system between $t_0 = (k + \Delta)T$ and $t = (k + 1)T$ is

$$\begin{bmatrix} x_1[(k+1)T] \\ x_2[(k+1)T] \end{bmatrix} = \begin{bmatrix} 2-(1-\Delta)T-e^{-(1-\Delta)T} & 1-e^{-(1-\Delta)T} \\ -1+e^{-(1-\Delta)T} & e^{-(1-\Delta)T} \end{bmatrix} \begin{bmatrix} x_1[(k+\Delta)T] \\ x_2[(k+\Delta)T] \end{bmatrix} + \begin{bmatrix} (1-\Delta)T-1+e^{-(1-\Delta)T} \\ 1-e^{-(1-\Delta)T} \end{bmatrix} r[(k+\Delta)T] \qquad (4\text{-}427)$$

4.21 ANALYSIS OF DISCRETE-TIME SYSTEMS WITH MULTIRATE, SKIP-RATE, AND NONSYNCHRONOUS SAMPLINGS

The state-variable method discussed in the preceding sections can readily be applied to digital systems with nonuniform sampling schemes. The analysis of digital systems with nonuniform sampling by means of the z-transform method usually becomes difficult and special equivalent models with uniform sampling must first be made. However, the state-variable method offers a unified approach to the analysis of a wide class of digital systems.

The sampling scheme shown in Fig. 4-21 is often referred to as a cyclic sampling operation. In general, the transition of state of a digital system may be considered to take place from t_0 to $t_0 + \tau$, where $0 < \tau \leqslant T$, and then from $t = \tau$ to T. For the cyclic sampling, we let $\tau = T_1$; the state transition equation of the system at $t = kT + T_1$ is written as

$$\mathbf{x}[(kT + T_1)] = \phi(T_1)\mathbf{x}(kT) + \theta(T_1)\mathbf{r}(kT) \qquad (4\text{-}428)$$

Then, at $t = (k+1)T$ the state transition equation is

$$\mathbf{x}[(k+1)T] = \phi(T - T_1)\mathbf{x}[(kT + T_1)] + \theta(T - T_1)\mathbf{r}[(kT + T_1)] \qquad (4\text{-}429)$$

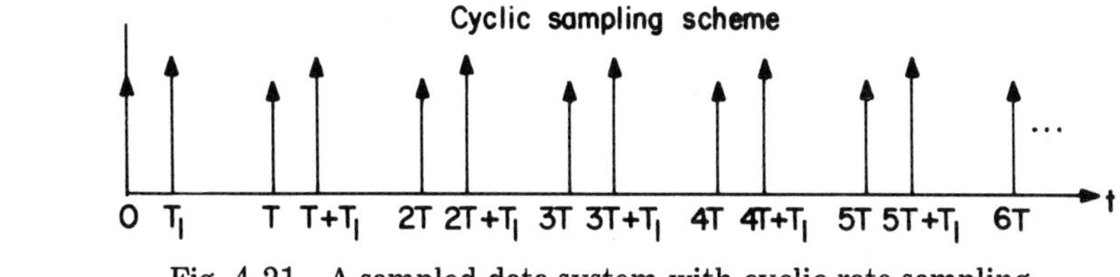

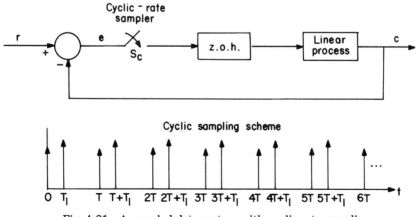

Fig. 4-21. A sampled-data system with cyclic-rate sampling.

Substituting Eq. (4-428) into Eq. (4-429) yields

$$\mathbf{x}[(k+1)T] = \phi(T-T_1)\phi(T_1)\mathbf{x}(kT) + \phi(T-T_1)\theta(T_1)\mathbf{r}(kT) + \theta(T-T_1)\mathbf{r}(kT+T_1) \qquad (4\text{-}430)$$

In general, if the cyclic sampler samples at $t = kT, kT + T_1, kT + T_1 + T_2, \ldots, (k+1)T, (k+1)T + T_1, (k+1)T + T_1 + T_2, \ldots, (k+1)T + T_1 + T_2 + \ldots + T_q, \ldots$ so that $T = T_1 + T_2 + \ldots + T_q$, then

$$\mathbf{x}(kT+T_1) = \phi(T_1)\mathbf{x}(kT) + \theta(T_1)\mathbf{r}(kT) \qquad (4\text{-}431)$$

$$\mathbf{x}(kT+T_1+T_2) = \phi(T_2)\mathbf{x}(kT+T_1) + \theta(T_2)\mathbf{r}(kT+T_1) \qquad (4\text{-}432)$$

$$\vdots$$

$$\mathbf{x}(kT+T_1+T_2+\ldots+T_{q-1}) = \phi(T_{q-1})\mathbf{x}(kT+T_1+T_2+\ldots+T_{q-2}) + \theta(T_{q-1})\mathbf{r}(kT+T_1+T_2+\ldots+T_{q-2}) \qquad (4\text{-}433)$$

At $t = (k+1)T$,

$$\mathbf{x}[(k+1)T] = \phi(T_q)\mathbf{x}(kT+T_1+T_2+\ldots+T_{q-1}) + \theta(T_q)\mathbf{r}(kT+T_1+T_2+\ldots+T_{q-1}) \qquad (4\text{-}434)$$

Substituting Eq. (4-431) into Eq. (4-432) and so on, we see that Eq. (4-433) can be written as

$$\begin{aligned}\mathbf{x}[(k+1)T] &= \phi(T_q)\phi(T_{q-1})\ldots\phi(T_2)\phi(T_1)\mathbf{x}(kT)\\ &+ \phi(T_q)\phi(T_{q-1})\ldots\phi(T_2)\theta(T_1)\mathbf{r}(kT)\\ &+ \phi(T_q)\phi(T_{q-1})\ldots\phi(T_3)\theta(T_2)\mathbf{r}(kT+T_1)\\ &+ \ldots + \phi(T_q)\theta(T_{q-1})\mathbf{r}(kT+T_1+T_2+\ldots+T_{q-2})\\ &+ \theta(T_q)\mathbf{r}(kT+T_1+T_2+\ldots+T_{q-1})\end{aligned} \qquad (4\text{-}435)$$

Example 4-17.

Consider that in the system of Fig. 4-21, the linear process is described by

$$G(s) = \frac{4}{s+1} \qquad (4\text{-}436)$$

and $T = 1$ sec, $T_1 = 0.25$ sec. The input signal is a unit-step function applied at $t = 0$. The system is initially at rest and $c(0) = 0$. The output response of the system for $t > 0$ is desired.

The periods of state transition in the case are $0 \leqslant t \leqslant T_1$ and $T_1 \leqslant t \leqslant T$, and, in general, $kT \leqslant t \leqslant kT + T_1$ and $kT + T_1 \leqslant t \leqslant (k + 1)T$. The state diagram of the system for $t \geqslant t_0$, where t_0 can be kT or $kT + T_1$, is drawn as shown in Fig. 4-22. The state transition equation in the Laplace domain is written as

$$X_1(s) = \frac{1}{s+1} x_1(t_0^+) - \frac{4}{s(s+1)} x_1(t_0^+) + \frac{4}{s(s+1)} r(t_0^+) \tag{4-437}$$

The inverse Laplace transform of $X_1(s)$ is

$$x_1(t) = e^{-(t-t_0)} x_1(t_0^+) - 4[1 - e^{-(t-t_0)}] x_1(t_0^+) + 4[1 - e^{-(t-t_0)}] r(t_0^+) \tag{4-438}$$

Therefore,

$$\phi(t - t_0) = -4 + 5e^{-(t-t_0)} \tag{4-439}$$

$$\theta(t - t_0) = 4[1 - e^{-(t-t_0)}] \tag{4-440}$$

From Eq. (4-438), with $t = kT + T_1$ and $t_0 = kT$, we have

$$x_1(kT + T_1) = \phi(T_1)x_1(kT) + \theta(T_1)r(kT)$$

$$= (-4 + 5e^{-T_1})x_1(kT) + 4(1 - e^{-T_1})r(kT) \tag{4-441}$$

In the same equation with $t = (k + 1)T$ and $t_0 = kT + T_1$, we get

$$x_1[(k+1)T] = [-4 + 5e^{-(T-T_1)}]x_1(kT + T_1) + 4[1 - e^{-(T-T_1)}]r(kT + T_1) \tag{4-442}$$

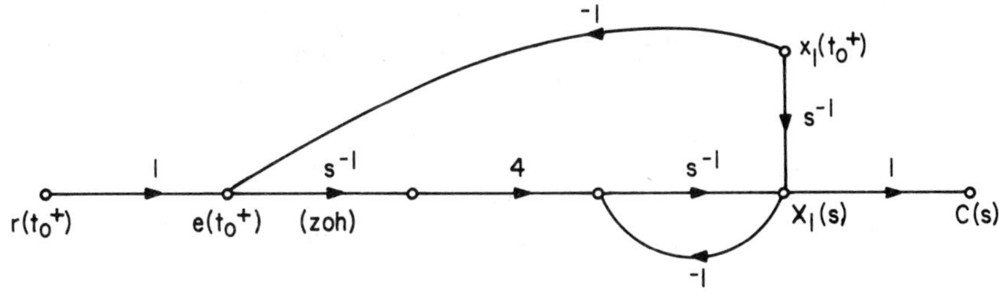

Fig. 4-22. State diagram for the sampled-data control system shown in Fig. 4-21.

For $T_1 = 0.25$ sec and $T = 1$ sec, the last two equations become

$$x_1(k + 0.25) = \phi(0.25)x_1(kT) + \theta(0.25)r(kT)$$

$$= -0.1x_1(kT) + 0.884r(kT) \tag{4-443}$$

and

$$x_1(k+1) = \phi(0.75)x_1(k+0.25) + \theta(0.75)r(kT+0.25)$$

$$= -1.64x_1(k+0.25) + 2.1r(k+0.25) \tag{4-444}$$

Equations (4-443) and (4-444) are two recursion relations from which the response of the system at $t = kT + T_1$ and $(k+1)T$ for $k = 0,1,2,\ldots$ can be computed. Note that the solutions are valid for any input r(t) whose values at t = kT and $kT + T_1$ are defined. For $r(t) = u_s(t)$ the following results are obtained.

t	0	0.25	1	1.25	2	2.25	3	3.25	4	4.25
$x_1(t)$	0	0.884	0.65	0.82	0.75	0.81	0.77	0.807	0.78	0.806

The unit-step response of the system is shown in Fig. 4-23.

In this case, it is shown that the system has a steady-state error of 0.2, or 20 percent.

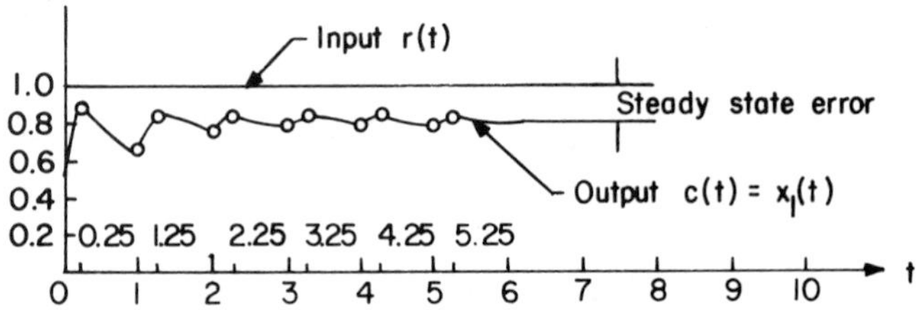

Fig. 4-23. Unit-step response of the digital control system of Example 4-17, Fig. 4-21.

Example 4-18.

A digital control system with nonsynchronous sampling is shown in Fig. 4-24. The sampling instants of the sampler S_2 lag behind that of S_1 by T_1 second. We shall write the state transition equations for the system using the same principles as discussed in this section.

Since S_1 and S_2 close at different sampling instants, the forward path of the system has a transition interval from t = kT to t = (k + 1)T, and the feedback path has a transition interval from $t = kT + T_1$ to $t = (k+1)T + T_1$. The state diagrams for the two paths with the corresponding transition intervals are shown in Fig. 4-25.

In the Laplace domain the state transition equations for the forward path and the feedback path of the system are written directly from Fig. 4-25.

$$kT \leqslant t \leqslant (k+1)T: \quad X_1(s) = \frac{1}{s}x_1(kT) - \frac{1}{s^2}x_2(kT) + \frac{1}{s^2}r(kT) \tag{4-445}$$

$$kT + T_1 \leqslant t \leqslant kT + T + T_1: \quad X_2(s) = \frac{1}{s(s+1)}x_1(kT + T_1)$$

$$+\frac{1}{s+1}x_2(kT+T_1) \tag{4-446}$$

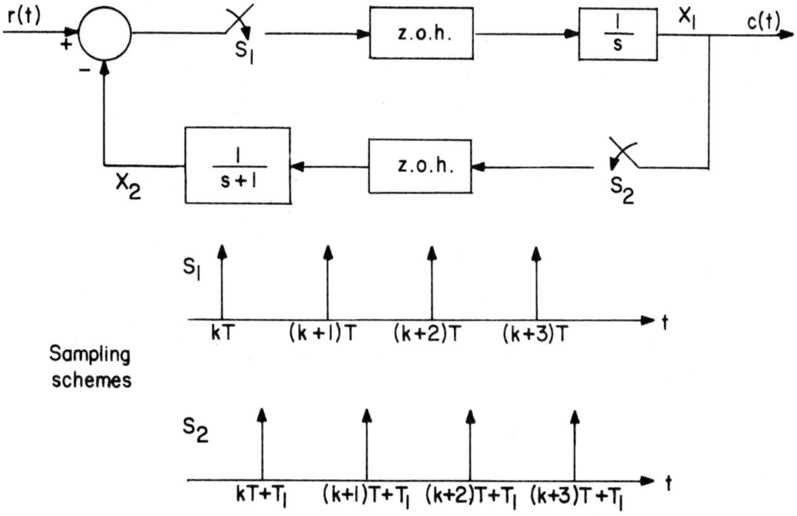

Fig. 4-24. A digital control system with nonsynchronous sampling.

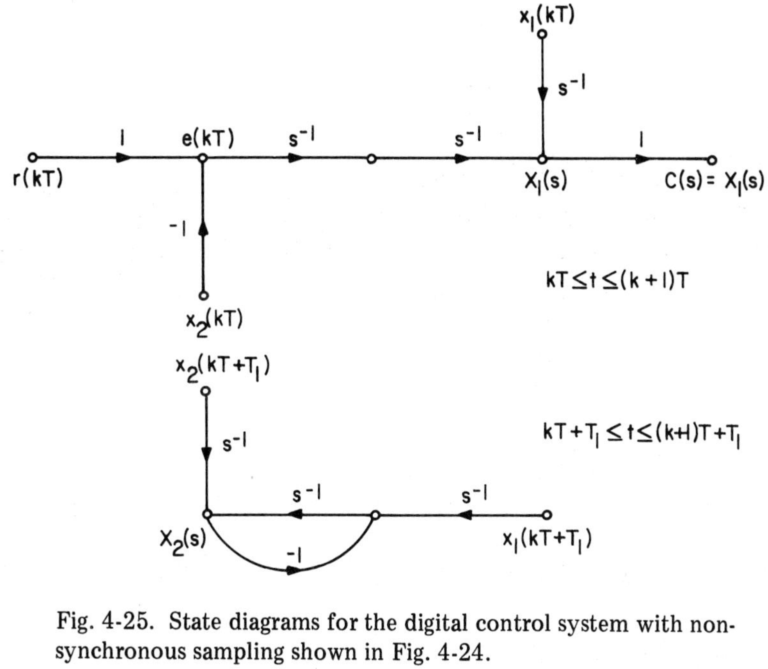

Fig. 4-25. State diagrams for the digital control system with nonsynchronous sampling shown in Fig. 4-24.

Taking the inverse Laplace transform of the last two equations yields

$$kT \leqslant t \leqslant (k+1)T: \quad x_1(t) = x_1(kT) - (t - kT)x_2(kT) + (t - kT)r(kT) \tag{4-447}$$

$$kT + T_1 \leqslant t \leqslant kT + T + T_1: \quad x_2(t) = [1 - e^{-(t-kT-T_1)}]x_1(kT + T_1) + e^{-(t-kT-T_1)}x_2(kT + T_1) \tag{4-448}$$

Since the initial states are $x_1(kT)$, $x_1(kT + T_1)$, $x_2(kT)$ and $x_2(kT + T_1)$, it is necessary to compute these quantities from the two transition equations. Therefore, setting $t = kT + T_1$ and $t = (k + 1)T$ in Eq. (4-447), we get

$$x_1(kT + T_1) = x_1(kT) - T_1x_2(kT) + T_1r(kT) \tag{4-449}$$

$$x_1(k + 1)T = x_1(kT) - Tx_2(kT) + Tr(kT) \tag{4-450}$$

Setting $t = (k + 1)T$ and $t = (k + 1)T + T_1$ in Eq. (4-448), we get

$$x_2(k + 1)T = [1 - e^{-(T-T_1)}]x_1(kT + T_1) + e^{-(T-T_1)}x_2(kT + T_1) \tag{4-451}$$

$$x_2[(k + 1)T + T_1] = (1 - e^{-T})x_1(kT + T_1) + e^{-T}x_2(kT + T_1) \tag{4-452}$$

Now by setting $k = 0,1,2,3,\ldots$ into these four transition equations, the states and the output of the system can be computed at the sampling instants of both samplers.

4.22 STATE-PLANE ANALYSIS

The phase-plane or the state-plane method is a well-known graphical technique for studying linear and nonlinear control systems. However, the method has been applied predominantly to the study of nonlinear second-order control systems for the following reasons:

1. Nonlinear systems are difficult to solve analytically, and very few nonlinear differential equations have closed-form solutions.
2. The phase-plane diagram can be used only for systems with an order less than or equal to two. For higher-order systems, the multidimensional "phase-space" diagram is very difficult to visualize and construct.

If a second-order system is characterized by the dependent variable $c(t)$ and its first derivative with respect to time $dc(t)/dt$, the phase-plane diagram of the system is a plot of $dc(t)/dt = c$ against $c(t)$ with time t as a variable parameter. At any time $t_1 \geqslant 0$, the coordinate $[c(t_1), c(t_1)]$ in the phase plane describes the "state" of the system at that time. As time progresses from zero to infinity, the representing point $(c,\dot{c})$ will trace out a continuous path which represents the complete history of the system in regard to c and $\dot{c}$ at all times. The path itself is called the *phase-plane trajectory*. For instance, for a second-order control system, the phase-plane diagram is usually a plot of the output velocity $\dot{c}$ against the output displacement c. Sometimes it may be more convenient to use the error function e and its derivative for the phase-plane plot.

In the state-variable analysis, a second-order system is characterized by its two state variables $x_1(t)$ and $x_2(t)$. A state-plane diagram is obtained when a trajectory of $x_2(t)$ versus $x_1(t)$ is plotted for $t \geqslant 0$. The trajectory is called the *state-plane trajectory*. It is apparent that the phase-plane and the state-plane diagrams are of the same nature; only the coordinate variables of the latter are state variables.

Phase-Plane Trajectories of a Second-Order System

In this section we shall demonstrate the derivation of the phase-plane trajectories of a second-order open-loop control system. The input to the system is assumed to be connected to the output of a sample-and-hold device. The block diagram of the system is shown in Fig. 4-26.

Let the linear process of the control system be described by

$$G(s) = \frac{C(s)}{H(s)} = \frac{K}{s(1 + \tau s)} \tag{4-453}$$

where K and τ are real constants. A state diagram of the process is drawn as shown in Fig. 4-27 for $kT \leqslant t \leqslant (kT + 1)T$. The input of the state diagram $h(t_0^+)$ is defined as

$$h(t_0^+) = e(kT)$$

for $kT \leqslant t < (k + 1)T$.

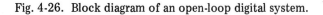

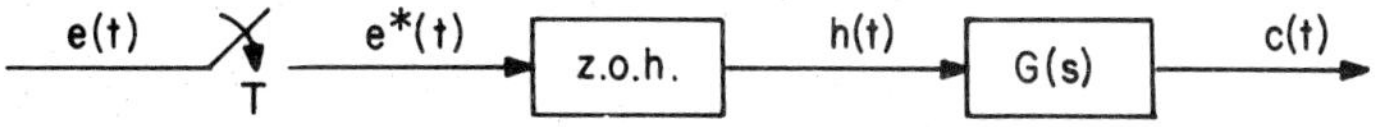

Fig. 4-26. Block diagram of an open-loop digital system.

The state transition equation of the process is written

$$\begin{bmatrix} x_1(t) \\ x_2(t) \end{bmatrix} = \begin{bmatrix} 1 & \tau(1 - e^{-(t-t_0)/\tau}) \\ 0 & e^{-(t-t_0)/\tau} \end{bmatrix} \begin{bmatrix} x_1(t_0^+) \\ x_2(t_0^+) \end{bmatrix}$$

$$+\begin{bmatrix} K(t - t_0 - \tau + \tau e^{-(t-t_0)/\tau}) \\ K(1 - e^{-(t-t_0)/\tau}) \end{bmatrix} h(t_0^+) \qquad (4\text{-}454)$$

for $kT \leqslant t \leqslant (k+1)T$, where $kT = t_0$.

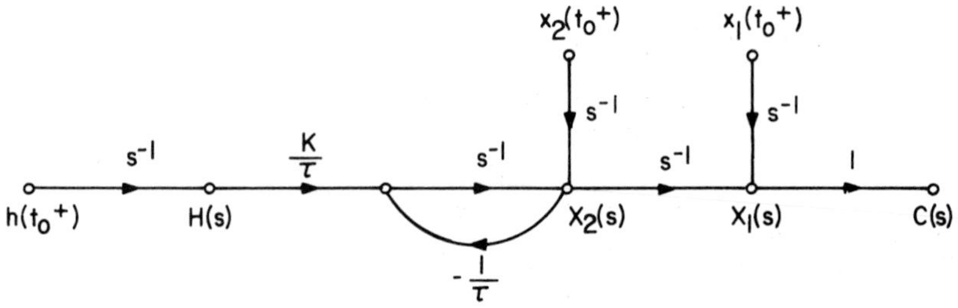

Fig. 4-27. A state diagram of the system shown in Fig. 4-26.

It is convenient to normalize the state variables by letting

$$x_{1n}(t) = x_1(t)$$

$$x_{2n}(t) = \tau x_2(t)$$

$$t_n = t/\tau$$

$$t_{0n} = t_0/\tau$$

$$h_n(t) = \tau K h(t)$$

Substituting these relationships into Eq. (4-454), the normalized state-transition equation is written

$$\begin{bmatrix} x_{1n}(t_n) \\ x_{2n}(t_n) \end{bmatrix} = \begin{bmatrix} 1 & 1 - e^{-(t_n - t_{0n})} \\ 0 & e^{-(t_n - t_{0n})} \end{bmatrix} \begin{bmatrix} x_{1n}(t_{0n}^+) \\ x_{2n}(t_{0n}^+) \end{bmatrix}$$

$$+ \begin{bmatrix} t_n - t_{0n} - 1 + e^{-(t_n - t_{0n})} \\ 1 - e^{-(t_n - t_{0n})} \end{bmatrix} h_n(t_{0n}^+) \qquad (4\text{-}455)$$

The equation of the state-plane trajectory in the x_{2n}-versus-x_{1n} plane is obtained by eliminating t_n from Eq. (4-455). Substituting the second equation in the first equation in Eq. (4-455), we get

$$x_{1n}(t_n) = x_{1n}(t_{0n}^+) + x_{2n}(t_{0n}^+) - x_{2n}(t_n) + (t_n - t_{0n})h_n(t_{0n}^+) \quad (4\text{-}456)$$

Also, solving for $t_n - t_{0n}$ from the second equation in Eq. (4-455), we have

$$t_n - t_{0n} = -\ln\left[\frac{h_n(t_{0n}^+) - x_{2n}(t_n)}{h_n(t_{0n}^+) - x_{2n}(t_{0n}^+)}\right] \quad (4\text{-}457)$$

Now substitution of Eq. (4-457) in Eq. (4-456) gives

$$x_{1n}(t_n) = x_{1n}(t_{0n}^+) + x_{2n}(t_{0n}^+) - x_{2n}(t_n) - h_n(t_{0n}^+) \times \ln\left[\frac{h_n(t_{0n}^+) - x_{2n}(t_n)}{h_n(t_{0n}^+) - x_{2n}(t_{0n}^+)}\right] \quad (4\text{-}458)$$

or

$$x_{1n}(t_n) - x_{1n}(t_{0n}^+) = -[x_{2n}(t_n) - x_{2n}(t_{0n}^+)] - h_n(t_{0n}^+) \times \ln\left[\frac{h_n(t_{0n}^+) - x_{2n}(t_n)}{h_n(t_{0n}^+) - x_{2n}(t_{0n}^+)}\right] \quad (4\text{-}459)$$

The last equation describes the normalized state-plane trajectory for the second-order transfer function in Eq. (4-453) during the time interval $t_0 = kT \leqslant t \leqslant (k + 1)T$ when the input is obtained from a zero-order hold. Since the normalized input $h_n(t_{0n}^+)$ is equal to the output of the zero-order hold at $t = t_0$, and it is held constant for one sampling duration, we need to consider only three values of $h_n(t_{0n}^+)$, namely, $+1$, -1, and 0. The state-plane trajectories described by Eq. (4-459) are plotted in Fig. 4-28 for $h_n(t_{0n}^+) = +1$ and -1 for various initial states. These are subsequently referred to as *the +1-trajectories* and *the −1-trajectories.*

The data of the zero-initial-state trajectory with $h_n(t_{0n}^+) = 1$ are tabulated in Table 4-2. These data correspond to the $h_n(t_{0n}^+)$ trajectory which passes through the origin in Fig. 4-28. It is shown that for nonzero initial states, the trajectory is simply shifted horizontally until it passes through the point $[x_{1n}(t_{0n}), x_{2n}(t_{0n})]$ in the normalized state plane. The state-plane trajectory for $h_n(t_{0n}^+) = -1$ has the same shape as that of the trajectory of $h_n(t_{0n}^+) = +1$ except that it is inverted.

When $h_n(t_{0n}^+) = 0$, Eq. (4-459) is reduced to

$$x_{1n}(t_n) - x_{1n}(t_{0n}^+) = -[x_{2n}(t_n) - x_{2n}(t_{0n}^+)] \quad (4\text{-}460)$$

which describes a straight line in the state plane (Fig. 4-29). These lines are subsequently referred to as the *0-trajectories.*

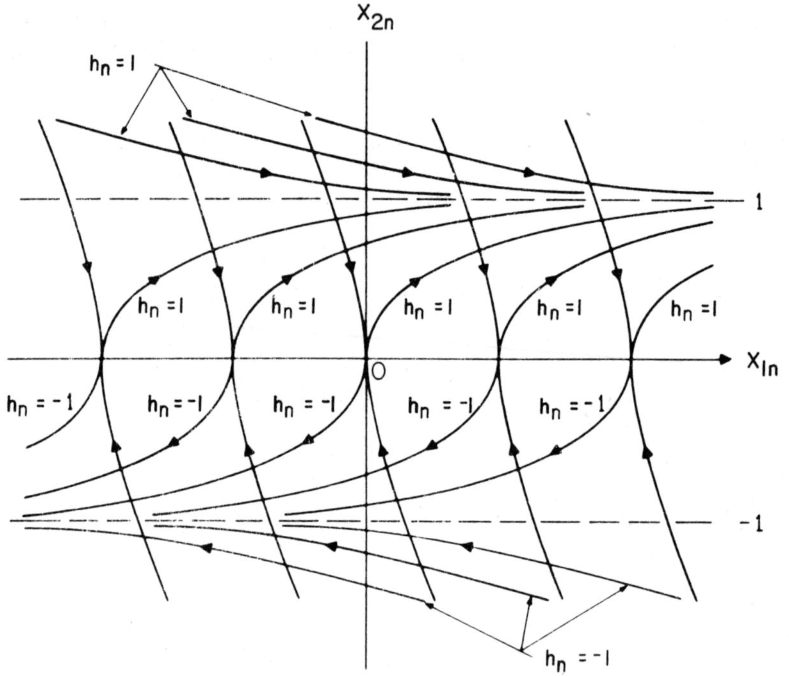

Fig. 4-28. State plane trajectories of a second-order system $G(s) = K/s(1 + \tau s)$ with input normalized to $h_n = +1$ and -1.

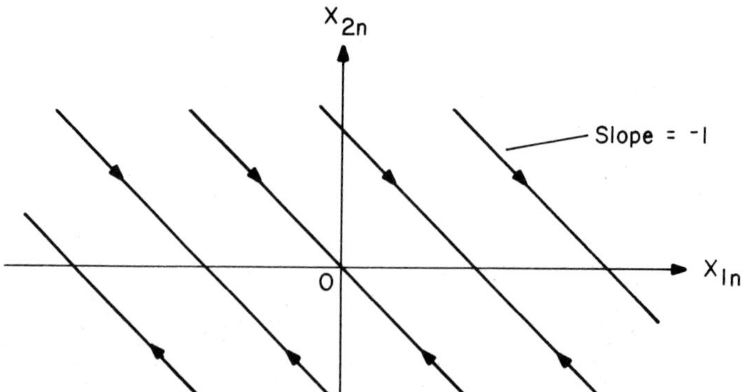

Fig. 4-29. State plane trajectories of a second-order system $G(s) = K/s(1 + \tau s)$ with $h_n = 0$.

TABLE 4-2.

x_{2n}	x_{1n}	x_{2n}	x_{1n}	x_{2n}	x_{1n}
0.1	0.0054	1.05	1.95	−0.2	0.018
0.2	0.023	1.10	1.20	−0.4	0.064
0.3	0.057	1.20	0.41	−0.5	0.095
0.4	0.112	1.28	0	−0.7	0.170
0.5	0.194	1.30	−0.10	−0.8	0.212
0.6	0.317	1.50	−0.80	−0.9	0.258
0.7	0.503			−1.0	0.307
0.8	0.810			−1.2	0.412
0.84	1.00			−1.4	0.525
0.90	1.40			−1.6	0.644
0.95	2.05			−1.8	0.770
0.98	2.93			−2.0	0.90
1.00	∞			−3.0	1.61

For digital control systems, if we are interested only in the response at the sampling instants, we let $t_n = (k+1)T_n$, $t_{0n}^+ = kT_n$, where $T_n = T/\tau$. Equations (4-459) and (4-460) become

$$x_{1n}[(k+1)T_n] - x_{1n}(kT_n) = -x_{2n}[(k+1)T_n] + x_{2n}(kT_n) - h_n(kT_n)\ln\left[\frac{h_n(kT_n) - x_{2n}[(k+1)T_n]}{h_n(kT_n) - x_{2n}(kT_n)}\right] \tag{4-461}$$

and

$$x_{1n}[(k+1)T_n] - x_{1n}(kT_n) = -x_{2n}[(k+1)T_n] + x_{2n}(kT_n) \tag{4-462}$$

respectively.

Thus far we have investigated only the state-plane trajectories of a second-order open-loop system when the input is constant. The state-plane trajectories in this case are shown to be those illustrated in Figs. 4-28 and 4-29 when normalized input of $h_n = +1, -1$, or 0 is applied. The state-plane analysis does not pose any advantage for the study of linear closed-loop digital systems since in a closed-loop system the magnitude of the input to the zero-order hold varies with time and, therefore, the magnitude of $h_n(t_{0n}^+)$ cannot be uniformly normalized to just the three values $+1, -1$, or 0. However, the state-plane method is convenient for the analysis and design of digital systems with relay-type or bang-bang control, since the output of a relay can assume values of only $+M$, $-M$, or 0 for any given input. Later we shall show that optimal digital control systems usually require a bang-bang type of control.

Now we shall demonstrate how the entire state-plane trajectory of a digital system with relay-type control can be constructed using the normalized trajectories of Figs. 4-28 and 4-29.

Consider the system shown in Fig. 4-30. The linear process is assumed to be of the second-order and is described by Eq. (4-378). The nonlinearity is a relay with dead zone and its input-output characteristics are defined by

$$u(t) = \begin{cases} +M & h(t) \geqslant D \\ -M & h(t) \leqslant -D \\ 0 & -D < h(t) < D \end{cases} \tag{4-463}$$

where M is a positive constant, and D denotes the magnitude of the dead zone. A pictorial representation of the relay characteristic is shown in Fig. 4-31. Since the input to the linear second-order process is a constant or zero, the state-plane trajectories described by Eqs. (4-461) and (4-462) can be applied.

It is well-known that for a continuous-data system with relay-type control, the switching of the relay from one mode to another depends on the magnitude of the error signal e(t). However, for a digital system, because of the sampling operation, the switching of the relay mode depends not only upon the magnitude of e(t) at the sampling instants, but also on the occurrence of sampling instants. The construction of the entire state-plane trajectory of the system shown in Fig. 4-30 may appear to be difficult, since time does not appear explicitly on the trajectory. However, for the second-order system under consideration, a set of switching lines can be derived in the state plane. The intersections of the state-plane trajectory and the switching lines indicate the possibility of relay switchings. The details of the method are described as follows.

With reference to Fig. 4-30, the system's variables are normalized as

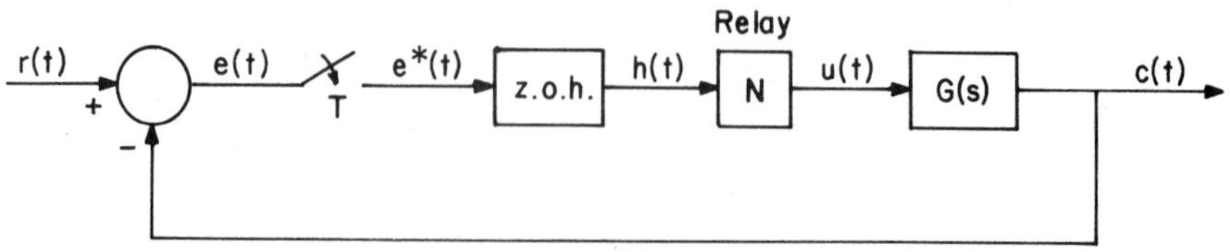

Fig. 4-30. A digital control system with relay control.

$$x_{1n}(t) = x_1(t) = c(t)$$

$$x_{2n}(t) = \tau x_2(t) = \tau \frac{dx_1}{dt}$$

$$t_n = t/\tau$$

$$t_{0n} = t_0/\tau = kT/\tau$$

$$T_n = [(k+1)T - t_0]/\tau$$

$$u_n(t) = \tau KMu(t)$$

where

$$u_n(t) = \begin{cases} +1 & h(t) \geqslant D \\ -1 & h(t) \leqslant -D \\ 0 & -D < h(t) < D \end{cases} \tag{4-464}$$

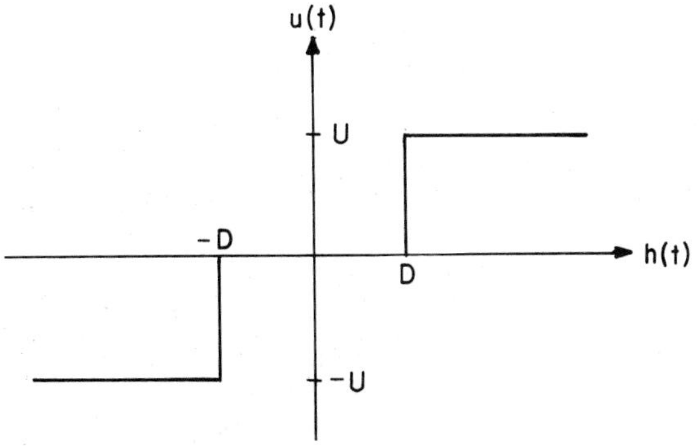

Fig. 4-31. Characteristic of relay with dead zone.

Then the state-plane trajectories of the second-order system are described by

$$x_{1n}[(k+1)T_n] - x_{1n}(kT_n) = -x_{2n}[(k+1)T_n] + x_{2n}(kT_n) - u_n(kT_n)\ln\left[\frac{u_n(kT_n) - x_{2n}[(k+1)T_n]}{u_n(kT_n) - x_{2n}(kT_n)}\right] \tag{4-465}$$

where $u_n(kT_n)$ can be $+1$, -1, or 0 only, and can change value only at the sampling instants.

At discrete sampling instants Eq. (4-456) is written

$$x_{1n}[(k+1)T_n] + x_{2n}[(k+1)T_n] = x_{1n}(kT_n) + x_{2n}(kT_n) + T_n u_n(kT_n) \tag{4-466}$$

The last equation gives the sum of the coordinates of the switching points as a function of the initial state $[x_{1n}(kT_n), x_{2n}(kT_n)]$, the input signal $m_n(kT_n)$, and the normalized sampling period T_n. In general, $u_n(kT_n)$ and T_n are known, so that, given the values of $x_{1n}(kT_n)$ and $x_{2n}(kT_n)$, the value of $x_{1n}[(k+1)T_n] + x_{2n}[(k+1)T_n]$ is determined from Eq. (4-466). However, to find the individual values of $x_{1n}[(k+1)T_n]$ and $x_{2n}[(k+1)T_n]$ another independent relation is needed. Since the state represented by $x_{1n}[(k+1)T_n], x_{2n}[(k+1)T_n]$ must also lie on the state-plane trajectory, it must satisfy Eq. (4-465).

Let the sum of $x_{1n}(kT_n)$ and $x_{2n}(kT_n)$ be designated by a constant A. Then, from Eq. (4-466),

$$x_{1n}[(k+1)T_n] + x_{2n}[(k+1)T_n] = A + T_n \qquad \text{for} \quad u_n(kT_n) = 1 \tag{4-467}$$

$$x_{1n}[(k+1)T_n] + x_{2n}[(k+1)T_n] = A - T_n \qquad \text{for} \quad u_n(kT_n) = -1 \tag{4-468}$$

$$x_{1n}[(k+1)T_n] + x_{2n}[(k+1)T_n] = A \qquad \text{for} \quad u_n(kT_n) = 0 \tag{4-469}$$

which means that, given the state $[x_{1n}(kT_n), x_{2n}(kT_n)]$, and $u_n(kT_n)$, the state $x_{1n}[(k+1)T_n], x_{2n}[(k+1)T_n]$ is determined from the intersection of the straight-line equations in (4-467), (4-468), or (4-469), and the corresponding state-plane trajectories. Figure 4-32 illustrates that once the initial state $[x_{1n}(kT_n), x_{2n}(kT_n)]$ is given, if $u_n(kT_n) = 0$, the state at t_n must satisfy Eqs. (4-465) and (4-469), and the latter is a straight line with a slope of -1 passing through the initial point in the state plane. When $u_n(kT_n) = 1$, the state point in the state plane moves from the initial state along the $+1$-trajectory shown in Fig. 4-28 until it reaches a point where the trajectory intersects the line described by Eq. (4-467); this intersection gives the point $x_{1n}[(k+1)T_n]$, $x_{2n}[(k+1)T_n]$. At this instant, $t_n = (k+1)T_n$, and if $u_n[(k+1)T_n]$ is again equal to $+1$, the state point continues to move on the same trajectory until it reaches the line described by the equation $x_{1n} + x_{2n} = A + 2T_n$. However, at $t_n = (k+1)T_n$, if $u_n[(k+1)T_n] = -1$, the operating point will move from $x_{1n}[(k+1)T_n], x_{2n}[(k+1)T_n]$ along the -1-trajectory until it intersects the line described by Eq. (4-469).

For $u_n(kT_n) = 0$, indicating that the input signal lies inside the relay dead zone, Eqs. (4-465) and (4-466) both become

$$x_{1n}[(k+1)T_n] + x_{2n}[(k+1)T_n] = x_{1n}(kT_n) + x_{2n}(kT_n) \tag{4-470}$$

Therefore, when $u_n(kT_n) = 0$, the switching lines and the state-plane trajectories (0-trajectories) are identical and are straight lines with slopes equal to -1. In Fig. 4-32 it is shown that when $u_n(kT_n) = 0$, the state point moves from the initial state along the line $x_{1n}[(k+1)T_n] + x_{2n}[(k+1)T_n] = A$. However, since time does not appear explicitly on the state-plane trajectory, the location of the $x_{1n}[(k+1)T_n], x_{2n}[(k+1)T_n]$ point on the switching line is uncertain. To find the switching point, we must turn to the second equation in Eq. (4-455) which leads to

$$x_{2n}[(k+1)T_n] = e^{-T_n} x_{2n}(kT_n) + u_n(kT_n)(1 - e^{-T_n}) \tag{4-471}$$

When $u_n(kT_n) = 0$, the last equation gives

$$x_{2n}[(k+1)T_n] = e^{-T_n} x_{2n}(kT_n) \tag{4-472}$$

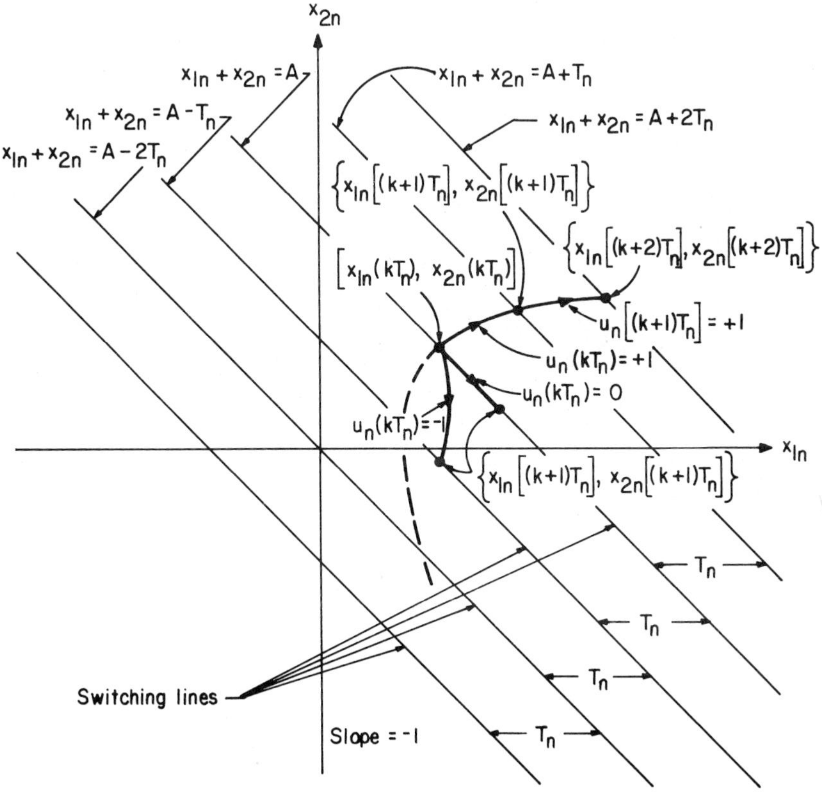

Fig. 4-32. Switching lines and state-plane trajectories of a second-order digital system with relay control.

Since T_n and $x_{2n}(kT_n)$ are known, Eq. (4-472) gives the switching point on the switching line described by Eq. (4-469) at $t_n = (k+1)T_n$.

The state-plane method of analyzing digital systems with relay control is illustrated by the following numerical examples.

Example 4-19.

Consider that for the system shown in Fig. 4-33,

$$G(s) = \frac{1}{s(s+1)} \tag{4-473}$$

$T = 1$ second, and the relay characteristic is that shown in Fig. 4-31 with $M = 1$ and $D = 0$ (ideal relay). The input to the system is a unit-step function, $r(t) = u_s(t)$. The output of the system is desired with the initial conditions of the system given as $c(0) = 0$ and $\dot{c}(0) = 0$.

The first step of the state-plane analysis is to define the normalized state variables and time variables. These are:

$$x_{1n}(t) = x_1(t) = c(t)$$

$$x_{2n}(t) = x_2(t) = \dot{c}(t)$$

$$t_n = t$$

$$T_n = T = 1 \text{ second}$$

$$e_n = r_n - x_{1n} = 1 - x_{1n}$$

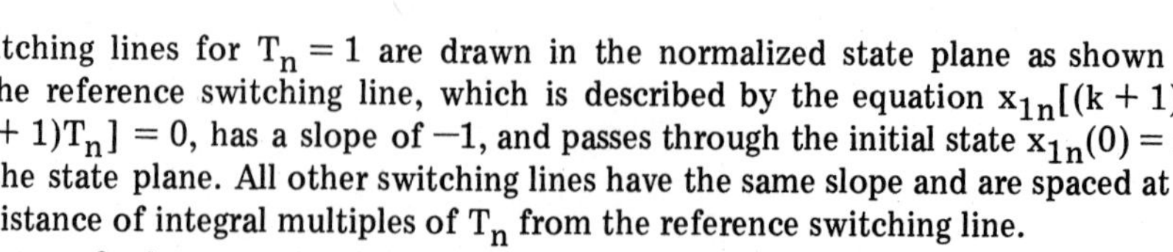

Fig. 4-33. A closed-loop digital system with relay control.

The switching lines for $T_n = 1$ are drawn in the normalized state plane as shown in Fig. 4-34. The reference switching line, which is described by the equation $x_{1n}[(k+1)T_n] + x_{2n}[(k+1)T_n] = 0$, has a slope of -1, and passes through the initial state $x_{1n}(0) = x_{2n}(0) = 0$ in the state plane. All other switching lines have the same slope and are spaced at a horizontal distance of integral multiples of T_n from the reference switching line.

At $t_n = 0$, the error signal e_n is equal to $+1.0$; hence $u_n(0) = +1$. The operating point will move from the origin along the $+1$-trajectory until it intersects the switching line $x_{1n}[(k+1)T_n] + x_{2n}[(k+1)T_n] = T_n$ and the intersection point is the $[x_{1n}(T_n), x_{2n}(T_n)]$ point. At this instant, $t_n = T_n$, since $x_{1n}(T_n)$ is less than unity, $e_n(T_n)$ is positive, and therefore $u_n(T_n) = +1$. The operating point continues to move along the same $+1$-trajectory

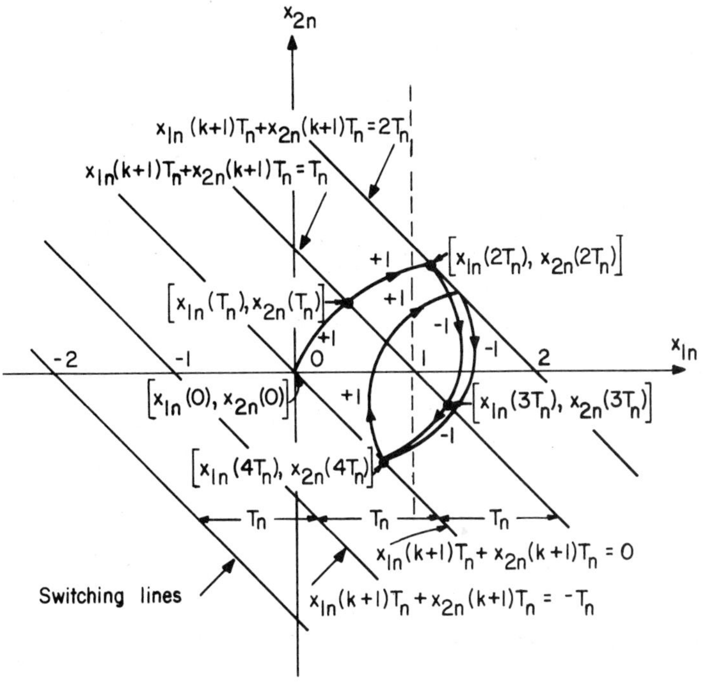

Fig. 4-34. State-plane diagram of the control system with relay control in Example 4-19.

until it intersects the next switching line $x_{1n}[(k+1)T_n] + x_{2n}[(k+1)T_n] = 2T_n$ at $t_n = 2T_n$, and the intersection represents $[x_{1n}(2T_n), x_{2n}(2T_n)]$. Now since $x_{1n}(2T_n)$ is greater than unity, the magnitude of the reference input $e_n(2T_n) < 0$, and $u_n(2T_n) = -1$. The operating point is now switched to and moves along the -1-trajectory which passes through the $[x_{1n}(2T_n), x_{2n}(2T_n)]$ point until the trajectory intersects the switching line $x_{1n}[(k+1)T_n] + x_{2n}[(k+1)T_n] = T_n$ and the intersect gives the coordinates for $[x_{1n}(3T_n), x_{2n}(3T_n)]$.

As is apparent, $x_{3n}(3T_n)$ is still greater than unity; therefore, $m_n(3T_n) = -1$, and the operating point continues to move along the -1-trajectory until it intersects the switching line $x_{1n}[(k+1)T_n] + x_{2n}[(k+1)T_n] = 0$ at $t_n = 4T_n$. At this instant, $x_{1n}(4T_n)$ is less than one, which gives $u_n(4T_n) = +1$, and the operating point is switched to a $+1$-trajectory. Figure 4-34 shows that as the switching process continues, the system eventually enters into sustained oscillation or a limit cycle. The period of the limit cycle as shown in Fig. 4-34 is 4 seconds, and the amplitude is approximately 0.85. The output response and the relay switching function $u_n(t_n)$ of the system are sketched as shown in Fig. 4-35.

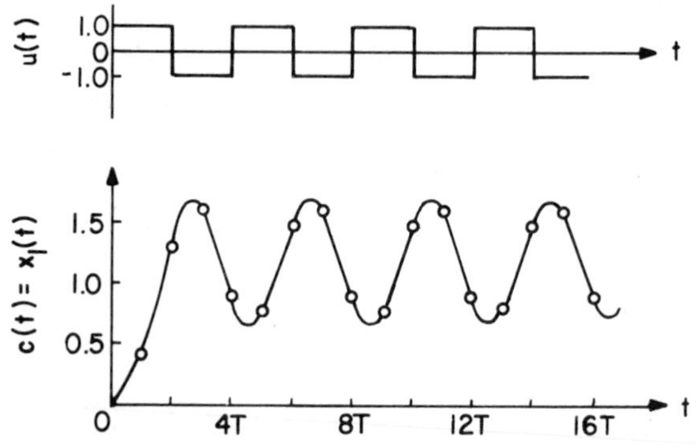

Fig. 4-35. Unit-step response of the relay control system in Example 4-19.

In this section we have discussed and illustrated the state-plane method of analyzing digital control systems. The discussion is not intended to be exhaustive since the graphical method is not emphasized here. In general, the method can be applied to second-order digital systems which have inputs defined at the sampling instants. Example 4-19 illustrates a case when the input is a step function. The method is readily applied to sinusoidal or any periodic inputs. For a ramp input, the state plane should be defined in terms of e(t) and $\dot{e}(t)$. Therefore, a new set of state variables $x_1(t) = e(t)$ and $x_2(t) = \dot{e}(t)$ should be defined.

REFERENCES

1. Kuo, B. C., *Automatic Control Systems*, 3rd edition. Prentice-Hall, Englewood Cliffs, N.J., 1975.
2. Zadeh, L. A., and Desoer, C. A., *Linear System Theory*. McGraw-Hill, New York, 1963.
3. Zadeh, L. A., "An Introduction to State Space Techniques," Workshop on State Space Techniques for Control Systems, *Proc., Joint Automatic Control Conference*, Boulder, Colorado, 1962.
4. Kalman, R. E., and Bertram, J. E., "A Unified Approach to the Theory of Sampling Systems," *Journal of Franklin Inst.*, 267, May 1959, pp. 405-436.
5. Kuo, B. C., *Analysis and Synthesis of Sampled-Data Control Systems*. Prentice-Hall, Englewood Cliffs, N.J., 1963.
6. Kuo, B. C., *Linear Networks and Systems*. McGraw-Hill, New York, 1967.
7. Tou, J. T., *Digital and Sampled-Data Control Systems*. McGraw-Hill, New York, 1959.
8. Jury, E. I., *Theory and Application of the z-Transform Method*. John Wiley & Sons, New York, 1964.

9. Gibson, J. E., *Nonlinear Automatic Control.* McGraw-Hill, New York, 1963.
10. Tou, J. T., *Modern Control Theory.* McGraw-Hill, New York, 1964.
11. Lindorff, D. P., *Theory of Sampled-Data Control Systems.* John Wiley & Sons, New York, 1964.
12. Brockett, R. W., *Finite Dimensional Linear Systems,* John Wiley & Sons, New York, 1970.

PROBLEMS

4-1 A digital control system is described by the following difference equation. Describe the system by a set of dynamic equations.

$$c(k+3) + 4c(k+2) + 3c(k+1) + c(k) = 2u(k+1) + u(k)$$

4-2 The state diagram of a digital control is shown in Fig. 4P-2. Write the dynamic equations for the system.

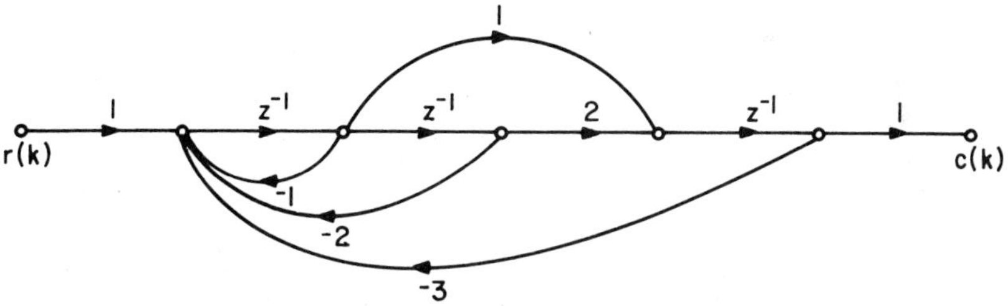

Fig. 4P-2.

4-3 Given the state equation

$$\mathbf{x}(k+1) = \mathbf{A}\mathbf{x}(k)$$

where

$$\mathbf{A} = \begin{bmatrix} 0.5 & 1 & 0 \\ 0 & 1 & 0 \\ 2 & 0.5 & 1 \end{bmatrix}$$

and the initial state $\mathbf{x}(0) = [1 \quad 1 \quad 0]'$. Find $\mathbf{x}(k)$ for $k > 0$.

4-4 Given the state equation

$$\mathbf{x}(k+1) = \mathbf{A}\mathbf{x}(k) + \mathbf{B}u(k)$$

where

$$\mathbf{A} = \begin{bmatrix} 0 & 1 & 0 \\ 0 & 0 & 1 \\ 0 & -0.5 & 1.5 \end{bmatrix} \qquad \mathbf{B} = \begin{bmatrix} 0 & 1 \\ 1 & 0 \\ 2 & 1 \end{bmatrix}$$

Find the state transition matrix $\phi(k)$.

4-5 Prove that if $\mathbf{A}$ is an $n \times n$ matrix with n distinct eigenvalues, then the set of n eigenvectors, $\mathbf{p}_i$, $i = 1,2,\ldots,n$, are linearly independent. [Hint: Let

$$a_1\mathbf{p}_1 + a_2\mathbf{p}_2 + \ldots + a_n\mathbf{p}_n = 0 \tag{4P-1}$$

where $a_1, a_2, \ldots, a_n$ are constants. If it can be shown that $a_1 = a_2 = \ldots = a_n = 0$ is the only condition that can satisfy Eq. (4P-1), then the eigenvectors are linearly independent.]

4-6 Control theory can be applied to the control of economic systems. A well-known dynamic model of supply-and-demand of economic systems is due to Leontief, and is given as follows*:

$$\dot{\mathbf{x}}(t) = -\mathbf{F}(\mathbf{I} - \mathbf{A})\mathbf{x}(t) + \mathbf{F}\mathbf{d}(t)$$

$$\mathbf{s}(t) = (\mathbf{I} - \mathbf{A})\mathbf{x}(t)$$

where $\mathbf{F}$ and $\mathbf{A}$ are $n \times n$ coefficient matrices; $\mathbf{I}$ is an $n \times n$ identity matrix; $\mathbf{x}(t)$ is the $n \times 1$ state vector, $\mathbf{d}(t)$ is the $n \times 1$ demand vector, and $\mathbf{s}(t)$ is the $n \times 1$ supply vector. The elements of these above matrices are defined as,

x_i = rate of production of the ith commodity

a_{ij} = rate at which the ith commodity is used to produce one unit of the jth commodity per unit time

s_i = supply rate of commodity i available for external consumption

d_i = external demand rate for commodity i

f_{ij} = matrix element giving the dynamics of the process

Assuming that the demand d_i $(i = 1,2,\ldots,n)$ is monitored periodically once every T (days), and that it is constant over two consecutive measures, devise a discrete-time model of the system in the form of

$$\mathbf{x}[(k+1)T] = \phi(T)\mathbf{x}(kT) + \theta(T)\mathbf{d}(kT)$$

*W. G. Vogt and M. H. Mickle, "A Stochastic Model For A Dynamic Leontief System," *Proc., Tenth Allerton Conference on Circuit and System Theory*, October 1972, pp. 351-356.

$$s(kT) = \psi(T)\mathbf{x}(kT)$$

given that

$$\mathbf{A} = \begin{bmatrix} 0 & 1 \\ 1 & 0 \end{bmatrix} \qquad \mathbf{F} = \begin{bmatrix} 1 & 0 \\ 0 & 2 \end{bmatrix} \qquad T = 1$$

Find the characteristic equation of the discrete-time system.

4-7 (a) Draw a state diagram for the digital system which is represented by the following dynamic equations:

$$\mathbf{x}(k+1) = \mathbf{A}\mathbf{x}(k) + \mathbf{B}u(k)$$

$$c(k) = x_1(k)$$

$$A = \begin{bmatrix} 0 & 2 & -1 \\ 0 & 1 & 2 \\ 3 & 5 & -1 \end{bmatrix} \qquad B = \begin{bmatrix} 0 \\ 0 \\ 1 \end{bmatrix}$$

(b) Find the characteristic equation of the system.

4-8 A digital control system is described by the following transfer function:

$$\frac{C(z)}{R(z)} = \frac{1 - 0.2z^{-1} - 0.5z^{-2}}{1 + 2z^{-1} + 2z^{-2}}$$

(a) Draw a state diagram for the system.
(b) Write the dynamic equations for the system.

4-9 The block diagram of a digital control system is shown in Fig. 4P-9.
(a) Draw a state diagram for the system.
(b) Write the state equations in normal form.
(c) Find the state transition matrix $\phi(kT)$.

$$G(s) = \frac{2(s + 0.5)}{s(s + 0.2)} \qquad T = 0.1 \text{ second}$$

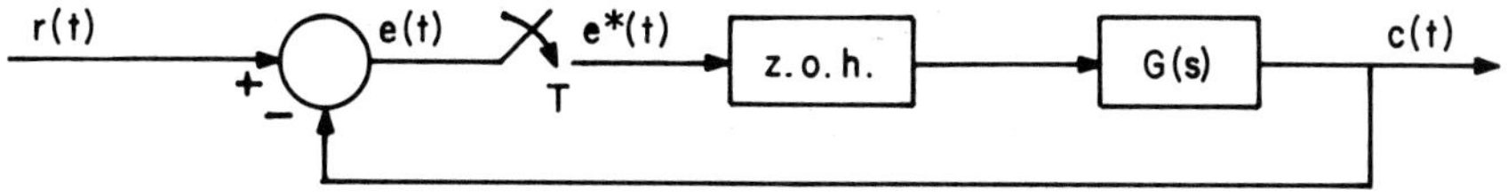

Fig. 4P-9.

4-10 The state diagram of a digital system is shown in Fig. 4P-10. Write the dynamic equations for the system.

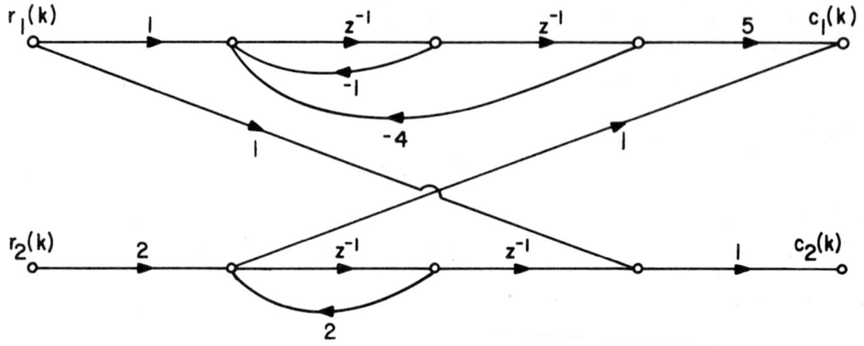

Fig. 4P-10.

4-11 Draw a state diagram for the first-order hold device.

4-12 Draw a state diagram for the system shown in Fig. 4P-12. Write the discrete state equation for the system in the form of

$$x_1[(k+1)T] = \phi(T)x_1(kT) + \theta_1(T)e(kT) + \theta_2(T)e[(k-1)T]$$

The sampling period is assumed to be one second.

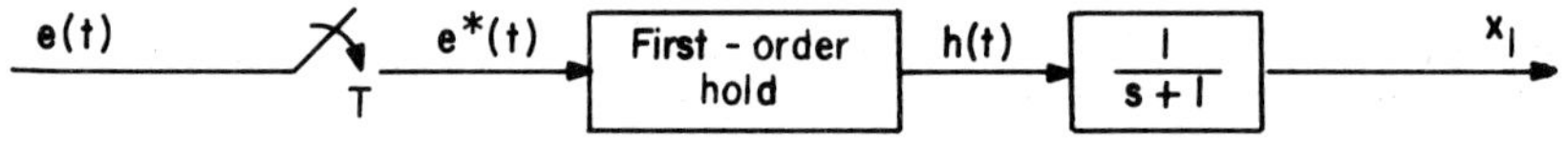

Fig. 4P-12.

4-13 The block diagram of a discrete-data system is shown in Fig. 4P-13. Find the state transition equations of the system at the sampling instants.

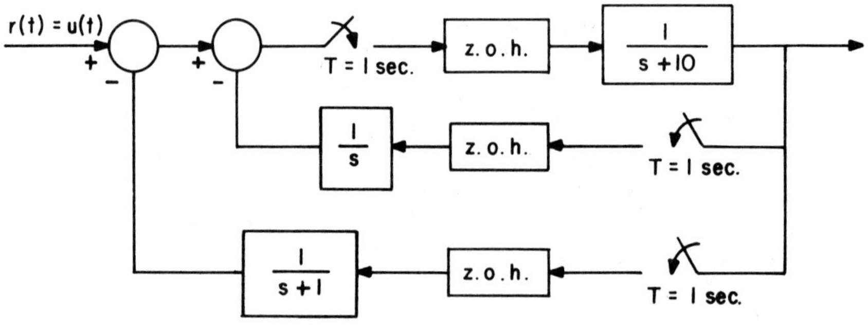

Fig. 4P-13.

4-14 The block diagram of a feedback control system with discrete-data is shown in Fig. 4P-14.

$$D(z) = \frac{1 + 0.5z^{-1}}{1 + 0.2z^{-1}} \qquad G(s) = \frac{10(s+5)}{s^2}$$

T = 0.1 second.

(a) Draw a state diagram for the system.

(b) Write the state equations in normal form for the system.

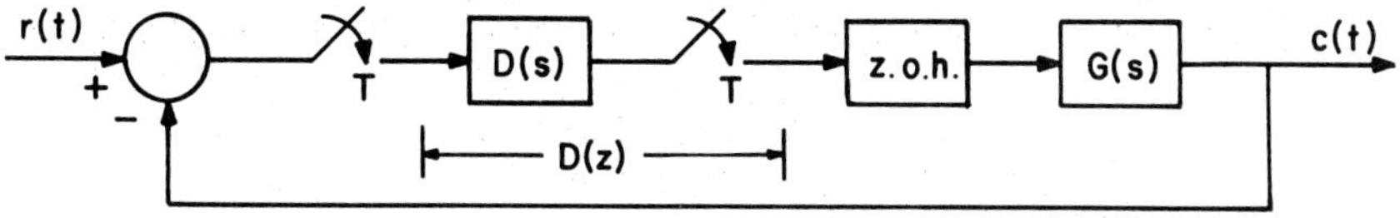

Fig. 4P-14.

4-15 Decompose the following transfer functions and draw the state diagrams.

(a) $$\frac{C(z)}{R(z)} = \frac{(1 - 0.1z^{-1})z^{-1}}{(1 - 0.5z^{-1})(1 - 0.8z^{-1})}$$

(b) $$\frac{C(z)}{R(z)} = \frac{z^{-1}}{(1 - 0.2z^{-1})(1 - z^{-1})}$$

(c) $$\frac{C(z)}{R(z)} = \frac{z + 0.8}{z^3 + 2z^2 + z + 0.5}$$

4-16 Given a linear homogeneous time-varying difference equation

$$\mathbf{x}(k+1) = \mathbf{A}(k)\mathbf{x}(k)$$

where $\mathbf{x}(k)$ is an n-vector, and $\mathbf{A}(k)$ is $n \times n$.

(a) Find the state transition matrix $\psi(i,j)$ of $\mathbf{A}(k)$ for $i > j$, $i < j$, and $i = j$.

(b) Prove the following properties of $\psi(i,j)$:

(i) $\psi(k,k) = \mathbf{I}$

(ii) $\psi(i,j) = \psi^{-1}(j,i)$ for *all* real integral i, j if $\mathbf{A}(k)$ is nonsingular for

$$\begin{aligned} k &= ? \quad \text{for} \quad j > i \\ k &= ? \quad \text{for} \quad i > j \end{aligned}$$

(iii) $\psi(i,j)\psi(j,m) = \psi(i,m)$ for *all* real integral i, j, m if $\mathbf{A}(k)$ is nonsingular for $\min(i,j,m) \leqslant k < \max(i,j,m)$.

(c) Find the range of i, j, m in which condition (iii) is valid if $\mathbf{A}(k)$ is singular for $k \geqslant p$, where p is a positive number.

4-17 Given a linear system with the state equation

$$\dot{\mathbf{x}}(t) = \mathbf{A}\mathbf{x}(t)$$

where

$$\mathbf{A} = \begin{bmatrix} 0 & 1 \\ -1 & 0 \end{bmatrix}$$

Let $t_{k+1} = t_k + \Delta t_k$ and $\dot{\mathbf{x}}(t_k) \cong \mathbf{x}(t_{k+1}) - \mathbf{x}(t_k)$ be a digital approximation of the system at $t = t_k$. Express the state equation of the digital-approximated system in the form of

$$\mathbf{x}(t_{k+1}) = \mathbf{A}_d\mathbf{x}(t_k)$$

Find the state transition equation of $\dot{\mathbf{x}}(t) = \mathbf{A}\mathbf{x}(t)$ and discretize it into the form of $\mathbf{x}[(k + 1)T] = \phi(T)\mathbf{x}(kT)$. Compare $\phi(T)$ with $\mathbf{A}_d$ and comment.

4-18 A digital control system is described by the state equation

$$x(k + 1) = e^{-1}x(k) + (1 - e^{-1})u(k)$$

It is desired that the equilibrium state of the system be at $x = x_d$. The control u(k) is given by

$$u(k) = K[x_d - x(k)]$$

where K is a constant. Determine the actual equilibrium state of the system.

4-19 The sampled-data system shown in Fig. 4P-19 has a triangular hold for data-reconstruction. Draw a state diagram for the system.

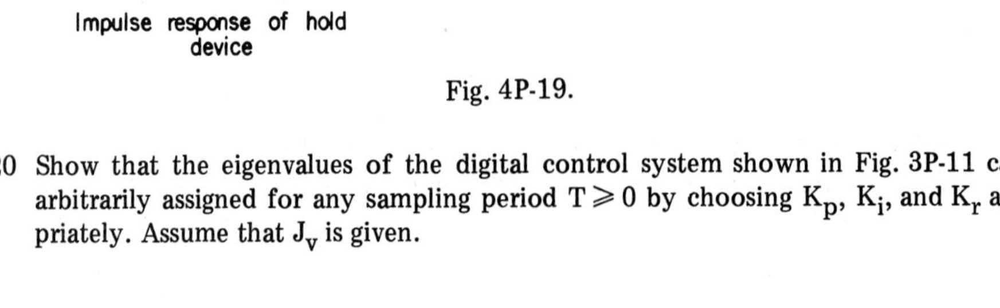

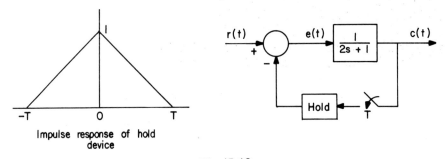

Fig. 4P-19.

4-20 Show that the eigenvalues of the digital control system shown in Fig. 3P-11 can be arbitrarily assigned for any sampling period $T \geqslant 0$ by choosing K_p, K_i, and K_r appropriately. Assume that J_v is given.

4-21 Given the state equations of a digital system,

$$x_1(k+1) = x_2(k)$$

$$x_2(k+1) = x_3(k)$$

$$x_3(k+1) = 0.4x_2(k) + 0.3x_3(k) + u(k)$$

(a) Find the eigenvalues and the corresponding eigenvectors.
(b) Find the transformation so that the state equations are transformed into a set of uncoupled equations.
(c) Consider that $u(k) = -[g_1 \quad g_2 \quad g_3]\mathbf{x}(k)$. Find the values of g_1, g_2, and g_3 so that the eigenvalues of the closed-loop system are all equal to zero.

4-22 The block diagram of an open-loop digital control system is shown in Fig. 4P-22. The input signal is a unit-step function. Write the state transition equations for the system at $t = kT_2$, $k = 0,1,2,\ldots$.

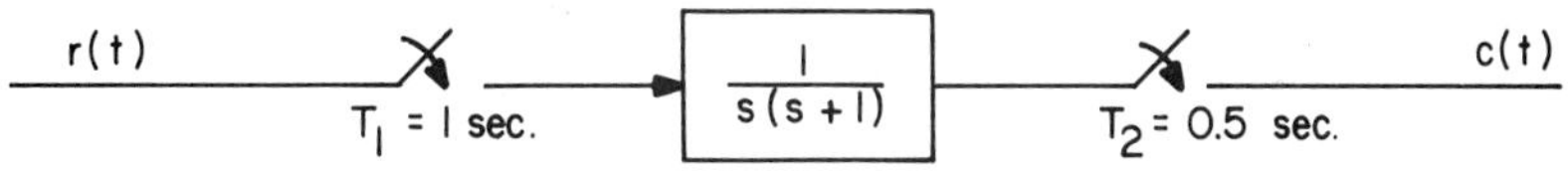

Fig. 4P-22.

4-23 The block diagram of a digital control system with multirate sampling is shown in Fig. 4P-23. It is assumed that the samplers are synchronized.
(a) Write the state equations for the system.
(b) Find the state transition equations for the system.

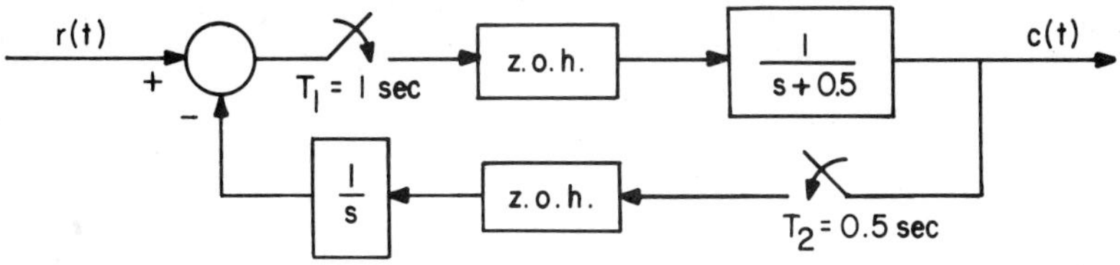

Fig. 4P-23.

4-24 Figure 4P-24 illustrates a linear process whose control signal is the output of a pulse-width modulator. The control signal u(t) is also shown. The pulsewidth at the ith sampling instant is denoted by a_i, where $a_i \leqslant T$ for all i. The magnitude of u(t) is designated by u(kT) which is +1 or −1 depending upon the sign of r(t). The linear process is described by the state equation

$$\frac{d\mathbf{x}(t)}{dt} = \mathbf{A}\mathbf{x}(t) + \mathbf{B}u(t)$$

Show that the system can be described at the sampling instants by the following discrete state equation:

$$\mathbf{x}[(k+1)T] = \Phi(T)\mathbf{x}(kT) + \Phi(T)\Phi(-a_k)\Theta(a_k)u(kT)$$

where

$$\Phi(T) = e^{\mathbf{A}T}$$

$$\Theta(T) = \int_0^T \Phi(T-\tau)\mathbf{B}d\tau$$

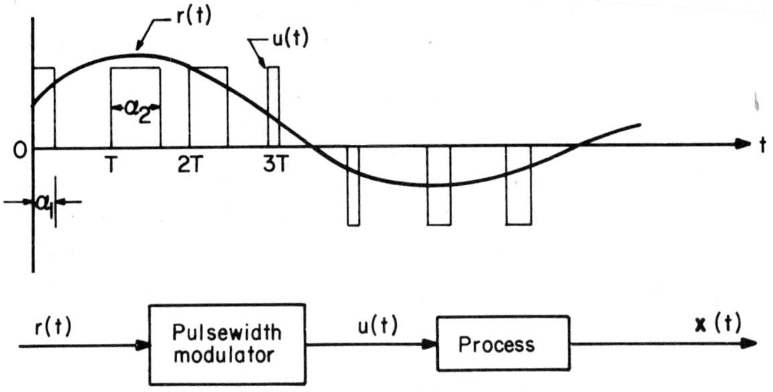

Fig. 4P-24.

4-25 The block diagram of a nonlinear digital control system with time delay is shown in Fig. 4P-25. Sketch the state-plane trajectory in x_2 versus x_1, where $x_1 = c$ and $x_2 = dc/dt$. Assume that a unit-step input is applied and the initial states are zero. The sampling period is one second.

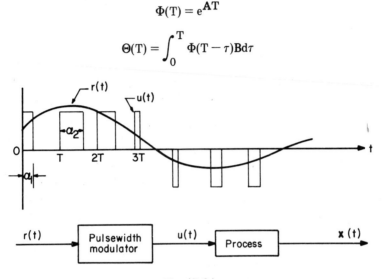

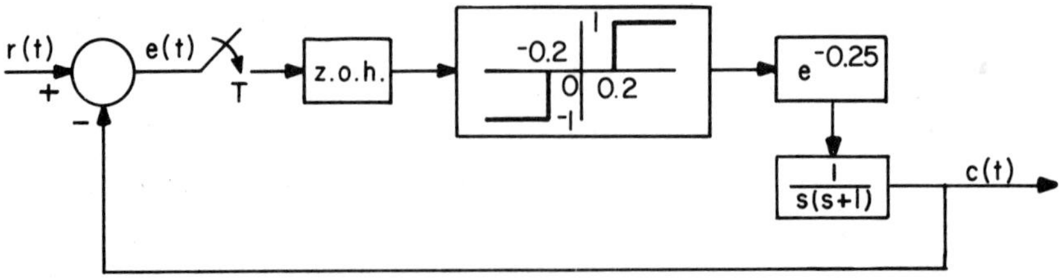

Fig. 4P-25.

4-26 The block diagram of a digital control system with quantization is shown in Fig. 4P-26. The sampling period is one second.

(a) Evaluate the output response by the state-variable method with $r(t) = 0.5u(t)$.

(b) Sketch the state-plane trajectory with $x_1 = c$ and $x_2 = dc/dt$.

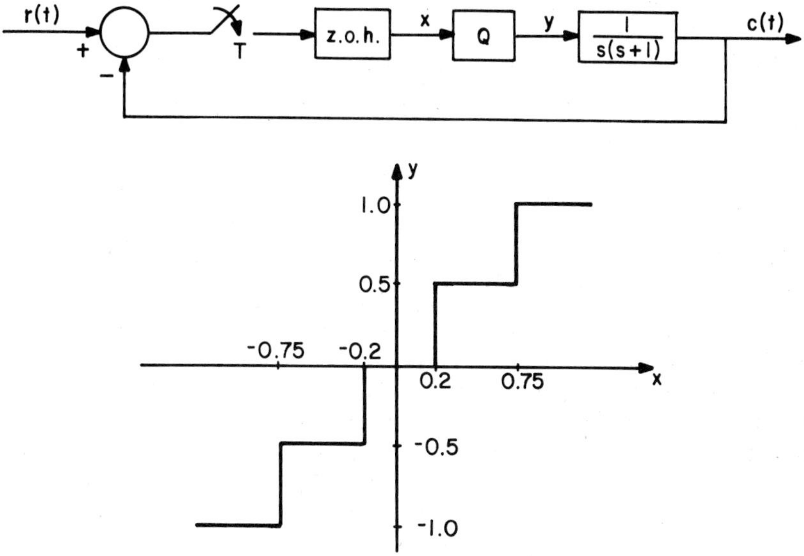

Fig. 4P-26.

4-27 The block diagram of a digital tracking loop of a missile system is shown in Fig. 4P-27. The sampling period is 1/12 second. Determine the maximum magnitude of the step input which will be just short of saturating the nonlinearity of the system. Let K = 8 and $\Delta = 0.4$.

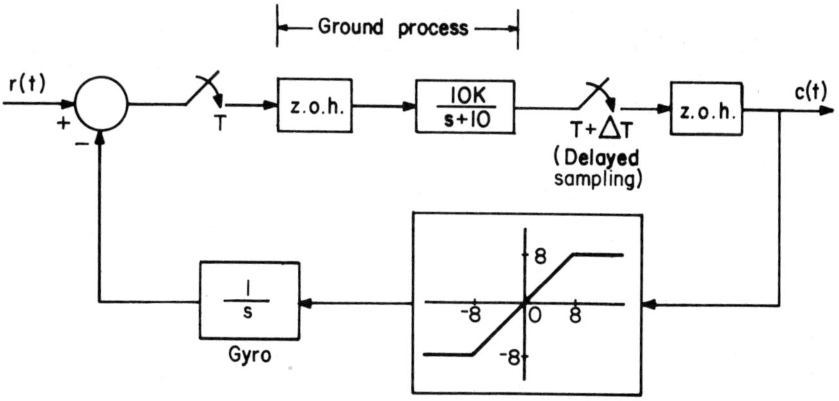

Fig. 4P-27.

4-28 A closed-loop sampled-data control system with a first-order hold is described by the following state equations

$$\dot{\mathbf{x}}(t) = \mathbf{A}\mathbf{x}(t) + \mathbf{B}\mathbf{u}(t)$$

where

$$\mathbf{u}(t) = [\mathbf{E}_0 + (t - kT)\mathbf{E}_1]\mathbf{r}(kT) - [\mathbf{G}_0 + (t - kT)\mathbf{G}_1]\mathbf{x}(kT)$$

for $kT \leqslant t < (k + 1)T$. The dimensions of the variables are: $\mathbf{x}(t)$ is $n \times 1$, $\mathbf{u}(t)$ is $p \times 1$, $\mathbf{r}(kT)$ is $p \times 1$. The matrices $\mathbf{E}_0$ and $\mathbf{E}_1$ are $p \times p$; $\mathbf{G}_0$ and $\mathbf{G}_1$ are $p \times n$.

Find the state transition equation of the system and discretize it in the form of

$$\mathbf{x}[(k + 1)T] = \phi(T)\mathbf{x}(kT) + \theta(T)\mathbf{r}(kT)$$

Express $\phi(T)$ and $\theta(T)$ in terms of $\mathbf{A}$, $\mathbf{B}$, $\mathbf{E}_0$, $\mathbf{E}_1$, $\mathbf{G}_0$ and $\mathbf{G}_1$.

4-29 A nonlinear sampled-data control system is shown in Fig. 4P-29.

(a) Determine the characteristics of the response of the system when $r(t) = 0$ and $c(0) = 1$, $\dot{c}(0) = -1$. The sampling period is 1 second. Sketch the state-plane trajectory of x_1 versus x_2 with $c(t) = x_1(t)$ and $\dot{c}(t) = x_2(t)$.

(b) Repeat part (a) when $r(t) = \sin \omega t$ and zero initial conditions. $\omega = 2\pi/T_s$ with $T_s = 4T = 4$ seconds.

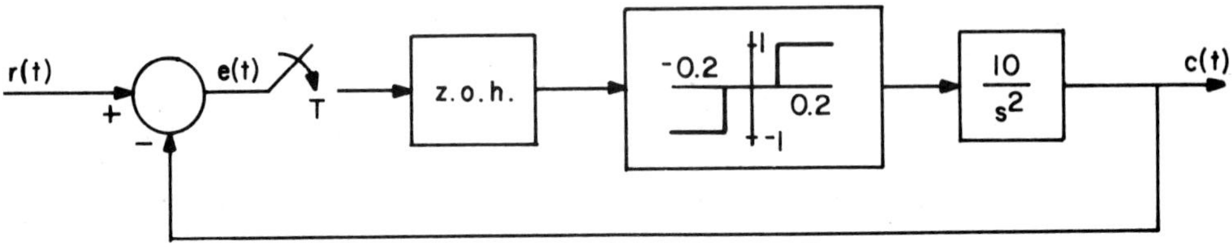

Fig. 4P-29.

5

STABILITY OF DIGITAL CONTROL SYSTEMS

5.1 INTRODUCTION

One of the most important requirements in the performance of control systems is stability. This is true whether the system has continuous data, digital data, or a combination of the two kinds of signals. In this chapter we shall investigate the methods of testing for the stability condition of digital systems. These methods are also applicable to continuous-data systems whose outputs are measured at discrete time intervals.

The following methods of stability analysis are well-known for continuous-data systems: (1) the Routh-Hurwitz criterion; (2) the Nyquist criterion; (3) the root-locus plot; (4) the Bode plot; (5) Liapunov's direct method. Most of these methods can be applied directly to the stability analysis of discrete-data systems with slight modifications. However, since the stability boundary in the z-plane is the unit circle $|z| = 1$, in contrast to the imaginary axis of the s-plane, the Routh-Hurwitz criterion cannot be applied without modification to the discrete case. Therefore, the Schur-Cohn criterion and Jury's criterion are devised as replacements of the Routh-Hurwitz criterion for digital systems. Finally, the Liapunov's direct method may be used to study the stability of linear as well as nonlinear systems with digital data.

Before embarking on the discussions of the aforementioned methods of stability analysis, let us collect some essential definitions and terminology on the stability of digital systems.

The digital system under consideration can be described by the state equation

$$\mathbf{x}(k_{i+1}) = \mathbf{f}[\mathbf{x}(k_i), \mathbf{u}(k_i), k_i] \tag{5-1}$$

where $\mathbf{x}(k_i)$ and $\mathbf{f}[\mathbf{x}(k_i), \mathbf{u}(k_i), k_i]$ are n-vectors. In general, $\mathbf{f}[\mathbf{x}(k_i), \mathbf{u}(k_i), k_i]$ may be linear or nonlinear, time-varying or time-invariant.

The Equilibrium State

The state $\mathbf{x}_e$ which is a *constant* solution of the state equation of Eq. (5-1) with $\mathbf{u}(k_i) = \mathbf{0}$, for all $k_i \geqslant k_0$, is defined as the equilibrium state of the system; that is,

$$\mathbf{x}_e = \mathbf{f}[\mathbf{x}_e(k_i), \mathbf{0}, k_i] \qquad k_i \geqslant k_0 \tag{5-2}$$

For a linear time-invariant system, the homogeneous state equation is written as

$$\mathbf{x}(k_{i+1}) = \mathbf{A}\mathbf{x}(k_i) \tag{5-3}$$

Then, Eq. (5-2) becomes

$$\mathbf{x}_e = \mathbf{A}\mathbf{x}_e \tag{5-4}$$

or

$$(\mathbf{A} - \mathbf{I})\mathbf{x}_e = \mathbf{0} \tag{5-5}$$

Since Eq. (5-5) corresponds to the condition for $\mathbf{x}_e$ as an eigenvector of $\mathbf{A}$ for a unity eigenvalue ($\lambda = 1$), we have

$$\mathbf{x}_e = \mathbf{0} \tag{5-6}$$

as the equilibrium state, if $\mathbf{A}$ does not have a unity eigenvalue. On the other hand, if any of the eigenvalues of $\mathbf{A}$ is unity, $\mathbf{x}_e$ is nonunique, as it can be any vector so long as it is an eigenvector that corresponds to $\lambda = 1$. Therefore, for unforced linear time-invariant systems, we can always select $\mathbf{x} = \mathbf{0}$ as the equilibrium state if $\mathbf{A}$ does not have an eigenvalue at $\lambda = 1$.

For nonlinear systems, the equilibrium state can be nonzero, and, in general, there may be more than one equilibrium state. However, we can always transform a nonzero equilibrium state into one which is at the origin of the state space. Consider that the equilibrium state of the system

$$\mathbf{x}(k_{i+1}) = \mathbf{f}[\mathbf{x}(k_i), k_i] \tag{5-7}$$

is at $\mathbf{x}_e \neq \mathbf{0}$. We let

$$\mathbf{y}(k_i) = \mathbf{x}(k_i) - \mathbf{x}_e \tag{5-8}$$

Then,

$$\mathbf{y}(k_{i+1}) = \mathbf{x}(k_{i+1}) - \mathbf{x}_e \tag{5-9}$$

Substituting Eq. (5-7) into Eq. (5-9) gives

$$\mathbf{y}(k_{i+1}) = \mathbf{f}[\mathbf{y}(k_i) + \mathbf{x}_e, k_i] - \mathbf{x}_e \tag{5-10}$$

or

$$\mathbf{y}(k_{i+1}) = \mathbf{g}[\mathbf{y}(k_i), k_i] \tag{5-11}$$

Thus, the system described by Eq. (5-11) will have an equilibrium state at $\mathbf{y}_e = \mathbf{0}$.

In the following sections we shall show that the definition of stability in the sense of Liapunov is related to the study of the behavior of the states in the neighborhood of the equilibrium state.

Definition of Norm of a Vector

The norm of an n-vector $\mathbf{x}$ is denoted by $||\mathbf{x}||$ and is defined as

$$||\mathbf{x}|| = \left[\sum_{k=1}^{n} x_k^2 \right]^{1/2} \tag{5-12}$$

Thus, the norm of $\mathbf{x}$ is interpreted as the distance measured from the origin to the vector $\mathbf{x}$. The properties of $||\mathbf{x}||$ are:

1. $||a\mathbf{x}|| = a||\mathbf{x}|| \qquad a = \text{scalar}$
2. $||\mathbf{x}|| \geqslant 0$
3. $||\mathbf{x} + \mathbf{y}|| \leqslant ||\mathbf{x}|| + ||\mathbf{y}||$
4. $||\mathbf{x}|| = 0 \quad \text{if and only if} \quad \mathbf{x} = \mathbf{0}$ (5-13)

Definition of Norm of a Matrix

The norm of a matrix $\mathbf{A}$, denoted by $||\mathbf{A}||$, is defined as the minimum value of a such that

$$||\mathbf{A}\mathbf{x}|| \leqslant a||\mathbf{x}|| \tag{5-14}$$

The properties of $||\mathbf{A}||$ are:

1. $||\mathbf{A}\mathbf{x}|| \leqslant ||\mathbf{A}|| \; ||\mathbf{x}||$
2. $||\mathbf{A}|| = \max ||\mathbf{A}\mathbf{x}|| = \max [\mathbf{x}'\mathbf{A}'\mathbf{A}\mathbf{x}] \quad \text{with} \quad ||\mathbf{x}|| = 1$
3. $||\mathbf{A}|| = 0 \quad \text{if and only if} \quad \mathbf{A} = \mathbf{0}$
4. $||\mathbf{A}\mathbf{B}|| \leqslant ||\mathbf{A}|| \; ||\mathbf{B}||$
5. $||\mathbf{A} + \mathbf{B}|| \leqslant ||\mathbf{A}|| + ||\mathbf{B}||$
6. $||\mathbf{A}||^2 \geqslant |\lambda_i|^2 \quad \text{where } \lambda_i \text{ is any eigenvalue of } \mathbf{A}$ (5-15)

5.2 DEFINITIONS OF STABILITY

Stability in the Sense of Liapunov

Let $\mathbf{x}_e$ denote an equilibrium state of the system

$$\mathbf{x}(k_{i+1}) = \mathbf{f}[\mathbf{x}(k_i), k_i] \tag{5-16}$$

Then $\mathbf{x}_e$ is said to be *stable* if, given any real number $\epsilon > 0$, there exists a real number $\delta(\epsilon, k_0) > 0$ such that for all initial states $\mathbf{x}(k_0)$ in the sphere of radius δ

$$||\mathbf{x}(k_0) - \mathbf{x}_e|| \leqslant \delta \tag{5-17}$$

the solution

$$\mathbf{x}(k_i) = \phi[\mathbf{x}(k_0), k_0] \tag{5-18}$$

remains within a sphere of radius ϵ for all $k_i \geqslant k_0$; that is,

$$||\mathbf{x}(k_i) - \mathbf{x}_e|| < \epsilon \tag{5-19}$$

This definition of stability is also referred to as *stability in the sense of Liapunov*, or *stability i.s.L.* for short.

The stability i.s.L. essentially states that if an equilibrium state $\mathbf{x}_e$ is stable, then every trajectory or solution starting in the neighborhood of $\mathbf{x}_e$ must remain arbitrarily close to the equilibrium state at all times (or stages). A graphical interpretation of the stability i.s.L. is illustrated in Fig. 5-1 for a second-order case.

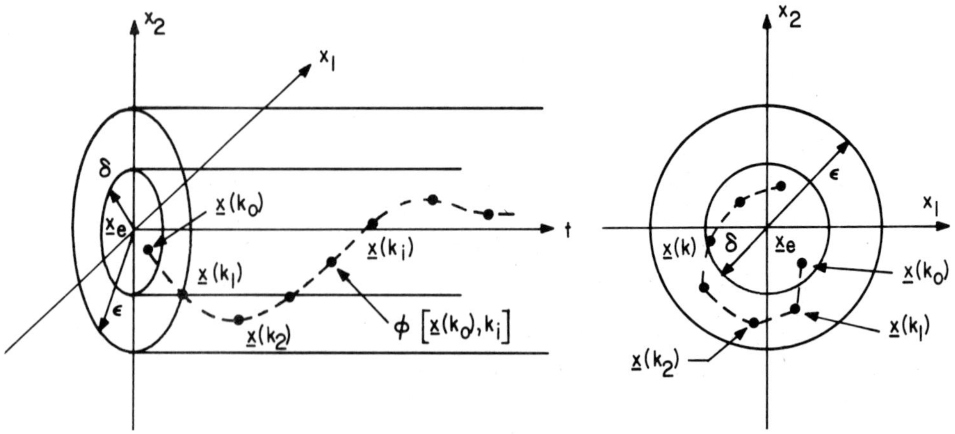

Fig. 5-1. Definition of stability.

The definitions of stability given in the following are all of the Liapunov type, and for simplicity they are all referred to as stability.

Uniform Stability

The equilibrium state $\mathbf{x}_e$ is said to be *uniformly stable* if it is stable and δ is independent of k_0.

Asymptotic Stability

The equilibrium state $\mathbf{x}_e$ is said to be *asymptotically stable* if

1. it is stable (in the sense of Liapunov),
2. for any k_0 and any $\mathbf{x}(k_0)$ sufficiently close to $\mathbf{x}_e$, the solution $\mathbf{x}(k_i) = \phi[\mathbf{x}(k_0),k_i]$ converges to $\mathbf{x}_e$ as k_i approaches infinity.

Mathematically, we can state that, given a number $\epsilon > 0$, there exists a number $\delta(k_0) > 0$ and $T(\epsilon,\delta,k_0)$ such that

$$||\mathbf{x}(k_0) - \mathbf{x}_e|| \leq \delta(k_0) \tag{5-20}$$

which implies that

$$||\phi[\mathbf{x}(k_0),k_i] - \mathbf{x}_e|| \leq \delta \qquad \text{for} \qquad k_i \geq k_0 \tag{5-21}$$

and

$$||\phi[\mathbf{x}(k_0),k_i] - \mathbf{x}_e|| \leq \epsilon \qquad \text{for} \qquad k_i \geq k_0 + T \tag{5-22}$$

The graphical interpretation of asymptotic stability is illustrated in Fig. 5-2.

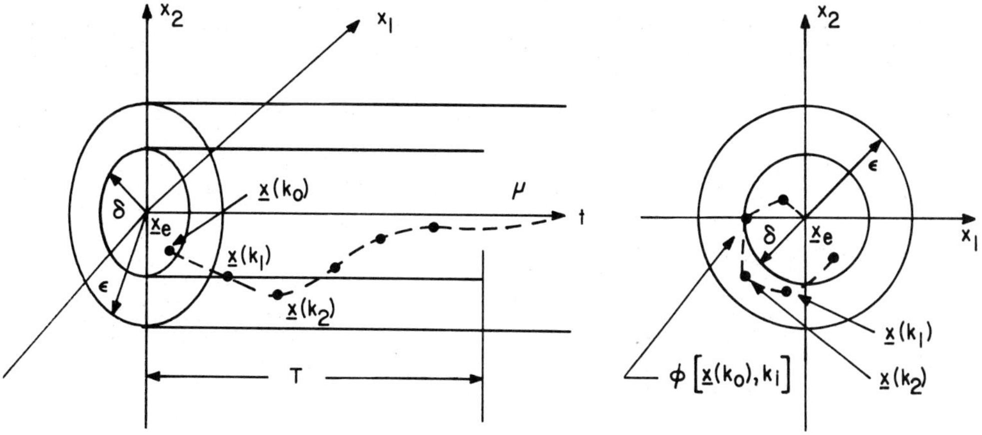

Fig. 5-2. Definition of asymptotic stability.

Uniform Asymptotic Stability

The equilibrium state $\mathbf{x}_e$ is *uniformly asymptotically stable* if it is asymptotically stable with δ independent of k_0.

Asymptotic Stability in the Large

The equilibrium state $\mathbf{x}_e$ is said to be *uniformly stable in the large* or *globally asymptotically stable* if δ, the domain in which the initial state $\mathbf{x}(k_0)$ lies, is arbitrarily large, that is, all the initial states can be made to converge to $\mathbf{x}_e$.

Instability or Unstable

The equilibrium state $\mathbf{x}_e$ is said to be *unstable* if it is neither stable nor asymptotically stable.

Condition of Asymptotic Stability for Linear Time-Invariant Systems

The linear time-invariant discrete-data system

$$\mathbf{x}(k + 1) = \mathbf{A}\mathbf{x}(k) \tag{5-23}$$

is asymptotically stable if and only if the eigenvalues of $\mathbf{A}$, λ_i, $i = 1,2,\ldots,n$, satisfy

$$|\lambda_i| < 1$$

Another way of stating the condition is that the roots of the characteristic equation

$$|z\mathbf{I} - \mathbf{A}| = 0 \tag{5-24}$$

must all lie inside the unit circle $|z| = 1$.

For the system of Eq. (5-23) the state-transition matrix is

$$\phi(k) = \mathbf{A}^k \tag{5-25}$$

Let the eigenvalues of $\mathbf{A}$ be denoted by λ_i, $i = 1,2,\ldots,n$. Then, the eigenvalues of $\phi(k)$ are λ_i^k, $i = 1,2,\ldots,n$.* For asymptotic stability,

$$\lim_{k\to\infty} ||\phi(k)|| = 0 \tag{5-26}$$

From the properties of the norm of a matrix, Eq. (5-15),

---

*This is from a theorem which states that if λ_i, $i = 1,2,\ldots,n$, are the eigenvalues (distinct or multiple order) of A, and if f(A) is any polynomial function of A, then the eigenvalues of f(A) are $f(\lambda_i)$, $i = 1,2,\ldots,n$. (See F. E. Hohn, *Elementary Matrix Algebra*, 2nd ed., The Macmillan Company, New York, 1964, p. 291.)

$$|\lambda_i^k|^2 \leqslant ||\phi(k)||^2 \tag{5-27}$$

Therefore, using Eqs. (5-26) and (5-27), it is seen that for the equilibrium state to be asymptotically stable, the eigenvalues of **A** must satisfy the condition

$$|\lambda_i| < 1 \qquad i = 1,2,\ldots,n$$

Bounded-Input-Bounded-State and Bounded-Output Stability of Linear Time-Invariant Systems

Thus far we have defined only the Liapunov-type stability which is concerned with an unforced system, i.e., $\mathbf{u}(k) = \mathbf{0}$. Also, the stability is defined with respect to the trajectories of the states of the system.

In this section we are concerned with a linear time-invariant system which is described by the following dynamic equations:

$$\mathbf{x}(k+1) = \mathbf{A}\mathbf{x}(k) + \mathbf{B}\mathbf{u}(k) \tag{5-28}$$

$$\mathbf{c}(k) = \mathbf{D}\mathbf{x}(k) + \mathbf{E}\mathbf{u}(k) \tag{5-29}$$

BIBS Stability

The system of Eqs. (5-28) and (5-29) is bounded-input-bounded-state stable (BIBS stable) if for any bounded input $\mathbf{u}(k)$, the state $\mathbf{x}(k)$ is also bounded.

For a bounded input it is implied that

$$||\mathbf{u}(k)|| \leqslant M \qquad \text{for all } k \tag{5-30}$$

where M is a finite number.

The state-transition equation of Eq. (5-28) is written

$$\mathbf{x}(k) = \phi(k)\mathbf{x}(0) + \sum_{i=0}^{k-1} \phi(k-i-1)\mathbf{B}\mathbf{u}(k) \tag{5-31}$$

where $\phi(k) = \mathbf{A}^k$. The BIBS stability requires that

$$||\mathbf{x}(k)|| \leqslant N \tag{5-32}$$

where N is some constant which may depend upon M and $\mathbf{x}(0)$. Taking the norm on both sides and applying some of the properties of the norm, we have

$$||\mathbf{x}(k)|| \leqslant ||\phi(k)||\ ||\mathbf{x}(0)|| + \sum_{i=0}^{k-1} ||\phi(k-i-1)\mathbf{B}||\ ||\mathbf{u}(k)|| \tag{5-33}$$

Since $\mathbf{u}(k)$ is bounded and $\mathbf{x}(0)$ is a constant vector, $\mathbf{x}(k)$ will be bounded if

$$||\phi(k)|| < P < \infty \qquad \text{for all } k \tag{5-34}$$

and

$$\sum_{i=0}^{k-1} ||\phi(k-i-1)\mathbf{B}|| < Q < \infty \qquad \text{for all k} \tag{5-35}$$

It can be proven that Eqs. (5-34) and (5-35) are the necessary and sufficient conditions for BIBS stability.

Since if a system is asymptotically stable $||\phi(k)||$ is necessarily bounded, as stated in Eq. (5-21), it follows that *if a linear system is asymptotically stable it is also BIBS stable.* Therefore, *the condition that the magnitudes of all the eigenvalues of* **A** *are less than one also assures BIBS stability.*

BIBO Stability

The system of Eqs. (5-28) and (5-29) is bounded-input-bounded-output (BIBO) stable if for any bounded input the output $\mathbf{c}(k)$ is also bounded. It should be noted that BIBS stability and BIBO stability are not related in any way. A system may be BIBO stable but not BIBS stable.

The following examples illustrate the concepts of the various types of stability defined in the preceding sections.

Example 5-1.

Consider that a second-order discrete-data system is described by the state equation

$$\mathbf{x}(k+1) = \mathbf{A}\mathbf{x}(k) \tag{5-36}$$

where

$$\mathbf{A} = \begin{bmatrix} 0 & 1 \\ -1 & 0 \end{bmatrix} \tag{5-37}$$

Taking the z-transform on both sides of Eq. (5-36), and solving for $\mathbf{X}(z)$, we get

$$\mathbf{X}(z) = (z\mathbf{I} - \mathbf{A})^{-1} z\mathbf{x}(0) \tag{5-38}$$

or

$$\begin{bmatrix} X_1(z) \\ X_2(z) \end{bmatrix} = \frac{1}{z^2+1} \begin{bmatrix} z^2 & z \\ -z & z^2 \end{bmatrix} \begin{bmatrix} x_1(0) \\ x_2(0) \end{bmatrix} \tag{5-39}$$

The responses of $x_1(k)$ and $x_2(k)$ are obtained by taking the inverse z-transform on both sides of Eq. (5-39). We have

$$\begin{bmatrix} x_1(k) \\ x_2(k) \end{bmatrix} = \begin{bmatrix} \cos\frac{k\pi}{2} & \sin\frac{k\pi}{2} \\ -\sin\frac{k\pi}{2} & \cos\frac{k\pi}{2} \end{bmatrix} \begin{bmatrix} x_1(0) \\ x_2(0) \end{bmatrix} \tag{5-40}$$

The state trajectories of $x_1(k)$ and $x_2(k)$ for two arbitrary initial states $\mathbf{x}(0)$ are shown in Fig. 5-3. The states at the instants of $k = 0,1,2,\ldots$ are represented by dots, and they are joined by dashed lines to indicate the transition of states from one instant to the next. In this case the trajectories form limit cycles which neither grow nor decay in amplitude. The corresponding time response is oscillatory as indicated by the sine and cosine terms in $\phi(k)$. Therefore, the system is stable but not asymptotically stable.

Another way of investigating the stability of the system is to find the eigenvalues of **A**. The characteristic equation of the system is

$$|z\mathbf{I} - \mathbf{A}| = z^2 + 1 = 0 \tag{5-41}$$

Thus, the roots of the characteristic equation are simple and are at $z = j$ and $z = -j$ which are on the unit circle in the z-plane.

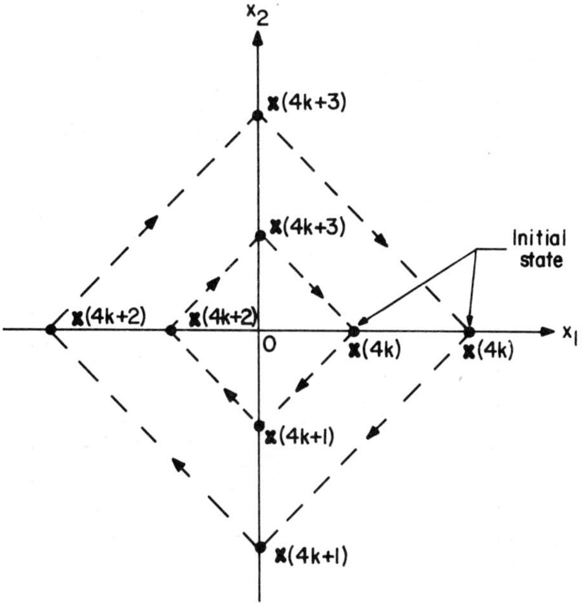

Fig. 5-3. State-plane trajectories of system in Example 5-1.

Example 5-2.

Consider that for the system of Eq. (5-36),

$$\mathbf{A} = \begin{bmatrix} -0.5 & 0 \\ 0 & 0.5 \end{bmatrix} \tag{5-42}$$

The solution of the state equation in the z-domain is

$$\mathbf{X}(z) = (z\mathbf{I} - \mathbf{A})^{-1} z\mathbf{x}(0) = \begin{bmatrix} \frac{z}{z+0.5} & 0 \\ 0 & \frac{z}{z-0.5} \end{bmatrix} \mathbf{x}(0) \tag{5-43}$$

The inverse z-transform of $\mathbf{X}(z)$ is

$$\mathbf{x}(k) = \begin{bmatrix} (-0.5)^k & 0 \\ 0 & (0.5)^k \end{bmatrix} \mathbf{x}(0) \tag{5-44}$$

For arbitrary initial states of $\mathbf{x}(0) = [1 \quad 1]'$ and $[-1 \quad -1]'$, the state trajectories of $\mathbf{x}(k)$ are shown in Fig. 5-4. In this case, the trajectories converge to the equilibrium state $\mathbf{x} = 0$ as k approaches infinity. Therefore, the system is asymptotically stable in the large.

The characteristic equation of the system is

$$|z\mathbf{I} - \mathbf{A}| = (z - 0.5)(z + 0.5) = 0 \tag{5-45}$$

which has two real roots at $z = -0.5, 0.5$. These roots are all inside the unit circle $|z| = 1$.

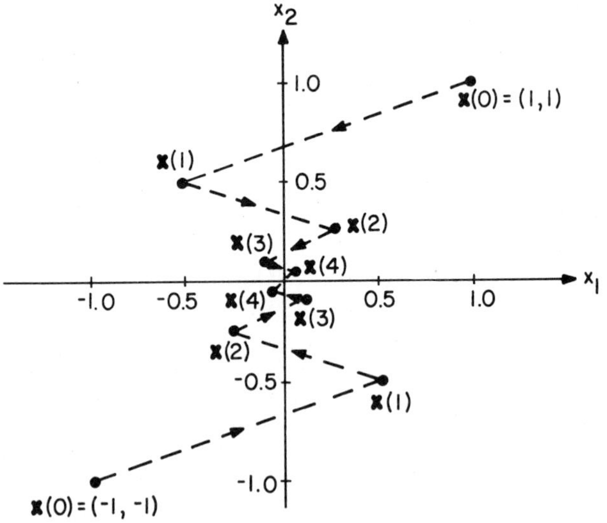

Fig. 5-4. State-plane trajectories of system in Example 5-2.

Example 5-3.

Now consider that the A matrix of the system of Eq. (5-36) is

$$\mathbf{A} = \begin{bmatrix} -1.5 & 0 \\ 0 & 0.5 \end{bmatrix} \tag{5-46}$$

Then,

$$X(z) = \begin{bmatrix} \dfrac{z}{z+1.5} & 0 \\ 0 & \dfrac{z}{z-0.5} \end{bmatrix} x(0) \tag{5-47}$$

In this case the characteristic equation of the system has a root at z = −1.5 which is outside the unit circle.

Therefore, the system is unstable, that is, it is neither stable nor asymptotically stable. The inverse z-transform of X(z) is

$$x(k) = \begin{bmatrix} (-1.5)^k & 0 \\ 0 & (0.5)^k \end{bmatrix} x(0) \tag{5-48}$$

For an initial state $x(0) = [1 \quad 1]'$, the state-plane trajectory of Fig. 5-5 shows that the response of $x_1(k)$ increases without bound as k approaches infinity.

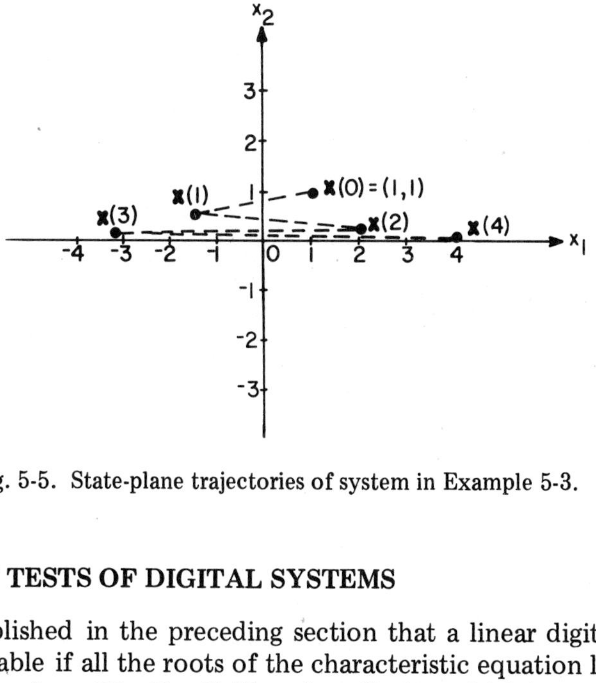

Fig. 5-5. State-plane trajectories of system in Example 5-3.

5.3 STABILITY TESTS OF DIGITAL SYSTEMS

It has been established in the preceding section that a linear digital system is asymptotically stable if all the roots of the characteristic equation lie inside the unit circle in the z-plane. The Routh-Hurwitz criterion which is normally useful for continuous-data systems cannot be applied directly to the z-domain since the stability boundary is now different.

One of the first methods which was devised for testing of the location of the roots of a polynomial in z with respect to the unit circle is the Schur-Cohn criterion. The criterion gives the necessary and sufficient conditions for the roots to lie inside the unit circle in terms of the signs of the "Schur-Cohn determinants." In this text we are not going to cover the Schur-Cohn criterion simply because it is very cumbersome for equations higher than the second order. In the following discussions the Jury's stability test and its variations are given.

Jury's Stability Test

A stability test which has some of the advantages of the Routh's test for continuous-data systems, and is simpler to apply than the Schur-Cohn criterion, is devised by Jury and Blanchard [7].

In general, given the polynomial in z,

$$F(z) = a_n z^n + a_{n-1} z^{n-1} + \ldots + a_2 z^2 + a_1 z + a_0 = 0 \tag{5-49}$$

where $a_0, a_1, \ldots, a_n$ are real coefficients. Assuming that a_n is positive, or that it can be made positive by changing the signs of all the coefficients, the following table is made:

Row	z^0	z^1	z^2	$\ldots$	z^{n-k}	$\ldots$	z^{n-1}	z^n
1	a_0	a_1	a_2	$\ldots$	a_{n-k}	$\ldots$	a_{n-1}	a_n
2	a_n	a_{n-1}	a_{n-2}	$\ldots$	a_k	$\ldots$	a_1	a_0
3	b_0	b_1	b_2	$\ldots$	b_{n-k}	$\ldots$	b_{n-1}	
4	b_{n-1}	b_{n-2}	b_{n-3}	$\ldots$	b_k	$\ldots$	b_0	
5	c_0	c_1	c_2	$\ldots$		c_{n-2}		
6	c_{n-2}	c_{n-3}	c_{n-4}	$\ldots$		c_0		
$\vdots$	$\vdots$	$\vdots$	$\vdots$	$\ldots$				
2n − 5	p_0	p_1	p_2	p_3				
2n − 4	p_3	p_2	p_1	p_0				
2n − 3	q_0	q_1	q_2					

Note that the elements of the (2k + 2)th row (k = 0,1,2,...) consist of the coefficients of the (2k + 1)th row written in the reverse order. The elements in the table are defined as

$$b_k = \begin{vmatrix} a_0 & a_{n-k} \\ a_n & a_k \end{vmatrix} \qquad c_k = \begin{vmatrix} b_0 & b_{n-1-k} \\ b_{n-1} & b_k \end{vmatrix}$$

$$d_k = \begin{vmatrix} c_0 & c_{n-2-k} \\ c_{n-2} & c_k \end{vmatrix} \qquad \cdots$$

$$\cdots \quad q_0 = \begin{vmatrix} p_0 & p_3 \\ p_3 & p_0 \end{vmatrix} \qquad q_2 = \begin{vmatrix} p_0 & p_1 \\ p_3 & p_2 \end{vmatrix}$$

The necessary and sufficient conditions for the polynomial F(z) to have no roots on and outside the unit circle in the z-plane (asymptotic stability) are:

$$F(1) > 0$$

$$F(-1) \begin{matrix} > 0 & \quad n \text{ even} \\ < 0 & \quad n \text{ odd} \end{matrix}$$

$$\left.\begin{matrix} |a_0| < a_n \\ |b_0| > |b_{n-1}| \\ |c_0| > |c_{n-2}| \\ |d_0| > |d_{n-3}| \\ \vdots \quad\quad \vdots \\ |q_0| > |q_2| \end{matrix}\right\} \quad (n-1) \text{ constraints} \tag{5-50}$$

For a second-order system, n = 2, Jury's tabulation contains only one row. Therefore, the requirements listed in Eq. (5-50) are reduced to

$$F(1) > 0$$

$$F(-1) > 0$$

and

$$|a_0| < a_n$$

It should be noted that the test of stability given in Eq. (5-50) is valid only if the inequality conditions provide conclusive results. As in the Routh-Hurwitz criterion which is used for stability testing of linear continuous-data systems, occasionally the first element of a row or a complete row of the tabulation may be zero before the tabulation is scheduled to terminate. These cases are referred to as *singular cases*. In Jury's tabulation a singular case is signified by either having the first and the last elements of a row be zero, or having a complete row of zeros.

The Singular Cases

When some or all of the elements of a row in the Jury's tabulation are zero, the tabulation ends prematurely. We refer to this situation as the *singular case*. The singular case can be eliminated by expanding and contracting the unit circle infinitesimally, which is equivalent to moving the roots off the unit circle. The transformation for this purpose is

$$z = (1 + \epsilon)z \qquad (5\text{-}51)$$

where ϵ is a very small real number. When ϵ is a positive number in Eq. (5-51), the radius of the unit circle is expanded to $1 + \epsilon$, and when ϵ is negative, the radius of the unit circle is reduced to $1 + \epsilon$. This is equivalent to moving the roots slightly. The difference between the number of roots found inside (or outside) the unit circle when the circle is expanded and contracted by ϵ is the number of roots on the circle.

The transformation in Eq. (5-51) is actually very easy to apply, since

$$(1 \pm \epsilon)^n z^n \cong (1 + n\epsilon)z^n \qquad (5\text{-}52)$$

This means that the coefficient of the z^n term is multiplied by $(1 \pm n\epsilon)$.

Raible's Tabular Form [30]

R. H. Raible devised a tabular form for Jury's stability test which simplifies the computation by hand. With reference to the equation of Eq. (5-49), Raible's tabulation is as follows:

$$
\begin{array}{cl}
 & \begin{array}{cccccccc} a_n & a_{n-1} & a_{n-2} & \cdots & a_2 & a_1 & a_0 & k_a \end{array} \\
\begin{array}{c} n \\ \text{calculated} \\ \text{rows} \end{array} &
\left\{ \begin{array}{cccccccc}
b_0 & b_1 & b_2 & \cdots & b_{n-2} & b_{n-1} & & k_b \\
c_0 & c_1 & c_2 & & c_{n-2} & & & k_c \\
\cdots & \cdots & \cdots & \cdots & \cdots & \cdots & \cdots & \cdots \\
p_0 & p_1 & p_2 & & & & & k_p \\
q_0 & q_1 & & & & & & k_q \\
r_0 & & & & & & &
\end{array} \right.
\end{array}
$$

where

$$k_a = a_0/a_n$$

$$k_b = b_{n-1}/b_0$$

$$k_c = c_{n-2}/c_0$$

$$\vdots$$

$$k_p = p_2/p_0$$

$$k_q = q_1/q_0$$

$$b_i = a_{n-i} - k_a a_i \qquad i = 0,1,2,\ldots,n-1$$

$$c_j = b_j - k_b b_{n-j-1} \qquad j = 0,1,2,\ldots,n-2$$

$$\vdots$$

$$q_0 = p_0 - k_p p_2$$

$$q_1 = p_1 - k_p p_1$$

$$r_0 = q_0 - k_q q_1$$

If Raible's tabulation is nonsingular, i.e., not one element in the first column is zero or one complete row of zeros is present, and if a_n is positive (or we can always have this arranged), the signs of the n-calculated elements in the first column indicate the number of roots of the equation $F(z) = 0$ that are inside (or outside) the unit circle. The conclusions are as follows:

Number of positive calculated elements in the first column

$(b_0, c_0, \ldots, p_0, q_0, r_0)$ = number of roots inside the unit circle

Or,

Number of negative calculated elements in the first column

$(b_0, c_0, \ldots, p_0, q_0, r_0)$ = number of roots outside the unit circle

If a singular case occurs in Raible's tabulation, it can be corrected by the transformation of Eq. (5-51). By setting $\epsilon > 0$ and $\epsilon < 0$ the difference between the positive (or negative) elements in the first column of Raible's tabulation gives the number of roots that are on the unit circle.

For the convenience of hand calculation, Raible's tabulation can also be formed as follows:

	a_n	a_{n-1}	a_{n-2}	$\cdots$	a_2	a_1	a_0	$k_a = a_0/a_n$
$-)$	$a_0 k_a$	$a_1 k_a$	$a_2 k_a$	$\cdots$	$a_{n-2} k_a$	$a_{n-1} k_a$		
	b_0	b_1	b_2	$\cdots$	b_{n-2}	b_{n-1}		$k_b = b_{n-1}/b_0$
$-)$	$b_{n-1} k_b$	$b_{n-2} k_b$	$b_{n-3} k_b$	$\cdots$	$b_1 k_b$			
	c_0	c_1	c_2	$\cdots$	c_{n-2}			$k_c = c_{n-2}/c_0$
$-)$	$\cdots$	$\cdots$	$\cdots$	$\cdots$	$\cdots$	$\cdots$		
	p_0	p_1	p_2					$k_p = p_2/p_0$
$-)$	$p_2 k_p$	$p_1 k_p$						
	q_0	q_1						$k_q = q_1/q_0$
$-)$	$q_1 k_q$							
	r_0							

Example 5-4.

Consider the same digital system of Example 5-2. The characteristic equation of the system is

$$F(z) = z^2 + z + 0.25 = 0 \tag{5-53}$$

The first two conditions of Jury's test in Eq. (5-50) lead to

$$F(1) = 2.25 > 0 \quad \text{and} \quad F(-1) = 0.25 > 0$$

Since $n = 2$ is even, these results satisfy the $F(1) > 0$ and $F(-1) > 0$ requirements for stability. Next, we tabulate the coefficients of $F(z)$ according to Jury's test; we have

Row	z^0	z^1	z^2
1	0.25	1	1

Since $2n - 3 = 1$, Jury's tabulation consists of only one row. The result is

$$|a_0| = 0.25 < a_2 = 1$$

and thus the system is asymptotically stable, and all the roots are inside the unit circle.

As an alternative we can also use Raible's tabulation, as follows:

	$a_2 = 1$	$a_1 = 1$	$a_0 = 0.25$	$k_a = 0.25$
−)	$a_0k_a = 0.0625$	$a_1k_a = 0.25$		
	$b_0 = 0.9375$	$b_1 = 0.75$		$k_b = 0.8$
−)	$b_1k_b = 0.6$			
	$c_0 = 0.3375$			

Since the calculated coefficients b_0 and c_0 in the first column are both positive, the two roots are all inside the unit circle.

Example 5-5.

Consider the equation

$$F(z) = z^3 + 3.3z^2 + 3z + 0.8 = 0 \tag{5-54}$$

which has roots at $z = -0.5, -0.8$, and -2.0.

From Jury's test, $F(1) = 8.1$ and $F(-1) = 0.1$. For odd n, since $F(-1)$ is not negative, F(z) has at least one root outside the unit circle. Let us apply Raible's tabulations to determine the number of roots that are outside the unit circle. Raible's tabulation is given as follows.

	1	3.3	3	0.8	$k_a = 0.8$
−)	0.64	2.4	2.64		
	0.36	0.9	0.36		$k_b = 1$
−)	0.36	0.9			
	0	0			

Thus we have a singular case. To alleviate the difficulty, we multiply the coefficient of z^k by $1 + k\epsilon$, $k = 1,2,3$, in Eq. (5-54). The result is

$$F[(1 + \epsilon)z] \cong (1 + 3\epsilon)z^3 + 3.3(1 + 2\epsilon)z^2 + 3(1 + \epsilon)z + 0.8 = 0 \tag{5-55}$$

Raible's tabulation for Eq. (5-55) is now made as follows, where the ϵ^2 terms have been neglected.

	$1 + 3\epsilon$	$3.3(1 + 2\epsilon)$	$3(1 + \epsilon)$	0.8	$k_a = \dfrac{0.8}{1 + 3\epsilon}$
−)	$\dfrac{0.64}{1 + 3\epsilon}$	$\dfrac{2.4(1 + \epsilon)}{1 + 3\epsilon}$	$\dfrac{2.64(1 + 2\epsilon)}{1 + 3\epsilon}$		

	$b_0 = \dfrac{0.36 + 6\epsilon}{1 + 3\epsilon}$	$\dfrac{0.9 + 14.1\epsilon}{1 + 3\epsilon}$	$\dfrac{0.36 + 6.72\epsilon}{1 + 3\epsilon}$	$k_b = \dfrac{0.36 + 6.72\epsilon}{0.36 + 6\epsilon}$
−)	$\dfrac{(0.36 + 6.72\epsilon)^2}{(0.36 + 6\epsilon)(1 + 3\epsilon)}$	$\dfrac{(0.9 + 14.1\epsilon)(0.36 + 6.72\epsilon)}{(0.36 + 6\epsilon)(1 + 3\epsilon)}$		
	$c_0 = \dfrac{-0.52\epsilon}{(0.36 + 6\epsilon)(1 + 3\epsilon)}$	$\dfrac{-0.64\epsilon}{(0.36 + 6\epsilon)(1 + 3\epsilon)}$		$k_c = 1.24$
−)	$\dfrac{-0.79\epsilon}{(0.36 + 6\epsilon)(1 + 3\epsilon)}$			
	$d_0 = \dfrac{0.27\epsilon}{(0.36 + 6\epsilon)(1 + 3\epsilon)}$			

From the above tabulation we observe that for

$$\epsilon > 0: \; b_0 > 0, c_0 < 0, d_0 > 0$$

$$\epsilon < 0: \; b_0 > 0, c_0 > 0, d_0 < 0$$

Since there is one negative element when ϵ is either positive or negative, there is one root outside the unit circle and none on it.

Example 5-6.

Consider the equation

$$F(z) = z^3 + z^2 + z + 1 = 0 \tag{5-56}$$

This is recognized as a singular case at the outset since all the coefficients of the equation are identical. Applying the transformation in Eq. (5-51), we have

$$F[(1 + \epsilon)z] \cong (1 + 3\epsilon)z^3 + (1 + 2\epsilon)z^2 + (1 + \epsilon)z + 1 = 0 \tag{5-57}$$

Raible's tabulation is given as follows (all ϵ^2 terms are neglected):

	$1 + 3\epsilon$	$1 + 2\epsilon$	$1 + \epsilon$	1	$k_a = \dfrac{1}{1 + 3\epsilon}$
−)	$\dfrac{1}{1 + 3\epsilon}$	$\dfrac{1 + \epsilon}{1 + 3\epsilon}$	$\dfrac{1 + 2\epsilon}{1 + 3\epsilon}$		
	$b_0 = \dfrac{6\epsilon}{1 + 3\epsilon}$	$\dfrac{4\epsilon}{1 + 3\epsilon}$	$\dfrac{2\epsilon}{1 + 3\epsilon}$		$k_b = \dfrac{1}{3}$
−)	$\dfrac{2\epsilon}{3(1 + 3\epsilon)}$	$\dfrac{4\epsilon}{3(1 + 3\epsilon)}$			

$$c_0 = \frac{16\epsilon}{3(1+3\epsilon)} \qquad \frac{8\epsilon}{3(1+3\epsilon)} \qquad\qquad k_c = \frac{1}{2}$$

$$-) \qquad \frac{4\epsilon}{3(1+3\epsilon)}$$

$$d_0 = \frac{4\epsilon}{1+3\epsilon}$$

From the first column of the tabulation we see that b_0, c_0 and d_0 are all directly proportional to ϵ. Thus, these coefficients are all positive when $\epsilon > 0$, and are negative when $\epsilon < 0$. When ϵ changes sign from positive to negative, the three roots move from the inside to the outside of the unit circle. The conclusion is that all three roots of Eq. (5-56) are on the unit circle. Of course, it is easy to show that the three roots are at $z = -1$, $+j$ and $-j$.

Example 5-7.

In this example we shall demonstrate the usefulness of Jury's stability test on the design of a digital system from the stability standpoint.

The characteristic equation of a digital control system is given as

$$F(z) = z^3 + (111.6T^2 + 16.74T - 3)z^2 + (3 - 33.48T + 1.395 \times 10^{-4}KT^3)z$$

$$+ (1.395 \times 10^{-4}KT^3 + 16.74T - 111.6T^2 - 1) = 0 \tag{5-58}$$

where T is the sampling period and K is a constant parameter. We want to determine the range of K and T so that the system is asymptotically stable.

Applying Jury's test to the last equation, we have

$$(1)\ \ F(1) > 0 \quad \text{or} \quad KT^3 > 0 \tag{5-59}$$

Since the sampling period T is necessarily positive, we have $K > 0$.

$$(2)\ \ F(-1) < 0 \quad \text{or} \quad -8 + 66.96T < 0 \tag{5-60}$$

which leads to $T < 0.12$ sec.

$$(3)\ \ |a_0| < a_n \quad \text{or} \quad |1.395 \times 10^{-4}KT^3 + 16.74T - 111.6T^2 - 1| < 1 \tag{5-61}$$

$$(4)\ \ |b_0| > |b_2| \tag{5-62}$$

where

$$b_0 = a_0^2 - a_3^2$$

$$b_2 = a_0a_2 - a_1a_3$$

$$a_0 = 1.395 \times 10^{-4}KT^3 + 16.74T - 111.6T^2 - 1$$

$$a_1 = 3 - 33.48T + 1.395 \times 10^{-4}KT^3$$

$$a_2 = 111.6T^2 + 16.74T - 3$$

$$a_3 = 1$$

The relations between K and T which satisfy Eqs. (5-61) and (5-62) are plotted as shown in Fig. 5-6, with the vertical axis in logarithmic scale. The figure shows that the inequality condition of Eq. (5-62) is the more stringent one.

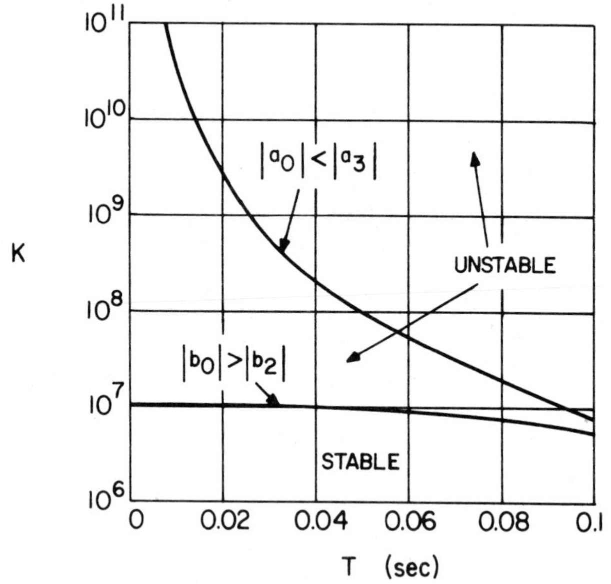

Fig. 5-6. Stable and unstable regions of the digital system described by the characteristic equation of Eq. (5-58).

Extension of Routh-Hurwitz Criterion to Discrete-Data Systems

As pointed out earlier, since the stability boundary in the z-plane is the unit circle $|z| = 1$, the Routh-Hurwitz criterion cannot be applied directly to the characteristic equation $F(z) = 0$. However, if we map the interior of the unit circle in the z-plane onto the left half of a complex-variable plane by a bilinear transformation, the Routh-Hurwitz criterion may again be applied directly to the equation in the new variable.

Let us consider the bilinear transformation

$$z = \frac{aw + b}{cw + d} \tag{5-63}$$

where w is a complex variable. We can show that the transformations

$$z = \frac{1 + w}{1 - w} \tag{5-64}$$

and

$$z = -\frac{1 + w}{1 - w} \tag{5-65}$$

will transform the interior of the unit circle onto the left half of the w-plane.

Once the characteristic equation $F(z) = 0$ is transformed into a polynomial of the same order in w, the Routh-Hurwitz criterion can be applied in a straightforward manner.

5.4 THE SECOND METHOD OF LIAPUNOV [2]

A powerful method of determining the stability of linear and nonlinear systems with continuous data is the second method of Liapunov. Much has been published about this subject, and it is readily extended into the study of stability of discrete-data systems. The second method of Liapunov is based on the determination of a V function called the *Liapunov function*. From the properties of the V function one is able to show stability or instability of the system. However, the main disadvantage of Liapunov's stability criterion is that it gives only the sufficient conditions for stability, not the necessary conditions. Furthermore, there are no unique methods of determining the V function for a wide class of systems.

The Liapunov Function

Any function $V = V[\mathbf{x}(k)]$ *of definite sign (positive definite or negative definite) is called a Liapunov function if* $V(\mathbf{0}) = 0$ *and* $\mathbf{x}(k)$ *is the solution of the state equation*

$$\mathbf{x}(k + 1) = \mathbf{f}[\mathbf{x}(k)] \tag{5-66}$$

We define a difference operation Δ so that $\Delta V[\mathbf{x}(k)]$ is defined as

$$\Delta V[\mathbf{x}(k)] = V[\mathbf{x}(k + 1)] - V[\mathbf{x}(k)] \tag{5-67}$$

Stability Theorem of Liapunov

Consider that a discrete-data system is described by

$$\mathbf{x}(k + 1) = \mathbf{f}[\mathbf{x}(k)] \tag{5-68}$$

where $\mathbf{x}(k)$ *is the* $n \times 1$ *state vector,* $\mathbf{f}[\mathbf{x}(k)]$ *is an* $n \times 1$ *function vector with the property that*

$$\mathbf{f}[\mathbf{x}(k) = \mathbf{0}] = \mathbf{0} \qquad \text{for all } k \tag{5-69}$$

Suppose that there exists a scalar function $V[\mathbf{x}(k)]$ *continuous in* $\mathbf{x}(k)$ *such that*

$$\begin{aligned} &1.\ V[\mathbf{x}(k) = \mathbf{0}] = V(\mathbf{0}) = 0 \\ &2.\ V[\mathbf{x}(k)] = V(\mathbf{x}) > 0 \quad \text{for} \quad \mathbf{x} \neq 0 \\ &3.\ V(\mathbf{x}) \textit{ approaches infinity as } ||\mathbf{x}|| \to \infty \\ &4.\ \Delta V(\mathbf{x}) < 0 \quad \text{for} \quad \mathbf{x} \neq \mathbf{0} \end{aligned} \tag{5-70}$$

Then the equilibrium state $\mathbf{x}(k) = \mathbf{0}$ *(for all* k*) is asymptotically stable in the large and* $V(\mathbf{x})$ *is a Liapunov function.*

Instability Theorem of Liapunov

For the system described by Eq. (5-66), if there exists a scalar function $V(\mathbf{x})$ *continuous in* $\mathbf{x}(k)$ *such that*

$$\Delta V(\mathbf{x}) < 0$$

then:

1. *The system is unstable in the finite region for which* V *is not positive semidefinite* ($\geqslant 0$).
2. *The response is unbounded as* k *approaches infinity if* V *is not positive semidefinite for all* $\mathbf{x}(k)$.

Example 5-8.

Consider the discrete-data system which is described by

$$x_1(k+1) = -0.5x_1(k) \tag{5-71}$$

$$x_2(k+1) = -0.5x_2(k) \tag{5-72}$$

Let us assign the Liapunov function to be

$$V(\mathbf{x}) = x_1^2(k) + x_2^2(k) \tag{5-73}$$

which is positive for all values of $x_1(k)$ and $x_2(k)$ not equal to zero. Then, the function $\Delta V(\mathbf{x})$ is evaluated using Eq. (5-67),

$$\begin{aligned} \Delta V(\mathbf{x}) &= V[\mathbf{x}(k+1)] - V[\mathbf{x}(k)] \\ &= x_1^2(k+1) + x_2^2(k+1) - x_1^2(k) - x_2^2(k) \\ &= -0.75x_1^2(k) - 0.75x_2^2(k) \end{aligned} \tag{5-74}$$

Since $\Delta V(\mathbf{x})$ is negative for all $\mathbf{x}(k) \neq 0$, the system is asymptotically stable.

Example 5-9.

Consider the digital system which is described by the state equations

$$x_1(k+1) = -1.5x_1(k) \tag{5-75}$$

$$x_2(k+1) = -0.5x_2(k) \tag{5-76}$$

It is simple to show that the eigenvalues of the **A** matrix are at -1.5 and -0.5 and that the system is unstable. However, without prior knowledge of the stability conditions of the system, the stability theorem of Liapunov is applied.

Let the Liapunov function be

$$V(\mathbf{x}) = x_1^2(k) + x_2^2(k) \tag{5-77}$$

Then,

$$\Delta V(\mathbf{x}) = V[\mathbf{x}(k+1)] - V[\mathbf{x}(k)]$$

$$= 1.25x_1^2(k) - 0.75x_2^2(k) \tag{5-78}$$

Since $\Delta V(\mathbf{x})$ is indefinite in sign, the test using the Liapunov function of Eq. (5-77) fails, and no conclusion on the stability condition of the system can be reached.

Now let us turn to the instability theorem of Liapunov. Let the Liapunov function be defined as

$$V(\mathbf{x}) = a_1x_1^2(k) + 2a_2x_1(k)x_2(k) + a_3x_2^2(k) \tag{5-79}$$

and let the function $\Delta V(\mathbf{x})$ be of the form

$$\Delta V(\mathbf{x}) = -x_1^2(k) - x_2^2(k) \tag{5-80}$$

so that it is negative definite for all $x_1(k)$ and $x_2(k) \neq 0$.

Forming $\Delta V(\mathbf{x})$ according to Eq. (5-67), we have

$$\Delta V(\mathbf{x}) = V[\mathbf{x}(k+1)] - V[\mathbf{x}(k)]$$

$$= 1.25a_1x_1^2(k) - 0.5a_2x_1(k)x_2(k) - 0.75a_3x_2^2(k) \tag{5-81}$$

Comparing Eqs. (5-80) and (5-81), we have

$$a_1 = -0.8,\ a_2 = 0 \text{ and } a_3 = 1.333$$

Thus, from Eq. (5-79),

$$V(\mathbf{x}) = -0.8x_1^2(k) + 1.333x_2^2(k) \tag{5-82}$$

Since $V(\mathbf{x})$ is indefinite there is again no conclusion of the stability condition.

Liapunov Stability Theorem for Linear Digital Systems

The stability and instability theorems of Liapunov are valid for both linear and nonlinear systems. The last two examples illustrate that a successful execution of the Liapunov tests depends upon the correct guess or assignment of $V(\mathbf{x})$ or $\Delta V(\mathbf{x})$, which is always a difficult task. However, for linear digital systems a simple test procedure is available.

Consider that a linear time-invariant digital system is described by the difference equation

$$\mathbf{x}(k+1) = \mathbf{A}\mathbf{x}(k) \tag{5-83}$$

where $\mathbf{x}(k)$ *is* $n \times 1$ *and* $\mathbf{A}$ *is an* $n \times n$ *matrix. The equilibrium state* $\mathbf{x}_e = \mathbf{0}$ *is asymptotically stable if and only if, given any positive-definite real symmetric matrix* $\mathbf{Q}$, *there exists a positive-definite real symmetric matrix* $\mathbf{P}$ *such that*

$$\mathbf{A}'\mathbf{P}\mathbf{A} - \mathbf{P} = -\mathbf{Q} \tag{5-84}$$

Then

$$V(\mathbf{x}) = \mathbf{x}'(k)\mathbf{P}\mathbf{x}(k) \tag{5-85}$$

is a Liapunov function for the system, and further,

$$\Delta V(\mathbf{x}) = -\mathbf{x}'(k)\mathbf{Q}\mathbf{x}(k) \tag{5-86}$$

where $\Delta V(\mathbf{x})$ *is as defined in Eq. (5-67).*

The proof of this theorem is based on Sylvester's theorem, which states that if $\mathbf{P}$ is a positive-definite matrix, then $V(\mathbf{x}) = \mathbf{x}'\mathbf{P}\mathbf{x}$ is positive definite. Using Eq. (5-85) as the Liapunov function,

$$\begin{aligned}\Delta V(\mathbf{x}) &= V[\mathbf{x}(k+1)] - V[\mathbf{x}(k)] \\ &= \mathbf{x}'(k+1)\mathbf{P}\mathbf{x}(k+1) - \mathbf{x}'(k)\mathbf{P}\mathbf{x}(k)\end{aligned} \tag{5-87}$$

Now substituting the state equation of Eq. (5-83) into Eq. (5-87), we have

$$\begin{aligned}\Delta V(\mathbf{x}) &= \mathbf{x}'(k)[\mathbf{A}'\mathbf{P}\mathbf{A} - \mathbf{P}]\mathbf{x}(k) \\ &= -\mathbf{x}'(k)\mathbf{Q}\mathbf{x}(k)\end{aligned} \tag{5-88}$$

Thus, from Sylvester's theorem, if $\Delta V(\mathbf{x})$ is to be negative definite, $\mathbf{Q}$ has to be positive definite.

Example 5-10.

Consider the same system in Example 5-8. The state equations are given in Eqs. (5-71) and (5-72). The coefficient matrix is

$$A = \begin{bmatrix} -0.5 & 0 \\ 0 & -0.5 \end{bmatrix} \tag{5-89}$$

The equilibrium state is $\mathbf{x}_e = \mathbf{0}$. Let $\mathbf{Q}$ be the identity matrix,

$$\mathbf{Q} = \begin{bmatrix} 1 & 0 \\ 0 & 1 \end{bmatrix}$$

and let $\mathbf{P}$ be of the form

$$\mathbf{P} = \begin{bmatrix} p_{11} & p_{12} \\ p_{21} & p_{22} \end{bmatrix}$$

Then Eq. (5-84) becomes

$$\begin{bmatrix} -0.5 & 0 \\ 0 & -0.5 \end{bmatrix}\begin{bmatrix} p_{11} & p_{12} \\ p_{21} & p_{22} \end{bmatrix}\begin{bmatrix} -0.5 & 0 \\ 0 & -0.5 \end{bmatrix} - \begin{bmatrix} p_{11} & p_{12} \\ p_{21} & p_{22} \end{bmatrix} = -\begin{bmatrix} 1 & 0 \\ 0 & 1 \end{bmatrix}$$

Solving for the elements of the $\mathbf{P}$ matrix from the last equation yields

$$\mathbf{P} = \begin{bmatrix} 1.33 & 0 \\ 0 & 1.33 \end{bmatrix}$$

which is positive definite. Therefore,

$$V(\mathbf{x}) = \mathbf{x}'(k)\mathbf{P}\mathbf{x}(k)$$

is the Liapunov function and is positive definite. The function $\Delta V(\mathbf{x})$ is given by Eq. (5-86), which is negative definite, and the equilibrium state is asymptotically stable.

Optimal State Feedback Design by the Liapunov Method

The stability criterion of Liapunov can be used for the design of a limited class of systems with state feedback. Figure 5-7 shows the block diagram of a linear digital system with state feedback. The state equations of the system are expressed as

$$\mathbf{x}(k + 1) = \mathbf{A}\mathbf{x}(k) + \mathbf{B}\mathbf{u}(k) \tag{5-90}$$

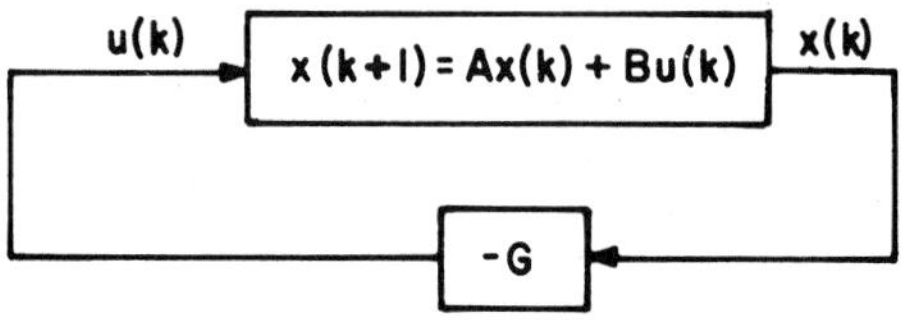

Fig. 5-7. A digital control system with state feedback.

The control $\mathbf{u}(k)$ is defined as

$$\mathbf{u}(k) = -\mathbf{G}\mathbf{x}(k)$$

where $\mathbf{x}(k)$ is an n-vector, $\mathbf{u}(k)$ an r-vector, and $\mathbf{G}$ is an $r \times n$ constant matrix.

The design objective is to find the feedback matrix $\mathbf{G}$ that will bring the state $\mathbf{x}(k)$ from any initial state $\mathbf{x}(0)$ to the equilibrium state $\mathbf{x} = \mathbf{0}$ in some optimal sense.

The basic assumption is that the system of Eq. (5-90) is asymptotically stable with $\mathbf{u}(k) = \mathbf{0}$. This guarantees that given a positive-definite real symmetric matrix $\mathbf{Q}$, there exists a positive-definite real symmetric matrix $\mathbf{P}$ such that

$$\mathbf{A}'\mathbf{P}\mathbf{A} - \mathbf{P} = -\mathbf{Q} \tag{5-91}$$

The Liapunov function is defined as

$$V[\mathbf{x}(k)] = \mathbf{x}'(k)\mathbf{P}\mathbf{x}(k) \tag{5-92}$$

and

$$\Delta V[\mathbf{x}(k)] = V[\mathbf{x}(k+1)] - V[\mathbf{x}(k)] = -\mathbf{x}(k)\mathbf{Q}\mathbf{x}(k) \tag{5-93}$$

as in Eqs. (5-85) and (5-86), respectively.

As an optimal control problem, we seek an optimal control $\mathbf{u}^{\circ}(k)$ which minimizes the performance index

$$J = \Delta V[\mathbf{x}(k)] \tag{5-94}$$

at the instant k.

Since $\Delta V[\mathbf{x}(k)]$ represents the discrete rate of change of $V[\mathbf{x}(k)]$, we can define $V[\mathbf{x}(k)]$ so that minimization of the performance index of Eq. (5-94) will carry a physical meaning in optimal control. For instance, $\Delta V[\mathbf{x}(k)]$ may represent the rate of change of distance or energy along the trajectory, $\mathbf{x}(k)$. For a second-order system, if we select $V[\mathbf{x}(k)]$ to be

$$V[\mathbf{x}(k)] = x_1^2(k) + x_2^2(k) \tag{5-95}$$

then

$$-\Delta V[\mathbf{x}(k)] = -x_1^2(k+1) - x_2^2(k+1) + x_1^2(k) + x_2^2(k) \tag{5-96}$$

which is in the form of a quadratic performance index over the period k to k + 1.

For the general form of $V[\mathbf{x}(k)]$ given in Eq. (5-92), we write

$$\begin{aligned}\Delta V[\mathbf{x}(k)] &= \mathbf{x}'(k+1)\mathbf{P}\mathbf{x}(k+1) - \mathbf{x}'(k)\mathbf{P}\mathbf{x}(k)\\ &= \mathbf{x}'(k)(\mathbf{A} - \mathbf{B}\mathbf{G})'\mathbf{P}(\mathbf{A} - \mathbf{B}\mathbf{G})\mathbf{x}(k) - \mathbf{x}'(k)\mathbf{P}\mathbf{x}(k)\end{aligned} \tag{5-97}$$

Using the state equation relationship in Eq. (5-91), the last equation is written

$$\Delta V[\mathbf{x}(k)] = \mathbf{x}'(k)\mathbf{A}'\mathbf{P}\mathbf{A}\mathbf{x}(k) + \mathbf{u}'\mathbf{B}'\mathbf{P}\mathbf{A}\mathbf{x}(k) + \mathbf{x}'(k)\mathbf{A}'\mathbf{P}\mathbf{B}\mathbf{u}(k) + \mathbf{u}'(k)\mathbf{B}'\mathbf{P}\mathbf{B}\mathbf{u}(k) - \mathbf{x}'(k)\mathbf{P}\mathbf{x}(k) \tag{5-98}$$

For minimum $\Delta V[\mathbf{x}(k)]$ with respect to $\mathbf{u}(k)$,

$$\frac{\partial \Delta V[\mathbf{x}(k)]}{\partial \mathbf{u}(k)} = \mathbf{0} \tag{5-99}$$

Substitution of Eq. (5-98) in Eq. (5-99) leads to

$$2\mathbf{B}'\mathbf{P}\mathbf{A}\mathbf{x}(k) + 2\mathbf{B}'\mathbf{P}\mathbf{B}\mathbf{u}(k) = \mathbf{0}$$

Therefore, the optimal control is

$$\mathbf{u}^{\circ}(k) = -(\mathbf{B}'\mathbf{P}\mathbf{B})^{-1}\mathbf{B}'\mathbf{P}\mathbf{A}\mathbf{x}(k) \tag{5-100}$$

which is in the form of *state feedback*.

It should be pointed out that this design technique is restricted by the requirement that the process of Eq. (5-90) must be asymptotically stable, which would exclude continuous-data systems with one or more poles at the origin of the s-plane.

Example 5-11.

Consider that the linear digital process shown in Fig. 5-7 is described by the state equations

$$\mathbf{x}(k+1) = \mathbf{A}\mathbf{x}(k) + \mathbf{B}u(k) \tag{5-101}$$

where

$$\mathbf{A} = \begin{bmatrix} 0.5 & 0 \\ 0 & 0.2 \end{bmatrix} \tag{5-102}$$

$$\mathbf{B} = \begin{bmatrix} 1 \\ 1 \end{bmatrix} \tag{5-103}$$

Find the optimal control $u^{\circ}(k)$ so that given $\mathbf{Q} = \mathbf{I}$ (identity matrix), with $V(\mathbf{x}) = \mathbf{x}'(k)\mathbf{P}\mathbf{x}(k)$, and $\mathbf{P}$ is the solution of

$$-\mathbf{Q} = \mathbf{A}'\mathbf{P}\mathbf{A} - \mathbf{P} \tag{5-104}$$

the performance index

$$\Delta V(\mathbf{x}) = V[\mathbf{x}(k+1)] - V[\mathbf{x}(k)] \tag{5-105}$$

is minimized.

It is simple to show that the process represented by Eq. (5-101) is asymptotically stable, since the eigenvalues of $\mathbf{A}$ are at 0.5 and 0.2.

Let **P** be written as

$$\mathbf{P} = \begin{bmatrix} p_{11} & p_{12} \\ p_{12} & p_{22} \end{bmatrix} \tag{5-106}$$

Substituting Eq. (5-106) into Eq. (5-104) and solving for the elements of **P**, we have

$$p_{11} = \frac{1}{1-(0.5)^2} = 1.333$$

$$p_{12} = 0$$

$$p_{22} = \frac{1}{1-(0.2)^2}$$

Thus,

$$\mathbf{P} = \begin{bmatrix} 1.333 & 0 \\ 0 & 1.042 \end{bmatrix}$$

which is positive definite.

The optimal control which minimizes the performance index of Eq. (5-105) is given by Eq. (5-100). The optimal feedback gain is

$$\begin{aligned} G &= (\mathbf{B}'\mathbf{P}\mathbf{B})^{-1}\mathbf{B}'\mathbf{P}\mathbf{A} \\ &= [0.28 \quad 0.0876] \end{aligned} \tag{5-107}$$

The optimal control is

$$u^{\circ}(k) = -0.28x_1(k) - 0.0876x_2(k) \tag{5-108}$$

The state diagram of the closed-loop system is shown in Fig. 5-8. It can be shown that the eigenvalues of the optimal closed-loop system are approximately 0 and 0.3324.

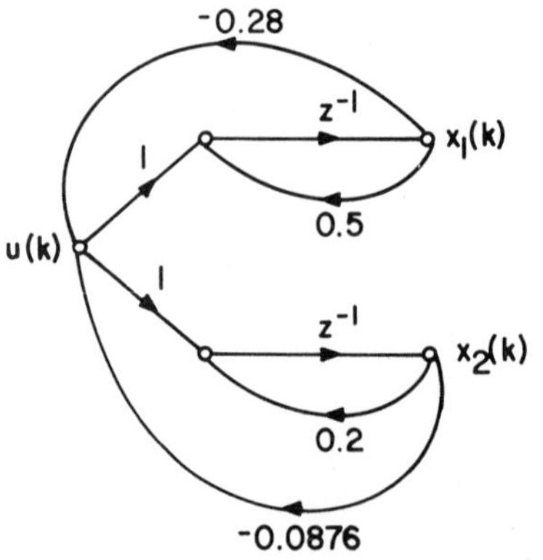

Fig. 5-8. Closed-loop system designed in Example 5-11.

REFERENCES

1. Kuo, B. C., *Automatic Control Systems*, 3rd edition. Prentice-Hall, Englewood Cliffs, N.J., 1975.
2. Kalman, R. E., and Bertram, J. E., "Control Systems Analysis and Design Via the Second Method of Liapunov: II, Discrete Time Systems," *Trans. ASME, J. Basic Eng.*, Series D, No. 3, June 1960, pp. 371-400.
3. Hahn, W., *Theory and Application of Liapunov's Direct Method.* Prentice-Hall, Englewood Cliffs, N.J., 1963.
4. Marden, M., *The Geometry of the Zeros of a Polynomial in a Complex Variable*, Chapter 10. American Math. Society, New York, 1949.
5. Kuo, B. C., *Analysis and Synthesis of Sampled-Data Control Systems.* Prentice-Hall, Englewood Cliffs, N.J., 1963.
6. Jury, E. I., and Bharucha, B. H., "Notes on the Stability Criterion for Linear Discrete Systems," *IRE Trans. on Automatic Control*, Vol. AC-6, February 1961, pp. 88-90.
7. Jury, E. I., and Blanchard, J., "A Stability Test for Linear Discrete Systems in Table Form," *IRE Proc.*, 49, No. 12, December 1961, pp. 1947-1948.
8. Ogata, K., *State Space Analysis of Control Systems.* Prentice-Hall, Englewood Cliffs, N.J., 1967.
9. Jury, E. I., "The Number of Roots of a Real Polynomial Inside (or Outside) the Unit Circle Using the Determinant Method," *IEEE Trans. on Automatic Control*, Vol. AC-10, July 1965, pp. 371-372.
10. Cohen, M. L., "A Set of Stability Constraints on the Denominator of a Sampled-Data Filter," *IEEE Trans. on Automatic Control*, Vol. AC-11, April 1966, pp. 327-328.
11. Pearson, J. B., Jr., "A Note on the Stability of a Class of Optimum Sampled-Data Systems," *IEEE Trans. on Automatic Control*, Vol. AC-10, January 1965, pp. 117-118.
12. Szego, G. P., and Pearson, J. B., Jr., "On the Absolute Stability of Sampled-Data Systems: the Indirect Control Case," *IEEE Trans. on Automatic Control*, Vol. AC-9, April 1964, pp. 160-163.
13. O'Shea, R. P. O., and Younis, M. I., "A Frequency-Time Domain Stability Criterion for Sampled-Data Systems," *IEEE Trans. on Automatic Control*, Vol. AC-12, December 1967, pp. 719-724.
14. Wu, S. H., "A Circle Stability Criterion for a Class of Discrete Systems," *IEEE Trans. on Automatic Control*, Vol. AC-12, February 1967, pp. 114-115.
15. Brockett, R. W., "The State of Stability Theory for Deterministic Systems," *IEEE Trans. on Automatic Control*, Vol. AC-11, July 1966, pp. 596-606.
16. Kodama, S., "Stability of a Class of Discrete Control Systems Containing a Nonlinear Gain Element," *IRE Trans. on Automatic Control*, Vol. AC-7, October 1962, pp. 102-109.
17. Kodama, S., "Stability on Nonlinear Sampled-Data Control Systems," *IRE Trans. on Automatic Control*, Vol. AC-7, January 1962, pp. 15-23.

18. Tsypkin, Y. S., "On the Stability in the Large of Nonlinear Sampled-Data Systems," *Dokl. Akad. Nank.*, Vol. 145, July 1962, pp. 52-55.
19. Tsypkin, Y. S., "Absolute Stability of Equilibrium Positions and of Responses in Nonlinear Sampled-Data, Automatic Systems," *Avtomat. i Telemekhan*, Vol. 24, December 1963, pp. 1601-1615.
20. Tsypkin, Y. S., "Frequency Criteria for the Absolute Stability of Nonlinear Sampled-Data Systems," *Avtomat. i Telemekhan*, Vol. 25, March 1964, pp. 281-290.
21. Iwens, R. P., and Bergen, A. R., "Frequency Criteria for Bounded-Input-Bounded-Output Stability of Nonlinear Sampled-Data Systems," *IEEE Trans. on Automatic Control*, Vol. AC-12, February 1967, pp. 46-53.
22. Iwens, R. P., and Bergen, A. R., "On Bounded-Input-Bounded-Output Stability of a Certain Class of Nonlinear Sampled-Data Systems," *J. Franklin Inst.*, Vol. 282, October 1966, pp. 193-205.
23. O'Shea, R. P., "Approximation of the Asymptotic Stability Boundary of Discrete-time Control Using an Inverse Transformation Approach," *IEEE Trans. on Automatic Control*, Vol. AC-9, October 1964, pp. 441-448.
24. Jury, E. I., and Lee, B. W., "On the Absolute Stability of Nonlinear Sampled-Data Systems," *IEEE Trans. on Automatic Control*, Vol. AC-9, October 1964, pp. 551-554.
25. Murphy, G. J., and Wu, S. H., "A Stability Criterion for Pulse-Width Modulated Feedback Control Systems," *IEEE Trans. on Automatic Control*, Vol. AC-9, October 1964, pp. 434-441.
26. Pearson, J. B., and Gibson, J. E., "On the Asymptotic Stability of a Class of Saturating Sampled-Data Systems," *IEEE Trans. on Application and Industry*, No. 71, March 1964, pp. 81-86.
27. Chen, C. T., "On the Stability of Nonlinear Sampled-Data Feedback Systems," *J. Franklin Inst.*, Vol. 280, October 1965, pp. 316-324.
28. Becker, P. W., "A Stability Criterion for Sampled-Data Control Systems," *IEEE Trans. on Automatic Control*, Vol. AC-8, January 1963, pp. 74-76.
29. Shortle, G. E., Jr., and Alexandro, F. J., Jr., "An Extended Frequency-Domain Stability Criterion for a Class of Sampled-Data Systems," *IEEE Trans. on Automatic Control*, Vol. AC-15, April 1970, pp. 232-234.
30. Raible, R. H., "A Simplification of Jury's Tabular Form," *IEEE Trans. on Automatic Control*, Vol. AC-19, June 1974, pp. 248-250.
31. Jury, E. I., *Theory and Application of the z-Transform Method.* John Wiley & Sons, New York, 1964.
32. Jury, E. I., and Gutman, S., "On the Stability of the A Matrix Inside the Unit Circle," *IEEE Trans. on Automatic Control*, Vol. AC-20, August 1975, pp. 533-535.
33. Jury, E. I., and Anderson, B. D. O., "A Simplified Schur-Cohen Test," *IEEE Trans. on Automatic Control*, Vol. AC-18, April 1973, pp. 157-163.
34. Seltzer, S. M., "A Launch Vehicle Stability Analysis Using the Parameter Plane Method," *IEEE Trans. on Automatic Control*, Vol. AC-15, October 1970, pp. 611-612.

PROBLEMS

5-1 Determine the stability conditions of the digital systems that are represented by the following characteristic equations:

(a) $z^3 + 5z^2 + 3z + 2 = 0$

(b) $z^4 + 9z^3 + 3z^2 + 9z + 1 = 0$

(c) $z^3 - 1.5z^2 - 2z + 3 = 0$

(d) $z^4 - 1.55z^3 + 0.5z^2 - 0.5z + 1 = 0$

(e) $z^4 - 2z^3 + z^2 - 2z + 1 = 0$

5-2 The characteristic equation of a linear digital system is

$$z^3 + Kz^2 + 1.5Kz - (K + 1) = 0$$

Determine the values of K for the system to be asymptotically stable.

5-3 The block diagram of a digital control system is shown in Fig. 5P-3. Determine the range of the sampling period T for the system to be asymptotically stable.

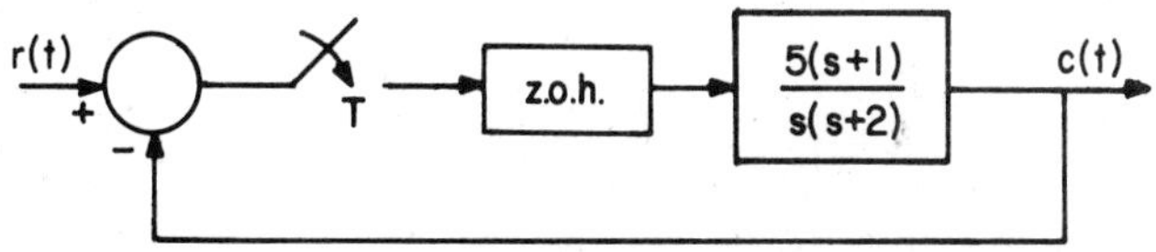

Fig. 5P-3.

5-4 Repeat Problem 5-3 when the hold device is a first-order hold.

5-5 A digital control system is described by the state equation

$$x(k + 1) = [0.368 - 0.632K]x(k) + 0.632(K + 1)u(k)$$

Determine the values of K for the equilibrium state to be asymptotically stable.

5-6 A digital control system is described by the state equation

$$\mathbf{x}(k + 1) = \mathbf{A}\mathbf{x}(k) + \mathbf{B}u(k)$$

where

$$\mathbf{A} = \begin{bmatrix} 0.5 & 1 \\ -1 & -1 \end{bmatrix} \qquad \mathbf{B} = \begin{bmatrix} 1 \\ 1 \end{bmatrix}$$

Determine the stability of the equilibrium state with $u(k) = 0$ by means of the second method of Liapunov. Find a Liapunov function $V(\mathbf{x})$.

5-7 A digital control system with state delays is described by the state equation

$$\mathbf{x}(k+1) = \mathbf{A}\mathbf{x}(k) + \mathbf{E}\mathbf{x}(k-1)$$

where

$$A = \begin{bmatrix} 0.5 & 1 \\ -1 & -1 \end{bmatrix} \qquad E = \begin{bmatrix} 0 & 0 \\ 0 & 1 \end{bmatrix}$$

Determine the stability of the equilibrium state.

5-8 For the digital control system with nonuniform sampling described in Problem 3-21, determine the region of asymptotic stability for the system in the K versus 1/N plane for $0 \leqslant K < \infty$ and $0 \leqslant 1/N \leqslant 1$.

5-9 For the digital control system with multirate sampling described in Problem 3-15, determine the maximum upper bound on the value of K for asymptotic stability as a function of N for $N \geqslant 1$.

5-10 A digital process is described by the state equation

$$\mathbf{x}(k+1) = \mathbf{A}\mathbf{x}(k) + \mathbf{B}u(k)$$

where

$$A = \begin{bmatrix} 0 & 0.81 \\ 1 & 0 \end{bmatrix} \qquad B = \begin{bmatrix} 0 \\ 1 \end{bmatrix}$$

Find the optimal state feedback control $u^{\circ}(k) = -\mathbf{G}\mathbf{x}(k)$ such that the performance index

$$J = [\mathbf{x}(k+1) - \mathbf{x}(k)]'\mathbf{P}[\mathbf{x}(k+1) - \mathbf{x}(k)]$$

is minimized, where $\mathbf{P}$ is the positive definite solution of the Liapunov equation $\mathbf{A}'\mathbf{P}\mathbf{A} - \mathbf{P} = \mathbf{I}$.

5-11 The state equations of a digital process are expressed as

$$\mathbf{x}(k+1) = \mathbf{A}\mathbf{x}(k) + \mathbf{B}u(k)$$

where

$$A = \begin{bmatrix} 0 & 1 \\ -1 & -2\zeta \end{bmatrix} \qquad B = \begin{bmatrix} 0 \\ 1 \end{bmatrix}$$

Find the optimal value of ζ so that the performance index

$$J = \mathbf{x}'(k)\mathbf{P}\mathbf{x}(k) = \text{minimum for all } k \geqslant 0$$

where **P** is the positive definite solution of the Liapunov equation $\mathbf{A}'\mathbf{P} + \mathbf{P}\mathbf{A} = -\mathbf{I}$, for $x_1(0) = x_0$ and $x_2(0) = 0$.

5-12 The block diagram of a digital tracking loop of a missile system is shown in Fig. 5P-12. The sampling period is 1/12 second.

(a) Determine the values of K for stability as a function of Δ $(0 \leqslant \Delta \leqslant 1)$. Plot the maximum value of K for stability as a function of Δ.

(b) With $K = 8$ and $\Delta = 0.4$, plot the output response of the system with a step input $r(t) = 2u(t)$.

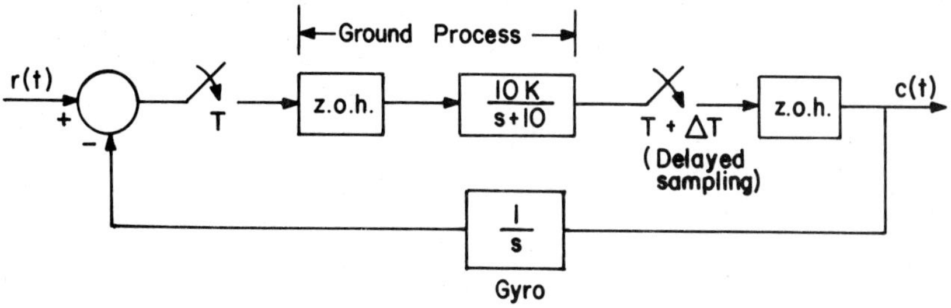

Fig. 5P-12.

5-13 The circuit diagram of an ac-to-dc SCR voltage-regulator system is shown in Fig. 5P-13(a). Systems of this type are generally difficult to analyze mathematically as the operations are nonlinear. For a sinusoidal input at e_i the signal at the output of the SCR bridge, e_1, is as shown in Fig. 5P-13(b). The conduction period, T_c, of the SCRs determines the average value of the output voltage e_0. The triggering angle α is directly related to the conduction period. As an approximation the SCRs and their phase angle control circuits can be represented by a sample-and-hold device, as shown in Fig. 5P-13(b). In this case the sample-and-hold has a delay in the sampling of α/ω sec and a partial holding period of T_c which is less than the sampling period T. The frequency of the input sinusoid is ω rad/sec, and, therefore, the sampling period is π/ω sec.

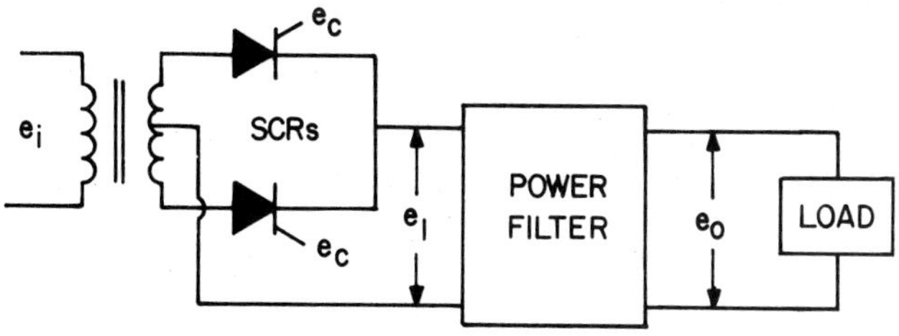

(a)

Fig. 5P-13.

Figure 5P-13(c) shows the block diagram of a closed-loop system which may be used as a steady-state model of an SCR system for the purpose of regulating the voltage e_0. The output voltage e_0 is fed back and compared with a reference voltage e_r, and the error $e = e_r - e_0$ is used to regulate the triggering angle a. Since the system shown in Fig. 5P-13(c) is still nonlinear, even with the use of the sample-and-hold approximation of Fig. 5P-13(b), for linear studies we can only consider the steady-state performance of the system about an equilibrium point for a.

Consider that the operating conditions are such that a nominal line voltage of 115 volts rms at 60 Hz is applied as e_i. The transfer function for the power filter is

$$G_f(s) = \frac{87}{(s+3)(s+29)}$$

(a) Determine the triggering angle a so that the average output voltage e_0 is 60 volts.
(b) For the closed-loop digital control system model shown in Fig. 5P-13(c), with the value of a as determined in (a), find the range of the gain of the phase-angle control circuit so that the system is asymptotically stable.
(c) Determine the steady-state error of e_0 in terms of K with a as determined in (a).

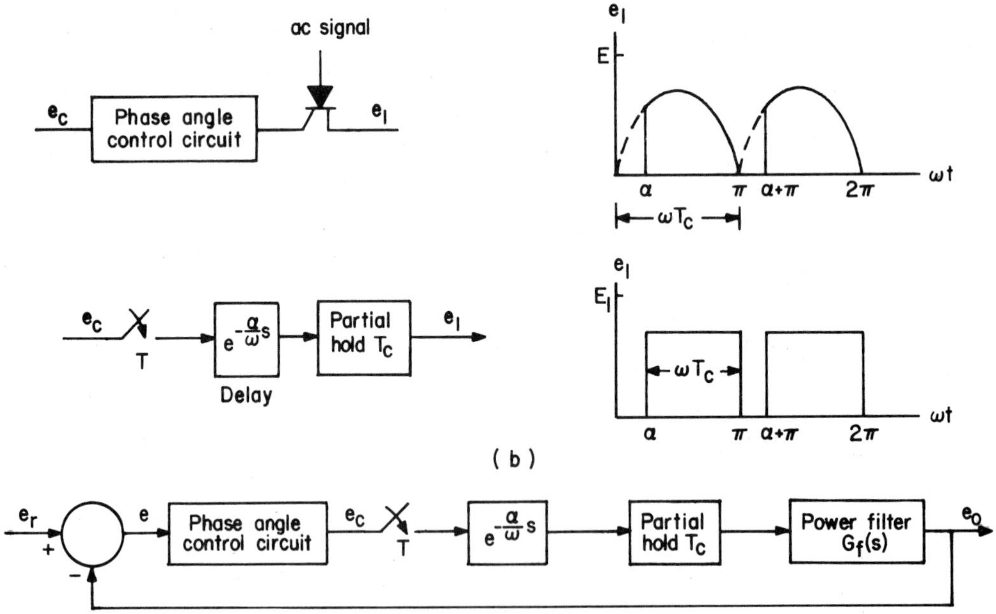

Fig. 5P-13.

5-14 Given a linear digital control system

$$\mathbf{x}(k+1) = \mathbf{A}\mathbf{x}(k) + \mathbf{B}u(k)$$

where

$$A = \begin{bmatrix} 0 & 1 \\ -1 & 1 \end{bmatrix} \qquad B = \begin{bmatrix} 0 \\ 1 \end{bmatrix}$$

The control u(k) is implemented through state feedback, u(k) = −Gx(k), where

$$G = [g_1 \quad g_2]$$

Find the ranges of g_1 and g_2 and express these in the g_2 versus g_1 plane so that the closed-loop system is asymptotically stable.

5-15 Figure 5P-15 shows the schematic diagram of a sampled-data control system for proportioning concentrate into a varying flow of forage for cattle feeding. The forage weigher converts the forage flow rate into an electrical signal via the load cell. The transfer function between the input of the amplifier u(t) and the concentrate flow rate c(t) is

$$\frac{C(s)}{U(s)} = \frac{36870}{s^2 + 83.3s + 7780}$$

Let the gain between the motor speed and the concentrate flow rate, and that of the speed transducer be unity. For T = 0.01 sec, find the closed-loop transfer function $C(z)/V_s(z)$ and determine if the system is stable.

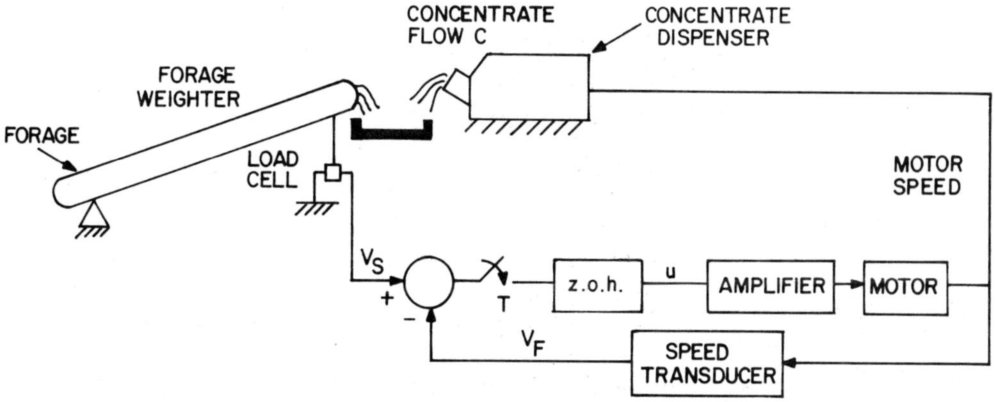

Fig. 5P-15.

5-16 Discrete pattern recognition and adaptive techniques require the use of prediction in digital control systems. Figure 5P-16 shows the block diagram of a digital control system with a digital predictor. The transfer function of the ideal predictor is given as $P(z) = z^p$, where p is a positive integer. For $G(s) = K/s^2$, sketch the root locus diagrams for the closed-loop system for p = 1 and 2, and discuss the stability conditions of the systems.

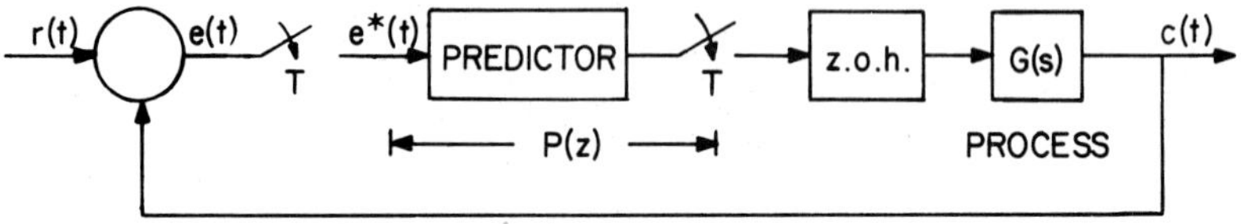

Fig. 5P-16.

6

DIGITAL SIMULATION AND DIGITAL REDESIGN

6.1 INTRODUCTION

It is well-known that digital computers play an increasingly important role in the analysis and design of feedback control systems. Digital computers are used not only in the computation and simulation of control systems performance, but also often in direct on-line control of systems. In addition, it is quite common to have airborne and shipboard computers for real-time and on-line controls.

In this chapter various methods of digital simulation and a digital redesign method are discussed.

6.2 DIGITAL SIMULATION — DIGITAL MODEL WITH SAMPLE-AND-HOLD

An analog system can be simulated on the digital computer once its system dynamics are approximated by a transfer function in the z-domain. The analysis usually consists of the following two steps:

1. Representation of the continuous-data system by a digital model.
2. Simulation of the digital model on a digital computer.

The second step has been discussed in Sec. 4.18.

There are many possible ways of representing a continuous-data system by a digital model. In general, the following three methods are used:

1. Insert sample-and-hold devices in the continuous-data system.
2. Numerical integration.
3. The z-form approximation.

The first method is discussed in this section. The simplest way of approximating a continuous-data system by a digital model is to insert fictitious sample-and-hold devices at strategic locations of the system. Then, the system can be described by z-transform transfer functions or difference state equations. For instance, the continuous-data control system shown in Fig. 6-1(a) is approximated by the digital model shown in Fig. 6-1(b). The hold device can be of any type at the discretion of the analyst. It can even be a polygonal hold which is not physically realizable but nevertheless quite suitable for digital simulation. In Fig. 6-1(b), let $G_h(s)$ denote the transfer function of the hold device. Then the discrete transfer function of the digital model is

$$\frac{C(z)}{R(z)} = \frac{G_h(z)G_hG(z)}{1 + G_hG(z)G_hH_1(z) + G_h(z)G_hG(z)G_hH_2(z)} \tag{6-1}$$

where

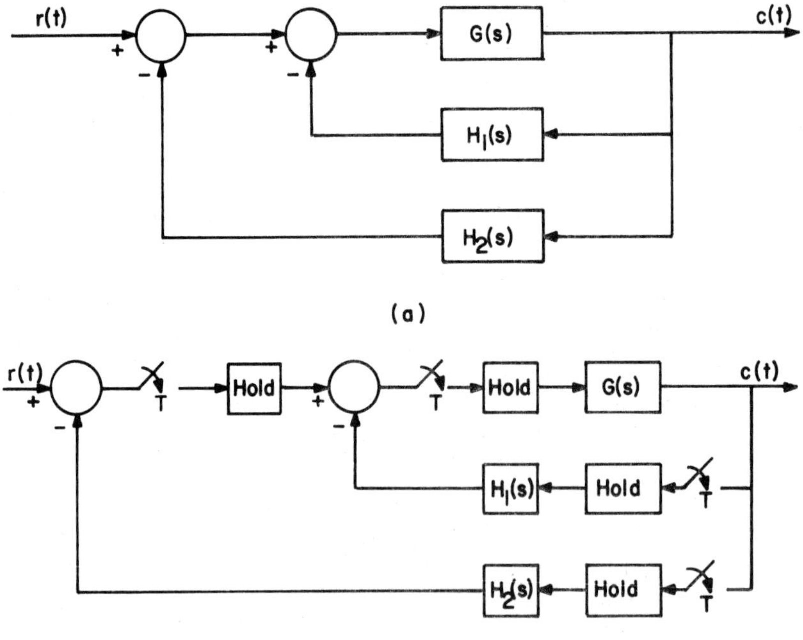

Fig. 6-1. (a) A continuous-data control system; (b) a digital model of the continuous-data system with fictitious sample-and-hold devices.

$$G_h(z) = \mathcal{Z}[G_h(s)] \tag{6-2}$$

$$G_hG(z) = \mathcal{Z}[G_h(s)G(s)] \tag{6-3}$$

$$G_hH_1(z) = \mathcal{Z}[G_h(s)H_1(s)] \tag{6-4}$$

$$G_hH_2(z) = \mathcal{Z}[G_h(s)H_2(s)] \tag{6-5}$$

Although in principle the digital modeling method just described is straightforward, it has two important considerations under actual applications. The first concerns the selection of the appropriate sampling period of the fictitious samplers. The sampling period is directly related to the accuracy as well as the time required of the digital simulation. The second consideration is that of stability. It was pointed out earlier that if the continuous-data system is stable, the digital model is not necessarily stable. In fact, sample-and-hold devices generally have an adverse effect on the system stability. Therefore, when inserting sample-and-hold devices in a stable continuous-data system, it is important that the digital model be a stable one.

As an illustrative example, let us refer to the continuous-data system shown in Fig. 6-2(a). A digital approximation of the system may be obtained by inserting a sample-and-hold in the feedback path as shown in Fig. 6-2(b). The z-transform of the output of the digital model is written

$$C(z) = \frac{RG(z)}{1 + G_{h0}G(z)} \tag{6-6}$$

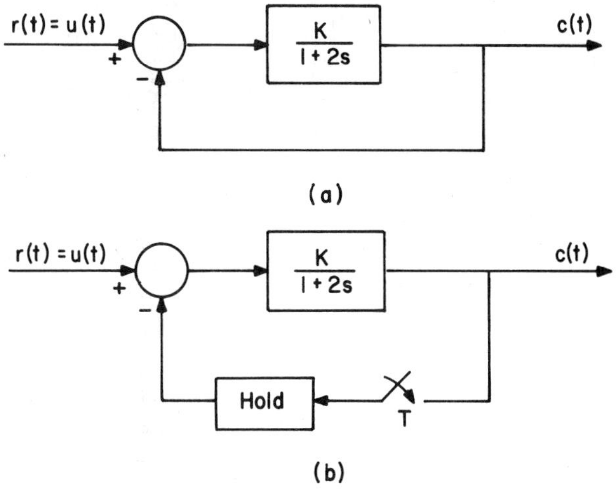

Fig. 6-2. (a) A continuous-data control system; (b) a digital model of the continuous-data system with sample-and-hold inserted in the feedback path.

where

$$RG(z) = \mathcal{Z}[R(s)G(s)] \tag{6-7}$$

and

$$G_{h0}G(z) = \mathcal{Z}[G_{h0}(s)G(s)] \tag{6-8}$$

For a unit-step input,

$$RG(z) = \mathcal{Z}\left[\frac{K}{s(1+2s)}\right] = \frac{K(1-e^{-0.5T})}{(z-1)(z-e^{-0.5T})} \tag{6-9}$$

Let us assume that the hold device is a zero-order hold. Then,

$$G_{h0}G(z) = \mathcal{Z}\left[\frac{1-e^{-Ts}}{s}\frac{K}{1+2s}\right] = \frac{K(1-e^{-0.5T})}{(z-e^{-0.5T})} \tag{6-10}$$

Substitution of Eqs. (6-9) and (6-10) into Eq. (6-6) yields

$$C(z) = \frac{K(1-e^{-0.5T})z}{(z-1)[z-e^{-0.5T}+K(1-e^{-0.5T})]} \tag{6-11}$$

Selecting $T = 0.25$ second and letting $K = 1$, the last equation becomes

$$C(z) = \frac{0.118z}{(z-1)(z-0.764)} \tag{6-12}$$

Expanding C(z) by continued division, we have

$$\begin{aligned} C(z) = {} & 0.118z^{-1} + 0.207z^{-2} + 0.276z^{-3} + 0.329z^{-4} \\ & + 0.369z^{-5} + 0.4z^{-6} + 0.423z^{-7} + 0.441z^{-8} \\ & + 0.455z^{-9} + 0.466z^{-10} + 0.473z^{-11} + 0.48z^{-12} \\ & + 0.485z^{-13} + 0.488z^{-14} + \dots \end{aligned} \tag{6-13}$$

The final value of c(kT) is 0.5. It is apparent that the digital model is a stable one.

Let us consider that the hold device is a polygonal hold, so that

$$G_h(s) = \frac{e^{Ts} + e^{-Ts} - 2}{Ts^2} \tag{6-14}$$

Then, for $T = 0.25$ second and $K = 1$, the z-transform of the output of the digital system is

$$C(z) = \frac{0.1108z}{(z-1)(z-0.778)} \tag{6-15}$$

which is expanded to

$$C(z) = 0.1108z^{-1} + 0.197z^{-2} + 0.264z^{-3} + 0.316z^{-4} + 0.357z^{-5} + 0.388z^{-6} + 0.413z^{-7} + 0.433z^{-8} + 0.447z^{-9} + 0.459z^{-10} + 0.468z^{-11} + 0.475z^{-12} + 0.481z^{-13} + 0.485z^{-14} + \dots \tag{6-16}$$

As a comparison, the responses of the continuous-data systems and the digital model with zero-order hold and polygonal hold, to a unit-step input, are computed and shown in Fig. 6-3(a). Figure 6-3(b) shows the responses when the sampling period on the digital simulation model is changed to one second. With this larger sampling period, the deviation between the actual response of the continuous-data system and the response of the digital model becomes greater as expected.

Since the continuous-data system is of the first order, it is stable for all finite positive values of K. However, the digital model may be stable or unstable depending on the values of K and T. For instance, with the zero-order hold and K = 2, the maximum value of the sampling time to insure stability in the digital system is 2.2 seconds. When K is increased to 10, the marginal value of T is 0.404 second. With T = 2 seconds, the z-transform of the digital system with a polygonal hold is

$$C(z) = \frac{0.632Kz}{(z - 1)[(1 + 0.368K)z + (0.264K - 0.368)]} \tag{6-17}$$

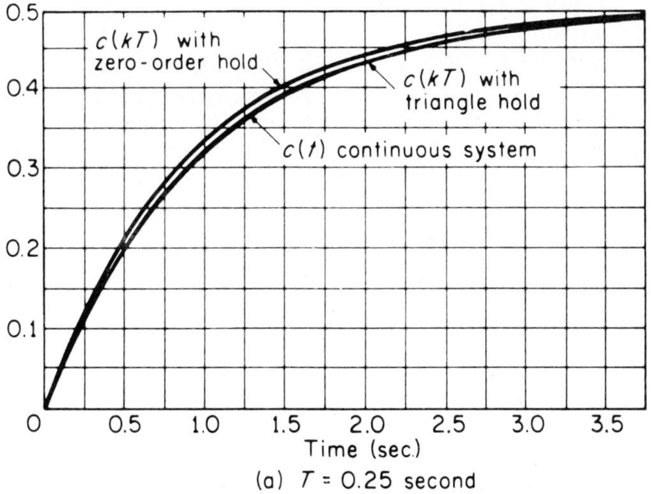

(a) T = 0.25 second

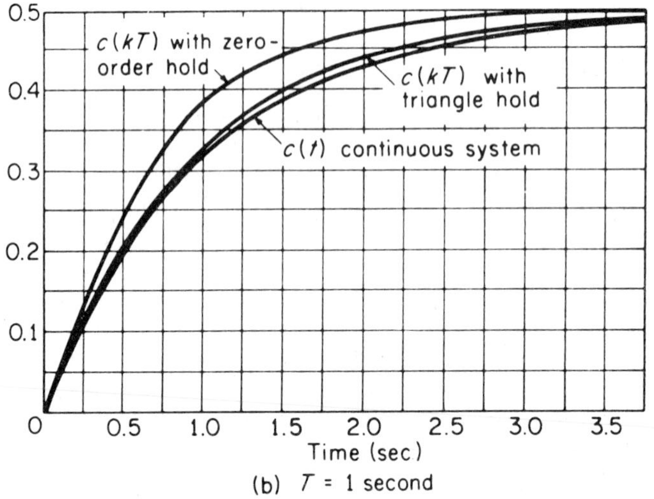

Fig. 6-3. Responses of systems by digital simulation.

It can be shown that the system is stable for all finite positive values of K. Therefore, the polygonal hold not only gives a better approximation but also a more stable system in this case.

6.3 DIGITAL SIMULATION — NUMERICAL INTEGRATION

Another popular method of digital simulation of continuous-data systems is the use of numerical integration. Since integration is the most time-consuming and difficult basic mathematical operation on a digital computer, its digital simulation plays an important role here. Instead of inserting sample-and-hold devices at strategic locations in a continuous-data system, the approach now is to approximate the continuous integration operation by numerical methods. The problem can also be stated as the simulation of the integrators s^{-1} in a continuous-data state diagram by digital models. Shown in Fig. 6-4 is the integrator element of a state diagram. The input-output relationships are written as

$$x(t) = \int_0^t r(\tau)d\tau \tag{6-18}$$

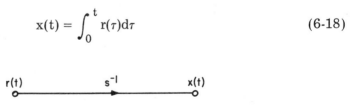

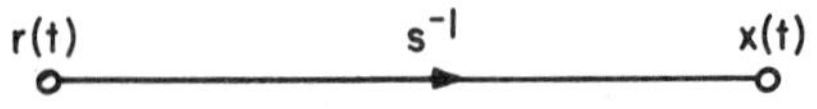

Fig. 6-4. State diagram representation of an integrator.

and

$$\frac{X(s)}{R(s)} = \frac{1}{s} \tag{6-19}$$

If the input $r(\tau)$ to the integrator is as shown in Fig. 6-5, the output $x(t)$ is equal to the area under the curve $r(\tau)$ between $\tau = 0$ and $\tau = t$.

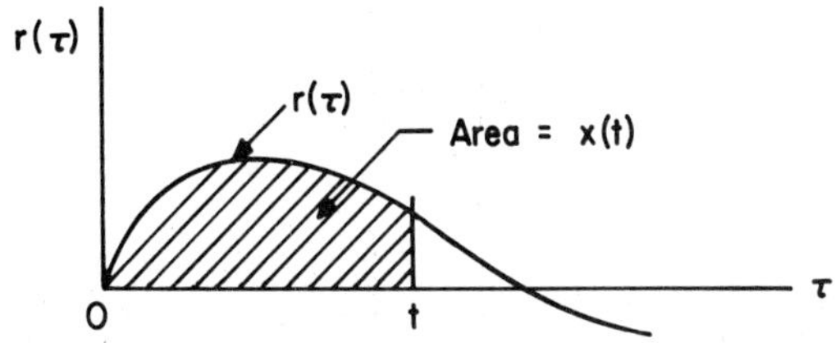

Fig. 6-5. Input and output relation of an integrator.

Rectangular Integration

One of the standard methods of numerical integration is the rectangular approximation, two types of which are shown in Fig. 6-6. The approximation is performed by summing the rectangular areas of width T under the staircase waveforms. These rectangular approximations are equivalent to inserting a sample-and-zero-order hold device before each integrator as illustrated in Fig. 6-6. The scheme shown in Fig. 6-6(a) is referred to as the rectangular approximation, whereas the one shown in Fig. 6-6(b) is the advanced rectangular approximation.

Referring to Fig. 6-6(a), the z-transfer function of the rectangular approxi-

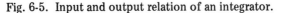

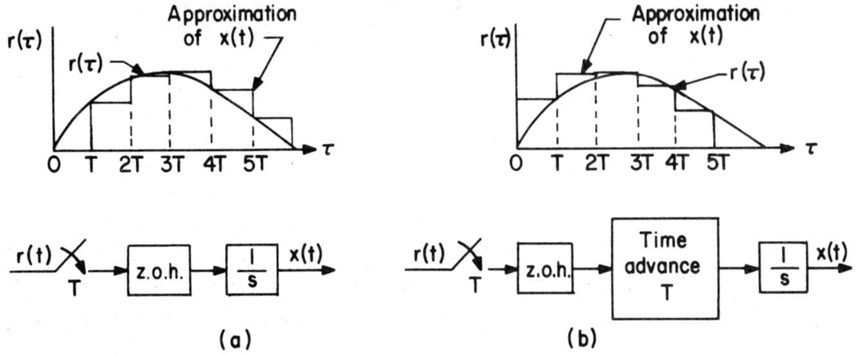

Fig. 6-6. (a) Rectangular approximation in numerical approximation; (b) advanced rectangular approximation in numerical approximation.

mated integrator is

$$\frac{X(z)}{R(z)} = (1 - z^{-1})\mathcal{Z}\left[\frac{1}{s^2}\right] = \frac{T}{z - 1} \tag{6-20}$$

The state equation is

$$x[(k + 1)] = x(kT) + Tr(kT) \tag{6-21}$$

Similarly, the z-transfer function of the advanced rectangular approximated integrator is written

$$\frac{X(z)}{R(z)} = z(1 - z^{-1})\mathcal{Z}\left[\frac{1}{s^2}\right] = \frac{Tz}{z - 1} \tag{6-22}$$

and the state equation is

$$x[(k + 1)T] = x(kT) + Tr(k + 1)T \tag{6-23}$$

As an illustrative example the continuous-data system of Fig. 6-2(a) is first approximated by rectangular integration. With the integrator 1/s replaced by the transfer function T/(z − 1), the state diagram of the digital model is shown in Fig. 6-7.

The z-transform of the output of the digital model when the input is a unit-step function is (K = 1, T = 0.25 sec)

$$C(z) = \frac{0.125z}{(z - 1)(z - 0.75)} \tag{6-24}$$

Expanding C(z), we get

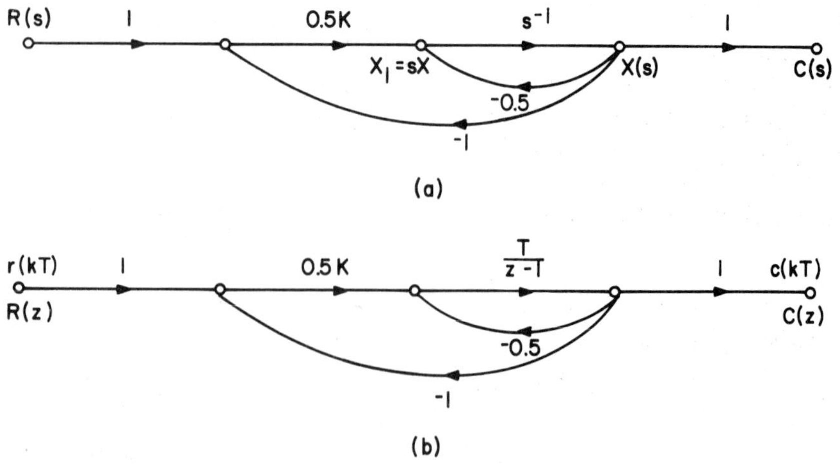

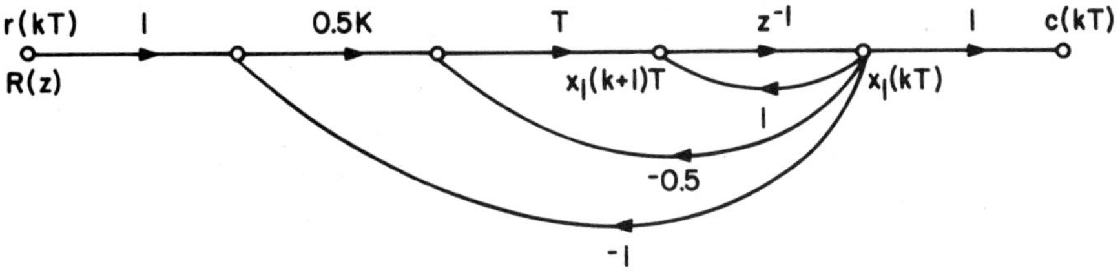

(c)

Fig. 6-7. State diagram of the control system in Fig. 6-2 with rectangular integration approximation: (a) state diagram of continuous-data system; (b) integration replaced by rectangular integrator; (c) state diagram of digital model with rectangular integration.

$$C(z) = 0.125z^{-1} + 0.218z^{-2} + 0.288z^{-3} + 0.34z^{-4}$$
$$+ 0.379z^{-5} + 0.408z^{-6} + 0.431z^{-7} + 0.45z^{-8}$$
$$+ 0.463z^{-9} + 0.473z^{-10} + \ldots$$

The response is plotted as shown in Fig. 6-8, together with the results of the sample-and-hold approximation scheme. Choosing the advanced rectangular ap-

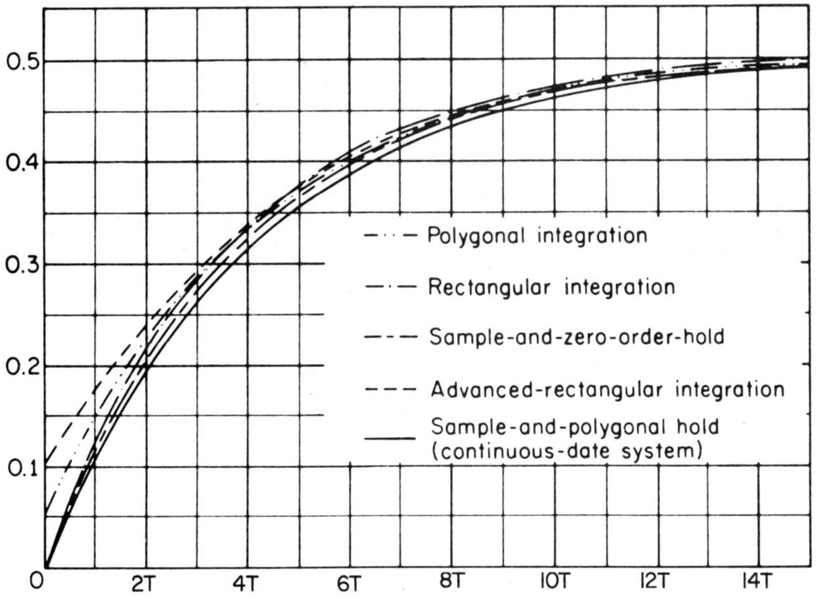

Fig. 6-8. Comparison of results of various methods of digital simulation.

proximation, the integrator transfer function 1/s in the continuous-data system is replaced by Tz/(z − 1). With K = 1 and T = 0.25 second, the output transfer function expression of the digital system with unit-step input is

$$C(z) = \frac{0.1z^2}{(z-1)(z-0.8)} \tag{6-25}$$

and the response is plotted as shown in Fig. 6-8.

Comparison of these results showed that both versions of the rectangular integration approximation give inferior results to those of the sample-and-hold approximation. The advanced rectangular approximation causes the output step response to jump at t = 0, although the response gets close to the actual values after the large errors for the first few sampling periods.

Polygonal Integration

A generally more accurate numerical integration scheme is achieved by using the polygonal hold concept. As shown in Fig. 6-9(a), the area under the curve $r(\tau)$ can be approximated by summing up the areas under the polygons of widths T. It is apparent that the approximation can be made as close as possible by reducing the sampling period T. This is often referred to as the polygonal or the trapezoidal integration approximation. This type of approximation is equivalent to inserting a sample-and-polygonal hold device before each integrator as illustrated in Fig. 6-9(b).

Since the transfer function of the polygonal hold is

$$G_h(s) = \frac{e^{Ts} + e^{-Ts} - 2}{Ts^2} \tag{6-26}$$

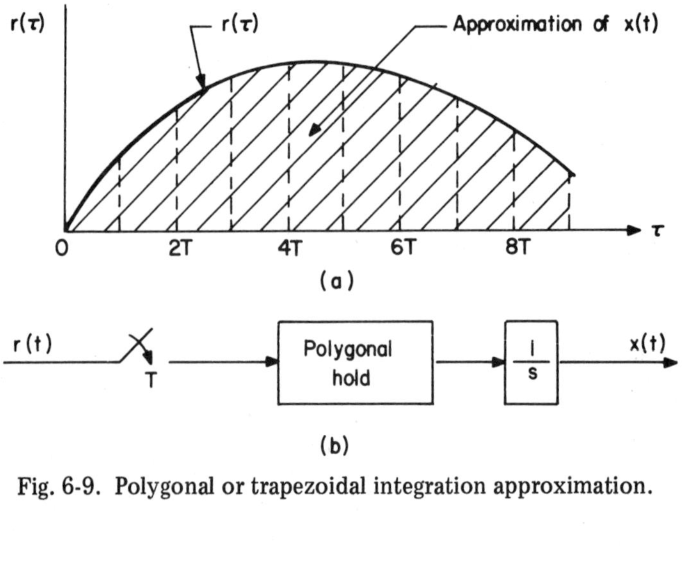

Fig. 6-9. Polygonal or trapezoidal integration approximation.

the transfer function of the polygonal integrator is written

$$\frac{X(z)}{R(z)} = \frac{z + z^{-1} - 2}{T} \mathcal{Z}\left[\frac{1}{s^3}\right] = \frac{T}{2}\left[\frac{z+1}{z-1}\right] \tag{6-27}$$

The state equation of the system is

$$x[(k+1)T] = x(kT) + \frac{T}{2} r[(k+1)T] + \frac{T}{2} r(kT) \tag{6-28}$$

The unit-step response of the system of Fig. 6-2(a) using the polygonal integration approximation is shown in Fig. 6-8. The fact that the polygonal integration gives an approximation inferior to that of the rectangular integration is due to the characteristic of this first-order system under consideration. Notice that the input to the integrator of the continuous-data system is written from Fig. 6-7(a).

$$X_1(s) = \frac{0.5s}{s+1} R(s) = \frac{0.5}{s+1} \tag{6-29}$$

Therefore,

$$X_1(t) = 0.5e^{-t} \qquad (t > 0) \tag{6-30}$$

The numerical integration of $x_1(t)$ by means of the three schemes described above is illustrated in Fig. 6-10. Since the signal $x_1(t)$ has a jump discontinuity at $t = 0$, the advanced rectangular and the polygonal integrations produce approximations which have areas prior to $t = 0$.

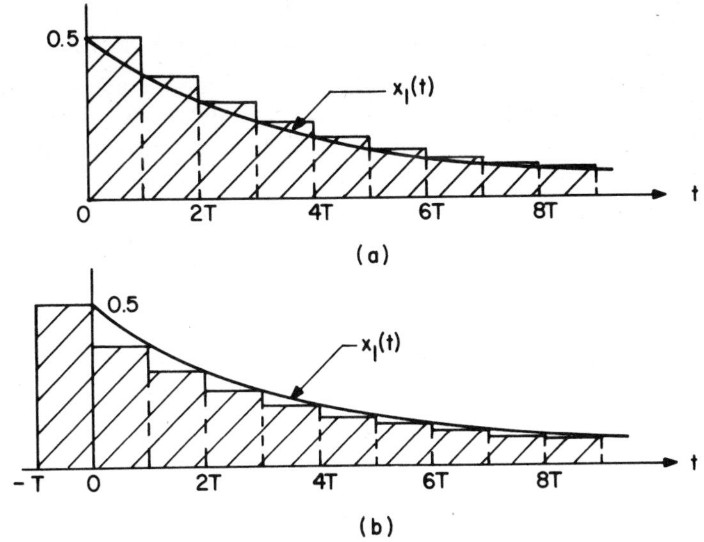

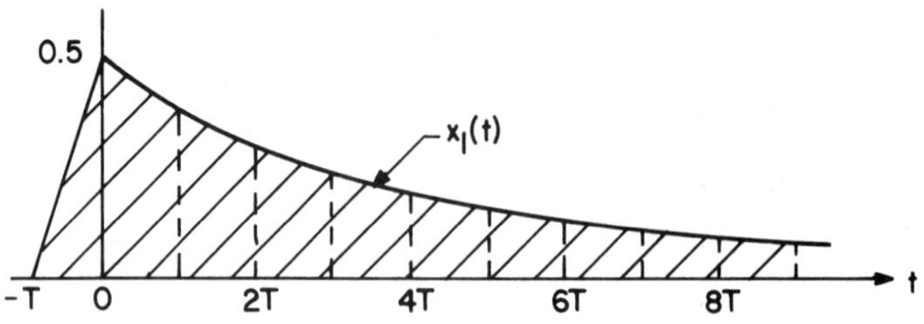

(c)

Fig. 6-10. Input to integrator $x_1(t)$ of system in Fig. 6-2(a) and areas showing the outputs of numerical integration by (a) rectangular integration, (b) advanced rectangular integration, (c) polygonal integration.

Stability Considerations

Since the digital model of the simulated system must be stable regardless of the method of approximation, it is necessary to investigate the stability effects of the integration schemes discussed in the preceding section. The root locus method is used here for its simplicity. Let us use the closed-loop system configuration of Fig. 6-11 for this study. Although in general the actual effects on stability due to the various numerical integration schemes will depend on the transfer function of the continuous-data system being simulated, it is believed that the simple integrator system of Fig. 6-11 will provide a qualitative indication of simulation stability in general. For the continuous-data system, the open-loop transfer function is simply

$$G(s) = \frac{1}{s} = \frac{T}{\ln z} \tag{6-31}$$

Therefore, when $T = 0$, the characteristic equation root is at $z = 1$, and when $T = \infty$, it is at $z = 0$. The roots stay inside the unit circle for all values of T between 0 and infinity. Therefore, the continuous-data model is always stable. For the rectangular integration, the transfer function of the system in Fig. 6-11 is replaced by $T/(z - 1)$. The system is unstable for $T \geqslant 2$.

For the advanced rectangular integration,

$$G(z) = \frac{Tz}{z - 1} \tag{6-32}$$

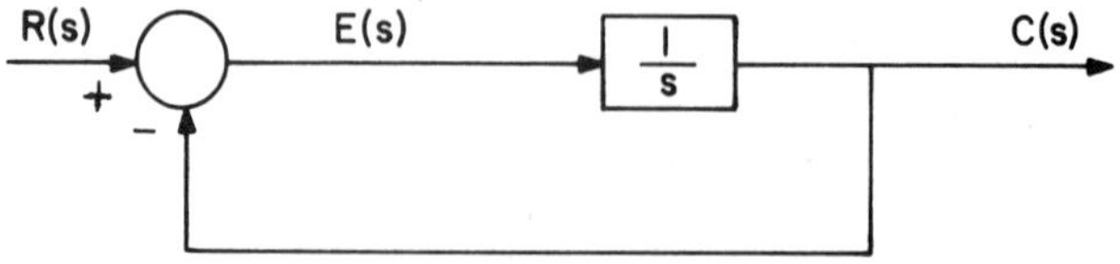

Fig. 6-11. A simple integrator system.

The system is again stable for all values of T between 0 and infinity. For the polygonal integration,

$$G(z) = \frac{T}{2}\left[\frac{z+1}{z-1}\right] \tag{6-33}$$

the digital simulation is stable for all finite T. Therefore, from this investigation, we conclude that, in general, the rectangular integration provides the worst effect on system stability.

In general, higher-order and more complex numerical integration methods such as the Simpson's rules are available. However, these schemes usually generate higher-order transfer functions which cause serious stability problems in the simulation models. Therefore, in control system applications, these higher-order integration methods are seldom used.

6.4 DIGITAL SIMULATION BY THE Z-FORMS

The disadvantage of the foregoing methods of numerical integration for digital simulation is that the system transfer function must first be decomposed so that the integrators are identified and then replaced by the approximating numerical integrator model. A simpler procedure is made possible by use of the z-forms. The z-form method is simpler because one can work directly with the continuous-data transfer function in the s-domain and transform it into an equivalent discrete-data transfer function in the z-domain. The method is described as follows:

Let us assume that a continuous-data signal is represented by g(t). Then, g(t) and its Laplace transform G(s) are related by

$$G(s) = \int_0^{\infty} g(t)e^{-st}dt \tag{6-34}$$

and

$$g(t) = \frac{1}{2\pi j}\int_{c-j\infty}^{c+j\infty} G(s)e^{st}ds \tag{6-35}$$

If g(t) is sampled once every T second, the discrete-data signal is represented by

$$g^*(t) = \sum_{k=0}^{\infty} g(kT)\delta(t-kT) \tag{6-36}$$

where

$$g(kT) = \frac{1}{2\pi j}\oint_{\Gamma} G(z)z^{k-1}dz \tag{6-37}$$

where the Γ path is a circle described by $|z| = e^{cT}$ in the z-plane. The line integral of Eq. (6-35) is performed along the path as shown in Fig. 6-12. Let us divide the path of integration between $c - j\infty$ and $c + j\infty$ in the s-plane into three parts as shown in Fig. 6-12. Then, Eq. (6-35) can be written as

$$g(t) = \frac{1}{2\pi j}\int_{c-j\infty}^{c-j\omega_s/2} G(s)e^{ts}ds + \frac{1}{2\pi j}\int_{c-j\omega_s/2}^{c+j\omega_s/2} G(s)e^{ts}ds + \frac{1}{2\pi j}\int_{c+j\omega_s/2}^{c+j\infty} G(s)e^{ts}ds \tag{6-38}$$

where $\omega_s = 2\pi/T$.

If the sampling period T is sufficiently small, g(t) can be approximated by just the second term on the right side of Eq. (6-38). Therefore,

$$g(t) \cong \frac{1}{2\pi j}\int_{c-j\omega_s/2}^{c+j\omega_s/2} G(s)e^{ts}ds \tag{6-39}$$

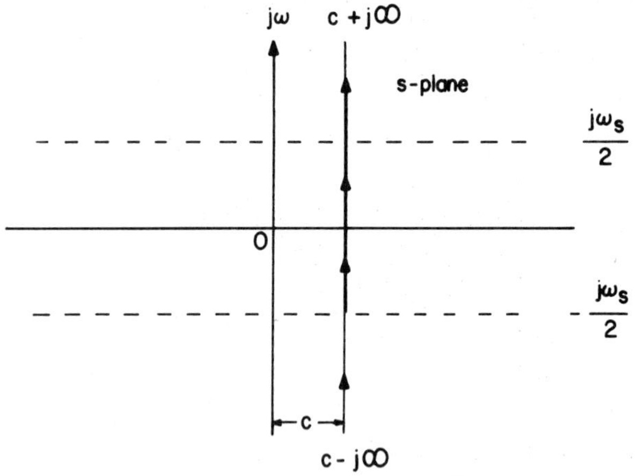

Fig. 6-12. Path of integration of the line integral in Eq. (6-35).

Replacing t by kT, and letting the right side of the last equation be represented by $g_A(t)$, we have

$$g(kT) \cong g_A(kT) = \frac{1}{2\pi j}\int_{c-j\omega_s/2}^{c+j\omega_s/2} G(s)e^{kTs}ds \tag{6-40}$$

Since s is related to z by

$$s = \frac{1}{T} \ln z \tag{6-41}$$

substituting this relation into Eq. (6-40) for s, we get

$$g_A(kT) = \frac{1}{2\pi j} \oint_\Gamma G(\frac{1}{T} \ln z) e^{kT(1/T)\ln z} d \frac{1}{T} \ln z \tag{6-42}$$

Or

$$g_A(kT) = \frac{1}{2\pi j} \oint_\Gamma \frac{1}{T} G(\frac{1}{T} \ln z) z^{k-1} dz \tag{6-43}$$

Comparing Eq. (6-43) with Eq. (6-37) we notice that the two integrals are similar except for the factor of 1/T in the former. This simply means that $g_A(kT)$, an approximation of g(kT), may be obtained by expanding G[(1/T)ln z] /T into a power series in z^{-1}. Therefore, we can initially summarize the z-form approximation method as follows:

1. Replace s^{-1} by $s^{-1} = T/\ln z$ in the transfer function G(s) of the continuous-data system. This gives G[(1/T)ln z].
2. The value of $g_A(kT)$ (k = 0,1,2,...) which approximates g(kT) is obtained by expanding G[(1/T)ln z] /T into a power series in z^{-1}. However, before G[(1/T)ln z] /T can be expanded it must be in the form of a rational function in z, which means that T/ln z must first be approximated by a truncated series.

Let ln z be represented by the following power series:

$$\ln z = 2\left\{ \frac{z-1}{z+1} + \frac{1}{3}\left[\frac{z-1}{z+1}\right]^3 + \frac{1}{5}\left[\frac{z-1}{z+1}\right]^5 + \dots \right\}$$
$$= 2\left\{\left[\frac{1-z^{-1}}{1+z^{-1}}\right] + \frac{1}{3}\left[\frac{1-z^{-1}}{1+z^{-1}}\right]^3 + \frac{1}{5}\left[\frac{1-z^{-1}}{1+z^{-1}}\right]^5 + \dots\right\} \qquad |z| > 0 \tag{6-44}$$

Then,

$$\frac{1}{s} = \frac{T}{\ln z} = \frac{T/2}{u + \frac{1}{3}u^3 + \frac{1}{5}u^5 + \dots} \tag{6-45}$$

where

$$u = \frac{1 - z^{-1}}{1 + z^{-1}} \tag{6-46}$$

By synthetic division, Eq. (6-45) is written

$$\frac{1}{s} = \frac{T}{\ln z} = \frac{T}{2}\left[\frac{1}{u} - \frac{1}{3}u - \frac{4}{45}u^3 - \frac{44}{945}u^5 + \dots\right] \tag{6-47}$$

In general, for positive integral n,

$$\frac{1}{s^n} = \left[\frac{T}{2}\right]^n \left[\frac{1}{u} - \frac{1}{3}u - \frac{4}{45}u^3 - \frac{44}{945}u^5 + \dots\right] \tag{6-48}$$

Notice that the right side of Eq. (6-47) as well as that of Eq. (6-48) is in the form of a Laurent series. If we take only the principal part and the constant term of the Laurent series, we get

$$\frac{1}{s^n} \cong \frac{N_n(z^{-1})}{(1 - z^{-1})^n} = G_n(z^{-1}) \tag{6-49}$$

where $N_n(z^{-1})$ is a polynomial in powers of z^{-1}, and $G_n(z^{-1})$ is called the *z-form* of s^{-n}. It should be pointed out that since $s = 0$ corresponds to $z = 1$, the poles on both sides of Eq. (6-49) correspond. This means that although the right side of Eq. (6-49) contains only the principal part and the constant term of the Laurent series of s^{-n}, inclusion of additional terms would introduce additional poles in the z-domain and, therefore, would actually lead to larger rather than smaller errors.

Let us now demonstrate the determination of the z-forms for $n = 1$ and $n = 2$.

For $n = 1$, taking the principal part of the right side of Eq. (6-47) yields

$$\frac{1}{s} = \frac{T}{\ln z} \cong G_1(z^{-1}) = \frac{T}{2}\left[\frac{1}{u}\right] = \frac{T}{2}\left[\frac{1 + z^{-1}}{1 - z^{-1}}\right] \tag{6-50}$$

For $n = 2$,

$$\frac{1}{s^2} = \left[\frac{T}{\ln z}\right]^2 \cong G_2(z^{-1}) = \left[\frac{T}{2}\right]^2 \left\{\left[\frac{1}{u}\right]^2 - \frac{2}{3}\right\} = \frac{T^2}{12}\left[\frac{1 + 10z^{-1} + z^{-2}}{(1 - z^{-1})^2}\right] \tag{6-51}$$

Following the same procedure, the z-forms for higher orders of s^{-1} are determined and are tabulated in Table 6-1.

Although the z-form for $1/s$ is identical to the expression for the polygonal integration, there is a basic difference between the present z-form method and the numerical integration method discussed earlier. In the polygonal and the rectangular integration approximation, the continuous-data system is first approximated by a digital model, and then the digital form of the input is applied to give the simulated output. However, in the z-form method, the Laplace transforms of the input signal are first multiplied by the transfer function of the continuous-data system, and then the z-forms are substituted into the expression of C(s) to give the approximated output $C_A(z)$.

Therefore, we can now summarize the steps of obtaining an approximation to the response of a continuous-data system by the z-form method as follows:

1. Express the Laplace transform of the output of the system, C(s), as a rational function in powers of s^{-1}.
2. Substitute for s^{-n} the corresponding z-forms using Table 6-1. This converts C(s) into a rational function in powers of z^{-1}.
3. Divide the expression obtained from the last step by T, the sampling period, to give $C_A(z)$.
4. Expand $C_A(z)$ by synthetic division into a power series of the form:

$$c_A(0) + c_A(T)z^{-1} + c_A(2T)z^{-2} + \ldots + c_A(kT)z^{-k} + \ldots$$

TABLE 6-1. z-FORMS OF s^{-1}

s^{-n}	$G_n(z^{-1})$
s^{-1}	$\frac{T}{2}\frac{1 + z^{-1}}{1 - z^{-1}}$
s^{-2}	$\frac{T^2}{12}\frac{1 + 10z^{-1} + z^{-2}}{(1 - z^{-1})^2}$
s^{-3}	$\frac{T^3}{2}\frac{z^{-1} + z^{-2}}{(1 - z^{-1})^3}$
s^{-4}	$\frac{T^4}{6}\frac{z^{-1} + 4z^{-2} + z^{-3}}{(1 - z^{-1})^4} - \frac{T^4}{720}$
s^{-5}	$\frac{T^5}{24}\frac{z^{-1} + 11z^{-2} + 11z^{-3} + z^{-4}}{(1 - z^{-1})^5}$

where $c_A(kT)$ is the approximate value of the time response $c(t)$ at $t = kT$.

Example 6-1.

As an illustrative example of the z-form method of digital approximation, let us consider that the open-loop transfer function of a feedback control system with unity feedback is given as

$$G(s) = \frac{K}{s(s+1)} \tag{6-52}$$

where K is a constant gain factor.

The closed-loop transfer function is written

$$\frac{C(s)}{R(s)} = \frac{K}{s^2 + s + K} \tag{6-53}$$

For a unit-step input, the output transform is

$$C(s) = \frac{K}{s(s^2 + s + K)} \tag{6-54}$$

Multiplying the numerator and the denominator of the right-hand side of the last equation by s^{-3}, we have

$$C(s) = \frac{Ks^{-3}}{1 + s^{-1} + Ks^{-2}} \tag{6-55}$$

Now substituting the corresponding forms from Table 6-1 and dividing the result by T, we get

$$C_A(z) = \frac{1}{T}\,\frac{\dfrac{T^3}{2}K\left[\dfrac{z^{-4} + z^{-2}}{(1 - z^{-1})^3}\right]}{1 + \dfrac{T}{2}\left[\dfrac{1 + z^{-1}}{1 - z^{-1}}\right] + \dfrac{T^2K}{12}\left[\dfrac{1 + 10z^{-1} + z^{-2}}{(1 - z^{-1})^2}\right]} \tag{6-56}$$

Rearranging and simplifying, the last equation is written

$$C_A(z) = \frac{6T^2K(z^{-1} + z^{-2})}{(1 - z^{-1})[(12 + 6T + T^2K) + (-24 + 10T^2K)z^{-1} + (12 - 6T + T^2K)z^{-2}]} \tag{6-57}$$

The two roots of the equation

$$(12 + 6T + T^2K)z^2 + (-24 + 10T^2K)z + (12 - 6T + T^2K) = 0 \tag{6-58}$$

determine the stability of the digital model obtained by use of the z-form. Applying the stability criterion, it can be shown that the values of K and T that correspond to a stable simulation are as illustrated in Fig. 6-13. We can further show that for all values of K and T which correspond to a stable digital system, the final value of $c_A(kT)$ as k approaches infinity is unity. That is,

$$\lim_{k\to\infty} c_A(kT) = \lim_{z\to 1} (1 - z^{-1})C_A(z) = 1 \tag{6-59}$$

The unit-step responses of the digital model for K = 2 and T = 1 and T = 2, together with the response of the actual continuous-data system, are drawn as shown in Fig. 6-14.

6.5 DIGITAL REDESIGN

Two of the most important subjects in digital control systems are simulation and design. Quite often in the analysis and design of control systems it is necessary to perform a digital simulation of a continuous-data system. Chapter 6 covers several methods of digital simulation, but the subject of digital redesign presented in this section will provide us with another method of digital simulation.

A large number of control systems in operation in the industry are continuous-data systems. From the performance standpoint these systems are quite satisfactory. However, as the technology of digital computers and digital

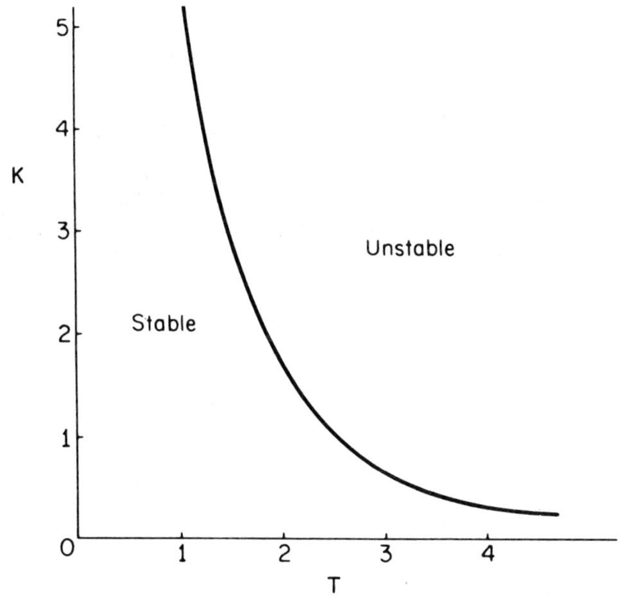

Fig. 6-13. Maximum values of K and T for a stable digital system.

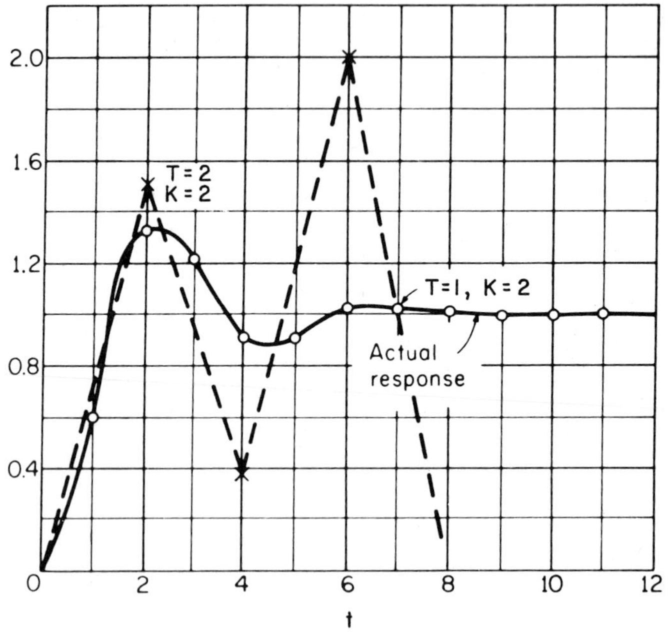

Fig. 6-14. Responses of system in Example 6-1.

processors becomes more advanced, it is often desirable to refit these systems with digital transducers and digital controllers. Rather than carrying out a completely new design using digital control theory, it is possible to apply the digital redesign technique to arrive at an equivalent digital system. The digital system is said to be equivalent to the continuous-data system if the responses of the two systems are closely matched for the same input and initial conditions.

The block diagram of the continuous-data system under consideration is shown in Fig. 6-15. As a matter of convenience the state-feedback model is used. It is conceivable that any linear feedback control system could be modeled in this form.

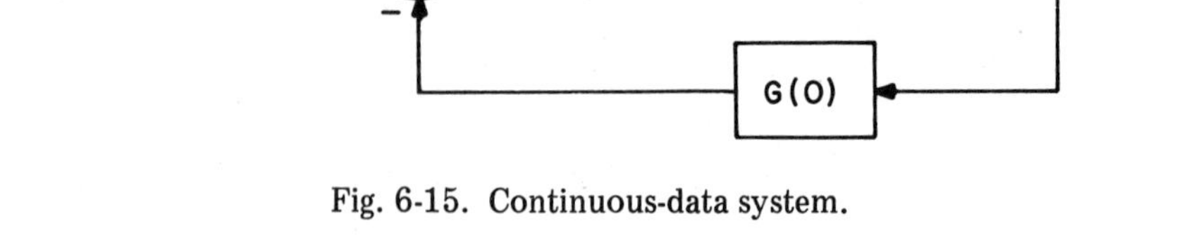

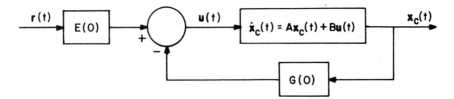

Fig. 6-15. Continuous-data system.

The controlled process of the continuous-data system is represented by the state equation

$$\dot{\mathbf{x}}_c(t) = \mathbf{A}\mathbf{x}_c(t) + \mathbf{B}\mathbf{u}(t) \tag{6-60}$$

The control vector is dependent upon the state vector $\mathbf{x}_c(t)$ and the input vector $\mathbf{r}(t)$ through the relation

$$\mathbf{u}(t) = \mathbf{E}(0)\mathbf{r}(t) - \mathbf{G}(0)\mathbf{x}_c(t) \tag{6-61}$$

The following vectors and matrices are defined:

$$\begin{aligned}
\mathbf{x}_c(t) &= \text{state vector} && (n \times 1)\\
\mathbf{u}(t) &= \text{control vector} && (m \times 1)\\
\mathbf{r}(t) &= \text{input vector} && (m \times 1)\\
\mathbf{A} &= \text{coefficient matrix} && (n \times n)\\
\mathbf{B} &= \text{coefficient matrix} && (n \times m)\\
\mathbf{E}(0) &= \text{input matrix} && (m \times m)\\
\mathbf{G}(0) &= \text{feedback matrix} && (m \times n)
\end{aligned}$$

Substituting Eq. (6-61) into Eq. (6-60) yields

$$\dot{\mathbf{x}}_c(t) = [\mathbf{A} - \mathbf{B}\mathbf{G}(0)]\mathbf{x}_c(t) + \mathbf{B}\mathbf{E}(0)\mathbf{r}(t) \tag{6-62}$$

The solution of the last equation for $t \geq t_0$ is

$$\mathbf{x}_c(t) = \phi_c(t - t_0)\mathbf{x}_c(t_0) + \int_{t_0}^{t} \phi_c(t - \tau)\mathbf{B}\mathbf{E}(0)\mathbf{r}(\tau)d\tau \tag{6-63}$$

where $\mathbf{x}_c(t_0)$ is the initial state of $\mathbf{x}_c(t)$ at $t = t_0$, and

$$\phi_c(t - t_0) = e^{[\mathbf{A}-\mathbf{B}\mathbf{G}(0)](t-t_0)} \tag{6-64}$$

As defined in Chapter 4, $\phi_c(t - t_0)$ is the state transition matrix of $\mathbf{A} - \mathbf{B}\mathbf{G}(0)$, and is represented by the power series

$$\phi_c(t - t_0) = \sum_{j=0}^{\infty} \frac{1}{j!} [\mathbf{A} - \mathbf{B}\mathbf{G}(0)]^j (t - t_0)^j \tag{6-65}$$

The block diagram of the digital control system which is to approximate the system of Fig. 6-15 is shown in Fig. 6-16. Here the digital system may be considered as a digital simulation model of the continuous-data system, or it

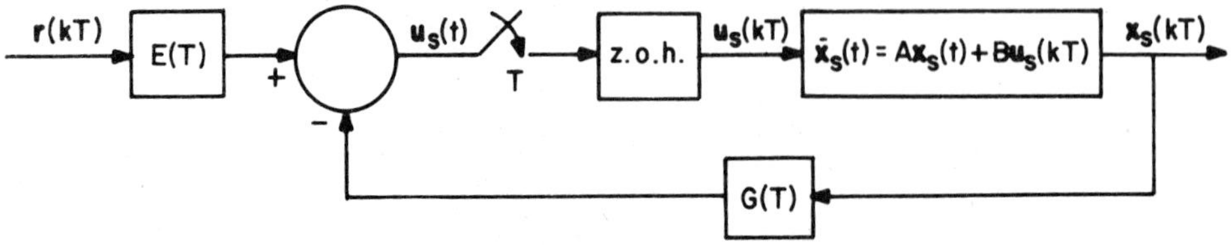

Fig. 6-16. Digital control system which approximates the continuous-data system in Fig. 6-15.

may be regarded as the digitally redesigned model of the continuous-data system.

The outputs of the sample-and-hold devices in Fig. 6-16 are a series of step functions with the amplitudes denoted by the elements of the vector $\mathbf{u}_s(kT)$ for $kT \leqslant t < (k+1)T$. The matrices $\mathbf{G}(T)$ and $\mathbf{E}(T)$ denote the feedback gain and the forward gain matrices of the digital system, respectively.

The state equations of the digital system are denoted by

$$\dot{\mathbf{x}}_s(t) = \mathbf{A}\mathbf{x}_s(t) + \mathbf{B}\mathbf{u}_s(kT) \tag{6-66}$$

for $kT \leqslant t \leqslant (k+1)T$, where

$$\mathbf{u}_s(kT) = \mathbf{E}(T)\mathbf{r}(kT) - \mathbf{G}(T)\mathbf{x}_s(kT) \tag{6-67}$$

It is important to note that the matrices **A** and **B** are identical to those of Eq. (6-60).

Substituting Eq. (6-67) into Eq. (6-66), we get

$$\dot{\mathbf{x}}_s(t) = \mathbf{A}\mathbf{x}_s(t) + \mathbf{B}[\mathbf{E}(T)\mathbf{r}(kT) - \mathbf{G}(T)\mathbf{x}_s(kT)] \tag{6-68}$$

for $kT \leqslant t \leqslant (k+1)T$.

The solution of Eq. (6-68) with $t = (k+1)T$ and $t_0 = kT$ is

$$\mathbf{x}_s[(k+1)T] = \left[\phi(T) - \int_{kT}^{(k+1)T} \phi(kT + T - \tau)d\tau\mathbf{B}\mathbf{G}(T)\right]\mathbf{x}_s(kT) + \int_{kT}^{(k+1)T} \phi(kT + T - \tau)d\tau\mathbf{B}\mathbf{E}(T)\mathbf{r}(kT) \tag{6-69}$$

where

$$\phi(T) = e^{\mathbf{A}T} = \sum_{j=0}^{\infty} \frac{(\mathbf{A}T)^j}{j!} \tag{6-70}$$

The problem is to find the matrices $\mathbf{E}(T)$ and $\mathbf{G}(T)$ so that the states of the digital model in Fig. 6-16 are as close as possible to those of the continuous-data system at the sampling instants, for a given input $\mathbf{r}(t)$.

So that the solution for $\mathbf{E}(T)$ is independent of $\mathbf{r}(t)$, it is necessary to assume that $\mathbf{r}(t) \cong \mathbf{r}(kT)$ for $kT \leqslant t < (k+1)T$. Therefore, effectively, the input of the continuous-data system of Fig. 6-15 is assumed to pass through sample-and-hold devices. The above assumption would not affect the solution if $\mathbf{r}(t)$ had step functions as its elements. If the inputs were other than step functions, the approximation would be a good one for small sampling periods.

Now letting $t_0 = kT$ and $t = (k+1)T$ in Eq. (6-63), and assuming $\mathbf{r}(\tau) \cong \mathbf{r}(kT)$ over one sampling period, we have

$$\mathbf{x}_c[(k+1)T] = \phi_c(T)\mathbf{x}_c(kT) + \int_{kT}^{(k+1)T} \phi_c(kT + T - \tau)\mathbf{BE}(0)d\tau\,\mathbf{r}(kT)$$

$$kT \leqslant t \leqslant (k+1)T \qquad (6\text{-}71)$$

The responses of Eq. (6-69) and Eq. (6-71) will match at $t = (k+1)T$ for an arbitrary initial state $\mathbf{x}_c(kT) = \mathbf{x}_s(kT)$ and an arbitrary input $\mathbf{r}(\tau)$ if and only if the following two equations are satisfied.

$$\phi_c(T) = \phi(T) - \int_{kT}^{(k+1)T} \phi(kT + T - \tau)d\tau\,\mathbf{BG}(T) \qquad (6\text{-}72)$$

and

$$\int_{kT}^{(k+1)T} \phi(kT + T - \tau)d\tau\,\mathbf{BE}(T) = \int_{kT}^{(k+1)T} \phi_c(kT + T - \tau)\mathbf{BE}(0)d\tau \qquad (6\text{-}73)$$

First referring to Eq. (6-72), and letting $\lambda = (k+1)T - \tau$, we have

$$\phi_c(T) = \phi(T) - \theta(T)\mathbf{G}(T) \qquad (6\text{-}74)$$

where

$$\theta(T) = \int_0^T e^{\mathbf{A}\lambda}\mathbf{B}d\lambda \qquad (6\text{-}75)$$

In principle, the feedback matrix, $\mathbf{G}(T)$, of the digital system can be determined from Eq. (6-74). However, it is subject to the limitations discussed in the following section.

A Closed-Form Solution For G(T)

In order that *all* n states of the digital system, $\mathbf{x}_s(kT)$, match those of the continuous-data system, $\mathbf{x}_c(kT)$, at each sampling instant, it is sufficient that

Eq. (6-74) be satisfied. However, Eq. (6-74) consists of n^2 scalar equations with mn unknown elements in **G**(T). If the number of state variables, n, is equal to the number of inputs, m, and the matrix $\theta(T)$ is nonsingular, the feedback matrix **G**(T) is solved from Eq. (6-74),

$$\mathbf{G}(T) = [\theta(T)]^{-1}[\phi(T) - \phi_c(T)] \tag{6-76}$$

For most control systems, $n > m$, that is, there are more states than inputs; Eq. (6-74) will generally not have a solution. However, if the following rank condition is satisfied, the system of equations in Eq. (6-74) is consistent, and a solution still exists.

Let

$$\mathbf{G}(T) = [\mathbf{g}_1 \quad \mathbf{g}_2 \quad \cdots \quad \mathbf{g}_n] \tag{6-77}$$

$$\phi(T) - \phi_c(T) = [\mathbf{d}_1 \quad \mathbf{d}_2 \quad \cdots \quad \mathbf{d}_n] \tag{6-78}$$

where $\mathbf{g}_i$, $i = 1,2,\ldots,n$, are m-dimensional vectors, and $\mathbf{d}_i$, $i = 1,2,\ldots,n$, are n-dimensional vectors; then if

$$\text{rank } [\theta] = \text{rank } [\theta, \mathbf{d}_i] \tag{6-79}$$

for all $i = 1,2,\ldots,n$, the system of equations in Eq. (6-74) has at least one solution. However, if the above rank condition is not satisfied, the equations are inconsistent and no solution exists.

Partial Matching of States

In general, the rank condition of Eq. (6-79) is rarely satisfied by the matrices $\theta(T)$ and $\phi(T) - \phi_c(T)$. Thus, when $n > m$, not all of the states of the continuous-data and the digital systems can be made to match at the sampling instants.

Although it is not possible to match all of the states exactly, it can be shown that it is possible to match some of the states or the algebraic sums of the states at each sampling instant.

We introduce a weighting matrix **H** which allows a partial matching of the states. Equation (6-74) is rewritten as

$$\mathbf{D}(T) = \phi(T) - \phi_c(T) = \theta(T)\mathbf{G}(T) \tag{6-80}$$

Both sides of the last equation are premultiplied by an $m \times n$ matrix **H** to give

$$\mathbf{HD}(T) = \mathbf{H}\theta(T)\mathbf{G}(T) \tag{6-81}$$

If **H** is chosen such that the $m \times m$ matrix $\mathbf{H}\theta(T)$ is nonsingular, Eq. (6-81) may be solved for **G**(T) to give a partial matching of the states. Let the feedback matrix for partial matching be designated by $\mathbf{G}_w(T)$. Then

$$\mathbf{G}_w(T) = [\mathbf{H}\theta(T)]^{-1}\mathbf{H}\mathbf{D}(T) \tag{6-82}$$

It should be noted that the solution of $\mathbf{G}_w(T)$ from Eq. (6-82) does not satisfy Eq. (6-80) except when **H** is an identity matrix. The reason is that when Eq. (6-80) is premultiplied by the matrix **H**, it reduces the system of n^2 equations to a system of mn equations. Therefore, for $n > m$, the solution will not satisfy the original equations.

To explore the physical meanings of the transformation of Eq. (6-81) and of the solution in Eq. (6-82), let us again equate Eq. (6-69) to Eq. (6-71), and then premultiply both sides of the equation by **H**. We get

$$\begin{aligned}\mathbf{H}\mathbf{x}_c[(k+1)T] = \mathbf{H}\mathbf{x}_s[(k+1)T] &= \mathbf{H}\phi_c(T)\mathbf{x}_c(kT) + \mathbf{H}\theta_c(T)\mathbf{E}(0)\mathbf{r}(kT)\\ &= \mathbf{H}[\phi(T) - \theta(T)\mathbf{G}(T)]\mathbf{x}_s(kT) + \mathbf{H}\theta(T)\mathbf{E}(T)\mathbf{r}(kT)\end{aligned} \tag{6-83}$$

where

$$\theta_c(T) = \int_0^T \phi_c(\lambda)\mathbf{B}d\lambda \tag{6-84}$$

For arbitrary $\mathbf{x}_c(kT)$, $\mathbf{x}_s(kT)$, and $\mathbf{r}(kT)$, Eq. (6-83) leads to

$$\mathbf{H}\phi_c(T)\mathbf{x}_c(kT) = \mathbf{H}[\phi(T) - \theta(T)\mathbf{G}(T)]\mathbf{x}_s(kT) \tag{6-85}$$

and

$$\mathbf{H}\theta_c(T)\mathbf{E}(0)\mathbf{r}(kT) = \mathbf{H}\theta(T)\mathbf{E}(T)\mathbf{r}(kT) \tag{6-86}$$

The significant point is that the solution $\mathbf{G}_w(T)$ of Eq. (6-82) satisfies Eq. (6-85) for any arbitrary initial state $\mathbf{x}_c(kT)$. Premultiplying both sides of the state transition equations of Eqs. (6-69) and (6-71) simply transforms the n-dimensional state vectors $\mathbf{x}_c[(k+1)T]$ and $\mathbf{x}_s[(k+1)T]$ into a new m-dimensional vector, $\mathbf{y}[(k+1)T]$, such that

$$\mathbf{y}[(k+1)T] = \mathbf{H}\mathbf{x}_c[(k+1)T] = \mathbf{H}\mathbf{x}_s[(k+1)T] \tag{6-87}$$

Equation (6-87) indicates that the m new states, $y_i[(k+1)T]$, $i = 1,2,\ldots,m$, are algebraic sums of the original n state variables; that is,

$$y_i[(k+1)T] = \sum_{j=1}^{n} h_{ij}x_j[(k+1)T] \qquad i = 1,2,\ldots,m \tag{6-88}$$

Therefore, the solution, $\mathbf{G}_w(T)$, from Eq. (6-82) represents a matching by the n states of the digital system, the weighted algebraic sum of the n states of the continuous-data system, at the sampling instants.

Solution of the Feedback Matrix by Series Expansion

Although Eq. (6-82) gives the exact solution of $\mathbf{G}_w(T)$, it is possible to simplify the procedure by expanding $\mathbf{G}(T)$ into a Taylor series about $T = 0$. In general, if the series converges, $\mathbf{G}(T)$ can be approximated by taking a finite number of terms of the series expansion.

Let $\mathbf{G}(T)$ be expanded into a Taylor series about $T = 0$,

$$\mathbf{G}(T) = \lim_{K \to \infty} \mathbf{G}_K(T) = \lim_{K \to \infty} \sum_{j=0}^{K-1} \frac{1}{j!} \mathbf{G}^{(j)}(T)T^j \tag{6-89}$$

where

$$\mathbf{G}^{(j)}(T) = \left. \frac{\partial^j \mathbf{G}(T)}{\partial T^j} \right|_{T=0} \tag{6-90}$$

Substituting $\mathbf{G}(T)$ from Eq. (6-89) into Eq. (6-74) gives

$$\boldsymbol{\phi}_c(T) = \boldsymbol{\phi}(T) - \boldsymbol{\theta}(T) \sum_{j=0}^{\infty} \frac{1}{j} \mathbf{G}^{(j)}(T)T^j \tag{6-91}$$

The last equation is written as

$$\sum_{j=0}^{\infty} \frac{[\mathbf{A} - \mathbf{BG}(0)]^j T^j}{j!} = \sum_{j=0}^{\infty} \frac{\mathbf{A}^j T^j}{j!} - \sum_{j=0}^{\infty} \frac{\mathbf{A}^j T^{j+1}}{(j+1)!} \mathbf{B} \sum_{k=0}^{\infty} \frac{\mathbf{G}^{(k)}(T)T^k}{k!} \tag{6-92}$$

or

$$\sum_{j=0}^{\infty} \left\{ \frac{[\mathbf{A} - \mathbf{BG}(0)]^j T^j}{j!} - \frac{\mathbf{A}^j T^j}{j!} + \frac{\mathbf{A}^j \mathbf{B}}{(j+1)!} \sum_{k=0}^{\infty} \frac{\mathbf{G}^{(k)}(T)T^{k+j+1}}{k!} \right\} = \mathbf{0} \tag{6-93}$$

Now equating the coefficients of T^i, $i = 1, 2, \ldots$, to zero, we have

$$\frac{[\mathbf{A} - \mathbf{BG}(0)]^i}{i!} - \frac{\mathbf{A}^i}{i!} + \sum_{j=0}^{i-1} \frac{\mathbf{A}^{i-j-1}\mathbf{BG}^{(j)}(T)}{(i-j)!j!} = \mathbf{0} \tag{6-94}$$

In general, it is possible to express $\mathbf{G}^{(i-1)}(T)$ in terms of $\mathbf{G}^{(i-2)}(T)$, $\mathbf{G}^{(i-3)}(T)$, $\ldots$, $\mathbf{G}^{(1)}(T)$, and $\mathbf{G}^{(0)}(T)$, where $\mathbf{G}^{(0)}(T) = \mathbf{G}(0)$.

Equation (6-84) is written

$$\frac{[\mathbf{A} - \mathbf{BG}(0)]^i}{i!} - \frac{\mathbf{A}^i}{i!} + \sum_{j=0}^{i-2} \frac{\mathbf{A}^{i-j-1}\mathbf{BG}^{(j)}(T)}{(i-j)!j!} + \frac{\mathbf{BG}^{(i-1)}(T)}{(i-1)!} = \mathbf{0} \tag{6-95}$$

$i = 1, 2, \ldots$. Solving for the last term in the last equation, we have

$$\frac{\mathbf{BG}^{(i-1)}(T)}{(i-1)!} = \frac{\mathbf{A}^i}{i!} - \frac{[\mathbf{A} - \mathbf{BG}(0)]^i}{i!} - \sum_{j=0}^{i-2} \frac{\mathbf{A}^{i-j-1}\mathbf{BG}^{(j)}(T)}{(i-j)!j!} \tag{6-96}$$

In order to solve for $\mathbf{G}^{(i-1)}(T)$ from the last equation, since **B** is generally not a square matrix, we again introduce the weighting matrix **H** (m X n) such that **HB** is nonsingular. Premultiplying both sides of Eq. (6-96) by **H** and solving for $\mathbf{G}^{(i-1)}(T)$, we have

$$\mathbf{G}_w^{(i-1)}(T) = (\mathbf{HB})^{-1}\mathbf{H}\left\{\frac{\mathbf{A}^i}{i} - \frac{[\mathbf{A} - \mathbf{BG}(0)]^i}{i} - (i-1)! \sum_{j=0}^{i-2} \frac{\mathbf{A}^{i-j-1}\mathbf{BG}^{(j)}(T)}{(i-j)!j!}\right\} \tag{6-97}$$

$i = 1, 2, \ldots$. Table 6-2 gives the expressions for $\mathbf{G}_w^{(i-1)}(T)$ for i = 1, 2, and 3.

TABLE 6-2.

i	$\mathbf{G}_w^{(i-1)}(T)$
1	$\mathbf{G}(0)$
2	$\frac{1}{2}\mathbf{G}(0)[\mathbf{A} - \mathbf{BG}(0)]$
3	$(\mathbf{HB})^{-1}\mathbf{H}\left[-\frac{1}{6}\mathbf{ABG}(0)[\mathbf{A} - \mathbf{BG}(0)] + \frac{1}{3}\mathbf{BG}(0)[\mathbf{A} - \mathbf{BG}(0)]^2\right]$

The results tabulated in Table 6-2 allow the approximation of $\mathbf{G}(T)$ by using up to three terms in the series expansion of Eq. (6-89). In Eq. (6-89),

$$K = 1 \qquad \mathbf{G}_1(T) = \mathbf{G}^{(0)}(T) = \mathbf{G}(0) \tag{6-98}$$

$$K = 2 \qquad \mathbf{G}_2(T) = \mathbf{G}(0) + T\mathbf{G}^{(1)}(T) \tag{6-99}$$

$$K = 3 \qquad \mathbf{G}_3(T) = \mathbf{G}(0) + T\mathbf{G}^{(1)}(T) + \frac{T^2}{2}\mathbf{G}^{(2)}(T) \tag{6-100}$$

In reality, we can only use $\mathbf{G}_w^{(0)}(T)$, $\mathbf{G}_w^{(1)}(T)$, and $\mathbf{G}_w^{(2)}(T)$ as approximations to their exact counterparts. However, it is interesting to note from Table 6-2 that $\mathbf{G}_w^{(0)}(T)$ and $\mathbf{G}_w^{(1)}(T)$ are not dependent upon **H**. Therefore, the one-term and the two-term approximations of $\mathbf{G}(T)$, $\mathbf{G}_1(T)$ and $\mathbf{G}_2(T)$, respectively, attempt to match all the states of the continuous-data and the digital systems. Beyond two terms, the weighting matrix **H** should be used, since $\mathbf{G}_w^{(2)}(T) \neq \mathbf{G}^{(2)}(T)$, and only certain states and combinations of states are matched depending on the **H** selected, when $\mathbf{G}_w^{(2)}(T)$ is used in place of $\mathbf{G}^{(2)}(T)$ in Eq. (6-100).

In general, as more terms are used in the series approximation of **G**(T) in Eq. (6-89), the solution will approach that of $\mathbf{G}_w(T)$ in Eq. (6-82).

An Exact Solution for E(T)

We now turn our attention to the determination of the feedforward matrix **E**(T). The condition of state matching in Eq. (6-73) is written as

$$\theta(T)\mathbf{E}(T) = \theta_c(T)\mathbf{E}(0) \tag{6-101}$$

As in the discussions on the closed-form solution of **G**(T), if m = n, and $\theta(T)$ is nonsingular, a unique solution of Eq. (6-101) exists and is given by

$$\mathbf{E}(T) = [\theta(T)]^{-1}\theta_c(T)\mathbf{E}(0) \tag{6-102}$$

In general, when n > m, the solution of **E**(T) that corresponds to a partial matching of states is

$$\mathbf{E}_w(T) = [\mathbf{H}\theta(T)]^{-1}\mathbf{H}\theta_c(T)\mathbf{E}(0) \tag{6-103}$$

where it is assumed that $\mathbf{H}\theta(T)$ is nonsingular.

Solution of E(T) by Series Expansion

As in the solution of **G**(T), the matrix **E**(T) may be expanded into a Taylor series about T = 0:

$$\mathbf{E}(T) = \lim_{K\to\infty} \mathbf{E}_K(T) = \lim_{K\to\infty} \sum_{j=0}^{K-1} \frac{1}{j!}\mathbf{E}^{(j)}(T)T^j \tag{6-104}$$

where

$$\mathbf{E}^{(j)}(T) = \left.\frac{\partial^j \mathbf{E}(T)}{\partial T^j}\right|_{T=0} \tag{6-105}$$

Substituting the series expansion of **E**(T) into Eq. (6-101) and carrying out steps similar to those in Eqs. (6-92) through (6-97), we have

$$\mathbf{E}_w^{(i-1)}(T) = (\mathbf{HB})^{-1}\mathbf{H}\left[\frac{[\mathbf{A} - \mathbf{BG}(0)]^{i-1}\mathbf{BE}(0)}{i} - (i-1)!\sum_{j=0}^{i-2}\frac{\mathbf{A}^{i-j-1}\mathbf{BE}^{(j)}(T)}{(i-j)!j!}\right] \tag{6-106}$$

i = 1,2,... . Table 6-3 gives the results of $\mathbf{E}_w^{(i-1)}(T)$ for i = 1, 2 and 3.

TABLE 6-3.

i	$E_w^{(i-1)}(T)$
1	$E(0)$
2	$-\frac{1}{2}G(0)BE(0)$
3	$(HB)^{-1}H\left[\frac{ABG(0)B}{6} - \frac{BG(0)[A - BG(0)]B}{3}\right]E(0)$

Like the situation for **G**(T), if only up to two terms are used for the series approximation for **E**(T), the weighting matrix **H** is not needed, and all the states of the digital and the continuous-data systems are attempted to be matched.

Stability Considerations and Constraints on the Selection of the Weighting Matrix H

In the preceding sections the weighting matrix **H** is introduced to effect a partial matching of states in the digital redesign problem. While the selection of the elements of **H** is not an exact science, one important requirement is that the closed-loop digital control system be asymptotically stable. The problem is to find the constraints on **H** such that the stability requirement is met.

If we solve the state equations in Eq. (6-66) and evaluate the states at the sampling instants, we get

$$\mathbf{x}_s[(k+1)T] = \phi(T)\mathbf{x}_s(kT) + \theta(T)\mathbf{u}_s(kT) \qquad (6\text{-}107)$$

where $\phi(T)$ and $\theta(T)$ are defined in Eqs. (6-70) and (6-75), respectively. For $\mathbf{r}(kT) = \mathbf{0}$, the state-feedback control is

$$\mathbf{u}_s(kT) = -\mathbf{G}(T)\mathbf{x}_s(kT) \qquad (6\text{-}108)$$

Substitution of Eq. (6-108) into Eq. (6-107) yields

$$\mathbf{x}_s[(k+1)T] = [\phi(T) - \theta(T)\mathbf{G}(T)]\mathbf{x}_s(kT) \qquad (6\text{-}109)$$

The digital closed-loop system described by Eq. (6-109) is asymptotically stable if all the eigenvalues of $[\phi(T) - \theta(T)\mathbf{G}(T)]$ are located inside the unit circle $|z| = 1$. Since $\phi(T)$ and $\theta(T)$ are known once the sampling period T is specified, the constraints on **G**(T) for stability can be established using standard stability test criteria.

Replacing **G**(T) by $\mathbf{G}_w(T)$ in Eq. (6-81), we have

$$\mathbf{HD}(T) = \mathbf{H}\theta(T)\mathbf{G}_w(T) \qquad (6\text{-}110)$$

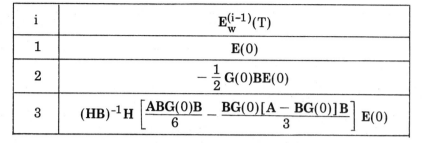

Taking the matrix transpose on both sides of the last equation, we get

$$\mathbf{D}'(\mathrm{T})\mathbf{H}' = \mathbf{G}'_{\mathrm{w}}(\mathrm{T})\theta'(\mathrm{T})\mathbf{H}' \tag{6-111}$$

or, rearranging,

$$[\mathbf{G}'_{\mathrm{w}}(\mathrm{T})\theta'(\mathrm{T}) - \mathbf{D}'(\mathrm{T})]\mathbf{H}' = \mathbf{0} \tag{6-112}$$

This matrix equation represents a set of n linear homogeneous equations which have nontrivial solutions if and only if the following condition is satisfied:

$$|\mathbf{G}'_{\mathrm{w}}(\mathrm{T})\theta'(\mathrm{T}) - \mathbf{D}'(\mathrm{T})| = 0 \tag{6-113}$$

which is equivalent to

$$|\theta(\mathrm{T})\mathbf{G}_{\mathrm{w}}(\mathrm{T}) - \mathbf{D}(\mathrm{T})| = 0 \tag{6-114}$$

Thus, if Eq. (6-114) is satisfied, there is always a nonzero **H** which will satisfy

$$\mathbf{G}_{\mathrm{w}}(\mathrm{T}) = [\mathbf{H}\theta(\mathrm{T})]^{-1}\mathbf{H}\mathbf{D}(\mathrm{T}) \tag{6-115}$$

Digital Redesign of the Simplified Skylab Satellite

An illustrative example of the digital redesign technique is presented in this section. The block diagram shown in Fig. 6-17 represents the dynamics of a simplified one-axis Skylab satellite system. It can be shown that the continuous-data Skylab system has a damping ratio of approximately 0.707.

The state equations of the system in Fig. 6-17 can be expressed in the form of Eq. (6-60) with

$$\mathbf{A} = \begin{bmatrix} 0 & 1 \\ 0 & 0 \end{bmatrix} \qquad \mathbf{B} = \begin{bmatrix} 0 \\ \dfrac{1}{970741} \end{bmatrix} \tag{6-116}$$

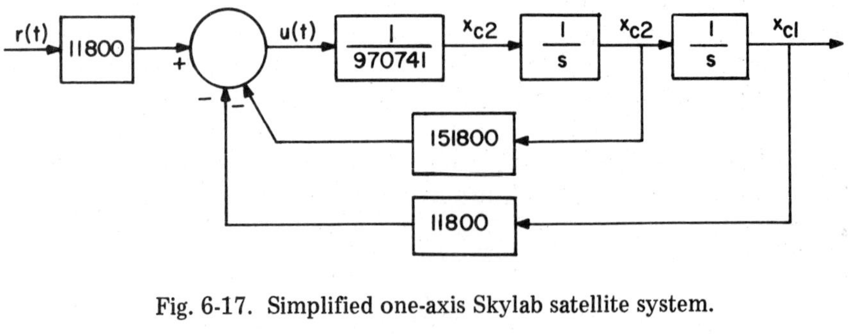

Fig. 6-17. Simplified one-axis Skylab satellite system.

The feedback matrix is

$$\mathbf{G}(0) = [11800 \quad 151800] \tag{6-117}$$

and the feedforward matrix is

$$\mathbf{E}(0) = 11800 \tag{6-118}$$

We now wish to find an equivalent digital system with the block diagram shown in Fig. 6-18 whose states will match those of the continuous-data Skylab at the sampling instants, for the same input and initial states.

The gain matrices $\mathbf{E}_w(T)$ and $\mathbf{G}_w(T) = [g_{w1}(T) \quad g_{w2}(T)]$ are to be determined using the digital redesign method discussed in the preceding sections.

When only one term is used in the series approximation of $\mathbf{G}(T)$ and $\mathbf{E}(T)$, the results are $\mathbf{G}(0)$ and $\mathbf{E}(0)$, respectively. In this case the sampling period of the digital system shown in Fig. 6-18 must be chosen so that the system is asymptotically stable. For a two-term approximation,

$$\mathbf{G}_w(T) \cong \mathbf{G}(0) + \frac{T}{2}\mathbf{G}(0)[\mathbf{A} - \mathbf{B}\mathbf{G}(0)]$$

$$= [11800 - 922.6T \quad 151800 - 5968.9T] \tag{6-119}$$

$$\mathbf{E}_w(T) \cong \mathbf{E}(0) - \frac{T}{2}\mathbf{G}(0)\mathbf{B}\mathbf{E}(0)$$

$$= 11800 - 922.6T \tag{6-120}$$

These results show that the gains of the feedback and feedforward matrices decrease as the sampling period increases. Table 6-4 gives the results of the elements of $\mathbf{G}_w(T)$ and $\mathbf{E}_w(T)$ with the two-term approximation for $T = 1$ through 5 seconds.

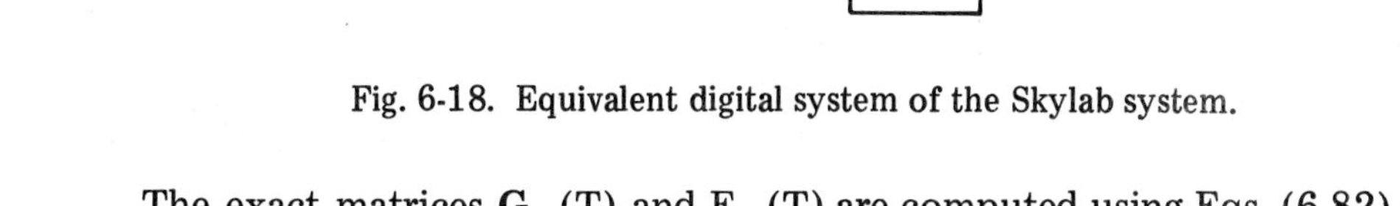

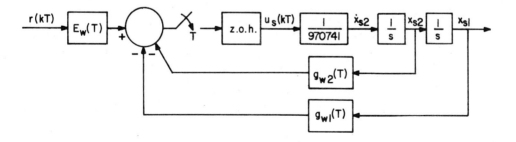

Fig. 6-18. Equivalent digital system of the Skylab system.

The exact matrices $\mathbf{G}_w(T)$ and $\mathbf{E}_w(T)$ are computed using Eqs. (6-82) and (6-103), respectively. Table 6-5 gives the results for the exact $\mathbf{G}_w(T)$ and $\mathbf{E}_w(T)$

TABLE 6-4.

T (sec)	$g_{w1}(T)$	$g_{w2}(T)$	$E_w(T)$
1	10877.4	145831	10877.4
2	9954.8	139862	9954.8
3	9032.2	133893	9032.2
4	8109.6	127924	8109.6
5	7187.0	121955	7187.0

for $T = 1$ through 5 seconds, and $\mathbf{H} = [0 \quad 1]$ and $\mathbf{H} = [1 \quad 0]$. With $\mathbf{H} = [0 \quad 1]$ the state x_{c2} is to be matched, and with $\mathbf{H} = [1 \quad 0]$ the state x_{c1} is to be matched.

The digital control system with the matrices tabulated in Table 6-5 is simulated on the digital computer with $T = 2$ seconds. The step responses of the continuous-data Skylab system are also computed, with the values of $\mathbf{G}(0)$ and $E(0)$ given in Eqs. (6-117) and (6-118), respectively. Figures 6-19 through 6-21 show the comparison for $T = 2$ seconds and $\mathbf{H} = [0 \quad 1]$. Figures 6-22 through 6-24 give the results for $\mathbf{H} = [1 \quad 0]$. These figures show that the digital redesign technique leads to a digital Skylab system with responses that match the continuous-data Skylab system very closely. The sampling period of 2 seconds appears to be quite adequate. The results are better with $\mathbf{H} = [0 \quad 1]$ than with $\mathbf{H} = [1 \quad 0]$. This means that the responses of the two systems are better matched if the matching of $x_{c2}(t)$ is emphasized rather than that of $x_{c1}(t)$.

We can investigate the general stability condition in the parameter plane of $g_{w2}(T)$ versus $g_{w1}(T)$. The characteristic equation of the closed-loop digital system is

$$F(z) = |z\mathbf{I} - \phi(T) + \theta(T)\mathbf{G}_w(T)|$$

$$= z^2 + 2\left[\frac{g_{w1}}{970741} + \frac{g_{w2}}{970741} - 1\right]z - \frac{2g_{w1}}{970741} - \frac{2g_{w2}}{970741} + 1 = 0 \quad (6\text{-}121)$$

Applying the stability criterion, we have the following three requirements:

$$F(0) < 1 \qquad g_{w1} < g_{w2}$$

$$F(1) > 0 \qquad g_{w1} > 0$$

$$F(-1) > 0 \qquad g_{w2} < 970741 \quad (6\text{-}122)$$

TABLE 6-5.

T (sec)	H	$g_{w1}(T)$	$g_{w2}(T)$	$E_w(T)$
1	[0 1]	10901.5	145840	10901.5
2	[0 1]	10051.2	139921	10051.2
3	[0 1]	9248.4	134071	9248.4
4	[0 1]	8492.5	128315	8492.5
5	[0 1]	7782.3	122674	7782.3
1	[1 0]	11197.0	147825	11197.0
2	[1 0]	10618.1	143867	10618.1
3	[1 0]	10063.1	139937	10063.1
4	[1 0]	9531.8	136048	9531.8
5	[1 0]	9023.7	132207	9023.7

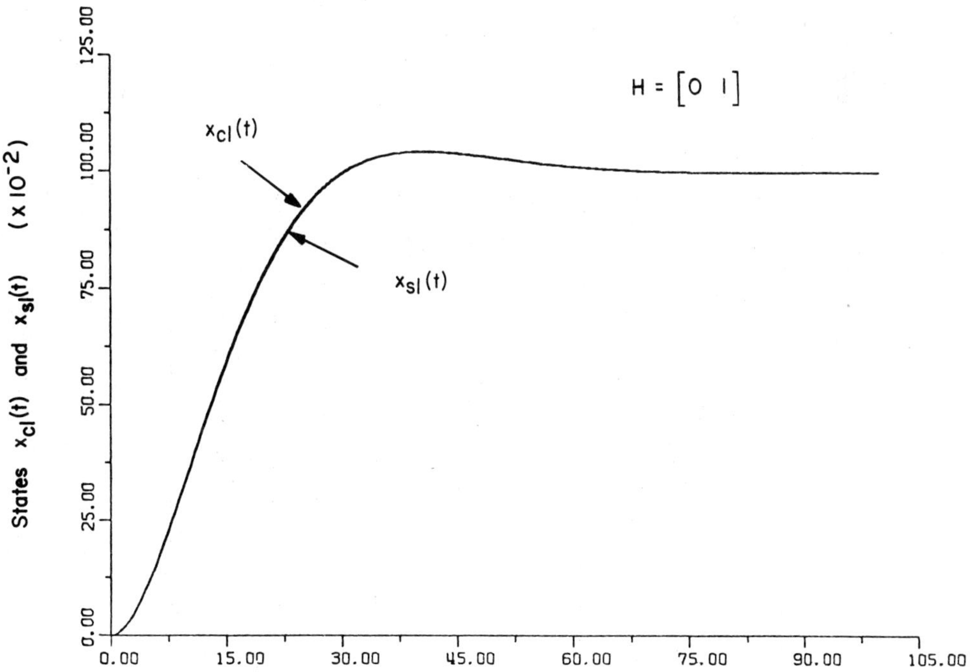

Fig. 6-19. State trajectories $x_{c1}(t)$ and $x_{s1}(t)$ for the simplified Skylab system.

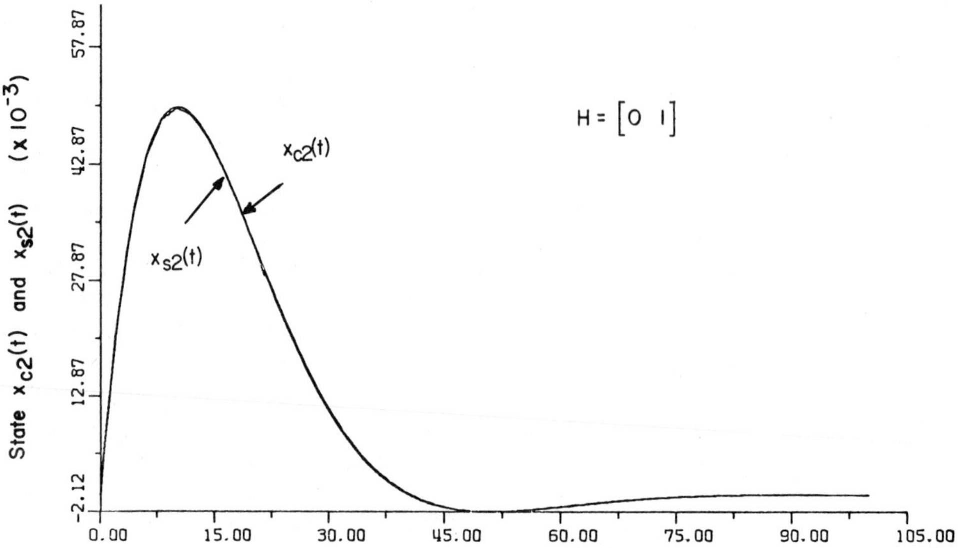

Fig. 6-20. State trajectories $x_{c2}(t)$ and $x_{s2}(t)$ for the simplified Skylab system.

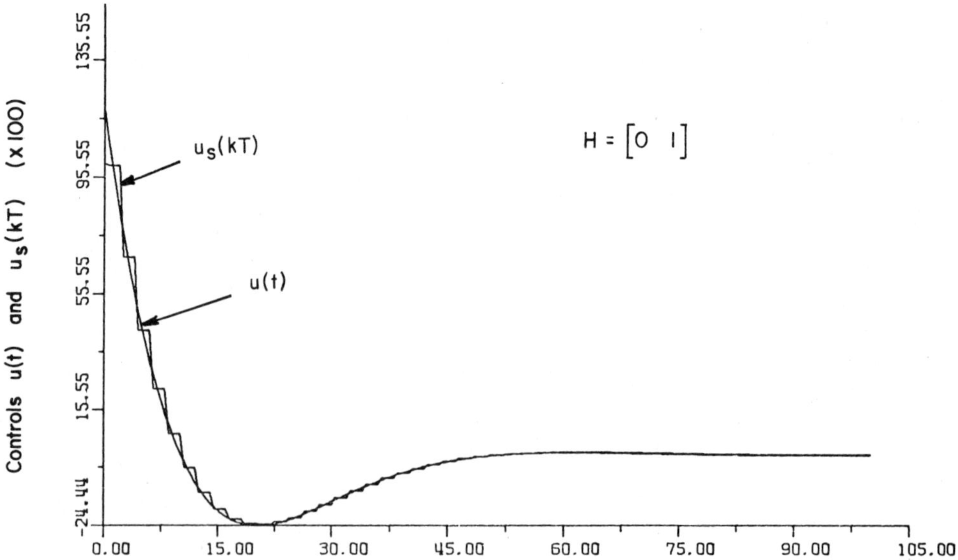

Fig. 6-21. Control signals u(t) and $u_s(kT)$ of the simplified Skylab system.

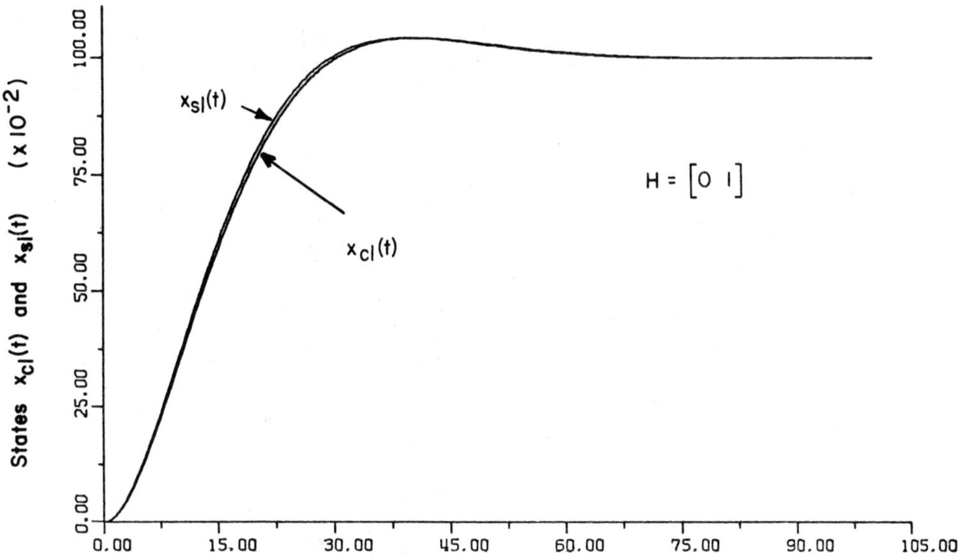

Fig. 6-22. State trajectories $x_{c1}(t)$ and $x_{s1}(t)$ of the simplified Skylab system.

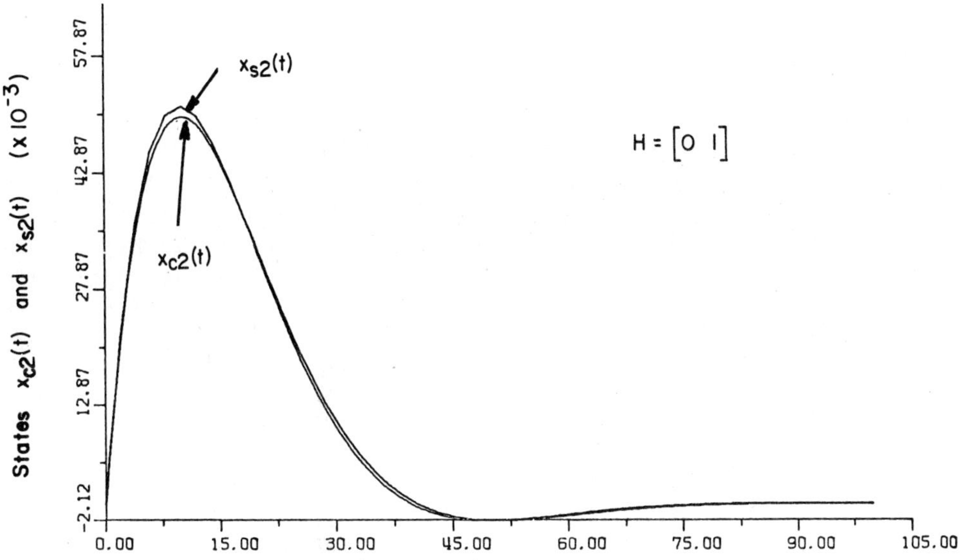

Fig. 6-23. State trajectories $x_{c2}(t)$ and $x_{s2}(t)$ of the simplified Skylab system.

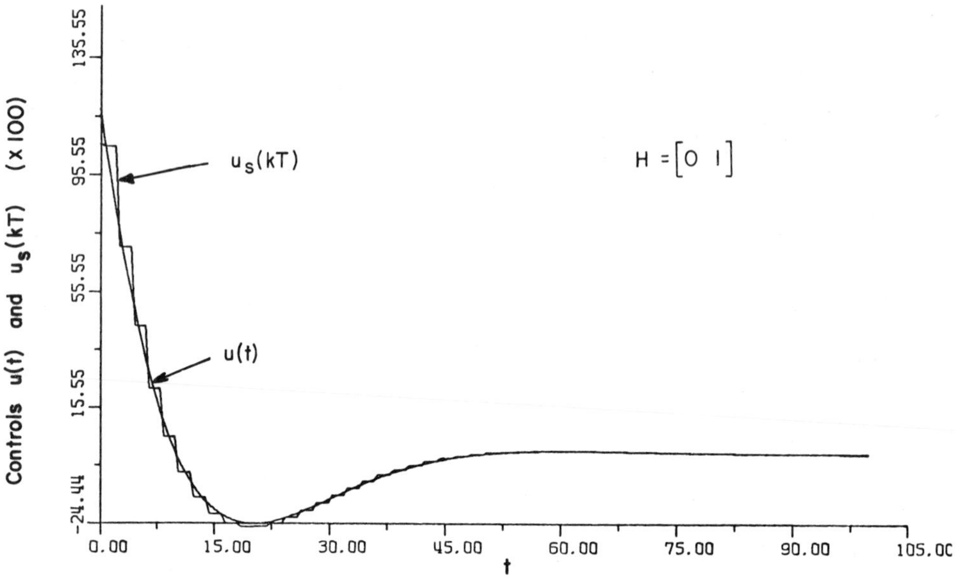

Fig. 6-24. Control signals u(t) and $u_s(kT)$ of the simplified Skylab system.

The stable region is bounded by these straight-line boundaries in the g_{w2} versus g_{w1} plane. It can be shown that all the results in Tables 6-4 and 6-5 lie inside the stable region.

In this section we have introduced the digital redesign method using the point-by-point comparison. In general, there are other realistic ways of matching the states between a continuous-data system and its equivalent digital model. For instance, it is possible to match the states at the sampling instants by use of a higher-order hold with the sampler, instead of the zero-order hold used in this section. Furthermore, it has been shown that states can be matched at multiples of sampling instants or by varying the feedback gain **G**(T) and feedforward gain **E**(T) between sampling instants, without using the weighting matrix **H**. These and other methods of digital redesign may be found in the literature listed in the reference section of this chapter.

REFERENCES

1. Salzer, J. M., "Frequency Analysis of Digital Computers Operating in Real Time," *Proc. IRE*, Vol. 42, No. 2, February 1954, pp. 457-466.
2. Fryer, W. D., and Schultz, W. C., "A Survey of Methods for Digital Simulation of Control Systems," Cornell Aeronautical Lab., Cornell University, New York.

3. Halijak, C. A., "Digital Approximation of the Solutions of Differential Equations Using Trapezoidal Convolution," Bendix Systems Divisions, Report ITM-64, August 1960.
4. Boxer, R., and Thaler, S., "A Simplified Method of Solving Linear and Nonlinear Systems," *Proc. IRE*, Vol. 44, January 1956, pp. 89-101.
5. Liff, A. I., and Wolf, J. K., "On the Optimum Sampling Rate for Discrete-Time Modeling of Continuous-Time Systems," *IEEE Trans. on Automatic Control*, Vol. AC-11, April 1966, pp. 288-290.
6. Widrow, B., "A Study of Rough Amplitude Quantization by Means of Nyquist Sampling Theory," *IRE Trans. on Circuit Theory*, Vol. CT-3, December 1964, pp. 266-276.
7. Beale, G. O., and Cook, G., "Frequency Domain Synthesis of Discrete Representations," *IEEE Trans. on Industrial Electronics and Control Instrumentation*, Vol. 23, No. 4, November 1976, pp. 438-443.
8. Rosko, J. S., *Digital Simulation of Physical Systems*, Addison Wesley, Reading, Mass., 1972.
9. Tustin, A., "A Method of Analyzing the Behavior of Linear Systems in Terms of Time Series," *Journal of the IEE*, Vol. 94, II-A, May 1947.
10. Sage, A. P., and Smith, S. L., "Real-Time Digital Simulation for Systems Control," *Proc. of the IEEE*, Vol. 54, December 1966, pp. 1802-1812.
11. Parrish, E. A., McVey, E. S., and Cook, G., "The Investigation of Optimal Discrete Approximations for Real-Time Flight Simulations," *NASA Technical Report EG-4041-102-76*, March 1976.
12. McVey, E. S., and Lee, Y. C., "Choice of Method for Discretization of Continuous Systems," *AIAA Journal of Guidance and Control*, Vol. 2, No. 1, January-February 1979, pp. 92-94.
13. Singh, G., Kuo, B. C., and Yackel, R. A., "Digital Approximation by Point-by-Point State Matching with Higher-Order Holds," *International J. on Control*, Vol. 20, No. 1, 1974, pp. 81-90.
14. Kuo, B. C., Singh, G., and Yackel, R. A., "Digital Approximation of Continuous-Data Control Systems by Point-by-Point State Comparison," *Computers & Elec. Eng.*, Vol. 1, Pergamon Press, 1973, pp. 155-170.
15. Kuo, B. C., and Peterson, D. W., "Optimal Discretization of Continuous-Data Control Systems," *IFAC Automatica*, Vol. 9, No. 1, 1973, pp. 125-129.
16. Fryer, W. D., and Schultz, W. C., "A Survey of Methods For Digital Simulation of Control Systems," Cornell Aeronautical Lab., Cornell University, New York.
17. Yackel, R. A., Kuo, B. C., and Singh, G., "Digital Redesign of Continuous Systems by Matching of States at Multiple Sampling Periods," *IFAC Automatica*, Vol. 10, Pergamon Press, 1974, pp. 105-111.
18. Winsor, C. A., and Roy, R. J., "The Application of Specific Optimal Control to the Design of Desensitized Model Following Control Systems," *Proc. JACC*, 1970, pp. 271-277.
19. Moore, B. C., and Silverman, L. M., "Model Matching by State Feedback and Dynamic

Compensation," *IEEE Trans. on Automatic Control,* Vol. AC-17, 1972, pp. 491-497.
20. Wang, S. H., and Desoer, C. A., "The Exact Model Matching of Linear Multivariable Systems," *IEEE Trans. on Automatic Control,* Vol. AC-17, 1972, pp. 347-349.
21. Wang, S. H., and Davison, E. J., "Solution of the Exact Model Matching Problem," *IEEE Trans. on Automatic Control,* Vol. AC-17, 1972, pp. 574.

PROBLEMS

6-1 The block diagram of a continuous-data control system is shown in Fig. 6P-1. Insert sample-and-zero-order-hold devices at appropriate locations in the system to simulate the system by a digital model. Select a suitable sampling period. Find the output of the digital model when the input is a unit-step function.

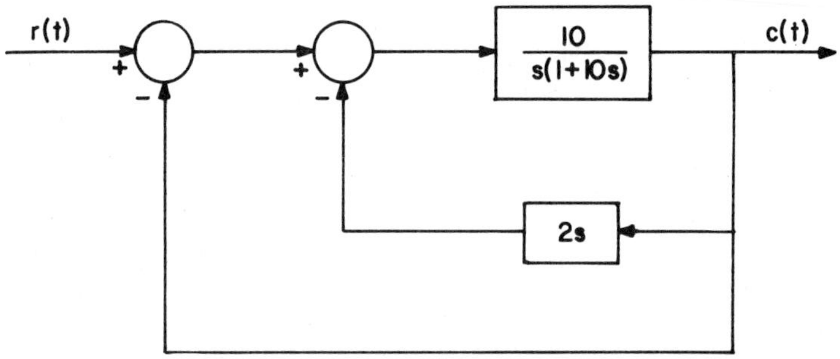

Fig. 6P-1.

6-2 The block diagram of a continuous-data system is shown in Fig. 6P-2. It is desired to simulate the system by a digital model.

(a) Insert sample-and-zero-hold devices at appropriate locations in the system to form the digital model. Select a suitable sampling period with considerations on stability and accuracy of the digital simulation. Evaluate the response of the digital model when the input is a unit-step function.

(b) Replace the zero-order hold devices in part (a) by an exponential hold device whose transfer function is $T/(1 + Ts)$, where T is the sampling period, and repeat the problem. Compare results.

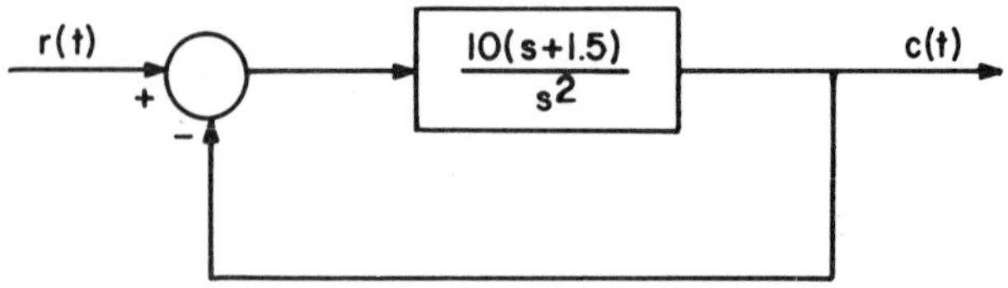

Fig. 6P-2.

6-3 Solve Problem 6-2 by means of the z-form method.

6-4 Solve the following differential equation using the z-form method. That is, find the output c(t) at discrete intervals. Select proper sampling periods with considerations on stability and accuracy of the simulation. In each case, determine $\lim_{k\to\infty} c(k)$.

$$\frac{d^2c}{dt^2} + 4\frac{dc}{dt} + c = 2u(t) \qquad c(0) = 1,\ \dot{c}(0) = 0$$

6-5 A continuous-data control system with state feedback is described by the following state equations:

$$\frac{d\mathbf{x}(t)}{dt} = \mathbf{A}\mathbf{x}(t) + \mathbf{B}u(t)$$

where

$$u(t) = E(0)r(t) - \mathbf{G}(0)\mathbf{x}(t)$$

$E(0) = 1$ and $\mathbf{G}(0) = [1 \quad 1]$.

$$\mathbf{A} = \begin{bmatrix} 0 & 0 \\ 0 & -0.6 \end{bmatrix} \qquad \mathbf{B} = \begin{bmatrix} 0 \\ 1 \end{bmatrix}$$

The system is to be approximated by a sampled-data model

$$\frac{d\mathbf{x}_s(t)}{dt} = \mathbf{A}\mathbf{x}_s(t) + \mathbf{B}u_s(kT)$$

where

$$u_s(kT) = E(T)r(kT) + \mathbf{G}(T)\mathbf{x}_s(kT)$$

Determine $E(T)$ and $\mathbf{G}(T)$ using the first two terms of the Taylor series expansion for $T = 0.3$ second and $T = 0.7$ second. Compare $\mathbf{x}_s(kT)$ with $\mathbf{x}(t)$ for $r(t) =$ unit-step function.

6-6 Given a continuous-data system

$$\frac{d\mathbf{x}(t)}{dt} = \mathbf{A}\mathbf{x}(t) + \mathbf{B}u(t)$$

where

$$u(t) = -\mathbf{G}(0)\mathbf{x}(t)$$

$$\mathbf{A} = \begin{bmatrix} 0 & 1 \\ 0 & 0 \end{bmatrix} \qquad \mathbf{B} = \begin{bmatrix} 0 \\ 1 \end{bmatrix} \qquad \mathbf{G}(0) = [2 \quad 3]$$

The following sampled-data system is used to approximate the continuous-data system by matching the responses at the sampling instants.

$$\frac{dx_s(t)}{dt} = Ax_s(t) + Bu_s(t)$$

$$u_s(kT) = -G(T)x_s(kT)$$

$$G(T) = [g_1(T) \quad g_2(T)]$$

The sampling period is 1 second. Using the weighted feedback gain matrix $G_w(T)$ for $G(T)$, find the stable region in the $g_2(T)$ versus $g_1(T)$ plane. Find the corresponding stable region in the h_2 versus h_1 plane, where $H = [h_1 \quad h_2]$ is the weighting matrix for $G_w(T)$.

6-7 The dynamics of a closed-loop continuous-data system are described by

$$\dot{x}(t) = Ax(t) + Bu(t)$$

where

$$A = \begin{bmatrix} 0 & 1 \\ 0 & 0 \end{bmatrix} \qquad B = \begin{bmatrix} 0 \\ 1 \end{bmatrix}$$

$$u(t) = -G_c x(t) + E_c r(t)$$

$$G_c = [100 \quad 20] \qquad E_c = -100$$

The system is to be approximated by a sampled-data model

$$\frac{dx_s(t)}{dt} = Ax_s(t) + Bu_s(kT) \qquad T = 0.01 \text{ sec}$$

where $u_s(kT)$ is the output of a first-order hold; i.e.,

$$u_s(kT) = [E_0 + (t - kT)E_1]r(kT) - [G_0 + (t - kT)G_1]x_s(kT)$$

E_0 and E_1 are scalars; G_0 and G_1 are 1×2 row matrices. For $r(t) \cong r(kT)$, $kT \leqslant t < (k + 1)T$, find E_0, E_1, G_0 and G_1 so that the responses $x_s(kT)$ and $x(t)$ are matched at the sampling instants. Compute and plot the responses of $x(t)$ and $x_s(t)$ for a unit-step input with zero initial states.

6-8 Given the continuous-data control system

$$\dot{x}_c(t) = Ax_c(t) + Bu_c(t)$$

where $x_c(t)$ is an n-vector, $u_c(t)$ is the scalar input, and

$$u_c(t) = E_c r(t) - G_c \mathbf{x}(t)$$

(a) Show that the response of the continuous-data system can be uniquely matched at every n = N sampling instants by that of the sampled-data system

$$\dot{\mathbf{x}}_s(t) = \mathbf{A}\mathbf{x}_s(t) + \mathbf{B}u_s(kT)$$

where

$$u_s(kT) = E(T)r(kT) - G(T)\mathbf{x}_s(kT) \qquad kT \leqslant t < (k+1)T$$

$\mathbf{x}_s(kT)$ is an n-vector, if the discretized system

$$\mathbf{x}_s[(k+1)T] = \phi(T)\mathbf{x}_s(kT) + \theta(T)u_s(kT)$$

is completely state controllable. Assume that $r(t) \cong r(kT)$ for $kT \leqslant t \leqslant (k+1)T$.

(b) For

$$A = \begin{bmatrix} 0 & 1 \\ 0 & 0 \end{bmatrix} \qquad B = \begin{bmatrix} 0 \\ 1 \end{bmatrix}$$

$$G_c = [100 \quad 20] \qquad E_c = -100$$

find G(T) and E(T) such that the state responses of the continuous-data system are matched by the sampled-data system every two sampling instants; T = 0.01 second.

7

TIME-DOMAIN ANALYSIS

7.1 INTRODUCTION

Since the outputs of digital control systems are usually functions of the continuous variable t, it is necessary to evaluate the performance of the system in the time domain. However, when the z-transform or the discrete-time state equation is used, the outputs of the system are measured only at the sampling instants. Depending on the sampling period and its relation to the time constants of the system, the discrete-time representation may or may not be accurate. In other words, there may be a large discrepency between the output c(t) and the sampled signal $c^*(t)$, so that the latter is not a valid representation of the system behavior.

As in the studies of continuous-data control systems, the time response of a digital control system may be characterized by such terms as the *overshoot*, *rise time*, *delay time*, *settling time*, *damping ratio*, *damping factor*, *natural undamped frequency*, etc.

The performance of a digital control system in the time domain is often measured by applying a test signal such as a *unit-step function* to the system input. For linear systems the unit-step input can provide valuable information on the transient and steady-state behavior of the system. In fact, *overshoot, rise time, delay time and settling time are all defined with respect to a unit-step input.* The response of a system to a unit-step input is called the *unit-step response.*

Figure 7-1 shows a typical unit-step response of a digital control system. Notice that although the system is called digital due to the fact that there are digital transducers and/or digital controllers in the system, the response or the output of the system is usually still a function of the continuous time variable t. Therefore, the parameters, maximum overshoot, delay time, settling time and peak time are still defined in the usual manner, as shown in Fig. 7-1.

When the z-transform or the discrete-time state equation is used for the analysis of digital control systems, the system responses are represented only at the sampling instants. Care must be taken in judging on the accuracy and validity of these discrete-time data as they may not be an accurate representation of the true responses of the digital system.

Figure 7-2 illustrates a typical output of a digital control system, c(t), which has a maximum overshoot c_m. The sampled signal is represented by $c^*(t)$. It is apparent that the maximum value of $c^*(t)$, c_m^*, is always less than or equal to c_m. The case illustrated in Fig. 7-2 is such that the sampling period is sufficiently small so that the sampled-data response $c^*(t)$ does give an adequate representation of the true response, and the discrepency between c_m and c_m^* is not significant. However, in general, if the sampling period is too large, the sampled-data representation may be entirely erroneous. It should be pointed out that the selection of the sampling period of a digital control system is usually not based on just the accuracy of representation of the system responses at the sampling instants, but more importantly on the overall system performance, stability, accuracy, and hardware considerations.

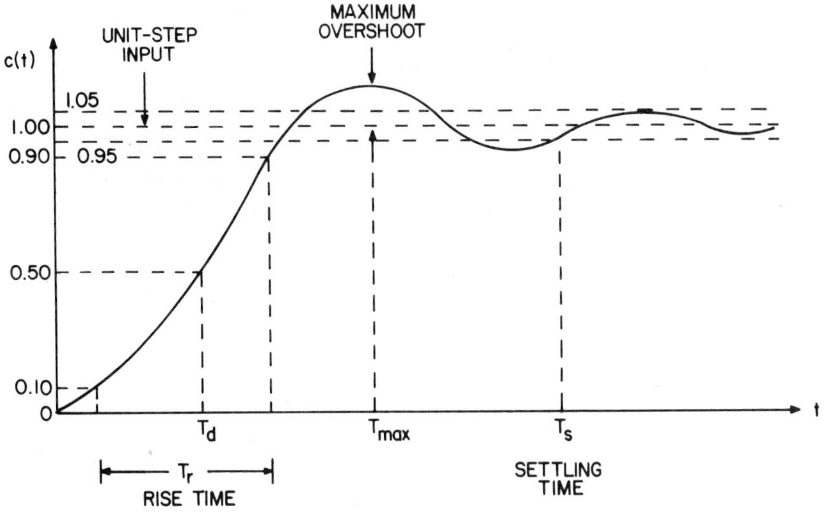

Fig. 7-1. Typical unit-step response of a digital control system and illustration of time-domain performance criteria.

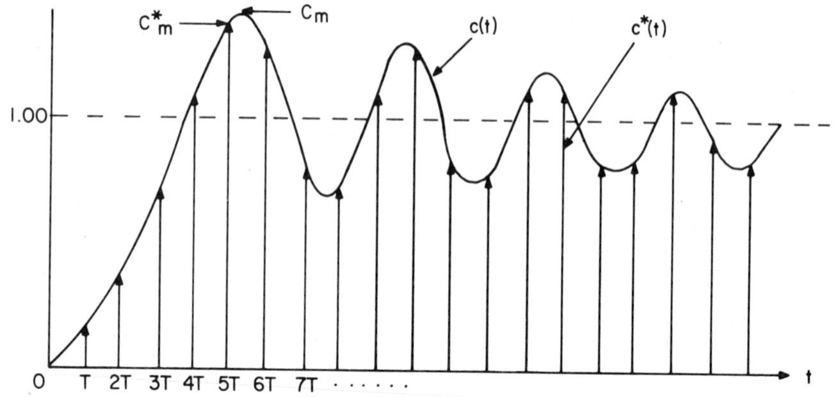

Fig. 7-2. Typical unit-step response of a digital control system and its sampled-data representation.

7.2 COMPARISON OF TIME RESPONSES OF CONTINUOUS-DATA AND DIGITAL CONTROL SYSTEMS

In this section we shall use a space-vehicle control system to compare the salient features of continuous-data and digital control systems. It should be borne in mind that the characteristics illustrated by this example are only typical, and generally there are exceptions which do not always conform to these results.

The block diagram of a simplified space-vehicle control system is shown in Fig. 7-3. The objective of the system is to control the attitude (position) of the space vehicle in one dimension. There may be two other similar systems which control the other two space variables if it is a three-coordinate system, and it has been assumed that the dynamics of these coordinates can be decoupled from each other. In most space-vehicle control problems, if the vehicle is a rigid structure (in practice it seldom is), then it can be modelled as a simple mass or inertia. In Fig. 7-3 the vehicle is represented by a pure inertia J_v, so that the transfer function between the applied torque and the output position

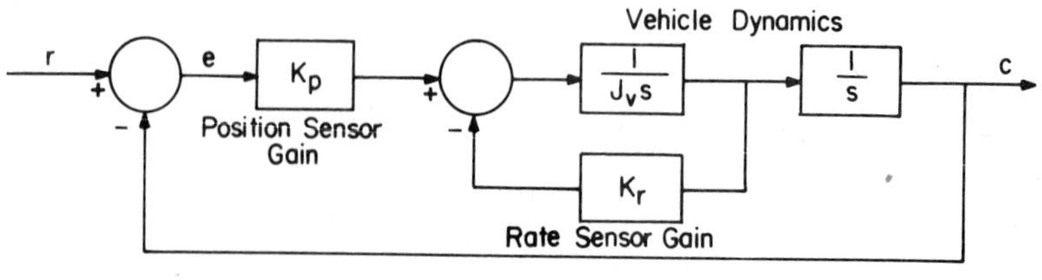

Fig. 7-3. A continuous-data space-vehicle control system.

is

$$\frac{C(s)}{T_m(s)} = \frac{1}{J_v s^2} \tag{7-1}$$

The position c(t) and its derivative, the velocity v(t), are fed back by the position and rate sensors, respectively, to form the closed-loop control. In continuous-data control theory, the rate feedback is used frequently for stabilization purposes.

The parameters of the simple continuous-data control system are given:

K_p = positional sensor gain = 1.65×10^6

K_r = rate sensor gain = 3.17×10^5

J_v = moment of inertia of vehicle = 41822

It is assumed that all the units are consistent so that for analytical purposes we intentionally left them to be unspecified.

From Fig. 7-3 the open-loop transfer function of the system is written,

$$G(s) = \frac{C(s)}{E(s)} = \frac{K_p}{s(J_v s + K_r)} \tag{7-2}$$

The closed-loop transfer function is

$$\frac{C(s)}{R(s)} = \frac{G(s)}{1 + G(s)} = \frac{K_p}{J_v s^2 + K_r s + K_p} \tag{7-3}$$

Substituting the system parameters in Eq. (7-3), we have

$$\frac{C(s)}{R(s)} = \frac{39.453}{s^2 + 8.871s + 39.453} \tag{7-4}$$

The characteristic equation of the system is obtained by setting the denominator of the closed-loop transfer function to zero, we get

$$s^2 + 8.871s + 39.453 = 0 \tag{7-5}$$

Comparing the last equation with the standard second-order characteristic equation,

$$s^2 + 2\zeta\omega_n s + \omega_n^2 = 0 \tag{7-6}$$

we have

Damping ratio $\zeta = 0.706$

Natural undamped frequency $\omega_n = 6.28$ rad/sec

The unit-step response of the system is shown in Fig. 7-4.

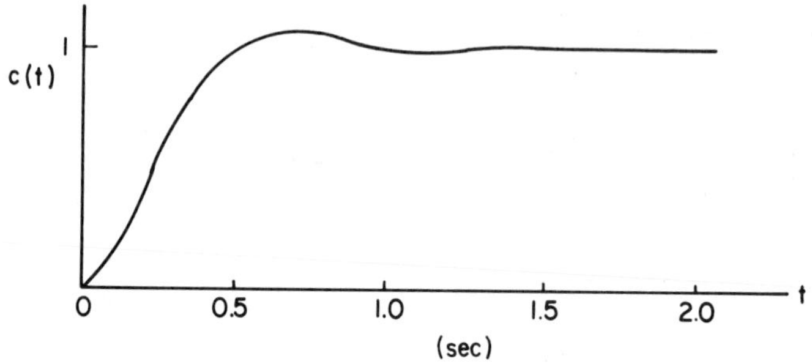

Fig. 7-4. Unit-step response of the continuous-data space-vehicle control system in Fig. 7-3.

Since the system is of the second order, the quadratic equation in Eq. (7-5) will always have roots in the left-half of the s-plane so long as all the parameters of the system are positive. Thus, the continuous-data system will always be asymptotically stable for all positive values of K_p, K_r and J_v.

Now let us consider that the control system in Fig. 7-3 is subject to digital control, which would be the case if the position and rate sensors are digital transducers. In practice, it is common to use a digital or incremental position transducer whose output is processed to give position as well as speed information.

For the digital control system the outputs of the position and rate sensors are all processed by sample-and-hold units which include A/D operations. Figure 7-5 illustrates the block diagram of the digital space vehicle control system; the sample-and-hold operations are represented by one sample-and-hold as shown. The sampling period is T seconds.

For the purpose of comparison we assume that the system parameters, K_p, K_r and J_v are the same as those of the continuous-data system. The open-

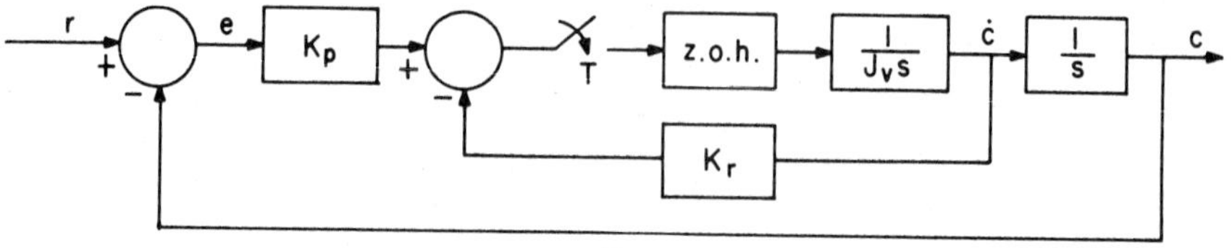

Fig. 7-5. A digital space-vehicle control system.

loop transfer function of the digital control system is written directly from Fig. 7-5.

$$G(z) = \frac{C(z)}{E(z)} = \frac{K_p \mathcal{Z}\left[G_h(s)\frac{1}{J_v s^2}\right]}{1 + K_r \mathcal{Z}\left[G_h(s)\frac{1}{J_v s}\right]} \tag{7-7}$$

where $G_h(s)$ represents the transfer function of the zero-order hold. Then,

$$\mathcal{Z}\left[G_h(s)\frac{1}{J_v s^2}\right] = (1 - z^{-1})\mathcal{Z}\left[\frac{1}{J_v s^3}\right] = \frac{T^2}{2J_v}\frac{z+1}{(z-1)^2} \tag{7-8}$$

$$\mathcal{Z}\left[G_h(s)\frac{1}{J_v s}\right] = (1 - z^{-1})\mathcal{Z}\left[\frac{1}{J_v s^2}\right] = \frac{T}{J_v(z-1)} \tag{7-9}$$

Substitution of the last two equations into Eq. (7-7) and simplifying, we get

$$G(z) = \frac{T^2K_p(z+1)}{2J_v z^2 + (2K_rT - 4J_v)z + 2J_v - 2K_rT} \tag{7-10}$$

The closed-loop transfer function of the digital system is

$$\frac{C(z)}{R(z)} = \frac{G(z)}{1 + G(z)} = \frac{T^2K_p(z+1)}{2J_v z^2 + (2K_rT - 4J_v + T^2K_p)z + (2J_v - 2K_rT + T^2K_p)} \tag{7-11}$$

Substituting the system parameters into the last two equations, we have

$$G(z) = \frac{1.65 \times 10^6 T^2(z+1)}{83644z^2 + (6.34 \times 10^5 T - 167288)z + 83644 - 6.34 \times 10^5 T} \tag{7-12}$$

and

$$\frac{C(z)}{R(z)} = \frac{1.65 \times 10^6 T^2(z+1)}{Az^2 + Bz + C} \tag{7-13}$$

where

$$A = 83644$$

$$B = 6.34 \times 10^5 T - 167288 + 1.65 \times 10^6 T^2$$

$$C = 83644 - 6.34 \times 10^5 T + 1.65 \times 10^6 T^2$$

The characteristic equation is obtained by equating the denominator of $C(z)/R(z)$ to zero.

$$Az^2 + Bz + C = 0 \tag{7-14}$$

where A, B and C are as given above.

Since now there is an additional system parameter in the sampling period T, the performance of the overall digital control system will depend on the values of K_p, K_r, J_v and T.

Since the roots of the characteristic equation of the digital control system must stay inside the unit circle $|z| = 1$ in the z-plane for the overall system to be asymptotically stable, we see that the second-order digital control system can be unstable for large values of T. Applying Jury's stability test to the characteristic equation of Eq. (7-14) we find that the stable range of T is $0 \leqslant T < 0.264$ second.

Figure 7-6 shows the plot of the roots of Eq. (7-14) when T varies from 0 to infinity. This plot is known as the *root locus plot* [1]. In the present situation the root locus plot of Fig. 7-6 is obtained by assigning a value to T and solving Eq. (7-14). Since T appears as a nonlinear parameter in Eq. (7-14) (T appearing as T^2) the rules of construction [1] of the conventional root locus diagram cannot be applied here.

Figure 7-7 shows the unit-step responses of the digital control system for several values of the sampling period T. Note that as the sampling period ap-

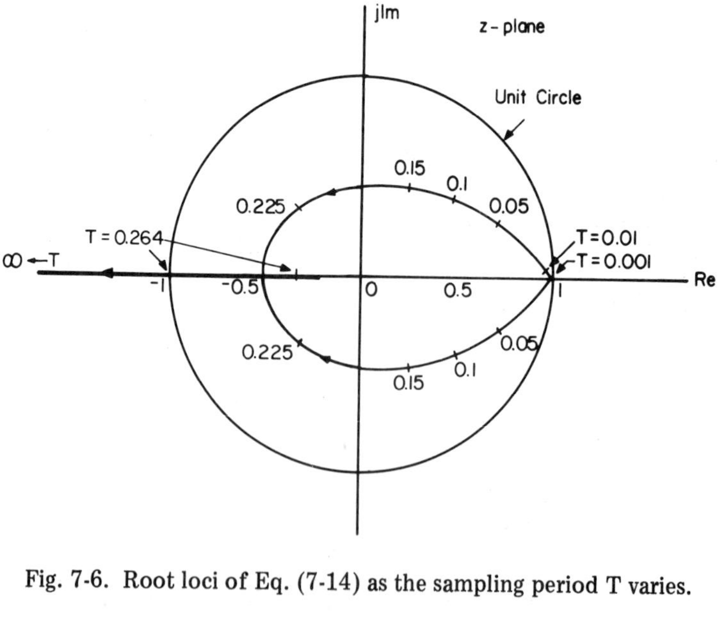

Fig. 7-6. Root loci of Eq. (7-14) as the sampling period T varies.

proaches zero, the response of the digital control system approaches that of the continuous-data system.

The steady-state performance of the digital control system can be determined simply by use of the final-value theorem. In general, the final-value theorem is the most straightforward method of evaluating the steady-state error as long as the theorem is valid for the given situation. When the closed-loop system is unstable or when the input is too fast that the output cannot follow, the error will be unbounded, and the final-value theorem cannot be applied.

The steady-state error of the digital control system is defined as

$$\lim_{k\to\infty} e(kT) = \lim_{z\to 1} (1 - z^{-1})E(z) \tag{7-15}$$

From Fig. 7-5,

$$E(z) = R(z) - C(z) = \frac{R(z)}{1 + G(z)} \tag{7-16}$$

For a unit-step input, $R(z) = z/(z-1)$, and substituting Eq. (7-16) into Eq. (7-15), we have

$$\lim_{k\to\infty} e(kT) = \lim_{z\to 1} \frac{1}{1 + G(z)} \tag{7-17}$$

Substituting Eq. (7-10) into the last equation, we get

$$\lim_{k\to\infty} e(kT) = 0 \tag{7-18}$$

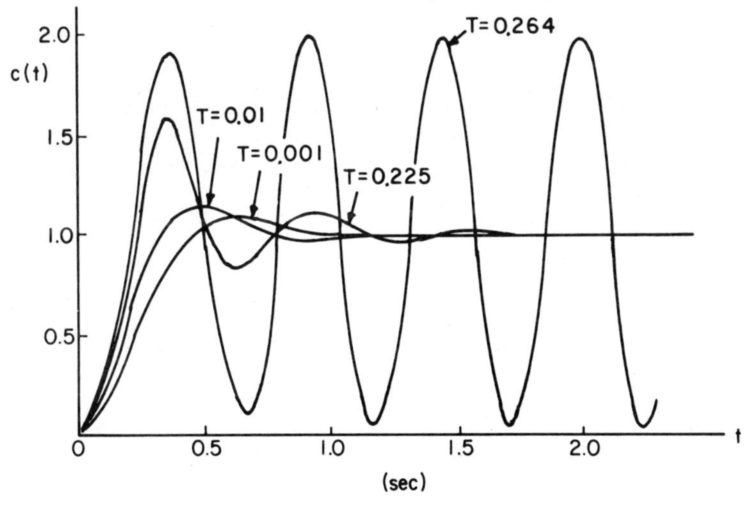

Fig. 7-7. Step responses of the digital control system in Fig. 7-5.

since G(z) approaches infinity as z approaches one. Thus, the digital space-vehicle control system is capable of tracking a step input signal without any steady-state error.

Now let us consider that the input r(t) is a unit-ramp function, $r(t) = tu_s(t)$. Then, the z-transform of r(t) is

$$R(z) = \frac{Tz}{(z-1)^2} \tag{7-19}$$

The steady-state error of the system becomes

$$\lim_{k\to\infty} e(kT) = \lim_{z\to 1} \frac{T}{(z-1)G(z)} \tag{7-20}$$

Substitution of Eq. (7-10) into Eq. (7-20) and carrying out the limiting process, we have

$$\lim_{k\to\infty} e(kT) = \frac{2K_r + TK_p}{2K_p} \tag{7-21}$$

Thus, the steady-state error of the digital system due to a ramp input is constant and is dependent on K_r, K_p and T. If we let T go to zero, we get

$$\lim_{t\to\infty} e(t) = K_r/K_p \tag{7-22}$$

which is the same as the steady-state error due to a unit-ramp input of the continuous-data space-vehicle control system.

We can summarize the findings from this simple illustrative example on the analysis of a continuous-data and a digital control system.

(1) For the same system configuration and parameters, the digital system is usually less stable than the continuous-data system. In Chapter 6 we have introduced the digital redesign method and have shown how the system parameters of the digital system can be altered to render a similar response as the continuous-data system.

(2) The performance of the digital system is dependent on the sampling period T. Larger sampling periods usually give rise to higher overshoots in the step response, and eventually may cause instability if the sampling period is too large.

(3) The root locus diagram in Fig. 7-6 reveals another important characteristic with regard to the analysis of digital control systems. The diagram shows that for small values of T the roots of the characteristic equation are located very close to the z = 1 point in the z-plane. The concentration of the dominant roots near the z = 1 point often pre-

sents two practical difficulties in analyzing digital control systems: one is that it becomes difficult to predict the closed-loop system behavior from the z-plane root configuration. Such parameters as the constant-damping-ratio loci, the constant-frequency loci, etc., become inaccurate when all the roots become cluttered together, and second, one may encounter accuracy problems when attempting to solve for these almost-identical roots on a digital computer.

7.3 CORRELATION BETWEEN TIME RESPONSE AND ROOT LOCATIONS IN THE S-PLANE AND THE Z-PLANE

The mapping between the domains of the s-plane and the z-plane was discussed in Sec. 3.3. In particular, the constant-damping, constant-damping ratio, and constant-frequency loci were established in the z-plane. These loci in the z-plane are useful in predicting the performance of a digital control system.

The correlation between the locations of the characteristic equation roots in the s-plane and the transient response is well-known for continuous-data systems. For instance, complex conjugate roots in the left-half s-plane give rise to exponentially damped sinusoidal responses; roots on the negative real axis correspond to monotonically decaying responses; and simple complex conjugate roots on the imaginary axis will give rise to undamped constant-amplitude sinusoidal oscillations. Multiple-order roots on the imaginary axis and roots in the right-half s-plane correspond to unstable responses.

Although the relationship between the s-plane and the z-plane has been well established, the sampling operation in digital systems does bring about conditions which require special attention. For instance, when the sampling theorem is not satisfied, the folding effect of sampling could distort the true response of the system. Figure 7-8(a) shows the poles of the transfer function of a second-order continuous-data system. The natural response of the system is shown in Fig. 7-8(b). Now if the system is subject to sampling with $2\omega_1 > \omega_s$, where ω_s is the sampling frequency, the sampling operation generates an infinite number of poles in the s-plane at $s = \sigma_1 \pm j\omega_1 + jn\omega_s$, $n = \pm 1, \pm 2, \ldots$. As shown in Fig. 7-9(a) the sampling operation folds the poles back into the primary strip, $-\omega_s/2 < \omega < \omega_s/2$, so that the net effect is equivalent to having a system originally with poles at $s = \sigma_1 \pm j(\omega_s - \omega_1)$.

The z-plane representation for both cases is shown in Fig. 7-9(b). Figure 7-9(c) illustrates the sampled signal which demonstrates the effect of frequency folding. The folding effect makes the sampled system to appear as if the frequency were equal to $\omega_s - \omega_1$ rather than ω_1.

Figure 7-10 illustrates several cases of root locations of a second-order system in the s-plane and the z-plane and their corresponding time responses.

Although only second-order characteristic roots are illustrated in the preceding discussion on the effects of root locations in the z-plane, the correlation between the time response and the second-order roots can generally be applied to high-order systems that can be approximated by second-order dominant roots.

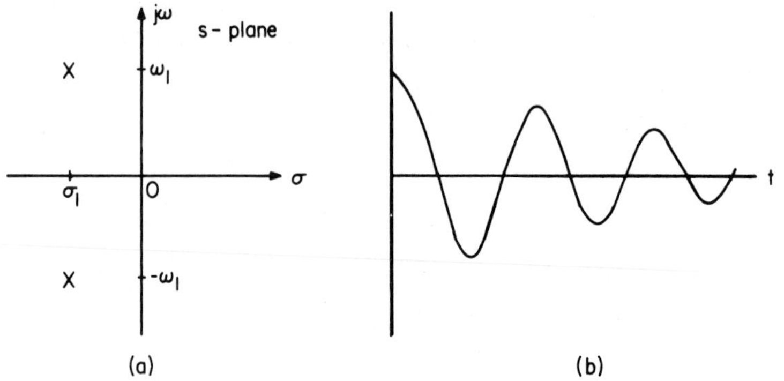

Fig. 7-8. (a) Poles of a second-order continuous-data system; (b) natural response of the system.

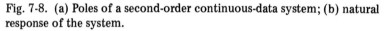

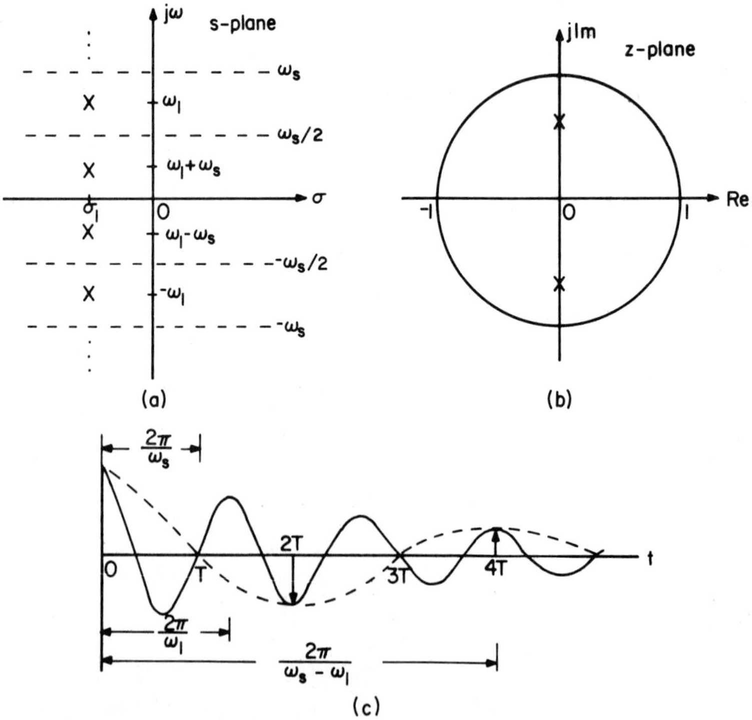

Fig. 7-9. Pole locations in the s-plane and the z-plane showing the effect of folding of frequency.

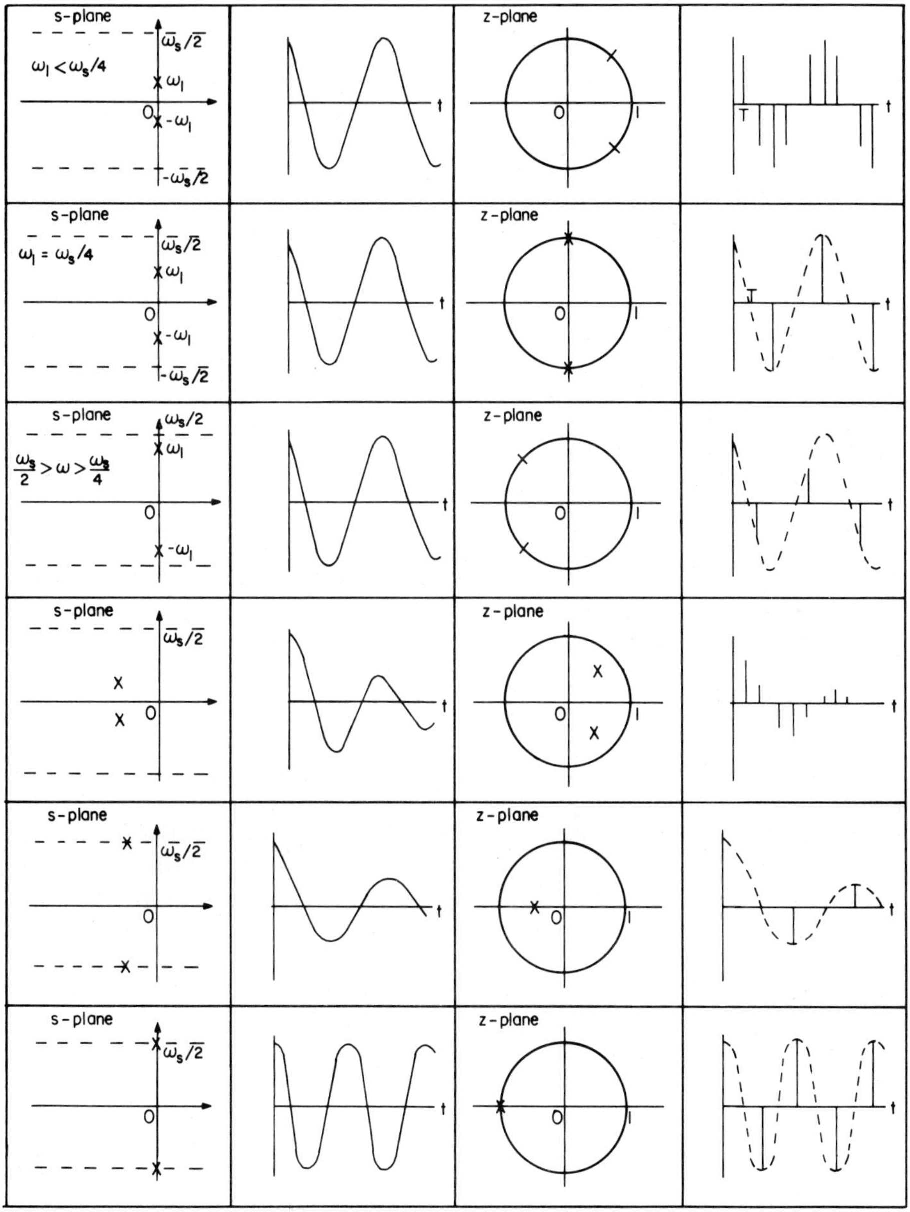
s-plane
z-plane
$\omega_1 < \omega_s/4$
$\omega_1 = \omega_s/4$
$\frac{\omega_s}{2} > \omega > \frac{\omega_s}{4}$
$\omega_s/2$
$-\omega_s/2$
ω_1
$-\omega_1$
T
t

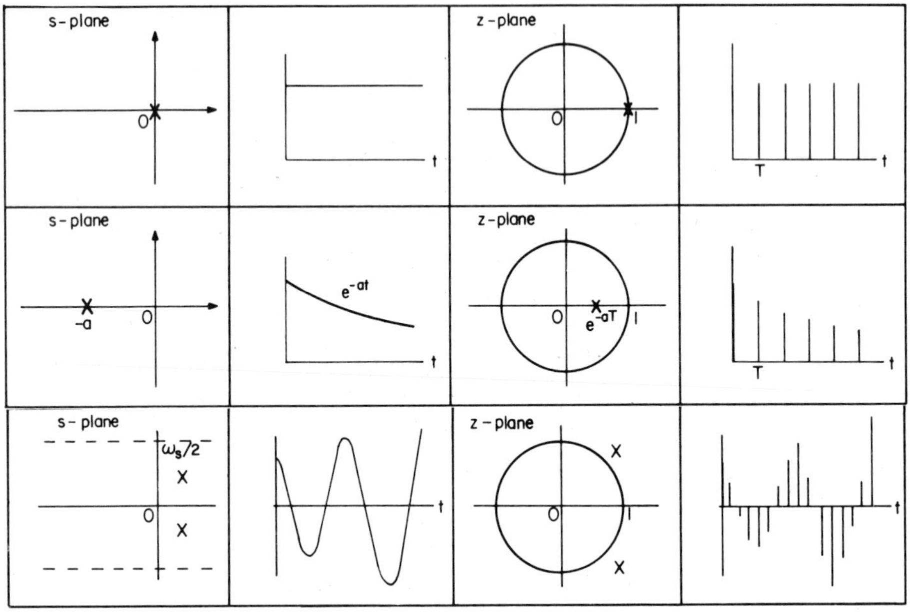

Fig. 7-10. Root locations in the s-plane and the z-plane and their corresponding time responses.

7.4 EFFECT OF THE POLE-ZERO CONFIGURATIONS IN THE Z-PLANE UPON THE MAXIMUM OVERSHOOT AND PEAK TIME OF TRANSIENT RESPONSE OF SAMPLED-DATA SYSTEMS

In the preceding section we have described the relationship between the characteristic equation roots in the z-plane and the transient response of a second-order digital control system. When the characteristic equation has complex roots inside the unit circle in the z-plane, the unit-step response of the system will be oscillatory with positive damping; that is, the response will decay with time. In general, the complex roots that are inside and are closest to the unit circle give rise to more oscillatory or less damped responses. It would be useful to determine the relationship between the pole-zero configuration of the closed-loop transfer function and the maximum overshoot and peak time T_{max} of digital control systems.

For a second-order continuous-data control system whose closed-loop transfer function is

$$\frac{C(s)}{R(s)} = \frac{\omega_n^2}{s^2 + 2\zeta\omega_n s + \omega_n^2} \tag{7-23}$$

the maximum overshoot and peak time are given, respectively, by [1]

$$c_{max} = 1 + e^{-\zeta\pi/\sqrt{1-\zeta^2}} \tag{7-24}$$

and

$$T_{max} = \frac{\pi}{\omega_n\sqrt{1-\zeta^2}} \tag{7-25}$$

For systems of the order higher than two, no simple relations can be obtained between c_{max}, T_{max}, and the pole-zero configuration of the closed-loop transfer function. However, if a system can be characterized by only a pair of dominant poles, that is, poles of the closed-loop transfer function that dominate the transient response, and the other poles and zeros are far to the left in the complex s-plane, the effects of these other poles and zeros on the transient response are negligible. Under this condition c_{max} and T_{max} can be approximated by using Eqs. (7-24) and (7-25) in conjunction with the second-order transfer function of Eq. (7-23). For example, consider the following transfer function of a fourth-order system,

$$\frac{C(s)}{R(s)} = \frac{K}{(s + p_1)(s + p_2)(s^2 + 2\zeta\omega_n s + \omega_n^2)} \tag{7-26}$$

with p_1 and p_2 being real constants. If p_1 and p_2 are at least five times greater than $\zeta\omega_n$, the two poles at $-p_1$ and $-p_2$ will contribute insignificantly to the transient response, and the poles at

$$s = -\zeta\omega_n + j\omega_n\sqrt{1-\zeta^2} \quad \text{and} \quad s = -\zeta\omega_n - j\omega_n\sqrt{1-\zeta^2}$$

are the dominant poles. However, we cannot simply throw away the two terms, $(s + p_1)$ and $(s + p_2)$ in Eq. (7-26) as they still affect the steady-state performance of the system.

For a digital control system, the problem of establishing the overshoot and peak-time-to-pole-zero configuration is more complex. This is because normally when the z-transform method or the discrete-data state equation is used, the output response is described only at the sampling instants.

Let us select the following model as the closed-loop transfer function of a typical second-order digital control system.

$$\frac{C(z)}{R(z)} = \frac{K(z - z_1)}{(z - p_1)(z - \bar{p}_1)} \tag{7-27}$$

where

where

$$K = \frac{(1 - p_1)(1 - \bar{p}_1)}{(1 - z_1)} \tag{7-28}$$

where z_1 is a real zero, and p_1 and $\bar{p}_1$ are complex conjugate poles. We have included the real zero at z_1 as a typical case, since most continuous-data transfer functions without zeros when sampled will produce at least one zero in the z-transform transfer function. A typical example is readily given by Eq. (7-11). We also intentionally consider only complex poles, since transfer functions with stable real poles in the z-plane would not produce any overshoot.

When the closed-loop digital control system is subject to a unit-step input, the output transform C(z) is given by

$$C(z) = \frac{zK(z - z_1)}{(z - 1)(z - p_1)(z - \bar{p}_1)} \tag{7-29}$$

The output response at the sampling instants is obtained by applying the inversion formula of z-transform, Eq. (3-60), to Eq. (7-29). We have

$$c(kT) = \frac{1}{2\pi j} \oint_{\Gamma} \frac{Kz(z - z_1)}{(z - 1)(z - p_1)(z - \bar{p}_1)} z^{k-1} dz \tag{7-30}$$

where Γ is a closed contour which encloses all the singularities of the integrand. Applying Cauchy's residue theorem to Eq. (7-30), the output sequence c(kT) is written

$$c(kT) = 1 + 2\left|\frac{K(p_1 - z_1)}{(p_1 - 1)(p_1 - \bar{p}_1)}\right| |p_1|^k \cos(k\phi_1 + \theta_1) \tag{7-31}$$

where

$$\phi_1 = \arg(p_1) \tag{7-32}$$

$$\theta_1 = \arg(p_1 - z_1) - \arg(p_1 - 1) - \pi/2 \tag{7-33}$$

and arg(·) represents the "argument of (·)" or the "angle of (·)".

The pole-zero configuration of Eq. (7-27) is shown in Fig. 7-11. From the locations of the poles and zero, we define the angle a, as shown in Fig. 7-11, as

$$\pm a = \arg(p_1 - z_1) - \arg(p_1 - 1) + \pi/2 \tag{7-34}$$

where the sign in front of a is selected according to the situations illustrated in Fig. 7-11. From Eq. (7-33), we see that θ_1 is related to a through

$$\theta_1 = \pm a - \pi \tag{7-35}$$

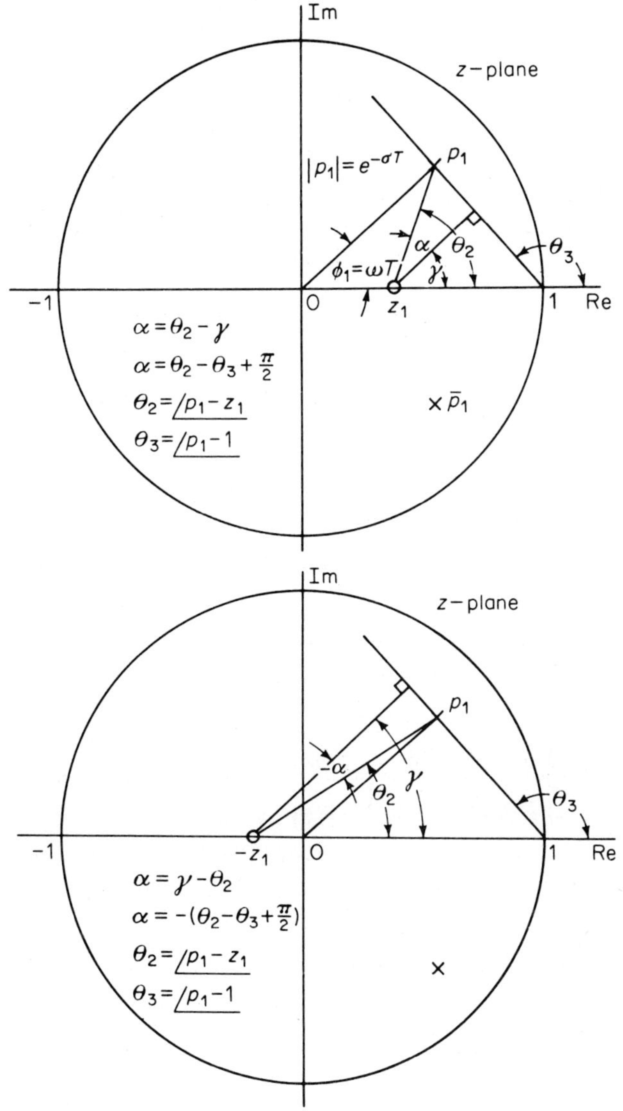

Fig. 7-11. Geometrical representation of α for a second-order sampled-data system with closed-loop transfer function given by Eq. (7-27).

Also, we can readily show that the following relationship holds between a and the closed-loop pole-zero locations:

$$|\sec a| = 2\left|\frac{K(p_1 - z_1)}{(p_1 - 1)(p_1 - \bar{p}_1)}\right| \tag{7-36}$$

where K is given in Eq. (7-28). Substituting Eqs. (7-35) and (7-36) into Eq. (7-31), the system output sequence is written

$$c(kT) = 1 + |\sec a| \times |p_1|^k \cos(k\phi_1 \pm a - \pi) \tag{7-37}$$

Equations (7-31) and (7-37) give the response of c(t) only at the sampling instants, and in principle, once c(t) is sampled, the information between the sampling instants is lost. In other words, we cannot reconstruct c(t) from c(kT). However, we can approximate c(t) by a function that passes through all the points of c(kT).

Letting t = kT, then

$$|p_1|^k = |p_1|^{t/T} = e^{-\zeta\omega_n t} \tag{7-38}$$

and

$$\phi_1 = \arg(p_1) = \omega T = \omega_n\sqrt{1 - \zeta^2}\, T \tag{7-39}$$

Therefore, a continuous-time function that passes through the points of c(kT) is

$$c(t) = 1 + |\sec a| e^{-\zeta\omega_n t} \cos(\omega_n\sqrt{1 - \zeta^2}\, t \pm a - \pi) \tag{7-40}$$

The overshoot of the time response c(t) in Eq. (7-40) is now used to approximate the overshoot of the unit-step response of the digital control system whose closed-loop transfer function is given by Eq. (7-27). The overshoot of the response of Eq. (7-40) is determined in the usual manner; that is, by taking the first derivative on both sides of Eq. (7-40) with respect to t and setting dc(t)/dt to zero. This step leads to the following relationship:

$$\tan(\omega_n\sqrt{1 - \zeta^2}\, t \pm a - \pi) = \frac{-\zeta}{\sqrt{1 - \zeta^2}} \tag{7-41}$$

Since the time t solved from the last equation is the peak time T_{max}, we have

$$T_{max} = \frac{1}{\omega_n\sqrt{1 - \zeta^2}}\left\{\tan^{-1}\frac{-\zeta}{\sqrt{1 - \zeta^2}} \mp a + \pi\right\} \tag{7-42}$$

Note that the signs of a are now reversed.

Substituting Eq. (7-42) into Eq. (7-40) and simplifying, the maximum value of c(t) is

$$c_{max} = 1 + \sqrt{1-\zeta^2}\ |\sec a|\ \exp \frac{-\zeta}{\sqrt{1-\zeta^2}}\left\{\tan^{-1}\frac{-\zeta}{\sqrt{1-\zeta^2}} \mp a + \pi\right\} \qquad (7\text{-}43)$$

Hence, the maximum overshoot of c(t), which is an approximation of c*(t), is

$$\text{max. overshoot} = \sqrt{1-\zeta^2}\ |\sec a|\ \exp \frac{-\zeta}{\sqrt{1-\zeta^2}}\left\{\tan^{-1}\frac{-\zeta}{\sqrt{1-\zeta^2}} \mp a + \pi\right\} \qquad (7\text{-}44)$$

We can readily see from Eq. (7-44) that the maximum overshoot is uniquely specified by the damping ratio ζ and the angle a (see Fig. 7-11).

In Fig. 7-12 the percent maximum overshoot given by Eq. (7-44) is plotted as a function of ζ and a. The angle a and its sign are determined from the pole-zero location of the closed-loop transfer function as described in Fig. 7-11. The value of ζ can be calculated by use of Eq. (7-38) once the complex pole p_1 is known. From Eq. (7-38) we write

$$|p_1| = e^{-\zeta\omega_n T} = \exp \frac{-\zeta\phi_1}{\sqrt{1-\zeta^2}} \qquad (7\text{-}45)$$

where ϕ_1 is measured in radians.

Thus, given p_1, ϕ_1 is determined from Eq. (7-39), and then ζ is computed from Eq. (7-45).

The peak time T_{max} is found by use of Eq. (7-42). Or,

$$T_{max} = \frac{T}{\phi_1}\left\{\tan^{-1}\frac{-\zeta}{\sqrt{1-\zeta^2}} \mp a + \pi\right\} \qquad (7\text{-}46)$$

where the sign of a is opposite to those shown in Fig. 7-11.

Figure 7-13 shows the curves of T_{max}, normalized in $T_{max}\phi_1/T$, plotted as functions of ζ and a.

It should be emphasized again that the time response of Eq. (7-40) is a close approximation to the actual discrete response c*(t) only if the sampling frequency of the sampler is sufficiently high. In general, for the approximation method described here to be reasonably accurate, the sampling theorem should be satisfied. In terms of pole location, the complex poles of the closed-loop transfer function, p_1 and $\bar{p}_1$ must be located in the first and the fourth quadrants in the z-plane, respectively.

As mentioned earlier, the method described above can be applied to high-order systems having two complex dominant poles and a zero located at any

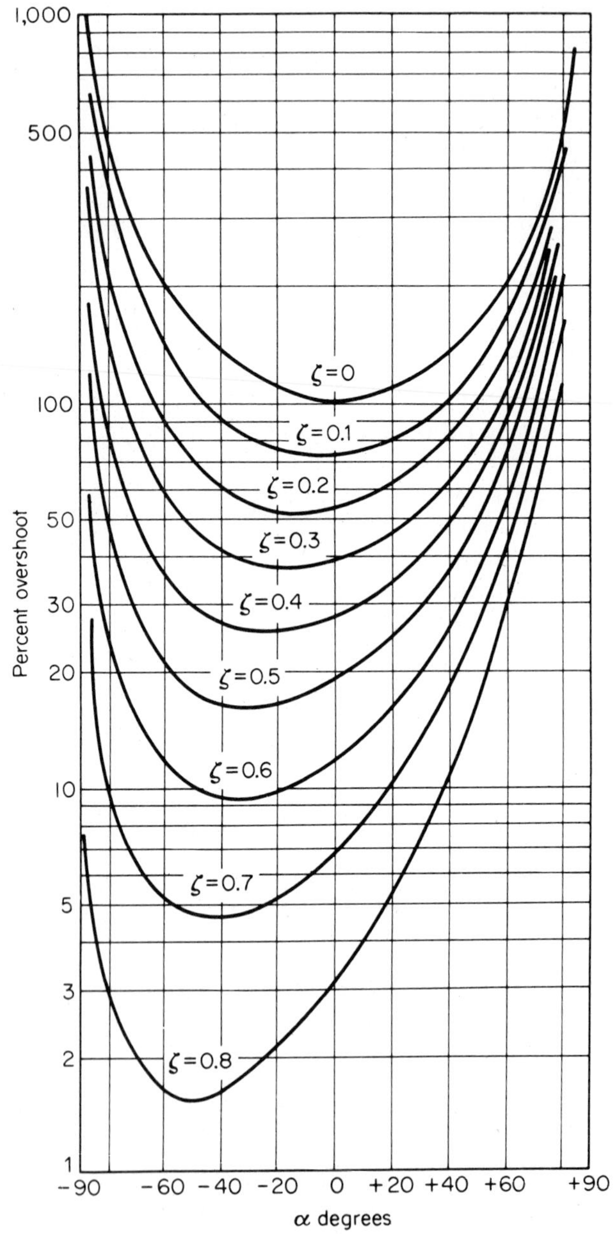

Fig. 7-12. Percent overshoot of second-order sampled-data control systems.

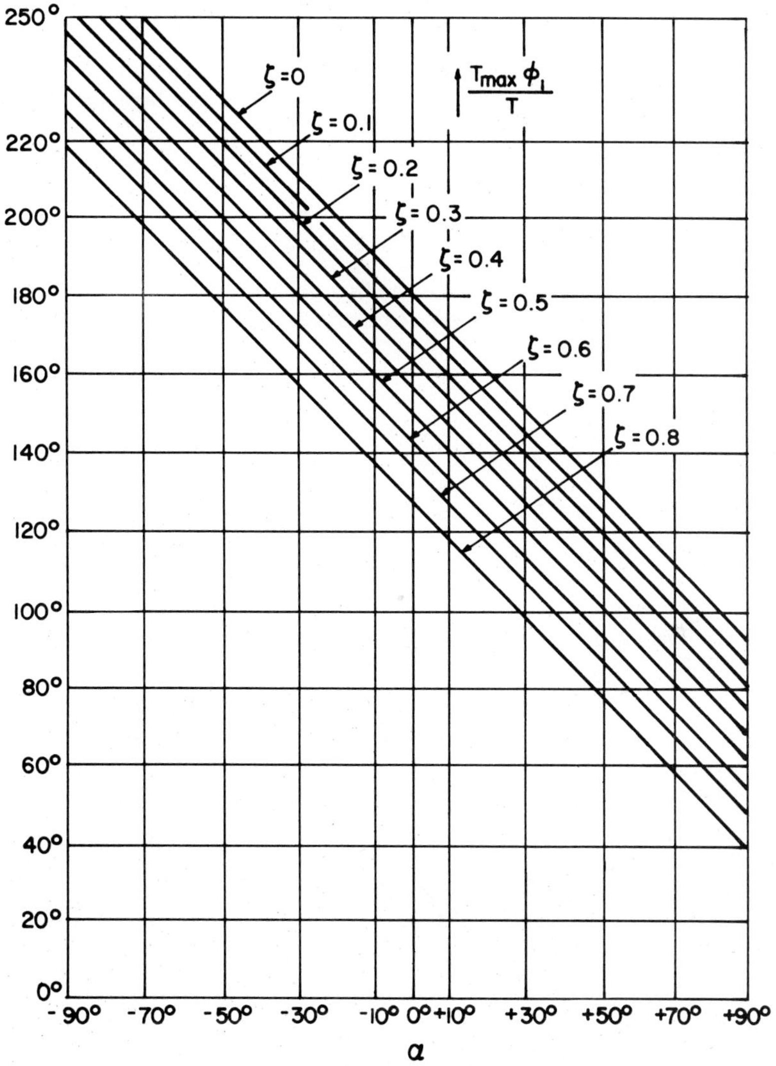

Fig. 7-13. Peak time of step response of second-order sampled-data systems versus a.

point on the real axis; all other poles and zeros are located near the origin in the z-plane.

Example 7-1.

Consider the digital space-vehicle control system shown in Fig. 7-5. The unit-step responses of the system are plotted in Fig. 7-7 for several values of the sampling period T. Now let us determine the maximum overshoot and the peak time of the system using the method presented in this section. We can see that the advantage of the present method is that the overshoot and peak-time properties of the unit-step response can be approximated without actually computing the response c*(t) numerically.

For T = 0.225 sec, the closed-loop transfer function of Eq. (7-13) becomes

$$\frac{C(z)}{R(z)} = \frac{83531.25(z+1)}{83644z^2 + 58893.25z + 24525.25} \tag{7-47}$$

The zero of C(z)/R(z) is at $z = -1$, and the two poles are at $z = p_1 = -0.352 + j0.4114$ and $z = \bar{p}_1 = -0.352 - j0.4114$. The pole-zero configuration of C(z)/R(z) is shown in Fig. 7-14.

From Fig. 7-14 we get

$$a = -40.66^\circ \tag{7-48}$$

and

$$\phi_1 = \omega T = 130.55^\circ = 2.278 \text{ rad}$$

$$|p_1| = 0.54$$

Substituting the values of ϕ_1 (in radians) and $|p_1|$ into Eq. (7-45), the damping ratio ζ is found to be

$$\zeta = 0.26$$

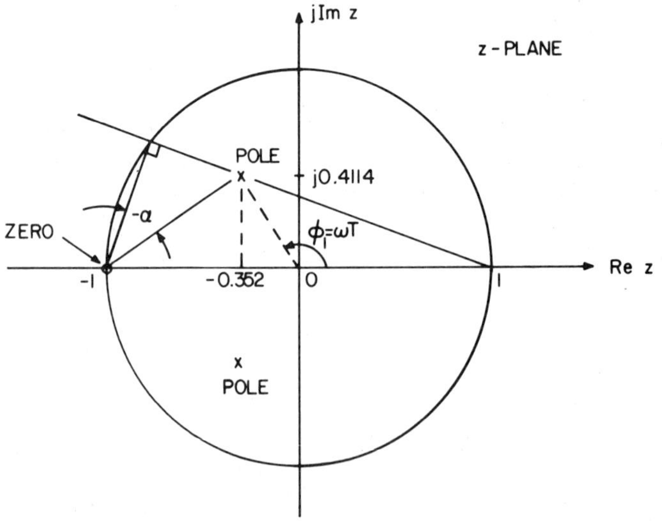

Fig. 7-14. Pole-zero configuration of the closed-loop transfer function in Eq. (7-47), T = 0.225 sec.

Once a and ζ are found, we go to Fig. 7-12 to determine the maximum overshoot. In this case, the interpolated value on the curves for the maximum overshoot is approximately 50 percent. This result agrees quite well with the computed response curve in Fig. 7-7 which shows a maximum overshoot of slightly over 50 percent.

Now let us consider the case for T = 0.01 sec. The closed-loop transfer function of Eq. (7-13) becomes

$$\frac{C(z)}{R(z)} = \frac{165(z+1)}{83644z^2 - 160783z + 77469} \tag{7-49}$$

The zero of this transfer function is at $z = -1$, and the two poles are at $z = p_1 = 0.96 + j0.0494$ and $z = \bar{p}_1 = 0.96 - j0.0494$. The pole-zero configuration of $C(z)/R(z)$ is shown in Fig. 7-15, from which we get

$$a = -37.56°$$

and

$$\phi_1 = 2.95° = 0.0514 \text{ rad}$$

$$|p_1| = 0.9613$$

Using Eq. (7-45), the damping ratio of the closed-loop digital control system is found to be

$$\zeta = 0.609$$

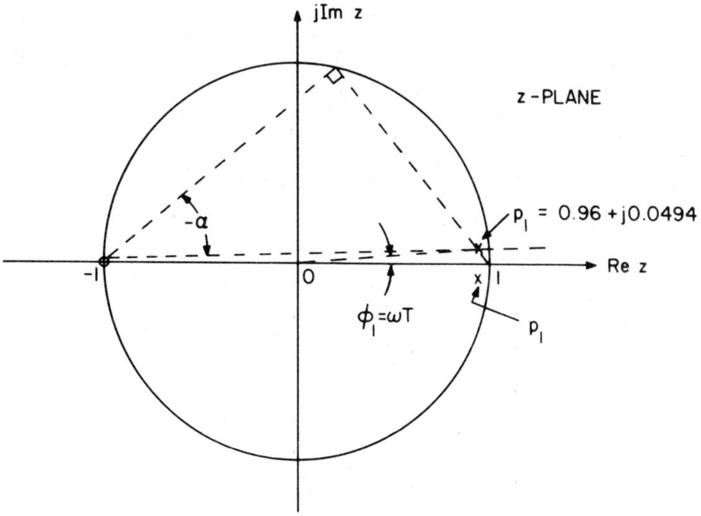

Fig. 7-15. Pole-zero configuration of the closed-loop transfer function in Eq. (7-47), T = 0.01 sec.

The maximum overshoot is now determined from Fig. 7-12, and the result is approximately 10 percent which agrees quite well with the exact value indicated by the response for T = 0.01 sec in Fig. 7-7.

The peak times for the two cases considered above can be computed directly from Eq. (7-46) or from Fig. 7-13. For T = 0.225 sec, Eq. (7-46) gives

$$T_{max} = \frac{0.225}{130.66^\circ}\left\{\tan^{-1}\frac{-0.26}{\sqrt{1-0.26^2}} + 40.66^\circ + 180^\circ\right\} = 0.354 \text{ sec} \tag{7-50}$$

For T = 0.01 sec,

$$T_{max} = \frac{0.01}{2.95^\circ}\left\{\tan^{-1}\frac{-0.609}{\sqrt{1-0.609^2}} + 37.56^\circ + 180^\circ\right\} = 0.61 \text{ sec} \tag{7-51}$$

The same results would have been obtained using Fig. 7-13, and these also agree very well with the results shown in Fig. 7-7.

7.5 ROOT LOCI FOR DIGITAL CONTROL SYSTEMS

The root locus method for the s-plane has been established as a useful tool for the analysis and design of continuous-data control systems. The root locus diagram of a continuous-data control system is essentially a plot of the loci of the roots of the characteristic equation as a function of some parameter K which varies from $-\infty$ to ∞. The root locus diagram gives indication on the absolute and relative stability of a control system with respect to the variation of one system parameter K.

Since the characteristic equation of a linear time-invariant digital control system is a rational polynomial in z, the same set of rules devised for the construction of root loci in the s-plane can be applied to the z-plane. Although the root locus diagram of a digital control system can be constructed in the s-plane using the characteristic equation obtained from the denominator polynomial of the closed-loop transfer function $C^*(s)/R^*(s)$, the root loci will contain infinite number of branches. To illustrate the difficulties of working in the s-plane for digital control systems, let us consider that the closed-loop pulse transfer function of a digital control system is given by

$$\frac{C^*(s)}{R^*(s)} = \frac{G^*(s)}{1 + G^*(s)} \tag{7-52}$$

where $G^*(s)$ is the sampled function of

$$G(s) = \frac{K}{s(s+1)} \tag{7-53}$$

Thus,

$$G^*(s) = \frac{1}{T} \sum_{n=-\infty}^{\infty} G(s + jn\omega_s)$$

$$= \frac{K}{T} \sum_{n=-\infty}^{\infty} \frac{1}{(s + jn\omega_s)(s + jn\omega_s + 1)} \qquad (7\text{-}54)$$

It is well-known [1] that the root loci of $1 + G^*(s) = 0$ can be constructed based on the poles and zeros of the open-loop transfer function $G^*(s)$. In the present case, the function $G^*(s)$ has an infinite number of poles as shown in Fig. 7-16(a). Therefore, the root loci of $1 + G^*(s) = 0$ contain infinite number of branches as shown in Fig. 7-16(b). It is apparent that for digital control systems with more complex transfer functions the construction of the root loci in the s-plane will be more tedious.

The use of z-transform folds the infinite number of poles and zero, thus the root loci, in the s-plane into a finite number of the same in the z-plane. For

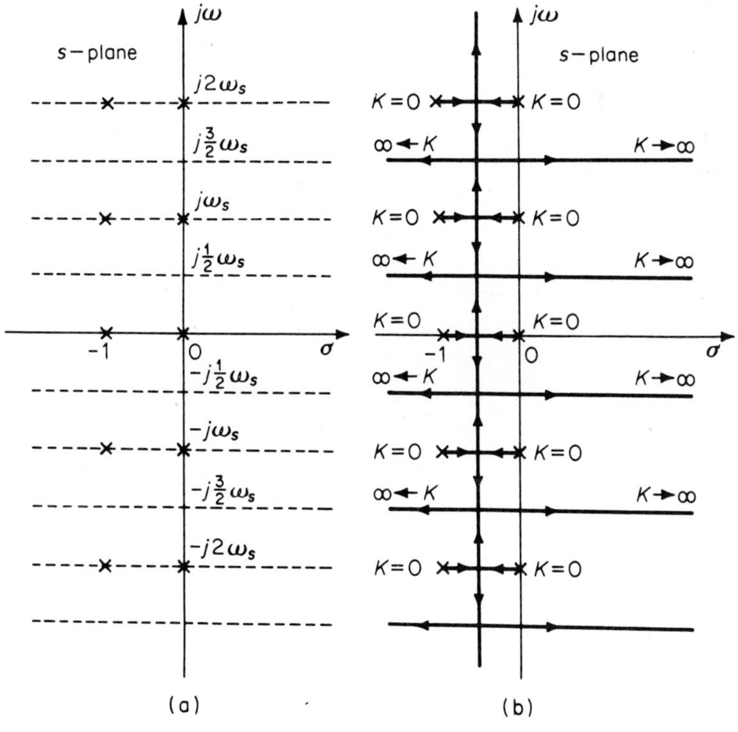

Fig. 7-16. Pole-zero configuration of $G^*(s)$ and the root locus diagram in the s-plane of the digital control system described by Eqs. (7-52) through (7-54). (a) Poles of $G^*(s)$. (b) Root loci.

the system described by Eq. (7-52), the z-transform expression is

$$\frac{C(z)}{R(z)} = \frac{G(z)}{1 + G(z)} \tag{7-55}$$

and the characteristic equation roots are obtained by solving

$$1 + G(z) = 0 \tag{7-56}$$

where corresponding to the G(s) given in Eq. (7-53),

$$G(z) = \mathcal{Z}[G(s)] = \frac{K(1 - e^{-T})z}{(z - 1)(z - e^{-T})} \tag{7-57}$$

Now the root loci of the characteristic equation

$$(z - 1)(z - e^{-T}) + K(1 - e^{-T})z = 0 \tag{7-58}$$

as K varies between 0 and ∞ are drawn based on the pole-zero configuration of G(z), as shown in Fig. 7-17. Since Eq. (7-58) is of the second order, there are only two root loci in Fig. 7-17.

We assume that the reader is already familiar with the conventional root locus technique which is discussed in standard texts on control systems [1]. We shall concentrate on the interpretation of the significance of the root loci in the z-plane with reference to the system performance, rather than how to construct root loci in the z-plane.

Once the root locus diagram of a digital control system is constructed in the z-plane, the absolute stability as well as the relative stability of the system can be observed from the root loci. Absolute stability requires that all the roots

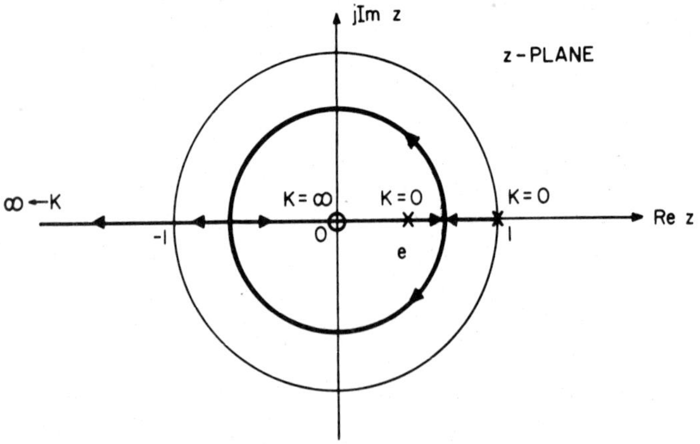

Fig. 7-17. Root locus diagram of the characteristic equation in Eq. (7-58) as K varies from 0 to ∞.

of the characteristic equation for a given set of system parameters lie inside the unit circle in the z-plane. Relative stability, on the other hand, is also indicated by the location of the roots. The questions that concern us are: If the system is stable, how stable and how good is it? Does the transient performance of the system satisfy the designer's specifications? Specifically, we want to obtain information of some performance criteria such as the maximum overshoot, peak time, which are closely related to damping ratio, damping factor and the natural undamped frequency. Thus, the relative stability problem in the z-plane is essentially that of studying the location of the characteristic equation roots with respect to the constant-damping, constant-damping-ratio and constant-frequency loci. These loci were introduced in Sec. 3.3. The constant-damping loci in the z-plane are a family of concentric circles, all centered at the origin; the radius of the circle corresponding to a given damping factor σ_1 is $e^{-\sigma_1 T}$. The constant-frequency loci are straight lines emanating from the origin of the z-plane at angles of $\theta = \omega T$ radians measured from the positive real axis. These loci and the corresponding loci in the s-plane are shown in Figs. 7-18 and 7-19. If the design requires the system to have a maximum damping factor σ_1 or a minimum time constant of $1/\sigma_1$, then all the characteristic roots of the system must lie to the left of the constant-damping locus $s = \sigma_1$ in the s-plane, and correspondingly, all the roots in the z-plane must lie inside the circle $|z| = e^{-\sigma_1 T}$ $(\sigma_1 > 0)$, as shown in Fig. 7-18.

The constant-damping ratio trajectories in the z-plane are a family of logarithmic spirals, except for $\zeta = 0$ and $\zeta = 1$. Typical constant-ζ loci for $\zeta = 0.5$ are shown in Fig. 7-19 for the s-plane and the z-plane. If a certain maximum damping ratio is specified in design, all the characteristic equation roots must lie to the left of the constant-damping-ratio line in the s-plane, or inside

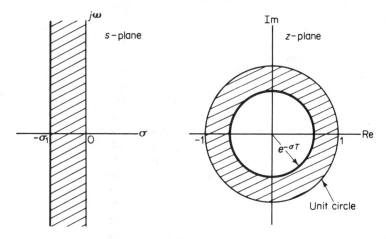

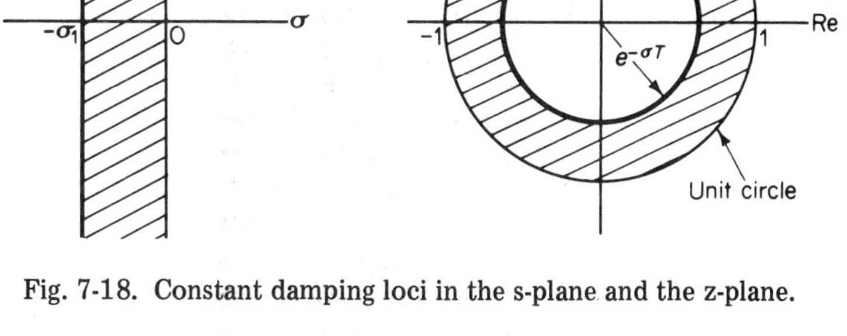

Fig. 7-18. Constant damping loci in the s-plane and the z-plane.

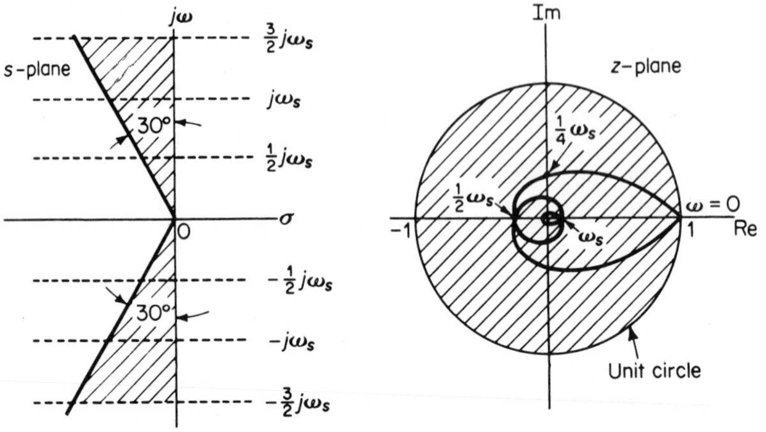

Fig. 7-19. Constant damping ratio in the s-plane and the z-plane for $\zeta = 0.5$.

the constant-damping-ratio logarithmic spiral in the z-plane, as shown by the shaded regions in Fig. 7-19.

Since most control systems have low-pass filter characteristics, it is sufficient in most practical situations to work with only the primary strip in the s-plane. This assumes that the sampling theorem is satisfied so that the highest frequency component in the system is less than $2\omega_s$. Then, Fig. 7-20 illustrates the constant-damping-ratio in the s-plane and the z-plane for various values of ζ for the positive half of the primary strip. The constant-ζ loci for $-\omega_s/2 \leqslant \omega \leqslant 0$ are just the mirror image of the curves shown in Fig. 7-20, with the mirror placed on the real axis.

The relative stability of a digital control system cannot be studied in the z-plane by superimposing the constant-ζ, σ, and ω_n loci on the root locus diagram.

Example 7-2.

Let us consider the space-vehicle control system shown in Fig. 7-5. The open-loop and the closed-loop transfer functions of the system are given in Eqs. (7-10) and (7-11), respectively. Let $T = 0.1$ sec, $K_r = 3.17 \times 10^5$, $J_v = 41822$, and K_p be the variable parameter, the open-loop transfer function becomes

$$G(z) = \frac{1.2 \times 10^{-7} K_p (z+1)}{(z-1)(z-0.242)} \tag{7-59}$$

and the characteristic equation of the closed-loop system is

$$z^2 + (1.2 \times 10^{-7} K_p - 1.242)z + 0.242 + 1.2 \times 10^{-7} K_p = 0 \tag{7-60}$$

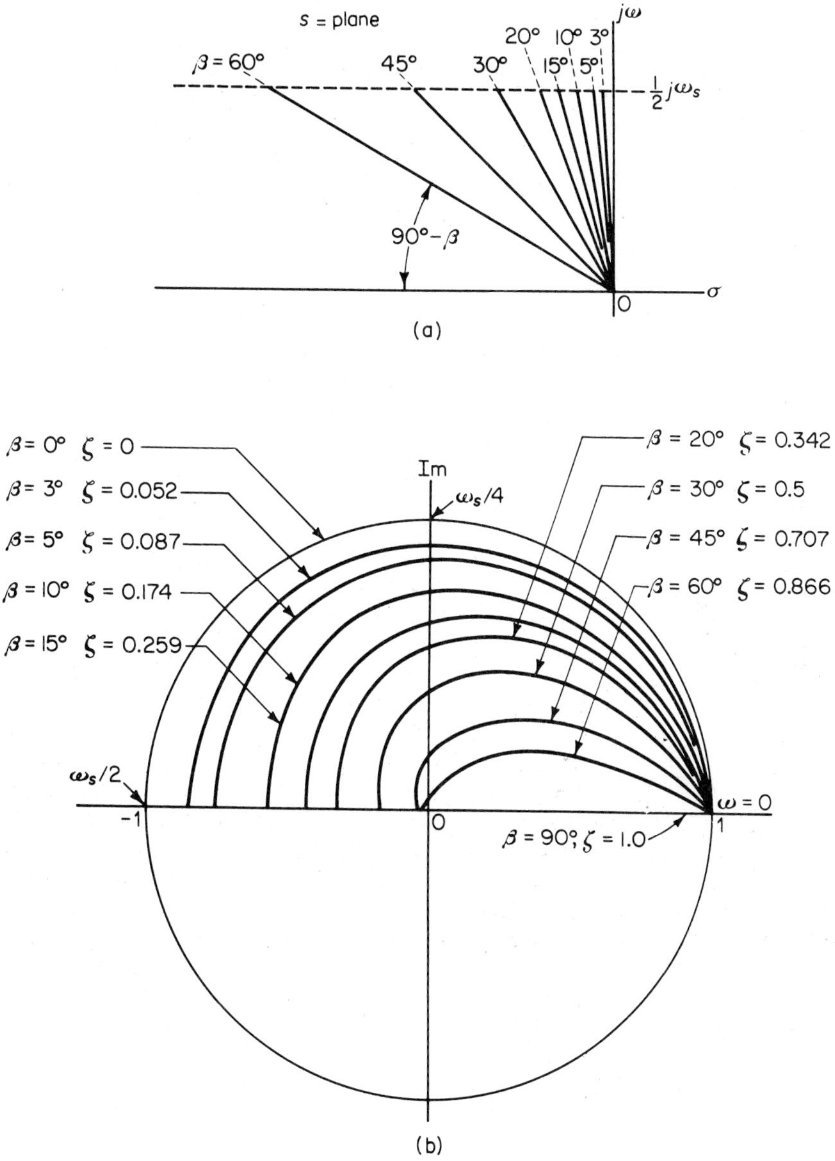

Fig. 7-20. (a) The constant-ζ loci in the s-plane for the periodic strip of 0 to $j\omega_s/2$. (b) The constant loci in the z-plane corresponding to the loci ζ in the s-plane in (a).

The root locus plot of the system when K_p varies between 0 and ∞ is constructed based on the pole-zero configuration of G(z), as shown in Fig. 7-21. From the root locus plot we find that when the root loci cross the unit circle in the z-plane, the value of K_p is 6.32×10^6. Thus, the critical value of K_p for stability is 6.32×10^6. If it is desired to realize a relative damping ratio of 50% for the system, we sketch the constant-damping-ratio locus of $\zeta = 0.5$ in Fig. 7-21. The intersect between the $\zeta = 0.5$ locus and the root loci gives the desired root location, and the corresponding value of K_p which is 1.65×10^6. The frequency ω that corresponds to $K_p = 1.65 \times 10^6$ is determined from the constant-frequency locus which has an angle of 38 degrees, as shown in Fig. 7-21. Thus,

$$\omega T = \theta = 38 \text{ deg} = 0.66 \text{ rad}$$

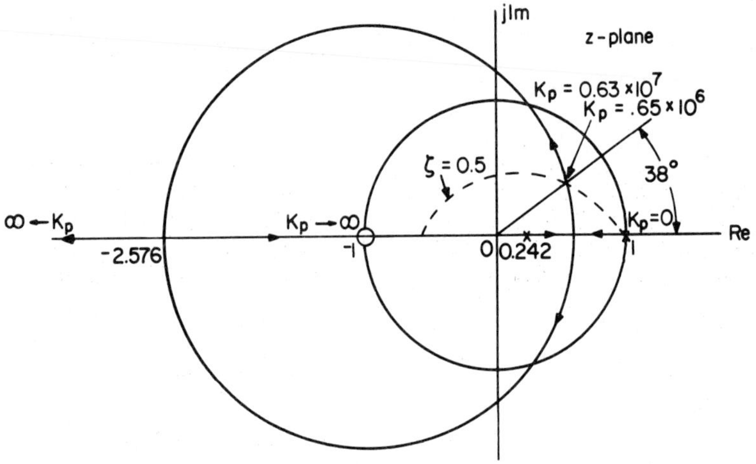

Fig. 7-21. Root loci of the digital system in Fig. 7-5.

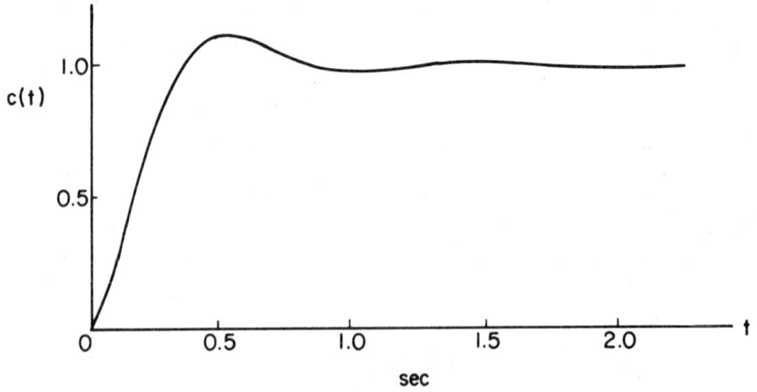

Fig. 7-22. Unit-step response of the system in Fig. 7-5.

The natural undamped frequency is determined as (for T = 0.1 sec)

$$\omega_n = \frac{\omega}{\sqrt{1-\zeta^2}} = \frac{6.6}{\sqrt{1-0.5^2}} = 7.62 \text{ rad/sec} \tag{7-61}$$

Using the values of ζ, and ω, and the pole-zero configuration of the closed-loop transfer function, we can again find the maximum overshoot and peak time from Figs. 7-12 and 7-13, respectively. Or, using a digital computer, the unit-step response of the system for $K_p = 1.65 \times 10^6$ can be computed and is plotted in Fig. 7-22.

7.6 STEADY-STATE ERROR ANALYSIS OF DIGITAL CONTROL SYSTEMS

In Sec. 7.2 we have defined and investigated the steady-state error of a digital control system. In this section we shall conduct a formal discussion on the subject.

The error signal of a control system may be defined in many different ways. It is usually the signal which is to be reduced to zero as quickly as possible; it is not always defined as the difference between the reference input and the output.

We may use the block diagram of Fig. 7-23 to conduct our study of the steady-state error of digital control systems. In this case the signal e(t) is defined as the error; that is,

$$e(t) = r(t) - b(t) \tag{7-62}$$

Since it is difficult to describe e(t) in a digital control system, the sampled error e*(t) is usually used. Thus, the steady-state error at the sampling instants is defined as

$$e^*_{ss} = \lim_{t\to\infty} e^*(t) = \lim_{k\to\infty} e(kT) \tag{7-63}$$

Using the z-transform, the final-value theorem leads to

$$e^*_{ss} = \lim_{t\to\infty} e^*(t) = \lim_{z\to 1} (1 - z^{-1})E(z) \tag{7-64}$$

provided that the function $(1 - z^{-1})E(z)$ does not have any pole on or outside the unit circle $|z| = 1$ in the z-plane.

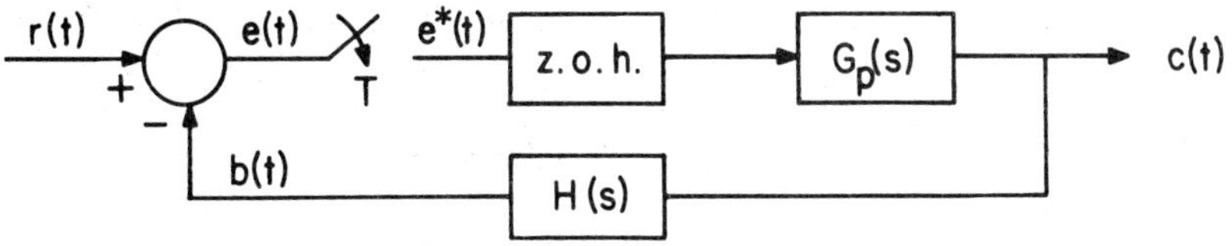

Fig. 7-23. A digital control system.

The steady-state error between the sampling instants can be determined by use of the modified z-transform. We can write

$$\lim_{\substack{k\to\infty \\ 0\leq m\leq 1}} e(kT,m) = \lim_{\substack{z\to 1 \\ 0\leq m\leq 1}} (1 - z^{-1})E(z,m) \tag{7-65}$$

where E(z,m) is the modified z-transform of e(t).

In the studies of linear continuous-data control systems, it is well-known that the steady-state error depends on the type of reference input applied to the system and also on the characteristics of the system. These properties are represented by the *error constants* [1]. We shall see that the error constants can also be applied to digital control systems.

For the system shown in Fig. 7-23, the z-transform of the error signal is written

$$E(z) = \frac{R(z)}{1 + GH(z)} \tag{7-66}$$

Substituting the last equation into Eq. (7-64), we have

$$e_{ss}^* = \lim_{t\to\infty} e^*(t) = \lim_{z\to 1} (1 - z^{-1}) \frac{R(z)}{1 + GH(z)} \tag{7-67}$$

This expression shows that the steady-state error depends on the reference input R(z), as well as the loop transfer function GH(z). In the following we shall consider three basic types of input signals: step function, ramp function and parabolic function.

Steady-State Error due to a Step Function Input

Let the reference input to the system of Fig. 7-23 be a step function of magnitude R. The z-transform of r(t) is

$$R(z) = \frac{Rz}{z - 1} \tag{7-68}$$

Substituting the last equation into Eq. (7-67), we have

$$e_{ss}^* = \lim_{z\to 1} \frac{R}{1 + GH(z)} = \frac{R}{1 + \lim_{z\to 1} GH(z)} \tag{7-69}$$

where

$$GH(z) = (1 - z^{-1})\mathcal{Z}\left[\frac{G_p(s)H(s)}{s}\right] \tag{7-70}$$

Let the *step-error constant* be defined as

$$K_p = \lim_{z \to 1} GH(z) \tag{7-71}$$

Eq. (7-69) becomes

$$e_{ss}^* = \frac{R}{1 + K_p} \tag{7-72}$$

Thus, we see that for the steady-state error due to a step function input to be zero, the step-error constant K_p must be infinite. This implies that the transfer function GH(z) must have at least one pole at $z = 1$. It should be pointed out that the step error constant K_p is meaningful only when the reference input is a step function.

Steady-State Error due to a Ramp Function Input

For a ramp function input, $r(t) = Rtu_s(t)$, the z-transform of r(t) is

$$R(z) = \frac{RTz}{(z-1)^2} \tag{7-73}$$

Substituting Eq. (7-73) into Eq. (7-67), we have

$$e_{ss}^* = \lim_{z \to 1} \frac{RT}{(z-1)[1 + GH(z)]}$$

$$= \frac{R}{\lim_{z \to 1} \frac{(z-1)}{T} GH(z)} \tag{7-74}$$

Let the *ramp-error constant* be defined as

$$K_v = \frac{1}{T} \lim_{z \to 1} [(z-1)GH(z)] \tag{7-75}$$

then, Eq. (7-74) becomes

$$e_{ss}^* = \frac{R}{K_v} \tag{7-76}$$

The ramp-error constant K_v is meaningful only when the input to the system is a ramp function. Again, Eq. (7-76) is valid only if the function after the limit sign in Eq. (7-67) does not have any poles on or outside the unit circle $|z| = 1$. This means that the closed-loop digital control system must *at least* be asymptotically stable.

Equation (7-76) shows that in order for e_{ss}^* due to a ramp function input be zero, K_v must equal infinity. From Eq. (7-75) we see that this is equivalent to the requirement of $(z-1)GH(z)$ having at least one pole at $z = 1$, or $GH(z)$ having two poles at $z = 1$.

Steady-State Error due to a Parabolic Function Input

For a parabolic input, $r(t) = Rtu_s(t)/2$, the z-transform of $r(t)$ is

$$R(z) = \frac{RT^2 z(z+1)}{2(z-1)^3} \tag{7-77}$$

From Eq. (7-67), the steady-state error at the sampling instants is written as

$$e^*_{ss} = \frac{T^2}{2} \lim_{z \to 1} \frac{R(z+1)}{(z-1)^2[1+GH(z)]} \tag{7-78}$$

or

$$e^*_{ss} = \frac{R}{\lim\limits_{z \to 1} \dfrac{(z-1)^2}{T^2} GH(z)} \tag{7-79}$$

Now, let the *parabolic-error constant* be defined as

$$K_a = \frac{1}{T^2} \lim_{z \to 1} [(z-1)^2 GH(z)] \tag{7-80}$$

Then, Eq. (7-79) becomes

$$e^*_{ss} = \frac{R}{K_a} \tag{7-81}$$

In a similar manner we must point out that the parabolic-error constant is associated only with the parabolic function input, and should not be used with any of the other types of inputs.

From the foregoing discussions, we see that when the reference input of a digital control system is of the step, ramp or parabolic type, the steady-state error at the sampling instants due to each input depends on the error constants K_p, K_v, and K_a, respectively. The error constants are summarized as follows:

Step-Error Constant
$$K_p = \lim_{z \to 1} GH(z) \tag{7-82}$$

Ramp-Error Constant
$$K_v = \frac{1}{T} \lim_{z \to 1} (z-1)GH(z) \tag{7-83}$$

Parabolic-Error Constant
$$K_a = \frac{1}{T^2} \lim_{z \to 1} (z-1)^2 GH(z) \tag{7-84}$$

In a similar way we can extend the definitions of the error constants to higher-order input functions if necessary.

Effects of Sampling on The Steady-State Error

Earlier in this chapter we have shown that sampling usually has a detrimental effect on the transient response and the relative stability of a control system. It is natural to ask what is the effect of sampling on the steady-state error of a closed-loop system? In other words, if we start out with a continuous-data system and then add sample-and-hold to form a digital control system, how would the steady-state errors of the two systems, when subject to the same type of input, compare?

Let us first consider the system shown in Fig. 7-23 without the sample-and-hold. The error constants for the continuous-data system are defined as

Step-Error Constant $$K_p = \lim_{s \to 0} G_p(s)H(s) \tag{7-85}$$

Ramp-Error Constant $$K_v = \lim_{s \to 0} sG_p(s)H(s) \tag{7-86}$$

Parabolic-Error Constant $$K_a = \lim_{s \to 0} s^2G_p(s)H(s) \tag{7-87}$$

In continuous-data control systems studies it is customary to classify the loop transfer function according to its pole property at $s = 0$; that is, assuming that $G_p(s)H(s)$ is of the following form:

$$G_p(s)H(s) = \frac{K(1 + T_a s)(1 + T_b s) \ldots (1 + T_m s)}{s^j(1 + T_1 s)(1 + T_2 s) \ldots (1 + T_n s)} \tag{7-88}$$

where the T's are nonzero real or complex constants, the type of the system is equal to j. Thus, referring to Eqs. (7-85), (7-86) and (7-87), we easily conclude that, for instance, a type-0 system will have a constant steady-state error due to a step function input, and infinite error due to all higher-order inputs. A type-1 system ($j = 1$) will have a zero steady-state error due to a step-function input, a constant error due to a ramp function input, and infinite errors due to all higher-order inputs, etc. Table 7-1 summarizes the relationships between the system type, and the error constants for the continuous-data systems.

In view of the error constants of the digital control system as given in Eqs. (7-83) and (7-84), it would seem that K_v and K_a all depend on the sampling period T. Since the digital control system still uses a continuous-data process, we may consider the same transfer function $G_p(s)H(s)$ given in Eq. (7-88). Let us consider only the cases of $j = 0$, 1 and 2, as follows:

Type-0 System

For a type-0 system, j = 0, Eq. (7-88) becomes

$$G_p(s)H(s) = \frac{K(1 + T_a s)(1 + T_b s) \ldots (1 + T_m s)}{(1 + T_1 s)(1 + T_2 s) \ldots (1 + T_n s)} \tag{7-89}$$

where we assume that the function has more poles than zeros.

Substitution of the last equation into Eq. (7-70), we get

$$GH(z) = (1 - z^{-1}) \mathcal{Z}\left\{\frac{K(1 + T_a s)(1 + T_b s) \ldots (1 + T_m s)}{s(1 + T_1 s)(1 + T_2 s) \ldots (1 + T_n s)}\right\} \tag{7-90}$$

TABLE 7-1. RELATIONSHIPS BETWEEN ERROR CONSTANTS AND TYPE OF SYSTEM FOR CONTINUOUS-DATA SYSTEMS

Type of System	K_p	K_v	K_a
0	K	0	0
1	∞	K	0
2	∞	∞	K

Performing partial-fraction expansion to the function inside the bracket in the last equation, we have

$$GH(z) = (1 - z^{-1}) \mathcal{Z}\left[\frac{K}{s} + \text{terms due to the nonzero poles}\right]$$

$$= (1 - z^{-1})\left[\frac{Kz}{z - 1} + \text{terms due to the nonzero poles}\right] \tag{7-91}$$

It is important to note that the terms due to the nonzero poles do not contain the term (z − 1) in the denominator. Thus, the step-error constant is

$$K_p = \lim_{z \to 1} GH(z) = \lim_{z \to 1} (1 - z^{-1}) \frac{Kz}{z - 1} = K \tag{7-92}$$

This result shows that for a type-0 process, in a closed-loop system with sample-and-hold the step error constant is identical to that of the closed-loop continuous-data system. Therefore, the sample-and-hold does not affect the steady-state behavior of this type of system.

Substituting Eq. (7-91) into the ramp-error constant of Eq. (7-83), we get

$$K_v = \frac{1}{T} \lim_{z \to 1} (z - 1)GH(z) = \frac{1}{T} \lim_{z \to 1} (z - 1)(1 - z^{-1}) \frac{Kz}{z - 1} = 0 \tag{7-93}$$

Similarly, $K_a = 0$ for a type-0 system.

Type-1 System

For a type-1 system, j = 1, Eq. (7-70) becomes

$$GH(z) = (1 - z^{-1})\mathcal{Z}\left\{\frac{K(1 + T_a s)(1 + T_b s)\dots(1 + T_m s)}{s^2(1 + T_1 s)(1 + T_2 s)\dots(1 + T_n s)}\right\}$$

$$= (1 - z^{-1})\mathcal{Z}\left\{\frac{K}{s^2} + \frac{K_1}{s} + \text{terms due to the nonzero poles}\right\} \qquad (7\text{-}94)$$

Then, the step-error constant is

$$K_p = \lim_{z \to 1} GH(z)$$

$$= \lim_{z \to 1} (1 - z^{-1})\left\{\frac{KTz}{(z-1)^2} + \frac{K_1 z}{z-1} + \text{terms due to the nonzero poles}\right\}$$

$$= \infty \qquad (7\text{-}95)$$

The ramp-error constant is

$$K_v = \frac{1}{T}\lim_{z \to 1} (z - 1)GH(z) = \frac{1}{T}\lim_{z \to 1} (z - 1)(1 - z^{-1})\left\{\frac{KTz}{(z-1)^2} + \frac{Kz}{z-1}\right\}$$

$$= K \qquad (7\text{-}96)$$

and we can easily show that $K_a = 0$.

Thus, for a type-1 system, the system with sample-and-hold has exactly the same steady-state error as the continuous-data system with the same process transfer function.

In the same fashion we can show that with a type-2 process, the digital control system will have $K_p = \infty$, $K_v = \infty$ and $K_a = K$. The conclusion is that the relationships tabulated in Table 7-1 can also be applied to a digital control system with the block diagram shown in Fig. 7-23. Equations (7-83) and (7-84) may purport to show that the ramp- and parabolic-error constants of a digital control system depend on the sampling period T; however, in the process of evaluation T gets cancelled, and the error depends only on the parameters of the process and the type of inputs.

It should be reiterated that when the system is designed to follow a ramp input, the response to a step input is no longer deadbeat.

REFERENCES

1. Kuo, B. C., *Automatic Control Systems*, 3rd ed. Prentice-Hall, Englewood Cliffs, N.J., 1975.
2. Kuo, B. C., *Analysis and Synthesis of Sampled-Data Control Systems*. Prentice-Hall, Englewood Cliffs, N.J., 1963.
3. Tou, J. T., *Digital and Sampled-Data Control Systems*. McGraw-Hill, New York, 1959.
4. Jury, E. I., *Theory and Applications of the z-Transform Method*. John Wiley & Sons, New York, 1964.
5. Ragazzini, J. R., and Franklin, G. F., *Sampled-Data Control Systems*. McGraw-Hill, New York, 1958.
6. Jury, E. I., *Sampled-Data Control Systems*. John Wiley & Sons, New York, 1958.
7. Lindorff, D. P., *Theory of Sampled-Data Control Systems*. John Wiley & Sons, New York, 1965.
8. Freeman, H., *Discrete-Time Systems*. John Wiley & Sons, New York, 1965.
9. Monroe, A. J., *Digital Processes For Sampled-Data Systems*. John Wiley & Sons, New York, 1962.
10. Tsypkin, Y. Z., *Theory of Pulse Systems*, (in Russian). State Press for Physics and Math. Lit., Moscow, 1958.
11. Phillips, C. L., "A Note on the Frequency-Response Design Technique for Multirate Digital Controllers," *IEEE Trans. on Automatic Control*, Vol. AC-15, April 1970, pp. 263-264.
12. Phillips, C. L., and Johnson, J. C., "Design of Multirate Controllers," *Proc. Houston Conference on Circuits, Systems, and Computers*, May 1969, pp. 301-309.
13. Pokoski, J. L., and Pierre, D. A., "Deadbeat Response to Parabolic Inputs With Minimum-Squared Error Restrictions on Ramp and Step Responses," *IEEE Trans. on Automatic Control*, Vol. AC-14, April 1969, pp. 199-200.
14. Pierre, D. A., Lorchirachoonkul, V., and Roos, M. E., "A Performance Limit for a Class of Linear Sampled-Data Control Systems," *IEEE Trans. on Automatic Control*, Vol. AC-12, February 1967, pp. 112-113.
15. Light, W. R., and McVey, E. S., "Analysis of Digital Predictive Compensation," *IEEE Trans. on Automatic Control*, Vol. AC-15, October 1970, pp. 604-606.
16. Pracht, C. P., and McVey, E. S., "Near Ideal Digital Predictive Compensation," *IEEE Trans. on Automatic Control*, Vol. AC-15, August 1970, pp. 471-474.

PROBLEMS

7-1 The schematic diagram of a steel rolling process is shown in Fig. 7P-1. The DC motor is characterized by the following parameters:

Armature resistance	$R = 5$ ohms
Armature inductance	negligible
Torque constant	$K_m = 0.721$ Nm/amp
Back emf constant	$K_b = 0.721$ volts/rad/sec
Lead of leadscrew	$L = 0.01$ m/rev $= 0.01/2\pi$ m/rad
Total inertia reflected to motor	$J = 0.001$ kg-m^2
Total viscous friction coefficient reflected to motor	$B = 0.01$ Nm/rad/sec
Gain of transducer for measurement of thickness	$K_s = 100$ volts/m
Sampling period	$T = 0.01$ sec

The motor shaft is coupled to a leadscrew or linear actuator whose output is connected to the roller to roll the steel plate to the desired thickness. The steel plate is fed through the rollers with a constant speed of v. The distance between the point where the plate is rolled and the point where the thickness is measured is d meters. Let $v = 10$ m/sec and $d = 1$ m. The following equations are written for the system:

$$e_a(t) = Ri_a(t) + e_b(t) \qquad T_m(t) = K_m i_a(t)$$

$$e_b(t) = K_b\omega(t) \qquad T_m(t) = J\dot{\omega}(t) + B\omega(t)$$

$$\frac{d\theta(t)}{dt} = \omega(t) \qquad y(t) = L\theta(t)$$

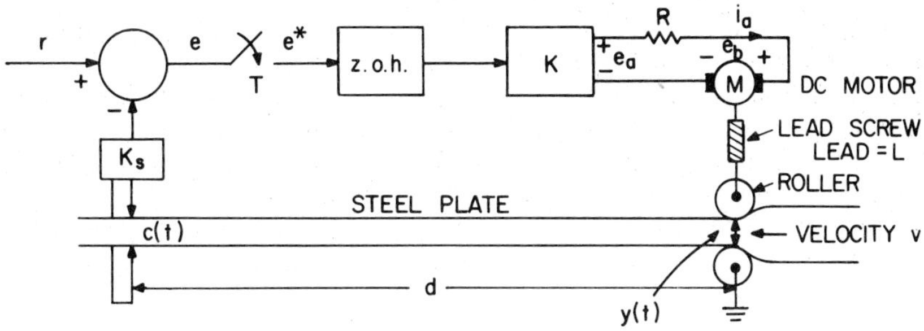

Fig. 7P-1.

$$c(t) = y(t - T_d) \qquad\qquad T_d = d/v$$

$$e(t) = r(t) - K_s c(t)$$

(a) Draw a block diagram to represent the transfer function relations of the digital steel-rolling control system.
(b) Find the closed-loop transfer function $Y(z)/R(z)$.
(c) Determine the range of K so that the system is asymptotically stable.

7-2 The block diagram of a microprocessor DC motor speed control system is shown in Fig. 7P-2(a). The speed of the motor-load is controlled to match the desired speed ω_r. The actual speed of the motor-load is measured by the microprocessor which processes the signal from the shaft encoder. The speed error is defined to be the difference between the desired speed and the actual speed. A controller is programmed into the microprocessor. The output of the microprocessor which is proportional to the speed error is transmitted to the pulse generator. This error signal controls the phase between two pulse trains generated by the pulse generator. The voltage applied to the DC motor is determined by the duty cycle which is a function of the phase between the two pulse trains.

The speed error is represented by an 8-bit binary number. Let the phase be from zero to 360 degrees. Then, the finest resolution of the digital representation of the phase shift is $2\pi/2^8 = 2\pi/256$ radians. The SCR chopper outputs a voltage to the DC motor which is determined by the phase between v_1 and v_2. Since the phase information between sampling instants is constant, we can use a sample-and-hold to model the input-output relation. Thus, the transfer function of the pulse-generator-phase relation is

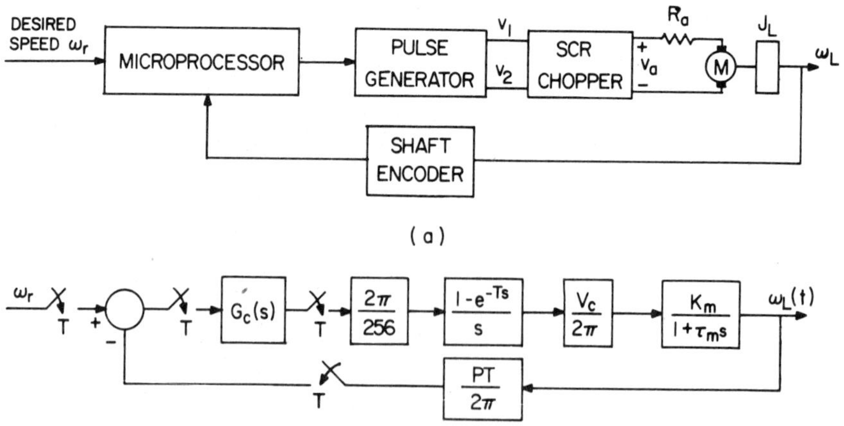

(b)

Fig. 7P-2.

$$G_p(s) = \frac{2\pi}{256} \cdot \frac{1 - e^{-Ts}}{s}$$

The transfer function of the SCR chopper is

$$G_d(s) = V_c/2\pi$$

where V_c = chopper output voltage (full scale). The block diagram of the overall system is given in Fig. 7P-2(b).

(a) Let $G_c(s) = K_c$ = constant; find the characteristic equation of the digital control system represented in Fig. 7P-2(b).
(b) Find the range of K_c so that the closed-loop digital control system is asymptotically stable.
(c) Find the closed-loop transfer function $\Omega_L(z)/\Omega_r(z)$.
(d) For $K_c = 1$ find the output response $\omega_L(k)$ when the input ω_r is a unit-step function. What is the steady-state value of $\omega_L(k)$ for this input?

Use the constants $V_c = 24$ v, $K_m = 5$, $\tau_m = 0.05$ sec, $T = 0.1$ sec, $P = 100$ pulses/rev.

7-3 Given the following closed-loop transfer functions for digital control systems, find the maximum overshoot and the normalized peak time T_{max}/T of the step response. First use the graphs in Figs. 7-12 and 7-13, then check the results by evaluating c(kT) for $k = 0, 1, 2, \ldots$.

(a) $$\frac{C(z)}{R(z)} = \frac{(z + 0.5)}{3(z^2 - z + 0.5)}$$

(b) $$\frac{C(z)}{R(z)} = \frac{0.5z}{(z^2 - z + 0.5)}$$

(c) $$\frac{C(z)}{R(z)} = \frac{z^{-2}(z + 0.5)}{3(z^2 - z + 0.5)}$$

(d) $$\frac{C(z)}{R(z)} = \frac{0.5z^{-2}}{(z^2 - z + 0.5)}$$

(e) $$\frac{C(z)}{R(z)} = \frac{0.316(z + 0.002)(z + 0.5)}{(z^2 - z + 0.5)(z - 0.05)}$$

7-4 The block diagram of a speed control system with a DC motor is shown in Fig. 7P-4. The system parameters are given as follows:

K_a = motor torque constant = 0.345

K_b = motor back emf constant = 0.345

R = motor armature resistance = 1 ohm

L = motor armature inductance = 1 mH

B = viscous friction coefficient of motor and load = 0.25

J = inertia of motor and load = 1.41×10^{-3}

T = sampling period = 1 millisecond

It is assumed that these parameters have the same consistent units so that no conversions are necessary.
The digital controller is implemented by a microprocessor and is modelled as shown in the block diagram. The transfer function of the digital controller is

$$G_c(z) = K_p + \frac{K_r T}{2}\left[\frac{z+1}{z-1}\right]$$

where $K_p = 1$ and $K_r = 295.276$.

(a) Find the open-loop transfer function $\Omega(z)/E(z)$.

(b) Find the closed-loop transfer function $\Omega(z)/\Omega_d(z)$.

(c) Find the characteristic equation of the overall system and the roots of the characteristic equation.

(d) For a unit-step function input, for ω_d, find the output response ω at the sampling instants.

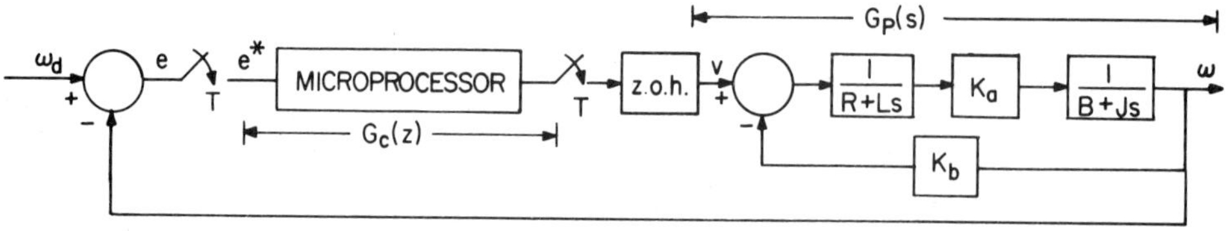

Fig. 7P-4.

7-5 For the automatic forage and concentrate-proportioning system described in Problem 5-15, let

$$\frac{C(s)}{U(s)} = \frac{36870K}{s^2 + 83.3s + 7780}$$

Sketch the root loci of the characteristic equation for $0 \leqslant K < \infty$ in the z-plane. Find the value of K when the damping ratio ζ is zero. Find the damping ratio ζ when K = 0.1. For K = 0.1 find the percent overshoot of the output c(kT) when the input v_s is a unit-step function.

7-6 The payload of a space-shuttle pointing control system is modelled as a pure mass M. It is suspended by magnetic bearings so that no friction is encountered in the control. The attitude in the y direction is controlled by magnetic actuators located at the base of the payload. The dynamic system model for the control of the y-axis motion is shown in Fig. 7P-6. The controls of the other degrees of motion are independent, and are not considered here. Since there are experiments located on the payload, electric power must be brought to the payload through wire cables. The linear spring with the

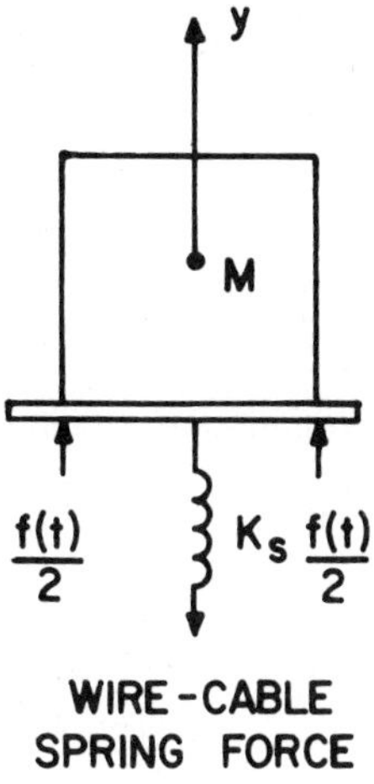

Fig. 7P-6.

spring constant K_s is used to model the wire-cable attachment. The total force produced by the magnetic actuators is denoted by f(t). The force equation of motion in the y direction is

$$f(t) - K_s y(t) = M\ddot{y}(t)$$

where $K_s = 0.35$ Newton/m, M = 600 Kg, and f(t) is in Newtons.
Let the magnetic actuators be controlled by sampled-data, so that

$$f(t) = f(kT) \qquad kT \leqslant t < (k+1)T$$

where

$$f(kT) = -K_p y(kT) - K_r \dot{y}(kT)$$

$K_p = 37.861$ and $K_r = 211$, and T is the sampling period in seconds.

(a) Draw a block diagram for the overall system.

(b) Find the characteristic equation of the overall system with the feedback control described above. Plot the root loci as a function of T. Find the range of T for the system to be asymptotically stable.

(c) Show the region in the parameter plane of K_r versus K_p in which the overall system is asymptotically stable. T = 1 second. Find the point in the parameter plane at which the system is able to follow a step input in a minimum number of sampling periods (deadbeat response).

7-7 The denominator of the closed-loop transfer function of a digital control system is given as 1 + G(z), where

$$G(z) = \frac{0.15Kz}{(z-1)(z-0.5)}$$

Sketch the root loci of the characteristic equation and determine the range of K for asymptotic stability.

7-8 The z-transform of the output which represents the positional response of a digital control system is given as C(z). Find an approximate z-transform representation of the output velocity v(kT) in terms of C(z). The sampling period is T and the initial conditions are c(0) and v(0) and are all zeros.

7-9 Given the closed-loop transfer function of a digital control system,

$$\frac{C(z)}{R(z)} = \frac{T^2(K_p z^2 + K_i Tz + K_i T - K_p)}{Az^3 + Bz^2 + Cz + D}$$

where

$$A = 2J_v$$

$$B = T^2 K_p + 2K_r T - 6J_v$$

$$C = 6J_v - 4K_r T + T^3 K_i$$

$$D = 2K_r T + K_i T^3 - 2J_v - K_p T^2$$

$J_v = 41822$. Find the values of K_p, K_i and K_r as functions of T so that the step response c(kT) reaches the step input in a minimum number of sampling periods. What is the maximum overshoot $c(kT)_{max}$?

7-10 The block diagram of a digital control system with sample-and-hold is shown in Fig. 7P-10.

(a) Sketch the root loci of the characteristic equation and determine the range of K for asymptotic stability. Find the value of K so that the damping ratio of the system is approximately 70.7%. T = 0.1 sec,

$$G(s) = \frac{K}{s(s+10)}$$

(b) Repeat part (a) for T = 1 sec.

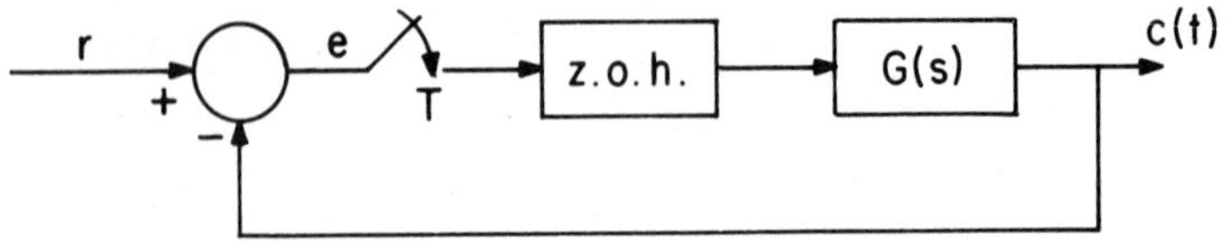

Fig. 7P-10.

7-11 The characteristic equation of a digital control system is

$$\Delta(z) = z^3 + (111.6T^2 + 16.74T - 3)z^2 + (3 - 33.48T + 1.395 \times 10^{-4}K_iT^3)z$$
$$+ (1.395 \times 10^{-4}K_iT^3 + 16.74T - 111.6T^2 - 1) = 0$$

Sketch the root loci of the characteristic equation for $T = 0.08$ sec and $0 \leqslant K_i < \infty$.

7-12 The schematic diagram of a DC motor controlling a load with inertia, friction and compliance is shown in Fig. 7P-12(a). The system parameters are defined as follows:

Amplifier gain	$K = 150$
Motor armature resistance	$R_a = 4$ ohms
Feedback resistance	$R_s = 0.18$ ohms
Torque constant	$K_a = 3.5$ oz-in/amp
Back emf constant	$K_b = 0.0247$ volts/rad/sec
Motor armature inductance	$L_a = 0.002$ H
Motor inertia	$J_m = 2.7 \times 10^{-4}$ oz-in-sec^2
Load inertia	$J_L = 12 \times 10^{-4}$ oz-in-sec^2
Motor friction	$B_s = 0.1045$ oz-in/rad/sec
Spring constant	$K_s = 3454$ oz-in/rad

(a) Show that the transfer function between the motor input voltage v_i and the load velocity ω_L is

$$\frac{\Omega_L(s)}{V_i(s)} = \frac{KK_a(B_s s + K_s)/L_aJ_mJ_L}{\Delta(s)}$$

where

$$\Delta(s) = s^4 + \left\{\frac{R_a + KR_s}{L_a} + \frac{B_s}{J_m} + \frac{B_s}{J_L}\right\}s^3 + \left\{\frac{K_aK_b}{J_mL_a} + \frac{(R_a + KR_s)B_s}{L_a}\left[\frac{1}{J_m} + \frac{1}{J_L}\right]\right.$$
$$\left. + K_s\left[\frac{1}{J_m} + \frac{1}{J_L}\right]\right\}s^2 + \left\{\frac{(R_a + KR_s)K_s}{L_a}\left[\frac{1}{J_m} + \frac{1}{J_L}\right] + \frac{K_aK_bB_s}{L_aJ_mJ_L}\right\}s$$
$$+ \frac{K_aK_bK_s}{L_aJ_mJ_L}$$

(b) The block diagram of a sampled-data system using the DC motor for speed control is shown in Fig. 7P-12(b). The encoder senses the motor speed and the output of the encoder is compared with the speed command. For analytical purposes, the pulsewidth-modulation amplifier and the encoder are modelled by constant gains. Find the transfer function $G_p(s)$ and $G_L(s)$. Let the gain of the encoder be unity. Based on the dynamics of the system, determine what is the largest sampling period that can be used without violating the sampling theorem. Let $T = 0.001$ sec. Sketch the root loci of the equation,

$$1 + K_m \mathcal{Z}[G_{h0}(s)G_p(s)] = 0$$

for $0 \leqslant K_m < \infty$. Comment on the stability of the overall sampled-data system. Repeat the root loci problem for $T = 0.0005$ sec.

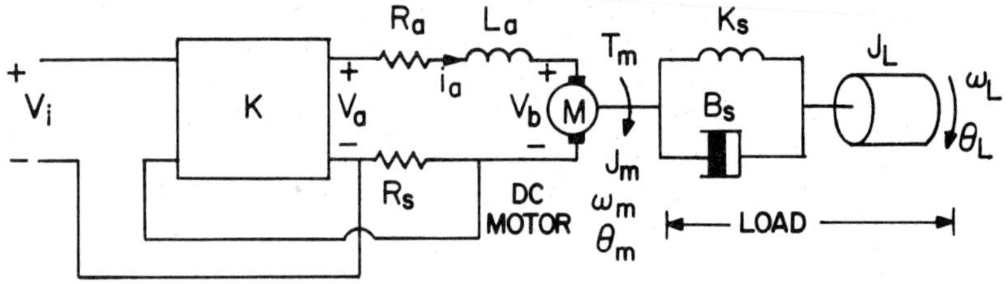

(a)

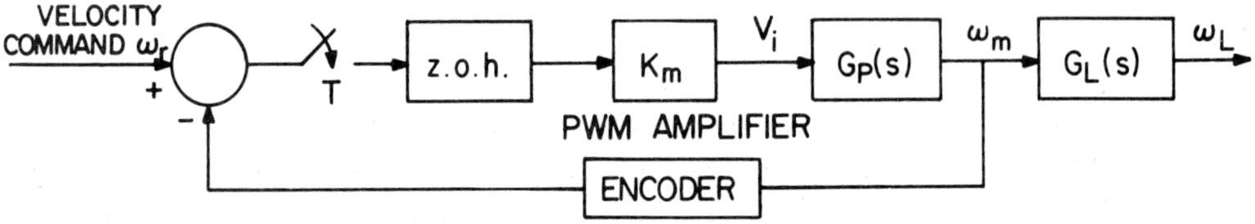

(b)

Fig. 7P-12.

7-13 The characteristic equation of the digital space-vehicle control system shown in Fig. 7-5 is

$$2J_v z^2 + (2K_r T - 4J_v + T^2 K_p)z + (2J_v - 2K_r T + T^2 K_p) = 0$$

For $T = 0.264$ sec, $J_v = 41822$ and $K_p = 1.65 \times 10^6$, sketch the root loci of the characteristic equation as a function of K_r. Find the range of K_r so that the system is asymptotically stable. Find the value of K_r so that the two roots are real and equal. Find the unit-step response of the system output for this case.

7-14 For the speed control system described in Problem 7-4, let $K_p = 1$. Sketch the root locus plot of the overall system characteristic equation for $0 \leqslant K_r < \infty$. Find the values of K_r so that the overall system is asymptotically stable.

7-15 The purpose of this problem is to apply digital control theory to the analysis of electronic control of the stoichiometric air-fuel-ratio control problem for automobiles. A considerable amount of effort is being spent by the automobile manufacturers to meet the exhaust emission performance standards of the various governmental agencies. Modern automotive powerplant systems consist of an internal combustion engine which has an internal cleanup device called the catalytic converter. Such a system requires control of such variables as engine air-fuel ratio A/F, ignition spark timing, exhaust gas recirculation, and injection air.

The control system problem considered in this exercise deals with the control of the air-to-fuel ratio A/F. In general, depending on fuel composition and other factors, a typical stoichiometric A/F is 14.7 to 1; that is, 14.7 g of air to each 1 g of fuel. An A/F greater or less than stoichiometry will cause high hydrocarbons, carbon monoxide and oxides of nitrogen in the tailpipe emission.

A control system is devised to control the air-to-fuel ratio so that a desired output variable is maintained for a given command signal. Figure 7P-15(a) shows the block diagram of such a system. The sensor senses the composition of the exhaust gas mixture entering the catalytic converter. The digital controller detects the difference or the error between the command and the sensor signals and computes the control signal necessary to achieve the desired exhaust gas composition. The disturbance signal is used to represent changing or unknown operating conditions such as variations in temperature, humidity, barometric pressure and fuel composition. The output variable c(t) denotes the effective air-fuel ratio A/F.

A sampled-data model of the system is shown in Fig. 7P-15(b). The transfer function of the engine is given by

$$G_p(s) = \frac{e^{-T_d s}}{(1 + \tau s)}$$

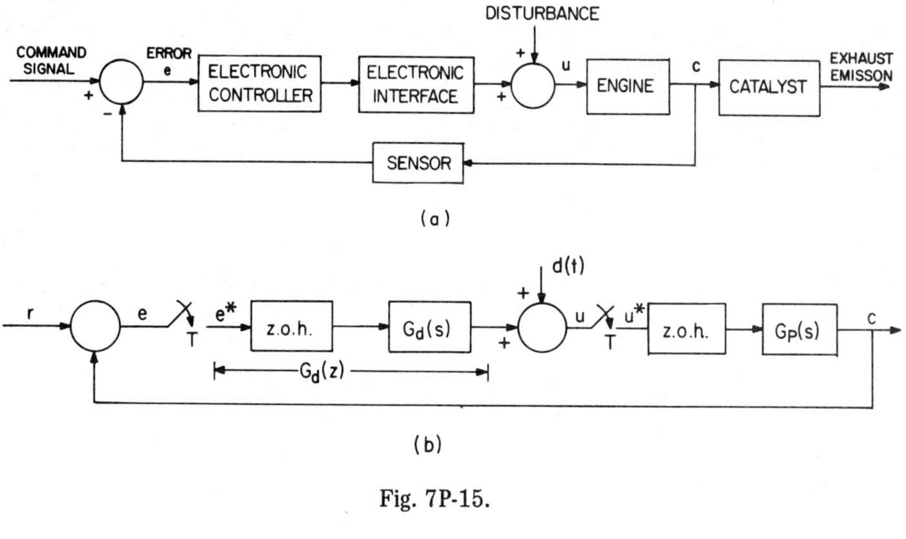

Fig. 7P-15.

where T_d is a time delay. Let $\tau = 0.25$ sec and $T_d = 1.0$ sec, and $T = 0.1$ sec.

(a) Let $d(t) = 0$ and $G_d(s) = K =$ constant. Find the characteristic equation of the closed-loop system and sketch the root loci as K varies from zero to infinity. Determine the marginal value of K for stability.

(b) Let $d(t) = 0$ and $G_d(s) = K =$ constant. Find the steady-state error between r(t) and c(t) when r(t) is a unit-step function. Evaluate the steady-state error at the sampling instants only (in terms of K).

(c) Under the conditions given above, obtain the time responses of c(t) at the sampling instants for $K = 1$ and $K = 0.5$ when the input r(t) is a unit-step input.

7-16 For the digital space vehicle control system shown in Fig. 7-5, let $J_v = 41822$ and $T = 0.1$ sec.

(a) Find the relation between K_p and K_r so that the ramp error constant K_v is equal to 10.

(b) Find the loci of the characteristic equation when K_p or K_r is varied from zero to infinity under the constraint that the ramp error constant K_v is 10. Find the ranges of K_r and K_v so that the closed-loop system is asymptotically stable.

7-17 The schematic diagram shown in Fig. 7P-17 represents a control system whose purpose is to hold the level of the liquid in the tank at a fixed level. The level of the liquid is controlled by a float whose position is connected to the wiper arm of a potentiometer error detector. The error between the reference level and the actual level of the liquid is fed into an A/D and then a digital controller. The transfer function of the A/D, the digital controller, and the D/A are lumped together and are represented by $G_c(z)$. The parameters of the system components are given as follows:

DC Motor:

Armature resistance	$R = 10$ ohms
Armature inductance	$L =$ negligible
Torque constant	$K_m = 10$ lb-ft/amp
Back emf constant	$K_b = 0.075$ volts/rad/sec
Rotor inertia	$J_m = 0.005$ lb-ft-sec^2
Load and rotor friction	negligible
Load inertia	$J_L = 10$ lb-ft-sec^2
Gear ratio	$n_1/n_2 = 1/100$

Motor Equations:

$$e_a = Ri + e_b$$

$e_b = K_b \omega_m$

$\omega_m = d\theta_m/dt$

$T_m = K_m i = J\dot{\omega}_m$ where $J = J_m + (n_1/n_2)^2 J_L$

$\theta_c = n_1\theta_m/n_2$

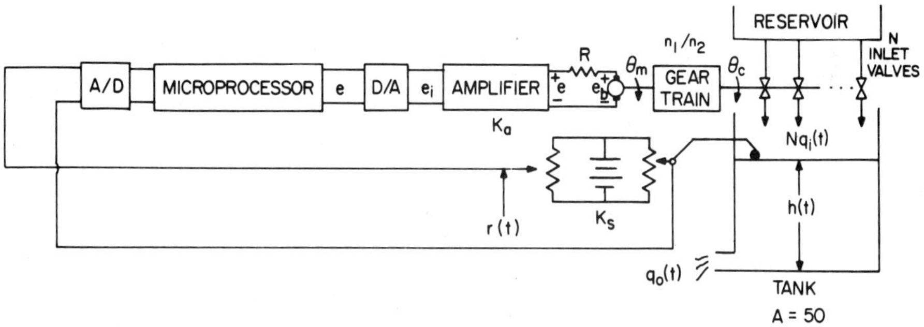

Fig. 7P-17.

Dynamics of Tank:

There are N inlets to the tank from the reservoir. All the inlet valves have the same characteristics, and are controlled simultaneously by θ_c. The equations that govern the volume of flow are:

Inlet: $q_i(t) = K_i N\theta_c(t)$ $K_i = 10\ \text{ft}^3/\text{sec-rad}$

$q_o(t) = K_o h(t)$ $K_o = 50\ \text{ft}^2/\text{sec}$

$$h(t) = \frac{\text{Volume of tank}}{\text{Area of tank}} = \frac{1}{A}\int [q_i(t) - q_o(t)]\,dt$$

Amplifier gain: $K_a = 50$

Error detector: $K_s = 1$ volt/ft

(a) Draw a block diagram of the control system with the A/D, D/A, and the digital controller represented by sample-and-hold and a digital controller.
(b) Find the maximum number of inlets, N, so that the closed-loop digital control system is stable. The sampling period is 0.05 seconds. $G_c(s) = 1$.
(c) Sketch the root locus of the closed-loop characteristic equation for $0 \leqslant N < \infty$ in the z-plane.

(d) For $N = 1$, find the time response of $h(t)$ at the sampling instants for a unit-step function input, $r(t) = u_s(t)$.
(e) Repeat part (d) for $N = 3$ and $N = 5$.

7-18 For the simplified digital space vehicle control system shown in Fig. 7-5,
(a) Find the step error constant.
(b) Find the ramp error constant.
(c) Find the parabolic error constant.
Express the results in terms of the system parameters.

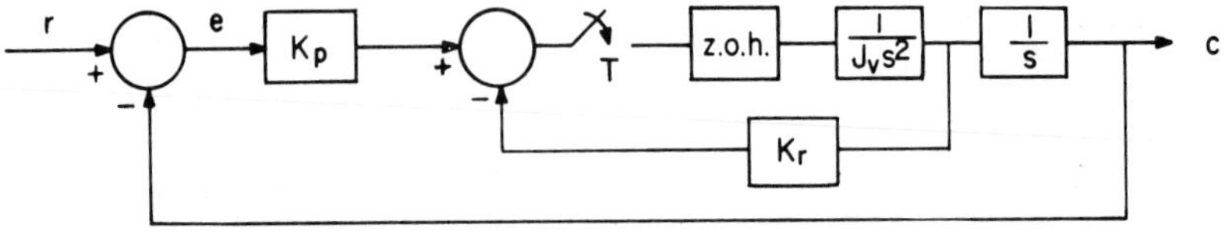

8

FREQUENCY-DOMAIN ANALYSIS

8.1 INTRODUCTION

In the preceding chapter time response of digital control systems was studied through the use of the pole-zero configuration of transfer functions in the z-plane. A direct method of time-domain analysis involves the computation of the system response analytically or by a computer. In general, there is a lack of a clear-cut or direct method of design of digital control systems in the time domain. An example of this difficulty is clearly illustrated by the restriction of the tractable correlation of maximum overshoot and peak time to the pole-zero configuration to only a second-order system.

The frequency-domain analysis and design, on the other hand, possess a wealth of graphical and semi-graphical techniques that can be applied to linear time-invariant control systems virtually of any complexity. Throughout the years the analysis and design of continuous-data control systems have been well developed, and practically all these methods can be extended to digital control systems. Such well-known methods as the Nyquist criterion for stability analysis, the Bode plot and the Nichols chart all can be applied to the analysis and design of digital control systems without modification.

The basic feature of the frequency-response method is that the description of performance of a linear time-invariant system is given in terms of its steady-state response to sinusoidally-varying input signals. The crux of the problem is that we are able to predict or project the time-domain characteristics of the performance of a system based on the sinusoidal-steady-state analysis information. For instance, the parameter, bandwidth, in the frequency-response analysis is linked directly to how "fast" and oscillatory the time response will be. Therefore, in many design problems instead of specifying the maximum overshoot, the designer often specifies what the desired bandwidth of the system should be.

It is well-known that in the Laplace-transform domain the sinusoidal-steady-state analysis is carried out by setting $s = j\omega$; thus, in the z-domain, $z = e^{j\omega T}$. However, some of the unique characteristics found in digital control systems make the frequency-domain study more interesting. For instance, the responses of a linear continuous-data system contain only the same frequency component as the sinusoidal input signal, and only in systems with nonlinear elements could harmonics or subharmonics of the input signal be found. However, in linear digital control systems, the sampler, whether it is an actual or a fictitious component, acts as a harmonic generator so that the system response in general may contain higher harmonics of the input sinusoid. Therefore, the main difficulty in studying digital control systems in the frequency domain stems from the fact that these high-frequency components often make the construction of the frequency loci plots more difficult.

The study of digital control systems in the frequency domain essentially relies on the extension of all the existing techniques devised for the analysis of continuous-data systems. Some of the well-known methods are described in the following:

1. The Nyquist Plot

The Nyquist plot of a transfer function $G(s)$ $[G(z)]$ is a mapping of the Nyquist path in the s-plane (z-plane) onto the $G(s)$ plane $[G(z)$ plane$]$. Usually, the Nyquist plot of the loop transfer function is made and information on the absolute and relative stability of the closed-loop system is obtained from the properties of the Nyquist plot with respect to the $(-1,j0)$ point.

2. The Bode Diagram (Gain and Phase versus Frequency)

The Bode diagram is a plot of the amplitude in decibels and phase angle of a transfer function (usually the loop transfer function) as functions of frequency ω (or log of frequency, $\log_{10}\omega$). The Bode diagram may be used to investigate the absolute and relative stabilities of a closed-loop system.

3. The Gain-Phase Plot

The gain-phase plot of the open-loop transfer function of a control system may be used to determine the absolute and relative stabilities of a closed-loop system. The gain-phase plot is normally a plot of amplitude in decibels versus phase shift in degrees. When the gain-phase plot is superimposed on the Nichols chart [1], relative stability and complete information on the closed-loop frequency response can be obtained.

We shall show in the following sections how these frequency-domain methods are applied to digital control systems.

8.2 THE NYQUIST PLOT [1]

The Nyquist criterion is a graphical study of the stability of a closed-loop system by plotting the frequency locus of the loop transfer function in polar coordinates. For the continuous-data system shown in Fig. 8-1(a), the closed-loop transfer function is

$$\frac{C(s)}{R(s)} = \frac{G(s)}{1 + G(s)H(s)} \tag{8-1}$$

The stability of the system is studied by investigating the zeros of the function $1 + G(s)H(s)$. The Nyquist path in the s-plane is defined as shown in Fig. 8-2(a). The path is described by four separate sections and virtually encloses the entire right-half of the s-plane, with the exception of the origin and points on the imaginary axis where there are poles and zeros of $G(s)H(s)$. The stability of the closed-loop system in Fig. 8-1(a) is investigated by first constructing the Nyquist plot of $G(s)H(s)$ which is a mapping of the Nyquist path Γ_s from the s-plane onto the $G(s)H(s)$, and then observing its behavior with respect to the $(-1,j0)$ point.

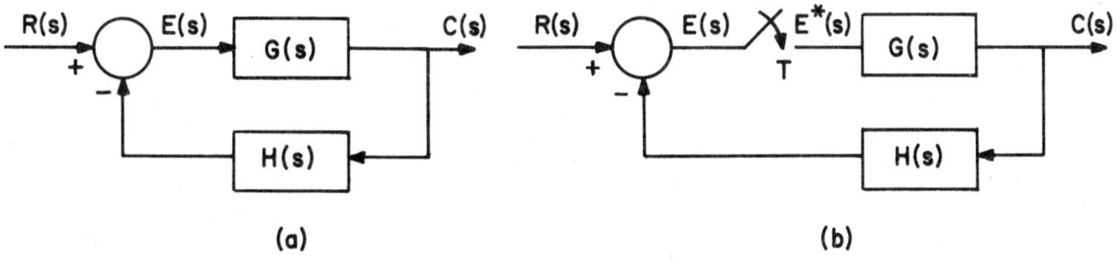

Fig. 8-1. (a) Block diagram of continuous-data control system.
(b) Block diagram of digital control system.

For the closed-loop digital control system shown in Fig. 8-1(b), the closed-loop pulse-transfer function is

$$\frac{C^*(s)}{R^*(s)} = \frac{G^*(s)}{1 + GH^*(s)} \tag{8-2}$$

Thus, the stability of the digital control system can be investigated from the Nyquist plot of $GH^*(s)$. Since $GH^*(s)$ is given by

$$GH^*(s) = \frac{1}{T} \sum_{n=-\infty}^{\infty} G(s + jn\omega_s)H(s + jn\omega_s) \tag{8-3}$$

the construction of GH*(s) actually depends on the properties of G(s)H(s). As an alternative approach, we can write the z-transfer function of Eq. (8-2) as

$$\frac{C(z)}{R(z)} = \frac{G(z)}{1 + GH(z)} \tag{8-4}$$

and then construct the Nyquist plot of GH(z) using the Nyquist path Γ_z in the z-plane. The Nyquist path in the z-plane may be obtained by mapping Γ_s onto the z-plane with the relation $z = e^{Ts}$. Figure 8-2(b) shows the Nyquist path Γ_z in the z-plane; the small semicircles at z = 1 and other points on the path correspond to the small indentations that may be needed on the s-plane Nyquist path. The large circle of infinite radius in the z-plane corresponds to the large semicircle (Sec. IV) in the s-plane. The Nyquist path in the z-plane is not shown as a continuous closed path because, corresponding to point (5) on the path in the s-plane, the z-plane path jumps abruptly from the unit circle to the infinite-radius circle, and at point (7) the z-plane path jumps abruptly back from the infinite circle to the unit circle. The exact place where the jump occurs is indeterminate. However, the area that is enclosed by the Nyquist path Γ_z is the

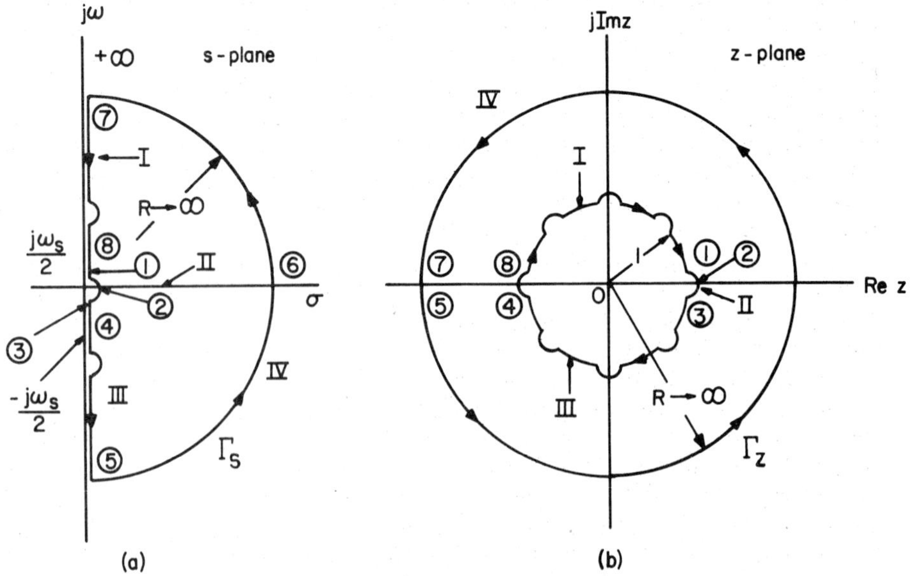

Fig. 8-2. (a) Nyquist path Γ_s in the s-plane.
(b) Nyquist path Γ_z in the z-plane.

region between the two circles. For transfer functions GH(z) that do not have poles and zeros outside the unit circle in the z-plane, it is sufficient to use only Sec. (I) of Γ_z which corresponds to the portion from $s = 0$ to $s = j\infty$ on Γ_s.

The procedure and statement of the Nyquist criterion for digital control systems are given as follows:

(1) The Nyquist path Γ_z in the z-plane is defined as shown in Fig. 8-2(b).

(2) The Nyquist plot of GH*(s) or GH(z) is constructed. This is a plot of GH*(s) or GH(z) in polar coordinates when z takes on values along Γ_z or s takes on values along Γ_s.

(3) The Nyquist criterion states that for a stable closed-loop digital system, the Nyquist plot of GH(z) [GH*(s)] must encircle the $(-1,j0)$ point of the GH(z)-plane [GH*(s)-plane] as many times as the number of poles of GH(z) [GH*(s)] that are found enclosed in the Nyquist path Γ_z (Γ_s), and the encirclement (if there is any), must be made in the clockwise direction.

In other words,

$$N = Z - P \tag{8-5}$$

where

N = number of encirclement of the $(-1,j0)$ point made by the GH(z) Nyquist plot in the GH(z)-plane. A positive sign for N corresponds to counterclockwise (positive) encirclement, and a negative sign for N means clockwise (negative) encirclement.

Z = number of zeros of $1 + GH(z)$ that are outside the unit circle (or "enclosed" by the Nyquist path, where enclosed means encircled in the counterclockwise direction) in the z-plane.

P = number of poles of $1 + GH(z)$ that are outside the unit circle (or enclosed by the Nyquist path) in the z-plane. [Note that the poles of $1 + GH(z)$ are the same as the poles of GH(z).]

For a closed-loop system to be stable $Z \equiv 0$; that is, all the roots of the characteristic equation must be found inside the unit circle. Thus, Eq. (8-5) becomes

$$N = -P \tag{8-6}$$

Usually, if GH(z) does not have poles outside the unit circle in the z-plane, the system is said to be "open-loop stable", and $P = 0$. The Nyquist criterion for closed-loop stability is then simplified to

$$N = 0 \tag{8-7}$$

which means that if the open-loop system is stable and Z = 0, the Nyquist plot of GH(z) should not encircle (or enclose) the (−1,j0) point at all to ensure that the closed-loop system is also stable.

As the above discussion indicates, the application of the Nyquist criterion to digital control systems involves essentially the study of the behavior of the Nyquist plot of GH(z) with respect to the (−1,j0) point. Therefore, the major effort in the application of the Nyquist criterion involves the construction of the Nyquist plot for GH(z). Once this is done, the stability condition of the closed-loop system is determined by mere inspection of the Nyquist plot.

If Eq. (8-7) is the criterion for stability, we need only to plot Sec. (I), ($z = e^{j\omega T}$, $\omega = 0$ to $\omega = \infty$) of the Nyquist plot.

Section (I) of the Nyquist plot of GH(z) is given by

$$GH^*(j\omega) = \frac{1}{T}\sum_{n=-\infty}^{\infty} G(j\omega + jn\omega_s)H(j\omega + jn\omega_s) \tag{8-8}$$

where ω takes on values between 0 and ∞. As an alternative, Sec. (I) of the Nyquist plot of GH(z) is given by $GH(z)\big|_{z=e^{j\omega T}}$ where ω varies from 0 to ∞.

The difficulty of working with Eq. (8-8) is that the expression contains an infinite number of terms, so that the Nyquist plot of GH(z) in general can no longer be sketched by inspection of the function. We present several methods of construction of the Nyquist plot of GH(z) in the following.

1. The z-Transform Method

For sinusoidal steady-state analysis of continuous-data systems, we set $s = j\omega$ in the transfer function. This way, only points on the positive imaginary axis of the s-plane are considered. For digital control systems, the corresponding analysis in the frequency domain is accomplished by setting

$$z = e^{j\omega T} \tag{8-9}$$

that is, only points on the unit circle, $|z| = 1$, in the z-plane are considered. Thus, given any transfer function G(z), we have the following identity:

$$G(z) = G^*(s)\Big|_{s=\frac{1}{T}\ln z} \tag{8-10}$$

Then

$$G(z)\Big|_{z=e^{j\omega T}} = G^*(s)\Big|_{s=j\omega} \tag{8-11}$$

A FORTRAN IV digital computer program for the computation of G(z) in the last equation is given in Sec. 8.8.

Example 8-1.

Consider that the loop transfer function of the digital control system shown in Fig. 8-1(b) is given by

$$G(s)H(s) = \frac{1.57}{s(s+1)} \tag{8-12}$$

and the sampling frequency of the system is $\omega_s = 4$ rad/sec, or the sampling period is $T = \pi/2$ sec.

The z-transform of G(s)H(s) is

$$GH(z) = 1.57 \frac{0.792z}{(z-1)(z-0.208)} \tag{8-13}$$

From Eq. (8-11), the frequency-response locus of GH(z) is described by

$$\begin{aligned} GH(e^{j\omega T}) &= \frac{1.243e^{j\omega T}}{(e^{j\omega T}-1)(e^{j\omega T}-0.208)} \\ &= \frac{1.243(\cos\omega T + j\sin\omega T)}{(\cos\omega T + j\sin\omega T - 1)(\cos\omega T + j\sin\omega T - 0.208)} \end{aligned} \tag{8-14}$$

Since the unit circle $|z| = 1$ is traversed once for every $-n\omega_s/2 \leqslant \omega \leqslant n\omega_s/2$ for $n = 1,2,\ldots$, the frequency locus of GH(z) repeats over the same frequency range. Figure 8-3 shows the plot of GH(z) when z takes on values along the unit circle. Only the values of ω corresponding to the primary strip on the $j\omega$-axis in the s-plane are shown on the locus. The effect of sampling on the stability of the closed-loop system is clearly illustrated by the Nyquist plots of G(s)H(s) and GH(z) shown in Fig. 8-3. Since the transfer function G(s)H(s) is of the second-order, a continuous-data control system with G(s)H(s) as its loop transfer function will always be stable, as shown by the G(s)H(s) plot in Fig. 8-3 not intersecting the negative real axis. For the digital control system, we see that the GH(z) plot does intersect the negative real axis at -0.515 with $K = 1.57$. For this situation, since the $(-1,j0)$ point is to the left of -0.515, it is not enclosed by the GH(z) plot. Thus, the closed-loop digital control system is stable for $K = 1.57$ and $T = \pi/2$ sec. However, for $K \geqslant 3.05$, the closed-loop digital control system is no longer asymptotically stable.

Example 8-2.

Consider the digital control system shown in Fig. 7-5; the open-loop transfer function of the system is given by Eq. (7-59), with $T = 0.1$ sec, $K_r = 3.17 \times 10^5$, $J_v = 41822$, and K_p is the variable parameter. The open-loop transfer function is

$$G(z) = \frac{1.2 \times 10^{-7} K_p (z+1)}{(z-1)(z-0.242)} \tag{8-15}$$

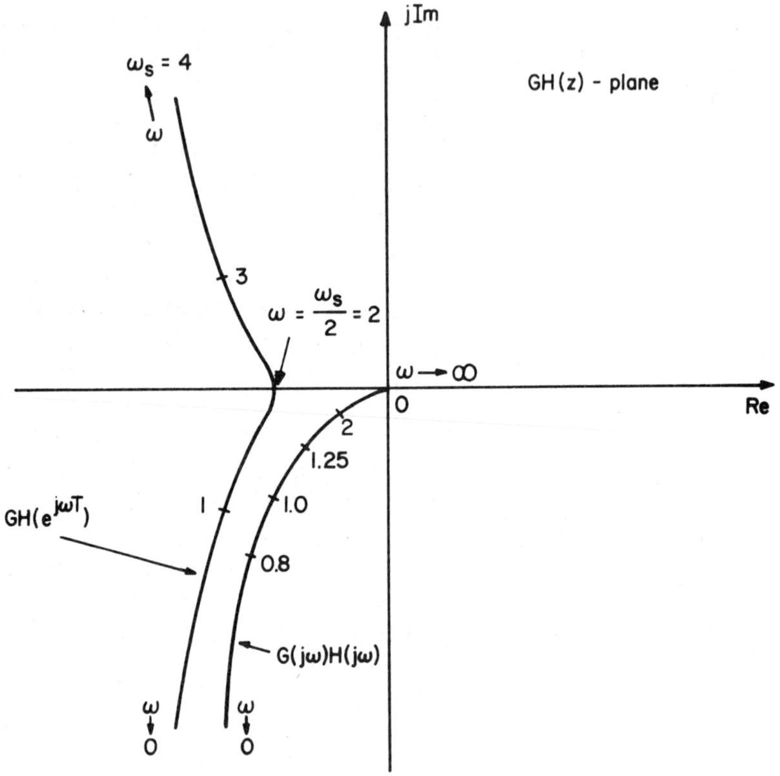

Fig. 8-3. Frequency-response loci of $G(s) = 1.57/[s(s+1)]$ and of $GH(z) = 0.792Kz/[(z-1)(z-0.208)]$, $K = 1.57$, $T = \pi/2$ sec.

Setting $z = e^{j\omega T}$ in Eq. (8-15), the magnitude and phase of G(z) are computed. Figure 8-4 shows the frequency-response or Nyquist plot of G(z) for $K_p = 1.65 \times 10^6$, 6.32×10^6 and 10^7. When $K_p = 1.65 \times 10^6$, the G(z) locus intersects the negative real axis at -0.26, and the frequency corresponding to that point is 13.2 rad/sec. Since the $(-1,j0)$ point is to the left of the crossover point, according to the Nyquist criterion, the closed-loop system is asymptotically stable. When $K_p = 6.32 \times 10^6$, the G(z) locus crosses over the negative real axis at the $(-1,j0)$ point, and the system is not asymptotically stable. Finally, for $K_p = 10^7$, the $(-1,j0)$ point is enclosed by the G(z) locus, and the system is unstable. The frequency at which the G(z) locus crosses the negative real axis corresponds to the value of ω where the root loci intersect the unit circle in the z-plane (Fig. 7-21).

It should be pointed out that only positive values of ω are considered in plotting the loci of Fig. 8-4. Furthermore, each locus represents a frequency range of $0 \leqslant \omega \leqslant \omega_s$, where ω_s is the sampling frequency. The locus for $0 \leqslant \omega \leqslant \omega_s/2$ is the mirror image of that for $\omega_s/2 \leqslant \omega \leqslant \omega_s$, with respect to the real axis. Therefore, the loci repeat for every $n\omega_s \leqslant \omega \leqslant (n+1)\omega_s$, $n = 0,1,2,\ldots$. This property of the frequency loci is due to the periodic strip in the s-plane and the sampling effect.

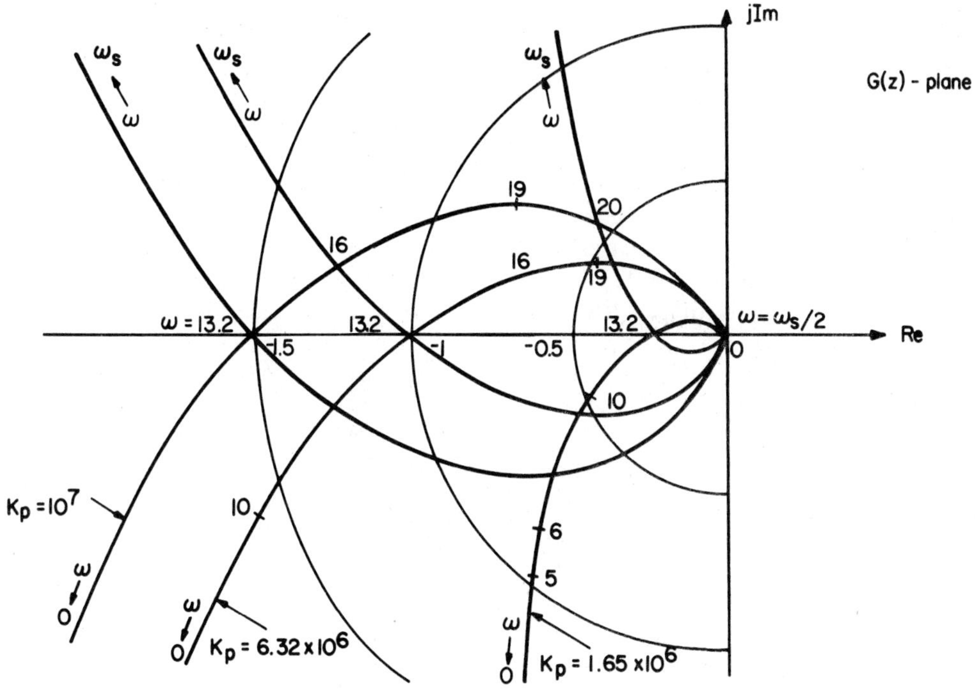

Fig. 8-4. Frequency-response loci of G(z) of the system in Fig. 7-5.

2. The Infinite-Series Method

The pulse transfer function of a typical digital control system is written as

$$GH^*(j\omega) = \frac{1}{T} \sum_{n=-\infty}^{\infty} G(j\omega + jn\omega_s)H(j\omega + jn\omega_s) \tag{8-16}$$

where we have substituted $s = j\omega$ for frequency-domain analysis. Since for most control systems $G(j\omega)H(j\omega)$ exhibits low-pass-filter characteristics, the magnitude of $G(j\omega)H(j\omega)$ decreases as ω increases. Thus, we can normally truncate the infinite series in Eq. (8-16) and use only a finite number of terms to approximate $GH^*(j\omega)$. For digital computer computation, an automatic check can be set up to terminate the series when a preset error criterion is met. In other words, let

$$GH^*_N(j\omega) = \frac{1}{T} \sum_{n=-N}^{N} G(j\omega + jn\omega_s)H(j\omega + jn\omega_s) \tag{8-17}$$

where the positive interger N is determined when the following error criterion is satisfied:

$$|G[j\omega + j(N+1)\omega_s]H[j\omega + j(N+1)\omega_s]| + |G[j\omega - j(N+1)\omega_s]H[j\omega - j(N+1)\omega_s]| \leqslant \Delta |GH_N^*(j\omega)| \quad (8\text{-}18)$$

where Δ is a predetermined small number.

The check in Eq. (8-18) essentially implies that if the contribution of the (N + 1)st terms in Eq. (8-17) are less than or equal to some small number Δ times the magnitude of the function due to $n = 0,1,2,\ldots,N$, the series can be truncated after $n = N$.

The advantage of using Eq. (8-17) as an approximation to $GH^*(j\omega)$ over the z-transform method of setting $z = e^{j\omega T}$ in GH(z) is, of course, the amount of work saved in not having to find GH(z) from G(s)H(s).

A FORTRAN IV digital computer program for the implementation of Eq. (8-17) using the error criterion of Eq. (8-18) is given in Sec. 8.8.

As an illustrative example, the digital computer program is used to compute $G_{h0}G^*(j\omega)$ with

$$G_{h0}(s)G(s) = \frac{1 - e^{-Ts}}{s}\left[\frac{s^2 + 2s + 10}{s^3 + 5s^2 + 5s + 1}\right] \quad (8\text{-}19)$$

Table 8-1 gives the results of the magnitude and phase of $G_{h0}G^*(j\omega)$ with $T = 0.2$, and an error of $\Delta = 0.01$. The first column in Table 8-1 is N, which is the upper and lower limits of the summation of Eq. (8-17) required to satisfy the specific error criterion. For instance, when $\omega = 0.1571$ rad/sec, $N = 1$. This means that $G_{h0}G^*(j0.1571)$ is approximated by

$$G_{h0}G^*(j0.1571) \cong G_{h0}G_1^*(j0.1571) = G_{h0}(j0.1571)G(j0.1571) + G_{h0}(j0.1571 + j\omega_s)G(j0.1571 + j\omega_s) + G_{h0}(j0.1571 - j\omega_s)G(j0.1571 - j\omega_s) \quad (8\text{-}20)$$

where ω_s = sampling frequency $= 2\pi/T = 10\pi$.

Apparently, the series is truncated at $N = 1$, since according to Eq. (8-18),

$$|G_{h0}(j0.1571 + j2\omega_s)G(j0.1571 + j2\omega_s)| + |G_{h0}(j0.1571 - j2\omega_s)G(j0.1571 - j2\omega_s)| \leqslant 0.01|G_{h0}G_1^*(j0.1571)| \quad (8\text{-}21)$$

For the function given in Eq. (8-19), N increases as ω becomes larger. When $\omega = 15.71$ rad/sec, $N = 5$.

Notice that in the computer program ω varies as a multiple of $\omega_s/2$, and

TABLE 8-1. $\Delta = 0.01$

N	ω	$\lvert G_{h0}G_N^*(j\omega)\rvert$	$\angle G_{h0}G_N^*(j\omega)$
1	0.1571	8.498	− 40.81
1	0.3142	6.120	− 69.96
1	0.4712	4.357	− 90.00
1	0.6283	3.172	−104.7
1	0.7854	2.363	−116.2
1	0.9425	1.794	−125.3
1	1.100	1.384	−132.6
1	1.257	1.081	−138.6
1	1.414	0.8526	−143.2
1	1.571	0.6786	−146.8
3	3.142	0.1301	−125.6
3	4.712	0.1211	−102.9
4	6.283	0.1197	−110.1
4	7.854	0.1118	−121.5
5	9.425	0.1044	−133.5
5	11.00	0.09806	−145.3
5	12.57	0.09357	−157.0
5	14.14	0.09093	−168.5
5	15.71	0.09005	−180.0

up to $\omega_s/2$. This is because the frequency locus of $GH^*(j\omega)$ repeats from $n\omega_s$ to $(n + 1)\omega_s$, $n = 0,1,2,\ldots$, and the locus from $\omega = 0$ to $\omega_s/2$ is symmetrical to that from $\omega_s/2$ to ω_s. Thus, the phase of $GH^*(j\omega)$ at $\omega = n\omega_s/2$ is always equal to an integral multiple of π.

Table 8-2 shows the results of the series approximation of Eq. (8-19) with $T = 0.2$ and $\Delta = 10^{-4}$. When $\omega = 0.1571$ apparently N is still one. However, as ω increases, more terms are needed in the approximating series to meet the error criterion.

TABLE 8-2. $\Delta = 10^{-4}$

N	ω	$\lvert G_{h0}G_N^*(j\omega)\rvert$	$\underline{/G_{h0}G_N^*(j\omega)}$ (deg)
1	0.1571	8.498	− 40.81
2	0.3142	6.120	− 69.96
2	0.4712	4.357	− 90.00
2	0.6283	3.172	−104.7
3	0.7854	2.364	−116.2
4	0.9425	1.795	−125.3
5	1.100	1.384	−132.6
5	1.257	1.081	−138.5
6	1.414	0.8536	−143.2
7	1.571	0.6797	−146.7
22	3.142	0.1315	−125.4
28	4.712	0.1233	−103.2
32	6.283	0.1219	−110.4
36	7.854	0.1145	−121.8
40	9.425	0.1070	−133.7
43	11.00	0.1009	−145.5
45	12.57	0.09665	−157.1
47	14.14	0.09415	−168.6
47	15.71	0.09333	−180.0

3. The Bilinear Transformation Method

The bilinear transformation introduced in Chapter 5, Eq. (5-64) or Eq. (5-65), can be used for the purpose of making frequency-domain plots. Consider the bilinear transformation

$$z = \frac{1 + w}{1 - w} \tag{8-22}$$

Solving for w from the last equation, we have

$$w = \frac{z-1}{z+1} \tag{8-23}$$

For $z = e^{j\omega T}$, and using the relation, $e^{j\omega T} = \cos\omega T + j\sin\omega T$, Eq. (8-23) becomes

$$w = j\frac{\sin\omega T}{1+\cos\omega T} = j\tan\left[\frac{\omega T}{2}\right] \tag{8-24}$$

Thus, we see that the unit circle in the z-plane is mapped onto the imaginary axis of the complex w-plane. Furthermore, the interior of the unit circle corresponds to the left-half of the w-plane, and the positive imaginary axis corresponds to the frequency range of $n\omega_s \leqslant \omega \leqslant (n+1)\omega_s$, $n = 0,1,2,\ldots$.

Let the complex variable w be represented by

$$w = \sigma_w + j\omega_w \tag{8-25}$$

Then from Eq. (8-24),

$$\omega_w = \tan\left[\frac{\omega T}{2}\right] \tag{8-26}$$

which establishes the relation between the frequency ω and the transformed "frequency" ω_w. Therefore, for frequency-domain analysis we can replace z in a transfer function by the relation

$$z = \frac{1+j\omega_w}{1-j\omega_w} \tag{8-27}$$

and then let ω_w take on values between zero and infinity. For example, the transfer function of Eq. (8-15) is transformed to

$$G(j\omega_w) = \frac{1.583 \times 10^{-7} K_p(1-j\omega_w)}{j\omega_w(1+j1.636\omega_w)} \tag{8-28}$$

through the use of Eq. (8-27) and after simplification. Notice that $G(j\omega_w)$ simply denotes that the transfer function is a function of $j\omega_w$, and not that z is replaced by $j\omega_w$. Once the expression of $G(j\omega_w)$ is obtained, the magnitude and phase of $G(j\omega_w)$ can be plotted by varying ω_w between zero and infinity.

8.3 THE BODE DIAGRAM [1]

It may seem that the bilinear transformation does not pose any advantage if the polar plot of a transfer function is desired. However, the bilinear transformation can condition the transfer function so that a Bode diagram can be sketched usually without tedious computation. For instance, the transformed transfer function in Eq. (8-28) represents a simple expression whose Bode diagram can be easily constructed without detailed plotting. The *corner frequencies* of the transfer function are at $\omega_w = 1$ and $\omega_w = 1/1.636 = 0.611$.

Figure 8-5 illustrates the Bode diagram of $G(j\omega_w)$ in Eq. (8-28) for $K_p = 1.65 \times 10^6$, 6.32×10^6, and 10^7. The magnitude curves are drawn using straight-line approximations of the component factors of $G(j\omega_w)$. For example, the magnitude of $(1 - j\omega_w)$, which is in the numerator of $G(j\omega_w)$, has an asymptotic line with zero slope at small values of ω_w, and a straight line with a slope of +20 db/decade of ω_w for large ω_w. The two lines meet at $\omega_w = 1.0$. The magnitude of the term $(1 + j1.636\omega_w)$, which is in the denominator of $G(j\omega_w)$, is represented by a straight line with zero slope at small ω_w, and a line with a slope of −20 db/decade of ω_w for large ω_w. The two straight lines meet at the corner frequency $\omega_w = 0.611$. The term $j\omega_w$ in the denominator of

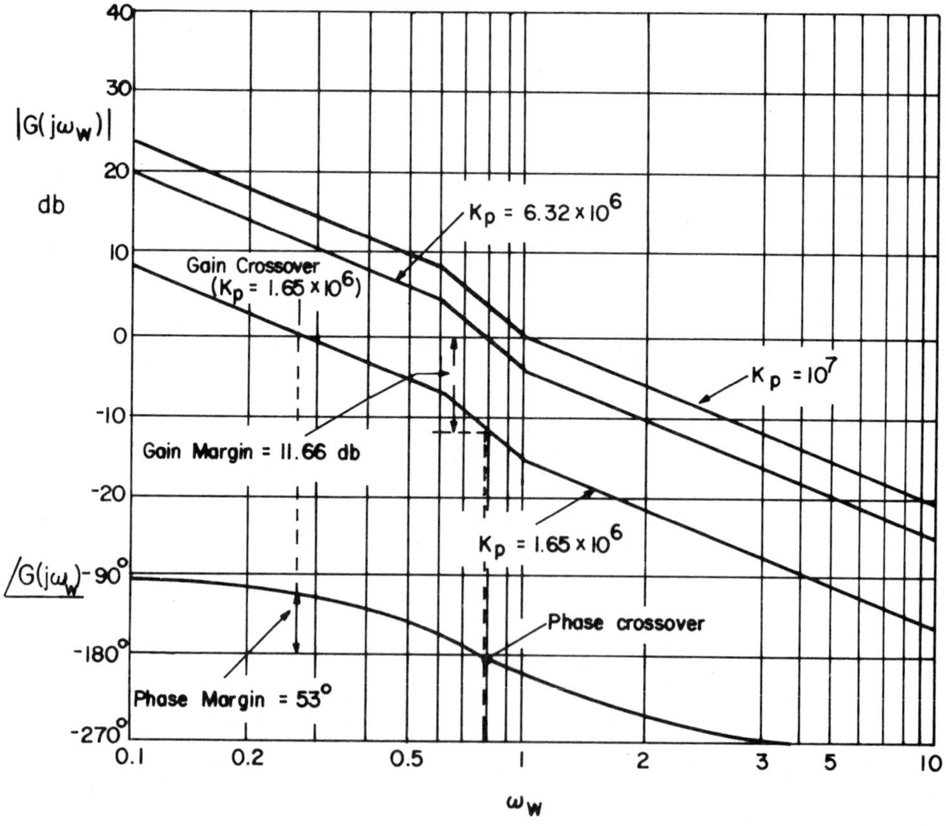

Fig. 8-5. Bode diagram of $G(j\omega_w)$ of the system in Fig. 7-5.

$G(j\omega_w)$ contributes to a straight line with a slope of -20 db/decade which passes through the point $\omega_w = 1$ on the horizontal axis. The sum of these three components gives the composite magnitude curves of $G(j\omega_w)$ shown in Fig. 8-5. When the gain K_p varies, the magnitude curve is simply shifted vertically up or down.

The phase of $(1 - j\omega_w)$ in the numerator of $G(j\omega_w)$ varies between 0 and -90 degrees as ω_w is varied between 0 and infinity. At $\omega_w = 1$, the phase is $-45°$. Similarly, the phase of the term $(1 + j1.636\omega_w)$ in the denominator has the same characteristics except that the $-45°$ point occurs at $\omega_w = 0.611$. The term $j\omega_w$ in the denominator contributes a constant phase of $-90°$ for all values of ω_w. The sum of all these phase curves is shown in Fig. 8-5.

It is informative to compare the Bode plots in Fig. 8-5 with the polar plots in Fig. 8-4, which are for the same transfer functions. The Bode plots correspond to the range of ω_w from 0 to infinity, or in terms of real frequency ω, it is from $n\omega_s$ to $(n+1)\omega_s$, $n = 0,1,2,\ldots$. For the polar plots, the frequency range is from $n\omega_s$ to $(n + 1/2)\omega_s$, $n = 0,1,2,\ldots$. The portion of $G(j\omega_w)$ from $\omega = (n + 1/2)\omega_s$ to $(n+1)\omega_s$ is simply the mirror image of the plot for $0 \leqslant \omega \leqslant (n + 1/2)\omega_s$, as shown in Fig. 8-4.

8.4 GAIN MARGIN AND PHASE MARGIN [1]

Gain Margin

Two of the quantities that give measure on the relative stability of control systems are the *gain margin* and the *phase margin*. These measures can be extended from the continuous-data control systems to digital control systems directly. With reference to Fig. 8-6 we have reconstructed the frequency-domain loci in Fig. 8-4. The $G(z)$ locus for $K_p = 1.65 \times 10^6$ is shown to intersect the negative real axis at -0.26. We can see that if K_p is increased to 6.32×10^6, or $1/0.26$ times 1.65×10^6, the system would be on the verge of being unstable. Therefore, for $K_p = 1.65 \times 10^6$, the number, $1/0.26 = 3.83$ is a "safety factor" which the gain K_p can be multiplied by to maintain marginal stability. The factor 3.83 is defined as the "gain margin" in linear scale.

It is customary to express the gain margin in decibels; then 3.83 corresponds to

$$20\log_{10} 3.83 = 11.66 \text{ db} \tag{8-29}$$

Clearly, the advantage of working with decibels becomes apparent if we use the Bode diagram.

Based upon the definition of the gain margin given above, we can express the gain margin analytically as

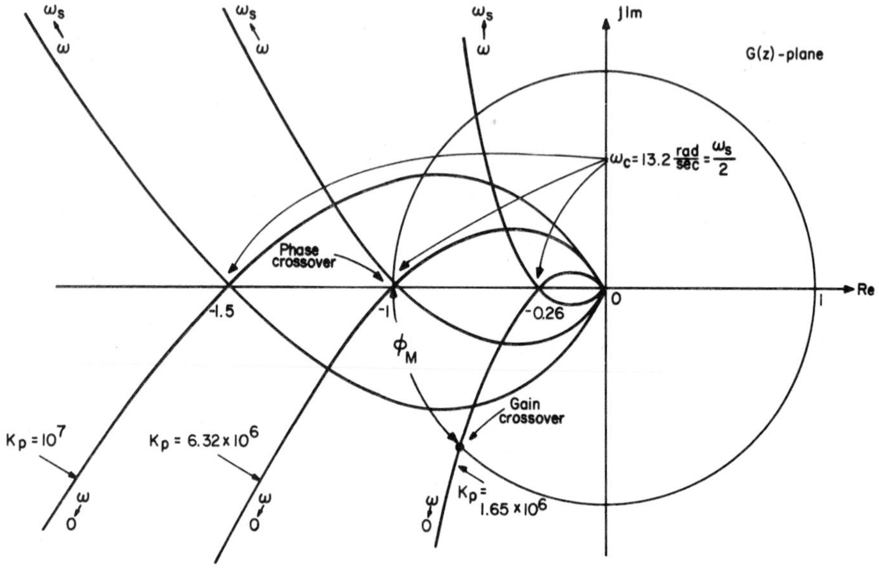

Fig. 8-6. Frequency-domain loci of G(z) of Eq. (8-15).

$$\text{Gain margin} = 20\log_{10} \frac{1}{\left|G(e^{j\omega_c T})\right|} \text{ db} \tag{8-30}$$

where $|G(e^{j\omega_c T})|$ is the magnitude of $G(e^{j\omega T})$ measured at the point where the locus crosses the real axis, and the frequency is denoted by ω_c. The frequency ω_c is also defined as the *phase-crossover frequency*, since at the crossover point, the phase of $G(e^{j\omega T})$ is 180 degrees.

Referring to Fig. 8-6, we see that for the case of $K_p = 6.32 \times 10^6$, $G(e^{j\omega_c T}) = -1$. Thus, the gain margin is given by

$$\text{Gain margin} = 20\log_{10} 1 = 0 \text{ db} \tag{8-31}$$

In this case it is apparent that the closed-loop system is on the verge of instability, and K_p can be increased no further.

For $K_p > 6.32 \times 10^6$, $|G(e^{j\omega_c T})|$ is greater than one, the gain margin is negative in db, and the system is unstable. Care must be taken in interpreting the correlation between negative-gain-margin and instability, as the validity is limited only to systems whose loop transfer functions have no poles and zeros outside the unit circle; that is, under the condition where Sec. (I) of the Nyquist plot is suffice to determine stability.

Phase Margin

Gain margin alone turns out to be inadequate to represent the relative stability of a closed-loop control system. Since the gain factor is not the only parameter that is subject to variation, a given system may have a large gain margin and yet still may have a low stability margin due to parameter variations that affect the phase of the transfer function. Referring to the loci in Fig. 8-6, we see that if the locus for $K_p = 1.65 \times 10^6$ is held rigid and rotates about the origin in the clockwise direction, it can intersect the $(-1,j0)$ point after a rotation of ϕ_M degrees. This angle of rotation, ϕ_M, is defined as the *phase margin.* Of course, the physical significance of the phase margin can only be interpreted for an ideal situation under which the parameter variation in the system causes a pure phase angle variation, which is only possible with an all-pass network in practice. On the other hand, it is common to vary only the loop gain of the system without affecting the phase.

Referring to the locus for $K_p = 6.32 \times 10^6$ in Fig. 8-6, we see that the phase margin is 0 degree, since the locus already passes through the $(-1,j0)$ point. Similarly, for $K_p > 6.32 \times 10^6$, the phase margin would be negative, indicating instability.

The point at which the unit circle intersects the G(z) locus is referred to as the gain-crossover point, due to the fact that the magnitude of $G(e^{j\omega T})$ at the point is equal to one. The frequency at the intersection is called the *gain-crossover frequency*.

From the above discussions we can state that *the gain margin is measured at the phase-crossover point*, whereas *the phase margin is measured at the gain-crossover point.*

The gain margin and the phase margin are more easily determined from the Bode diagram. Again using Fig. 8-5 as an illustration, we can identify the gain-crossover and the phase-crossover points as shown in the figure. Then, the gain margin in db is the distance the magnitude curve of $G(e^{j\omega T})$ lies below the 0-db axis at the phase crossover. The phase margin is measured as the difference between the phase of $G(e^{j\omega T})$ and $-180°$ at the gain crossover. It is readily seen from the Bode diagram that for $K_p = 1.65 \times 10^6$ if the gain K_p is increased by the gain margin, 11.66 db, the closed-loop system would be on the verge of instability. Similarly, if we could hold the magnitude curve invariant and lower the phase shift curve down by an amount equal to the phase margin, 53°, then the gain crossover and the phase crossover would occur at the same frequency, and the same stability condition would exist.

8.5 THE GAIN-PHASE DIAGRAM AND THE NICHOLS CHART

Still another way of representing frequency-domain characteristics is the gain-versus-phase plot. In this case, the magnitude of G(z) or $G(j\omega_w)$ is plotted as a function of the phase of the same function with frequency ω as a varying parameter on the locus. One advantage of using the gain-phase plot is that when superposed with the Nichols chart [1] the plot gives information on the frequency response of the closed-loop system. Information on the gain and phase margins, and the bandwidth of the closed-loop system, are also easily obtained from the gain-phase plot.

The bode plots of Fig. 8-5 are transferred to the gain-phase coordinates in Fig. 8-7. Notice that the gain and phase margins are easily measured from the gain-phase plot, as illustrated by the case for $K_p = 1.65 \times 10^6$. The Nichols chart provides information on the magnitude of the closed-loop transfer function C(z)/R(z) in Eq. (8-4) as a function of frequency. If desired, the values of

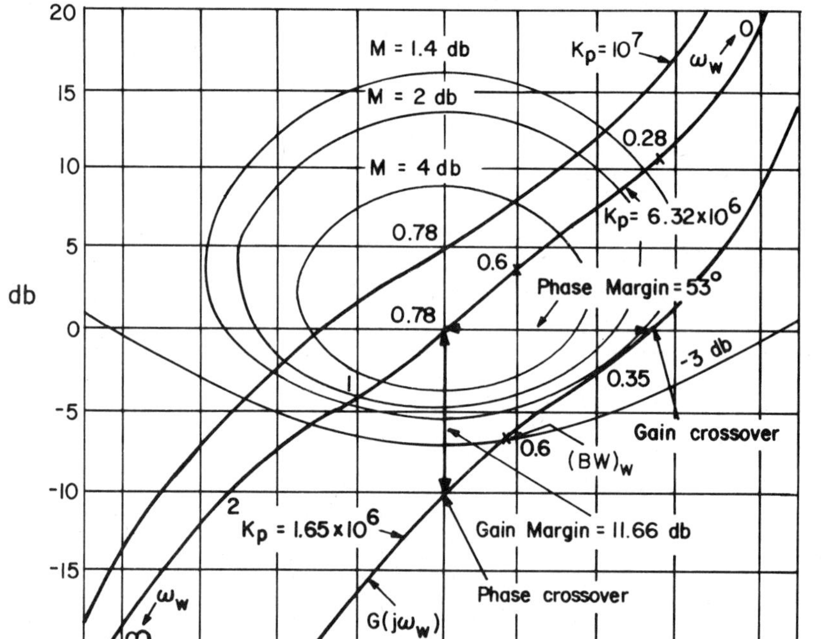

Fig. 8-7. Frequency loci of $G(j\omega_w)$ in the gain-phase coordinates for the system in Fig. 7-5, together with the Nichols chart.

$M = |C(j\omega_w)/R(j\omega_w)|$ at various frequencies can be obtained by observing the intersects between the $G(j\omega_w)$ locus and the Nichols chart trajectories. The Nichols trajectory which is tangent to the $G(j\omega_w)$ curve represents the maximum value of the closed-loop gain, M_p, and the frequency at which it occurs is termed the resonant frequency. From Fig. 8-7, it is observed that for $K_p =$

1.65×10^6, $M_p \cong 1.4$ db or 1.175. The resonant frequency is $\omega_w = 0.35$, or using Eq. (8-26),

$$\omega = \frac{2}{T} \tan^{-1} \omega_w = 6.73 \text{ rad/sec} \tag{8-32}$$

for T = 0.1 sec.

8.6 BANDWIDTH CONSIDERATIONS

In the design of control systems in the frequency domain it is common to use bandwidth as a qualitative measure of the system performance. For instance, the practicing engineer often specifies that the bandwidth of a designed system should be, say, 2 Hz. The use of bandwidth as a system specification can be applied to continuous-data as well as digital control systems, since from the input-output performance standpoint it does not matter what type of system it is.

It is important to clarify what bandwidth means in general terms. As it turns out, for a second-order continuous-data control system there is a direct correlation between the bandwidth, BW, and ζ, and ω_n. Specifically, for a second-order system that is described by the closed-loop transfer function

$$\frac{C(s)}{R(s)} = \frac{\omega_n^2}{s^2 + 2\zeta\omega_n s + \omega_n^2} \tag{8-33}$$

bandwidth is given by [1]

$$BW = \omega_n \left[(1 - 2\zeta^2) + \sqrt{4\zeta^4 - 4\zeta^2 + 2} \right]^{1/2} \tag{8-34}$$

which is determined by finding the frequency ω at which the magnitude of $C(j\omega)/R(j\omega)$ is equal to 0.707.

For digital control systems, the simplest second-order transfer function usually has a first-order term, $(z - z_1)$, in the numerator, plus the fact that z is related to ω through the relation $z = e^{j\omega T}$ in the frequency domain. It becomes difficult to arrive at a clean analytical expression such as Eq. (8-34) which relates the bandwidth to the pole-zero location of $C(z)/R(z)$. However, a bandwidth expression does exist for a continuous-data system with the following transfer function:

$$\frac{C(s)}{R(s)} = \frac{\omega_n^2(1 + Ts)}{s^2 + (2\zeta\omega_n + T\omega_n^2)s + \omega_n^2} \tag{8-35}$$

In this case the bandwidth is given as

$$BW = \left[-a + \frac{1}{2}\sqrt{a^2 + 4\omega_n^4} \right]^{1/2} \tag{8-36}$$

where

$$a = 4\zeta^2\omega_n^2 + 4\zeta\omega_n^3 T - 2\omega_n^2 - \omega_n^4 T^2 \tag{8-37}$$

We can use these results and the w-transformation to find the bandwidth relation for a second-order digital control system.

Consider that we start with the closed-loop transfer function

$$\frac{C(z)}{R(z)} = \frac{K(z - z_1)}{(z - p_1)(z - \bar{p}_1)} \tag{8-38}$$

where z_1 is a real constant, p_1 and $\bar{p}_1$ are either real or in complex-conjugate pair. Using the bilinear transformation of Eq. (8-22), Eq. (8-38) is transformed to the following form:

$$\frac{C(w)}{R(w)} = \frac{\omega_{wn}^2(1 + T_w w)}{w^2 + (2\zeta_w\omega_{wn} + T_w\omega_{wn}^2)w + \omega_{wn}^2} \tag{8-39}$$

where ω_{wn}, ζ_w, and T_w do not necessarily have any direct physical meaning, and are merely working parameters in the w-domain to conform with the format of Eq. (8-35). Now the bandwidth in ω_w, where $w = j\omega_n$, in the w-domain for the system in Eq. (8-39) is determined from

$$(BW)_w = \left[-a_w + \frac{1}{2}\sqrt{a_w^2 + 4\omega_{wn}^4} \right]^{1/2} \tag{8-40}$$

where

$$a_w = 4\zeta_w^2\omega_{wn}^2 + 4\zeta_w\omega_{wn}^3 T_w - 2\omega_{wn}^2 - 4\omega_{wn}^4 T_w^2 \tag{8-41}$$

Once $(BW)_w$ is found from Eq. (8-40), the true bandwidth in the real frequency domain is determined using Eq. (8-26), or

$$BW = \frac{2}{T}\tan^{-1}(BW_w) \tag{8-42}$$

Example 8-3.

Consider again the digital control system shown in Fig. 7-5. The open-loop transfer function of the system is given in Eq. (8-15), and the closed-loop transfer function in the z domain is

$$\frac{C(z)}{R(z)} = \frac{0.198(z+1)}{z^2 - 1.044z + 0.44} \tag{8-43}$$

for $K_p = 1.65 \times 10^6$. We wish to determine the bandwidth of the closed-loop digital system. Substituting $z = (1+w)/(1-w)$ into Eq. (8-43), and simplifying, we get

$$\frac{C(w)}{R(w)} = \frac{0.159(1-w)}{w^2 - 0.451w + 0.159} \tag{8-44}$$

Comparing Eq. (8-44) with Eq. (8-39), we get

$$\omega_{wn} = 0.4$$

$$T_w = -1$$

$$\zeta_w = -0.365$$

Thus, using Eqs. (8-40) and (8-41), we have

$$(BW)_w = 0.586 \tag{8-45}$$

and from Eq. (8-42) the real bandwidth is

$$BW = \frac{2}{T}\tan^{-1}(0.586) = 10.61 \text{ rad} \tag{8-46}$$

In the gain-phase plot bandwidth is easily found from the intersect between the $G(e^{j\omega T})$ locus and the -3 db constant-M locus of the Nichols chart. For instance, for the system considered in the last example, the gain-phase plot of $G(z)$ is drawn in Fig. 8-7 using the w-transform. The value of ω_w at the intersect between $G(j\omega_w)$ and the $M = -3$ db locus of the Nichols chart is shown to be 0.6. This result is quite close to that found in the last example in Eq. (8-45) which is 0.586.

The gain-phase-plot method of determining bandwidth has the advantage that the method is applicable to systems of any order. The analytical method involving Eqs. (8-39) through (8-42) is applicable only to systems of the second order, or those that can be approximated by a second-order model.

As mentioned at the start of this section, the bandwidth requirement is often specified in the design of control systems. For a continuous-data control system, once the bandwidth of the overall closed-loop system is specified, we can use a pair of dominant characteristic roots to realize the bandwidth requirements. The system dynamics are thus approximated by the second-order transfer function of Eq. (8-33). For high-order systems all the other poles and zeros are placed far to the left relative to the dominant roots. For instance, given BW, Eq. (8-34) can be used for the determination of ζ and ω_n, although the solutions are not unique. Typically, we may assign a value to ζ, say 0.707, then ω_n is uniquely determined.

For a digital control system, given a desired bandwidth BW, the w-domain equivalent is determined from

$$(BW)_w = \tan\left[\frac{(BW)T}{2}\right] \tag{8-47}$$

We can again designate a pair of dominant characteristic roots in the w-plane. If we use the second-order transfer function in Eq. (8-39) as the dominant model we would have three unknown parameters in ζ_w, ω_{wn} and T_w, but with only one known quantity in $(BW)_w$. However, similar to the continuous-data system case, we can set the dominant-mode closed-loop transfer function as

$$\frac{C(w)}{R(w)} = \frac{\omega_{wn}^2}{w^2 + 2\zeta_w \omega_{wn} w + \omega_{wn}^2} \tag{8-48}$$

which has no dominant zero. Then we may use the bandwidth relation

$$(BW)_w = \omega_{wn}\left[(1 - 2\zeta_w^2) + \sqrt{4\zeta_w^4 - 4\zeta_w^2 + 2}\,\right]^{1/2} \tag{8-49}$$

to determine ζ_w and ω_{wn}.

8.7 COMPUTER PROGRAMS FOR FREQUENCY-DOMAIN PLOTS

Two digital computer programs are presented in this section to facilitate the computation of the frequency-response plot of G(z) or G*(s).

The first FORTRAN program, listed in Table 8-3, gives the magnitude and phase of G(z) as functions of ω, when $z = e^{j\omega T}$. The magnitude is given in terms of real values and in decibels, and the phase is in degrees. The user must input the following parameters:

T	sampling period (sec)
NP	integer which represents the total number of data points
J	integer which determines the resolution of ω.
	$\omega = (\omega_s/2)J/NP$
GN	numerator polynomial of G(z)
GD	denominator polynomial of G(z)
GZ	G(z) = GN/GD

Table 8-4 gives the results of $|G(z)|$, $|G(z)|$ in db, and the phase of G(z), with $z = e^{j\omega T}$, for the function

$$G(z) = \frac{1.2(z+1)}{(z-1)(z-0.242)} \tag{8-50}$$

with ω varying up to $\omega_s/2$. The sampling period is $T = 0.1$ sec, so $\omega_s = 2\pi/T = 62.8$ rad/sec. NP = 50, so there are 50 data points.

The computer program listing given in Table 8-5 is for the computation of $G^*(s)$ with $s = j\omega$, using the truncated infinite series

$$G^*(j\omega) \cong \frac{1}{T} \sum_{n=-N}^{N} G(j\omega + jn\omega_s) \tag{8-51}$$

where N is a positive integer which is determined by an error criterion, DELTA. Let DELTA be a small number, say 10^{-4}. For a given value of ω, let $G^*(j\omega)$ be represented by Eq. (8-51), and is denoted

$$G^*(j\omega) = \text{GSTAR} \tag{8-52}$$

and

$$\text{GSTAR1} = G(j\omega + jN\omega_s) + G(j\omega - jN\omega_s) \tag{8-53}$$

Then N is determined to be the termination of the series for $G^*(j\omega)$ when the following condition is met:

$$|\text{GSTAR1}| \leqslant \text{DELTA} * |\text{GSTAR}| \tag{8-54}$$

The user must input the following parameters in the computer program:

T	sampling period (sec)
NP	integer which represents the total number of data points
J	integer which determines the resolution of ω.
DELTA	error (small number)
GN	numerator polynomial of G(s)
GD	s [denominator polynomial of G(s)]
GFUN	GFUN = $(1 - e^{-Ts})$GAIN*GN/GD (for zero-order hold)
GAIN	constant gain of G(s)

Tables 8-1 and 8-2 already show the sample results of using the digital computer program listed in Table 8-5 for the transfer function in Eq. (8-19).

TABLE 8-3.

```
      COMPLEX CWT,Z,GZ,GD,GN
      REAL GZR,GZI,GZMAG,GZDB,GZPH
      REAL K,T,OMEGA,WT
      PI=3.14159
      RAD=180./PI
      WRITE(5,101)
      T=0.1
      NP=50
      DO 1 J=1,NP
      OMEGAS=2.*PI/T
      OMEGA=(OMEGAS/2.)*FLOAT(J)/NP
      WT=OMEGA*T
      CWT=CMPLX(0.,WT)
      Z=CEXP(CWT)
      K=1.2
      GN=K*(Z+1.0)
      GD=(Z-1.)*(Z-0.242)
      GZ=GN/GD
      GZR=REAL(GZ)
      GZI=AIMAG(GZ)
      GZMAG=CABS(GZ)
      GZDB=20.*ALOG10(GZMAG)
      GZPH=RAD*ATAN2(GZI,GZR)
      WRITE(5,102)OMEGA,GZMAG,GZDB,GZPH
1     CONTINUE
100   FORMAT(25X,'FREQUENCY RESPONSE'/)
101   FORMAT(4X,'OMEGA',7X,'GZMAG',7X,'GZDB',7X,'PHASE'/)
102   FORMAT(1P4E12.3)
      STOP
      END
```

TABLE 8-4.

OMEGA	GZMAG	GZDB	PHASE
6.283E-01	5.033E+01	3.404E+01	-9.475E+01
1.257E+00	2.508E+01	2.799E+01	-9.949E+01
1.885E+00	1.662E+01	2.441E+01	-1.042E+02
2.513E+00	1.237E+01	2.185E+01	-1.089E+02
3.142E+00	9.796E+00	1.982E+01	-1.135E+02
3.770E+00	8.064E+00	1.813E+01	-1.182E+02
4.398E+00	6.815E+00	1.667E+01	-1.227E+02
5.027E+00	5.868E+00	1.537E+01	-1.272E+02
5.655E+00	5.124E+00	1.419E+01	-1.317E+02
6.283E+00	4.522E+00	1.311E+01	-1.360E+02
6.911E+00	4.025E+00	1.210E+01	-1.403E+02
7.540E+00	3.608E+00	1.114E+01	-1.446E+02
8.168E+00	3.252E+00	1.024E+01	-1.487E+02
8.796E+00	2.945E+00	9.380E+00	-1.528E+02
9.425E+00	2.677E+00	8.552E+00	-1.569E+02
1.005E+01	2.442E+00	7.754E+00	-1.608E+02

1.068E+01	2.233E+00	6.979E+00	-1.647E+02
1.131E+01	2.048E+00	6.226E+00	-1.685E+02
1.194E+01	1.882E+00	5.492E+00	-1.723E+02
1.257E+01	1.732E+00	4.773E+00	-1.760E+02
1.319E+01	1.597E+00	4.067E+00	-1.796E+02
1.382E+01	1.474E+00	3.372E+00	1.768E+02
1.445E+01	1.363E+00	2.687E+00	1.733E+02
1.508E+01	1.260E+00	2.009E+00	1.698E+02
1.571E+01	1.166E+00	1.336E+00	1.664E+02
1.634E+01	1.080E+00	6.674E-01	1.630E+02
1.696E+01	1.000E+00	8.041E-05	1.597E+02
1.759E+01	9.260E-01	-6.676E-01	1.564E+02
1.822E+01	8.573E-01	-1.338E+00	1.531E+02
1.885E+01	7.932E-01	-2.012E+00	1.499E+02
1.948E+01	7.333E-01	-2.694E+00	1.467E+02
2.011E+01	6.772E-01	-3.386E+00	1.436E+02
2.073E+01	6.244E-01	-4.090E+00	1.404E+02
2.136E+01	5.747E-01	-4.812E+00	1.373E+02
2.199E+01	5.276E-01	-5.554E+00	1.343E+02
2.262E+01	4.830E-01	-6.322E+00	1.312E+02
2.325E+01	4.405E-01	-7.122E+00	1.282E+02
2.388E+01	3.999E-01	-7.960E+00	1.252E+02
2.450E+01	3.611E-01	-8.848E+00	1.222E+02
2.513E+01	3.238E-01	-9.795E+00	1.192E+02
2.576E+01	2.878E-01	-1.082E+01	1.163E+02
2.639E+01	2.530E-01	-1.194E+01	1.133E+02
2.702E+01	2.193E-01	-1.318E+01	1.104E+02
2.765E+01	1.864E-01	-1.459E+01	1.074E+02
2.827E+01	1.542E-01	-1.624E+01	1.045E+02
2.890E+01	1.227E-01	-1.823E+01	1.016E+02
2.953E+01	9.159E-02	-2.076E+01	9.870E+01
3.016E+01	6.086E-02	-2.431E+01	9.580E+01
3.079E+01	3.037E-02	-3.035E+01	9.290E+01
3.142E+01	1.289E-06	-1.178E+02	9.000E+01

TABLE 8-5.

```
      DIMENSION DB(200),PHASE(200)
      COMPLEX GFUN,GSTAR,GSTAR1,GSTAT
      DELTA=1.E-4
      TOPI=2.*3.14159
      RAD=180./3.14159
      WRITE(5,101)
101   FORMAT(5X,'GSTAR PROGRAM')
      T=0.2
      WRITE(5,100)T
100   FORMAT(//5X,'T= ',1PE12.5/)
      WRITE(5,102)
102   FORMAT(3X,'NN',4X,'OMEGA',4X,'MAGNITUDE',6X,'DB',6X,'PHASE',
     15X,'REAL PART',2X,'IMAG PART'/)
      OMEGAS=TOPI/T
      NP=100
      DO 1 J=1,NP
      OMEGA=(OMEGAS/2.)*FLOAT(J)/NP
```

```
          GSTAR=GFUN(T,OMEGA)
          DO 4 NN=1,1000
          OMEGT1=OMEGA+OMEGAS*FLOAT(NN)
          OMEGT2=OMEGA-OMEGAS*FLOAT(NN)
          GSTAR1=GFUN(T,OMEGT1)+GFUN(T,
         1OMEGT2)
          GSTAR=GSTAR+GSTAR1
          IF(CABS(GSTAR1).LE.DELTA*CABS(GSTAR))GO TO 5
4         CONTINUE
          GO TO 1000
5         CONTINUE
          GSTAT=GSTAR/T
          GMAG=CABS(GSTAT)
          GREAL=REAL(GSTAT)
          GIMAG=AIMAG(GSTAT)
          GPHASE=RAD*ATAN2(GIMAG,GREAL)
          GDB=20.*ALOG10(GMAG)
          WRITE(5,103)NN,OMEGA, GMAG,GDB,GPHASE,GREAL,GIMAG
103       FORMAT(' ',I4,1P6E11.3)
          GO TO 1
1000      WRITE(5,104)
104       FORMAT(5X,'NO CONVERGENCE IN 1000ITERATIONS'/)
1         CONTINUE
1001      CALL EXIT
          END
          COMPLEX FUNCTION GFUN(T,OMEGA)
          COMPLEX S,GN,GD
          S=CMPLX(0.,OMEGA)
          GN=S*S+2.*S+10.
          GAIN=1.0
          GD=S*(S*S*S+5.*S*S+5.*S+1.)
          GFUN=GAIN*(1.-CEXP(-S*T))*GN/GD
          RETURN
          END
```

REFERENCES

1. Kuo, B. C., *Automatic Control Systems*, 3rd ed., Prentice-Hall, Inc., Englewood Cliffs, N.J., 1975.

2. Whitbeck, R. F., *Analysis of Digital Flight Control Systems with Flying Qualities Applications*, Vol. I — Executive Summary, Tech. Report AFFDL-TR-78-115, Air Force Flight Dynamics Lab., Wright-Patterson Air Force Base, Ohio, Sept. 1978.

3. Whitbeck, R. F., and Hoffmann, L. G., *Analysis of Digital Flight Control Systems with Flying Qualities Applications*, Vol. II — Executive Summary, Tech. Report AFFDL-TR-78-115, Air Force Flight Dynamics Lab., Wright-Patterson Air Force Base, Ohio, Sept. 1978.

PROBLEMS

8-1 The block diagram of a sampled-data control system is shown in Fig. 8P-1. Plot the Nyquist diagram of the open-loop transfer function for $\omega = 0$ to $\omega_s/2$, when
(a) the hold device is a zero-order hold,
(b) the hold device is the first-order hold described in Problem 2-18.
Use the truncated series approximation of $G^*(s) = C^*(s)/E^*(s)$ with an error criterion of $\Delta = 10^{-4}$. The sampling period T is one second. The transfer function G(s) is

$$G(s) = \frac{2}{s(1 + 0.2s)}$$

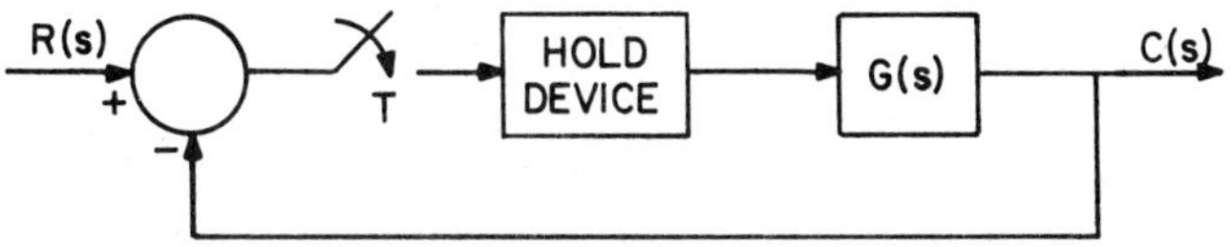

Fig. 8P-1.

8-2 Repeat Problem 8-1 with

$$G(s) = \frac{2.4}{s(1 + 0.5s)}$$

8-3 For the stoichiometric air-fuel ratio control system described in Problem 7-15, the open-loop transfer function of the unity-feedback system is

$$G(z) = \frac{0.33Kz^{-10}}{(z - 0.67)}$$

The sampling period is 0.1 sec. Sketch the Nyquist plot of G(z) with $z = e^{j\omega T}$, and determine the value of K for the closed-loop system to be asymptotically stable.

8-4 Sketch the Nyquist plot of

$$G(z) = \frac{1.2 \times 10^{-7} K_p (z + 1)}{(z - 1)(z - 0.242)} \qquad T = 0.1 \text{ sec}$$

for $\omega = 0$ to $\omega = \omega_s/2$, and determine the critical value of K_p for the closed-loop system to be stable.

8-5 The open-loop transfer function of the level-control system described in Problem 7-17 is represented as $G^*(s)$, where

$$G(s) = \frac{1 - e^{-Ts}}{s} \left[\frac{16.67N}{s(s + 12.5)(s + 1)} \right] \qquad T = 0.05 \text{ sec}$$

Construct a Nyquist plot of G(z)/N with $z = e^{j\omega T}$ for $\omega = 0$ to $\omega = \omega_s/2$. Determine the values of N for the closed-loop system to be stable.

8-6 For the digital control system shown in Fig. 3P-14, Problem 3-14, let $K_r = 731885$, $K_p = 10455500$ and $K_i = 41822000$, $T = 0.1$ sec. Plot the frequency response of the open-loop transfer function C(z)/E(z) for $z = e^{j\omega T}$ from $\omega = 0.1$ to 20 rad/sec. Find the gain margin and the phase margin of the closed-loop system.

8-7 Consider the multirate sampled-data control system described in Problem 3-18, Fig. 3P-18. Sketch the Nyquist plots of the open-loop transfer function, $G(z) = C(z)/E(z)$, for $\omega = 0$ to $\omega = \omega_s/2$, $N = 1, 2, 3, 5, 10$, and $N = \infty$; with $K = 1$. Find the critical values of K for the closed-loop system to be stable for these cases.

8-8 Consider the speed control system described in Problem 7-4.

(a) Find the open-loop transfer function $G(z) = \Omega(z)/E(z)$. First set $K_p = 1$ and $K_r = 0$. Construct a Bode plot of G(z) using the w-transform. Find the gain margin and the phase margin of the closed-loop system. Sketch the Nyquist plot of G(z) for $\omega = 0$ to $\omega = \omega_s/2$. $T = 0.001$ sec.

(b) Repeat part (a) with $K_p = 1$ and $K_r = 295.276$.

8-9 The closed-loop transfer function of a unity-feedback digital control system is

$$\frac{C(z)}{R(z)} = \frac{z + 0.5}{3(z^2 - z + 0.5)} \qquad T = 1 \text{ sec}$$

Find the open-loop transfer function G(z) and construct a Bode diagram for G(w) using the w-transform. Find the gain margin, phase margin, and M_p, ω_p, BW (rad/sec) from the Nichols chart.

8-10 For the open-loop transfer function given in Eq. (7-59),

$$G(z) = \frac{1.2 \times 10^{-7} K_p(z + 1)}{(z - 1)(z - 0.242)} \qquad T = 0.1 \text{ sec}$$

Construct a Bode plot in normalized magnitude for G(w), using the w-transform and with $z = e^{j\omega T}$.

(a) Find the value of K_p so that the gain margin is 20 db.

(b) Find the value of K_p so that the phase margin is 45 degrees.

(c) Sketch the normalized G(w) in the Nichols chart for the case in (b) and find the value of M_p. Find the bandwidth in rad/sec for the closed-loop system.

8-11 For the closed-loop digital control system which has the open-loop transfer function given by

$$G(z) = \frac{1.2 \times 10^{-7} K_p(z + 1)}{(z - 1)(z - 0.242)} \qquad T = 0.1 \text{ sec}$$

Find the bandwidth of the closed-loop system in rad/sec using Eqs. (8-39) through (8-42). $K_p = 3.165 \times 10^6$.

9

CONTROLLABILITY AND OBSERVABILITY

9.1 INTRODUCTION

The concepts of controllability and observability introduced first by Kalman [2,3] play an important role in modern control system design theory. The significance of controllability can be stated with reference to the block diagram of Fig. 9-1. *The controlled process G is said to be controllable if every state variable of G can be affected or controlled in finite time by some unconstrained control signal u(t).* Intuitively, we understand that if any one of the state variables is independent of the control **u**(t), then there is no way of driving this particular state variable to a desired state in finite time by means of some control effort. Therefore, this particular state variable is said to be *uncontrollable*, and the system is said to be uncontrollable.

The concept of observability is the dual of controllability. Essentially, the process G is said to be observable if every state variable of the process eventually affects some of the outputs of the process. In other words, it is often desirable to obtain information on the state variables from measurements of the

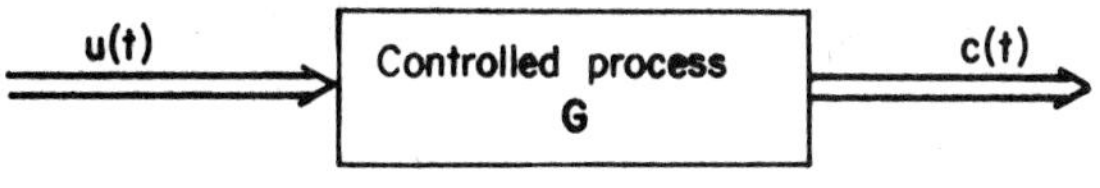

Fig. 9-1. Block diagram of a controlled process.

outputs and the inputs. For example, we may want to "estimate" those state variables that are not accessible for feedback control from measurements of the outputs and inputs. If any one of the states cannot be observed from the measurements of the outputs, then the state is said to be *unobservable*, and the system is unobservable.

The descriptions of controllability and observability given above are general in nature, and they are for systems with continuous or digital data. Before giving formal definitions of the subjects, let us first consider some simple illustrative examples of the ideas. Figure 9-2 shows the state diagram of a controlled process with two state variables. Since the input u(t) affects only the state variable $x_1(t)$, we say that $x_2(t)$ is uncontrollable, and the process is not completely controllable or is simply uncontrollable.

In Figure 9-3 the state diagram of another controlled process is shown. It is observed that the state variable $x_2(t)$ is not connected to the output c(t) in any way. Therefore, if we have measured c(t) once, we can observe from the measurement the state variable $x_1(t)$, since $c(t) = x_1(t)$. However, the state variable $x_2(t)$ cannot be observed from the information on c(t). Thus, the process is described as not completely observable or simply unobservable. It should be noted that the two examples given in Figs. 9-2 and 9-3 are special cases which are devised to illustrate the concepts of controllability and observability, respectively. In general, the controllability and observability of a linear system usually cannot be detected by mere inspection, and some mathematical test criteria must be applied.

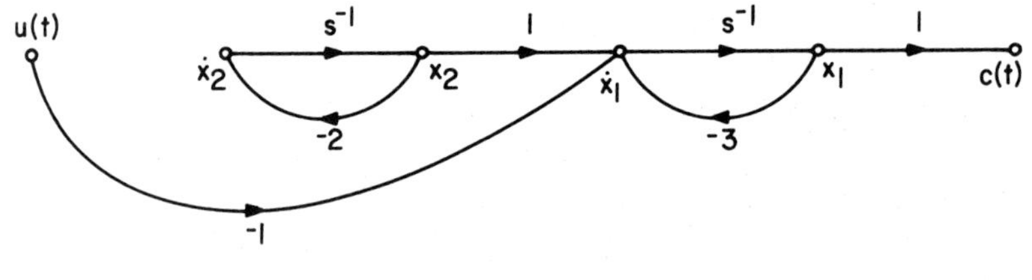

Fig. 9-2. State diagram of a process which is not completely controllable.

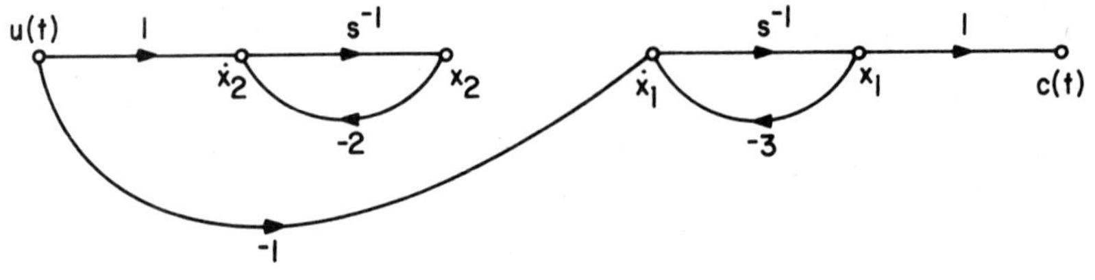

Fig. 9-3. State diagram of a process which is not completely observable.

9.2 DEFINITIONS OF CONTROLLABILITY

The concept of controllability of linear digital systems is similar to that of continuous-data systems except that the state equations are linear first-order difference equations.

Consider that a digital system is described by the state equation

$$\mathbf{x}(k_{i+1}) = \mathbf{A}(k_i)\mathbf{x}(k_i) + \mathbf{B}(k_i)\mathbf{u}(k_i) \tag{9-1}$$

$$\mathbf{c}(k_i) = \mathbf{D}(k_i)\mathbf{x}(k_i) + \mathbf{E}(k_i)\mathbf{u}(k_i) \tag{9-2}$$

where $\mathbf{x}(k_i)$ is an n-vector, $\mathbf{u}(k_i)$ is an r-vector, $\mathbf{c}(k_i)$ is a p-vector; $\mathbf{A}(k_i)$, $\mathbf{B}(k_i)$, $\mathbf{D}(k_i)$, and $\mathbf{E}(k_i)$ are coefficient matrices with appropriate dimensions.

Definition 9-1. Complete State Controllability

The system described by Eq. (9-1) is said to be completely state controllable if for any initial time (stage) k_0 there exists a set of unconstrained controls $\mathbf{u}(k_i)$, $i = 0,1,2,\ldots,N-1$, which transfers each initial state $\mathbf{x}(k_0)$ to any final state $\mathbf{x}(k_N)$ for some finite $k_N \geqslant k_0$.

Definition 9-2. Complete Output Controllability

The system described by Eqs. (9-1) and (9-2) is said to be completely output controllable if for any initial time (stage) $k = k_0$ there exists a set of unconstrained controls $\mathbf{u}(k_i)$, $i = 0,1,2,\ldots,N-1$, such that any final output $\mathbf{c}(k_N)$ can be reached from arbitrary initial states in the system in finite time (stage) $k_N \geqslant k_0$.

Definition 9-3. Total Controllability

A system is said to be *totally* (state or output) *controllable* if it is completely (state or output) controllable for every k_0 and every $k_N \geqslant k_0$.

Note that *complete* controllability refers to the requirement that *every* initial condition be controllable. *Total* controllability is a stronger condition in that it requires that, in addition to complete controllability, every k_0 and k_N must be included. If $\mathbf{A}(k)$, $\mathbf{B}(k)$, $\mathbf{D}(k)$, and $\mathbf{E}(k)$ contain analytical functions of k, then complete controllability also implies total (state or output) controllability. Since the coefficient matrices of a linear time-invariant system are constant, total and complete controllability are the same.

Definition 9-4. Strongly Versus Weakly Controllable

A system is said to be *strongly controllable* (in the sense of total, complete; state or output) if it is controllable by each control separately while all others are zero; otherwise it is *weakly controllable*, or simply controllable.

9.3 THEOREMS ON CONTROLLABILITY (TIME-VARYING SYSTEMS)

Theorem 9-1. Complete State Controllability

The linear digital system of Eq. (9-1) is completely state controllable if and only if the following matrix is of rank n:

$$\mathbf{Q} = [\mathbf{Q}_0 \quad \mathbf{Q}_1 \quad \cdots \quad \mathbf{Q}_{N-1}] \qquad (n \times Nr) \tag{9-3}$$

where

$$\mathbf{Q}_i = \psi(k_N, k_{i+1})\mathbf{B}(k_i) \qquad (n \times r) \tag{9-4}$$

$$\psi(k_N, k_{i+1}) = \mathbf{A}(k_{N-1})\mathbf{A}(k_{N-2}) \cdots \mathbf{A}(k_{i+1}) \qquad (n \times n)$$

$$i = 0, 1, 2, \ldots, N-2 \tag{9-5}$$

$$\psi(k_N, k_{i+1}) = \mathbf{I}$$

$$i = N-1 \tag{9-6}$$

Proof: The state transition equation of Eq. (9-1) is written

$$\mathbf{x}(k_N) = \psi(k_N, k_0)\mathbf{x}(k_0) + \sum_{i=0}^{N-1} \psi(k_N, k_{i+1})\mathbf{B}(k_i)\mathbf{u}(k_i) \tag{9-7}$$

or

$$\mathbf{x}(k_N) - \psi(k_N, k_0)\mathbf{x}(k_0) = \sum_{i=0}^{N-1} \psi(k_N, k_{i+1})\mathbf{B}(k_i)\mathbf{u}(k_i) \tag{9-8}$$

The left side of Eq. (9-8) can be represented by an $n \times 1$ vector $\mathbf{X}(k_N, k_0)$; then the equation is written as

$$\mathbf{X}(k_N, k_0) = \mathbf{QU} \tag{9-9}$$

where $\mathbf{Q}$ is an $n \times Nr$ matrix which is defined by Eqs. (9-3) and (9-4), and $\mathbf{U}$ is an $Nr \times 1$ vector, that is,

$$\mathbf{U} = \begin{bmatrix} \mathbf{u}(k_0) \\ \mathbf{u}(k_1) \\ \vdots \\ \mathbf{u}(k_{N-1}) \end{bmatrix} \qquad (Nr \times 1) \tag{9-10}$$

For complete state controllability, every initial state $\mathbf{x}(k_0)$ for any k_0

must be transferred by unconstrained controls $\mathbf{u}(k_i)$, $i = 0,1,\ldots,N-1$, to any final state $\mathbf{x}(k_N)$ for some finite $k_N \geqslant k_0$. Therefore, the problem is that of given $\mathbf{Q}$ and every vector $\mathbf{X}(k_N,k_0)$ in the n-dimensional state space, solve for the controls from Eq. (9-9). Since Eq. (9-9) represents n simultaneous linear equations, from the theory of linear equations these equations must be linearly independent in order for solutions to exist, and the necessary and sufficient condition for this to be true is that the matrix $\mathbf{Q}$ be of n. In other words, $\mathbf{Q}$ must have n linearly independent columns.

If the system in Eq. (9-1) has only one input, $r = 1$, $\mathbf{S}$ is an $n \times n$ square matrix; then the condition of complete state controllability is that $\mathbf{Q}$ must be nonsingular.

An alternate way of stating that the matrix $\mathbf{Q}$ must be of rank n is that the $n \times n$ matrix $\mathbf{QQ}'$

$$\mathbf{QQ}' = \sum_{i=0}^{N-1} \mathbf{Q}_i\mathbf{Q}_i' \qquad n \times n \tag{9-11}$$

is nonsingular. Since $\mathbf{Q}$ is of rank n, its transpose, $\mathbf{Q}'$, has the same rank. Thus, $\mathbf{QQ}'$ must also have a rank of n or is nonsingular.

The matrix $\mathbf{QQ}'$ in Eq. (9-11) is also known as the Gramian matrix. Thus, an alternate way of stating the theorem is that the system of Eq. (9-1) is completely state controllable if and only if the Gramian matrix

$$\mathbf{W}(k_0,k_N) = \mathbf{QQ}' = \sum_{i=0}^{N-1} \psi(k_N,k_{i+1})\mathbf{B}(k_i)\mathbf{B}'(k_i)\psi'(k_N,k_{i+1}) \tag{9-12}$$

is nonsingular.

Theorem 9-2. Complete Output Controllability

The linear digital system of Eqs. (9-1) and (9-2) is completely output controllable if and only if the following matrix is of rank p:

$$\mathbf{T} = [\mathbf{T}_0 \quad \mathbf{T}_1 \quad \cdots \quad \mathbf{T}_N] \qquad [p \times (N+1)r] \tag{9-13}$$

where

$$\begin{aligned} \mathbf{T}_i &= \mathbf{D}(k_N)\psi(k_N,k_{i+1})\mathbf{B}(k_i) \qquad & i = 0,1,\ldots,N-1 \\ &= \mathbf{E}(k_N) \qquad & i = N \end{aligned} \tag{9-14}$$

Proof: Substituting the state transition equation of Eq. (9-7) into the output equation, Eq. (9-2), with $k_i = k_N$, we have, after rearranging,

$$\mathbf{c}(k_N) - \mathbf{D}(k_N)\psi(k_N,k_0)\mathbf{x}(k_0) = \sum_{i=0}^{N-1} \mathbf{D}(k_N)\psi(k_N,k_{i+1})\mathbf{B}(k_i)\mathbf{u}(k_i)$$

$$+ \mathbf{E}(k_N)\mathbf{u}(k_N) \tag{9-15}$$

Using Eq. (9-14), and representing the left side of Eq. (9-15) by a $p \times 1$ vector $\mathbf{C}(k_N, k_0)$, Eq. (9-15) is written

$$\mathbf{C}(k_N, k_0) = \sum_{i=0}^{N} \mathbf{T}_i \mathbf{u}(k_i)$$

$$= \mathbf{TV} \tag{9-16}$$

where

$$\mathbf{V} = \begin{bmatrix} \mathbf{u}(k_0) \\ \mathbf{u}(k_1) \\ \cdot \\ \cdot \\ \cdot \\ \mathbf{u}(k_N) \end{bmatrix} \qquad (N+1)r \times 1 \tag{9-17}$$

Equation (9-16) represents p linear simultaneous equations, and for arbitrary initial state $\mathbf{x}(k_0)$ and any final state $\mathbf{c}(k_N)$, and given $\mathbf{T}$, the solution of these equations for the controls requires that the rank of $\mathbf{T}$ be p. In other words, the $p \times (N+1)r$ matrix $\mathbf{T}$ must have p independent columns in order for the system to be completely output controllable.

Alternately, for complete output controllability, the $p \times p$ Gramian matrix

$$\mathbf{TT}' = \sum_{i=0}^{N} \mathbf{T}_i \mathbf{T}_i' \tag{9-18}$$

must be nonsingular.

9.4 THEOREMS ON CONTROLLABILITY (TIME-INVARIANT SYSTEMS)

The following theorems are given on state and output controllability of linear digital time-invariant systems. Some of these theorems are stated without proofs.

Theorem 9-3. Complete State Controllability

The linear time-invariant digital system

$$\mathbf{x}(k_{i+1}) = \mathbf{Ax}(k_i) + \mathbf{Bu}(k_i) \tag{9-19}$$

is completely state controllable if and only if the $n \times Nr$ matrix

$$\mathbf{Q} = [\mathbf{B} \quad \mathbf{AB} \quad \mathbf{A}^2\mathbf{B} \quad \ldots \quad \mathbf{A}^{N-1}\mathbf{B}] \tag{9-20}$$

is of rank n, or the $n \times n$ matrix $\mathbf{QQ}'$ is nonsingular.

Proof: The proof of this theorem follows directly from the condition of state controllability for time-varying systems. From Eq. (9-4), when $\mathbf{A}$ is a constant matrix,

$$\mathbf{Q}_i = \mathbf{A}^i\mathbf{B} \tag{9-21}$$

Note that the condition of state controllability of a linear time-invariant digital system is identical to that of a time-invariant continuous-data system which has $\mathbf{A}$ and $\mathbf{B}$ as its coefficient matrices. However, if the digital system were formed by sampling in a continuous-data system, the discrete state equation would be of the form:

$$\mathbf{x}[(k+1)T] = \phi(T)\mathbf{x}(kT) + \theta(T)\mathbf{u}(kT) \tag{9-22}$$

Then, the digital system would be completely state controllable if and only if the matrix

$$\mathbf{Q} = [\mathbf{Q}_0 \quad \mathbf{Q}_1 \quad \cdots \quad \mathbf{Q}_{N-1}] \tag{9-23}$$

were of rank n, where

$$\mathbf{Q}_i = \phi(iT)\theta(T) \qquad i = 0,1,\ldots,N-1 \tag{9-24}$$

Theorem 9-4. Complete State Controllability

The linear time-invariant digital system of Eq. (9-19) is completely state controllable if and only if the $n \times n$ Gramian matrix

$$\mathbf{W} = \sum_{i=0}^{N-1} \mathbf{A}^{N-i-1}\mathbf{BB}'(\mathbf{A}^{N-i-1})' \tag{9-25}$$

is nonsingular, where $N \geqslant n$.

This theorem follows directly from the time-varying case in Theorem 9-1.

Theorem 9-5. Complete State Controllability

The linear time-invariant digital system of Eq. (9-19) is completely state controllable if and only if the rows of the $n \times m$ matrix

$$(z\mathbf{I} - \mathbf{A})^{-1}\mathbf{B}$$

are linearly independent.

Theorem 9-6. Complete State Controllability

The linear time-invariant digital system of Eq. (9-19) is completely state controllable if and only if the $n \times (n + m)$ matrix

$$[\lambda\mathbf{I} - \mathbf{A} \vdots \mathbf{B}]$$

has rank n for $\lambda =$ *every* eigenvalue of **A**. If **A** is diagonal or in the Jordan canonical form, then the rows of $[\lambda\mathbf{I} - \mathbf{A} \vdots \mathbf{B}]$ that correspond to the last rows of the Jordan blocks for $\lambda =$ every eigenvalue of **A** cannot contain all zeros.

Theorem 9-7. Complete Output Controllability

The linear time-invariant digital system

$$\mathbf{x}(k_{i+1}) = \mathbf{A}\mathbf{x}(k_i) + \mathbf{B}\mathbf{u}(k_i) \tag{9-26}$$

$$\mathbf{c}(k_i) = \mathbf{D}\mathbf{x}(k_i) + \mathbf{E}\mathbf{u}(k_i) \tag{9-27}$$

is completely output controllable if and only if the following matrix is of rank p:

$$\mathbf{T} = [\mathbf{D}\mathbf{A}^{N-1}\mathbf{B} \quad \mathbf{D}\mathbf{A}^{N-2}\mathbf{B} \quad \ldots \quad \mathbf{D}\mathbf{A}\mathbf{B} \quad \mathbf{D}\mathbf{B} \quad \mathbf{E}] \tag{9-28}$$

Proof: The proof of this theorem follows directly from that of the time-varying case, since in Eq. (9-13),

$$\begin{aligned} \mathbf{T}_i &= \mathbf{D}\mathbf{A}^{N-i-1}\mathbf{B} \qquad & i &= 0,1,\ldots,N-1 \\ &= \mathbf{E} & i &= N \end{aligned} \tag{9-29}$$

Theorem 9-8. Complete State Controllability of Systems with Distinct Eigenvalues

For the time-invariant digital system

$$\mathbf{x}(k+1) = \mathbf{A}\mathbf{x}(k) + \mathbf{B}\mathbf{u}(k) \tag{9-30}$$

consider that **A** has distinct eigenvalues in λ_i, $i = 1,2,\ldots,n$. Let **P** be the nonsingular transformation which transforms $\mathbf{x}(k)$ into

$$\mathbf{y}(k) = \mathbf{P}^{-1}\mathbf{x}(k) \tag{9-31}$$

and such that

$$\Lambda = \mathbf{P}^{-1}\mathbf{A}\mathbf{P} = \text{diag}[\lambda_i] \tag{9-32}$$

$$\Gamma = \mathbf{P}^{-1}\mathbf{B} \tag{9-33}$$

The new state equations are expressed as

$$\mathbf{y}(k+1) = \Lambda\mathbf{y}(k) + \Gamma\mathbf{u}(k) \tag{9-34}$$

The system is completely (totally) state controllable if and only if Γ does not have any *rows* containing all zeros.

Proof: Since Λ is a diagonal matrix, the system of state equations in Eq. (9-34) is decoupled. Therefore, if any one row of Γ contains all zeros, the corresponding state variable will not be affected by any one of the inputs and that state is uncontrollable.

Theorem 9-9. Complete State Controllability of Systems with Multiple-Order Eigenvalues

Consider that the digital system of Eq. (9-30) has distinct as well as multiple-order eigenvalues. Then, there is a nonsingular transformation **P** which transforms Eq. (9-30) into the state equations of Eq. (9-34), with Λ being a Jordan canonical form. In general, Λ will be of the following form:

$$\Lambda = \begin{bmatrix} \lambda_1 & 0 & 0 & 0 & 0 \\ 0 & \lambda_2 & 0 & 0 & 0 \\ 0 & 0 & \lambda_3 & 0 & 0 \\ 0 & 0 & 0 & \lambda_4 & 1 \\ 0 & 0 & 0 & 0 & \lambda_4 \end{bmatrix} \tag{9-35}$$

where λ_1, λ_2, and λ_3 are distinct eigenvalues, and λ_4 denotes an eigenvalue of multiplicity two. The elements in each block bounded by dotted lines form the Jordan blocks.

The system is completely state controllable if and only if

(1) each Jordan block corresponds to one distinct eigenvalue, and
(2) all elements of Γ that correspond to the *last row* of each Jordan block are nonzero.

Proof: Since the last row of each Jordan block corresponds to the state equation that is completely uncoupled from the other state equations, if the elements of the rows of Γ which correspond to these uncoupled states were all zero, these states would not be controllable by any of the controls. The elements in the rows of Γ which correspond to the other rows of the Jordan blocks could indeed contain all zeros, as the ones in the off-diagonal positions of Λ form coupling between the states.

Thus, for the Λ given in Eq. (9-35), for complete state controllability, the first three rows and the last row of Γ cannot be all zeros.

9.5 DEFINITIONS OF OBSERVABILITY

Definition 9-5. Complete Observability

The digital system described in Eqs. (9-1) and (9-2) is said to be *completely observable* if for any k_0 any state $\mathbf{x}(k_0)$ can be determined from knowledge of the output $\mathbf{c}(k)$ and input $\mathbf{u}(k)$ for $k_0 \leq k < k_N$, where k_N is some finite time or stage.

Definition 9-6. Total Observability

If a system is completely observable for every k_0 and every $k_f > k_0$, it is said to be *totally observable.*

9.6 THEOREM ON OBSERVABILITY (TIME-VARYING SYSTEMS)

Theorem 9-10. Complete Observability

The linear digital system of Eqs. (9-1) and (9-2) is completely observable if and only if the following $n \times pN$ matrix is of rank n:

$$\mathbf{L}(k_0,k_{N-1}) = [\mathbf{D}'(k_0) \quad \psi'(k_1,k_0)\mathbf{D}'(k_1) \quad \cdots \quad \psi'(k_{N-1},k_0)\mathbf{D}'(k_{N-1})] \tag{9-36}$$

or the $n \times n$ matrix

$$\mathbf{L}(k_0,k_{N-1})\mathbf{L}'(k_0,k_{N-1}) = \sum_{i=0}^{N-1} \psi'(k_i,k_0)\mathbf{D}'(k_i)\mathbf{D}(k_i)\psi(k_i,k_0) \tag{9-37}$$

should be nonsingular.

Proof: The state transition equation of Eq. (9-1) is written

$$\mathbf{x}(k_i) = \psi(k_i,k_0)\mathbf{x}(k_0) + \sum_{j=0}^{i-1} \psi(k_i,k_{j+1})\mathbf{B}(k_j)\mathbf{u}(k_j) \tag{9-38}$$

where $i = 1,2,\ldots,N-1$.

Substitution of Eq. (9-38) into the output equation of Eq. (9-2) gives

$$\mathbf{c}(k_i) = \mathbf{D}(k_i)\psi(k_i,k_0)\mathbf{x}(k_0) + \sum_{j=0}^{i-1} \mathbf{D}(k_i)\psi(k_i,k_{j+1})\mathbf{B}(k_j)\mathbf{u}(k_j) + \mathbf{E}(k_i)\mathbf{u}(k_i) \tag{9-39}$$

for $i = 1,2,\ldots,N-1$.

When i assumes values from 0 through $N-1$, Eq. (9-39) together with Eq. (9-2) represents pN linear algebraic equations that can be put in a matrix form as follows:

$$\begin{bmatrix} \mathbf{c}(k_0) \\ \mathbf{c}(k_1) \\ \vdots \\ \mathbf{c}(k_{N-1}) \end{bmatrix} = \begin{bmatrix} \mathbf{D}(k_0) \\ \mathbf{D}(k_1)\psi(k_1,k_0) \\ \vdots \\ \mathbf{D}(k_{N-1})\psi(k_{N-1},k_0) \end{bmatrix} \mathbf{x}(k_0)$$

$$+ \begin{bmatrix} \mathbf{E}(k_0) & \mathbf{0} & \mathbf{0} & \cdots & \mathbf{0} \\ \mathbf{D}(k_1)\mathbf{B}(k_0) & \mathbf{E}(k_1) & \mathbf{0} & & \mathbf{0} \\ \mathbf{D}(k_2)\psi(k_2,k_1)\mathbf{B}(k_0) & \mathbf{D}(k_2)\mathbf{B}(k_1) & \mathbf{E}(k_2) & & \mathbf{0} \\ \vdots & \vdots & \vdots & & \vdots \\ \mathbf{D}(k_{N-1})\psi(k_{N-1},k_1)\mathbf{B}(k_0) & \mathbf{D}(k_{N-1})\psi(k_{N-1},k_2)\mathbf{B}(k_1) & \cdots & \cdots & \mathbf{E}(k_{N-1}) \end{bmatrix}$$

$$\times \begin{bmatrix} \mathbf{u}(k_0) \\ \mathbf{u}(k_1) \\ \mathbf{u}(k_2) \\ \vdots \\ \mathbf{u}(k_{N-1}) \end{bmatrix} \qquad (9\text{-}40)$$

In order to determine $\mathbf{x}(k_0)$ from the last equation, given $\mathbf{c}(k_i)$ and $\mathbf{u}(k_i)$, $i = 0,1,2,\ldots,N-1$, it is apparent that the $pN \times n$ matrix

$$\mathbf{L}'(k_0,k_{N-1}) = \begin{bmatrix} \mathbf{D}(k_0) \\ \mathbf{D}(k_1)\psi(k_1,k_0) \\ \vdots \\ \mathbf{D}(k_{N-1})\psi(k_{N-1},k_0) \end{bmatrix} \qquad (9\text{-}41)$$

must have n independent rows (assuming that $pN \geqslant n$), or the matrix $\mathbf{L}(k_0,k_{N-1})$ given by Eq. (9-36) must have n independent columns or be of rank n.

The condition in Eq. (9-37) follows naturally since if $\mathbf{L}(k_0,k_{N-1})$ is of rank n, then the square $n \times n$ matrix $\mathbf{L}(k_0,k_{N-1})\mathbf{L}'(k_0,k_{N-1})$ must also be of rank n.

9.7 THEOREMS ON OBSERVABILITY (TIME-INVARIANT SYSTEMS)

Theorem 9-11. Complete Observability

The linear time-invariant digital system

$$\mathbf{x}(k_{i+1}) = \mathbf{A}\mathbf{x}(k_i) + \mathbf{B}\mathbf{u}(k_i) \tag{9-42}$$

$$\mathbf{c}(k_i) = \mathbf{D}\mathbf{x}(k_i) + \mathbf{E}\mathbf{u}(k_i) \tag{9-43}$$

is completely (totally) observable if and only if the following matrix is of rank n:

$$\mathbf{L} = [\mathbf{D}' \quad \mathbf{A}'\mathbf{D}' \quad (\mathbf{A}')^2\mathbf{D}' \quad \ldots \quad (\mathbf{A}')^{N-1}\mathbf{D}'] \qquad (n \times Np) \tag{9-44}$$

where n is the dimension of $\mathbf{x}(k_i)$, and p is the dimension of $\mathbf{c}(k_i)$.

Proof: The proof of this theorem follows directly from that of the time-varying case. In general, we need only information on $\mathbf{c}(k_i)$ and $\mathbf{u}(k_i)$ for $i = 0,1,\ldots,N-1$, or any N intervals. For the time-invariant system, if the system is observable for any interval $[k_0, k_N]$, then it is also observable on any other interval. Therefore, for a time-invariant system complete observability also implies total observability.

For the sampled-data system modeled by the state equation of Eq. (9-22), the condition of complete observability is that the matrix

$$\mathbf{L} = [\mathbf{D}' \quad \phi(T)'\mathbf{D}' \quad \phi(2T)'\mathbf{D}' \quad \ldots \quad \phi[(N-1)T]'\mathbf{D}'] \qquad (n \times Np) \tag{9-45}$$

must be of rank n.

Theorem 9-12. Complete Observability

The linear time-invariant digital system described by Eqs. (9-42) and (9-43) is completely observable if and only if the Gramian matrix

$$\mathbf{V} = \sum_{i=0}^{N-1} (\mathbf{A}^{N-i-1})'\mathbf{D}'\mathbf{D}\mathbf{A}^{N-i-1} \tag{9-46}$$

is nonsingular.

Theorem 9-13. Complete Observability

The linear time-invariant digital system described by Eqs. (9-42) and (9-43) is completely observable if and only if the columns of the $p \times n$ matrix

$$\mathbf{D}(z\mathbf{I} - \mathbf{A})^{-1}$$

are linearly independent. If $\mathbf{A}$ is diagonal or is in the Jordan canonical form, then the columns of $\mathbf{D}(z\mathbf{I} - \mathbf{A})^{-1}$ that correspond to the first column of each Jordan block cannot contain all zeros.

Theorem 9-14. Complete Observability

The linear time-invariant digital system described by Eqs. (9-42) and (9-43) is completely observable if and only if the $n \times (n + p)$ matrix

$$[(\lambda \mathbf{I} - \mathbf{A})' \vdots \mathbf{D}']$$

has rank n for λ = every eigenvalue of **A**. If **A** is diagonal or in the Jordan canonical form, the columns of $[(\lambda \mathbf{I} - \mathbf{A})' \vdots \mathbf{D}']$ that correspond to the first columns of the Jordan blocks for λ = every eigenvalue of **A** cannot contain all zeros.

Theorem 9-15. Complete Observability of Systems with Distinct Eigenvalues

For a digital time-invariant system,

$$\mathbf{x}(k + 1) = \mathbf{A}\mathbf{x}(k) + \mathbf{B}\mathbf{u}(k) \tag{9-47}$$

$$\mathbf{c}(k) = \mathbf{D}\mathbf{x}(k) + \mathbf{E}\mathbf{u}(k) \tag{9-48}$$

if **A** has distinct eigenvalues, the dynamic equations are transformed by the similarity transformation $\mathbf{x}(k) = \mathbf{P}\mathbf{y}(k)$ to

$$\mathbf{y}(k + 1) = \Lambda\mathbf{y}(k) + \Gamma\mathbf{u}(k) \tag{9-49}$$

$$\mathbf{c}(k) = \mathbf{F}\mathbf{y}(k) + \mathbf{E}\mathbf{u}(k) \tag{9-50}$$

where Λ and Γ are given in Eq. (9-32) and Eq. (9-33), respectively, and

$$\mathbf{F} = \mathbf{D}\mathbf{P} \tag{9-51}$$

The system is completely observable if and only if the matrix **F** does not have any *columns* containing all zeros.

Proof: The reasoning behind this theorem is that, since Λ is a diagonal matrix, the system of state equations in Eq. (9-49) is decoupled. Therefore, if any one column of **F** contains all zeros, the corresponding state variable will not be observed by any one of the outputs.

Theorem 9-16. Complete State Observability of Systems with Multiple Eigenvalues

For the system described by Eqs. (9-47) and (9-48), if **A** has multiple eigenvalues, then the similarity transformation $\mathbf{x}(k) = \mathbf{P}\mathbf{y}(k)$ transforms **A** into a Jordan canonical form Λ, Eq. (9-35). The conditions of complete observability are:

(1) each Jordan block corresponds to one distinct eigenvalue, and
(2) all elements of columns of **F** that correspond to the *first column* of each Jordan block are nonzero.

The proof of this theorem is similar to that of the theorem on controllability of systems with multiple eigenvalues.

9.8 DUAL RELATION BETWEEN OBSERVABILITY AND CONTROLLABILITY

Just as in the case of continuous-data systems, a dual relation exists between the conditions of state controllability and observability of linear digital systems.

Consider that the dynamic equations of a linear digital system (System 1) are given as

System 1:
$$\mathbf{x}(k_{i+1}) = \mathbf{A}(k_i)\mathbf{x}(k_i) + \mathbf{B}(k_i)\mathbf{u}(k_i) \tag{9-52}$$
$$\mathbf{c}(k_i) = \mathbf{D}(k_i)\mathbf{x}(k_i) + \mathbf{E}(k_i)\mathbf{u}(k_i) \tag{9-53}$$

The adjoint system (System 2) of System 1 is represented by

System 2:
$$\mathbf{y}(k_{i+1}) = \mathbf{A}^*(k_i)\mathbf{y}(k_i) + \mathbf{D}'(k_i)\mathbf{u}(k_i) \tag{9-54}$$
$$\mathbf{z}(k_i) = \mathbf{B}'(k_i)\mathbf{y}(k_i) + \mathbf{E}(k_i)\mathbf{u}(k_i) \tag{9-55}$$

where

$$\mathbf{A}^*(k_i) = [\mathbf{A}^{-1}(k_{i+1})]' \tag{9-56}$$

Notice that the adjoint system in this case is defined in a slightly different manner than in Sec. 4.11, so that the dual relation between controllability and observability will be exact. It is assumed that $\mathbf{A}(k_i)$ is nonsingular for all $k_i = k_0, k_1, \ldots, k_N$.

Let the state transition matrix of $\mathbf{A}(k_i)$ be represented by

$$\begin{aligned} \psi_1(k_{j+1}, k_1) &= \mathbf{A}(k_j)\mathbf{A}(k_{j-1}) \ldots \mathbf{A}(k_1) \qquad & k_{j+1} > k_1 \\ &= \mathbf{I} & k_{j+1} = k_1 \end{aligned} \tag{9-57}$$

and the state transition matrix of $\mathbf{A}^*(k_i)$ be written as

$$\begin{aligned} \psi_2(k_j, k_0) &= \mathbf{A}^*(k_{j-1})\mathbf{A}^*(k_{j-2}) \ldots \mathbf{A}(k_0) \qquad & k_j > k_0 \\ &= \mathbf{I} & k_j = k_0 \end{aligned} \tag{9-58}$$

Using Eq. (9-56), $\psi_2(k_j, k_0)$ becomes

$$\begin{aligned} \psi_2(k_j, k_0) &= [\mathbf{A}^{-1}(k_j)]'[\mathbf{A}^{-1}(k_{j-1})]' \ldots [\mathbf{A}^{-1}(k_1)]' \\ &= [\mathbf{A}^{-1}(k_1)\mathbf{A}^{-1}(k_2) \ldots \mathbf{A}^{-1}(k_j)]' \end{aligned} \tag{9-59}$$

The transpose of $\psi_2(k_j, k_0)$ is

$$\psi_2'(k_j, k_0) = [\mathbf{A}^{-1}(k_1)\mathbf{A}^{-1}(k_2) \ldots \mathbf{A}^{-1}(k_j)]$$

$$= [\mathbf{A}(k_j)\mathbf{A}(k_{j-1}) \ldots \mathbf{A}(k_1)]^{-1}$$

$$= \psi_1(k_1, k_{j+1}) \qquad k_{j+1} > k_0 \qquad (9\text{-}60)$$

Notice that Eq. (9-60) is similar to the relation between the state transition matrices of a continuous-data system and its adjoint system.

The dual relations between the state controllability and observability of digital systems, System 1 and System 2, are now tabulated as follows:

System 1

State Controllability

$$[\psi_1(k_N, k_1)\mathbf{B}(k_0) \quad \psi_1(k_N, k_2)\mathbf{B}(k_1) \quad \ldots \quad \mathbf{B}(k_{N-1})]$$

Observability

$$[\mathbf{D}'(k_0) \quad \psi_1'(k_1, k_0)\mathbf{D}'(k_1) \quad \ldots \quad \psi_1'(k_{N-1}, k_0)\mathbf{D}'(k_{N-1})]$$

System 2

Observability

$$[\mathbf{B}(k_0) \quad \psi_2'(k_1, k_0)\mathbf{B}(k_1) \quad \ldots \quad \psi_2'(k_{N-1}, k_0)\mathbf{B}(k_{N-1})]$$

State Controllability

$$[\psi_2(k_N, k_1)\mathbf{D}'(k_0) \quad \psi_2(k_N, k_2)\mathbf{D}'(k_1) \quad \ldots \quad \mathbf{D}'(k_{N-1})]$$

In view of Eq. (9-60), the matrix for observability for System 2 is written

$$\mathbf{L}_2(k_0, k_{N-1}) = [\mathbf{B}(k_0) \quad \psi_2'(k_1, k_0)\mathbf{B}(k_1) \quad \ldots \quad \psi_2'(k_{N-1}, k_0)\mathbf{B}(k_{N-1})]$$

$$= [\mathbf{B}(k_0) \quad \psi_1(k_1, k_2)\mathbf{B}(k_1) \quad \ldots \quad \psi_1(k_1, k_N)\mathbf{B}(k_{N-1})] \qquad (9\text{-}61)$$

Therefore, for nonsingular $\psi_1(k_N, k_1)$, the rank of $\mathbf{L}_2(k_0, k_{N-1})$ is the same as the rank of

$$\psi_1(k_N, k_1)\mathbf{L}_2(k_0, k_{N-1}) = [\psi_1(k_N, k_1)\mathbf{B}(k_0) \quad \psi_1(k_N, k_2)\mathbf{B}(k_1) \quad \ldots \quad \mathbf{B}(k_{N-1})] \qquad (9\text{-}62)$$

We have shown that the state controllability of System 1 implies the observability of its adjoint system (slightly modified), System 2. Using the same approach, we can easily show that the observability condition of System 1 is the same as the condition of state controllability of System 2.

It is useful to investigate why the adjoint system has to be slightly modified [Eq. (9-56)] in order to establish a dual relationship between state controllability and observability between the two systems. It should be noted that when deriving the condition of controllability, the state is transferred from $\mathbf{x}(k_0)$ to $\mathbf{x}(k_N)$, whereas in observing $\mathbf{x}(k_0)$, only the states from $\mathbf{x}(k_1)$ to $\mathbf{x}(k_{N-1})$ are used in Eq. (9-38). This is one way of explaining why the indices in the state transition matrices of the two systems are shifted.

For digital time-invariant systems, the adjoint relation can be defined in the usual manner; that is,

System 1:

$$\mathbf{x}(k_{i+1}) = \mathbf{A}\mathbf{x}(k_i) + \mathbf{B}\mathbf{u}(k_i) \tag{9-63}$$

$$\mathbf{c}(k_i) = \mathbf{D}\mathbf{x}(k_i) + \mathbf{E}\mathbf{u}(k_i) \tag{9-64}$$

System 2:

$$\mathbf{y}(k_{i+1}) = \mathbf{A}^*\mathbf{y}(k_i) + \mathbf{D}'\mathbf{u}(k_i) \tag{9-65}$$

$$\mathbf{z}(k_i) = \mathbf{B}'\mathbf{y}(k_i) + \mathbf{E}\mathbf{u}(k_i) \tag{9-66}$$

where

$$\mathbf{A}^* = (\mathbf{A}^{-1})' \tag{9-67}$$

The relation between the state transition matrices of the two systems is

$$\psi_1(k_i) = [\psi_2^{-1}(k_i)]^{-1} = \sum_{j=1}^{i} \mathbf{A}^j \tag{9-68}$$

The dual relation between the state controllability and observability of the two systems becomes straightforward.

9.9 RELATIONSHIP BETWEEN CONTROLLABILITY, OBSERVABILITY AND TRANSFER FUNCTIONS

Classical analysis and design of control systems relies greatly on the use of transfer functions for the modeling of linear time-invariant systems. One advantage of using transfer functions is that state controllability and observability are directly related to the minimum order of the transfer function. The following theorem gives the relation between controllability and observability and the pole-zero cancellation of a transfer function.

Theorem 9-17. Controllability, Observability and Transfer Functions

If the input-output transfer function of a linear time-invariant digital system has pole-zero cancellation, the system will be either not state controllable, unobservable, or both, depending on how the state variables are defined. If the input-output transfer function does not have pole-zero cancellation, the system can always be represented by discrete dynamic equations as a completely controllable and observable system.

Proof: Consider that an nth-order digital system with a single input and single output and distinct eigenvalues is represented by the dynamic equations

$$\mathbf{x}(k+1) = \mathbf{A}\mathbf{x}(k) + \mathbf{B}u(k) \tag{9-69}$$

$$c(k) = \mathbf{D}\mathbf{x}(k) \tag{9-70}$$

Let the **A** matrix be diagonalized by an $n \times n$ Vandermonde matrix **P** of Eq. (4-203). The new state equation is

$$\mathbf{y}(k+1) = \Lambda\mathbf{y}(k) + \Gamma u(k) \tag{9-71}$$

where $\Lambda = \mathbf{P}^{-1}\mathbf{A}\mathbf{P}$. The output equation is transformed into

$$c(k) = \mathbf{F}\mathbf{y}(k) \tag{9-72}$$

where $\mathbf{F} = \mathbf{D}\mathbf{P}$. The state vectors $\mathbf{x}(k)$ and $\mathbf{y}(k)$ are related by

$$\mathbf{x}(k) = \mathbf{P}\mathbf{y}(k) \tag{9-73}$$

Since Λ is a diagonal matrix, the ith ($i = 1,2,\ldots,n$) equation of Eq. (9-71) is

$$y_i(k+1) = \lambda_i y_i(k) + \gamma_i u(k) \tag{9-74}$$

where λ_i is the eigenvalue of **A** and γ_i is the ith element of Γ, where Γ is an $n \times 1$ matrix in the present case. Taking the z-transform on both sides of Eq. (9-74) and assuming zero initial conditions, the transfer function relation between $Y_i(z)$ and $U(z)$ is obtained as

$$Y_i(z) = \frac{\gamma_i}{z - \lambda_i} U(z) \tag{9-75}$$

The z-transform of Eq. (9-72) is

$$C(z) = \mathbf{F}\mathbf{Y}(z) = \mathbf{D}\mathbf{P}\mathbf{Y}(z) \tag{9-76}$$

Let the $1 \times n$ matrix **D** be represented by

$$\mathbf{D} = [d_1 \quad d_2 \quad \ldots \quad d_n] \tag{9-77}$$

Then

$$\mathbf{F} = \mathbf{D}\mathbf{P} = [f_1 \quad f_2 \quad \ldots \quad f_n] \tag{9-78}$$

where

$$f_i = d_1 + d_2\lambda_i + \ldots + d_n\lambda_i^{n-1} \tag{9-79}$$

for i = 1,2,...,n. Using Eqs. (9-75) and (9-78), Eq. (9-76) is written

$$C(z) = [f_1 \quad f_2 \quad \ldots \quad f_n] \begin{bmatrix} \dfrac{\gamma_1}{z - \lambda_1} \\ \dfrac{\gamma_2}{z - \lambda_2} \\ \vdots \\ \dfrac{\gamma_n}{z - \lambda_n} \end{bmatrix} U(z) \tag{9-80}$$

or

$$\frac{C(z)}{U(z)} = \sum_{i=1}^{n} \frac{f_i\gamma_i}{z - \lambda_i} \tag{9-81}$$

which is the transfer function of the system in the form of partial-fraction expansion. If the transfer function in Eq. (9-81) had a cancellation of pole and zero, then the corresponding coefficient on the right side of the equation would be zero. Assuming that the pole at $z = \lambda_j$ is cancelled, then

$$f_j\gamma_j = 0$$

which implies that $f_j = 0$, *or* $\gamma_j = 0$, *or* both. Since γ_j is the jth element of the matrix $\boldsymbol{\Gamma}$, $\gamma_j = 0$ would mean that the system is uncontrollable. On the other hand, if $f_j = 0$, where f_j is the jth element of **F**, the system is unobservable.

9.10 CONTROLLABILITY AND OBSERVABILITY VERSUS SAMPLING PERIOD

With a digital control system containing sampling operations, the condition of controllability and observability can be dependent on the sampling period T. For the sampled-data system modeled by the state equations in Eq. (9-22), the condition of state controllability is that the **Q** matrix of Eq. (9-23) is of rank n. This means that if

$$\mathbf{Q}_i = \phi(iT)\theta(T) = \mathbf{Q}_j = \phi(jT)\theta(T) \tag{9-82}$$

for i,j = 1,2,...,N − 1, $i \neq j$, and $T \neq 0$, the system will be uncontrollable. Similarly, Eq. (9-45) indicates that if

$$\phi(iT)'D' = \phi(jT)'D' \tag{9-83}$$

for $i,j = 1,2,\ldots,N-1$, $i \neq j$, and $T \neq 0$, the system will be unobservable. Therefore, the last two conditions lead to the conclusion that if for some sampling period T ($\neq 0$),

$$\phi(iT) = \phi(jT)$$

$i,j = 1,2,\ldots,N-1$, $i \neq j$, the system is neither controllable nor observable. The following example illustrates a digital system whose controllability and observability depend on the choice of the sampling period.

Example 9-1.

Consider that the transfer function of a linear process is described by

$$G(s) = \frac{\omega}{s^2 + \omega^2} \tag{9-84}$$

It is apparent that the continuous-data system is controllable and observable since G(s) has no pole-zero cancellation.

The transfer function G(s) is decomposed to the following state equations

$$\dot{x}_1(t) = x_2(t) \tag{9-85}$$

$$\dot{x}_2(t) = -\omega^2 x_1(t) + u(t) \tag{9-86}$$

The state transition matrix with t replaced by T is

$$\phi(T) = \begin{bmatrix} \cos\omega T & \frac{1}{\omega}\sin\omega T \\ -\omega\sin\omega T & \cos\omega T \end{bmatrix} \tag{9-87}$$

Since $\cos\omega T = \pm 1$ and $\sin\omega T = 0$ for $\omega T = n\pi$, $n = 0,\pm 1,\pm 2,\ldots$, the system is uncontrollable and unobservable for $\omega = n\pi/T$.

9.11 ILLUSTRATIVE EXAMPLES

In this section several illustrative examples are given to demonstrate the applications of the methods of checking controllability and observability discussed in the preceding sections.

Example 9-2.

One of the methods of testing for controllability of linear systems with distinct eigenvalues is to diagonalize the **A** matrix and then observe the properties of the rows of the matrix **Γ** in Eq. (9-33). However, certain **A** matrices with multiple eigenvalues can still be diagonalized, so that just because **Λ** is a diagonal matrix and **Γ** does not have all zero rows does not mean that the system is completely controllable. Consider the digital system that is described by the following equations:

$$x_1(k+1) = ax_1(k) + b_1 u \tag{9-88}$$

$$x_2(k+1) = ax_2(k) + b_2 u \tag{9-89}$$

$b_1 \neq 0$, $b_2 \neq 0$. This system is clearly uncontrollable although Λ is diagonal and $\Gamma = [b_1 \;\; b_2]'$ contains no zero rows. We simply have to form the matrix $[\Gamma \;\; \Lambda\Gamma]$ and show that it is singular. From the physical standpoint, the two states in Eqs. (9-88) and (9-89) are decoupled. However, since the two states have the same dynamics, it is impossible to control them independently through the same input.

We shall verify that the system is not state controllable by means of Theorems 9-3 through 9-6.

(Theorem 9-3.)

$$\mathbf{Q} = [\mathbf{B} \quad \mathbf{AB}] = \begin{bmatrix} b_1 & ab_1 \\ b_2 & ab_2 \end{bmatrix} \tag{9-90}$$

Since $\mathbf{Q}$ is singular, the pair $[\mathbf{A}, \mathbf{B}]$ is uncontrollable.

(Theorem 9-4.)

For $N = 2$,

$$\mathbf{W} = \mathbf{ABB'A'} + \mathbf{BB'}$$

$$= \begin{bmatrix} b_1^2 & b_1 b_2 \\ b_1 b_2 & b_2^2 \end{bmatrix} (1 + a^2) \tag{9-91}$$

Since $\mathbf{W}$ is singular, the pair $[\mathbf{A}, \mathbf{B}]$ is uncontrollable.

(Theorem 9-5.)

$$(z\mathbf{I} - \mathbf{A})^{-1}\mathbf{B} = \begin{bmatrix} z-a & 0 \\ 0 & z-a \end{bmatrix}^{-1} \begin{bmatrix} b_1 \\ b_2 \end{bmatrix}$$

$$= \begin{bmatrix} \dfrac{b_1}{z-a} \\ \dfrac{b_2}{z-a} \end{bmatrix} \tag{9-92}$$

Since the rows of $(z\mathbf{I} - \mathbf{A})^{-1}\mathbf{B}$ are dependent, the pair $[\mathbf{A}, \mathbf{B}]$ is uncontrollable.

(Theorem 9-6.)

$$[\lambda I - A \vdots B] = \begin{bmatrix} \lambda - a & 0 & \vdots & b_1 \\ 0 & \lambda - a & \vdots & b_2 \end{bmatrix} \tag{9-93}$$

Then,

$$[\lambda_1 I - A \vdots B] = \begin{bmatrix} 0 & 0 & \vdots & b_1 \\ 0 & 0 & \vdots & b_2 \end{bmatrix} \tag{9-94}$$

$$[\lambda_2 I - A \vdots B] = \begin{bmatrix} 0 & 0 & \vdots & b_1 \\ 0 & 0 & \vdots & b_2 \end{bmatrix} \tag{9-95}$$

Since $[\lambda_1 I - A \vdots B]$ and $[\lambda_2 I - A \vdots B]$ do not have a rank of 2, the pair $[A,B]$ is uncontrollable.

Example 9-3.

Consider a linear digital control system whose input-output relation is described by the difference equation

$$c(k+2) + 2c(k+1) + c(k) = u(k+1) + u(k) \tag{9-96}$$

This difference equation is decomposed into the following dynamic equations:

$$\mathbf{x}(k+1) = \mathbf{A}\mathbf{x}(k) + \mathbf{B}u(k) \tag{9-97}$$

$$c(k) = \mathbf{D}\mathbf{x}(k) \tag{9-98}$$

where

$$\mathbf{A} = \begin{bmatrix} 0 & 1 \\ -1 & -2 \end{bmatrix} \qquad \mathbf{B} = \begin{bmatrix} 0 \\ 1 \end{bmatrix}$$

$$\mathbf{D} = [1 \quad 1]$$

Since

$$\mathbf{Q} = [\mathbf{B} \quad \mathbf{AB}] = \begin{bmatrix} 0 & 1 \\ 1 & -2 \end{bmatrix} \tag{9-99}$$

is nonsingular, the system is completely state controllable for the state variables $x_1(k)$ and $x_2(k)$. The observability of the system is investigated by forming the matrix

$$\mathbf{L} = [\mathbf{D}' \quad \mathbf{A}'\mathbf{D}'] = \begin{bmatrix} 1 & -1 \\ 1 & -1 \end{bmatrix} \tag{9-100}$$

Since $\mathbf{L}$ is singular, the system is unobservable, i.e., not all the states $x_1(k)$ and $x_2(k)$ can be determined from the knowledge of the output c(0) over the finite interval (0,N).

The system modeled by Eqs. (9-97) and (9-98) is output controllable since the matrix

$$\mathbf{T} = [\mathbf{DAB} \quad \mathbf{DB}] = [-1 \quad 1] \tag{9-101}$$

is of rank 1.

The conclusion is that the choice of state variables which results in Eqs. (9-97) and (9-98) produces a system which is state controllable and output controllable, but unobservable.

Now let us redefine the state variables of the system so that the dynamic equations are as in Eqs. (9-97) and (9-98), but with

$$\mathbf{A} = \begin{bmatrix} 0 & 1 \\ -1 & -2 \end{bmatrix} \qquad \mathbf{B} = \begin{bmatrix} 1 \\ -1 \end{bmatrix}$$

$$\mathbf{D} = [1 \quad 0]$$

Carrying out the exercise similar to Eqs. (9-99), (9-100) and (9-101), we find that

$$\mathbf{Q} = [\mathbf{B} \quad \mathbf{AB}] = \begin{bmatrix} 1 & -1 \\ -1 & 1 \end{bmatrix} \qquad \text{not state controllable} \tag{9-102}$$

$$\mathbf{L} = [\mathbf{D}' \quad \mathbf{A}'\mathbf{D}'] = \begin{bmatrix} 1 & 0 \\ 0 & 1 \end{bmatrix} \qquad \text{observable} \tag{9-103}$$

$$\mathbf{T} = [\mathbf{DAB} \quad \mathbf{DB}] = [1 \quad 1] \qquad \text{output controllable} \tag{9-104}$$

The variations in the conditions of controllability and observability depending on the state variable assignments are due to the fact that the system in Eq. (9-96) has a pole-zero cancellation in its transfer function. The output controllability is not affected by the choice of state variables.

Example 9-4.

In this illustrative example we shall demonstrate certain practical considerations in the application of the criteria on controllability and observability. Most practical control systems found in the process industry and the aerospace industry are of high order, so that the application of the criteria on controllability and observability would not be as straightforward as that demonstrated in the previous two examples.

The digital model of the dynamics of a space vehicle is represented by the vector-matrix state equation

$$\mathbf{x}(k+1) = \mathbf{Ax}(k) + \mathbf{Bu}(k) \tag{9-105}$$

where **A** is an 11×11 matrix, and **B** is an 11×1 matrix. For convenience we express **A** as a partitioned matrix as follows:

$$\mathbf{A} = \begin{bmatrix} \mathbf{A}_{11} & \mathbf{A}_{12} \\ \mathbf{A}_{21} & \mathbf{A}_{22} \end{bmatrix} \tag{9-106}$$

where $A_{11} = 0$ (5 X 5),

$$A_{12} = \begin{bmatrix} 1 & 0 & 0 & 0 & 0 & 0 \\ 0 & 1 & 0 & 0 & 0 & 0 \\ 0 & 0 & 0 & 1 & 0 & 0 \\ 0 & 0 & 0 & 0 & 1 & 0 \\ 0 & 0 & 0 & 0 & 0 & 1 \end{bmatrix} \tag{9-107}$$

$$A_{21} = \begin{bmatrix} 0.176 & 0 & 0 & -85.26 & -2.56 \\ 0 & -134.6 & -1.06 & 0 & 0 \\ 0 & -0.69 & 10.17 & 0 & 0 \\ 0 & 19.9 & -293.0 & 0 & 0 \\ -0.366 & 0 & 0 & -52.86 & -0.168 \\ 6.02 & 0 & 0 & -85.4 & -18.84 \end{bmatrix} \tag{9-108}$$

$$A_{22} = \begin{bmatrix} 0 & -6.1 & -0.188 & 0 & -0.113 & -0.077 \\ -24.0 & 0 & 0 & 0.02 & 0 & 0 \\ 0.69 & 0 & 0 & 0.304 & 0 & 0 \\ -19.9 & 0 & 0 & -8.76 & 2.0 & 0 \\ 0 & -3.66 & 2.36 & -11.67 & -7.0 & -0.005 \\ 0 & -6.1 & -0.188 & 0 & -0.113 & -0.388 \end{bmatrix} \tag{9-109}$$

$$B = [0 \quad 0 \quad 0 \quad 0 \quad 0 \quad -7.28 \quad 0 \quad 0 \quad 0 \quad -0.478 \quad -7.28] \tag{9-110}$$

Since the system has only one input, we can investigate the controllability of the system by evaluating the determinant of the 11 X 11 matrix

$$Q = [B \quad AB \quad A^2B \quad \ldots \quad A^{10}B]' \tag{9-111}$$

and the result is $|Q| = 4.46 \times 10^{18}$. Therefore, the system is completely state controllable. However, during the process of designing a state-feedback control to place the eigenvalues of the closed-loop system at appropriate locations, numerical difficulties are encountered. We would suspect that certain peculiarities of the system may not have been uncovered by the controllability matrix in Eq. (9-111). As an alternate way of testing controllability, we transform A into a diagonal matrix by a similarity transformation. The eigenvalues of A are distinct and are tabulated as follows:

$$-0.603 \pm j30.15$$
$$-0.0328 \pm j1.76$$
$$-0.0276 \pm j1.36$$

$$\pm j1.0$$
$$-0.00056 \pm j0.29$$
$$0$$

The Γ matrix of the decoupled system is

$$\Gamma = \begin{bmatrix} -8.16 \times 10^{-2} \\ 1.67 \times 10^{-3} \\ -9.32 \times 10^{-1} \\ 2.99 \times 10^{-2} \\ -1.32 \times 10^{-3} \\ 9.00 \times 10^{-2} \\ -1.76 \times 10^{-1} \\ 1.54 \times 10^{-10} \\ 1.97 \times 10^{-3} \\ -3.19 \\ -2.79 \times 10^{-16} \end{bmatrix} \tag{9-112}$$

Notice that, theoretically, since the matrix **Q** in Eq. (9-111) has a rank of eleven, and none of the elements of Γ are zero, the system should be completely state controllable. However, Eq. (9-112) shows that the last element of Γ is nearly zero, and we may consider the system to be "nearly uncontrollable." From the practical standpoint, since the elements of Γ do correspond to the coupling of the decoupled states to the input, the magnitudes of these elements do give indication of the degree of controllability of the states. In the present case, the last element of Γ corresponds to the eigenvalue of **A** at $z = 0$, and as it turns out it is practically impossible to move this eigenvalue by state feedback.

The purpose of this example is to illustrate that, while the alternate methods of controllability and observability testing may seem to be equivalent, in solving practical problems the numerical properties of a system may render one method more informative than the others. In the present case, while the check on the rank of the **Q** matrix indicates that the system is completely state controllable, the similarity transformation method reveals the difficulty of controlling certain states.

9.12 INVARIANT THEOREM ON CONTROLLABILITY

In Chapter 4 similarity transformation and the phase-variable canonical transformation were introduced to facilitate the analysis and design of digital control systems for certain purposes. We shall now investigate the effects of these nonsingular transformations on the properties of controllability and observability. The effects on controllability and observability due to state and output feedbacks will also be studied.

Theorem 9-18. Invariant Theorem on Nonsingular Transformation, Controllability

Given the n-th order digital control system

$$\mathbf{x}(k+1) = \mathbf{A}\mathbf{x}(k) + \mathbf{B}\mathbf{u}(k) \tag{9-113}$$

where the pair $[\mathbf{A},\mathbf{B}]$ is completely controllable. The transformation $\mathbf{x}(k) = \mathbf{P}\mathbf{y}(k)$, where $\mathbf{P}$ is nonsingular, transforms the system of Eq. (9-113) to

$$\mathbf{y}(k+1) = \Lambda\mathbf{y}(k) + \Gamma\mathbf{u}(k) \tag{9-114}$$

where

$$\Lambda = \mathbf{P}^{-1}\mathbf{A}\mathbf{P}$$

$$\Gamma = \mathbf{P}^{-1}\mathbf{B}$$

Then, the pair $[\Lambda,\Gamma]$ is also controllable.

Proof: The controllability matrix of the system of Eq. (9-114) is

$$\begin{aligned}
\mathbf{S}_1 &= [\Gamma \quad \Lambda\Gamma \quad \Lambda^2\Gamma \quad \ldots \quad \Lambda^{n-1}\Gamma] \\
&= [\mathbf{P}^{-1}\mathbf{B} \quad \mathbf{P}^{-1}\mathbf{A}\mathbf{P}\mathbf{P}^{-1}\mathbf{B} \quad (\mathbf{P}^{-1}\mathbf{A}\mathbf{P})(\mathbf{P}^{-1}\mathbf{A}\mathbf{P})\mathbf{P}^{-1}\mathbf{B} \quad \ldots \quad (\mathbf{P}^{-1}\mathbf{A}\mathbf{P})^{n-1}\mathbf{P}^{-1}\mathbf{B}] \\
&= \mathbf{P}^{-1}[\mathbf{B} \quad \mathbf{A}\mathbf{B} \quad \mathbf{A}^2\mathbf{B} \quad \ldots \quad \mathbf{A}^{n-1}\mathbf{B}] \\
&= \mathbf{P}^{-1}\mathbf{S}
\end{aligned} \tag{9-115}$$

where

$$\mathbf{S} = [\mathbf{B} \quad \mathbf{A}\mathbf{B} \quad \mathbf{A}^2\mathbf{B} \quad \ldots \quad \mathbf{A}^{n-1}\mathbf{B}] \tag{9-116}$$

since the pair $[\mathbf{A},\mathbf{B}]$ is controllable, and $\mathbf{S}$ is of rank n. Thus, since $\mathbf{P}^{-1}$ is nonsingular, the rank of $\mathbf{S}_1$ is the same as that of $\mathbf{S}$, and the pair $[\Lambda,\Gamma]$ is completely controllable.

The above theorem actually covers both the similarity and the phase-variable canonical transformation by proper choice of the nonsingular matrix $\mathbf{P}$.

Theorem 9-19. Invariant Theorem on Nonsingular Transformation, Observability

Given the n-th order digital control system

$$\mathbf{x}(k+1) = \mathbf{A}\mathbf{x}(k) + \mathbf{B}\mathbf{u}(k) \tag{9-117}$$

$$\mathbf{c}(k) = \mathbf{D}\mathbf{x}(k) \tag{9-118}$$

where the pair $[\mathbf{A},\mathbf{D}]$ is completely observable. The transformation $\mathbf{x}(k) = \mathbf{P}\mathbf{y}(k)$, where $\mathbf{P}$ is nonsingular, transforms the system equations to

$$\mathbf{y}(k+1) = \Lambda\mathbf{y}(k) + \Gamma\mathbf{u}(k) \tag{9-119}$$

and

$$\mathbf{c}(k) = \mathbf{D}\mathbf{P}\mathbf{y}(k) \tag{9-120}$$

Then, the pair $[\Lambda,\mathbf{DP}]$ is also observable.

Proof: The proof of this theorem is similar to that of Theorem 9-18. We form the observability matrix

$$\mathbf{V}_1 = [(\mathbf{DP})' \quad \Lambda'(\mathbf{DP})' \quad (\Lambda^2)'(\mathbf{DP})' \quad \ldots \quad (\Lambda^{n-1})'(\mathbf{DP})'] \tag{9-121}$$

Substituting $\Lambda = \mathbf{P}^{-1}\mathbf{A}\mathbf{P}$ into the last equation, and simplifying, we can show that $\mathbf{V}_1$ is of rank n if $[\mathbf{A},\mathbf{D}]$ is observable.

Theorem 9-20. Theorem on Controllability of Closed-Loop Systems With State Feedback

If the digital control system

$$\mathbf{x}(k+1) = \mathbf{A}\mathbf{x}(k) + \mathbf{B}\mathbf{u}(k) \tag{9-122}$$

is completely controllable, then the closed-loop system derived through state feedback,

$$\mathbf{u}(k) = \mathbf{r}(k) - \mathbf{G}\mathbf{x}(k) \tag{9-123}$$

so that the state equation becomes

$$\mathbf{x}(k+1) = (\mathbf{A} - \mathbf{B}\mathbf{G})\mathbf{x}(k) + \mathbf{B}\mathbf{r}(k) \tag{9-124}$$

is also completely controllable. On the other hand, if the pair $[\mathbf{A},\mathbf{B}]$ is uncontrollable, then there is no $\mathbf{G}$ which will make the pair $[\mathbf{A} - \mathbf{BG},\mathbf{B}]$ controllable. In other words, if an open-loop system is uncontrollable, it cannot be made controllable through state feedback.

Proof: By controllable of the pair $[\mathbf{A},\mathbf{B}]$ we mean that there exists a control $\mathbf{u}(k)$ over the interval $[0,N]$ such that the initial state $\mathbf{x}(0)$ is driven to $\mathbf{x}(N)$ for the finite time interval N. We can write Eq. (9-123) as

$$\mathbf{r}(k) = \mathbf{u}(k) + \mathbf{G}\mathbf{x}(k) \tag{9-125}$$

which is the control of the closed-loop system. Thus, if $\mathbf{u}(k)$ exists which can drive $\mathbf{x}(0)$ to $\mathbf{x}(N)$ in finite time, Eq. (9-125) implies that $\mathbf{r}(k)$ also exists, and the closed-loop system is also controllable.

Conversely, if the pair $[\mathbf{A},\mathbf{B}]$ is uncontrollable, which means that no $\mathbf{u}(k)$ exists which will drive any $\mathbf{x}(0)$ to any $\mathbf{x}(N)$ in finite N, then we cannot find an input $\mathbf{r}(k)$ which will do the same to $\mathbf{x}(k)$, since otherwise we can set $\mathbf{u}(k)$ as in Eq. (9-123) to control the open-loop system.

Theorem 9-21. Theorem on Observability of Closed-Loop Systems With State Feedback

If the system described by Eqs. (9-117) and (9-118) is controllable and observable, then state feedback of the form of Eq. (9-123) could destroy observability. In other words, the observability of open-loop and closed-loop systems due to state feedback are unrelated.

The following example will illustrate relation between observability and state feedback.

Example 9-5.

For the system described by Eqs. (9-117) and (9-118), let

$$\mathbf{A} = \begin{bmatrix} 0 & 1 \\ -2 & -3 \end{bmatrix} \qquad \mathbf{B} = \begin{bmatrix} 1 \\ 1 \end{bmatrix}$$

$$\mathbf{D} = [1 \quad 2]$$

We can show that the pair $[\mathbf{A},\mathbf{D}]$...

We can show that the pair $[\mathbf{A},\mathbf{B}]$ is controllable and the pair $[\mathbf{A},\mathbf{D}]$ is observable.

Let the state feedback be defined by

$$u(k) = r(k) - \mathbf{G}\mathbf{x}(k) \tag{9-126}$$

where

$$\mathbf{G} = [g_1 \quad g_2] \tag{9-127}$$

Then, the closed-loop system is described by the state equation

$$\mathbf{x}(k+1) = (\mathbf{A} - \mathbf{B}\mathbf{G})\mathbf{x}(k) + \mathbf{B}r(k) \tag{9-128}$$

$$\mathbf{A} - \mathbf{B}\mathbf{G} = \begin{bmatrix} -g_1 & 1 - g_2 \\ -2 - g_1 & -3 - g_2 \end{bmatrix} \tag{9-129}$$

The observability matrix of the closed-loop system is

$$\mathbf{V} = [\mathbf{D}' \quad (\mathbf{A} - \mathbf{B}\mathbf{G})'\mathbf{D}'] = \begin{bmatrix} 1 & -3g_1 - 4 \\ 2 & -3g_2 - 5 \end{bmatrix} \tag{9-130}$$

The determinant of V is

$$|\mathbf{V}| = 6g_1 - 3g_2 + 3 \tag{9-131}$$

Therefore, if g_1 and g_2 are chosen so that $|\mathbf{V}| = 0$, the closed-loop system would be unobservable.

REFERENCES

1. Kreindler, E., and Sarachik, P. E., "On the Concepts of Controllability and Observability of Linear Systems," *IEEE Trans. on Automatic Control,* Vol. AC-9, April 1964, pp. 129-136.
2. Kalman, R. E., "Contributions to the Theory of Optimal Control," *Bol. Soc. Mat. Mexicana,* 5, 1960, pp. 102-119.
3. Kalman, R. E., "On the General Theory of Control Systems," *Proc. IFAC,* Vol. 1, Butterworths, London, 1961, pp. 481-492.
4. Locatelli, A., and Rinaldi, S., "Controllability versus Sensitivity in Linear Discrete Systems," *IEEE Trans. on Automatic Control,* Vol. AC-15, April 1970, pp. 254-255.
5. Weiss, L., "Controllability, Realization and Stability of Discrete-Time Systems," *SIAM Journal on Control,* Vol. 10, May 1972, pp. 230-251.
6. Mullis, C. T., "On the Controllability of Discrete Linear Systems with Output Feedback," *IEEE Trans. on Automatic Control,* Vol. AC-18, December 1973, pp. 608-615.
7. Kwon, W. H., and Pearson, A. E., "On the Stabilization of a Discrete Constant Linear System," *IEEE Trans. on Automatic Control,* Vol. AC-20, December 1975, pp. 800-801.
8. Hautus, M. L. J., "Controllability and Stabilizability of Sampled Systems," *IEEE Trans. on Automatic Control,* Vol. AC-17, August 1972, pp. 528-531.
9. Brockett, R. W., "Poles, Zeros, and Feedback: State Space Interpretation," *IEEE Trans. on Automatic Control,* Vol. AC-10, April 1965, pp. 129-135.

PROBLEMS

9-1 Consider that a digital control system is described by the state equation

$$\mathbf{x}(k+1) = \mathbf{A}\mathbf{x}(k) + \mathbf{B}\mathbf{u}(k)$$

where

$$\mathbf{A} = \begin{bmatrix} 1 & -2 \\ 1 & -1 \end{bmatrix} \qquad \mathbf{B} = \begin{bmatrix} 1 & 0 \\ 0 & -1 \end{bmatrix}$$

(a) Determine the controllability of the system.
(b) Find a **P** matrix which will diagonalize **A** in the form of $\mathbf{A} = \mathbf{P}^{-1}\mathbf{AP}$. Verify the results of part (a) with the uncoupled state equations.

9-2 Consider that a digital control system is described by the state equation

$$\mathbf{x}(k+1) = \mathbf{Ax}(k) + \mathbf{Bu}(k)$$

where

$$\mathbf{A} = \begin{bmatrix} 1 & -2 & 0 \\ 3 & 2 & 1 \\ -1 & 1 & 4 \end{bmatrix} \qquad \mathbf{B} = \begin{bmatrix} 1 & 0 \\ -1 & 1 \\ 0 & 1 \end{bmatrix}$$

Determine the controllability of the system.

9-3 The block diagram of a digital control system is shown in Fig. 9P-3. Determine the controllability of the system.

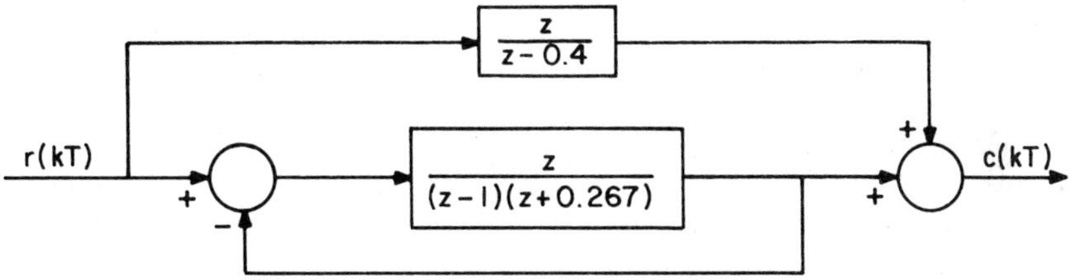

Fig. 9P-3.

9-4 The block diagram of a digital control system is shown in Fig. 9P-4. Determine what values of T must be avoided so that the system will be assured of complete state controllability.

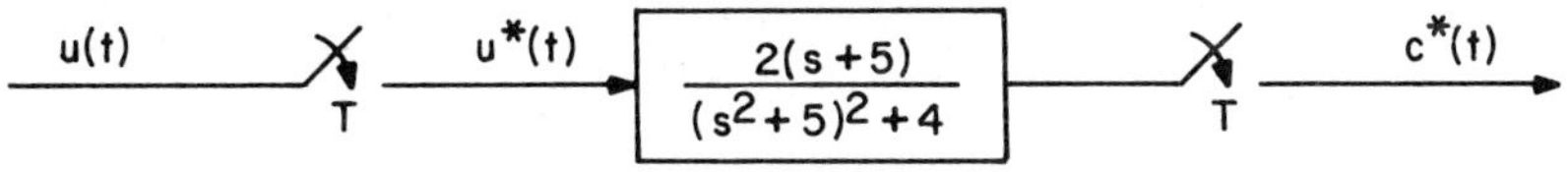

Fig. 9P-4.

9-5 The linear process of a digital control system is described by the following state equation

$$\begin{bmatrix} \dfrac{dx_1}{dt} \\ \dfrac{dx_2}{dt} \end{bmatrix} = \begin{bmatrix} 0 & 1 \\ -1 & 0 \end{bmatrix} \begin{bmatrix} x_1 \\ x_2 \end{bmatrix} + \begin{bmatrix} 0 \\ 1 \end{bmatrix} u(t)$$

where $u(t) = u(kT)$ for $kT \leqslant t < (k+1)T$. Determine the values of the sampling period T which make the system uncontrollable; that is, the initial state $\mathbf{x}(0)$ cannot be driven to the equilibrium state $\mathbf{x}(NT) = \mathbf{0}$ for any finite N.

9-6 Figure 9P-6 shows a stick-balancing system in which the objective is to control the attitude of the stick with the force u(t) applied to the car. The force u(t) is sampled and is described by $u(t) = u(kT)$, $kT \leqslant t < (k+1)T$, where T is the sampling period. The linearized equations that approximate the motion of the stick are

$$\ddot{\theta}(t) = \theta(t) + u(t)$$

$$\ddot{y}(t) = \theta(t) - u(t)$$

(a) Let the state variables be defined as $x_1(t) = \theta(t)$, $x_2(t) = \dot{\theta}(t)$, $x_3(t) = y(t)$ and $x_4(t) = \dot{y}(t)$. Discretize the system equations and express the state equations in the following form:

$$\mathbf{x}[(k+1)T] = \phi(T)\mathbf{x}(kT) + \theta(T)\mathbf{u}(kT)$$

(b) For $T = 1$ sec determine if the discrete-data system in (a) is completely state controllable.

(c) If only one of the state variables can be regarded as the output, that is, $c(k) = x_i(k)$, $i = 1, 2, 3$ or 4, determine which one corresponds to a completely observable system.

(d) Find the state feedback $\mathbf{u}(k) = -\mathbf{G}\mathbf{x}(k)$ where $\mathbf{G} = [g_1 \quad g_2 \quad g_3 \quad g_4]$ such that the eigenvalues of the closed-loop system are all zero.

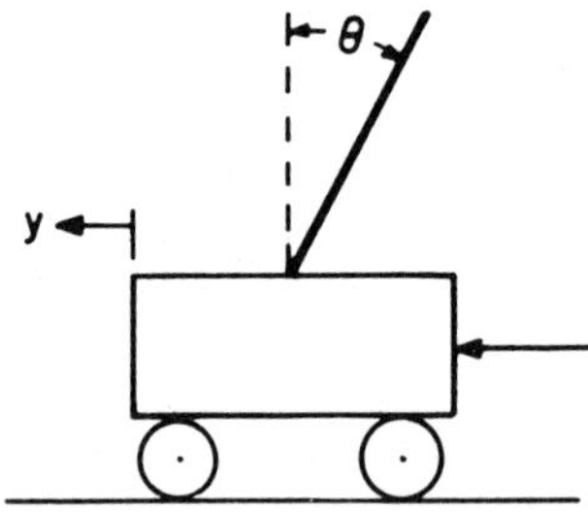

Fig. 9P-6.

9-7 A digital control process is described by the state equation

$$\mathbf{x}(k+1) = \mathbf{A}\mathbf{x}(k) + \mathbf{B}u(k)$$

where

$$\mathbf{A} = \begin{bmatrix} 0 & 0 & 0 \\ 0 & 0.5 & 0 \\ 0 & 0 & 2 \end{bmatrix} \qquad \mathbf{B} = \begin{bmatrix} 1 \\ 0 \\ 1 \end{bmatrix}$$

(a) Determine the state controllability of the system.
(b) Can the system be stabilized by state feedback of the form:

$$u(k) = -[g_1 \quad g_2 \quad g_3]\mathbf{x}(k)$$

where g_1, g_2 and g_3 are constants?

9-8 The state equations of a time-varying digital control system are described by

$$\mathbf{x}[(k+1)T] = \mathbf{A}(kT)\mathbf{x}(kT) + \mathbf{B}(kT)u(kT)$$

where

$$\mathbf{A}(kT) = \begin{bmatrix} kT & 2 \\ 1 & (k+1)T \end{bmatrix} \qquad \mathbf{B}(kT) = \begin{bmatrix} 1 \\ 1 \end{bmatrix}$$

Determine the state controllability of the system as a function of the sampling period T.

9-9 The state equations of a time-varying system are given as

$$\begin{bmatrix} \dot{x}_1(t) \\ \dot{x}_2(t) \end{bmatrix} = \begin{bmatrix} 0 & 2 \\ 0 & 0 \end{bmatrix} \begin{bmatrix} x_1(t) \\ x_2(t) \end{bmatrix} + \begin{bmatrix} 0 \\ 1-t \end{bmatrix} u(t)$$

where $u(t) = u(kT)$, $kT \leqslant t < (k+1)T$. Discretize the equations by approximating $\dot{x}_i(t)$ by $\left[x_i[(k+1)T] - x_i(kT)\right]/T$, $i = 1, 2$, and letting $t = kT$. Determine the controllability of the discretized system in terms of the sampling period T.

9-10 The input-output transfer function of a digital control system is

$$\frac{C(z)}{U(z)} = \frac{1.65(z+0.1)}{z^3 + 0.7z^2 + 0.11z + 0.005}$$

(a) Assign state variables to the system so that the system is state controllable but not observable.
(b) Assign state variables to the system in such a way that the system is observable but not state controllable.

9-11 Given the digital control system

$$\mathbf{x}(k+1) = \mathbf{A}\mathbf{x}(k) + \mathbf{B}\mathbf{u}(k)$$

$$\mathbf{c}(k) = \mathbf{D}\mathbf{x}(k)$$

Prove that if the pair $[\mathbf{A}, \mathbf{D}]$ is observable, then the closed-loop system realized through output feedback,

$$\mathbf{u}(k) = -\mathbf{G}\mathbf{c}(k) + \mathbf{r}(k)$$

is also observable.

10

DESIGN OF DIGITAL CONTROL SYSTEMS

10.1 INTRODUCTION

The design problems encountered in digital control systems are essentially similar to those found in the design of continuous-data control systems. Basically, we have a process which needs to be controlled so that the outputs of the process will behave according to some preset performance specifications. In the conventional design philosophy we decide at the outset that there should be feedback from the output to the input reference, so that an error between the two signals can be formed. Then, in general, we find that a controller is needed to operate on the error signal in such a way that the design specifications are all satisfied. In digital control systems, the problem has more variations and flexibility. For instance, the digital or discrete data may be due to the usage of certain digital or incremental transducers which naturally put out digital or discrete data, or it may simply be a case that the designer decides he wants to use a digital controller. On the other hand, it is also feasible to use a continuous-data controller on the digital signal after it has been smoothed out by a data hold. Thus, we see that there are a great variety of possible configurations and schemes when it comes to the design of a digital control system, and the final choice is entirely at the hands of the designer.

In this chapter we shall introduce several methods of designs of digital control systems, or perhaps, more appropriately, these should be called systems with digital or discrete data. Some of these methods are classified as conventional and the others rely on the state-variable formulation.

The conventional design of control systems is characterized by the fixed-configuration concept, in that the designer first fixes the composition of the overall system, including the controlled process and the controller. Figures 10-1 through 10-4 illustrate some of the schemes most frequently encountered in practice. Figure 10-1 shows the block diagram of a sampled-data system in which the controller is analog. The sampler may represent the fact that digital or sampled data exist at the input and the feedback channels, due to the use of digital transducers. However, we may still elect to use a continuous-data controller to act on the output of the sampler after the digital signal is decoded and smoothed out by a data hold.

Figure 10-2 shows the classical case of a digital control system in which a digital controller is located in the forward path.

Figure 10-3 shows the case with the continuous-data controller located in the minor feedback loop. Figure 10-4 gives the counterpart of the system in Fig. 10-3, with the digital controller replacing the analog controller.

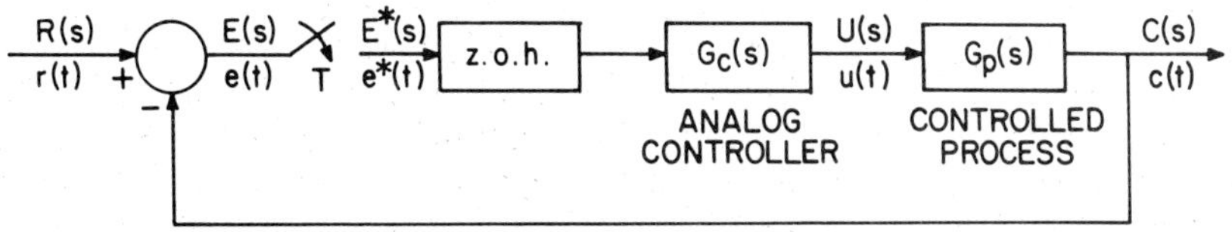

Fig. 10-1. Digital control system with cascade analog controller.

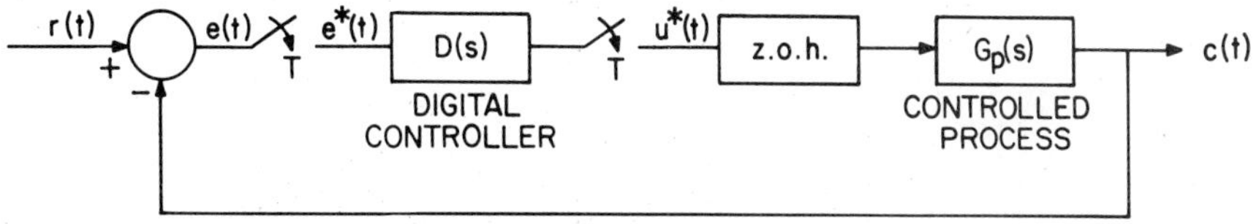

Fig. 10-2. Digital control system with cascade digital controller.

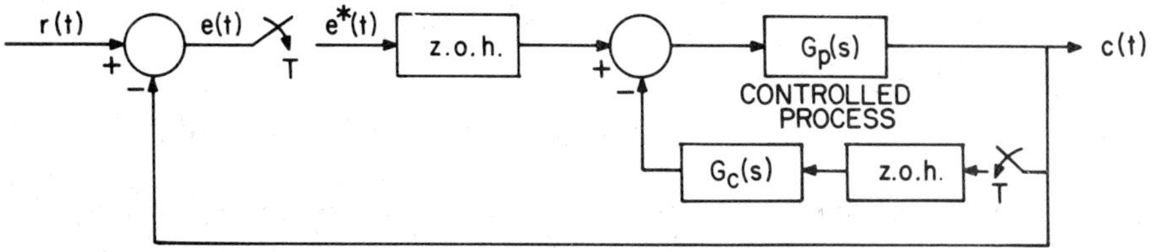

Fig. 10-3. Digital control system with analog controller in minor feedback path.

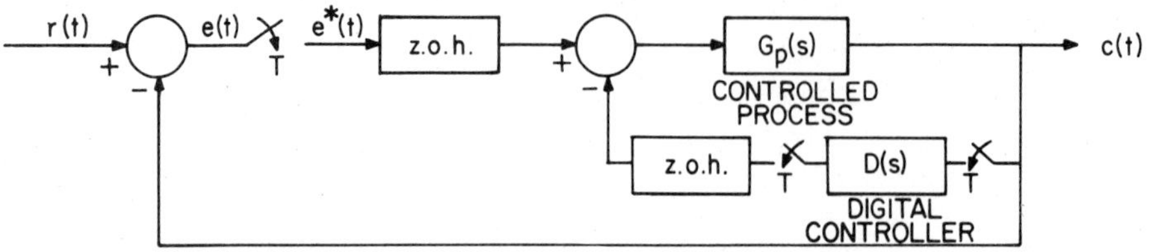

Fig. 10-4. Digital control system with digital controller in minor feedback path.

A powerful design method in the state-variable domain is the state-feedback or output-feedback design. Figure 10-5 illustrates the block diagram of a multivariable digital control system with state-variable feedback, assuming that all the state variables are accessible. In practice, not all the state variables are accessible, so that we either have to use an observer to estimate some or all of the state variables, or simply use output feedback. Figure 10-6 shows the block diagram of a multivariable digital control system with state feedback and an observer, and Fig. 10-7 gives the system configuration with output feedback. In the system of Fig. 10-6, the output vector $\mathbf{c}(k)$ is assumed to be accessible. The observer estimates the state variables from information it receives from $\mathbf{c}(k)$. The output of the observer is the estimated state variable $\mathbf{x}(k)$. In the case of Fig. 10-7 the output $\mathbf{c}(k)$ is fed directly into the feedback gain matrix G. Since in general there are fewer output variables than state variables, the output feedback would be more restricted than the state feedback design.

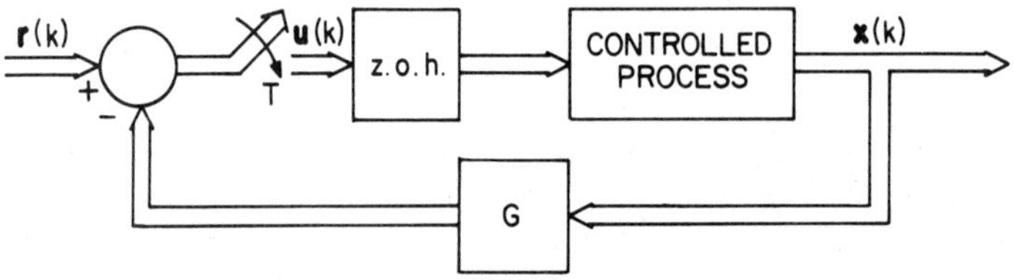

Fig. 10-5. Digital control system with state feedback.

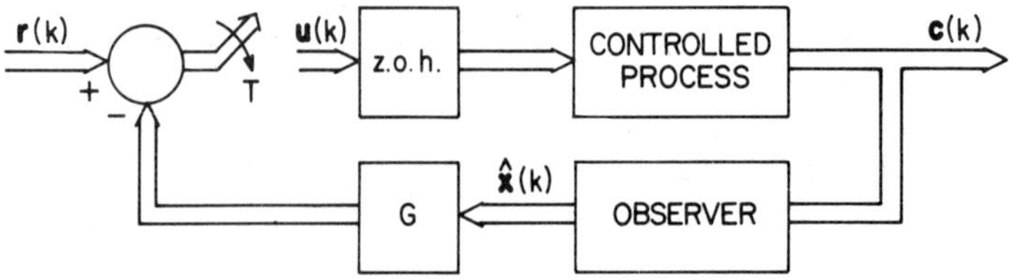

Fig. 10-6. Digital control system with state-feedback and observer.

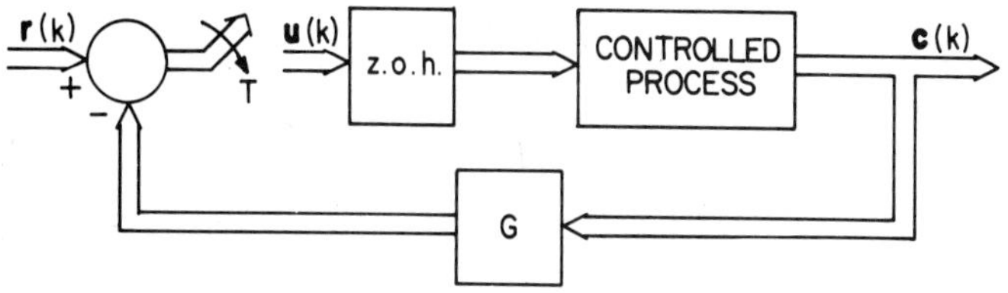

Fig. 10-7. Digital control system with output feedback.

10.2 CASCADE DECOMPENSATION BY CONTINUOUS-DATA CONTROLLERS

In this section we shall consider the design of a digital control system with a cascade continuous-data controller, as shown in Fig. 10-1. Since the controller is located between the zero-order hold and the controlled process, the z-transform of the open-loop transform is written

$$\frac{C(z)}{E(z)} = \mathcal{Z}\left[G_{h0}(s)G_c(s)G_p(s)\right] = (1 - z^{-1})\,\mathcal{Z}\left[\frac{G_c(s)G_p(s)}{s}\right] \tag{10-1}$$

The design objective is to find a physically realizable transfer function $G_c(s)$ of the continuous-data controller which will cause the digital control system to perform according to specifications. Unfortunately, as shown by Eq. (10-1), the transfer function $G_c(s)$ is imbedded with the transfer function of the process, $G_p(s)$, so that we cannot easily investigate the effect of the controller independently. In the following section we shall introduce a simple approximation method and show why it must be applied with great care. A design method using the w-transformation is also introduced.

Time-Delay Approximation of Sample-and-Hold

A method which has been frequently used by the practicing engineer to approximate a sampled-data system by a continuous-data system relies on the approximation of the sample-and-hold by a pure time delay. Although we are presenting this method here, the opinion is that the approximation is not accurate in general, and great care must be exercised when applying the method to design problems.

Let us refer to the sinusoidal steady-state transfer function of the zero-order hold given in Eq. (2-108),

$$G_{h0}(j\omega) = T\,\frac{\sin \pi(\omega/\omega_s)}{(\omega\pi/\omega_s)}\,e^{-j\pi(\omega/\omega_s)} \tag{10-2}$$

Using this expression, the pulse transfer function of $G_{h0}(s)G_c(s)G_p(s)$ is written

$$G_{h0}G_cG_p^*(s) = \frac{1}{T}\sum_{n=-\infty}^{\infty} G_{h0}(j\omega + jn\omega_s)G_c(j\omega + jn\omega_s)G_p(j\omega + jn\omega_s)$$

$$= \frac{1}{T}\sum_{n=-\infty}^{\infty} T\frac{\sin\frac{\omega + n\omega_s}{2}T}{\frac{\omega + n\omega_s}{2}T} e^{-j(\omega+n\omega_s)T/2}G_c(j\omega + jn\omega_s)G_p(j\omega + jn\omega_s) \qquad (10\text{-}3)$$

Since in most control systems $G_p(j\omega)$ has a low-pass filter characteristic, and at low frequencies $\sin(1/2)(\omega + n\omega_s)T/(1/2)(\omega + n\omega_s)T$ is approximately equal to one, we can approximate $G_{h0}G_cG_p^*(j\omega)$ by just the first term $(n = 0)$ of its series expansion. Thus, we have

$$G_{h0}G_cG_p^*(j\omega) \cong G_c(j\omega)G_p(j\omega)e^{-j\omega T/2} \qquad (10\text{-}4)$$

which means that the sample-and-hold can be approximated by a pure time delay of one-half the sampling period T. Figure 10-8 shows the continuous-data system with the pure time delay approximating the sample-and-hold. A more convincing evidence of this pure time-delay approximation of the sample-and-hold is illustrated by the signal waveforms shown in Fig. 10-9. It seems that if we approximate the output of the zero-order hold by a smooth curve passing through the center of the flat-top signal during each sampling period, we have a continuous-data signal which is approximately the same as the input signal

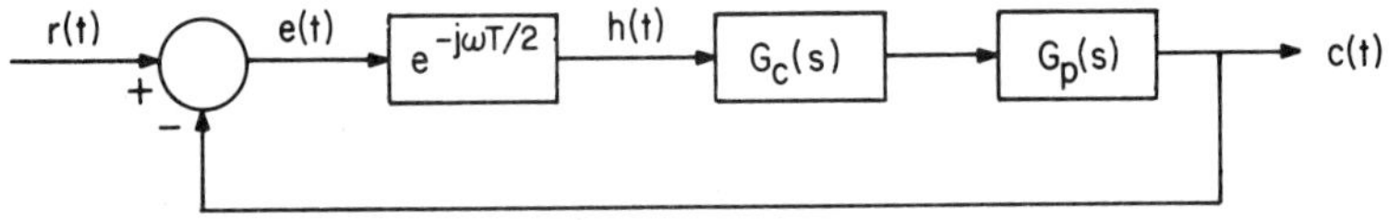

Fig. 10-8. Continuous-data system approximation of the sampled-data system in Fig. 10-1.

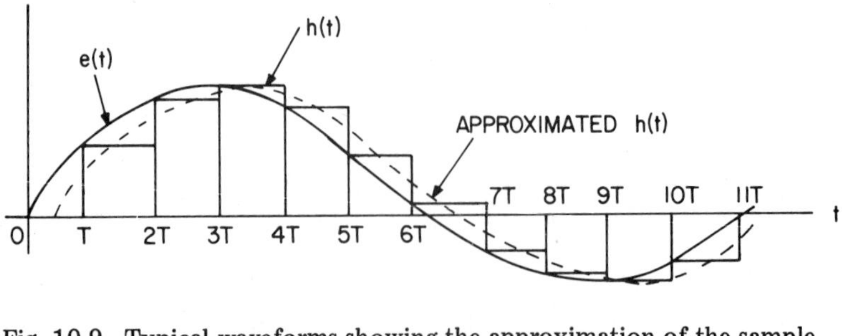

Fig. 10-9. Typical waveforms showing the approximation of the sample-and-hold by a pure time delay of T/2.

of the sampler, except that it is delayed by one-half the sampling period. However, it should be cautioned that this type of approximation is not applicable to a broad class of sampled-data control systems. In general, it is not certain if the one-term approximation of the infinite series in Eq. (10-3) is adequate, and the ramifications of the approximation as illustrated by Fig. 10-9 are not at all clear. Therefore, the best we can say is that the continuous-data system with pure time delay shown in Fig. 10-8 can be used to approximate the sampled-data system of Fig. 10-1 only if the sampling period is *relatively very small.* Otherwise, if all sampled-data control systems could be approximated by continuous-data systems with time delays, there would be no need to devise and study the analysis and design techniques of digital control systems. However, if and when the time-delay approximation is valid, the design methods established for continuous-data control systems can be applied directly to the approximating system.

We shall use the following example to illustrate the pitfall of the time-delay approximation method.

Example 10-1.

Consider that the transfer function of the controlled process of the system shown in Fig. 10-1 is

$$G_p(s) = \frac{K}{s(s+1)} \tag{10-5}$$

where $K = 1.57$ and $T = 1.57$ sec. The open-loop pulse transfer function of the approximating continuous-data system with time delay shown in Fig. 10-8 is

$$G_{h0}G_p^*(s) \cong e^{-Ts/2}G_p(s) \tag{10-6}$$

where the controller $G_c(s)$ is absent for the moment; i.e., $G_c(s) = 1$. Substituting Eq. (10-5) into Eq. (10-6), we have

$$G_{h0}G_p^*(s) \cong e^{-Ts/2}\frac{K}{s(s+1)} \tag{10-7}$$

The Bode plot of the last transfer function is shown in Fig. 10-10 for $K = 1.57$ and $T = 1.57$ sec. From this plot we see that the phase margin of the uncompensated system with $G_c(s) = 1$ is approximately 0 degree. If we use the exact magnitude plot of $G_{h0}G_p^*(j\omega)$ instead of the time-delay approximation as shown in Fig. 10-10, the phase margin would be approximately 15 degrees. Thus, the approximation of a sampled-data system using Eq. (10-4) usually gives a pessimistic representation of the system stability. Furthermore, the actual phase curve of $G_{h0}G_p^*(j\omega)$ is discontinuous at $\omega = \omega_s$ which is 4 rad/sec in the present case, since the term $\sin(1/2)(\omega T)/(1/2)(\omega T)$ changes sign at $\omega = \omega_s$. The phase relation resulting from Eq. (10-4) would be continuous for all values of ω.

Let us assume that the design specification calls for a phase margin of at least 45 degrees. We may, at this point, investigate the possibility of using either a phase-lead (high-pass filter) or a phase-lag (low-pass filter) controller for the continuous-data system. From Fig. 10-10 we see that the phase of $G_{h0}G_p^*(j\omega)$ decreases rapidly beyond 1 rad/sec. Therefore, a phase-lead controller may not be effective in improving the phase margin of the system. As is well-known [1], a phase-lead controller would increase the gain-crossover frequency [1] and, therefore, would make the addition of more phase lead less effective if the phase of the uncompensated system decreases rapidly in the vicinity of the new gain crossover.* Therefore, a phase-lag controller would seem to be more appropriate for this system.

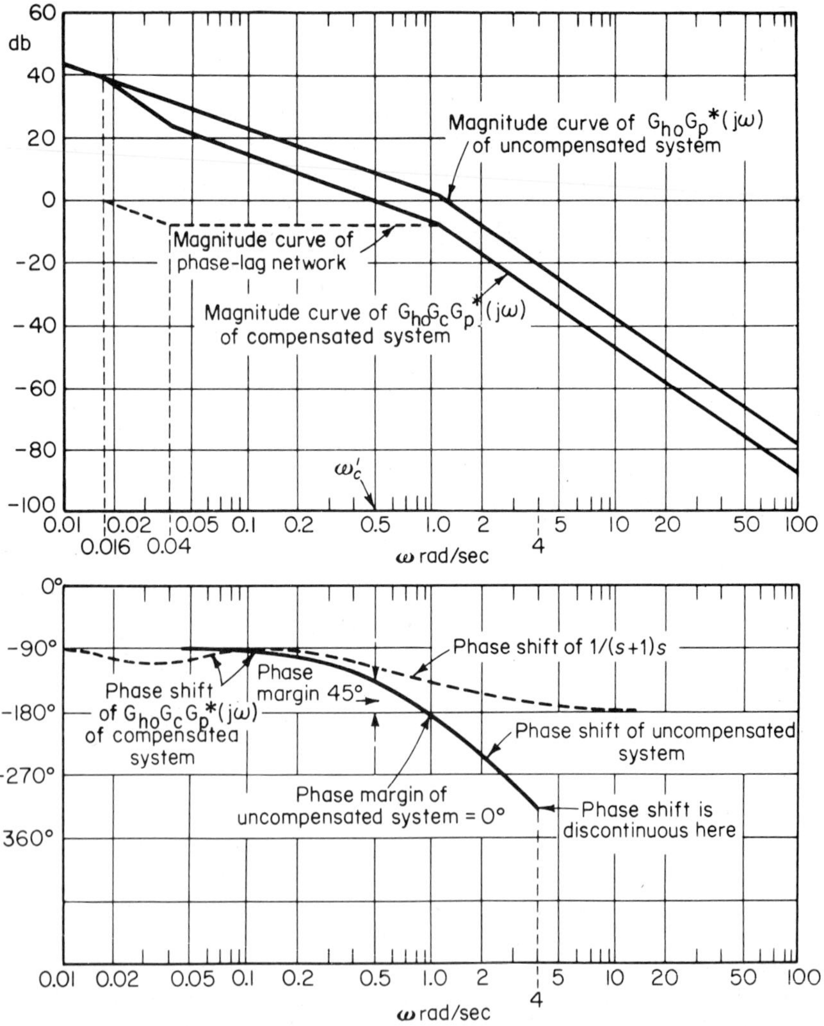

Fig. 10-10. Bode plot of the sampled-data control system in Example 10-1.

*The gain crossover is the point at which the gain or magnitude of the Bode plot is 0 db.

A typical phase-lag controller is described by the following transfer function:

$$G_c(s) = \frac{1 + a\tau s}{1 + \tau s} \tag{10-8}$$

where a and τ are constants and a is always less than unity. From the Bode plot of the uncompensated system shown in Fig. 10-10, we see that the phase margin of 45 degrees may be realized if the gain-crossover frequency is moved from the present value of approximately 1 rad/sec to a frequency of 0.5 rad/sec. Since the phase lag of the compensating network will inevitably have some effect on the final phase curve of the compensated system, as a standard practice, the new gain crossover should be selected at a frequency somewhat lower than 0.5 rad/sec, say $\omega_c' = 0.4$ rad/sec. From the Bode plot in Fig. 10-10, the gain of $G_{h0}G_p^*(j\omega)$ at this new gain crossover is found to be approximately 8 db. Thus, the phase-lag controller must produce 8 db of attenuation at the new gain-crossover frequency. Hence,

$$20\log_{10} a = -8 \text{ db} \tag{10-9}$$

Solving for a from the last equation yields

$$a = -10^{-8/20} = 10^{-0.4} = 0.4 \tag{10-10}$$

This fixes the distance between the two corner frequencies of the phase-lag controller. To determine the exact location of the two corners, $1/a\tau$ and $1/\tau$, it is necessary to first locate the upper corner frequency $\omega = 1/a\tau$ at such a place that the phase lag contributed by the phase-lag controller is negligible at the new gain crossover. This is done by placing the upper corner at a frequency which is equal to 1/10 times of the new gain-crossover frequency. Therefore,

$$\frac{1}{a\tau} = \frac{\omega_c'}{10} = \frac{0.4}{10} = 0.04 \text{ rad/sec} \tag{10-11}$$

and the lower corner frequency of the controller is determined readily from

$$\frac{1}{\tau} = 0.04a = 0.016 \text{ rad/sec} \tag{10-12}$$

Substituting the values of a and τ determined from above into Eq. (10-8), the transfer function of the phase-lag controller is determined as

$$G_c(s) = \frac{1 + 25s}{1 + 62.5s} \tag{10-13}$$

The magnitude and phase plots of the compensated system transfer function $G_{h0}G_cG_p^*(j\omega)$ are sketched in Fig. 10-10. The phase margin of the compensated system is shown to be about 45 degrees as required.

At the outset of the design we had specified that the phase margin of the compensated system should be 45 degrees. This requirement was selected because from design experience a phase margin of 45 degrees would correspond to adequate relative stability for this type of simple system. However, since the continuous-data system is only an approximation of the sampled-data system, the significance of the 45-degree phase margin becomes questionable.

The only valid way to judge the designed system is to substitute the phase-lag controller of Eq. (10-13) into the sampled-data system in Fig. 10-1, and evaluate the performance of the system.

For $K = 1.57$, the open-loop z-transfer function of the compensated sampled-data system is written

$$\mathscr{Z}[G_{h0}(s)G_c(s)G_p(s)] = (1 - z^{-1})\,\mathscr{Z}\left[\frac{0.628(s + 0.04)}{s^2(s + 1)(s + 0.016)}\right] \tag{10-14}$$

Or

$$\mathscr{Z}[G_{h0}(s)G_c(s)G_p(s)] = \frac{2.667z^2 - 0.815z - 0.172}{(z - 1)(z - 0.208)(z - 0.984)} \tag{10-15}$$

The closed-loop transfer function of the sampled-data system is

$$\frac{C(z)}{R(z)} = \frac{2.667z^2 - 0.815z - 0.172}{z^3 + 0.475z^2 + 0.582z - 0.384} \tag{10-16}$$

Without detailed computation we can readily see from the first term of the numerator of the closed-loop transfer function that the peak overshoot of the unit-step response is at least 2.667. Figure 10-11 shows the unit-step response of the compensated sampled-data system with the controller as described by Eq. (10-13). Since the response is highly oscillatory, though stable, the design criterion of a 45-degree phase margin is no longer an accurate measure of performance for this case. This means that the time-delay approximation could cause serious inaccuracies in the design of sampled-data control systems.

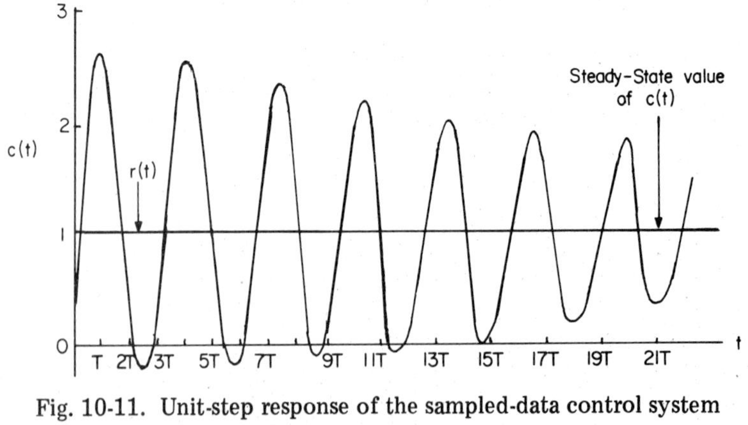

Fig. 10-11. Unit-step response of the sampled-data control system designed in Example 10-1.

Design by Bilinear Transformation

In the preceding chapters the bilinear transformation, the w-transform, given in Eq. (8-22) was used to transform the z-transfer function G(z) of a digital control system into a new function G(w), which is a rational function in the complex variable w. This bilinear w-transformation maps the unit circle in the z-plane onto the imaginary axis in the w-plane (see Sec. 8.2). As shown in Chapter 8, the Bode diagram of G(w) can be drawn in db of magnitude and phase in degrees versus ω_w which is the imaginary part of w. Then, information on stability and relative stability in terms of gain margin and phase margin can be determined directly from the Bode plot. The data presented on the Bode plot can be transferred to the Nichols chart [1] so that information on such performance criteria as the resonance peak [1], resonant frequency and the bandwidth, can be obtained.

In this section we present a straightforward method of designing digital control systems with continuous-data controllers utilizing the w-transform.* The method does not rely on any approximation of the sample-and-hold, and has all the advantages of the frequency-domain design.

Referring to the block diagram in Fig. 10-1, the proposed steps of the design are given as follows:

(1) Determine the w-transform function $G_{h0}G_p(w)$ of the system without compensation, where

$$G_{h0}G_p(w) = G_{h0}G_p(z)\Big|_{z=\frac{1+w}{1-w}} \tag{10-17}$$

(2) Construct the Bode diagram of $G_{h0}G_p(w)$ by setting $w = j\omega_w$, and if necessary, transfer the db magnitude and phase versus ω_w loci to the Nichols chart. Predict the dynamic behavior of the uncompensated system by evaluating the phase margin, gain margin, peak resonance, resonant frequency, and bandwidth from the Nichols chart.

(3) Should the system need improvement, a cascade controller with the transfer function $G_c'(w)$ is multiplied to $G_{h0}G_p(w)$. The objective of the controller is to reshape the transfer function loci of $G_{h0}G_p(w)$ in the Bode plot or the Nichols chart. The transfer function of the controller in the w-domain can be classified as phase-lead, phase-lag, lead-lag or lag-lead configurations. It is important to point out that the filtering characteristics referred to in the w-domain may have no direct significant correlation in the s-domain. The purpose of identifying $G_c'(w)$ in the w-domain according to lead or lag is purely for the purpose of making use of the established knowledge on the effects of lead or lag controllers on the compensation of continuous-data systems with regard to the transfer function loci in the s-domain.

Once the transfer function $G_c'(w)G_{h0}G_p(w)$ is determined, the

*Design of digital control systems is also carried out in the w′-domain. For details of the w′-transform, the reader may refer to [4].

transfer function of the actual controller in the s-domain, $G_c(s)$ is found by first transforming $G_c'(w)G_{h0}G_p(w)$ to $G_c(s)G_{h0}(s)G_p(s)$, and then solving $G_c(s)$ from the latter expression, since $G_{h0}(s)G_p(s)$ is already known.

One of the important requirements is that $G_c(s)$ must be a physically realizable transfer function. Furthermore, if possible, $G_c(s)$ should be realized by a network with only R and C elements. In order for $G_c(s)$ to be RC realizable, the transfer function must satisfy the following requirements:

(a) The poles of $G_c(s)$ must lie in the left-half of the s-plane and must be simple and real.
(b) The number of poles of $G_c(s)$ must be greater than or at least equal to the number of zeros of $G_c(s)$.

In general, the zeros of $G_c(s)$ may lie anywhere in the s-plane. From (a), the poles of $G_c(s)$ must be restricted to the negative real axis of the s-plane. Since the poles of $G_c(s)$ are generated by the poles of $G_c'(w)$, and the negative real axis of the s-plane corresponds to the portion, $-1 \leqslant w \leqslant 0$, in the w-plane (see Fig. 10-12), it follows that $G_c'(w)$ can have only simple poles that lie in the range of -1 and 0 on the real axis of the w-plane.

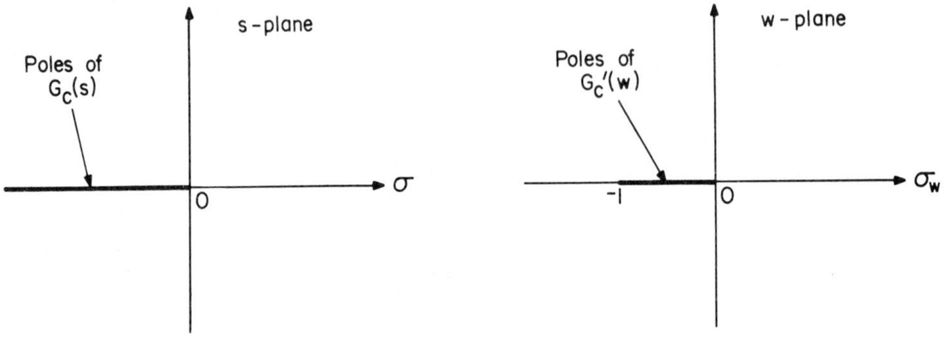

Fig. 10-12. Locations of the poles of $G_c(s)$ in the s-plane and $G_c'(w)$ in the w-plane for RC realizability.

(4) Once $G_c'(w)G_{h0}G_p(w)$ is determined, take the inverse transform from w to s to obtain $G_c(s)G_{h0}(s)G_p(s)$. It should be noted that since

$$G_c'(w)G_{h0}G_p(w) = [G_cG_{h0}G_p(z)]\Big|_{z=\frac{1+w}{1-w}} \tag{10-18}$$

$$G_c'(w) \neq \mathcal{Z}[G_p(s)]\Big|_{z=\frac{1+w}{1-w}} \tag{10-19}$$

The z-transform of the uncompensated open-loop transfer function is written as

$$G_{h0}G_p(z) = (1 - z^{-1})\mathcal{Z}\left[\frac{G_p(s)}{s}\right] \tag{10-20}$$

In terms of the w-transform, we have

$$\begin{aligned} G_{h0}G_p(w) &= (1 - z^{-1})\mathcal{Z}\left[\frac{G_p(s)}{s}\right]\Bigg|_{z=\frac{1+w}{1-w}} \\ &= \frac{2w}{1+w}\mathcal{Z}\left[\frac{G_p(s)}{s}\right]\Bigg|_{z=\frac{1+w}{1-w}} \end{aligned} \tag{10-21}$$

With the cascade controller $G_c'(w)$, the w-transform open-loop transfer function is written

$$\begin{aligned} G_c'(w)G_{h0}G_p(w) &= G_{h0}G_cG_p(w) \\ &= \frac{2w}{w+1}\mathcal{Z}\left[\frac{G_c(s)G_p(s)}{s}\right]\Bigg|_{z=\frac{1+w}{1-w}} \end{aligned} \tag{10-22}$$

or

$$\mathcal{Z}\left[\frac{G_c(s)G_p(s)}{s}\right]\Bigg|_{z=\frac{1+w}{1-w}} = \frac{w+1}{2w}G_c'(w)G_{h0}G_p(w) \tag{10-23}$$

The transfer function $G_c(s)G_p(s)/s$ is found from Eq. (10-23) by making the partial-fraction expansion of $G_c'(w)G_{h0}G_p(w)$ and finding the corresponding pairs of transfer functions in s. A table of the Laplace- z- w-transform functions is given in Table 10-1 for this purpose. Since in order to find $G_c(s)$ the function $G_p(s)/s$ must eventually be factored out of $G_c(s)G_p(s)/s$, care must be taken in performing the partial-fraction expansion of $G_c'(w)G_{h0}G_p(w)$.

(5) Once $G_c(s)G_p(s)/s$ is determined, $G_c(s)$ is readily obtained. However, because of the lack of correspondence between the number of s-plane zeros and w-plane zeros, the function $G_c(s)$ obtained from the previous steps will in many cases have more zeros than poles. This can be circumvented by adding one or more remote poles to $G_c(s)$ on the negative real axis in the s-plane, in order to make $G_c(s)$ physically

realizable. These additional poles should have negligible effect on the overall transient and steady-state performance of the system.

TABLE 10-1.

Laplace transform G(s)	z-transform G(z)	w-transform G(w)
$\dfrac{1}{s}$	$\dfrac{z}{z-1}$	$\dfrac{w+1}{2w}$
$\dfrac{1}{s^2}$	$\dfrac{Tz}{(z-1)^2}$	$\dfrac{T(1+w)(1-w)}{4w^2}$
$\dfrac{1}{s^3}$	$\dfrac{T^2z(z+1)}{2(z-1)^3}$	$\dfrac{T^2(1+w)(1-w)}{8w^3}$
$\dfrac{1}{s+a}$	$\dfrac{z}{z-e^{-aT}}$	$\dfrac{1+w}{(1-e^{-aT})\left[1+\dfrac{1+e^{-aT}}{1-e^{-aT}}w\right]}$
$\dfrac{1}{(s+a)^2}$	$\dfrac{Tze^{-aT}}{(z-e^{-aT})^2}$	$\dfrac{(1+w)(1-w)Te^{-aT}}{(1-e^{-aT})^2\left[1+\dfrac{1+e^{-aT}}{1-e^{-aT}}w\right]^2}$
$\dfrac{a}{s(s+a)}$	$\dfrac{(1-e^{-aT})z}{(z-1)(z-e^{-aT})}$	$\dfrac{(1+w)(1-w)}{2w\left[1+\dfrac{1+e^{-aT}}{1-e^{-aT}}w\right]}$
$\dfrac{a}{s^2(s+a)}$	$\dfrac{Tz}{(z-1)^2}-\dfrac{1-e^{-aT}}{a(z-1)(z-e^{-aT})}$	$\dfrac{T(1-w)(1+w)}{4w^2}-\dfrac{(1-w)(1+w)}{2aw\left[1+\dfrac{1+e^{-aT}}{1-e^{-aT}}w\right]}$
$\dfrac{\omega}{s^2+\omega^2}$	$\dfrac{z\sin\omega T}{z^2-2z\cos\omega T+1}$	$\dfrac{(1-w)(1+w)\sin\omega T}{2[(1+w^2)-(1-w^2)\cos\omega T]}$
$\dfrac{\omega}{(s+a)^2+\omega^2}$	$\dfrac{ze^{-aT}\sin\omega T}{z^2-2ze^{-aT}\cos\omega T+e^{-2aT}}$	$\dfrac{w\,\dfrac{(1+w)(1-w)e^{-aT}\sin\omega T}{(1+w)^2-2(1+w)(1-w)}}{e^{-aT}\cos\omega T+(1-w)e^{-2aT}}$

Example 10-2.

In this example we shall repeat the same design problem as in Example 10-1, using the bilinear-transformation method. The transfer function of the controlled process is given in Eq. (10-5), with K = 1.57, and the sampling period is T = 1.57 sec.

The z-transform of the open-loop transfer function is

$$\mathcal{Z}[G_{h0}(s)G_p(s)] = G_{h0}G_p(z) = (1 - z^{-1})\mathcal{Z}\left[\frac{K}{s^2(s+1)}\right]$$

$$= \frac{1.22(z + 0.598)}{(z-1)(z-0.208)} \tag{10-24}$$

Taking the w-transform of the last equation, we have

$$G_{h0}G_p(w) = G_{h0}G_p(z)\Big|_{z=\frac{1+w}{1-w}} = \frac{1.232(1+0.251w)(1-w)}{w(1+1.525w)} \tag{10-25}$$

The Bode diagram of $G_{h0}G(w)$ is constructed in Fig. 10-13. The phase-crossover frequency is at $\omega_w = 1$, and we see that the uncompensated system has very low margin of stability. In order to achieve a phase margin of 45°, the gain crossover must be moved to $\omega_w = 0.4$ at which point the magnitude of $G_{h0}G_p(w)$ is approximately 8 db. This means that in order to realize the gain crossover of $\omega_w = 0.4$, the magnitude curve must be attenuated by 8 db in the vicinity of $\omega_w = 0.4$. This must be done without reducing the gain of $G_{h0}G_p(w)$ at $\omega_w = 0$.

Let us select the "phase-lag" model for the continuous-data controller $G_c'(w)$; i.e.,

$$G_c'(w) = \frac{1 + a\tau w}{1 + \tau w} \qquad (a < 1) \tag{10-26}$$

Since the attenuation required is 8 db, we set

$$20\log_{10} a = -8 \text{ db}$$

Thus,

$$a = 10^{-8/20} = 0.398$$

In order that the phase lag of the controller produces negligible effect on the phase characteristics of the original system, we set the value of $1/a\tau$ to be one tenth of the new gain-crossover frequency. Thus,

$$\frac{1}{a\tau} = 0.04$$

or

$$\frac{1}{\tau} = 0.016$$

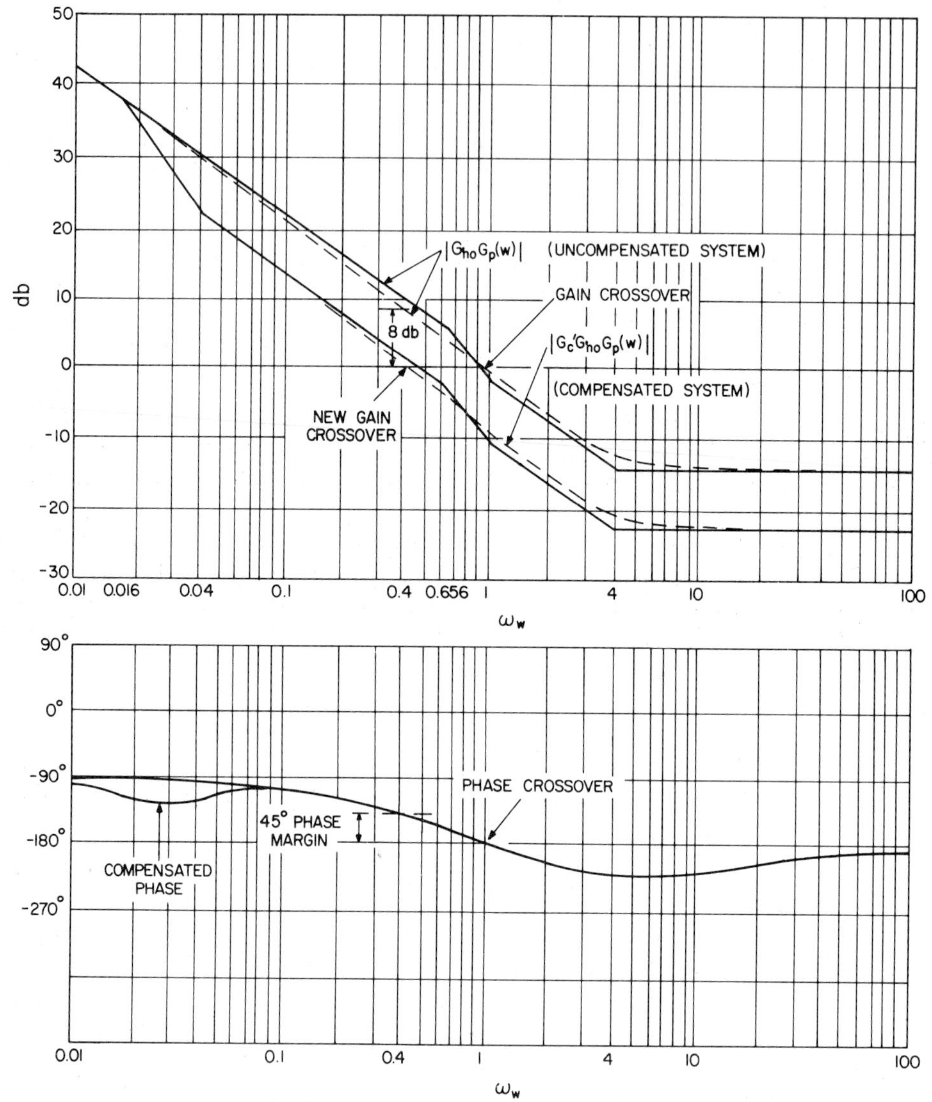

Fig. 10-13. Bode diagram of the digital control system in Example 10-2.

The transfer function of the controller is determined as

$$G_c'(w) = \frac{1 + 25w}{1 + 62.8w} \tag{10-27}$$

Thus,

$$G_c'(w)G_{h0}G_p(w) = \frac{1.232(1 + 0.251w)(1 - w)(1 + 25w)}{w(1 + 1.525w)(1 + 62.8w)} \tag{10-28}$$

The z-transform of the open-loop transfer function of the compensated system is obtained by substituting $w = (z - 1)/(z + 1)$ into Eq. (10-28). We have,

$$G_{h0}G_cG_p(z) = G_c'(w)G_{h0}G_p(w)\Big|_{w=\frac{z-1}{z+1}}$$

$$= \frac{0.497(z - 0.92)(z + 0.599)}{(z - 0.969)(z - 1)(z - 0.208)} \tag{10-29}$$

The closed-loop transfer function of the compensated system is

$$\frac{C(z)}{R(z)} = \frac{G_{h0}G_cG_p(z)}{1 + G_{h0}G_cG_p(z)} = \frac{0.497z^2 - 0.16z - 0.274}{z^3 - 1.68z^2 + 1.219z - 0.476} \tag{10-30}$$

The unit-step response of the compensated system is shown in Fig. 10-14. In this case we see that the designed system has a much better unit-step response than the system designed in Example 10-1.

The remaining task in this design problem involves the determination of the transfer function $G_c(s)$ of the continuous-data controller. From Eq. (10-23),

$$\mathcal{Z}\left[\frac{G_c(s)G_p(s)}{s}\right]\Bigg|_{z=\frac{1+w}{1-w}} = \frac{w+1}{2w}\,\frac{1.232(1 + 0.251w)(1 - w)(1 + 25w)}{w(1 + 1.525w)(1 + 62.8w)} \tag{10-31}$$

The function inside the bracket on the right-hand side of the last equation is to be expanded by partial-fraction expansion. At this point, it is important to investigate the form of the transfer function $G_p(s)/s$. Since $G_p(s)/s$ has two poles at $s = 0$, and Table 10-1 shows that corresponding to $G(s) = 1/s^2$, $G(w)$ has the term $(1 + w)(1 - w)$ in its numerator, we should perform the partial-fraction expansion on $G_c'(w)G_{h0}G_p(w)/(1 - w)$, so that the term $(1 + w)(1 - w)$ is preserved. Thus,

$$\frac{G_c'(w)G_{h0}G_p(w)}{1 - w} = \frac{1.232(1 + 0.251w)(1 + 25w)}{w(1 + 1.525w)(1 + 62.8w)}$$

$$= \frac{0.0807(w + 3.984)(w + 0.04)}{w(w + 0.656)(w + 0.016)}$$

$$= \frac{7.6886}{w} + \frac{0.1922}{w + 0.656} - \frac{7.881}{w + 0.016} \tag{10-32}$$

Substitution of the result of the last equation into Eq. (10-31), we have

$$\mathcal{Z}\left[\frac{G_c(s)G_p(s)}{s}\right]\Bigg|_{z=\frac{1+w}{1-w}} = \frac{7.6886(1 + w)(1 - w)}{2w^2} + \frac{0.293(1 + w)(1 - w)}{2w(1 + 1.525w)}$$

$$- \frac{492.56(1 + w)(1 - w)}{2w(1 + 62.5w)} \tag{10-33}$$

The transform pairs for the terms in Eq. (10-33) are identified in Table 10-1, and we have

$$\frac{G_c(s)G_p(s)}{s} = \frac{7.6886}{s^2} + \frac{0.293}{s(s+1)} - \frac{10.04}{s(s+0.0204)}$$

$$= \frac{0.1568 - 2.1886s - 2.0584s^2}{s^2(s+1)(s+0.0204)} \tag{10-34}$$

from which we get

$$G_c(s) = \frac{0.1 - 1.394s - 1.31s^2}{s + 0.0204} \tag{10-35}$$

In order for $G_c(s)$ to be physically realizable, we add a remote pole at $s = -10$. Thus,

$$G_c(s) = \frac{1 - 13.94s - 13.1s^2}{(s + 0.0204)(s + 10)} \tag{10-36}$$

Note that the factor $10/(s + 10)$ is actually multiplied to $G_c(s)$, since the added pole must not affect the zero-frequency gain of the system.

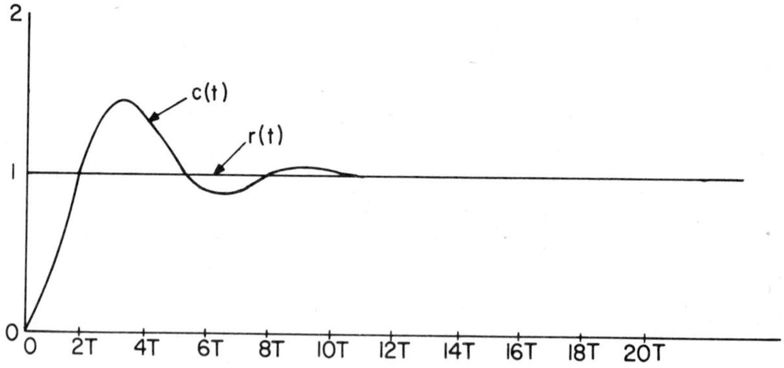

Fig. 10-14. Unit-step response of the sampled-data control system designed in Example 10-2.

10.3 FEEDBACK COMPENSATION WITH CONTINUOUS-DATA CONTROLLERS

In general, continuous-data controllers can also be placed in the feedback paths of a control system for the improvement on the performance of the system.

For example, the rate sensor used in the feedback path of the digital space vehicle control system shown in Fig. 7-5 is aimed primarily at the improvement on the damping of the system. A design problem may involve the determination of the sensor gain K_r so that the overall system will perform in a specific way. In general, we may consider that the transfer function H(s) of the feedback element is unknown and is to be determined.

To illustrate how feedback compensation with continuous-data controllers may be handled, we consider the system configuration shown in Fig. 10-15. In general, of course, we may encounter situations in which controllers are located in both the forward path as well as the feedback path.

A digital control system with the digital controller located in the forward path is shown in Fig. 10-16. Since the transfer function of the digital controller, $G_c(z)$, is isolated from the zero-order hold and the controlled process, it is generally simple to investigate the effects of varying the elements of $G_c(z)$. Thus, assuming that we can determine the transfer function of a desired $G_c(z)$ that will satisfy the performance criteria of the system in Fig. 10-16, we shall show that the feedback controllers in the systems of Fig. 10-15 can be determined by finding the relation between H(s) and $G_c(s)$ so that the systems are equivalent. The closed-loop transfer function of the system in Fig. 10-16 is

$$\frac{C(z)}{R(z)} = \frac{G_c(z)G_{h0}G_p(z)}{1 + G_c(z)G_{h0}G_p(z)} \tag{10-37}$$

and the closed-loop transfer function of the system in Fig. 10-15 is

$$\frac{C(z)}{R(z)} = \frac{G_{h0}G_p(z)}{1 + G_{h0}H(z) + G_{h0}G_p(z)} \tag{10-38}$$

Multiplying the numerator and the denominator of the last equation by $G_c(z)$, we have

$$\frac{C(z)}{R(z)} = \frac{G_c(z)G_{h0}G_p(z)}{G_c(z) + G_c(z)G_{h0}H(z) + G_c(z)G_{h0}G_p(z)} \tag{10-39}$$

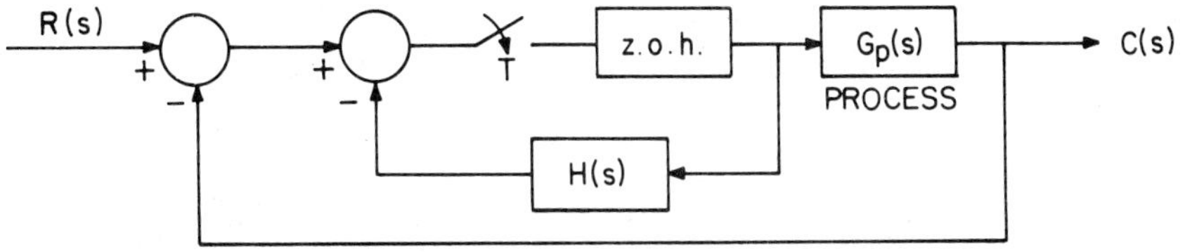

Fig. 10-15. Digital control system with continuous-data feedback controller.

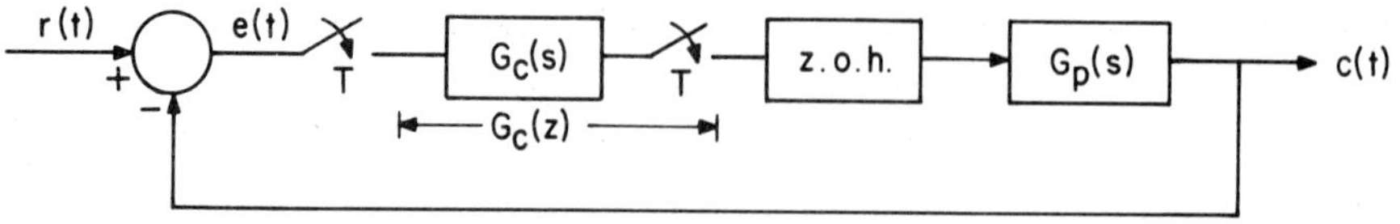

Fig. 10-16. Digital control system with cascade digital controller.

Comparing the two transfer functions in Eqs. (10-37) and (10-39) we see that the two expressions are equal when the following relation holds:

$$1 + G_c(z)G_{h0}G_p(z) = G_c(z) + G_c(z)G_{h0}H(z) + G_c(z)G_{h0}G_p(z) \tag{10-40}$$

Solving for $G_{h0}H(z)$ from the last equation, we get

$$G_{h0}H(z) = \frac{1 - G_c(z)}{G_c(z)} \tag{10-41}$$

or

$$\mathcal{Z}\left[\frac{H(s)}{s}\right] = \frac{z}{z-1}\,\frac{1 - G_c(z)}{G_c(z)} \tag{10-42}$$

Thus, once the digital controller transfer function $G_c(z)$ for the system of Fig. 10-16 is determined, the transfer function H(s) of the equivalent system in Fig. 10-15 can be obtained from Eq. (10-42).

However, in determining $G_c(z)$ it is important to keep in mind the physical realizability requirement on H(s). In other words, what are the restrictions on $G_c(z)$ so that H(s) is a physically realizable transfer function? For a transfer function to be physically realizable, the system represented by the transfer function must not produce an output prior to the application of an input. This is true for digital as well as continuous-data systems. Since if H(s) is physically realizable, H(s)/s also is, the power-series expansion of $\mathcal{Z}$[H(s)/s] should have no positive powers in z. Let $\mathcal{Z}$[H(s)/s] be represented by the following form:

$$\mathcal{Z}\left[\frac{H(s)}{s}\right] = \frac{\sum_{k=0}^{n} a_k z^{-k}}{\sum_{k=0}^{m} b_k z^{-k}} \tag{10-43}$$

The transfer function in the last equation is physically realizable if $b_0 \neq 0$, and m and n can be any positive integer.

Substituting Eq. (10-43) into Eq. (10-42) and solving for $G_c(z)$, we have

$$G_c(z) = \frac{\sum_{k=0}^{m} b_k z^{-k}}{\sum_{k=0}^{m} b_k z^{-k} + (1 - z^{-1}) \sum_{k=0}^{n} a_k z^{-k}} \tag{10-44}$$

Thus, the physical realizability of H(s)/s requires that $b_0 \neq 0$ in $G_c(z)$, or the constant term in the numerator polynomial in z^{-1} of $G_c(z)$ cannot be zero.

For H(s) to be an RC realizable transfer function, all the poles of H(s) must be simple and must lie on the negative real axis of the s-plane, excluding the origin and infinity. The zeros of H(s) can be anywhere in the s-plane. Thus, H(s)/s can be expanded into the following form by partial-fraction expansion:

$$\frac{H(s)}{s} = \frac{A_0}{s} + \sum_{k=1}^{m} \frac{A_k}{s + s_k} \tag{10-45}$$

where A_0 and A_k are constants and $-s_k$ $(k = 1,2,\ldots,m)$ are simple negative poles. The z-transform of H(s)/s is written

$$\mathcal{Z}\left[\frac{H(s)}{s}\right] = \frac{A_0 z}{z - 1} + \sum_{k=1}^{m} \frac{A_k z}{z - e^{-s_k T}} \tag{10-46}$$

where there is only one pole at $z = 1$ and the rest of the poles are all inside the unit circle $|z| = 1$. The form of $G_c(z)$ represented by Eq. (10-44) implies that $G_c(z)$ has the same number of poles and zeros. Furthermore, comparing Eqs. (10-43) and (10-44), we see that the poles of [H(s)/s] are generated from the zeros of $G_c(z)$. Therefore, the RC realizability of H(s) requires in addition that $G_c(z)$ be of the form of Eq. (10-44), $G_c(z)$ must have an equal number of poles and zeros, and the zeros of $G_c(z)$ must be simple and lie inside the unit circle in the z-plane. There are no restrictions on the location of the poles of $G_c(z)$, and, therefore, an RC transfer function for H(s) can be derived from even an unstable $G_c(z)$ that satisfies all the above requirements. To illustrate the realization of a digital controller by an equivalent continuous-data feedback controller configured in Fig. 10-15, the following illustrative example is given.

Example 10-3.

Consider that the transfer function of the digital controller of the system in Fig. 10-16 is given as

$$G_c(z) = \frac{(1 - 0.2z^{-1})(1 - 0.1z^{-1})}{(1 - 0.5z^{-1})(1 - 0.8z^{-1})} \tag{10-47}$$

We do not have to be concerned with how $G_c(z)$ is determined and what its effects are on the system in Fig. 10-16.

Since $G_c(z)$ has an equal number of poles and zeros, and all the zeros are simple and lie inside the unit circle $|z| = 1$, H(s) can be realized by an RC network. Substituting Eq. (10-47) into Eq. (10-42), we have

$$\mathcal{Z}\left[\frac{H(s)}{s}\right] = \frac{-z^{-1} + 0.38z^{-2}}{(1 - z^{-1})(1 - 0.2z^{-1})(1 - 0.1z^{-1})}$$

$$= \frac{-0.86z}{z-1} - \frac{2.25z}{z-0.2} + \frac{3.11z}{z-0.1} \tag{10-48}$$

The corresponding Laplace transform of each of the terms of the last equation is found, and we get (T = 1 sec)

$$\frac{H(s)}{s} = \frac{-0.86}{s} - \frac{2.25}{s+1.61} + \frac{3.11}{s+2.3} \tag{10-49}$$

from which

$$H(s) = \frac{-3.53s - 3.18}{(s+1.61)(s+2.3)} \tag{10-50}$$

In this section we have outlined a method of using continuous-data feedback controllers in digital control systems. In designing the feedback controller, an equivalent cascade-digital-controller controlled system is used. Since the transfer function of the digital controller $G_c(z)$ is separated from the transfer function of the controlled process $G_{h0}G_p(z)$, the design of $G_c(z)$ can be carried out more easily. The design of digital control systems with digital controllers is discussed in the remaining sections of this chapter.

10.4 THE DIGITAL CONTROLLER

The most versatile way of compensating a digital control system is by use of a digital controller. In general, digital controllers can be implemented by digital networks or digital computers such as microcomputers. Compared with the continuous-data controller, far better performance can be realized with a given control system by using a digital controller. Another advantage of using a digital-computer-controlled system is that the control algorithm can be easily changed by changing the program of the controller, especially if the controller is implemented by a microprocessor, whereas for a continuous-data controller this is rather difficult to do once the controller has been implemented.

Before entering into the design of digital controllers, we shall discuss the physical realizability and the composition of the digital controller. We can represent the digital controller by the block diagram notation of Fig. 10-17. The input of the digital controller, $e_1^*(t)$ is in the form of a sequence of number $e_1(kT)$, analytically represented as the samples of $e_1(t)$. The digital controller performs certain linear operations on the sequence $e_1(kT)$ and delivers

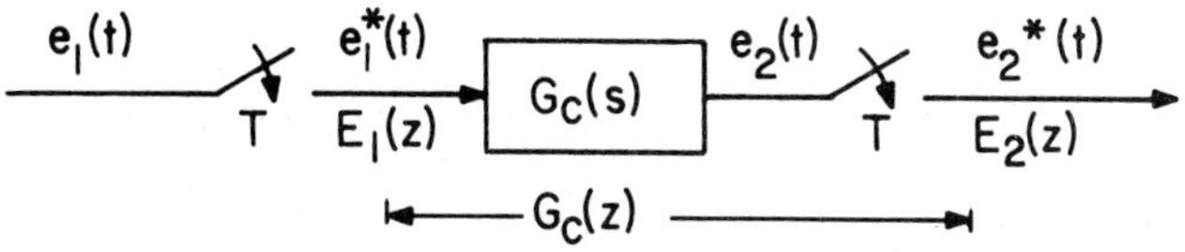

Fig. 10-17. The block diagram of a digital controller.

the output sequence $e_2(kT)$ in the form of the sampled signal $e_2^*(t)$ at the output. The transfer function of the digital controller is defined as

$$G_c(z) = \frac{E_2(z)}{E_1(z)} \tag{10-51}$$

There are many practical ways of implementing a digital controller. A digital controller may be implemented by a passive network preceded and followed by sample-and-hold devices. It may also be realized by a digital computer, such as a microcomputer. When using a microcomputer as a digital controller we must be aware of the limitation of the computer in terms of its wordlength, memory capacity and computation speed. We do not impose any constraint on the design of digital controllers in this chapter. However, the constraints on the design due to the limitations of a microcomputer will be discussed in Chapter 14. The reader simply has to keep in mind that in practice not all the controller configurations and parameter values determined by analytical means can be implemented exactly by physical components.

Physical Realizability

The a priori requirement on the design of the digital controller is that the transfer function $G_c(z)$ be physically realizable. The condition of physical realizability of a system implies that no output signal of the system will appear before an input signal is applied. We can express the transfer function $G_c(z)$ as a quotient of two polynomials in z,

$$G_c(z) = \frac{b_m z^m + b_{m-1} z^{m-1} + \ldots + b_0}{a_n z^n + a_{n-1} z^{n-1} + \ldots + a_0} \tag{10-52}$$

Expanding $G_c(z)$ in a power series in z^{-1}, the coefficients of the series represent the values of the weighting sequence of the digital controller. The coefficient of the z^{-k} term $(k = 0,1,2,\ldots)$ corresponds to the value of the weighting sequence at $t = kT$. Clearly, for the digital controller to be physically realizable, the power-series expansion of $G_c(z)$ must not contain any positive power in z. Any positive power in z in the series expansion of $G_c(z)$ indicates "prediction" or simply that the output precedes the input. Therefore, for the transfer function given in Eq. (10-52) to be physically realizable, the highest power of the denominator must be equal to or greater than that of the numerator, or simply $n \geqslant m$.

If the digital controller has the same number of poles and zeros, it is expressed as

$$G_c(z) = \frac{b_0 + b_1 z^{-1} + b_m z^{-m}}{a_0 + a_1 z^{-1} + a_n z^{-n}} \tag{10-53}$$

where n and m are any positive integers. In this case, in order that the power-series expansion of $G_c(z)$ will not contain positive powers of z, the denominator must not contain any factor of z^{-1}, if $b_0 \neq 0$. Therefore, if $b_0 \neq 0$, the condition for $G_c(z)$ to be physically realizable is that $a_0 \neq 0$.

Realization of Digital Controllers by Digital Networks

A digital network is defined as an electric network preceded and followed by a sample-and-hold. We show in the following how a digital controller may be realized by a *series digital network*, a *feedback digital network*, or a *series-feedback digital network*.

1. *Series Digital Network*

The block diagrams showing the implementation of a digital controller by a series digital network are shown in Fig. 10-18. The zero-order hold at the output of the digital controller is included to show that in general the digital data are processed by a D/A before being sent to the analog controlled process. This zero-order hold at the output of the digital controller is absorbed by the sample-and-hold at the output of the series digital network shown in Fig. 10-18(b). The transfer function $G_d(s)$ in Fig. 10-18(b) represents any electric network; however, for simplicity and economy, a network with only resistors and capacitors is preferred (RC network).

From Fig. 10-18(b), we have

$$G_c(z) = G_{h0}G_d(z) = \mathscr{Z}[G_{h0}(s)G_d(s)]$$
$$= (1 - z^{-1})\mathscr{Z}\left[\frac{G_d(s)}{s}\right] \tag{10-54}$$

Thus,

$$\mathscr{Z}\left[\frac{G_d(s)}{s}\right] = \frac{1}{1 - z^{-1}} G_c(z) \tag{10-55}$$

This relation shows that given the transfer function $G_c(z)$ of a digital controller,

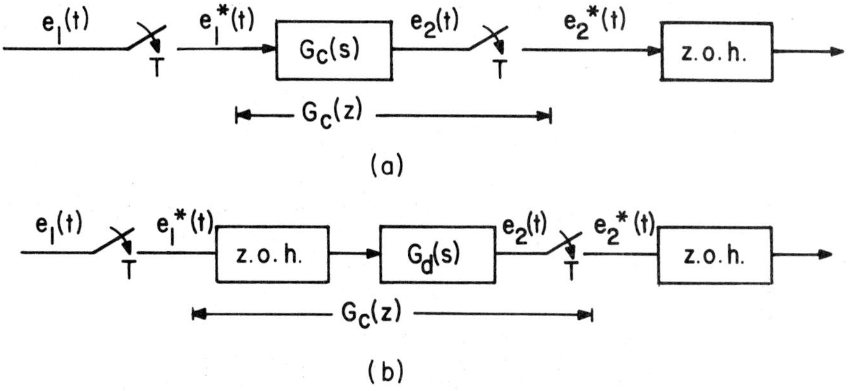

Fig. 10-18. Realization of a digital controller by a series digital network. (a) Digital controller, (b) Series digital network.

the transfer function of the series digital network can be determined from Eq. (10-55). If $G_d(s)$ is to be an RC-realizable transfer function, all the poles of $G_d(s)$ must be simple and lie on the negative real axis of the s-plane with the exception of the origin and infinity. The zeros of $G_d(s)$ can be located anywhere in the s-plane. Let $G_d(s)/s$ be expanded into the following form by partial-fraction expansion:

$$\frac{G_d(s)}{s} = \frac{A_0}{s} + \sum_{k=1}^{n} \frac{A_k}{s + s_k} \tag{10-56}$$

where A_0 and A_k are constants, and $-s_k$ $(k = 1,2,\ldots,n)$ are simple negative real poles. The z-transform of Eq. (10-56) is

$$\mathcal{Z}\left[\frac{G_d(s)}{s}\right] = \frac{A_0}{1 - z^{-1}} + \sum_{k=1}^{n} \frac{A_k}{1 - e^{-s_k T} z^{-1}} \tag{10-57}$$

which has simple positive real poles inside the unit circle $|z| = 1$, with only one pole at $z = 1$. Comparing Eq. (10-57) with Eq. (10-55) we see that in order for $G_d(s)$ to represent an RC network, the transfer function $G_c(z)$ must have the following properties:

(i) The number of poles of $G_c(z)$ must be equal to or greater than the number of zeros.
(ii) The zeros of $G_c(z)$ are arbitrary.
(iii) The poles of $G_c(z)$ must be simple, real and positive and less than unity.

As an illustration on the realizability of a digital controller by a digital RC network, an example is presented below.

Example 10-4.

Given the transfer function of a digital controller

$$G_c(z) = \frac{1 - 0.5z^{-1}}{1 - 0.2z^{-1}} \tag{10-58}$$

The sampling period is assumed to be 1 sec.

The problem is to realize $G_c(z)$ by the series digital network configuration shown in Fig. 10-18(b). Since $G_c(z)$ meets the three requirements of RC realizability listed above, we can proceed to substitute $G_c(z)$ from Eq. (10-58) in Eq. (10-55).

$$\mathcal{Z}\left[\frac{G_d(s)}{s}\right] = \frac{1 - 0.5z^{-1}}{(1 - z^{-1})(1 - 0.2z^{-1})} \tag{10-59}$$

Expanding the last equation by partial fraction, we have

$$\mathcal{Z}\left[\frac{G_d(s)}{s}\right] = \frac{0.625}{1 - z^{-1}} + \frac{0.375}{1 - 0.2z^{-1}} \tag{10-60}$$

Thus,

$$\frac{G_d(s)}{s} = \frac{0.625}{s} + \frac{0.375}{s + 1.61} = \frac{s + 1}{s(s + 1.61)} \tag{10-61}$$

and

$$G_d(s) = \frac{s + 1}{s + 1.61} \tag{10-62}$$

The last equation is realized by the RC network shown in Fig. 10-19. The digital network is complete when a sample-and-hold is applied at the input and the output of the RC network of Fig. 10-19, as in Fig. 10-18(b).

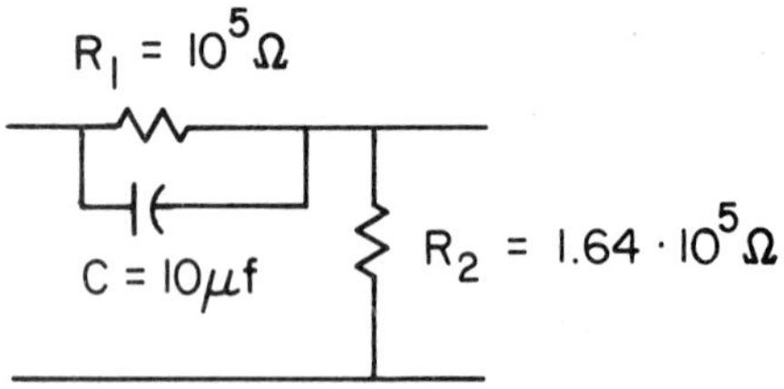

Fig. 10-19. Network realization of the transfer function $G_d(s) = (s + 1)/(s + 1.61)$.

2. *Feedback Digital Network*

Figure 10-20(b) shows the block diagram of a feedback digital network realization of the digital controller. The continuous-data transfer function H(s) represents an electric network, preferably an RC network. For the transfer functions of the two systems in Fig. 10-20 to be identical, the following condition must hold:

$$G_c(z) = \frac{1}{1 + G_{h0}H(z)} \tag{10-63}$$

Thus,

$$\mathcal{Z}\left[\frac{H(s)}{s}\right] = \frac{1}{1 - z^{-1}}\left[\frac{1 - G_c(z)}{G_c(z)}\right] \tag{10-64}$$

We see that the problem we have at hand is identical to the feedback-compensation problem discussed in Sec. 10.3. In fact, Eq. (10-64) is the same as Eq. (10-42). Thus, the results obtained earlier can all be applied here. The conclusion is that in order for H(s) to be realizable by an RC network, the digital controller transfer function, $G_c(z)$, must have the following properties:

(i) $G_c(z)$ must have the same number of poles and zeros.
(ii) The poles of $G_c(z)$ are arbitrary.
(iii) The zeros of $G_c(z)$ must be simple, real, positive, and lie inside the unit circle of the z-plane.

Note that in the feedback structure, restrictions are placed on the location of the zeros rather than on the poles of $G_c(z)$. In fact, H(s) is RC realizable

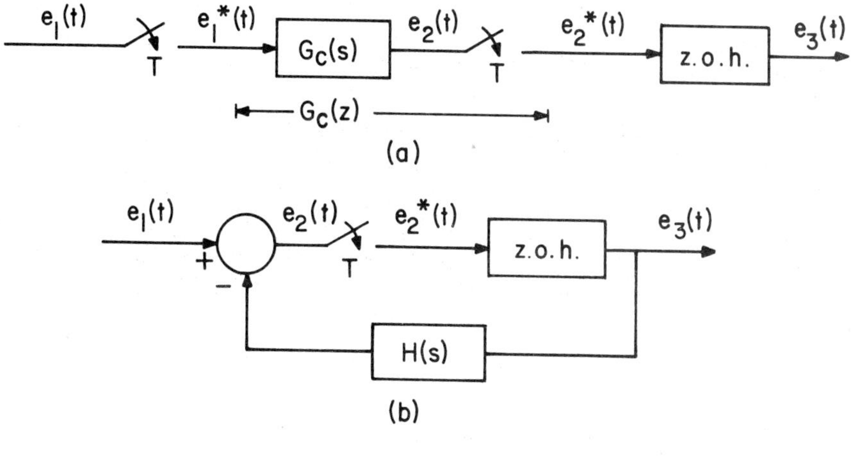

Fig. 10-20. Realization of digital controller by a feedback digital network. (a) Digital controller, (b) Equivalent feedback digital network.

even if $G_c(z)$ has poles outside the unit circle and is thus an unstable controller. Example 10-3 can be regarded as an illustration to the feedback digital network realization of a digital controller.

3. *Series-Feedback Digital Network*

Both the series digital network and the feedback digital network discussed above have restrictions in the realization of digital controller transfer functions. If the digital controller transfer function has complex poles and zeros or poles and zeros that are on or outside the unit circle, it can be realized neither by a series nor a feedback digital network. In the series structure the zeros of $G_c(z)$ are arbitrary, whereas in the feedback structure the poles of $G_c(z)$ are arbitrary. This means that certain combinations of the series and feedback configurations may relieve all the restrictions on the poles and zeros of $G_c(z)$ with the exception of course that $G_c(z)$ must be physically realizable. Many such combinations are conceivable, but let us consider the simple feedback-series digital network shown in Fig. 10-21. It can be shown that any physically realizable digital controller transfer function $G_c(z)$ in Fig. 10-21(a) can be realized by the feedback-series digital network configuration shown in Fig. 10-21(b), where H(s) and $G_d(s)$ represent RC networks.

From Fig. 10-21(b) the transfer function of the feedback-series digital network is written

$$\frac{E_2(z)}{E_1(z)} = \frac{G_{h0}G_d(z)}{1 + G_{h0}H(z)} \tag{10-65}$$

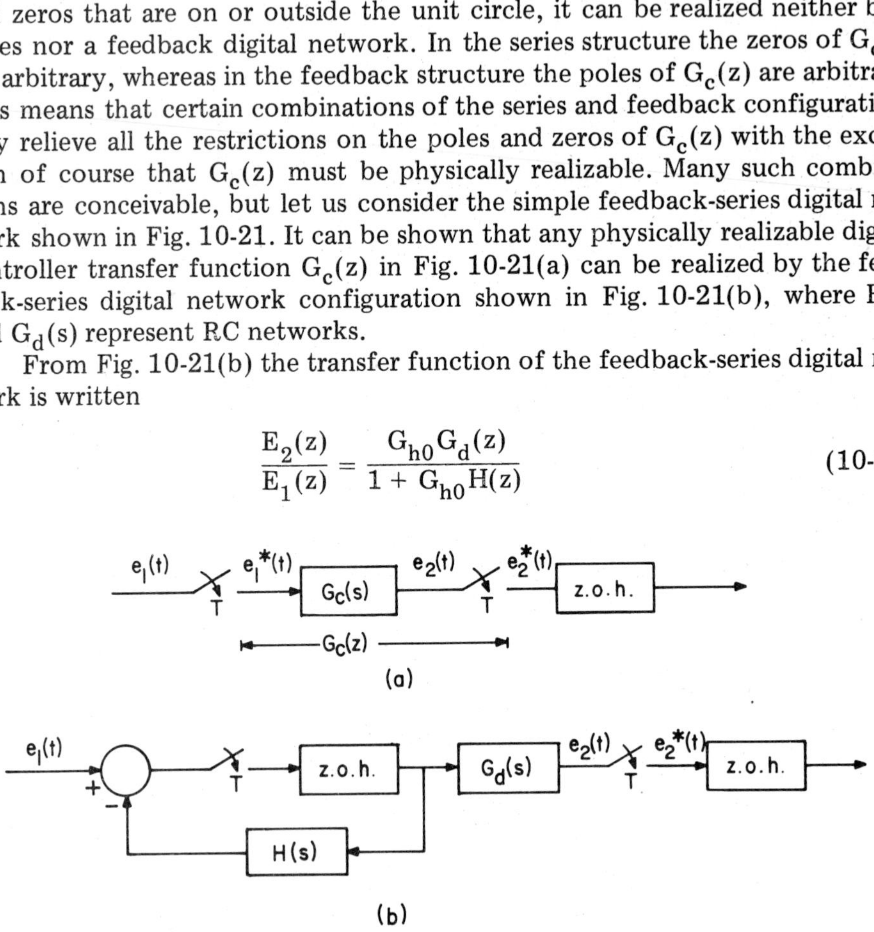

Fig. 10-21. Realization of digital controller by a feedback-series digital network. (a) Digital controller, (b) Equivalent feedback-series digital network.

For the two systems in Fig. 10-21 to be equivalent,

$$G_c(z) = \frac{G_{h0}G_d(z)}{1 + G_{h0}H(z)} \tag{10-66}$$

Consider that $G_c(z)$ may have complex conjugate poles and zeros and/or poles and zeros that are on or outside the unit circle. Then, we can write $G_c(z)$ as the product of two transfer functions, $G_{cf}(z)$ and $G_{cs}(z)$; that is,

$$G_c(z) = G_{cf}(z)G_{cs}(z) \tag{10-67}$$

where

$$G_{cf}(z) = \frac{1}{1 + G_{h0}H(z)}$$

$$G_{cs}(z) = G_{h0}G_d(z) \tag{10-68}$$

It is clear that $G_{cf}(z)$ and $G_{cs}(z)$ are the transfer functions of the feedback digital network and the series digital network in Eq. (10-63) and Eq. (10-54), respectively. Therefore, $G_{cf}(z)$ should have all the properties of a transfer function that can be realized by a feedback digital network with RC elements. This means that $G_{cf}(z)$ should contain all the poles of $G_c(z)$ that are complex, on or outside the unit circle. The zeros of $G_{cf}(z)$ must be simple, real, positive, and less than unity. The number of poles and zeros of $G_{cf}(z)$ must be equal. Similarly, $G_{cs}(z)$ must satisfy all the requirements for the series digital network, and in addition, it must contain all the zeros of $G_c(z)$ that are complex, on or outside the unit circle. The number of poles and zeros of $G_{cs}(z)$ must also be equal.

To illustrate the formulation of $G_{cf}(z)$ and $G_{cs}(z)$ for a given digital controller transfer function $G_c(z)$, let us consider that $G_c(z)$ is given by the following expression:

$$G_c(z) = \frac{\overbrace{(1 - 0.5z^{-1})(1 - 0.6z^{-1})(1 - 0.8z^{-1})(1 + 0.2z^{-1})}^{\text{realizable zeros}}\overbrace{(1 - z^{-1} + z^{-2})}^{\text{unrealizable zeros}}}{\underbrace{(1 - 0.2z^{-1})(1 - 0.37z^{-1})(1 - 0.9z^{-1})}_{\text{realizable poles}}\underbrace{(1 - z^{-1})(1 - 0.5z^{-1} + z^{-2})}_{\text{unrealizable poles}}} \tag{10-69}$$

We have referred to the zeros and poles of $G_c(z)$ that are real, simple, positive, and less than one, as the *realizable* zeros and poles, respectively. The complex, non-positive (including $z = 0$) or multiple-order zeros and poles as well as those that are on or outside the unit circle are referred to as the *unrealizable* zeros and poles, respectively.

The transfer functions $G_{cf}(z)$ and $G_{cs}(z)$ are formulated as follows:

$$G_{cf}(z) = \frac{\text{Realizable zeros of } G_c(z)}{\text{Unrealizable poles of } G_c(z)}$$

$$= \frac{(1 - 0.5z^{-1})(1 - 0.6z^{-1})(1 - 0.8z^{-1})}{(1 - z^{-1})(1 - 0.5z^{-1} + z^{-2})} \qquad (10\text{-}70)$$

$$G_{cs}(z) = \frac{\text{Unrealizable zeros of } G_c(z)}{\text{Realizable poles of } G_c(z)}$$

$$= \frac{(1 + 0.2z^{-1})(1 - z^{-1} + z^{-2})}{(1 - 0.2z^{-1})(1 - 0.37z^{-1})(1 - 0.9z^{-1})} \qquad (10\text{-}71)$$

Since $G_{cs}(z)$ and $G_{cf}(z)$ now satisfy their respective requirements to be realized by digital networks, the digital controller transfer function in Eq. (10-69) can be realized by the feedback-series configuration of Fig. 10-21(b) with RC network elements.

The illustrative example given above is fabricated in such a manner that the transfer functions $G_{cf}(z)$ and $G_{cs}(z)$ that are formulated from $G_c(z)$ by properly assigning the realizable and the unrealizable poles and zeros of $G_c(z)$ to the two functions are directly realizable by the feedback and the series networks, respectively. Careful inspection of the transfer function in Eq. (10-69) reveals that $G_{cf}(z)$ and $G_{cs}(z)$ are directly realizable by their corresponding digital networks only if the following conditions are met:

(i) The number of realizable poles of $G_c(z)$ is equal to the number of unrealizable zeros of $G_c(z)$.

(ii) The number of realizable zeros of $G_c(z)$ is equal to the number of unrealizable poles of $G_c(z)$.

Let us investigate a general case for which the above conditions are not met. Consider the transfer function of a digital controller,

→| realizable zero |← unrealizable zeros →|

$$G_c(z) = \frac{(1 - 0.5z^{-1})(1 + 0.2z^{-1})(1 - z^{-1} + z^{-2})}{(1 - 0.2z^{-1})(1 - z^{-1})(1 - 0.5z^{-1} + z^{-2})} \qquad (10\text{-}72)$$

→| realizable pole |← unrealizable poles →|

We assign $G_{cf}(z)$ and $G_{cs}(z)$ as

$$G_{cf}(z) = \frac{\text{Realizable zero of } G_c(z)}{\text{Unrealizable poles of } G_c(z)} = \frac{z - 0.5}{(z - 1)(z^2 - 0.5z + 1)} \qquad (10\text{-}73)$$

$$G_{cs}(z) = \frac{\text{Unrealizable zeros of } G_c(z)}{\text{Realizable pole of } G_c(z)} = \frac{(z+0.2)(z^2-z+1)}{(z-0.2)} \tag{10-74}$$

Since these transfer functions do not have the same number of poles and zeros, they are not RC realizable. To overcome this difficulty, we can modify $G_{cf}(z)$ and $G_{cs}(z)$ as follows:

$$G_{cf}(z) = \frac{(z-0.5)(z-a)(z-b)}{(z-1)(z^2-0.5z+1)} \tag{10-75}$$

$$G_{cs}(z) = \frac{(z+0.2)(z^2-z+1)}{(z-0.2)(z-a)(z-b)} \tag{10-76}$$

where a and b are real numbers, $0 < a < 1$, $0 < b < 1$, and $a \neq b$; a and b are not equal to any other poles and zeros of $G_c(z)$.

Realization of Digital Controllers by Digital Computers

The most versatile way of implementing a digital controller is by use of a digital computer or digital-data processing unit. Because of the advantages in computing speed, storage capacity, and flexibility, the use of digital computers in control systems as controllers has become increasingly important. In recent years the advancements made in the field of microcomputers have made the applications of this type of computer very attractive for control systems.

In general, the transfer function of a digital controller can be realized by a program of a digital computer. Three basic methods of programming are known. These are: *direct programming*, *cascade programming*, and *parallel programming*. From the analytical standpoint these programming methods are very similar to the decomposition techniques presented in Chapter 4.

We shall consider that a digital computer is capable of performing specific arithmetic operations of addition, multiplication, subtraction, storage, shifting, etc. Some microcomputers cannot multiply two numbers directly, and a subprogram must be written to perform the simple multiplication.

1. *Direct Digital Programming*

A physically realizable discrete-data transfer function $G_c(z)$ of a digital controller may be written as

$$G_c(z) = \frac{E_2(z)}{E_1(z)} = \frac{b_0 + b_1 z^{-1} + b_2 z^{-2} + \dots + b_m z^{-m}}{a_0 + a_1 z^{-1} + a_2 z^{-2} + \dots + a_n z^{-n}} \tag{10-77}$$

where $a_0 \neq 0$ if $b_0 \neq 0$; m and n are positive integers, and $E_1(z)$ and $E_2(z)$ represent the z-transforms of the input and the output of the controller, respectively. To conduct a direct digital programming of $G_c(z)$, we perform cross-multiplication in Eq. (10-77), and then take the inverse z-transform; we get the following equation:

$$a_0 e_2^*(t) + \sum_{k=1}^{n} a_k e_2^*(t - kT) = \sum_{k=0}^{m} b_k e_1^*(t - kT) \tag{10-78}$$

Solving for $e_2^*(t)$ from the last equation yields

$$e_2^*(t) = \frac{1}{a_0} \sum_{k=0}^{m} b_k e_1^*(t - kT) - \frac{1}{a_0} \sum_{k=1}^{n} a_k e_2^*(t - kT) \tag{10-79}$$

The last equation shows that the present value of the output $e_2^*(t)$ depends on the present and the past values of the input $e_1^*(t)$ as well as the past information of the output. To implement the digital computer program using Eq. (10-79), two basic mathematical operations are required. The first operation is the *data storage*. The data storage stores the past samples of the output and the input so that they may be used in the computation of the output $e_2^*(t)$. The second operation involves arithmetic manipulations including multiplication of the stored output and input data by constants, addition and subtraction.

Let

$$x^*(t) = \frac{1}{a_0} \sum_{k=0}^{m} b_k e_1^*(t - kT) \tag{10-80}$$

and

$$y^*(t) = \frac{1}{a_0} \sum_{k=1}^{n} a_k e_2^*(t - kT) \tag{10-81}$$

Equation (10-79) becomes

$$e_2^*(t) = x^*(t) - y^*(t) \tag{10-82}$$

The block diagram representation of the direct digital programming of Eqs. (10-80), (10-81) and (10-82) is given in Fig. 10-22. Note that the data storages are simply time-delay units which delay the input data for one sampling period.

The direct digital program implemented in Fig. 10-22 requires a total of n + m data storages. As an alternate method, we can use the direct-decomposition scheme discussed in Chapter 4 as a direct digital program implementation of Eq. (10-77).

Applying direct decomposition to Eq. (10-77), we have the following equations:

$$E_2(z) = \frac{1}{a_0}(b_0 + b_1 z^{-1} + \ldots + b_m z^{-m})X(z) \tag{10-83}$$

$$X(z) = \frac{1}{a_0} E_1(z) - \frac{1}{a_0}(a_1 z^{-1} + a_2 z^{-2} + \ldots + a_n z^{-n})X(z) \tag{10-84}$$

where X(z) is simply a dummy variable. Figure 10-23 shows the block diagram of the direct digital programming by direct decomposition. It is assumed in the illustrated case that n = m. The block diagram can be easily modified if $n \neq m$. In general, the number of data-storage (time-delay) units is equal to the greater of n and m.

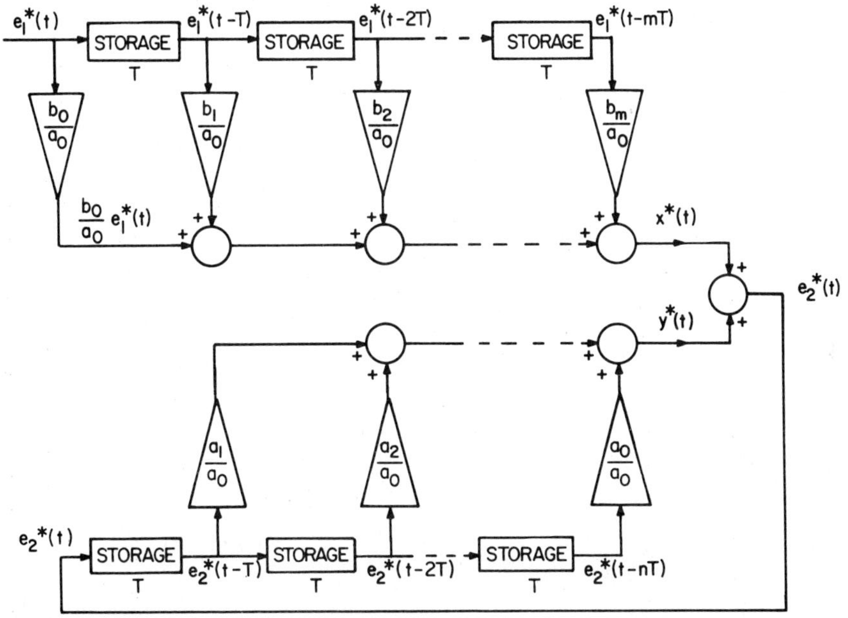

Fig. 10-22. Block diagram for direct digital programming of $G_c(z)$ in Eq. (10-77).

2. *Cascade Digital Programming*

The transfer function $G_c(z)$ may be written as the product of a number of simple transfer functions each realizable by a simple digital program. Then, the digital programming of $G_c(z)$ may be represented by the series of cascaded digital programs of the simple transfer functions. Writing Eq. (10-77) in factored form, we have

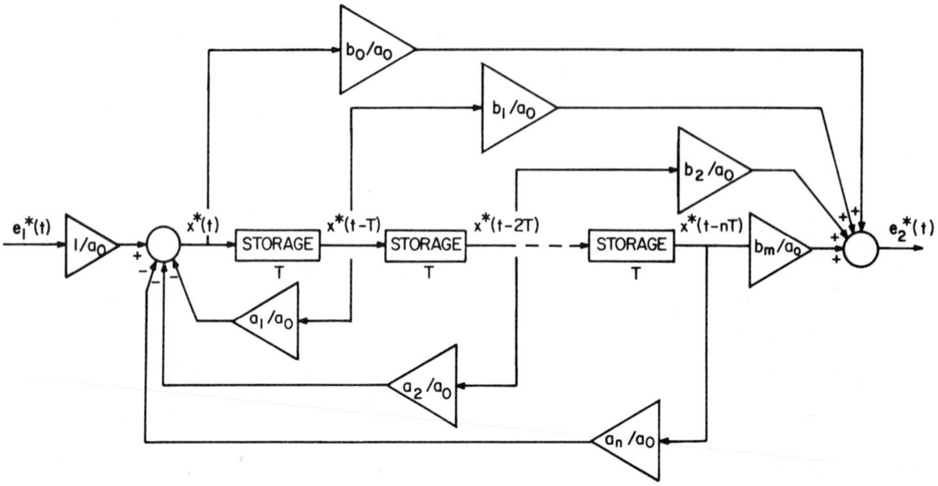

Fig. 10-23. Block diagram for direct digital programming of $G_c(z)$ in Eq. (10-77) by direct decomposition, n = m.

$$G_c(z) = \prod_{k=1}^{p} G_{ck}(z) \tag{10-85}$$

where p is the greater of n and m. Figure 10-24 shows the block diagram representation of the cascade digital programming of $G_c(z)$. In general, the transfer function $G_{ck}(z)$ may assume the following forms depending on the poles and zeros of $G_c(z)$, and the relative magnitude of m and n.

(i) $G_{ck}(z) = K_k \dfrac{1 + c_k z^{-1}}{1 + d_k z^{-1}}$ (real pole and zero)

(ii) $G_{ck}(z) = K_k \dfrac{1}{1 + d_k z^{-1} + f_k z^{-2}}$ (two complex conjugate poles)

(iii) $G_{ck}(z) = K_k \dfrac{1 + c_k z^{-1}}{1 + d_k z^{-1} + f_k z^{-2}}$ (one real zero and two complex conjugate poles)

(iv) $G_{ck}(z) = K_k \dfrac{1 + g_k z^{-1} + h_k z^{-2}}{1 + d_k z^{-1} + f_k z^{-2}}$ (complex conjugate poles and zeros)

(v) $G_{ck}(z) = K_k(1 + c_k z^{-1})$ (real zero, m > n)

(vi) $G_{ck}(z) = K_k(1 + g_k z^{-1} + h_k z^{-2})$ (complex conjugate zeros, $m > n$)

(vii) $G_{ck}(z) = K_k \dfrac{1 + g_k z^{-1} + h_k z^{-2}}{1 + d_k z^{-1}}$ (one real pole and two complex conjugate zeros, $m > n$)

These transfer functions can all be realized by the direct digital programming method discussed in the last section. The case illustrated in (i) is perhaps the most common as the real pole and zero of $G_c(z)$ are represented. The case in (ii) is for two complex conjugate poles of $G_c(z)$, and we do not want to use complex numbers in the digital program. The cases in (iii) through (vii) are just some other possible configurations depending on the relative magnitude of n and m.

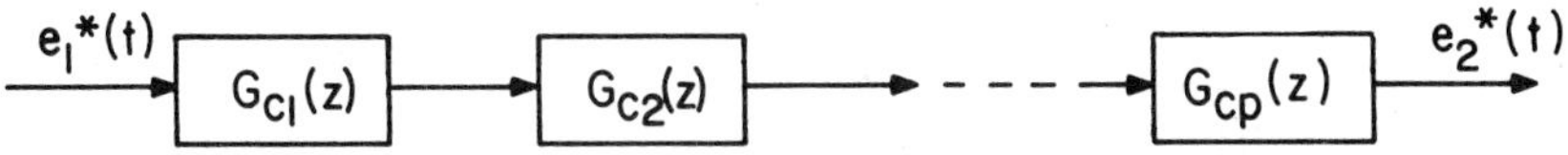

Fig. 10-24. Block diagram of cascade digital programming of $G_c(z)$.

3. *Parallel Digital Programming*

Another method of implementing digital programming is called the parallel programming, in which case the transfer function $G_c(z)$ is first expanded by partial-fraction expansion into a sum of simple transfer functions.

In general, the transfer function $G_c(z)$ in Eq. (10-77) can be written as

$$G_c(z) = \sum_{k=1}^{p} G_{ck}(z) \tag{10-86}$$

where p is the greater of m and n. Depending on the nature of $G_c(z)$, $G_{ck}(z)$ can be of the following forms:

(a) $G_{ck} = \dfrac{A_k}{1 + d_k z^{-1}}$ (simple real pole)

(b) $G_{ck} = \dfrac{A_k}{(1 + d_k z^{-1})^j}$ ($j = 1, 2, \ldots, N$; real pole of multiplicity N)

(c) $G_{ck} = \dfrac{A_k(1 + c_k z^{-1})}{(1 + d_k z^{-1} + f_k z^{-2})}$ (simple complex conjugate poles)

(d) $G_{ck} = \dfrac{A_k(1 + c_k z^{-1})}{(1 + d_k z^{-1} + f_k z^{-2})^j}$ (j = 1,2,...,N; complex poles of multiplicity N)

(e) $G_{ck} = \dfrac{A_k}{z^j}$ (j = 1,2,...,N; m − n = N > 0)

Each of these above-listed transfer functions can again be realized by the direct digital programming method. The block diagram of the parallel digital programming is illustrated in Fig. 10-25.

We present the following example to illustrate the three methods of digital programming presented in this section.

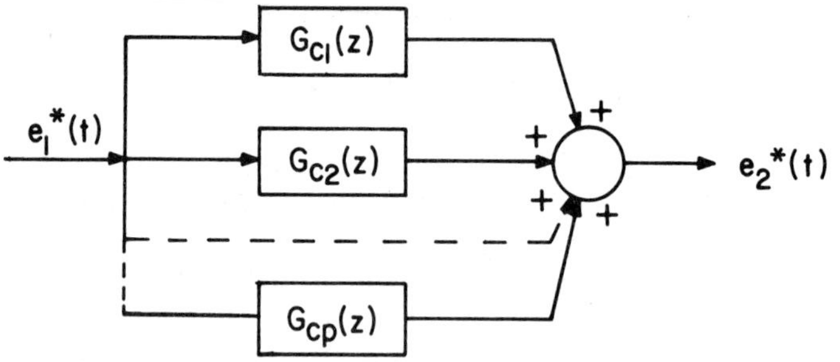

Fig. 10-25. Block diagram of parallel digital programming of $G_c(z)$.

Example 10-5.

Consider the following transfer function of a digital controller.

$$\frac{E_2(z)}{E_1(z)} = G_c(z) = \frac{5(1 + 0.25z^{-1})}{(1 - 0.5z^{-1})(1 - 0.1z^{-1})} \tag{10-87}$$

It is apparent that the transfer function is physically realizable. We shall realize the transfer function by the direct programming, cascade programming and parallel programming methods.

1. *Direct Digital Programming*

Expanding the denominator factors and cross-multiplying, Eq. (10-87) is written

$$(1 - 0.6z^{-1} + 0.05z^{-2})E_2(z) = 5(1 + 0.25z^{-1})E_1(z) \tag{10-88}$$

Taking the inverse z-transform on both sides of the last equation and solving for $e_2^*(t)$, we get

$$e_2^*(t) = 5e_1^*(t) + 1.25e_1^*(t - T) + 0.6e_2^*(t - T) - 0.05e_2^*(t - 2T) \tag{10-89}$$

where T is the sampling period. The block diagram of the direct digital programming of Eq. (10-87) according to Eq. (10-89) is shown in Fig. 10-26.

Alternately, we may apply the direct decomposition to Eq. (10-87), and get

$$E_2(z) = (5 + 1.25z^{-1})X(z) \tag{10-90}$$

$$X(z) = E_1(z) + 0.6z^{-1}X(z) - 0.05z^{-2}X(z) \tag{10-91}$$

These last two equations are realized by the digital program shown in Fig. 10-27.

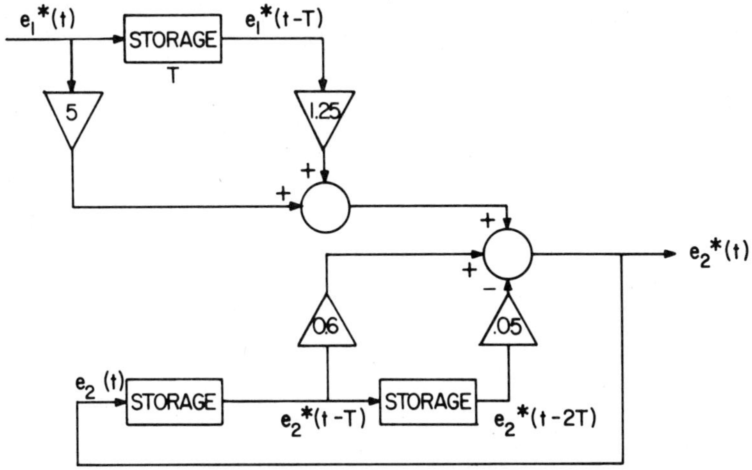

Fig. 10-26. A direct digital program of the transfer function $G_c(z)$ in Eq. (10-87).

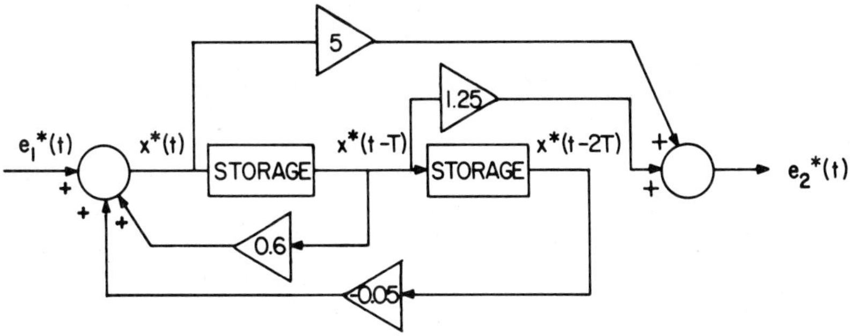

Fig. 10-27. A direct digital program of the transfer function $G_c(z)$ in Eq. (10-87).

2. *Cascade Digital Programming*

The right-hand side of Eq. (10-87) is arbitrarily divided and written as the product of two functions:

$$\frac{E_2(z)}{E_1(z)} = \frac{1 + 0.25z^{-1}}{1 - 0.5z^{-1}} \; \frac{5}{1 - 0.1z^{-1}} \tag{10-92}$$

The digital program shown in Fig. 10-28 is drawn according to the division of the transfer function as shown in Eq. (10-92).

3. *Parallel Digital Programming*

The right-hand side of Eq. (10-87) is expanded by partial-fraction expansion into the following form:

$$\frac{E_2(z)}{E_1(z)} = \frac{9.375}{1 - 0.5z^{-1}} - \frac{4.375}{1 - 0.1z^{-1}} \tag{10-93}$$

Figure 10-29 shows that the transfer function is realized by the parallel connection of two first-order programs, according to Eq. (10-93).

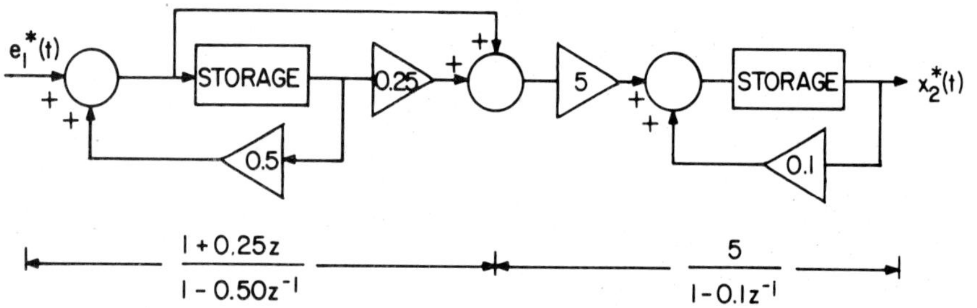

Fig. 10-28. A cascade digital program of the transfer function $G_c(z)$ in Eq. (10-87).

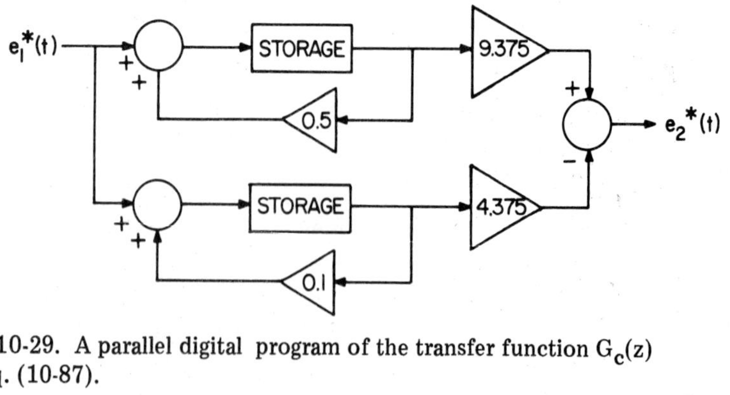

Fig. 10-29. A parallel digital program of the transfer function $G_c(z)$ in Eq. (10-87).

10.5 DESIGN OF DIGITAL CONTROL SYSTEMS WITH DIGITAL CONTROLLERS THROUGH THE BILINEAR TRANSFORMATION

In this section we shall consider the design of a digital control system with a digital controller, using the frequency-domain technique. This type of design problem is generally simpler to carry out than the design of digital control systems with continuous-data controllers. The reason is because the transfer function of the digital controller is isolated from that of the controlled process, so that the effects of varying the controller parameters may be investigated by means of the Bode diagram. With reference to the block diagram of Fig. 10-30, the design procedure may be outlined as follows:

1. Evaluate the z-transform of the zero-order hold and controlled process combination, $G_{h0}G_p(z)$. Apply the w-transform, $z = (1 + w)/(1 - w)$, to obtain $G_{h0}G_p(w)$.
2. Construct the Bode diagram for $G_{h0}G_p(w)$ in db magnitude and phase versus ω_w. Transfer the data from the Bode plot to the Nichols chart if necessary. Determine the performance characteristics of the uncompensated system by finding the gain margin, phase margin, bandwidth, resonance peak and resonant frequency from the Bode plot and the Nichols chart.
3. Should the system need compensation, the open-loop transfer function of the system with digital controller becomes $G_c(z)G_{h0}G_p(z)$, or in the w-domain, $G_c(w)G_{h0}G_p(w)$. The digital controller transfer function $G_c(w)$ is to be determined so that the desired system performance specifications are satisfied. The selection of $G_c(w)$ may follow the principle of design of continuous-data control systems, which involves a trial-and-error procedure, and to some extent based on the experience and imagination of the designer.

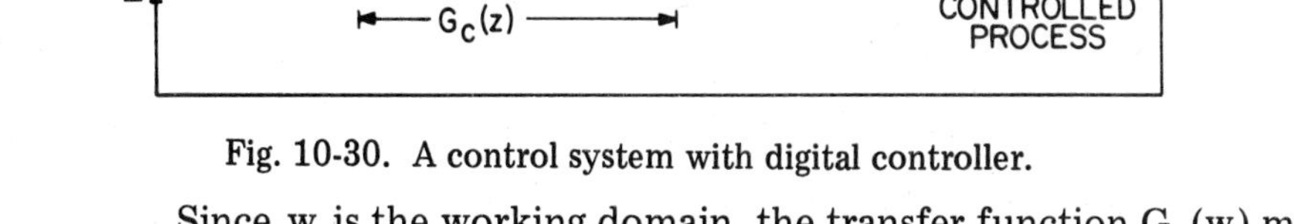

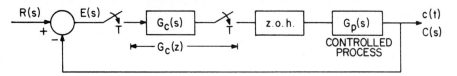

Fig. 10-30. A control system with digital controller.

Since w is the working domain, the transfer function $G_c(w)$ must be so chosen that the corresponding z-domain transfer function of the digital controller, $G_c(z)$, is physically realizable. Let $G_c(w)$ be of the following form:

$$G_c(w) = \frac{c_m w^m + c_{m-1} w^{m-1} + \ldots + c_1 w + c_0}{d_n w^n + d_{n-1} w^{n-1} + \ldots + d_1 w + d_0} \tag{10-94}$$

where n and m are positive integers. Let us investigate what are the constraints on the relative magnitudes of n and m and the coefficients of $G_c(w)$ so that the corresponding transfer function $G_c(z)$, obtained through the transformation, $w = (z - 1)/(z + 1)$, is physically realizable. Substituting $w = (z - 1)/(z + 1)$ into Eq. (10-94) and simplifying, we have

$$G_c(z) = \frac{c_m(z-1)^m + c_{m-1}(z-1)^{m-1}(z+1) + \ldots + c_1(z-1)(z+1)^{m-1} + c_0(z+1)^m}{d_n(z-1)^n + d_{n-1}(z-1)^{n-1}(z+1) + \ldots + d_1(z-1)(z+1)^{n-1} + d_0(z+1)^n}(z+1)^{n-m} \tag{10-95}$$

This expression for $G_c(z)$ indicates that if $G_c(w)$ is as given in Eq. (10-94), regardless of the relative magnitudes of m and n, $G_c(z)$ will always have the same number of poles and zeros. This means that $G_c(z)$ will always be physically realizable so long as $G_c(w)$ is of the form of Eq. (10-94). However, if it is desired that the digital controller is stable, then all the poles of $G_c(z)$ must lie inside the unit circle $|z| = 1$. This means that all the poles of $G_c(w)$ must lie inside the left-half of the w-plane. In addition, we note from Eq. (10-95) that if $m > n$ $G_c(z)$ will have $m - n$ poles at $z = -1$ which correspond to an unstable condition. Therefore, for a stable digital controller we must also require that $G_c(w)$ does not have more zeros than poles. In practice, it is permissible to have an unstable digital controller so long as the overall system is stable, although, if possible, it is desirable to keep both the open-loop as well as the closed-loop systems stable.

4. Once $G_c(w)$ is determined, $G_c(z)$ is obtained by substituting $w = (z - 1)/(z + 1)$ in $G_c(w)$. The final step in the design involves the realization of $G_c(z)$ by one of the methods discussed in Sec. 10.4. If $G_c(z)$ is to be implemented by a microcomputer, then the designer should be aware of the limitations and the constraints of the microcomputer, and take these into consideration when carrying out the design.

To illustrate the design method outlined above, an illustrative example follows.

Example 10-6.

Consider that the controller process of the digital control system shown in Fig. 10-30 is described by the transfer function

$$G_p(s) = \frac{K}{s(1 + 0.1s)(1 + 0.5s)} \tag{10-96}$$

The sampling period of the system is fixed at 0.5 second. In practice the sampling period is one of the system parameters that is subject to determination by the designer. However, in the present case, we assume that the sampling period has already been selected by some criterion that is not directly related to the design problem concerned here. The problem involves the design of the digital controller so that the following set of performance specifications are met:

ramp error constant $K_v \geqslant 1.4$

phase margin $\geqslant 50$ deg.

resonance peak $M_p \leqslant 1.3$

We would like to mention again that these specifications on relative stability are selected only as a qualitative means of creating an effective and convenient way for the designer to carry out the design using the available analytical tools. In the end one still has to check and see if the phase margin of 50 deg. and the resonance peak of 1.3 are indeed adequate for this system.

The z-transform of the uncompensated open-loop system transfer function is (for T = 0.5)

$$G_{h0}G_p(z) = \mathcal{Z}\left[\frac{1 - e^{-Ts}}{s}\,\frac{K}{s(1 + 0.1s)(1 + 0.2s)}\right]$$

$$= \frac{K(0.13z^2 + 0.177z + 0.0092)}{z(z-1)(z-0.368)}$$

$$= \frac{0.13K(z + 1.31)(z + 0.054)}{z(z-1)(z-0.368)} \tag{10-97}$$

Let us set the ramp error constant K_v to be at 1.5, from Eq. (7-75),

$$K_v = 1.4 = \frac{1}{T}\lim_{z \to 1}[(z-1)G_{h0}G_p(z)]$$

$$= 0.46\ K \tag{10-98}$$

Thus, K = 3. Substitution of $z = (1 + w)/(1 - w)$ and K = 3 into Eq. (10-97) yields

$$G_{h0}G_p(w) = \frac{0.75(1 - w)(1 + 0.9w)(1 - 0.134w)}{w(1 + w)(1 + 2.17w)} \tag{10-99}$$

The Bode plot of the transfer function in the last equation is sketched as shown in Fig. 10-31. From the magnitude and phase curves of the Bode diagram we see that for K = 3 the uncompensated system is only marginally stable. The gain margin and the phase margin are both approximately zero. It can be shown that the actual marginal value of K for stability is 3.3. To realize a phase margin of 50 deg. while maintaining the ramp error constant at 1.4, a phase-lag model is suggested for $G_c(w)$, since a phase-lead model will be ineffective for the present problem, due to the rapid fall-off of the phase curve beyond the phase-crossover frequency.

Let the transfer function of $G_c(w)$ be of the form,

$$G_c(w) = \frac{1 + a\tau w}{1 + \tau w} \qquad (a < 1) \tag{10-100}$$

From Fig. 10-31 we see that in order to realize a phase margin of 50 deg, the gain crossover of the system should be shifted from $\omega_w = 0.6$ to $\omega_w = 0.2$, provided that the phase at the new gain crossover is not significantly affected by the phase-lag characteristics of $G_c(w)$. The Bode plot indicates that the gain of $G_c(w)$ at $\omega_w = 0.2$ is approximately 12 db. Therefore, to make $\omega_w = 0.2$ the new gain crossover, the controller $G_c(w)$ must contribute 12 db of attenuation at this frequency. This allows us to determine the value of a in Eq. (10-100) by setting

$$20\log_{10} a = -12 \text{ db} \tag{10-101}$$

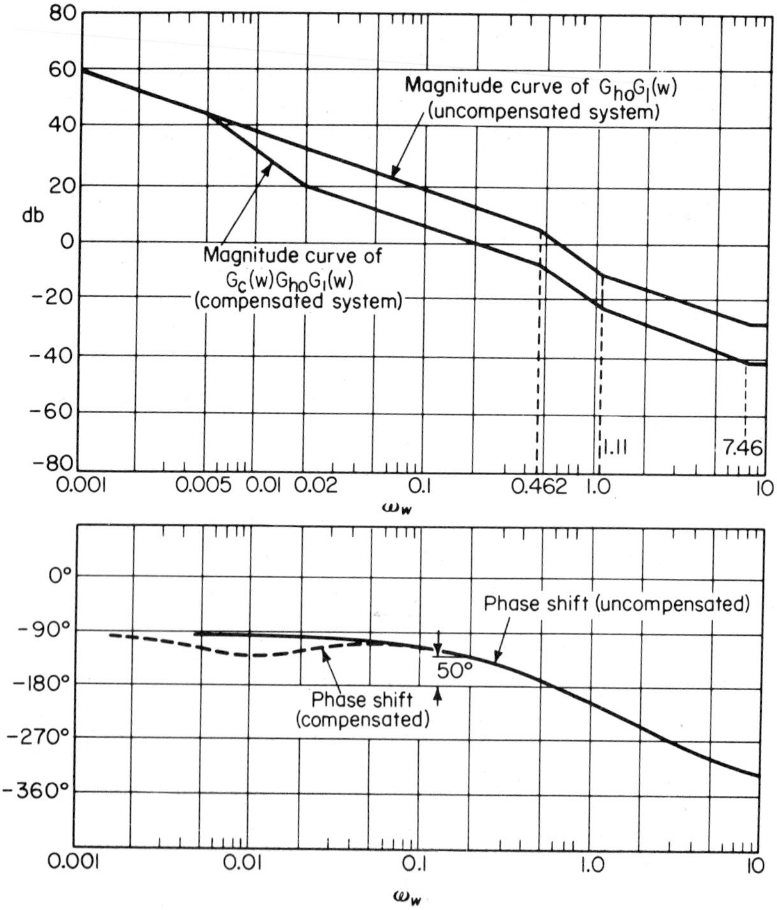

Fig. 10-31. Bode diagram of the sampled-data system with digital controller in Example 10-6.

Thus, $a = 0.25$. In order that the phase characteristic of $G_c(w)$ does not greatly affect the phase of the compensated system at $\omega_w = 0.2$, we set the upper corner frequency of $G_c(w)$ at one decade below $\omega_w = 0.2$. We set

$$1/a\tau = 0.02 \tag{10-102}$$

and thus,

$$1/\tau = 0.005 \tag{10-103}$$

Now the transfer function of the digital controller in the w-domain is

$$G_c(w) = \frac{1 + 50w}{1 + 100w} \tag{10-104}$$

The open-loop transfer function of the compensated system is given by

$$G_c(w)G_{h0}G_p(w) = \frac{0.75(1 + 50w)(1 - w)(1 + 0.9w)(1 - 0.134w)}{w(1 + 100w)(1 + w)(1 + 2.17w)} \tag{10-105}$$

The Bode plot of the last transfer function is plotted as shown in Fig. 10-31. Note that the phase margin is now approximately equal to 50 deg.

The magnitude and phase curves of $G_{h0}G_p(w)$ and $G_c(w)G_{h0}G_p(w)$ are transferred from the Bode diagram to the Nichols chart shown in Fig. 10-32. From these plots the following quantities on the system's performance are measured and comparisons are made between the compensated and the uncompensated system.

	Uncompensated System	Compensated System
Resonance peak M_p	almost infinite	1.2
Resonant frequency ω_{wr} (ω_r)	0.6 (2.1 rad/sec)	0.2 (0.8 rad/sec)
Phase margin	$\cong$ 0 deg.	50 deg.
Gain margin	$\cong$ 0 db	12 db
Bandwidth	0.9 (3 rad/sec)	0.4 (1.5 rad/sec)

The transfer function $G_c(z)$ is obtained by setting $w = (z - 1)/(z + 1)$ in Eq. (10-104). We have

$$G_c(z) = 0.25 \frac{z - 0.96}{z - 0.99} \tag{10-106}$$

To evaluate the true effectiveness of the design, we write the closed-loop transfer function of the compensated system.

$$\frac{C(z)}{R(z)} = \frac{G_c(z)G_{h0}G_p(z)}{1 + G_c(z)G_{h0}G_p(z)}$$

$$= \frac{0.0975z^3 + 0.0289z^2 - 0.1245z + 0.0066}{z^4 - 2.2605z^3 + 1.7512z^2 - 0.4888z + 0.0066} \qquad (10\text{-}107)$$

The closed-loop poles are at $z = 0.14213$, 0.955863, $0.64521 + j0.26364$ and $0.64521 - j0.26364$. The unit-step response of the system is plotted in Fig. 10-33. Note that the maximum overshoot is less than 13 percent indicating that the choice of the phase margin of 50 deg. is effective for this system.

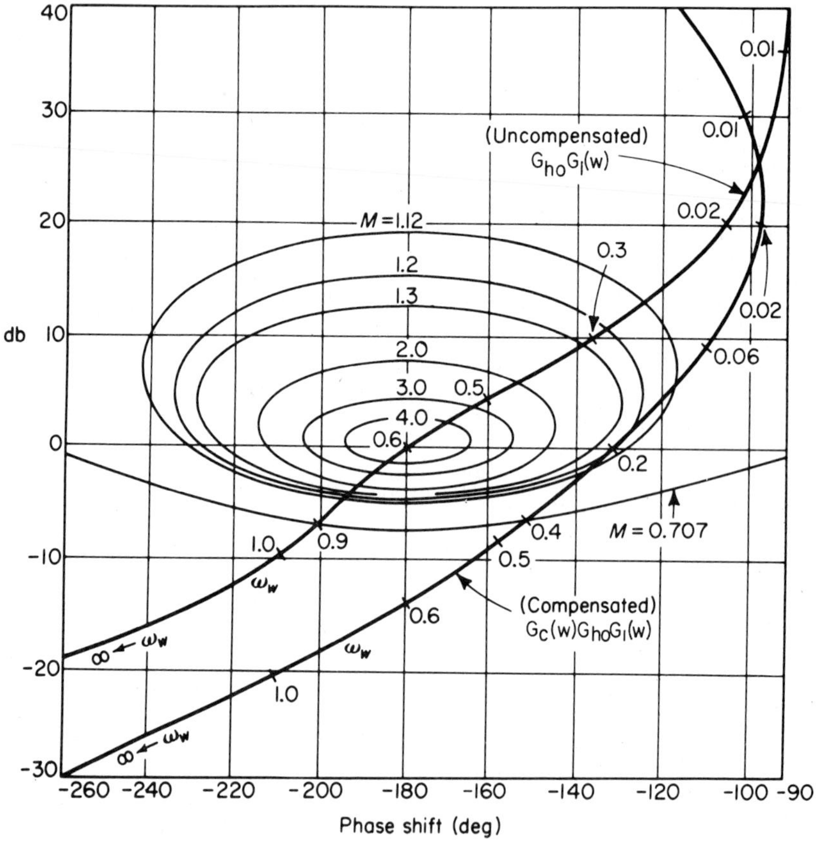

Fig. 10-32. Nichols chart and the plots of the open-loop transfer functions for the sampled-data system in Example 10-6.

The final step in the design may involve the implementation of $G_c(z)$ by any one of the methods described in Sec. 10.4. Figure 10-34 gives a direct-digital-program realization of $G_c(z)$.

Substituting Eq. (10-106) into Eq. (10-55), we get

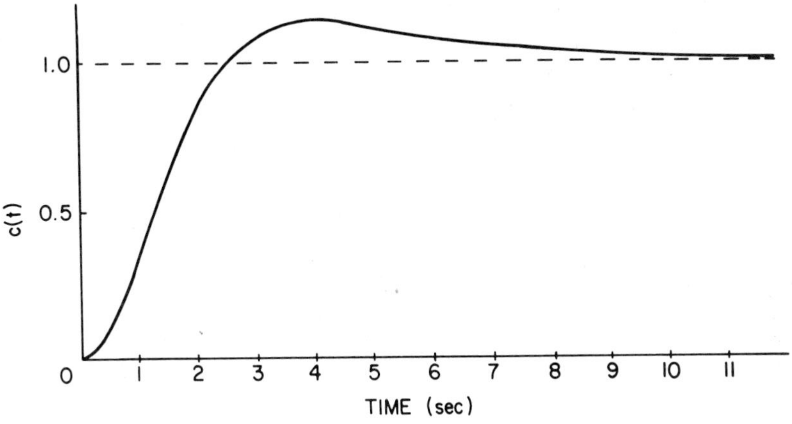

Fig. 10-33. Unit-step response of the designed system in Example 10-6.

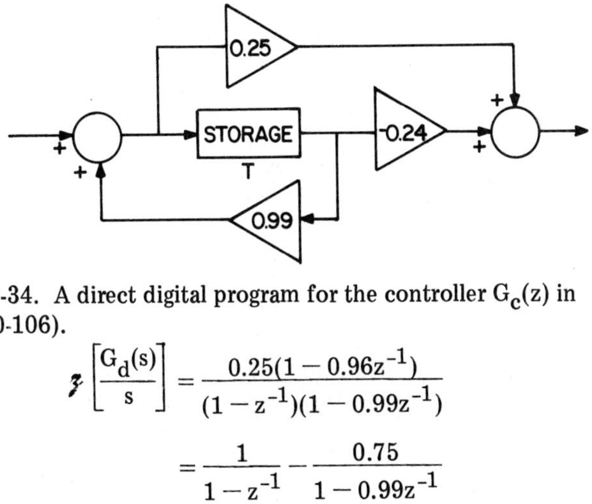

Fig. 10-34. A direct digital program for the controller $G_c(z)$ in Eq. (10-106).

$$\mathcal{Z}\left[\frac{G_d(s)}{s}\right] = \frac{0.25(1-0.96z^{-1})}{(1-z^{-1})(1-0.99z^{-1})}$$

$$= \frac{1}{1-z^{-1}} - \frac{0.75}{1-0.99z^{-1}} \tag{10-108}$$

Taking the inverse z-transform on both sides of the last equation, and solving for $G_d(s)$, we have

$$G_d(s) = \frac{1+12.5s}{1+50s} \tag{10-109}$$

which is the transfer function of the series-digital-network realization of $G_c(z)$. We can also use the feedback-digital-network realization of $G_c(z)$ by using Eq. (10-64). The result is

$$H(s) = \frac{3s}{s+0.0816} \tag{10-110}$$

Figure 10-35 shows the series-digital-program realization of $G_c(z)$, and Fig. 10-36 shows the feedback-digital-program realization of $G_c(z)$.

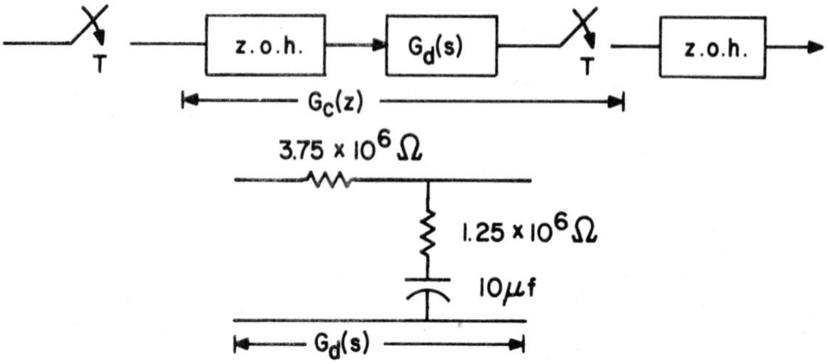

Fig. 10 35. A series-digital-network realization of $G_c(z)$ in Eq. (10-106).

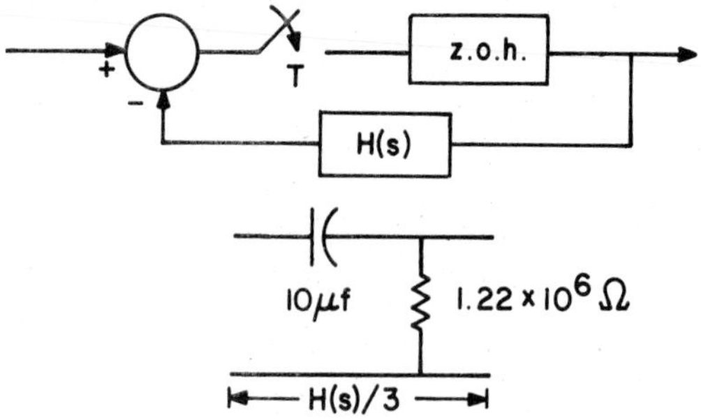

Fig. 10-36 A feedback-digital-program realization of $G_c(z)$ in Eq. (10-106).

10.6 DESIGN OF THE Z-PLANE USING THE ROOT LOCUS DIAGRAM

In Chapter 7 the root locus technique was extended to the analysis of digital control systems. It was shown that the properties of the construction of the root loci in the z-plane are essentially the same as those for the root loci of continuous-data systems which are drawn in the s-plane. However, in interpreting the system performance from the z-plane roots, the unit-circle $|z| = 1$ is the stability boundary, and all information on relative stability should be evaluated accordingly. One may regard the design of a control system by the root locus method as a pole-placement problem solved by trial-and-error. In other words, the root-locus design essentially involves the determination of the system and controller parameters so that the roots of the characteristic equation are at the desired locations. For systems with an order higher than third, it is generally very difficult to establish relation between the controller parameters and the characteristic equation roots. Furthermore, the conventional root locus diagram only allows one parameter to vary at a given time. Therefore, the design of a digital control system in the z-plane using a root locus diagram is essentially a trial-and-error method. Or, the designer may rely on a digital computer to plot out a large number of root loci by scanning through a wide range of possible values of the controller parameters, and select the best solution. However, the experienced designer can still make proper and intelligent initial "guesses' so that the amount of trial-and-error effort is kept to a minimum. Therefore, it is useful to investigate the effects of the various pole-zero configurations of the digital controller on the overall system performance and the characteristic equation roots.

Phase-Lead and Phase-Lag Controllers

Since we are now working in the z-plane, we should investigate digital controllers that are described by $G_c(z)$. Let us consider the first-order controller transfer function

$$G_c(z) = K_c \frac{z - z_1}{z - p_1} \tag{10-111}$$

where z_1 is a real zero, and p_1 is a real pole. In general, if the digital controller is to not affect the steady-state performance of the system, we set

$$\lim_{z \to 1} G_c(z) = 1 \tag{10-112}$$

Then K_c in Eq. (10-111) is set to

$$K_c = \frac{1 - p_1}{1 - z_1} \tag{10-113}$$

We can classify $G_c(z)$ as a low-pass controller or a high-pass controller depending on the relative magnitudes of z_1 and p_1. Substituting $z = e^{Ts}$ in Eq. (10-111), we get

$$G_c^*(s) = K_c \frac{e^{Ts} - z_1}{e^{Ts} - p_1} \tag{10-114}$$

Apparently $G_c^*(s)$ has an infinite number of poles and zeros. However, if we consider only the poles and zeros that lie in the primary strip in the s-plane, the zero of $G_c^*(s)$ is given by

$$s = \frac{1}{T} \ln(z_1) \tag{10-115}$$

and the pole of $G_c^*(s)$ is

$$s = \frac{1}{T} \ln(p_1) \tag{10-116}$$

where z_1 and p_1 are real numbers. Equation (10-114) is approximated by the following rational function in the primary strip of the s-plane.

$$G_c^*(s) \cong K_c \frac{s + \frac{1}{T}\ln(z_1)}{s + \frac{1}{T}\ln(p_1)} \tag{10-117}$$

As an alternative we can approximate e^{Ts} by the first two terms of its power-series expansion. Then Eq. (10-114) is approximated by

$$G_c^*(s) \cong K_c \frac{Ts + 1 - z_1}{Ts + 1 - p_1} \tag{10-118}$$

This is a better choice since Eq. (10-117) does not allow negative values of p_1 and z_1. As an illustration of the approximations made above, consider the transfer function of the digital controller in Eq. (10-106),

$$G_c(z) = 0.25 \frac{z - 0.96}{z - 0.99} \tag{10-119}$$

Using Eq. (10-117), we have

$$G_c^*(s) \cong 0.25 \frac{s + 0.0816}{s + 0.0201} \qquad (T = 0.5 \text{ sec}) \tag{10-120}$$

and from Eq. (10-118),

$$G_c^*(s) \cong 0.25 \frac{s + 0.08}{s + 0.02} \qquad (T = 0.5 \text{ sec}) \tag{10-121}$$

Thus, we see that $G_c^*(s)$ can be classified as a low-pass filter, since the pole is to the right of the zero in the s-plane. For this reason we refer to the transfer function $G_c(z)$ in Eq. (10-119) as a low-pass controller. The pole-zero configurations of Eq. (10-121) and Eq. (10-119) are shown in Fig. 10-37. Therefore, we may regard the pole-zero configuration of Fig. 10-37(b) as that of a typical first-order phase-lag digital controller. It should be pointed out that the approximation of $G_c(z)$ by $G_c^*(s)$ used above is solely for the purpose of identifying the filter characteristics of $G_c(z)$.

In general, we may also have a phase-lag controller with one pole in the right-half and one zero in the left-half of the z-plane, all inside the unit circle; or both the pole and the zero on the negative real axis inside the unit circle, with the pole located to the right of the zero. However, these configurations are generally not as effective in stabilizing a system as the "dipole" arrangement shown in Fig. 10-37.

Using a similar approach, we can show that since for a high-pass filter in the s-domain the zero is always to the right of the pole on the negative real axis in the s-plane, the same is true for the relative pole-zero location of $G_c(z)$ in the z-plane. Figure 10-38 illustrates three possible pole-zero configurations of $G_c(z)$ as a first-order high-pass filter.

The following examples illustrate typical digital control system design problems carried out in the z-plane using the root locus technique.

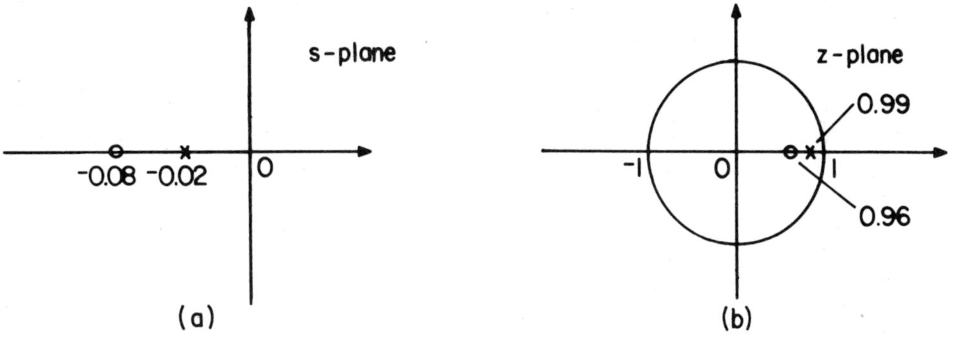

Fig. 10-37. (a) Pole-zero configuration of the phase-lag $G_c^*(s)$ in Eq. (10-121).
(b) Pole-zero configuration of the phase-lag $G_c(z)$ in Eq. (10-119).

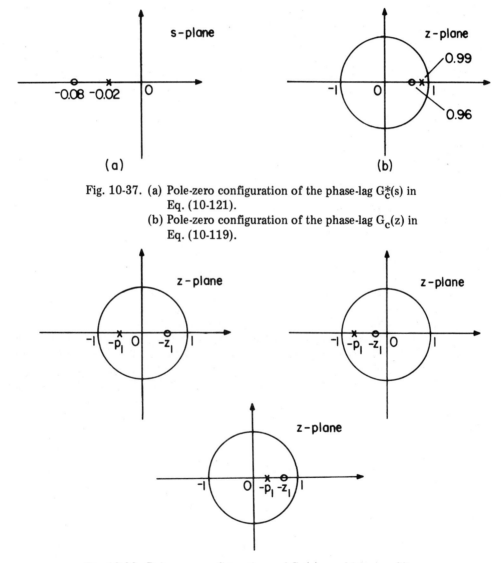

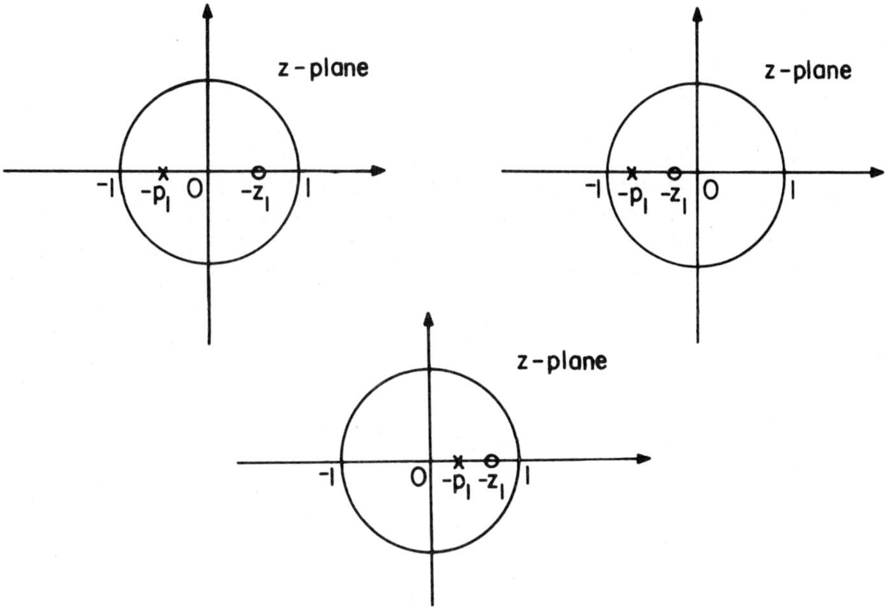

Fig. 10-38. Pole-zero configurations of $G_c(z)$ as a high-pass filter.

Example 10-7.

Let us consider the same design problem stated in Example 10-6. The controlled process is described by Eq. (10-96), and the sampling period is 0.5 sec. Since the design is to be carried out in the z-plane using the root locus, it is no longer convenient to use the gain or phase margin or any other frequency-domain measures as the performance specification. Let the performance specifications be:

ramp-error constant $K_v \geqslant 1.4$

relative damping ratio of the dominant roots $\zeta \cong 0.707$

The open-loop transfer function of the uncompensated system is given by Eq. (10-97), and is repeated below.

$$G_{h0}G_p(z) = \frac{0.13K(z + 1.31)(z + 0.054)}{z(z - 1)(z - 0.368)} \tag{10-122}$$

The characteristic equation of the closed-loop system is obtained by setting the numerator of $1 + G_{h0}G_p(z)$ to zero. Thus,

$$z(z - 1)(z - 0.368) + 0.13K(z + 1.31)(z + 0.054) = 0 \tag{10-123}$$

The root locus diagram of the last equation is shown in Fig. 10-39. Since the open-loop zero at $z = -0.054$ is very close to the open-loop pole at $z = 0$, the two nearly cancel each other, so that the root loci are essentially formed by the open-loop poles at $z = 1$, 0.368, and the zero at -1.31. It has been established in Example 10-6 that for $K_v = 1.4$, K should equal 3.0. The root locus diagram in Fig. 10-39 shows that the marginal value of K for stability is 3.3. Thus, setting $K = 3$ would yield a very low margin of stability.

A great deal has been learned from the frequency-domain design of the system. The digital controller arrived at in Eq. (10-119) is of the phase-lag type. Furthermore, the zero at $z = 0.96$ and the pole at $z = 0.99$ are very close to each other; they are also very close to the $z = 1$ point. These characteristics are by no means coincidental. The digital controller represented by Eq. (10-119) is known as a "dipole". We shall show that we can arrive at a similar result using the root-locus design method.

The constant-damping locus in the z-plane for $\zeta = 0.707$ is shown in Fig. 10-39. The intersect between the root loci and the $\zeta = 0.707$ constant-damping locus corresponds to $K \cong 0.7$. The problem is to find a digital controller which will realize $K = 3$ and $\zeta = 0.707$ simultaneously.

Let us introduce the controller with the transfer function

$$G_c(z) = K_c \frac{z - z_1}{z - p_1} \tag{10-124}$$

where

$$K_c = \frac{1 - p_1}{1 - z_1} \tag{10-125}$$

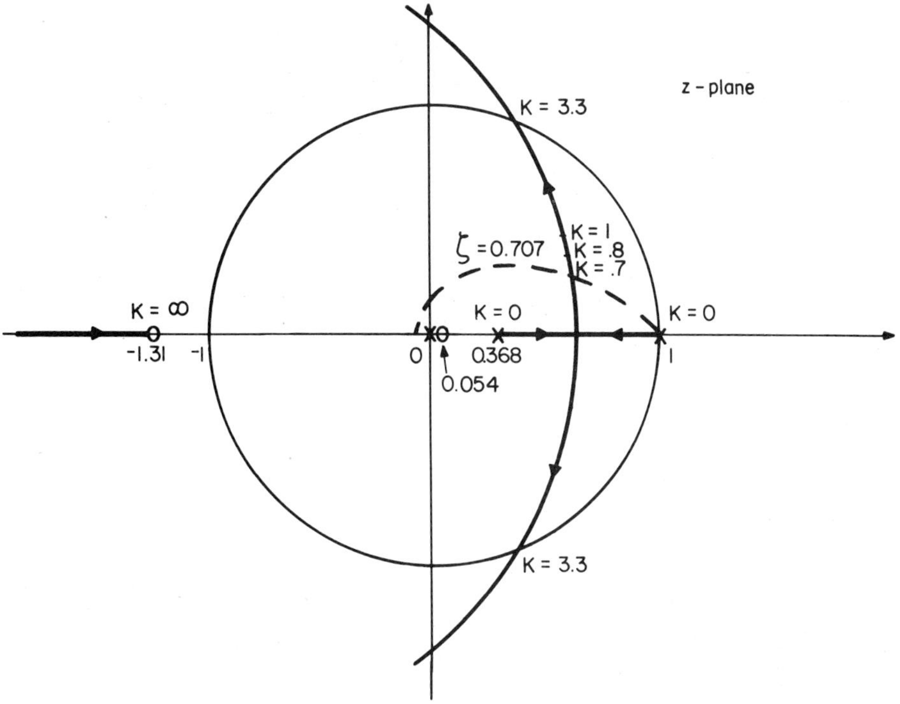

Fig. 10-39. Root loci of the uncompensated digital control system in Example 10-7.

Most important, we stipulate that the values of z_1 and p_1 should be very close together, and that they are also very close to 1 (but less than 1). Also, z_1 is less than p_1, so that the controller is of the phase-lag type. The motivation for selecting this type of controller is explained as follows.

Figure 10-39 shows that the damping ratio requirement can be satisfied if K is set to 0.7. This means that we do not have to reshape the root loci, and only the system gain has to be modified. But we cannot simply just lower the value of K. Therefore, we need a controller which will raise the loop gain effectively, but at the same time does not alter the pole-zero configuration of the overall system appreciably. The controller described by Eq. (10-122) with the properties stated above will satisfy these requirements. The reason is that since z_1 and p_1 are very close, the factor $(z - z_1)/(z - p_1)$ is effectively equal to one. Thus, the main purpose of $G_c(z)$ is to introduce the attenuation $K_c = (1 - p_1)/(1 - z_1)$. Once this principle is established, the design becomes extremely simple, because all we have to do is set K_c to the ratio of 0.7/3.0, which is the ratio of (the gain to realize $\zeta = 0.707$)/(the gain to realize $K_v = 1.4$); i.e.,

$$K_c = \frac{1 - p_1}{1 - z_1} = \frac{0.7}{3.0} \tag{10-126}$$

However, we have one equation with two unknowns in the last equation. We have yet to apply the condition that $p_1 \cong z_1 \cong 1$ and $z_1 < p_1 < 1$. Setting p_1 arbitrarily to 0.99 in Eq. (10-126), we have $z_1 = 0.957$. Thus, Eq. (10-124) becomes

$$G_c(z) = 0.233 \frac{z - 0.957}{z - 0.99} \tag{10-127}$$

Since the "dipole" is located very near the $z = 1$ point, the root loci at a distance relatively far from $z = 1$ are not affected in any significant way by the addition of the "dipole", and the only net effect is that the loop gain of the system is changed by K_c. Figure 10-40 shows the important portions of the root loci of the compensated system and the uncompensated. It is important to emphasize that the absolute values of p_1 and z_1 are not important, so long as they are very close to one. However, the ratio of $(1 - p_1)/(1 - z_1)$ is important and must be set according to the amount of attenuation desired in the loop in order to achieve the desired damping. This is the same principle as in the phase-lag design conducted in Example 10-6, in that the upper-corner frequencies of the controller transfer function are placed far below the desired new gain-crossover frequency. The exact values of these two corner frequencies are unimportant as long as their ratio is equal to the desired

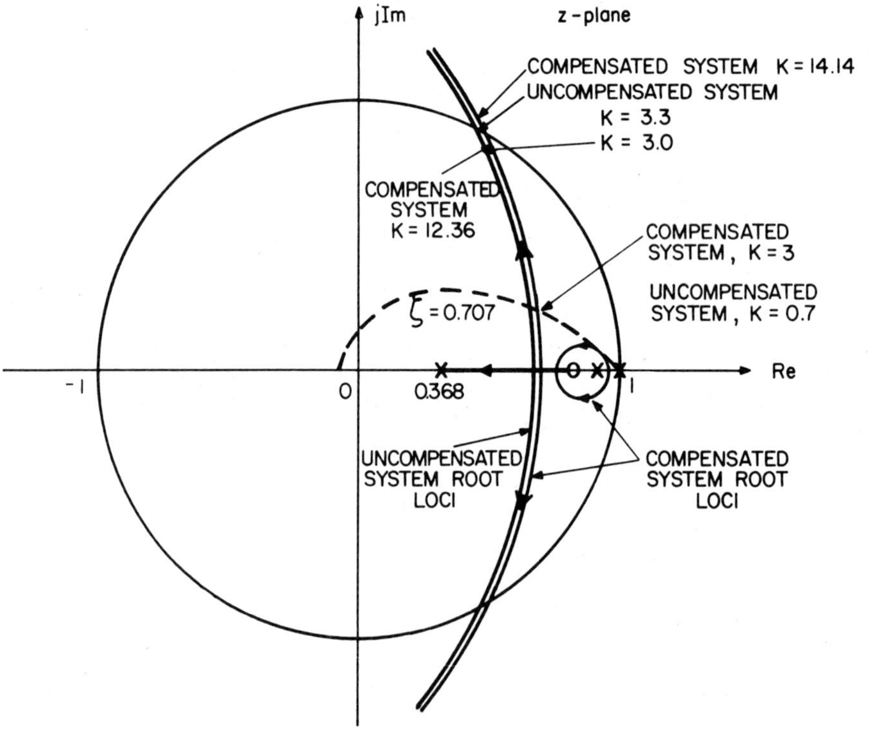

Fig. 10-40. Root loci of the uncompensated and the compensated digital control system in Example 10-7.

attenuation for the realization of the specified phase or gain margin. Therefore, we could have selected p_1 to be 0.98 for the present problem, and the corresponding z_1 would have to be 0.9143, and the root loci shown in Fig. 10-40 would not be significantly altered. The transfer function of the digital controller would be

$$G_c(z) = 0.233 \frac{z - 0.9143}{z - 0.98} \tag{10-128}$$

where the value of K_c is not altered.

Example 10-8.

The block diagram of a digital control system is shown in Fig. 10-41. The controlled process is described by the transfer function

$$G_p(s) = \frac{K}{s^2} \tag{10-129}$$

which may represent a pure inertial load. The open-loop transfer function of the uncompensated system is

$$G_{h0}G_p(z) = (1 - z^{-1})\mathcal{Z}\left[\frac{K}{s^3}\right] = \frac{KT^2(z+1)}{2(z-1)^2} \tag{10-130}$$

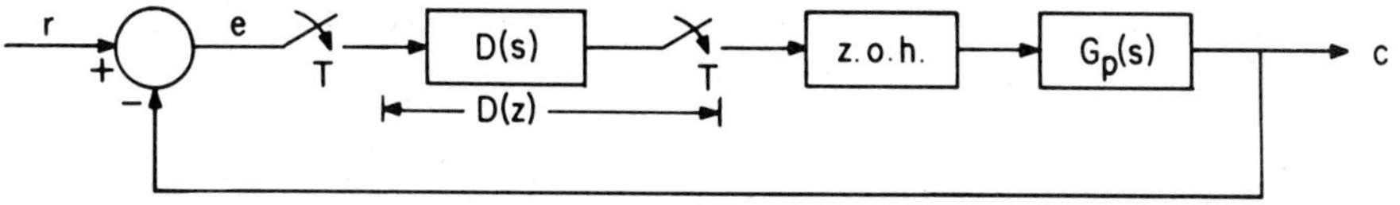

Fig. 10-41. A digital control system.

For this particular process, since T appears only as a multiplying factor in the open-loop transfer function, the root locus method can be applied to study the effects of variable K and T simultaneously. Figure 10-42 shows the root loci of the system based on the pole-zero configuration of the transfer function in Eq. (10-130). The variable parameter on the loci is $K' = KT^2$. It is apparent that the system without a controller is unstable for any value of KT^2.

Since $G_{h0}G_p(z)$ has two poles at $z = 1$, a phase-lag controller with a pole located to the right of its zero would simply push the loci further toward the right near $z = 1$. Thus, it would seem logical to try a phase-lead controller. Let the transfer function of the phase-lead controller be

$$G_c(z) = K_c \frac{z - z_1}{z - p_1} \tag{10-131}$$

where $z_1 > p_1$, and $K_c = (1 - p_1)/(1 - z_1)$. To select the values of p_1 and z_1, we notice that it would be desirable to place the zero near the two poles of $G_{h0}G_p(z)$ at $z = 1$. Figure

10-42 shows the root loci of the compensated system with $z_1 = 0.9$ and $p_1 = 0.5$. The closed-loop system is now stable for $KT^2 < 0.2$. Thus, if the sampling period is 0.1 sec, the marginal value of K for stability would be 20.

It would be ideal if the zero z_1 could be placed at $z = 1$ to cancel one of the poles of $G_{h0}G_p(z)$; however, this would correspond to an infinite K_c.

In the above design, the pole p_1 was arbitrarily placed at 0.5. What would be the effect of moving p_1 along the real axis inside the unit circle? It is apparent that p_1 should not be placed too close to the zero at z_1, or the phase-lead controller would not be too effective in improving the stability of the system. On the other hand, the pole can be placed on the negative real axis. This corresponds to more phase lead, since the distance between z_1 and p_1 is increased. Figure 10-43 shows the root loci when the digital controller has the transfer function

$$G_c(z) = 15 \frac{z - 0.9}{z + 0.5} \tag{10-132}$$

Now when $K' = 0.05$, the closed-loop system has three real poles at $z = -0.322$, 0.584 and 0.864. For higher values of K' which corresponds to two complex conjugate roots, the natural frequency of the system is higher. Thus, the rise time of the system would be faster than

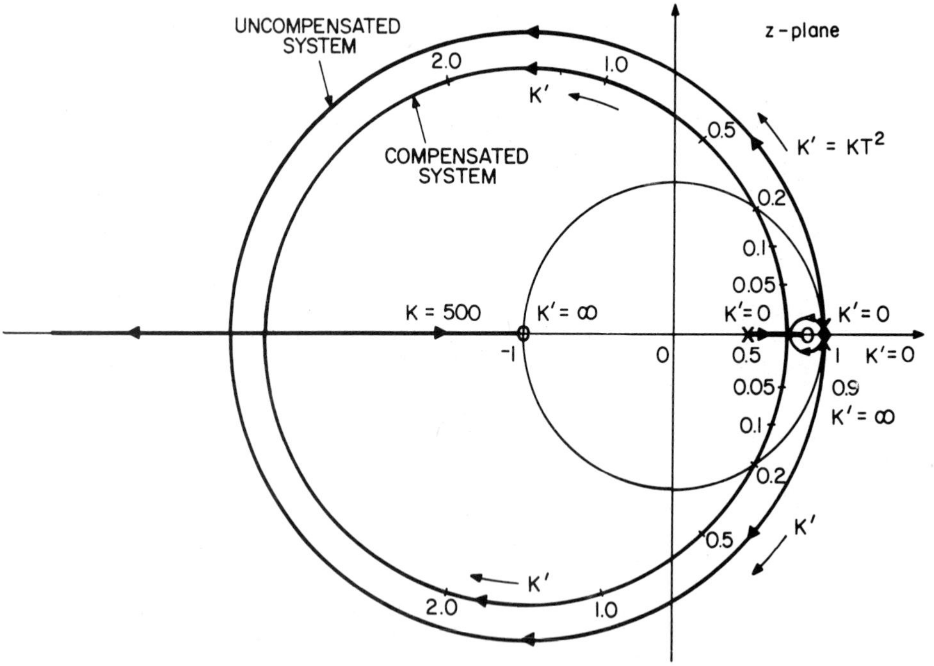

Fig. 10-42. Root loci of the uncompensated digital control system and the compensated system with the controller $G_c(z) = 5(z - 0.9)/(z - 0.5)$, in Example 10-8.

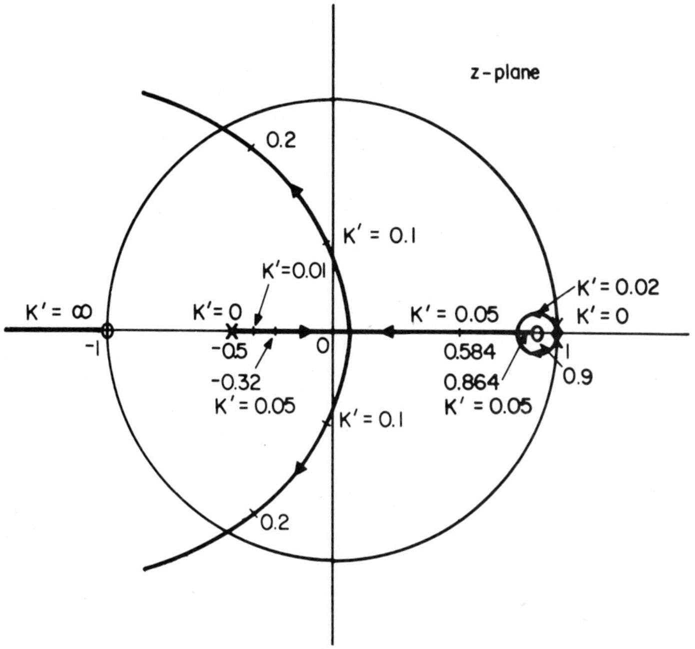

Fig. 10-43. Root loci of the compensated digital control system in Example 10-8 with $G_c(z) = 15(z - 0.9)/(z + 0.5)$.

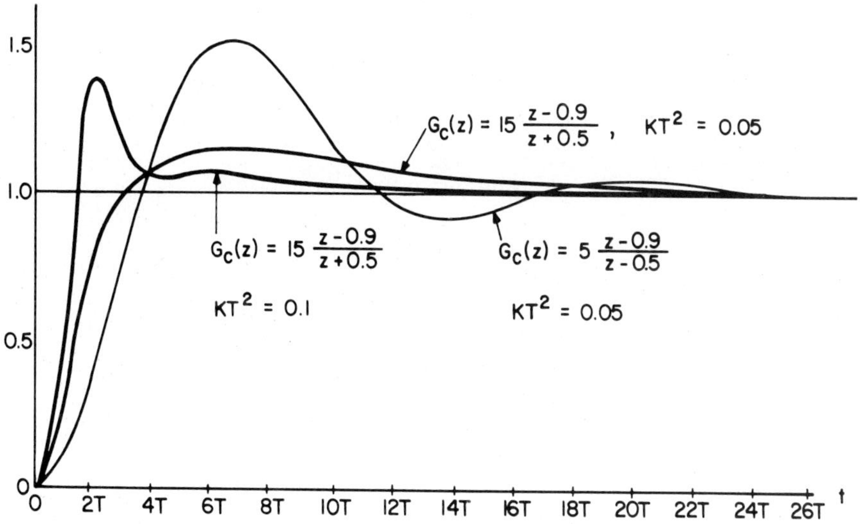

Fig. 10-44. Unit-step responses of the digital control system designed in Example 10-8.

the case with $p_1 = 0.5$. Figure 10-44 shows the unit-step responses of the designed system with the two different controllers and two values of K'. With the controller as described by Eq. (10-131) with $z_1 = 0.9$ and $p_1 = 0.5$, the unit-step response has a maximum overshoot in excess of 50%, although the relative damping ratio of 0.707 was achieved. With the controller as described by Eq. (10-132), and $KT^2 = K' = 0.05$, the unit-step response has a maximum overshoot of only 15%. For the same controller but with K' increased to 0.1, the response is much faster although the maximum overshoot is increased to 38%.

The Root Contour Method [1]

The conventional root locus diagram of the characteristic equation of a closed-loop control system is often defined with the loop gain K as the variable parameter, and all the other system parameters are held constant. Often, it is desirable to investigate the effects of varying other parameters than just the loop gain K. The root contour is simply defined for this purpose; it is still a root locus as far as all its properties are concerned, and the only difference is that the variable parameter along the root contour can be any parameter of the system other than the loop gain. For instance, in the design problem carried out in Example 10-8, a phase-lead digital controller was used. In that problem, we set the zero of the controller at $z_1 = 0.9$, and investigated the effects on the root loci when the pole was placed at $p_1 = 0.5$ and then at -0.5. It would be desirable to investigate the effect of various values of p_1 by varying p_1 continuously from $-\infty$ to $+\infty$, while all the other parameters are fixed.

Let us illustrate the root contour technique by using the same system as described in Example 10-8. The open-loop transfer function of the system with digital controller is

$$G_c(z)\, G_{h0}G_p(z) = \frac{1 - p_1}{1 - z_1}\,\frac{z - z_1}{z - p_1}\,\frac{KT^2(z + 1)}{2(z - 1)^2} \tag{10-133}$$

Let $z_1 = 0.9$ and $KT^2 = 0.05$; the last equation becomes

$$G_c(z)G_{h0}G_p(z) = \frac{1 - p_1}{0.1}\,\frac{z - 0.9}{z - p_1}\,\frac{0.05(z + 1)}{2(z - 1)^2} \tag{10-134}$$

The characteristic equation of the closed-loop system is obtained by setting the numerator of $1 + G_c(z)G_{h0}G_p(z)$ to zero. Thus, the characteristic equation is

$$(z - p_1)(z - 1)^2 + 0.25(1 - p_1)(z + 1)(z - 0.9) = 0 \tag{10-135}$$

Since the last equation contains only one parameter in p_1, the rules of construction of the root locus diagram can be applied, and the loci of the roots of the equation when p_1 varies are called the *root contours*. Multiplying out the terms and collecting the terms that are associated with p_1 in Eq. (10-135), we have

$$z^3 - 1.75z^2 + 1.025z - 0.225 - 1.25p_1(z^2 - 1.58z + 0.62) = 0 \qquad (10\text{-}136)$$

Dividing both sides of the last equation by the terms that do not contain p_1 yields

$$1 + \frac{-1.25p_1(z^2 - 1.58z + 0.62)}{z^3 - 1.75z^2 - 1.025z - 0.225} = 0 \qquad (10\text{-}137)$$

Now Eq. (10-137) is of the form of $1 + G_e(z) = 0$ where p_1 appears in $G_e(z)$ only as a multiplying constant. This means that we can construct the root contour of Eq. (10-136) from the pole-zero configuration of $G_e(z)$ which is written

$$G_e(z) = \frac{-1.25p_1(z - 0.726)(z - 0.854)}{(z - 0.867)(z - 0.441 + j0.254)(z - 0.441 - j0.254)} \qquad (10\text{-}138)$$

The root contours of Eq. (10-136) when p_1 varies from $-\infty$ to ∞ are sketched as shown in Fig. 10-45. Since p_1 covers the entire range of real values, both the phase-lag controller and the phase-lead controller with $z_1 = 0.9$ are covered on

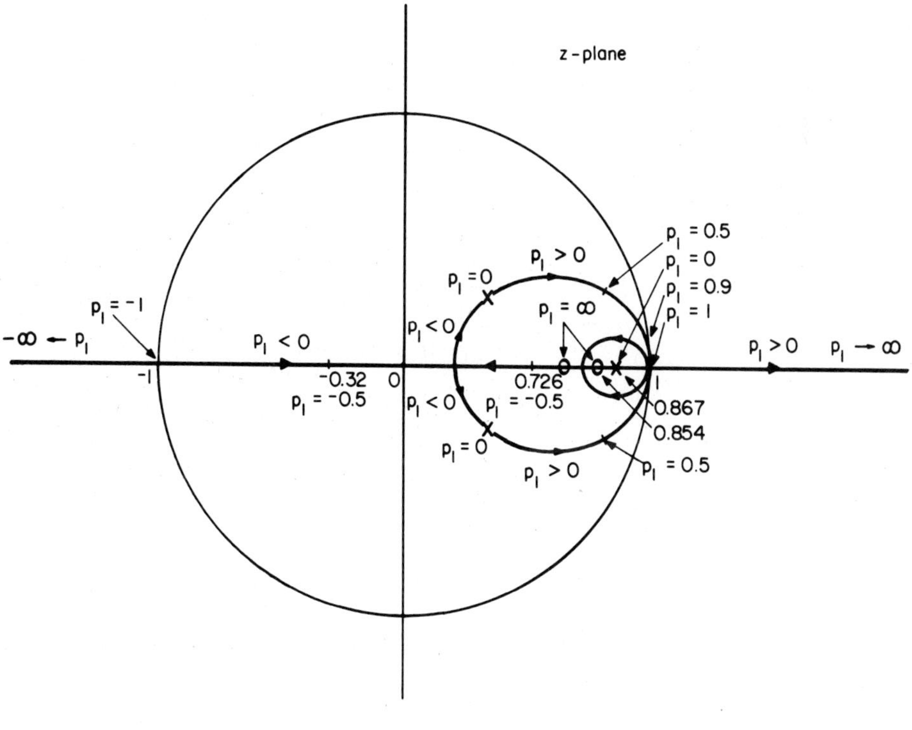

Fig. 10-45. Root contours of Eq. (10-136).

the root contour diagram. It is clear that for the phase-lag controller, $p_1 > z_1 = 0.9$, at least one closed-loop characteristic root is outside the unit circle, and the system is unstable. For the phase-lead controller, $p_1 < z_1 = 0.9$, the root contours show that it is more effective to use a negative value of p_1. However, p_1 cannot be less than -1 or one of the roots will again move outside the unit circle. For negative p_1 one of the three roots will always lie between 0.854 and 0.867.

Pole-Zero Cancellation Design

In the design of control systems using the s-plane or the z-plane it has been a practical practice to attempt to cancel the undesirable poles and zeros of the controlled process transfer function by poles and zeros of the controller, and new open-loop poles and zeros are added at more advantageous locations to satisfy the design specifications. The cancellation compensation suggested above is used often in systems that have complex poles in the controlled process and the stability problem is more severe. It would seem that canceling the undesired poles and adding new ones at any place one wishes is the simplest strategy to solve any problem. However, it should be kept in mind that the pole-zero cancellation compensation scheme does not always furnish a satisfactory solution to a broad class of design problems. It is possible that the resulting controller may involve excessive complexity; and if the undesirable poles of the controlled process are near the unit circle in the z-plane, inexact cancellation, which is almost always true in practice, could result in a conditionally stable system. As a simple illustrative example to the effect of inexact cancellation, let us refer to the root locus diagram shown in Fig. 10-46. In Fig. 10-46(a) the pole-zero configuration of the open-loop transfer function of a certain digital control system is shown. A digital controller with the transfer function

$$G_c(z) = K\frac{(z - p_1)(z - p_2)}{(z - a)(z - b)} \qquad (10\text{-}139)$$

is devised to cancel the complex poles p_1 and p_2 and to add new poles at $z = a$ and $z = b$. Figure 10-46(b) shows an assumed case of inexact cancellation, where the complex zeros of the controller are not exactly equal to the poles p_1 and p_2 of the controlled process. In this case, the relative positions of the poles and zeros involved are such that the inexact cancellation does not have any adverse effect on the system performance, since the closed-loop poles near p_1 and p_2 are stable and are very close to the two zeros of the controller. On the other hand, if the poles and zeros of the inexact cancellation are arranged as shown in Fig. 10-46(c), a portion of the root loci between the uncanceled pole and zero pairs may lie outside the unit circle. The system is called a conditionally stable system since it is stable only for low and high values of the loop

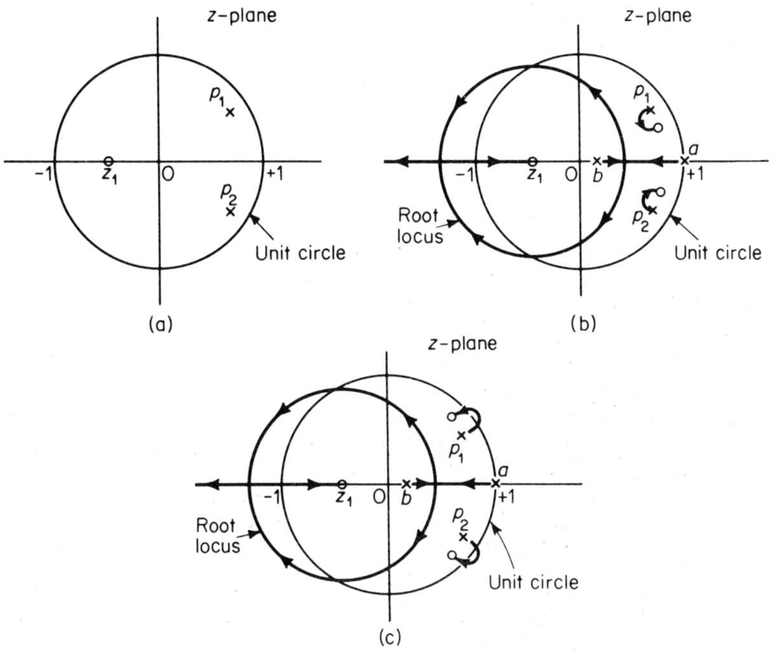

Fig. 10-46. Root locus diagrams illustrating the effects of inexact cancellation of undesirable poles.

gain. Thus, in applying pole-zero cancellation in control system design, inexact cancellation is always inevitable, and we must make certain that the inexact cancellation does not cause stability problems.

10.7 THE DIGITAL PID CONTROLLER

One of the most widely used controllers in the design of continuous-data control systems is the PID controller, where PID stands for Proportional-Integral-Derivative control. Figure 10-47 shows the block diagram of a continuous-data PID controller acting on an error signal e(t). The proportional control simply multiplies the error signal by a constant K_p; the integral control multiplies the integral of e(t) by K_1, and the derivative control generates a signal which is proportional to the time derivative of the error signal. The function of the integral control is to provide action to reduce the steady-state error, whereas the derivative control provides an anticipatory action to reduce the overshoots in the response. The same principle of the PID control can be applied to digital control. In digital control, the proportional control is still implemented by a proportional constant K_p. In general, there are a number of ways to implement integration and derivatives digitally. For example, the integral operation

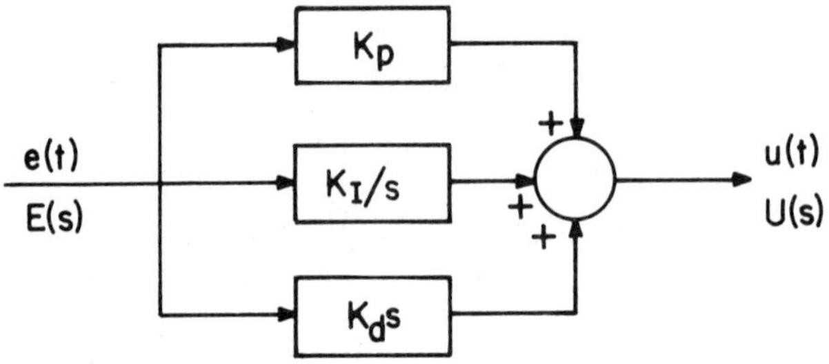

Fig. 10-47. A continuous-data PID controller.

K_1/s can be approximated by the z-form on the polygonal integration, $K_1 T(z+1)/2(z-1)$. The derivative of e(t) at t = T can be approximated by the following equation:

$$\left.\frac{de(t)}{dt}\right|_{t=T} \cong \frac{e(kT) - e[(k-1)T]}{T} \qquad (10\text{-}140)$$

Taking the z-transform on both sides of the last equation, we have the transfer function of the digital derivative controller.

$$G_d(z) = K_d \frac{z-1}{Tz} \qquad (10\text{-}141)$$

where K_d is the derivative control constant.

Now when the complete PID controller is applied, we have the block diagram shown in Fig. 10-48.

The design of the PID controller essentially involves the determination of the values of K_p, K_I and K_d so that the controlled system performs as specified. All the conventional design methods discussed in the preceding sections can be applied to the design of the PID controller since the latter is simply a specific

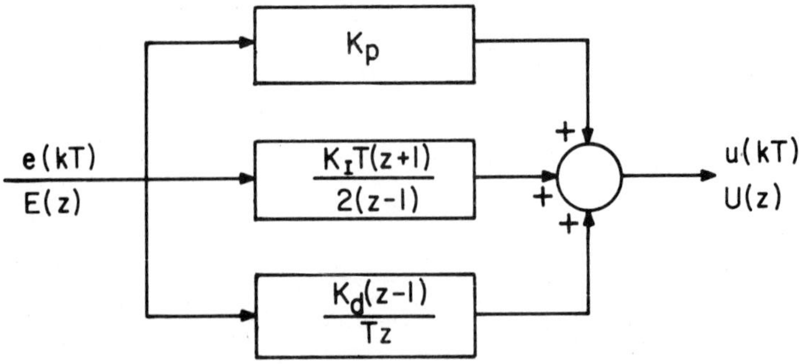

Fig. 10-48. A digital PID controller.

type of controller. We shall use the following example to illustrate the effects and the properties of the PID controller.

Example 10-9.

The block diagram of a digital control system is shown in Fig. 10-49. The design objective is to determine the transfer function of the digital controller so that the system will meet certain performance specifications. The controlled process is represented by the transfer function

$$G_p(s) = \frac{10}{(s+1)(s+2)} \tag{10-142}$$

The open-loop transfer function of the uncompensated system is

$$G_{h0}G_p(z) = (1 - z^{-1})\mathcal{Z}\left[\frac{10}{s(s+1)(s+2)}\right]$$

$$= \frac{0.0453(z+0.904)}{(z-0.905)(z-0.819)} \tag{10-143}$$

where the sampling period T is equal to 0.1 second. The closed-loop transfer function of the system is

$$\frac{C(z)}{R(z)} = \frac{0.0453(z+0.904)}{z^2 - 1.679z + 0.782} \tag{10-144}$$

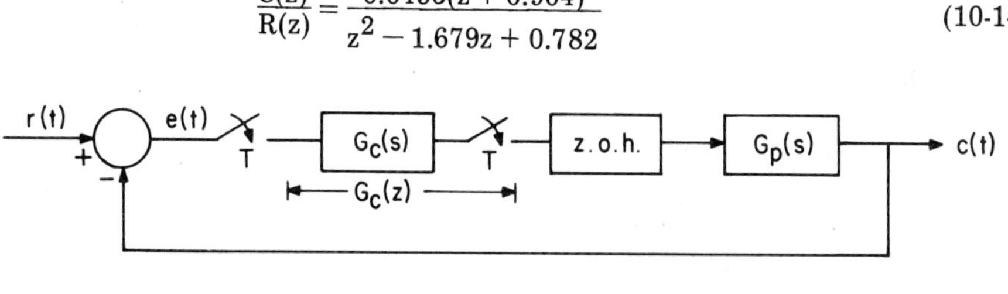

Fig. 10-49. A digital control system with a digital controller.

The roots of the characteristic equation are at $z = 0.84 + j0.278$ and $z = 0.84 - j0.278$. Therefore, the system is stable. However, for a step input, the steady-state error of the system is not zero. This property of the system is revealed by the fact that the open-loop transfer function $G_{h0}G_p(z)$ does not have at least one pole at $z = 1$.

For a unit-step input, the steady-state value of the output is determined from

$$\lim_{k\to\infty} c(kT) = \lim_{z=1} (1 - z^{-1})C(z)$$

$$= \lim_{z=1} \frac{0.0453(z+0.904)}{z^2 - 1.679z + 0.782} = 0.837 \tag{10-145}$$

Thus, the steady-state error of the system to a unit-step input is 0.163. The unit-step response of the uncompensated system is shown in Fig. 10-50.

In order to eliminate the steady-state error, we introduce the integral control, since the integral of the constant steady-state error will produce a constant signal which may be used to eliminate this error. Let us first apply only the PI controller ($K_d = 0$). The open-loop transfer function of the system with the PI controller is written

$$G_c(z)G_{h0}G_p(z) = \frac{(2K_p + K_I T)\left[z + \dfrac{K_I T - 2K_p}{K_I T + 2K_p}\right]}{2(z-1)} \frac{0.0453(z + 0.904)}{(z - 0.905)(z - 0.819)} \tag{10-146}$$

Since the choice of K_p and K_I is arbitrary, we can design the controller in such a way that its zero cancels one of the poles of the process dynamics. Let

$$\frac{K_I T - 2K_p}{K_I T + 2K_p} = -0.905 \tag{10-147}$$

Therefore,

$$K_p/K_I = 1.00263 \tag{10-148}$$

For $K_p = 1$ and $K_I = 0.997$, the transfer function of the digital controller is

$$D(z) = 1.0499 \frac{z - 0.905}{z - 1} \tag{10-149}$$

The open-loop transfer function of the compensated system is now

$$G_c(z)G_{h0}G_p(z) = \frac{0.0453(z + 0.904)}{(z - 1)(z - 0.819)} \tag{10-150}$$

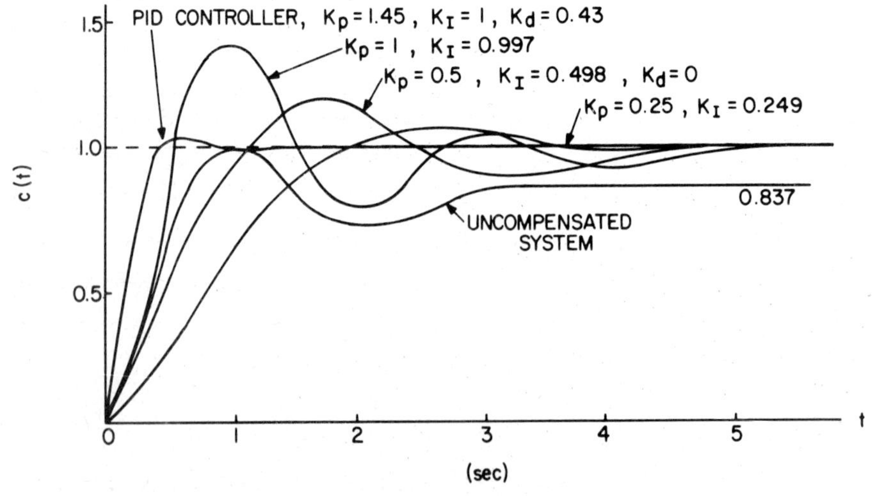

Fig. 10-50. Unit-step responses of the digital control system in Example 10-9.

Since the open-loop transfer function has a pole at $z = 1$, the steady-state error of the closed-loop system to a step input is zero. The unit-step response of the system with the digital PI controller with $K_p = 1$ and $K_I = 0.997$ is shown in Fig. 10-50. Notice that although the steady-state error is completely eliminated by the PI controller, the maximum overshoot is increased to 45%. The maximum overshoot can be reduced at the expense of longer rise time. Figure 10-50 illustrates the step responses of the system with two other sets of values for K_p and K_I.

In order to eliminate the steady-state error due to a step input and simultaneously realizing a good transient response, we must use a PID controller. Let the digital PID controller be described by the transfer function

$$G_c(z) = K_p + \frac{K_I T(z+1)}{2(z-1)} + \frac{K_d(z-1)}{Tz}$$

$$= \frac{(K_I T^2 + 2K_d + 2K_p T)z^2 + (K_p T^2 - 2K_p T - 4K_d)z + 2K_d}{2Tz(z-1)} \tag{10-151}$$

Then, the open-loop transfer function of the compensated system becomes

$$G_c(z)G_{h0}G_p(z) = \frac{(K_I T^2 + 2K_d + 2K_p T)z^2 + (K_p T^2 - 2K_p T - 4K_d)z + 2K_d}{2Tz(z-1)}$$

$$\times \frac{0.0453(z+0.904)}{(z-0.905)(z-0.819)} \tag{10-152}$$

Now there are three unknown parameters in K_p, K_I and K_d which must be determined in the design, depending on the specifications of performance.

Let us require that the ramp-error constant K_v should equal 5, and, in addition, the two zeros of the PID controller should cancel the two poles of the controlled process at $z = 0.905$ and $z = 0.819$. These specifications should give us three linearly independent equations to solve the three unknown parameters of the digital PID controller.

The ramp-error constant is written

$$K_v = \frac{1}{T} \lim_{z \to 1} (z-1)G_c(z)G_{h0}G_p(z) = 5K_I \tag{10-153}$$

and it is interesting to note that K_v is affected by only K_I and the parameters of the controlled process, but not by K_p, and K_d. Since it is required that K_v be equal to 5, K_I must be equal to 1.

Setting the two zeros of the controller so that they cancel the poles of the process, we get

$$z^2 + \frac{K_I T^2 - 2K_p T - 4K_d}{2K_p T + K_I T^2 + 2K_d} z + \frac{2K_d}{2K_p T + K_I T^2 + 2K_d} = (z-0.905)(z-0.819) \tag{10-154}$$

With $K_I = 1$ and $T = 0.1$ sec, the values of K_p and K_d are solved from the last equation,

$$K_p = 1.45$$

$$K_d = 0.43$$

Substitution of the controller parameters in Eq. (10-152), the open-loop transfer function of the compensated system is simply

$$G_c(z)G_{h0}G_p(z) = \frac{0.263(z + 0.904)}{z(z - 1)} \tag{10-155}$$

The closed-loop transfer function is

$$\frac{C(z)}{R(z)} = \frac{0.263(z + 0.904)}{z^2 - 0.737 + 0.238} \tag{10-156}$$

The characteristic equation roots are at $z = 0.369 + j0.319$ and $z = 0.369 - j0.319$. The unit-step response of the system with the PID controller just designed is shown in Fig. 10-50. It is evident that the addition of the derivative control not only cut down the overshoot but also reduced the rise time. The maximum overshoot in this case is approximately 4%.

10.8 DESIGN OF DIGITAL CONTROL SYSTEMS WITH THE DEADBEAT RESPONSE

A large number of control systems are designed with the objective that the responses of the systems should reach their respective desired values as quickly as possible. This class of control systems is called *minimum-time control systems*, or *time-optimal control systems*, as described in Chapter 9. With reference to the design methods discussed thus far in this chapter, again one of the design objectives is to have a small maximum overshoot and a fast rise time in the step response. In reality, the design principles discussed in the preceding sections all involve the extension of the design experience acquired in the design of continuous-data control systems; e.g., phase-lag and phase-lead controllers, and the PID controllers. However, since the digital controller has a great deal of flexibility in its configuration, we should be able to come up with independent methods not relying completely on the principles of design of continuous-data control systems. We were perhaps amazed by what the PID controller could accomplish in the improvement on the system response for the digital control system in Example 10-9. Can we do better? Let us repeat the transfer function of the controlled process given in Eq. (10-143) as follows.

$$G_{h0}G_p(z) = \frac{0.0453(z + 0.904)}{(z - 0.905)(z - 0.819)} \tag{10-157}$$

Let the transfer function of the cascade digital controller be

$$G_c(z) = \frac{(z - 0.905)(z - 0.819)}{0.0453(z - 1)(z + 0.904)} \tag{10-158}$$

We see that the objective of the controller is to cancel all the poles and the zeros of the process and then add a pole at z = 1. The motivation for doing this will become immediately clear. The open-loop transfer function of the compensated system now simply becomes

$$G_c(z)G_{h0}G_p(z) = \frac{1}{z-1} \tag{10-159}$$

The corresponding closed-loop transfer function is

$$\frac{C(z)}{R(z)} = \frac{1}{z} \tag{10-160}$$

Thus, for a unit-step input, the output transform is

$$C(z) = \frac{1}{z-1} = z^{-1} + z^{-2} + z^{-3} + \ldots \tag{10-161}$$

The significance of this result is that the output response c(kT) reaches the desired steady-state value in one sampling period and stays at that value thereafter. The overshoot of c(kT) is *zero*. In reality, however, it must be kept in mind that the true judgement on the system performance should be based on the behavior of c(t). In general, although c(kT) may exhibit little or no overshoot, the actual response c(t) may have oscillations between the sampling instants. For the present system, since the sampling period T = 0.1 sec is much smaller than the time constants of the controlled process, it is expected that C(z) or c(kT) gives a fairly accurate description of c(t). Thus, it is expected that the digital controller given in Eq. (10-158) will produce a unit-step response that reaches its steady-state value of unity in 0.1 sec, and there should be little or no ripple in between the sampling instants. This type of response is generally referred to as the *deadbeat response*.

It is important to point out that the deadbeat response is obtainable only under the ideal condition that the cancellation of poles and zeros as required by the design is exact. In practice, the uncertainty of the poles and zeros of the controlled process, due to the approximations required in the modelling of the process, and the restrictions in the realization of the controller transfer function by a digital computer or processor, would make a deadbeat response almost impossible to achieve exactly.

Now let us look at the deadbeat response design as a general design tool. The system configuration considered is shown in Fig. 10-51, where G(z) is the transfer function of the controlled process which could include a zero-order hold. Once we have established the criteria for the deadbeat-response design it will be easy to see that the system configuration is not limited to that shown in Fig. 10-51.

The closed-loop transfer function of the system is written

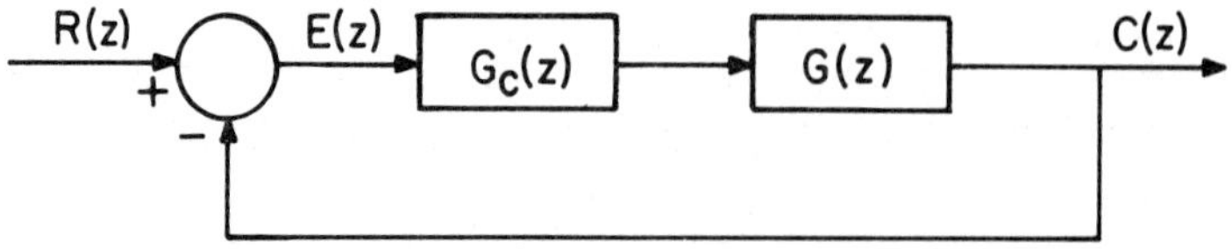

Fig. 10-51. Digital control system.

$$\frac{C(z)}{R(z)} = M(z) = \frac{G_c(z)G_p(z)}{1 + G_c(z)G_p(z)} \tag{10-162}$$

The deadbeat-response design is characterized by the following design criteria:

1. The system must have zero steady-state error at the sampling instants for the specified reference input signal.
2. The response time defined as the time required to reach the steady state should be a minimum.
3. The digital controller $G_c(z)$ must be physically realizable.

Solving for $G_c(z)$ from Eq. (10-162), we have

$$G_c(z) = \frac{1}{G(z)} \frac{M(z)}{1 - M(z)} \tag{10-163}$$

The z-transform of the error signal is written

$$\begin{aligned} E(z) &= R(z) - C(z) \\ &= R(z)[1 - M(z)] = \frac{R(z)}{1 + G_c(z)G(z)} \end{aligned} \tag{10-164}$$

Let the z-transform of the input be described by the function

$$R(z) = \frac{A(z)}{(1 - z^{-1})^N} \tag{10-165}$$

where N is a positive integer, and A(z) is a polynomial in z^{-1} with no zeros at $z = 1$. For example, for a unit-step function input, $A(z) = 1$ and $N = 1$; for a unit-ramp function input, $A(z) = Tz^{-1}$ and $N = 2$, etc. In general, R(z) of Eq. (10-165) represents inputs of the type t^{N-1}. For zero steady-state error,

$$\begin{aligned} \lim_{k\to\infty} e(kT) &= \lim_{z\to 1} (1 - z^{-1})E(z) \\ &= \lim_{z\to 1} (1 - z^{-1}) \frac{A(z)[1 - M(z)]}{(1 - z^{-1})^N} = 0 \end{aligned} \tag{10-166}$$

Since the polynomial A(z) does not contain any zeros at z = 1, the necessary condition for the steady-state error to be zero is that 1 − M(z) must contain the factor $(1 - z^{-1})^N$. Thus, 1 − M(z) should have the form

$$1 - M(z) = (1 - z^{-1})^N F(z) \tag{10-167}$$

where F(z) is a polynomial of z^{-1}. Solving for M(z) in the last equation, we have

$$M(z) = \frac{z^N - (z-1)^N F(z)}{z^N} \tag{10-168}$$

Since F(z) is a polynomial in z^{-1}, it has only poles at z = 0. Therefore, Eq. (10-168) clearly indicates that the characteristic equation of the system with zero steady-state error is of the form

$$z^p = 0 \tag{10-169}$$

where p is a positive integer $\geqslant$ N.

Substitution of Eq. (10-167) in Eq. (10-164), the z-transform of the error is written

$$E(z) = A(z)F(z) \tag{10-170}$$

Since A(z) and F(z) are both polynomials of z^{-1}, E(z) in Eq. (10-170) will have a *finite* number of terms in its power-series expansion in inverse powers of z. Thus, *when the characteristic equation of a digital control system is of the form of Eq. (10-169), that is, when the characteristic equation roots are all at z = 0, the error signal will go to zero in finite number of sampling periods.*

In general, F(z) can be a rational function of z, that is, F(z) can be of the form

$$F(z) = \frac{F_n(z)}{F_d(z)} \tag{10-171}$$

where $F_n(z)$ and $F_d(z)$ are polynomials of z, but with no poles and zeros at z = 1 and with no poles with $|z| > 1$. Then, the characteristic equation becomes

$$z^p F_d(z) = 0 \tag{10-172}$$

Physical Realizability Considerations

Equation (10-168) indicates that the design of a digital control system with the deadbeat response for a given input requires first the selection of the function F(z). Once M(z) is determined from Eq. (10-168), the transfer function of the digital controller is obtained from Eq. (10-163). However, the physical realizability requirement on $G_c(z)$ and the fact that $G(z)$ is the transfer function of a physical process put constraints on the closed-loop transfer function $M(z)$. In general, let $G(z)$ and $M(z)$ be expressed by the following series expansions:

$$G(z) = g_n z^{-n} + g_{n+1} z^{-n-1} + \ldots \tag{10-173}$$

$$M(z) = m_k z^{-k} + m_{k+1} z^{-k-1} + \ldots \tag{10-174}$$

where $n \geqslant 0$ and $k \geqslant 0$. Substituting the last two equations in Eq. (10-163), we have

$$G_c(z) = \frac{m_k z^{-k} + m_{k+1} z^{-k-1} + \ldots}{(g_n z^{-n} + g_{n+1} z^{-n-1} + \ldots)(1 - m_k z^{-k} - m_{k-1} z^{-k-1} - \ldots)}$$

$$= d_{k-n} z^{-(k-n)} + d_{k-n+1} z^{-(k-n+1)} + \ldots \tag{10-175}$$

Thus, for $G_c(z)$ to be physically realizable, $k \geqslant n$; i.e., the lowest power of the series expansion of $M(z)$ in inverse powers of z must be at least as large as that of $G(z)$. Once the minimum requirement on $M(z)$ is established, for a specific input, $F(z)$ must be chosen according to Eq. (10-167) to satisfy this requirement.

Referring to the input transform of Eq. (10-165), the value of N is determined by the type of input, i.e., for a step input, $N = 1$; ramp input, $N = 2$; parabolic input, $N = 3$, etc. The relations between the basic form of $M(z)$ and the type of input for a deadbeat response are determined from Eq. (10-168) and are tabulated in Table 10-2.

TABLE 10-2.

Input function	N	M(z)
Step input $u_s(t)$	1	$1 - (1 - z^{-1})F(z)$
Ramp input $tu_s(t)$	2	$1 - (1 - z^{-1})^2 F(z)$
Parabolic input $t^2 u_s(t)$	3	$1 - (1 - z^{-1})^3 F(z)$

Investigating the relations between $M(z)$ and $F(z)$ given in Table 10-2, we see that $F(z)$ must contain at least the term 1, or $M(z)$ will contain it, and would not be a physically realizable function. In fact, there does not seem to be any objection to selecting $F(z) = 1$ for all inputs that are described by Eq. (10-165). Then, for the three basic inputs considered in Table 10-2, the closed-loop transfer functions of the systems with deadbeat responses are tabulated in Table 10-3.

TABLE 10-3.

Step input	$M(z) = z^{-1}$
Ramp input	$M(z) = 2z^{-1} - z^{-2}$
Parabolic input	$M(z) = 3z^{-1} - 3z^{-2} + z^{-3}$

Thus, the results in Table 10-3 indicate that when the input is a step function, the minimum time for the error to go to zero is one sampling period, for a ramp function input, it takes two sampling periods for the error to be reduced to zero, and the minimum number of sampling periods for the error due to a parabolic input to diminish is three.

As illustrated by the example given at the beginning of this section, the deadbeat-response design depends essentially on the cancellation of the poles and zeros of the controlled process by the zeros and poles of the digital controller, and then new ones are added at appropriate places in the z-plane. One difficulty may arise when $G(z)$ has one or more zeros that are on or outside the unit circle, which would require an unstable controller to cancel these zeros. Another difficulty is revealed by referring to Eq. (10-163), when $M(z)$ assumes any one of the forms given in Table 10-3. Since the highest power term in $M(z)$ is z^{-1}, $M(z)/[1 - M(z)]$ will always have one more pole than zero. Then, in order that $G_c(z)$ is a physically realizable transfer function, $G(z)$ must have at most one more pole than zero. Of course, $G(z)$ cannot have more zeros than poles. For example, for a step input, $M(z) = z^{-1}$, Eq. (10-163) gives

$$G_c(z) = \frac{1}{G(z)} \frac{1}{z - 1} \tag{10-176}$$

Thus, the condition on $G(z)$ given above is arrived at. The conclusion is that when $G(z)$ has one or more zeros on or outside the unit circle $|z| = 1$, or has more than one pole than zeros, $F(z)$ cannot be simply 1. The following example illustrates the case when $G(z)$ has an excess of poles over zeros greater than one.

Example 10-10.

Consider that the controlled process of the digital control system shown in Fig. 10-51 is described by

$$G(z) = \frac{z^{-2}}{1 - z^{-1} - z^{-2}} \tag{10-177}$$

A digital controller is to be designed so that a deadbeat response is obtained when the input is a unit-step function.

Since the transfer function G(z) has two more poles than zeros we cannot chose M(z) to be z^{-1}, since it will lead to a physically unrealizable $G_c(z)$. Let us try $M(z) = z^{-2}$. Then, Eq. (10-163) gives

$$G_c(z) = \frac{1 - z^{-1} - z^{-2}}{z^{-2}} \frac{z^{-2}}{1 - z^{-2}}$$

$$= \frac{1 - z^{-1} - z^{-2}}{1 - z^{-2}} \tag{10-178}$$

which is a physically realizable transfer function. In this case, the function F(z) is given by

$$F(z) = \frac{1 - M(z)}{1 - z^{-1}} = 1 + z^{-1} \tag{10-179}$$

In general it can be shown that for a given type of input which determines the value of N, the minimum number of sampling periods for the system to reach zero is N + M, where the controlled process G(z) has M + 1 more poles than zeros.

Deadbeat-Response Design of Systems With Process Poles or Zeros On or Outside the Unit Circle

It has been established that the deadbeat-response design of a digital control system depends entirely on the cancellation of the poles and zeros of the controlled process by the zeros and poles of the digital controller. However, if the controlled process has poles or zeros that are on or outside the unit circle in the z-plane, imperfect cancellation, which is very likely to occur in practice, will result in an unstable closed-loop system. Thus, for all practical purposes no attempt is made to cancel the poles and zeros of the process that are on or outside the unit circle with the zeros and poles of the digital controller. This simply imposes additional constraints on the selection of the closed-loop transfer function M(z).

Let the transfer function of the controlled process of Fig. 10-51 be written as

$$G(z) = \frac{\prod_{i=1}^{K} (1 - z_i z^{-1})}{\prod_{j=1}^{L} (1 - p_j z^{-1})} A_g(z) \qquad (10\text{-}180)$$

where z_i ($i = 1,2,\ldots,K$) and p_j ($j = 1,2,\ldots,L$) represent the zeros and poles, respectively, of G(z) that lie on or outside the unit circle, and $A_g(z)$ is a rational function in z^{-1} with poles and zeros only inside the unit circle. Substitution of Eq. (10-180) in Eq. (10-163), we get

$$G_c(z) = \frac{\prod_{j=1}^{L} (1 - p_j z^{-1})}{\prod_{i=1}^{K} (1 - z_i z^{-1})} \frac{M(z)}{A_g(z)[1 - M(z)]} \qquad (10\text{-}181)$$

Since $G_c(z)$ cannot contain the poles p_j and zeros z_i as its zeros and poles, respectively, they must be cancelled by the zeros and poles of $1 - M(z)$ and M(z). In other words, M(z) must contain

$$\prod_{i=1}^{K} (1 - z_i z^{-1})$$

and $1 - M(z)$ must contain

$$\prod_{j=1}^{L} (1 - p_j z^{-1})$$

Thus, in general, M(z) and $1 - M(z)$ have the following forms:

$$M(z) = \prod_{i=1}^{K} (1 - z_i z^{-1})(M_m z^{-m} + M_{m+1} z^{-m-1} + \ldots) \qquad (10\text{-}182)$$

$$1 - M(z) = \prod_{j=1}^{L} (1 - p_j z^{-1})(1 - z^{-1})^N (1 + a_1 z^{-1} + a_2 z^{-2} + \ldots) \qquad (10\text{-}183)$$

where m should equal or be greater than the lowest power in z^{-1} in the series expansion of G(z), and N is an integer which depends on the order of the input signal. Note that if some of the poles of G(z) are at $z = 1$, then $1 - M(z)$ is raised to the power of $(1 - z^{-1})$ equal to either the order of the input poles or the order of the poles of G(z) at $z = 1$, whichever is higher.

To illustrate the design of a digital control system which has a controlled process with poles or zeros outside or on the unit circle, the following example is given.

Example 10-11.

For the system in Fig. 10-51, the transfer function of the controlled process is given by

$$G(z) = \frac{0.000392z^{-1}(1 + 2.78z^{-1})(1 + 0.2z^{-1})}{(1 - z^{-1})^2(1 - 0.286z^{-1})} \tag{10-184}$$

The problem is to design a digital controller which will produce an output response with zero steady-state error and minimum finite settling time in response to a unit-ramp input.

Since G(z) has a zero outside the unit circle at $z = -2.78$ and two poles on the unit circle at $z = 1$, M(z) and $1 - M(z)$ should be of the forms given in Eqs. (10-182) and (10-183), respectively. Thus,

$$M(z) = (1 + 2.78z^{-1})(M_1 z^{-1} + M_2 z^{-2}) \tag{10-185}$$

Since the first term in the series expansion of G(z) is $0.000392z^{-1}$, the first term in the series expansion of M(z) should be $M_1 z^{-1}$, where the coefficient M_1 is yet to be determined. For a unit-ramp function input, $1 - M(z)$ should contain the factor $(1 - z^{-1})^2$, as well as all the poles of G(z) that are on or outside the unit circle. In this case since G(z) has two poles at $z = 1$, having the term $(1 - z^{-1})^2$ in $1 - M(z)$ would be sufficient to satisfy both requirements. Thus, $1 - M(z)$ should be of the form

$$1 - M(z) = (1 - z^{-1})^2 F(z) = (1 - z^{-1})^2 (1 + a_1 z^{-1}) \tag{10-186}$$

where in order for the response to be deadbeat and simultaneously result in a physically realizable $G_c(z)$, $1 - M(z)$ must contain the constant term 1, so that $a_1 \neq 0$. Since M(z) now has a minimum order of three, the transient response will settle in three sampling periods, the shortest possible settling time for this system. Referring to Table 10-3, we see that if G(z) does not have poles or zeros on or outside the unit circle, the minimum settling time for a ramp input would be two sampling periods. Because of the zero of G(z) at $z = -2.78$, F(z) can no longer be chosen to be one, and the settling time is increased by one sampling period. In general, the more zeros and poles of G(z) which lie on or outside the unit circle, the longer will be the minimum settling time of the deadbeat response.

Now with three unknowns and three independent equations, M_1, M_2, and a_1 are solved from Eqs. (10-185) and (10-186). Substitution of Eq. (10-185) into Eq. (10-186) yields the following equations:

$$M_1 z^{-1} + (2.78M_1 + M_2)z^{-2} + 2.78M_2 z^{-3} = (2 - a_1)z^{-1} + (2a_1 - 1)z^{-2} - a_1 z^{-2} \tag{10-187}$$

Equating the coefficients of the corresponding terms on both sides of the last equation, we have

$$M_1 = 2 - a_1 \tag{10-188}$$

$$2.78M_1 + M_2 = 2a_1 - 1 \tag{10-189}$$

$$2.78M_2 = -a_1 \tag{10-190}$$

Solving the last three equations yields the coefficients

$$M_1 = 0.723$$

$$M_2 = -0.46$$

$$a_1 = 1.277$$

The closed-loop transfer function is

$$M(z) = 0.723z^{-1} + 1.554z^{-2} - 1.277z^{-3}$$

$$= \frac{0.723z^2 + 1.554z - 1.277}{z^3} \tag{10-191}$$

For a unit-ramp input, the z-transform of the output is

$$C(z) = M(z)\frac{Tz^{-1}}{(1 - z^{-1})^2} = 0.05(0.723z^{-2} + 3z^{-3} + 4z^{-4} + 5z^{-5} + \ldots) \tag{10-192}$$

The output response at the sampling instants is shown in Fig. 10-52(a). When a unit step function input is applied, the output z-transform is

$$C(z) = M(z)\frac{1}{1 - z^{-1}} = 0.723z^{-1} + 2.277z^{-2} + z^{-3} + z^{-4} + \ldots \tag{10-193}$$

Thus, the unit-step response of the system has a maximum overshoot of 127.7%, as shown in Fig. 10-52(b).

Substitution of Eqs. (10-191) and (10-184) in Eq. (10-163) yields the transfer function of the digital controller

$$G_c(z) = \frac{1840(1 - 0.286z^{-1})(1 - 0.636z^{-1})}{(1 + 0.2z^{-1})(1 + 1.277z^{-1})} \tag{10-194}$$

Comments of the Deadbeat-Response Design

Although the deadbeat-response design leads to a digital control system which gives an "ideal" output response to a specific type of input, the system is really "tuned" only for the input it is designed for, and would give inferior or unacceptable performance to other inputs. For example, the system designed in Example 10-11 is for a ramp-function input. The output reaches the input ramp and follows it after three sampling periods. However, when the same system is subject to a unit-step input, the peak overshoot is 2.277, or 127.7 percent.

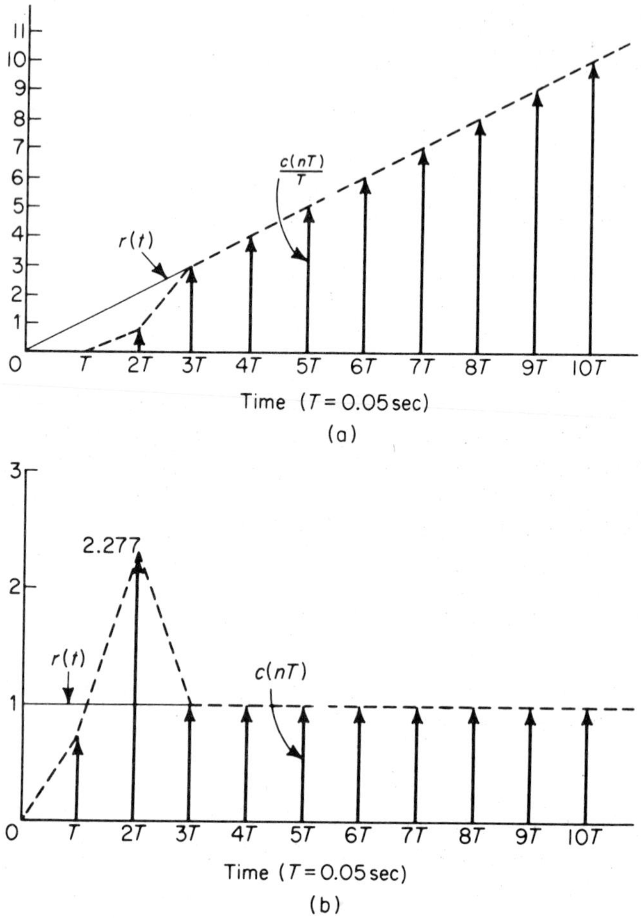

Fig. 10-52. Ramp and step responses of the system in Example 10-11.

Another problem with the deadbeat-response design is due to the multiple-order pole of the closed-loop transfer function M(z), since the characteristic equation is of the form: $z^N = 0$. It can be shown that the sensitivity of a multiple-order characteristic root due to a change in any system parameter is theoretically equal to infinity. The sensitivity of the roots of the characteristic equation of a digital control system with respect to the incremental change of a system parameter k is defined as

$$S_k^z = \frac{dz/z}{dk/k} = \frac{dz}{dk}\frac{k}{z} \tag{10-195}$$

Let the characteristic equation be written as

$$P(z) + kQ(z) = 0 \tag{10-196}$$

where k is any system parameter; P(z) and Q(z) are polynomials of z with constant coefficients, and they do not contain k. Consider that k is varied by an increment Δk; Eq. (10-196) becomes

$$P(z) + (k + \Delta k)Q(z) = 0 \tag{10-197}$$

Dividing both sides of the last equation by P(z) + kQ(z), we have

$$1 + \frac{\Delta k Q(z)}{P(z) + kQ(z)} = 0 \tag{10-198}$$

which can be written as

$$1 + \Delta k B(z) = 0 \tag{10-199}$$

where

$$B(z) = \frac{Q(z)}{P(z) + kQ(z)} \tag{10-200}$$

Notice that the denominator of B(z) contains the original characteristic equation. Let us assume that the characteristic equation in Eq. (10-196) has a root z_i with multiplicity n (> 1). Expanding B(z) by partial-fraction expansion in the neighborhood of z_i, we have

$$B(z) \cong \frac{A_i}{(z - z_i)^n} = \frac{A_i}{(\Delta z)^n} \tag{10-201}$$

Substitution of Eq. (10-201) into Eq. (10-199) yields

$$1 + \frac{\Delta k A_i}{(\Delta z)^n} \cong 0 \tag{10-202}$$

from which we obtain

$$\frac{\Delta k}{\Delta z} = \frac{-(\Delta z)^{n-1}}{A_i} \tag{10-203}$$

Now taking the limit on both sides of Eq. (10-203) as Δk and Δz approach zero, we have

$$\lim_{\substack{\Delta k \to 0 \\ \Delta z \to 0}} \frac{\Delta k}{\Delta z} = \frac{dk}{dz} = \lim_{\substack{\Delta k \to 0 \\ \Delta z \to 0}} \frac{-(\Delta z)^{n-1}}{A_i} = 0 \tag{10-204}$$

Thus, we have shown that at a multiple-order root of the characteristic equation, dk/dz is equal to zero, and from Eq. (10-195), this corresponds to infinite S_k^z. This result simply shows that the deadbeat-response design is very sensitive to parameter variations.

Design Using The Staleness Weighting Factor

Thus far in the deadbeat-response design it was pointed out that a system which is designed for a deadbeat response for a particular input does not provide satisfactory performance when other types of inputs are applied. A comprised design may be devised to lead to a system that will respond satisfactorily to a number of different types of inputs. Of course, none of the responses can be described to be deadbeat, and the design based on the deadbeat-response principle is used simply as an established design procedure.

The design is effected by introducing a *weighting factor* c into the desired closed-loop transfer function M(z). Let us modify the function $1 - M(z)$ by dividing it by $(1 - cz^{-1})$. We write

$$1 - M_w(z) = \frac{1 - M(z)}{1 - cz^{-1}} \tag{10-205}$$

Since c must now appear as a pole of $M_w(z)$, we must limit the value of c to between -1 and $+1$ so that $M_w(z)$ is stable. When the *weighting factor* c is introduced into the denominator of $1 - M(z)$, the characteristic equation of the compensated system can no longer be of the form $z^N = 0$, and the error of the system will not settle to zero in a finite period of time. However, the use of the weighting factor will give a better all-around performance to several different types of inputs, and the sensitivity of the system response to parameter variations is reduced. With a proper choice of the value of c, the maximum overshoot of the step response can be reduced when the system is designed for a ramp input, and the transient response can be reduced to a tolerable amount within a reasonable time interval. The use of the weighting factor in the design of digital control systems is illustrated by the following example.

Example 10-12.

Consider that the controlled process of a digital control system is given by

$$G(z) = \frac{0.005z^{-1}(1 - 0.9z^{-1})}{(1 - z^{-1})(1 - 0.905z^{-1})} \tag{10-206}$$

For a unit-ramp input, the deadbeat-response design gives the closed-loop transfer function as (from Table 10-2)

$$M(z) = 2z^{-1} - z^{-2} \tag{10-207}$$

The unit-step response and the unit-ramp response of the designed system are shown in Figs. 10-53(a) and (b), respectively (c = 0). Notice that while the ramp response reaches the input in two sampling instants, the step response exhibits a peak overshoot of 100 percent.

Now using Eq. (10-205), we have

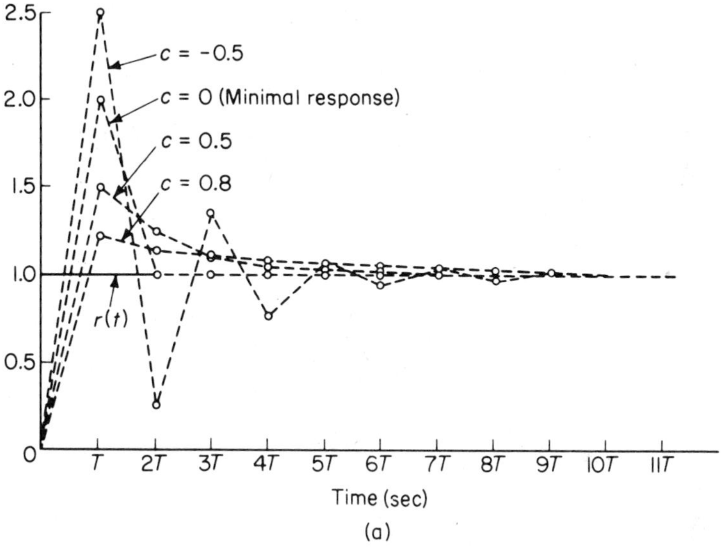

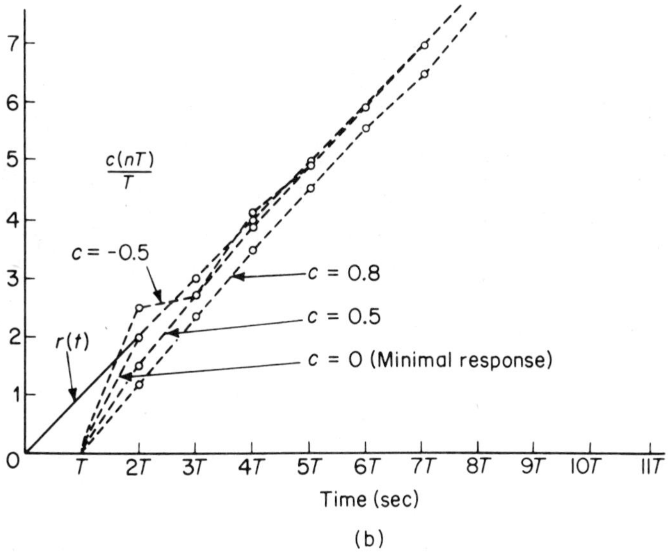

Fig. 10-53. Step and ramp responses of the system in Example 10-12.

$$1 - M_w(z) = \frac{1 - M(z)}{1 - cz^{-1}} = \frac{(1 - z^{-1})^2}{1 - cz^{-1}} \tag{10-208}$$

which gives

$$M_w(z) = \frac{(2 - c)z^{-1} - z^{-2}}{1 - cz^{-1}} \tag{10-209}$$

To illustrate the effects of the weighting factor c, and to assist the selection of an optimum value of c, the unit-step and unit-ramp responses of the system at the sampling instants are plotted as shown in Figs. 10-53(a) and 10-53(b), respectively, for several values of c between -1 and 1. It is seen that for negative values of c the overshoot of the step response is actually increased from the deadbeat design. For $c = 0.8$, the peak overshoot in the step response is only 20 percent, but the transient in the step and the ramp responses settles very slowly. It would seem that a best all-around choice for c in this case is 0.5.

10.9 POLE PLACEMENT DESIGN BY STATE FEEDBACK (SINGLE INPUT)

In Chapter 4 we have considered the design of a single-input digital control system by state feedback. It was shown by means of the phase-variable-canonical-form transformation that for a single-input system

$$\mathbf{x}[(k + 1)T] = \mathbf{Ax}(kT) + \mathbf{B}u(kT) \tag{10-210}$$

with the state feedback control

$$u(kT) = -\mathbf{Gx}(kT) \tag{10-211}$$

where $\mathbf{G}$ is the $1 \times n$ constant feedback matrix, the eigenvalues of $\mathbf{A} - \mathbf{BG}$ can be arbitrarily assigned if and only if the pair $[\mathbf{A}, \mathbf{B}]$ is completely controllable. As a matter of fact, the property of arbitrary pole-placement by state feedback for a controllable system is also true for a system with multiple inputs.

In Chapter 4 the subject of pole-placement design by state feedback was touched upon using the phase-variable canonical form. In this section we shall present more general methods of pole-placement design for single-input systems with state feedback. First we shall establish certain functional relationships as design tools. We define the following functional relations:

$$T_0(z) = -\mathbf{G}(z\mathbf{I} - \mathbf{A})^{-1}\mathbf{B} = \text{open-loop return signal matrix} \tag{10-212}$$

$$T_c(z) = -\mathbf{G}(z\mathbf{I} - \mathbf{A} + \mathbf{BG})^{-1}\mathbf{B} = \text{closed-loop return signal matrix} \tag{10-213}$$

$$\mathbf{T}(z) = \mathbf{I} - T_0(z) = \text{return difference matrix} \tag{10-214}$$

$$\Delta_0(z) = |z\mathbf{I} - \mathbf{A}| = \text{characteristic equation of } \mathbf{A} \text{ (open-loop system)} \tag{10-215}$$

$$\Delta_c(z) = |z\mathbf{I} - \mathbf{A} + \mathbf{BG}| = \text{characteristic equation of } \mathbf{A} - \mathbf{BG} \text{ (closed-loop system)} \tag{10-216}$$

$$\Delta(z) = |\mathbf{I} - \mathbf{T}_0(z)| = |\mathbf{T}(z)| \tag{10-217}$$

In the preceding equations $\mathbf{I}$ denotes the identity matrix of the appropriate dimension in a given equation.

First we shall show that

$$\Delta(z) = \frac{\Delta_c(z)}{\Delta_0(z)} \tag{10-218}$$

This is accomplished by writing

$$z\mathbf{I} - \mathbf{A} + \mathbf{BG} = (z\mathbf{I} - \mathbf{A})[\mathbf{I} + (z\mathbf{I} - \mathbf{A})^{-1}\mathbf{BG}] \tag{10-219}$$

Taking the determinant on both sides of the last equation, we get

$$\Delta_c(z) = |z\mathbf{I} - \mathbf{A} + \mathbf{BG}| = \Delta_0(z)|\mathbf{I} + (z\mathbf{I} - \mathbf{A})^{-1}\mathbf{BG}| \tag{10-220}$$

Since

$$|\mathbf{I} + (z\mathbf{I} - \mathbf{A})^{-1}\mathbf{BG}| = |\mathbf{I} + \mathbf{BG}(z\mathbf{I} - \mathbf{A})^{-1}| = |\mathbf{I} + \mathbf{G}(z\mathbf{I} - \mathbf{A})^{-1}\mathbf{B}| \tag{10-221}$$

where the identity matrices are of different dimensions, Eq. (10-220) becomes

$$\Delta_c(z) = \Delta_0(z)\Delta(z) \tag{10-222}$$

Thus, Eq. (10-218) is proven.

The next functional relation of importance is

$$\mathbf{T}_0(z) = \mathbf{T}(z)\mathbf{T}_c(z) \tag{10-223}$$

or

$$\mathbf{G}(z\mathbf{I} - \mathbf{A})^{-1}\mathbf{B} = [\mathbf{I} + \mathbf{G}(z\mathbf{I} - \mathbf{A})^{-1}\mathbf{B}]\mathbf{G}(z\mathbf{I} - \mathbf{A} + \mathbf{BG})^{-1}\mathbf{B} \tag{10-224}$$

Taking the matrix inverse on both sides of Eq. (10-219), we get

$$(z\mathbf{I} - \mathbf{A} + \mathbf{BG})^{-1} = [\mathbf{I} + (z\mathbf{I} - \mathbf{A})^{-1}\mathbf{BG}]^{-1}(z\mathbf{I} - \mathbf{A})^{-1} \tag{10-225}$$

Premultiplication on both sides of the last equation by $\mathbf{I} + (z\mathbf{I} - \mathbf{A})^{-1}\mathbf{BG}$ yields

$$[\mathbf{I} + (z\mathbf{I} - \mathbf{A})^{-1}\mathbf{BG}](z\mathbf{I} - \mathbf{A} + \mathbf{BG})^{-1} = (z\mathbf{I} - \mathbf{A})^{-1} \tag{10-226}$$

Then we premultiply both sides by **G** and post-multiply by **B**; the last equation becomes

$$\mathbf{G}[\mathbf{I} + (z\mathbf{I} - \mathbf{A})^{-1}\mathbf{B}\mathbf{G}](z\mathbf{I} - \mathbf{A} + \mathbf{B}\mathbf{G})^{-1}\mathbf{B} = \mathbf{G}(z\mathbf{I} - \mathbf{A})^{-1}\mathbf{B} \tag{10-227}$$

The last equation is written

$$[\mathbf{I} + \mathbf{G}(z\mathbf{I} - \mathbf{A})^{-1}\mathbf{B}]\mathbf{G}(z\mathbf{I} - \mathbf{A} + \mathbf{B}\mathbf{G})^{-1}\mathbf{B} = \mathbf{G}(z\mathbf{I} - \mathbf{A})^{-1}\mathbf{B} \tag{10-228}$$

Thus, the relation in Eq. (10-223) is proven.

The last relation we need is derived by use of Eqs. (10-218) and (10-225). Equation (10-214) is written as

$$\begin{aligned} \mathbf{T}(z) &= \mathbf{I} + \mathbf{G}(z\mathbf{I} - \mathbf{A})^{-1}\mathbf{B} \\ &= \mathbf{I} + \mathbf{G}\,\frac{\mathrm{Adj}(z\mathbf{I} - \mathbf{A})\mathbf{B}}{\Delta_0(z)} \end{aligned} \tag{10-229}$$

where Adj$(z\mathbf{I} - \mathbf{A})$ denotes the adjoint of the matrix $z\mathbf{I} - \mathbf{A}$.

Let

$$\mathbf{k}(z) = [\mathrm{Adj}(z\mathbf{I} - \mathbf{A})]\mathbf{B} \qquad (n \times 1) \tag{10-230}$$

Then, Eq. (10-229) is written as

$$\mathbf{T}(z) = \frac{\Delta_0(z) + \mathbf{G}\mathbf{k}(z)}{\Delta_0(z)} = \Delta(z) \tag{10-231}$$

where $\mathbf{T}(z)$ is shown to be a scalar.

Using Eq. (10-218), the last equation leads to

$$\mathbf{G}\mathbf{k}(z) = \Delta_c(z) - \Delta_0(z) \tag{10-232}$$

Thus, knowing $\mathbf{k}(z)$, $\Delta_c(z)$ and $\Delta_0(z)$, we should be able to solve for the feedback gain matrix **G** from Eq. (10-232) if the pair (**A**,**B**) is completely controllable.

Two expressions for **G** can be obtained from Eq. (10-232). Let

$$\Delta_c(z) = z^n + \alpha_n z^{n-1} + \alpha_{n-1} z^{n-2} + \ldots + \alpha_2 z + \alpha_1 \tag{10-233}$$

$$\Delta_0(z) = z^n + a_n z^{n-1} + a_{n-1} z^{n-2} + \ldots + a_2 z + a_1 \tag{10-234}$$

From Eq. (4-247), $\mathbf{k}(z)$ is written

$$\mathbf{k}(z) = [\mathrm{Adj}(z\mathbf{I} - \mathbf{A})]\mathbf{B} = \sum_{j=1}^{n} z^{j-1} \sum_{i=j}^{n} a_{i+1}\mathbf{A}^{i-j}\mathbf{B} \tag{10-235}$$

Then, Eq. (10-232) is written

$$\mathbf{G}\sum_{j=1}^{n} z^{j-1} \sum_{i=j}^{n} a_{i+1}\mathbf{A}^{i-j}\mathbf{B} = \sum_{i=0}^{n-1} (\alpha_{i+1} - a_{i+1})z^{i} \tag{10-236}$$

Equating the coefficients of the like powers of z on both sides of the last equation, we get

$$\begin{aligned}
&\mathbf{G}(a_{n+1}\mathbf{B}) = \alpha_n - a_n \qquad\qquad (a_{n+1} = 1)\\
&\mathbf{G}(a_n + a_{n+1}\mathbf{A})\mathbf{B} = \alpha_{n-1} - a_{n-1}\\
&\qquad\vdots\\
&\mathbf{G}\sum_{i=2}^{n} a_{i+1}\mathbf{A}^{i-2}\mathbf{B} = \alpha_2 - a_2\\
&\mathbf{G}\sum_{i=1}^{n} a_{i+1}\mathbf{A}^{i-1}\mathbf{B} = \alpha_1 - a_1
\end{aligned} \tag{10-237}$$

Or, in matrix form

$$\begin{bmatrix} a_{n+1} & 0 & 0 & \cdots & 0 \\ a_n & a_{n+1} & 0 & \cdots & 0 \\ a_{n-1} & a_n & a_{n+1} & \cdots & 0 \\ \vdots & \vdots & \vdots & \cdots & \vdots \\ a_2 & a_3 & a_4 & \cdots & a_{n+1} \end{bmatrix} \begin{bmatrix} \mathbf{B}' \\ \mathbf{B}'\mathbf{A}' \\ \mathbf{B}'(\mathbf{A}')^2 \\ \vdots \\ \mathbf{B}'(\mathbf{A}')^{n-1} \end{bmatrix} \mathbf{G}' = \begin{bmatrix} \alpha_n - a_n \\ \alpha_{n-1} - a_{n-1} \\ \alpha_{n-2} - a_{n-2} \\ \vdots \\ \alpha_1 - a_1 \end{bmatrix} \tag{10-238}$$

Let

$$\mathbf{M} = \begin{bmatrix} a_{n+1} & 0 & 0 & \cdots & 0 \\ a_n & a_{n+1} & 0 & \cdots & 0 \\ a_{n-1} & a_n & a_{n+1} & \cdots & 0 \\ \vdots & \vdots & \vdots & & \vdots \\ a_1 & a_2 & a_3 & \cdots & a_{n+1} \end{bmatrix} \tag{10-239}$$

$$\mathbf{S} = [\mathbf{B} \quad \mathbf{AB} \quad \mathbf{A}^2\mathbf{B} \quad \ldots \quad \mathbf{A}^{n-1}\mathbf{B}] \tag{10-240}$$

$$\alpha = [\alpha_n \quad \alpha_{n-1} \quad \cdots \quad \alpha_1]' \tag{10-241}$$

$$\mathbf{a} = [a_n \quad a_{n-1} \quad \cdots \quad a_1]' \tag{10-242}$$

Thus, Eq. (10-238) is written

$$\mathbf{MS}'\mathbf{G}' = \alpha - \mathbf{a} \tag{10-243}$$

Solving for $\mathbf{G}$ from the last equation, we get

$$\mathbf{G} = [(\mathbf{MS}')^{-1}(\alpha - \mathbf{a})]' \tag{10-244)*$$

Since $\mathbf{M}$ is a triangular matrix with 1's on the main diagonal, it is nonsingular. Thus, for $\mathbf{G}$ to have the solution given by Eq. (10-244), the controllability matrix $\mathbf{S}$ must be of rank n, or the pair $(\mathbf{A},\mathbf{B})$ must be controllable.

The feedback matrix $\mathbf{G}$ is expressed as a function of the coefficients of the closed-loop characteristic equation, α_i, $i = 1,2,\ldots,n$, in Eq. (10-244). An alternate expression for $\mathbf{G}$ can be obtained in terms of the desired closed-loop eigenvalues. Let these eigenvalues be $z_1, z_2, z_3, \ldots, z_m$, all distinct, and the remaining ones are of multiple order. Then,

$$\Delta_c(z_i) = 0 \qquad i = 1,2,\ldots,n \tag{10-245}$$

and, thus, Eq. (10-232) leads to

$$\mathbf{Gk}(z_i) = -\Delta_0(z_i) \qquad i = 1,2,\ldots,m \tag{10-246}$$

Letting

$$\mathbf{k}_i = \mathbf{k}(z_i) \tag{10-247}$$

and

$$\Delta_0(z_i) = \Delta_{0i} \tag{10-248}$$

for $i = 1,2,\ldots,m$, Eq. (10-246) becomes

$$\mathbf{Gk}_i = -\Delta_{0i} \tag{10-249}$$

For a multiple-order eigenvalue of multiplicity q, we take the derivatives on both sides of Eq. (10-232) with respect to z, and then setting $z = z_{m+j}$, $j = 1,2,\ldots,q$, we have

$$\mathbf{Gk}_{m+j} = -\Delta_{m+j} \tag{10-250}$$

*The expression for $\mathbf{G}$ is simplified and can be expressed in terms of the closed-loop eigenvalues if $\mathbf{A}$ and $\mathbf{B}$ are in phase-variable canonical form. (See Problem 10-11).

where

$$\mathbf{k}_{m+j} = \frac{d^j}{dz^j}\mathbf{k}(z)\Big|_{z=z_{m+j}} \qquad (j = 1,2,\ldots,q) \qquad (10\text{-}251)$$

and

$$\Delta_{m+j} = \frac{d^j}{dz^j}\Delta_0(z)\Big|_{z=z_{m+j}} \qquad (j = 1,2,\ldots,q) \qquad (10\text{-}252)$$

Now for all n eigenvalues, we have

$$\mathbf{G}[\mathbf{k}_1 \quad \mathbf{k}_2 \quad \ldots \quad \mathbf{k}_n] = -[\Delta_{01} \quad \Delta_{02} \quad \cdots \quad \Delta_{0n}] \qquad (10\text{-}253)$$

Thus,

$$\mathbf{G} = -[\Delta_{01} \quad \Delta_{02} \quad \cdots \quad \Delta_{0n}]\mathbf{K}^{-1} \qquad (10\text{-}254)$$

where

$$\mathbf{K} = [\mathbf{k}_1 \quad \mathbf{k}_2 \quad \ldots \quad \mathbf{k}_n] \qquad (10\text{-}255)$$

Since if $(\mathbf{A},\mathbf{B})$ is controllable, we are assured of a solution for $\mathbf{G}$ from Eq. (10-254); and, under this condition, $\mathbf{K}^{-1}$ exists.

Example 10-13.

This example is aimed at illustrating the design of a digital control system with pole placement and state feedback. The controlled process is a dc motor whose block diagram is shown in Fig. 10-54. The dc motor is used to control a pure inertia load in such a way that any non-zero initial values in the armature current i_a, and the motor-load velocity ω, should be reduced to zero as quickly as possible. This type of system belongs to a well-known class of control systems called *regulators*. The system parameters of the dc motor are given as follows:

Armature resistance $R_a = 1$ ohm

Armature inductance L_a = negligible

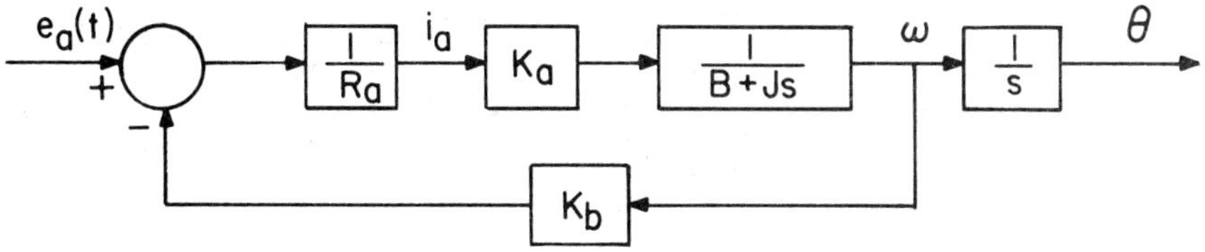

Fig. 10-54. Block diagram of a dc-motor system.

Torque constant	$K_a = 0.345$ Nm/amp
Back emf constant	$K_b = 0.345$ Nm/amp
Motor and load inertia	$J = 1.41 \times 10^{-3}$ kg-m^2
Viscous friction coefficient	$B = 0.25$ kg-m

Note that the parameters are given in the SI units and are consistent. Figure 10-55 shows a state diagram of the dc motor. The input is the armature voltage $e_a(t)$. From the state diagram it is apparent that the state variables are $\theta(t)$ and $\omega(t)$. The state equations of the dc motor system are written as

$$\mathbf{x}(t) = \mathbf{A}\mathbf{x}(t) + \mathbf{B}u(t) \tag{10-256}$$

where

$$\mathbf{x}(t) = \begin{bmatrix} \theta(t) \\ \omega(t) \end{bmatrix} \tag{10-257}$$

$$u(t) = e_a(t) \tag{10-258}$$

$$\mathbf{A} = \begin{bmatrix} 0 & 1 \\ 0 & -\dfrac{(BR_a + K_aK_b)}{JR_a} \end{bmatrix} \tag{10-259}$$

$$\mathbf{B} = \begin{bmatrix} 0 \\ \dfrac{K_a}{JR_a} \end{bmatrix} \tag{10-260}$$

Substitution of the system parameters in the matrices **A** and **B**, we have

$$\mathbf{A} = \begin{bmatrix} 0 & 1 \\ 0 & -267.19 \end{bmatrix} \tag{10-261}$$

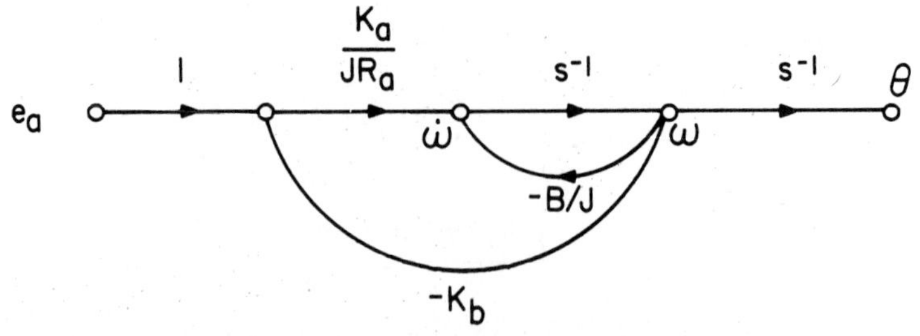

Fig. 10-55. A state diagram of the dc-motor system shown in Fig. 10-54.

$$\mathbf{B} = \begin{bmatrix} 0 \\ 244.68 \end{bmatrix} \tag{10-262}$$

Let us incorporate state feedback by sensing the displacement θ and the velocity ω, and the control is derived through sample-and-hold; i.e.,

$$u(kT) = -\mathbf{G}\mathbf{x}(kT) \tag{10-263}$$

where

$$\mathbf{G} = [g_1 \quad g_2] \tag{10-264}$$

and the sampling period T is 0.005 sec.

The design problem is to find g_1 and g_2 such that the eigenvalues of the closed-loop digital control system are $\lambda_1 = \lambda_2 = 0$. It is important to point out that the eigenvalues of the closed-loop digital control system are the eigenvalues of $\phi(T) - \theta(T)\mathbf{G}$, where

$$\phi(T) = e^{\mathbf{A}T} \tag{10-265}$$

and

$$\theta(T) = \int_0^T \phi(\lambda)\mathbf{B}d\lambda \tag{10-266}$$

The state transition matrix $\phi(T)$ for $T = 0.005$ sec is determined by use of the **A** matrix in Eq. (10-261). Thus,

$$\phi(T) = \mathcal{L}^{-1}[(s\mathbf{I} - \mathbf{A})^{-1}]\Big|_{t=T=0.005} = \begin{bmatrix} 1 & 0.00276 \\ 0 & 0.263 \end{bmatrix} \tag{10-267}$$

$$\theta(T) = \mathcal{L}^{-1}[(s\mathbf{I} - \mathbf{A})^{-1}\mathbf{B}s^{-1}]\Big|_{t=T=0.005} = \begin{bmatrix} 0.00205 \\ 0.675 \end{bmatrix} \tag{10-268}$$

The controlled process is now discretized, and the state equations in the difference equation form are written as

$$\mathbf{x}[(k+1)T] = \phi(T)\mathbf{x}(kT) + \theta(T)u(kT) \tag{10-269}$$

where $\phi(T)$ and $\theta(T)$ are given by Eqs. (10-267) and (10-268), respectively. The state feedback control is described by

$$u(kT) = -\mathbf{G}\mathbf{x}(kT) \tag{10-270}$$

The characteristic equation of the open-loop system, or of $\phi(T)$, is

$$\Delta_0(z) = |z\mathbf{I} - \phi(T)| = \begin{vmatrix} z-1 & -0.00276 \\ 0 & z-0.263 \end{vmatrix}$$

$$= z^2 - 1.263z + 0.263 = 0 \tag{10-271}$$

Since the pair $[\phi(T), \theta(T)]$ is completely controllable, the eigenvalues of $\phi(T) - \theta(T)G$ can be arbitrarily placed by state feedback. Let the desired closed-loop characteristic equation be

$$\Delta_c(z) = z^2 \tag{10-272}$$

From Eq. (10-230),

$$\mathbf{k}(z) = \mathrm{Adj}[z\mathbf{I} - \phi(T)] \cdot \theta(T)$$

$$= \begin{bmatrix} 0.00205z + 0.00132 \\ 0.675(z-1) \end{bmatrix} \tag{10-273}$$

The feedback matrix G is determined from Eq. (10-254),

$$G = -[\Delta_{01} \quad \Delta_{02}]K^{-1} \tag{10-274}$$

where

$$\Delta_{01} = \Delta_0(z)\Big|_{z=0} = 0.263 \tag{10-275}$$

$$\Delta_{02} = \frac{d}{dz}\Delta_0(z)\Big|_{z=0} = (2z - 1.263)\Big|_{z=0} = -1.263 \tag{10-276}$$

$$K = [\mathbf{k}_1 \quad \mathbf{k}_2]$$

where

$$\mathbf{k}_1 = \mathrm{Adj}[z\mathbf{I} - \phi(T)] \cdot \theta(T)\Big|_{z=0}$$

$$= \begin{bmatrix} 0.001323 \\ -0.675 \end{bmatrix} \tag{10-277}$$

$$\mathbf{k}_2 = \frac{d}{dz}\,\mathrm{Adj}[z\mathbf{I} - \phi(T)] \cdot \theta(T)\Big|_{z=0}$$

$$= \begin{bmatrix} 0.00205 \\ 0.675 \end{bmatrix} \tag{10-278}$$

Substitution of Eqs. (10-275) through (10-278) in Eq. (10-274), we get

$$G = [296.3 \quad 0.970] \tag{10-279}$$

Substitution of the feedback gain matrix and the control u(kT) in Eq. (10-269), we get the state equation of the closed-loop system as

$$\begin{bmatrix} x_1[(k+1)T] \\ x_2[(k+1)T] \end{bmatrix} = \begin{bmatrix} 0.395 & 0.00078 \\ -200 & -0.392 \end{bmatrix} \begin{bmatrix} x_1(kT) \\ x_2(kT) \end{bmatrix} \tag{10-280}$$

The solution of the last equation in the z-domain is

$$\mathbf{X}(z) = \begin{bmatrix} z^{-1} + 0.395z^{-2} & 0.00078z^{-2} \\ -200z^{-2} & z^{-1} - 0.392z^{-2} \end{bmatrix} \mathbf{x}(0) \tag{10-281}$$

Thus, we see that the natural responses at the sampling instants of the designed system with state feedback are deadbeat responses, since the two eigenvalues of the closed-loop system are all placed at $z = 0$. For any initial conditions for $x_1(kT)$ and $x_2(kT)$, the responses will become zero after two sampling periods, as indicated by Eq. (10-281).

As an alternate method the feedback matrix **G** can also be found by use of Eq. (10-244).

10.10 POLE PLACEMENT DESIGN BY STATE FEEDBACK (MULTIPLE INPUTS)

With a slight modification the pole-placement design method for systems with single input can be applied to systems with multiple inputs. Consider the system

$$\mathbf{x}[(k+1)T] = \mathbf{A}\mathbf{x}(kT) + \mathbf{B}\mathbf{u}(kT) \tag{10-282}$$

where $\mathbf{x}(kT)$ is an n-vector, and $\mathbf{u}(kT)$ is an r-vector. The pair $(\mathbf{A},\mathbf{B})$ is assumed to be completely controllable. The design problem is stated as: Find the feedback matrix $\mathbf{G}$ $(r \times n)$ such that the control

$$\mathbf{u}(kT) = -\mathbf{G}\mathbf{x}(kT) \tag{10-283}$$

places the eigenvalues of $\mathbf{A} - \mathbf{B}\mathbf{G}$ at arbitrarily assigned positions in the z-plane.

Let us construct the single-input system

$$\mathbf{x}[(k+1)T] = \mathbf{A}\mathbf{x}(kT) + \mathbf{B}^*u(kT) \tag{10-284}$$

and let the $n \times 1$ matrix $\mathbf{B}^*$ be defined as

$$\mathbf{B}^* = \mathbf{B}\mathbf{w} \tag{10-285}$$

where $\mathbf{w}$ is $r \times 1$. The matrix $\mathbf{w}$ must be so chosen that the pair $(\mathbf{A},\mathbf{B}^*)$ is controllable. Then, we can apply the feedback

$$u(kT) = -\mathbf{G}^*\mathbf{x}(kT) \tag{10-286}$$

to place the eigenvalues of $\mathbf{A} - \mathbf{B}^*\mathbf{G}^*$ at the same locations as those of $\mathbf{A} - \mathbf{B}\mathbf{G}$. Then, the problem becomes that of designing the state feedback for the single-input system of Eq. (10-284). Once the feedback matrix $\mathbf{G}^*$ is determined, $\mathbf{G}$ is given by

$$\mathbf{G} = \mathbf{w}\mathbf{G}^* \tag{10-287}$$

since $\mathbf{BG} = \mathbf{B}^*\mathbf{G}^*$.

It is apparent that in general $\mathbf{w}$ is not unique, and it only has to satisfy the condition that $(\mathbf{A}, \mathbf{Bw})$ is controllable. The feedback gain $\mathbf{G}^*$ of the single-input model can be determined using either Eq. (10-244) or Eq. (10-254).

Example 10-14.

Consider the multiple-input digital control system

$$\mathbf{x}[(k+1)T] = \mathbf{A}\mathbf{x}(kT) + \mathbf{B}\mathbf{u}(kT) \tag{10-288}$$

where

$$\mathbf{A} = \begin{bmatrix} 0 & 1 \\ -1 & -2 \end{bmatrix} \qquad \mathbf{B} = \begin{bmatrix} 1 & 0 \\ 0 & 1 \end{bmatrix} \tag{10-289}$$

The pair $(\mathbf{A}, \mathbf{B})$ is controllable. The design problem involves the determination of the feedback matrix $\mathbf{G}$ such that the state feedback

$$\mathbf{u}(kT) = -\mathbf{G}\mathbf{x}(kT) \tag{10-290}$$

places the closed-loop eigenvalues at $z_1 = 0.1$ and $z_2 = 0.2$.

Let us define

$$\mathbf{B}^* = \mathbf{B}\mathbf{w} = \begin{bmatrix} 1 & 0 \\ 0 & 1 \end{bmatrix} \begin{bmatrix} w_1 \\ w_2 \end{bmatrix} = \begin{bmatrix} w_1 \\ w_2 \end{bmatrix} \tag{10-291}$$

For the pair $(\mathbf{A}, \mathbf{B}^*)$ to be controllable, we require that the matrix $[\mathbf{B}^* \quad \mathbf{AB}^*]$ be nonsingular. Thus,

$$|\mathbf{B}^* \quad \mathbf{AB}^*| = \begin{vmatrix} w_1 & w_2 \\ w_2 & -w_1 - 2w_2 \end{vmatrix} = -(w_1 + w_2)^2 \neq 0 \tag{10-292}$$

or $w_1 \neq -w_2$.

The feedback matrix $\mathbf{G}^*$ of the single-input system is first determined by use of Eq. (10-254). We have

$$\mathbf{G}^* = -[\Delta_{01} \quad \Delta_{02}]\mathbf{K}^{-1} \tag{10-293}$$

where

$$\Delta_{01} = |z\mathbf{I} - \mathbf{A}|_{z=0.1} = 1.21 \tag{10-294}$$

$$\Delta_{02} = |z\mathbf{I} - \mathbf{A}|_{z=0.2} = 1.44 \tag{10-295}$$

$$\mathbf{k}(z) = \text{Adj}(z\mathbf{I} - \mathbf{A}) \cdot \mathbf{B}^*$$

$$= \text{Adj}\begin{bmatrix} z & -1 \\ 1 & z+2 \end{bmatrix}\begin{bmatrix} w_1 \\ w_2 \end{bmatrix} = \begin{bmatrix} w_1(z+2) + w_2 \\ -w_1 + w_2 z \end{bmatrix} \tag{10-296}$$

Then,

$$\mathbf{k}_1 = \mathbf{k}(z_1) = \begin{bmatrix} 2.1w_1 + w_2 \\ -w_1 + 0.1w_2 \end{bmatrix} \tag{10-297}$$

$$\mathbf{k}_2 = \mathbf{k}(z_2) = \begin{bmatrix} 2.2w_1 + w_2 \\ -w_1 + 0.2w_2 \end{bmatrix} \tag{10-298}$$

$$\mathbf{K} = [\mathbf{k}_1 \quad \mathbf{k}_2] = \begin{bmatrix} 2.1w_1 + w_2 & 2.2w_1 + w_2 \\ -w_1 + 0.1w_2 & -w_1 + 0.2w_2 \end{bmatrix} \tag{10-299}$$

From the last equation we can again show that in order for $\mathbf{K}$ to be nonsingular, the condition in Eq. (10-280) must be satisfied. Let us arbitrarily choose $\mathbf{w} = [1 \quad 1]'$, although w_1 and w_2 satisfy the condition $w_1 \neq -w_2$. Then,

$$\mathbf{K} = \begin{bmatrix} 3.1 & 3.2 \\ -0.9 & -0.8 \end{bmatrix} \tag{10-300}$$

Substitution of Eqs. (10-294), (10-295) and (10-299) in Eq. (10-293), we get

$$\mathbf{G}^* = -[0.82 \quad 1.48] \tag{10-301}$$

The feedback matrix of the multiple-input system is determined by using Eq. (10-287),

$$\mathbf{G} = \mathbf{w}\mathbf{G}^* = -\begin{bmatrix} 0.82 & 1.48 \\ 0.82 & 1.48 \end{bmatrix} \tag{10-302}$$

Now it is simple to show that

$$|z\mathbf{I} - \mathbf{A} + \mathbf{B}^*\mathbf{G}^*| = |z\mathbf{I} - \mathbf{A} + \mathbf{B}\mathbf{G}| = z^2 - 0.3z + 0.02 \tag{10-303}$$

which has the desired roots at $z = 0.1$ and 0.2.

An alternate method of finding $\mathbf{G}^*$ is to use Eq. (10-244). In the present case,

$$\mathbf{M} = \begin{bmatrix} 1 & 0 \\ 2 & 1 \end{bmatrix} \tag{10-304}$$

$$S = [B^* \quad AB^*] = \begin{bmatrix} w_1 & w_2 \\ w_2 & -w_1 - 2w_2 \end{bmatrix} \tag{10-305}$$

For the closed-loop eigenvalues to be at 0.1 and 0.2,

$$\alpha = [-0.3 \quad 0.02]' \tag{10-306}$$

Also,

$$\mathbf{a} = [2 \quad 1]' \tag{10-307}$$

Then, Eq. (10-244) gives

$$G^* = [(MS')^{-1}(\alpha - \mathbf{a})]'$$

$$= \frac{1}{(w_1 + w_2)^2} [-2.3w_1 - 0.98w_2 \quad -3.62w_1 - 2.3w_2] \tag{10-308}$$

which shows that w_1 cannot be equal to $-w_2$. If as in the above we choose $w_1 = w_2 = 1$, we have

$$G^* = -[0.82 \quad 1.48] \tag{10-309}$$

which is the same answer as obtained in Eq. (10-301).

Weighted- or Incomplete-Input Feedback

The method introduced in the above discussions showed that the pole-placement design of multiple-input systems can be conducted in lieu of an "equivalent" single-input system. *The two systems are "equivalent" only in the sense that they have the same eigenvalues.*

The matrix **w** used to convert the multiple-input system to a single-input system has only the constraint that the latter must be controllable. However, although **w** can otherwise be arbitrarily chosen, in practice, we can impose other useful constraints on the selection of its elements. Since **w** is multiplied to the input vector **u**(kT), the significance of **w** is that its elements place various weights on the state feedback control. For instance, if we choose $w_1 = 2w_2$, this means that there would be twice as much gain in the feedback to u_1 as compared to that for u_2.

10.11 POLE PLACEMENT DESIGN BY INCOMPLETE STATE FEEDBACK OR OUTPUT FEEDBACK

In practice not all the state variables are accessible. For economical reasons, it may not be feasible to feed back all the state variables, especially for high-order systems. Thus, it is necessary to consider incomplete state feedback or output feedback. The design of digital control systems with incomplete state feedback and output feedback for pole placement is discussed in the following.

Incomplete State Feedback

Consider the digital control system

$$\mathbf{x}[(k+1)T] = \mathbf{A}\mathbf{x}(kT) + \mathbf{B}\mathbf{u}(kT) \tag{10-310}$$

where $\mathbf{x}(kT)$ is an n-vector and $\mathbf{u}(kT)$ is an r-vector. The state feedback is described by

$$\mathbf{u}(kT) = -\mathbf{G}\mathbf{x}(kT) \tag{10-311}$$

Let us assume that $x_i(kT)$ is not available for feedback, where i can be one or more values from 1 to n. *This requires that the corresponding columns of G must contain all zeros.* Using the design procedure introduced in the last section, we let

$$\mathbf{G} = \mathbf{w}\mathbf{G}^* \tag{10-312}$$

where $\mathbf{w}$ is $r \times 1$ and $\mathbf{G}^*$ is $1 \times r$. The matrix $\mathbf{w}$ should be so chosen that $(\mathbf{A}, \mathbf{B}\mathbf{w})$ is completely controllable. *With incomplete feedback, the columns of G* that correspond to the zero columns of G must equal zero.* Since the feedback matrix $\mathbf{G}^*$ for the single-input model is related to the desired closed-loop eigenvalue, the system parameters, and $\mathbf{w}$, through the following relationship,

$$\mathbf{G}^* = -[\Delta_{01} \quad \Delta_{02} \quad \cdots \quad \Delta_{0n}]\mathbf{K}^{-1} \tag{10-313}$$

with one or more columns of $\mathbf{G}^*$ forced to be zero, constraints are placed on the values of the desired closed-loop eigenvalues. The following example illustrates the design of a digital control system with incomplete state feedback.

Example 10-15.

Consider the system

$$\mathbf{x}[(k+1)T] = \mathbf{A}\mathbf{x}(kT) + \mathbf{B}\mathbf{u}(kT) \tag{10-314}$$

where

$$\mathbf{A} = \begin{bmatrix} 0 & 1 \\ -1 & -2 \end{bmatrix} \qquad \mathbf{B} = \begin{bmatrix} 0 \\ 1 \end{bmatrix}$$

The state feedback is defined as

$$u(kT) = -\mathbf{G}\mathbf{x}(kT) \tag{10-315}$$

where

$$G = [g_1 \quad g_2] \tag{10-316}$$

First let us assume that only $x_1(kT)$ is available for feedback; thus, $g_2 = 0$. The characteristic equation of the closed-loop system is

$$|z\mathbf{I} - \mathbf{A} + \mathbf{B}\mathbf{G}| = z^2 + (g_1 + 2)z + (1 - g_1) = 0 \tag{10-317}$$

Since we have only one parameter in g_1, the two eigenvalues of the closed-loop system cannot be arbitrarily assigned. Dividing both sides of Eq. (10-317) by the terms that do not contain g_1, we have

$$1 + \frac{g_1(z-1)}{z^2 + 2z + 1} = 0 \tag{10-318}$$

The root loci of Eq. (10-317) are constructed based on the pole-zero configuration of $(z-1)/(z^2 + 2z + 1)$, as shown in Fig. 10-56(a). Notice that for negative values of g_1 both roots are outside the unit circle, and for positive values of g_1 one root always stays to the left of the -1 point in the z-plane. Thus, the conclusion is that with only $x_1(kT)$ available for feedback, the trajectories of the roots of Eq. (10-317) are restricted to those shown in Fig. 10-56(a), and the system cannot be stabilized for any value of g_1.

Now let us consider that only $x_2(kT)$ is available for feedback, so that

$$G = [0 \quad g_2] \tag{10-319}$$

The characteristic equation of the closed-loop system is

$$|z\mathbf{I} - \mathbf{A} + \mathbf{B}\mathbf{G}| = z^2 + (g_2 + 2)z + (1 + g_2) = 0 \tag{10-320}$$

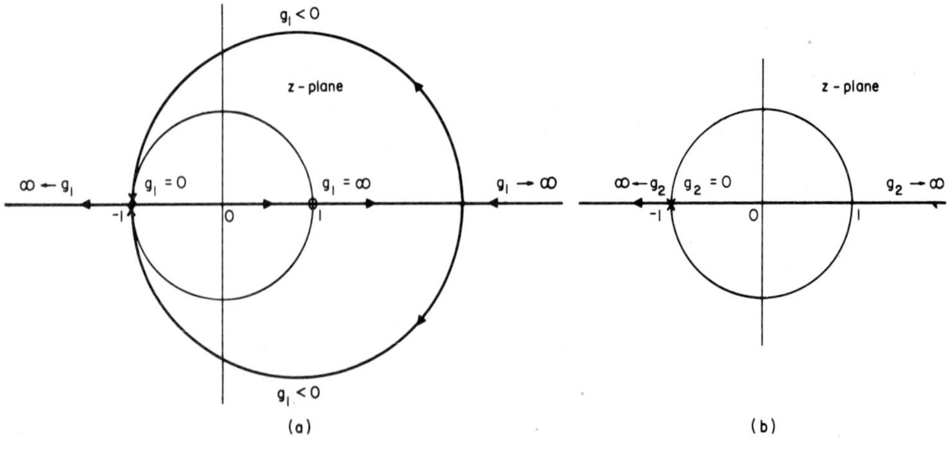

Fig. 10-56. Root loci of the system in Example 10-15.

The root loci of the last equation are constructed with the help of the following equation,

$$1 + \frac{g_2(z+1)}{z^2 + 2z + 1} = 0 \tag{10-321}$$

as shown in Fig. 10-56(b). In this case, the open-loop pole at $z = -1$ cannot be moved. As g_2 varies between $-\infty$ and ∞, the other root moves from $z = \infty$ to $z = -\infty$ on the real axis in the z-plane. Thus, we see that for the single-input system not only the eigenvalues of the closed-loop system cannot be arbitrarily placed by incomplete state feedback, but the system is not stabilizable.

Let us modify the system by changing **B** to

$$\mathbf{B} = \begin{bmatrix} 1 & 0 \\ 0 & 1 \end{bmatrix}$$

The feedback matrix is

$$\mathbf{G} = \begin{bmatrix} g_{11} & g_{12} \\ g_{21} & g_{22} \end{bmatrix}$$

Then,

$$|z\mathbf{I} - \mathbf{A} + \mathbf{BG}| = z^2 + (2 + g_{11} + g_{22})z + g_{11}(2 + g_{22}) + (1 - g_{12})(1 + g_{21}) = 0 \tag{10-322}$$

When $x_1(kT)$ is not available for feedback, $g_{11} = g_{21} = 0$. Equation (10-322) becomes

$$z^2 + (2 + g_{22})z + (1 - g_{12}) = 0 \tag{10-323}$$

Since we have two independent parameters in g_{12} and g_{22} in the last equation, the two eigenvalues of $\mathbf{A} - \mathbf{BG}$ can be arbitrarily assigned. Similarly, for $g_{12} = g_{22} = 0$, we have

$$z^2 + (2 + g_{11})z + 2g_{11} + 1 + g_{21} = 0 \tag{10-324}$$

Again, the eigenvalues can be arbitrarily assigned by choosing the values of g_{11} and g_{21}.

Output Feedback

Since the outputs of a system are always measurable, we can always count on feeding back these signals through constant gains for control purposes. Thus, output feedback may be considered as an alternate to incomplete feedback.

Consider the system

$$\mathbf{x}[(k+1)T] = \mathbf{Ax}(kT) + \mathbf{Bu}(kT) \tag{10-325}$$

$$\mathbf{c}(kT) = \mathbf{Dx}(kT) \tag{10-326}$$

where $\mathbf{x}(kT)$ is an n-vector, $\mathbf{u}(kT)$ is an r-vector, and $\mathbf{c}(kT)$ is a p-vector. The output feedback is defined as

$$\mathbf{u}(kT) = -\mathbf{G}\mathbf{c}(kT) \tag{10-327}$$

where $\mathbf{G}$ is the $r \times p$ output-feedback matrix. The design objective is to find $\mathbf{G}$ so that the eigenvalues of the closed-loop system are at the desired values. However, since in general $p \leq r \leq n$, not all the n eigenvalues can be arbitrarily assigned. We shall show that the number of eigenvalues that can be arbitrarily assigned depends on the ranks of $\mathbf{D}$ and $\mathbf{B}$.

The design of the output-feedback control can be carried out similar to that of the state-feedback control. Let us consider first the single-input case. Substitution of Eq. (10-326) in Eq. (10-327) and then in Eq. (10-325), we get

$$\mathbf{x}[(k+1)T] = (\mathbf{A} - \mathbf{BGD})\mathbf{x}(kT) \tag{10-328}$$

Since the last equation is equivalent to the state equation of a closed-loop system with state-feedback gain $\mathbf{GD}$, the latter can be solved directly from Eq. (10-244) or Eq. (10-254) if the pair $(\mathbf{A},\mathbf{B})$ is completely controllable. Thus, we have

$$\mathbf{GD} = [(\mathbf{MS}')^{-1}(\alpha - \mathbf{a})]' \tag{10-329}$$

or

$$\mathbf{GD} = [\Delta_{01} \quad \Delta_{02} \quad \cdots \quad \Delta_{0n}]\mathbf{K}^{-1} \tag{10-330}$$

However, since $\mathbf{D}$ and $\mathbf{DK}$ are generally not square matrices, we cannot solve for $\mathbf{G}$ directly from the last two equations.

For the single-input case, $\mathbf{G}$ is $1 \times p$, $\mathbf{D}$ is $p \times n$, and $\mathbf{B}$ is $n \times 1$; so $\mathbf{GD}$ is always a $1 \times n$ row matrix. There are p gain elements in $\mathbf{G}$, but only m of these are free or independent parameters which may be used for design purposes, where m is the rank of $\mathbf{D}$, and $m \leq p$. For example, if

$$\mathbf{D} = \begin{bmatrix} 1 & 0 & 0 \\ 0 & 1 & 0 \\ 0 & 2 & 0 \end{bmatrix}$$

which has a rank of two, and let $\mathbf{G} = [g_1 \quad g_2 \quad g_3]$. Then,

$$\mathbf{GD} = [g_1 \quad g_2 \quad 2g_2]$$

which has only two independent gain parameters. This means that only two of the three eigenvalues of the system can be placed arbitrarily by the output feedback. For the single-input case, if the rank of $\mathbf{D}$ is equal to n, the order of the system, then output feedback is the same as complete state feedback, and if (A,B) is controllable, all the eigenvalues can be arbitrarily assigned.

For multiple-input systems, $\mathbf{B}$ is n × r; we again form

$$\mathbf{B}^* = \mathbf{B}\mathbf{w} \tag{10-331}$$

where $\mathbf{w}$ is an r × 1 matrix with r parameters, thus, $\mathbf{B}^*$ is n × 1. Similarly,

$$\mathbf{G} = \mathbf{w}\mathbf{G}^* \tag{10-332}$$

where

$$\mathbf{G}^* = [g_1^* \quad g_2^* \quad \dots \quad g_n^*] \qquad (1 \times n) \tag{10-333}$$

Then,

$$\mathbf{BGD} = \mathbf{BwG}^*\mathbf{D} = \mathbf{B}^*\mathbf{G}^*\mathbf{D} \tag{10-334}$$

and the characteristic equation of the closed-loop system is

$$|z\mathbf{I} - \mathbf{A} + \mathbf{BGD}| = |z\mathbf{I} - \mathbf{A} + \mathbf{B}^*\mathbf{G}^*\mathbf{D}| = 0 \tag{10-335}$$

Thus, $\mathbf{G}^*\mathbf{D}$ can be determined using Eq. (10-244) or Eq. (10-254).

Unlike in the single-input case with output feedback, the solution of the feedback gain now depends on the ranks of **D** and **B**. In general, if the rank of **D** is greater than or equal to that of **B**, the elements of **w** can be arbitrarily selected, of course, subject to the controllability of the pair (**A**,**B**). However, if the rank of **B** is greater than that of **D**, we cannot arbitrarily assign all the elements of **w** if we wish to arbitrarily assign a maximum number of eigenvalues of the closed-loop system. For the design of systems with state feedback we do not have this problem, since the rank of **D**, where **D** is a unity matrix of dimension n, is always n, and the rank of **B** cannot exceed n. The following example will illustrate the design of output feedback and the constraints mentioned above.

Example 10-16.

Given the digital control system that is described by the following dynamic equations,

$$\mathbf{x}[(k+1)T] = \mathbf{A}\mathbf{x}(kT) + \mathbf{B}\mathbf{u}(kT) \tag{10-336}$$

$$\mathbf{c}(kT) = \mathbf{D}\mathbf{x}(kT) \tag{10-337}$$

where

$$\mathbf{A} = \begin{bmatrix} 0 & 1 & 0 \\ 0 & 0 & 1 \\ -1 & 0 & 0 \end{bmatrix} \qquad \mathbf{B} = \begin{bmatrix} 0 & 1 \\ 1 & 0 \\ 0 & 0 \end{bmatrix}$$

$$D = \begin{bmatrix} 1 & 0 & 0 \\ 1 & 1 & 0 \end{bmatrix}$$

The problem is to find the feedback gain matrix G such that the output feedback

$$\mathbf{u}(kT) = -\mathbf{Gc}(kT) \tag{10-338}$$

places the eigenvalues of $\mathbf{A} - \mathbf{BGD}$ at certain desired values. Since D has a rank of two, and the rank of B is also two, a maximum of two of the eigenvalues can be arbitrarily assigned. Let these two eigenvalues be at $z_1 = 0.1$ and $z_2 = 0.2$.

The characteristic equation of A is

$$|z\mathbf{I} - \mathbf{A}| = z^3 + 1 \tag{10-339}$$

Thus,

$$\mathbf{M} = \begin{bmatrix} a_4 & 0 & 0 \\ a_3 & a_4 & 0 \\ a_2 & a_3 & a_4 \end{bmatrix} = \begin{bmatrix} 1 & 0 & 0 \\ 0 & 1 & 0 \\ 0 & 0 & 1 \end{bmatrix} \tag{10-340}$$

From Eq. (10-331), B* is written

$$\mathbf{B}^* = \begin{bmatrix} 0 & 1 \\ 1 & 0 \\ 0 & 0 \end{bmatrix} \begin{bmatrix} w_1 \\ w_2 \end{bmatrix} = \begin{bmatrix} w_2 \\ w_1 \\ 0 \end{bmatrix} \tag{10-341}$$

which has two independent parameters in w_1 and w_2, due to the fact that B has rank two. The controllability matrix for the pair $(\mathbf{A}, \mathbf{B}^*)$ is

$$\mathbf{S} = [\mathbf{B}^* \quad \mathbf{AB}^* \quad \mathbf{A}^2\mathbf{B}^*] = \begin{bmatrix} w_2 & w_1 & 0 \\ w_1 & 0 & -w_2 \\ 0 & -w_2 & -w_1 \end{bmatrix} \tag{10-342}$$

The matrix S is nonsingular if $w_1^3 - w_2^3 \neq 0$.

Let

$$\mathbf{G}^* = [g_1^* \quad g_2^*] \tag{10-343}$$

Then

$$\mathbf{G}^*\mathbf{D} = [g_1^* + g_2^* \quad g_2^* \quad 0] \tag{10-344}$$

Since **D** is of rank two, **G*D** has two independent parameters in g_1^* and g_2^*. Now using Eq. (10-329), with G replaced by G*, we have

$$\mathbf{G^*D} = [(\mathbf{MS}')^{-1}(\alpha - \mathbf{a})]' \tag{10-345}$$

or

$$\begin{bmatrix} g_1^* + g_2^* \\ g_2^* \\ 0 \end{bmatrix}' = \frac{1}{w_1^3 - w_2^3} \begin{bmatrix} -w_2^2\alpha_3 + \alpha_2 w_1^2 - w_1 w_2(\alpha_1 - 1) \\ \alpha_3 w_1^2 - w_1 w_2 \alpha_2 + w_2^2(\alpha_1 - 1) \\ -w_1 w_2 \alpha_3 + \alpha_2 w_2^2 - w_1^2(\alpha_1 - 1) \end{bmatrix}' \tag{10-346}$$

The last row in Eq. (10-346) corresponds to the constraint equation

$$-w_1 w_2 \alpha_3 + \alpha_2 w_2^2 - w_1^2(\alpha_1 - 1) = 0 \tag{10-347}$$

Since only two of the three eigenvalues, or, correspondingly, two of the three coefficients of the closed-loop characteristic equation can be arbitrarily assigned, although w_1 and w_2 can be arbitrary ($w_1^3 \neq w_2^3$) they should be selected so that the third eigenvalue is stable. This requirement on stability does impose additional constraints on the values of w_1 and w_2. For example, the necessary condition for the closed-loop system to be stable is that $|\alpha_1| < 1$. From Eq. (10-347), we see that w_2 cannot be zero.

In order for $z = 0.1$ and 0.2 to be roots of the characteristic equation

$$z^3 + \alpha_3 z^2 + \alpha_2 z + \alpha_1 = 0 \tag{10-348}$$

the following two equations must be satisfied:

$$\alpha_2 + 0.3\alpha_3 + 0.07 = 0 \tag{10-349}$$

$$\alpha_1 - 0.02\alpha_3 - 0.006 = 0 \tag{10-350}$$

Solving the last two equations together with Eq. (10-347) gives

$$\alpha_1 = \frac{0.02w_1^2 + 0.0004w_2^2 + 0.006w_1 w_2}{0.3w_2^2 + w_1 w_2 + 0.02w_1^2} \tag{10-351}$$

$$\alpha_2 = \frac{-0.2996w_1^2 - 0.07w_1 w_2}{0.3w_2^2 + w_1 w_2 + 0.02w_1^2} \tag{10-352}$$

$$\alpha_3 = \frac{0.994w_1^2 - 0.07w_2^2}{0.3w_2^2 + w_1 w_2 + 0.02w_1^2} \tag{10-353}$$

Setting $w_1 = 0$ and $w_2 = 1$, we have

$$a_1 = 0.001333 \quad , \quad a_2 = 0 \quad , \quad a_3 = -0.23333 \quad .$$

With these coefficients, Eq. (10-348) gives the following three roots:

$$z_1 = 0.1 \quad , \quad z_2 = 0.2 \quad \text{and} \quad z_3 = -0.06667 \quad .$$

Thus, with the choice of w_1 and w_2, the third root which we cannot place arbitrarily is at $z = -0.06667$, and the closed-loop system is stable. Apparently, there are other combinations of w_1 and w_2 that will correspond to a stable system with $z = 0.1$ and 0.2.

Substitution of the values of $w_1 = 0$, $w_2 = 1$, and the corresponding values of a_1, a_2 and a_3 in Eq. (10-346), we get

$$\begin{bmatrix} g_1^* + g_2^* \\ g_2^* \\ 0 \end{bmatrix}' = \begin{bmatrix} a_3 \\ -a_1 + 1 \\ a_2 \end{bmatrix}' = \begin{bmatrix} -0.23333 \\ 0.99866 \\ 0 \end{bmatrix}' \tag{10-354}$$

The feedback matrix G^* is written

$$G^* = [-1.232 \quad 0.99866] \tag{10-355}$$

The feedback matrix G is

$$G = wG^* = \begin{bmatrix} 0 \\ 1 \end{bmatrix} G^* = \begin{bmatrix} 0 & 0 \\ -1.232 & 0.99866 \end{bmatrix} \tag{10-356}$$

Example 10-17.

Consider the same system as given in Example 10-16, except that

$$D = \begin{bmatrix} 1 & 0 & 0 \\ 1 & 0 & 0 \end{bmatrix} \tag{10-357}$$

which has rank 1. This means that

$$G^*D = [\ g_1^* + g_2^* \quad 0 \quad 0\ ] \tag{10-358}$$

which has only one independent parameter in $g_1^* + g_2^*$. Now Eq. (10-345) reads

$$\begin{bmatrix} g_1^* + g_2^* \\ 0 \\ 0 \end{bmatrix}' = \frac{1}{w_1^3 - w_2^3} \begin{bmatrix} -w_2^2 a_3 + a_2 w_1^2 - w_1 w_2 (a_1 - 1) \\ a_3 w_1^2 - w_1 w_2 a_2 + w_2^2 (a_1 - 1) \\ -w_1 w_2 a_3 + a_2 w_2^2 - w_1^2 (a_1 - 1) \end{bmatrix}' \tag{10-359}$$

Since the last two rows in the last equation are constrained to be zero, we can only assign either w_1 or w_2 arbitrarily, but not both. Thus, from Eq. (10-359),

$$\alpha_3 w_1^2 - w_1 w_2 \alpha_2 + w_2^2(\alpha_1 - 1) = 0 \tag{10-360}$$

$$-w_1 w_2 \alpha_3 + \alpha_2 w_2^2 - w_1^2(\alpha_1 - 1) = 0 \tag{10-361}$$

In order for two of the closed-loop eigenvalues to be at $z_1 = 0.1$ and $z_2 = 0.2$, Eqs. (10-349) and (10-350) must also be satisfied. These two equations together with Eqs. (10-360) and (10-361) form a collection of four equations with five unknowns in $\alpha_1, \alpha_2, \alpha_3, w_1$ and w_2. Thus, only one of these unknowns can be arbitrarily assigned. Unfortunately, these four equations are nonlinear in w_1 and w_2 so that they are difficult to solve. In this case, it would be simpler to use the brute force method by writing

$$|z\mathbf{I} - \mathbf{A} + \mathbf{BGD}| = z^3 + (g_{21} + g_{22})z^2 + (g_{11} + g_{12})z + 1 = 0 \tag{10-362}$$

Hence, it is clear that only two of the three coefficients of the last equation can be arbitrarily assigned. Since the constant term is equal to one, the system cannot be stabilized by output feedback with the $\mathbf{D}$ matrix given.

10.12 DESIGN OF DIGITAL CONTROL SYSTEMS WITH STATE FEEDBACK AND DYNAMIC OUTPUT FEEDBACK

The state-feedback and output-feedback designs discussed in the preceding sections are suitable for the design of digital regulators. By properly selecting the eigenvalues of the closed-loop system, the natural responses of the state variables of the system can be properly controlled. When the design objective is to have the system track a certain input, the state-feedback and the output-feedback schemes should be properly modified. Since state feedback and output feedback do not increase the order of a system, they do not guarantee in general that the output or states of the system will follow the input in the steady state. In the conventional design technique, the PID controller has been used extensively in the industry for the compensation of a control process so that the transient response and the steady-state response behave in a prescribed manner. In fact, in Sec. 10.7 the design of a control system with a digital PID controller has been investigated. Since the PID controller always increases the order of the overall system (assuming that no cancellation of poles and zeros exist), whereas the state or output feedback through constant gains does not affect the system order, the control effects of the two schemes are not equivalent. It may be stated that the state or output feedback through constant gains cannot achieve the same control objectives as the PID controller or any dynamic controllers in general.

In this section we present a method which will lead to a constant-gain feedback from the state variables in order to control the dynamics of the system, and simultaneously feedback the system output through a dynamic controller for output regulation. In particular, the dynamic controller turns out to be a digital approximation of an integrator.

The digital control system under consideration is described by the following set of dynamic equations:

$$\mathbf{x}(k+1) = \mathbf{A}\mathbf{x}(k) + \mathbf{B}\mathbf{u}(k) + \mathbf{F}\mathbf{w} \tag{10-363}$$

$$\mathbf{c}(k) = \mathbf{D}\mathbf{x}(k) + \mathbf{E}\mathbf{u}(k) + \mathbf{H}\mathbf{w} \tag{10-364}$$

where

$$\begin{aligned} \mathbf{x}(k) &= \text{n-vector} \quad &&\text{(state)} \\ \mathbf{u}(k) &= \text{r-vector} \quad &&\text{(input)} \\ \mathbf{c}(k) &= \text{p-vector} \quad &&\text{(output)} \\ \mathbf{w} &= \text{q-vector} \quad &&\text{(disturbance)} \end{aligned}$$

The dimensions of the matrices **A**, **B**, **D**, **E**, **F**, and **H** are defined according to that of the variables. The disturbance vector **w** is assumed to be constant. The elements of **w** may also include set points or reference inputs for the states or outputs of the system to follow. The true disturbance components of **w** are generally unknown, although their magnitudes are assumed to be constant.

The design objectives of the digital control system described by Eqs. (10-363) and (10-364) may be stated as: Find the control $\mathbf{u}(k)$ such that

$$\lim_{k\to\infty} \mathbf{x}(k+1) = \lim_{k\to\infty} \mathbf{x}(k) \tag{10-365}$$

and

$$\lim_{k\to\infty} \mathbf{c}(k) = \mathbf{0} \tag{10-366}$$

The condition in Eq. (10-365) is equivalent to requiring that the system be asymptotically stable, and Eq. (10-366) implies output regulation. The output vector $\mathbf{c}(k)$ need not be the true output of the system. In fact, by constructing $\mathbf{c}(k)$ appropriately, a great variety of regulating and tracking problems can be defined.

Let us define the augmented state and control vectors as

$$\mathbf{y}(k) = \begin{bmatrix} \mathbf{x}(k+1) - \mathbf{x}(k) \\ \mathbf{c}(k) \end{bmatrix} \qquad (n+p) \times 1 \tag{10-367}$$

$$\mathbf{v}(k) = \mathbf{u}(k+1) - \mathbf{u}(k) \tag{10-368}$$

Then, from Eq. (10-367),

$$\mathbf{y}(k+1) = \begin{bmatrix} \mathbf{x}(k+2) - \mathbf{x}(k+1) \\ \mathbf{c}(k+1) \end{bmatrix} \tag{10-369}$$

From Eqs. (10-363) and (10-364), we have

$$\mathbf{x}(k+2) = \mathbf{A}\mathbf{x}(k+1) + \mathbf{B}\mathbf{u}(k+1) + \mathbf{F}\mathbf{w} \tag{10-370}$$

$$\mathbf{c}(k+1) = \mathbf{D}\mathbf{x}(k+1) + \mathbf{E}\mathbf{u}(k+1) + \mathbf{H}\mathbf{w} \tag{10-371}$$

Forming the difference vector between $\mathbf{y}(k+1)$ and $\mathbf{y}(k)$, we have

$$\mathbf{y}(k+1) - \mathbf{y}(k) = \begin{bmatrix} \mathbf{A}[\mathbf{x}(k+1) - \mathbf{x}(k)] + \mathbf{B}[\mathbf{u}(k+1) - \mathbf{u}(k)] - [\mathbf{x}(k+1) - \mathbf{x}(k)] \\ \mathbf{D}[\mathbf{x}(k+1) - \mathbf{x}(k)] + \mathbf{E}[\mathbf{u}(k+1) - \mathbf{u}(k)] \end{bmatrix}$$

$$= \begin{bmatrix} \mathbf{A} - \mathbf{I}_n & \mathbf{0} \\ \mathbf{D} & \mathbf{0} \end{bmatrix} \mathbf{y}(k) + \begin{bmatrix} \mathbf{B} \\ \mathbf{E} \end{bmatrix} \mathbf{v}(k) \tag{10-372}$$

where $\mathbf{I}_n$ is an $n \times n$ identity matrix.

After rearranging, Eq. (10-372) is written as

$$\mathbf{y}(k+1) = \begin{bmatrix} \mathbf{A} & \mathbf{0} \\ \mathbf{D} & \mathbf{I}_p \end{bmatrix} \mathbf{y}(k) + \begin{bmatrix} \mathbf{B} \\ \mathbf{E} \end{bmatrix} \mathbf{v}(k) \tag{10-373}$$

Thus, the design objectives stated in Eqs. (10-365) and (10-366) are equivalent to driving the system in Eq. (10-373) from any initial state $\mathbf{y}(0)$ to $\mathbf{y}(k) \to 0$ as $k \to \infty$.

Condition on Controllability

In order to achieve the design objectives stated above, we must first investigate the controllability of the system described by Eq. (10-373).

Let

$$\hat{\mathbf{A}} = \begin{bmatrix} \mathbf{A} & \mathbf{0} \\ \mathbf{D} & \mathbf{I}_p \end{bmatrix} \qquad \hat{\mathbf{B}} = \begin{bmatrix} \mathbf{B} \\ \mathbf{E} \end{bmatrix}$$

Equation (10-373) is written

$$\mathbf{y}(k+1) = \hat{\mathbf{A}}\mathbf{y}(k) + \hat{\mathbf{B}}\mathbf{v}(k) \tag{10-374}$$

The necessary and sufficient conditions for the pair $(\hat{\mathbf{A}}, \hat{\mathbf{B}})$ to be completely controllable are that the $(n + p) \times (n + p + r)$ matrix

$$[\lambda\mathbf{I} - \hat{\mathbf{A}} \vdots \hat{\mathbf{B}}]$$

has rank (n + p) for λ equals every eigenvalue of $\hat{\mathbf{A}}$. We have

$$[\lambda\mathbf{I} - \hat{\mathbf{A}} \vdots \hat{\mathbf{B}}] = \begin{bmatrix} \lambda\mathbf{I}_n - \mathbf{A} & \mathbf{0} & \mathbf{B} \\ -\mathbf{D} & (\lambda - 1)\mathbf{I}_m & \mathbf{E} \end{bmatrix} \tag{10-375}$$

where $\mathbf{I}_m$ is an m $\times$ m identity matrix. From the last equation we know that $\hat{\mathbf{A}}$ has at least p eigenvalues at $\lambda = 1$. When $\lambda = 1$, Eq. (10-375) becomes

$$[\lambda\mathbf{I} - \hat{\mathbf{A}} \vdots \hat{\mathbf{B}}] = \begin{bmatrix} \mathbf{I}_n - \mathbf{A} & \mathbf{0} & \mathbf{B} \\ -\mathbf{D} & \mathbf{0} & \mathbf{E} \end{bmatrix} \tag{10-376}$$

When $\lambda \neq 1$, $(\lambda - 1)\mathbf{I}_m$ is of rank p, in order that $[\lambda\mathbf{I}_n - \hat{\mathbf{A}} \vdots \hat{\mathbf{B}}]$ has a rank of n + p, the matrix $[\lambda\mathbf{I}_n - \hat{\mathbf{A}} \vdots \hat{\mathbf{B}}]$ must be of rank n, which implies that the pair $(\mathbf{A},\mathbf{B})$ is controllable. Thus, the controllability of $(\hat{\mathbf{A}},\hat{\mathbf{B}})$ requires that

(1) $(\hat{\mathbf{A}},\hat{\mathbf{B}})$ is controllable, and

(2) $\begin{bmatrix} \mathbf{A} - \mathbf{I}_n & \mathbf{B} \\ \mathbf{D} & \mathbf{E} \end{bmatrix}$ is of rank n + p (10-377)

Thus, we have established the controllability condition of the augmented system in Eq. (10-374) in terms of the coefficient matrices of the original system.

Now let us assume that the control vector $\mathbf{v}(k)$ is effected by state feedback, i.e.,

$$\mathbf{v}(k) = -\mathbf{G}\mathbf{y}(k) \tag{10-378}$$

where $\mathbf{G}$ is the r $\times$ (n + p) feedback matrix with constant-gain elements.

In view of the definitions of $\mathbf{v}(k)$ and $\mathbf{y}(k)$, Eq. (10-378) is written

$$\mathbf{v}(k) = \mathbf{u}(k+1) - \mathbf{u}(k) = -\mathbf{G}_1[\mathbf{x}(k+1) - \mathbf{x}(k)] - \mathbf{G}_2\mathbf{c}(k) \tag{10-379}$$

where $\mathbf{G}_1$ is an r $\times$ n matrix and $\mathbf{G}_2$ is an r $\times$ p matrix.

Taking the z-transform on both sides of Eq. (10-379) and rearranging, we get

$$\mathbf{U}(z) = -\mathbf{G}_1\mathbf{X}(z) - \frac{1}{z-1}\mathbf{G}_2\mathbf{C}(z) \tag{10-380}$$

The significance of the last equation is that the control $\mathbf{u}(k)$ is obtained as a combination of constant-gain state feedback and dynamic output feedback. The transfer function $1/(z-1)$ may be interpreted as a digital approximation of integration. Figure 10-57 shows the block diagram of the closed-loop system.

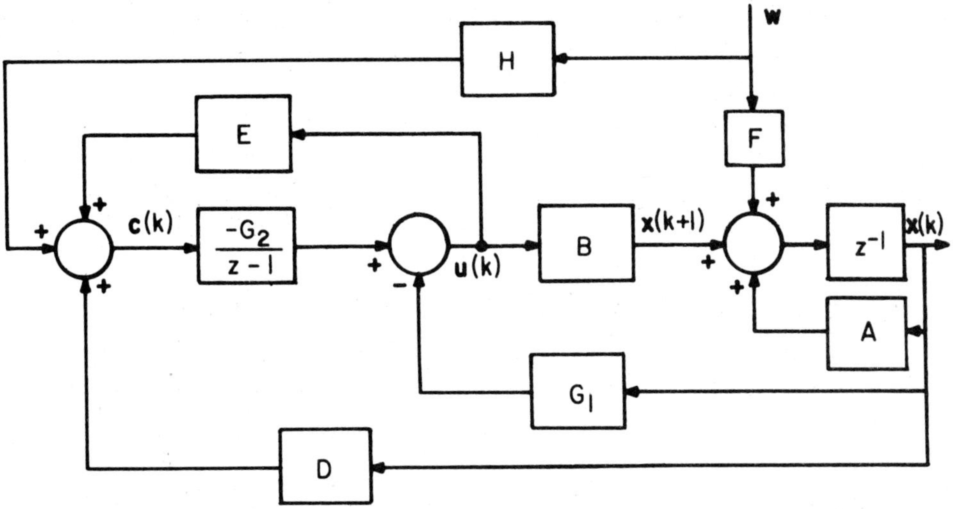

Fig. 10-57. Block diagram of closed-loop digital control system with dynamic controller and state feedback.

Example 10-18.

Consider the digital control system

$$\mathbf{x}(k+1) = \mathbf{A}\mathbf{x}(k) + \mathbf{B}u(k) + \mathbf{w}_2 \tag{10-381}$$

where

$$\mathbf{A} = \begin{bmatrix} 0 & 1 \\ -1 & 0 \end{bmatrix} \qquad \mathbf{B} = \begin{bmatrix} 0 \\ 1 \end{bmatrix}$$

w_2 denotes a constant disturbance whose amplitude is unknown.

The design objective is to find the control scheme such that

1. x_1 reaches the reference input $r = w_1$ as $k \to \infty$, and
2. the eigenvalues of the closed-loop system should be at certain specified values.

Let us define the output variable as

$$c(k) = w_1 - x_1(k) \tag{10-382}$$

Then, we want c(k) to go to zero as k goes to infinity. Using the output equation model given in Eq. (10-364), we have

$$\mathbf{D} = [-1 \quad 0] \qquad \mathbf{E} = 0$$

$$\mathbf{H} = [1 \quad 0] \qquad \mathbf{F} = \begin{bmatrix} 0 & 1 \\ 0 & 0 \end{bmatrix}$$

We can see that the pair (A, B) is completely controllable, and

$$\begin{bmatrix} A - I_n & B \\ D & E \end{bmatrix} = \begin{bmatrix} -1 & 1 & 0 \\ -1 & -1 & 1 \\ -1 & 0 & 0 \end{bmatrix} \quad \text{has rank three} \tag{10-383}$$

Thus, the pair $(\hat{A}, \hat{B})$ is completely controllable, where

$$\hat{A} = \begin{bmatrix} 0 & 1 & 0 \\ -1 & 0 & 0 \\ -1 & 0 & 1 \end{bmatrix} \qquad \hat{B} = \begin{bmatrix} 0 \\ 1 \\ 0 \end{bmatrix}$$

The control in the z-domain is given by Eq. (10-380), and is written

$$U(z) = -g_1 X_1(z) - g_2 X_2(z) - \frac{g_3}{z-1} C(z) \tag{10-384}$$

where g_1, g_2 and g_3 are feedback gain constants. The characteristic equation of the closed-loop system is

$$|zI - \hat{A} + \hat{B}G| = \begin{bmatrix} z & -1 & 0 \\ 1 + g_1 & z + g_2 & g_3 \\ 1 & 0 & z-1 \end{bmatrix}$$

$$= z^3 + (g_2 - 1)z^2 + (1 + g_1 - g_2)z - (1 + g_1 + g_3) = 0 \tag{10-385}$$

Let the closed-loop eigenvalues be at $z = 0.5$, $0.3 + j0.3$ and $0.3 - j0.3$. Then the characteristic equation becomes

$$z^3 - 1.1z^2 + 0.39z - 0.045 = 0 \tag{10-386}$$

Equating the coefficients of Eqs. (10-385) and (10-386), we get

$$g_1 = -0.71 \ , \quad g_2 = -0.1 \ , \quad g_3 = -0.245$$

The block diagram of the closed-loop system is shown in Fig. 10-58, from which the z-transforms of $x_1(k)$ and $x_2(k)$ are written

$$\begin{bmatrix} X_1(z) \\ X_2(z) \end{bmatrix} = \frac{1}{\Delta} \begin{bmatrix} \dfrac{-z^{-2}}{z-1} g_3 & z^{-1}(1 + g_2 z^{-1}) \\ \dfrac{-z^{-1}}{z-1} g_3 & -z^{-2} - g_1 z^{-2} + \dfrac{g_3}{z-1} z^{-2} \end{bmatrix} \begin{bmatrix} w_1 \\ w_2 \end{bmatrix} \frac{z}{z-1} \tag{10-387}$$

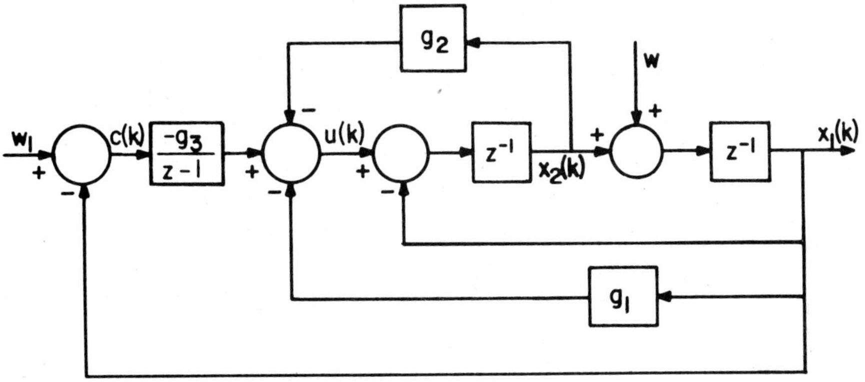

Fig. 10-58. Block diagram of closed-loop digital control system.

where

$$\Delta = 1 + g_2 z^{-1} + z^{-2} + g_1 z^{-2} - \frac{g_3 z^{-2}}{z-1} \tag{10-388}$$

Applying the final-value theorem of the z-transform to both sides of Eq. (10-387), we get

$$\lim_{k\to\infty} \begin{bmatrix} x_1(k) \\ x_2(k) \end{bmatrix} = \begin{bmatrix} 1 & 0 \\ 0 & -1 \end{bmatrix} \begin{bmatrix} w_1 \\ w_2 \end{bmatrix} = \begin{bmatrix} w_1 \\ w_1 - w_2 \end{bmatrix} \tag{10-389}$$

Therefore, the final value of $x_1(k)$ is equal to the reference input w_1 as desired.

Figure 10-59 illustrates the responses of $x_1(k)$ for $w_1 = 1$ and two values of w_2, the true disturbance. The actual responses are defined only at discrete intervals but are represented by smooth curves passing through the data points. When the true disturbance is zero, the response of $x_1(k)$ has a delay time of two sampling periods, and approaches the reference input $w_1 = r = 1$ without overshoot. When $w_2 = 1$, which was chosen arbitrarily, the response of $x_1(k)$ actually approaches its final value a great deal faster. However, this result is better than the response for $w_2 = 0$ only because when $w_2 = 1$, the condition is most favorable. Apparently, when w_2 is of any other values, the response would not be as impressive.

10.13 REALIZATION OF STATE FEEDBACK BY DYNAMIC CONTROLLERS

The design of control systems by state feedback often encounters one practical problem in that not all the state variables are accessible. Furthermore, for high-order systems, the implementation of state feedback would require a large

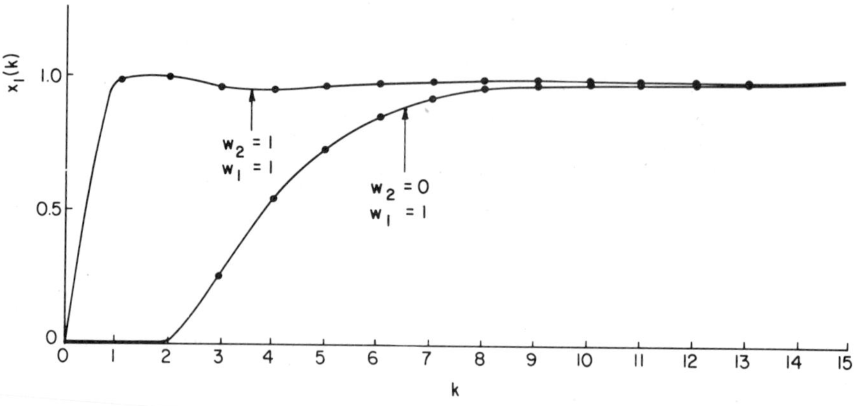

Fig. 10-59. Responses of $x_1(k)$ for the system designed in Example 10-17.

number of transducers for the measurements of the state variables, a costly proposition. In Chapter 13 we shall see that if the process to be controlled is observable, then an "observer" can be designed which utilizes the input and the output signal measurements to generate a set of "observed" states. These observed states, which are estimates of the true state variables, are then used to control the process through state feedback.

In the conventional design principles, feedback from the output signal is usually used. This is natural, because the output is always measurable. The conventional design is also characterized in such a way that the controller configuration is selected a priori at the outset of the design. In this section we shall present a method which allows the approximation of state feedback by a cascade controller or a feedback controller. In a general sense the equivalent controller discussed here can be regarded as a *dynamic observer-controller* of the system.

Consider the digital system described by

$$\mathbf{x}(k + 1) = \mathbf{A}\mathbf{x}(k) + \mathbf{B}\mathbf{u}(k) \tag{10-390}$$

$$\mathbf{c}(k) = \mathbf{D}\mathbf{x}(k) + \mathbf{E}\mathbf{u}(k) \tag{10-391}$$

where $\mathbf{x}(k)$ = n-vector, $\mathbf{c}(k)$ = q-vector, and $\mathbf{u}(k)$ = p-vector; **A**, **B**, **D** and **E** are coefficients of the appropriate dimensions.

Let the control be realized by state feedback such that

$$\mathbf{u}(k) = -\mathbf{G}\mathbf{x}(k) \tag{10-392}$$

where **G** is the p $\times$ q feedback gain matrix.

The objective is to approximate the digital control system with state feedback shown in Fig. 10-60 by the system of Fig. 10-61 which has a feedback controller with input coming from the output $\mathbf{c}(k)$. Let the transfer function

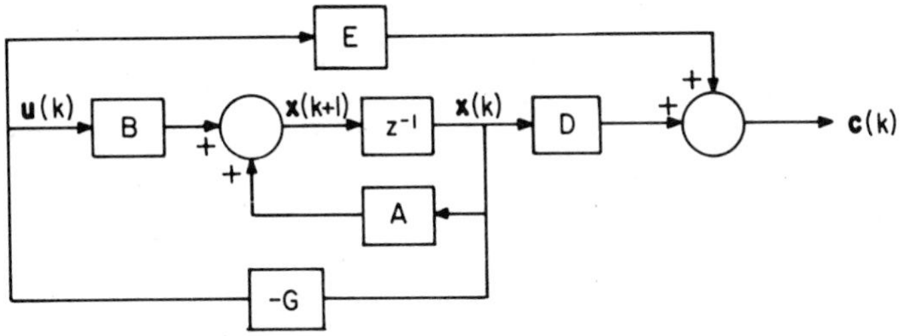

Fig. 10-60. Digital control system with state feedback.

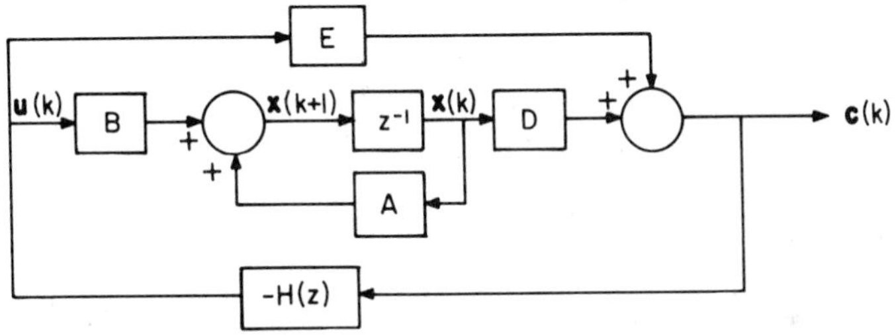

Fig. 10-61. Digital control system with dynamic feedback controller from output.

relation of the feedback controller be represented by

$$\mathbf{U}(z) = -\mathbf{H}(z)\mathbf{C}(z) \tag{10-393}$$

where $\mathbf{H}(z)$ is the transfer function matrix. Let us express $\mathbf{H}(z)$ as

$$\mathbf{H}(z) = \begin{bmatrix} H_{11}(z) & H_{12}(z) & \cdots & H_{1q}(z) \\ H_{21}(z) & H_{22}(z) & \cdots & H_{2q}(z) \\ \vdots & \vdots & & \vdots \\ H_{p1}(z) & H_{p2}(z) & \cdots & H_{pq}(z) \end{bmatrix} \quad (p \times q) \tag{10-394}$$

Let $H_{ij}(z)$ $(i = 1,2,\ldots,p; j = 1,2,\ldots,q)$ be an mth-order transfer function,

$$H_{ij}(z) = \frac{K_{ij}(z^m + a_{ij1}z^{m-1} + a_{ij2}z^{m-2} + \ldots + a_{ijm})}{z^m + \beta_{ij1}z^{m-1} + \beta_{ij2}z^{m-2} + \ldots + \beta_{ijm}} \tag{10-395}$$

Expanding $H_{ij}(z)$ into a Laurent's series about $z = 0$, we get

$$H_{ij}(z) = K_{ij} \sum_{k=0}^{\infty} d_{ijk} z^{-k} \tag{10-396}$$

where

$$d_{ij0} = 1$$

$$d_{ij1} = \alpha_{ij1} - \beta_{ij1}$$

$$d_{ijk} = \alpha_{ijk} - \beta_{ijk} - \sum_{v=1}^{k-1} \beta_{k-v} d_{ijv} \qquad (k > 1) \tag{10-397}$$

Truncating $H_{ij}(z)$ at m terms, we have

$$H_{ij}(z) \cong K_{ij} \sum_{k=0}^{m-1} d_{ijk} z^{-k} \tag{10-398}$$

where we have assumed that the infinite series converges. The significance of Eq. (10-393) and the truncated series representation of $H_{ij}(z)$ is that the control $\mathbf{u}(k)$ is implemented by feeding back the output $\mathbf{c}(k)$ and all the delayed outputs up to $\mathbf{c}(k - m + 1)$, where m is not yet specified. Let

$$H_{ijm}(z) = K_{ij} \sum_{k=0}^{m-1} d_{ijk} z^{-k} \tag{10-399}$$

Substituting the last equation into Eq. (10-393) as an approximation for $H_{ij}(z)$, we have

This equation is displayed on page 559. (10-400)

The elements of the last equation are rearranged to give

This equation is displayed on page 559. (10-401)

The time-domain equivalence of the last equation is

$$\mathbf{u}(k) = -\mathbf{P} \begin{bmatrix} \mathbf{c}(k) \\ \mathbf{c}(k-1) \\ \vdots \\ \mathbf{c}(k-m+1) \end{bmatrix} \tag{10-402}$$

$$
U(z) \cong -\begin{bmatrix} K_{11}[d_{110} \cdots d_{11(m-1)}] & K_{12}[d_{120} \cdots d_{12(m-1)}] & \cdots & K_{1q}[d_{1q0} \cdots d_{1q(m-1)}] \\ K_{21}[d_{210} \cdots d_{21(m-1)}] & K_{22}[d_{220} \cdots d_{22(m-1)}] & \cdots & K_{2q}[d_{2q0} \cdots d_{2q(m-1)}] \\ \vdots & \vdots & & \vdots \\ K_{p1}[d_{p10} \cdots d_{p1(m-1)}] & K_{p2}[d_{p20} \cdots d_{p2(m-1)}] & \cdots & K_{pq}[d_{pq1} \cdots d_{pq(m-1)}] \end{bmatrix} \begin{bmatrix} C_1(z) \\ z^{-1}C_1(z) \\ \vdots \\ z^{-m+1}C_1(z) \\ C_2(z) \\ z^{-1}C_2(z) \\ \vdots \\ z^{-m+1}C_2(z) \\ \vdots \\ C_q(z) \\ z^{-1}C_q(z) \\ \vdots \\ z^{-m+1}C_q(z) \end{bmatrix} \tag{10-400}
$$

$$
U(z) \cong -\begin{bmatrix} [K_{11}d_{110} \cdots K_{1q}d_{1q0}] & [K_{11}d_{111} \cdots K_{1q}d_{1q1}] & \cdots & [K_{11}d_{11(m-1)} \cdots K_{1q}d_{1q(m-1)}] \\ [K_{21}d_{210} \cdots K_{2q}d_{2q0}] & [K_{21}d_{211} \cdots K_{2q}d_{2q1}] & \cdots & [K_{21}d_{21(m-1)} \cdots K_{2q}d_{2q(m-1)}] \\ \vdots & \vdots & & \vdots \\ [K_{p1}d_{p10} \cdots K_{pq}d_{pq0}] & [K_{p1}d_{p11} \cdots K_{pq}d_{pq1}] & \cdots & [K_{p1}d_{p1(m-1)} \cdots K_{pq}d_{pq(m-1)}] \end{bmatrix} \begin{bmatrix} \mathbf{C}(z) \\ z^{-1}\mathbf{C}(z) \\ z^{-2}\mathbf{C}(z) \\ \vdots \\ z^{-m+1}\mathbf{C}(z) \end{bmatrix} \tag{10-401}
$$

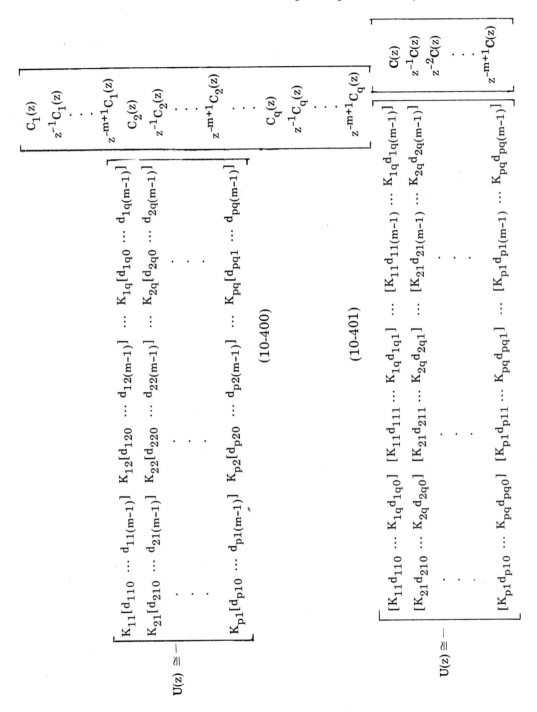

where **P** denotes the $p \times qm$ coefficient matrix in Eq. (10-401).

Substitution of Eq. (10-392) in Eq. (10-391), we have

$$\mathbf{c}(k) = (\mathbf{D} - \mathbf{EG})\mathbf{x}(k) \tag{10-403}$$

Thus,

$$\mathbf{c}(k-1) = (\mathbf{D} - \mathbf{EG})\mathbf{x}(k-1) \tag{10-404}$$

Also, from Eq. (10-390) we write

$$\begin{aligned}\mathbf{Ax}(k-1) &= \mathbf{x}(k) - \mathbf{Bu}(k-1)\\ &= \mathbf{x}(k) + \mathbf{BGx}(k-1)\end{aligned} \tag{10-405}$$

Solving $\mathbf{x}(k-1)$ from the last equation, we get

$$\mathbf{x}(k-1) = (\mathbf{A} - \mathbf{BG})^{-1}\mathbf{x}(k) \tag{10-406}$$

Now substituting the last equation in Eq. (10-404), we have

$$\mathbf{c}(k-1) = (\mathbf{D} - \mathbf{EG})(\mathbf{A} - \mathbf{BG})^{-1}\mathbf{x}(k) \tag{10-407}$$

Recursively, we can write the following relationships:

$$\mathbf{c}(k-2) = (\mathbf{D} - \mathbf{EG})(\mathbf{A} - \mathbf{BG})^{-2}\mathbf{x}(k) \tag{10-408}$$

$$\vdots$$

$$\mathbf{c}(k-m+1) = (\mathbf{D} - \mathbf{EG})(\mathbf{A} - \mathbf{BG})^{-m+1}\mathbf{x}(k) \tag{10-409}$$

Substitution of Eqs. (10-403) and (10-407) through (10-409) in Eq. (10-402), the state-feedback control is written

$$\mathbf{u}(k) \cong -\mathbf{P}\begin{bmatrix}\mathbf{D} - \mathbf{EG}\\ (\mathbf{D} - \mathbf{EG})(\mathbf{A} - \mathbf{BG})^{-1}\\ (\mathbf{D} - \mathbf{EG})(\mathbf{A} - \mathbf{BG})^{-2}\\ \vdots\\ (\mathbf{D} - \mathbf{EG})(\mathbf{A} - \mathbf{BG})^{-m+1}\end{bmatrix}\mathbf{x}(k) \tag{10-410}$$

$$(p \times qm) \qquad (qm \times n) \qquad (n \times 1)$$

Now comparing Eq. (10-410) with Eq. (10-392), we have

$$\mathbf{P}\begin{bmatrix} \mathbf{D}-\mathbf{EG} \\ (\mathbf{D}-\mathbf{EG})(\mathbf{A}-\mathbf{BG})^{-1} \\ (\mathbf{D}-\mathbf{EG})(\mathbf{A}-\mathbf{BG})^{-2} \\ \vdots \\ (\mathbf{D}-\mathbf{EG})(\mathbf{A}-\mathbf{BG})^{-m+1} \end{bmatrix} = \mathbf{G} \qquad (10\text{-}411)$$

$$(p \times qm) \qquad\qquad (qm \times n) \qquad\qquad (p \times n)$$

In order to solve for **P** from the last equation, qm must equal n, or $m = n/q$. This means that n/q must be an integer, and the series expansion of $H_{ij}(z)$ should be truncated at $m = n/q$ terms. Solving for **P** from Eq. (10-411), we get

$$\mathbf{P} = \mathbf{G}\begin{bmatrix} \mathbf{D}-\mathbf{EG} \\ (\mathbf{D}-\mathbf{EG})(\mathbf{A}-\mathbf{BG})^{-1} \\ (\mathbf{D}-\mathbf{EG})(\mathbf{A}-\mathbf{BG})^{-2} \\ \vdots \\ (\mathbf{D}-\mathbf{EG})(\mathbf{A}-\mathbf{BG})^{-m+1} \end{bmatrix}^{-1} \qquad (10\text{-}412)$$

where the indicated matrix inverse must exist.

Since **P** is $p \times qm$, there are (pqm) unknowns in **P**. However, there are only pn equations in Eq. (10-419). Thus, $p(qm - n)$ of the elements of **P** may be assigned arbitrarily. It should also be noted that solution of the elements of the matrix **P** from Eq. (10-412) gives only the values of the coefficients K_{ij} and d_{ijk} in Eq. (10-398). The coefficients of the transfer function of Eq. (10-395) still have to be determined using Eq. (10-397). In general, there are more unknowns than equations in Eq. (10-397). This means that ideally we can simply set

$$d_{ij0} = 1 \qquad (10\text{-}413)$$

and

$$d_{ijk} = \alpha_{ijk} \qquad k \geq 1 \qquad (10\text{-}414)$$

and set all β_{ijk} to 0, for $k = 1,2,\ldots,q$. However, for a physically-realizable transfer function, $H_{ijm}(z)$ must not have more zeros than poles. Therefore, the values of β_{ijk} should be assigned such that the dynamic behavior of the overall system is not appreciably affected by the presence of β_{ijk}, $k = 1,2,\ldots,q$. This is similar to the classical design practice of designing the zeros of the transfer function of the controller, $H_{ijm}(z)$ to control the dynamics of the system, while placing the poles of $H_{ijm}(z)$ so that they do not have appreciable effects on the system performance. In the case of digital control systems the poles of $H_{ijm}(z)$ should be placed near the origin in the z-plane.

In the following we shall investigate the single-variable case of the dynamic feedback controller realization of state-feedback problems.

Single-Variable Systems

Consider that the digital control system described by Eqs. (10-390) through (10-392) has one input and one output; i.e., $p = q = 1$. Then, the feedback gain matrix is written as

$$\mathbf{G} = [g_1 \quad g_2 \quad \cdots \quad g_n] \qquad (1 \times n) \tag{10-415}$$

The dynamic feedback controller is described by the scalar transfer function

$$H(z) = \frac{K(1 + \alpha_1 z + \alpha_2 z^2 + \ldots + \alpha_n z^n)}{(1 + \beta_1 z + \beta_2 z^2 + \ldots + \beta_n z^n} \tag{10-416}$$

The n-term Laurent's series expansion of H(z) is

$$H_n(z) = K(1 + d_1 z^{-1} + d_2 z^{-2} + \ldots + d_{n-1} z^{-n+1}) \tag{10-417}$$

where

$$d_k = \alpha_k - \beta_k - \sum_{v=1}^{k-1} \beta_{k-v} d_v \tag{10-418}$$

for $k = 1, 2, \ldots, n - 1$. Equation (10-414) gives

$$\mathbf{P} = \mathbf{K}[1 \quad d_1 \quad d_2 \quad \cdots \quad d_{n-1}] = \mathbf{G} \begin{bmatrix} \mathbf{D} - \mathbf{EG} \\ (\mathbf{D} - \mathbf{EG})(\mathbf{A} - \mathbf{BG})^{-1} \\ (\mathbf{D} - \mathbf{EG})(\mathbf{A} - \mathbf{BG})^{-2} \\ \vdots \\ (\mathbf{D} - \mathbf{EG})(\mathbf{A} - \mathbf{BG})^{-n+1} \end{bmatrix}^{-1} \tag{10-419}$$

Equivalent Forward Cascade Controller

The development carried out in the preceding section is based on a dynamic controller $\mathbf{H}(z)$ being located in the feedback path of the system as shown in Fig. 10-61. When the reference input $\mathbf{r}(k)$ is zero, that is, when the system is designed as a regulator, it does not matter whether the dynamic controller is placed in the forward path or the feedback path. However, when the system is designed to track an input **r**(k), it may be desirable to place the dynamic controller in the forward path, as shown in Fig. 10-62. For a single-variable system, we can show that the equivalent forward cascade controller $G_c(z)$ can be determined from H(z).

$$G_c(z) = \frac{1}{1 + \mathbf{D}(z\mathbf{I} - \mathbf{A})^{-1}\mathbf{B}[H(z) - 1]} \tag{10-420}$$

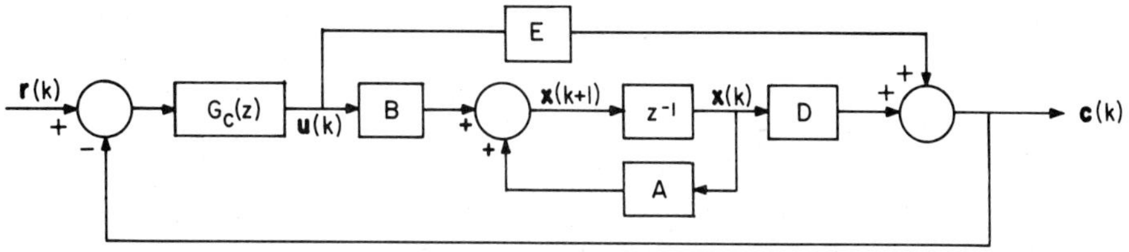

Fig. 10-62. Digital control system with dynamic cascade controller.

Example 10-19.

Figure 10-63 shows the state diagram of a digital control process which is subject to state feedback. The feedback gains are chosen so that the eigenvalues of the closed-loop system are at 0.5 + j0.5 and 0.5 − j0.5.

The coefficient matrices of the process are

$$\mathbf{A} = \begin{bmatrix} 0 & 1 \\ -0.368 & 1.368 \end{bmatrix} \qquad \mathbf{B} = \begin{bmatrix} 0 \\ 1 \end{bmatrix}$$

$$\mathbf{D} = [0.264 \quad 0.368] \qquad \mathbf{E} = 0$$

The feedback gain matrix is

$$\mathbf{G} = [0.132 \quad 0.368]$$

The equivalent feedback dynamic controller which approximates the state feedback is described by

$$H(z) = K\frac{z + \alpha_1}{z + \beta_1} \cong K(1 + d_1 z^{-1}) \tag{10-421}$$

Equation (10-419) gives

$$P = K[1 \quad d_1] = G \begin{bmatrix} D \\ D(A - BG)^{-1} \end{bmatrix}^{-1}$$

$$= [0.132 \quad 0.368] \begin{bmatrix} 0.264 & 0.368 \\ 0.896 & -0.528 \end{bmatrix}^{-1}$$

$$= 0.8514[1 \quad -0.1216] \tag{10-422}$$

Thus,

$$K = 0.8514 \quad \text{and} \quad d_1 = -0.1216$$

In order that the dynamic controller is a physically realizable one, we arbitrarily set β_1 to be a very small number compared with d_1, say, $\beta_1 = 0.0005$. Then

$$a_1 = d_1 + \beta_1 = -0.1211 \tag{10-423}$$

The transfer function of the feedback dynamic controller is

$$H(z) = 0.8514 \frac{z - 0.1211}{z - 0.0005} \tag{10-424}$$

The block diagram of the closed-loop system with the dynamic controller is shown in Fig. 10-64.

Now let us compare the characteristics and the performance of the two systems shown in Figs. 10-63 and 10-64. The characteristic equation of the system with the state feedback is

$$|zI - A + BG| = z^2 - z + 0.5 = 0 \tag{10-425}$$

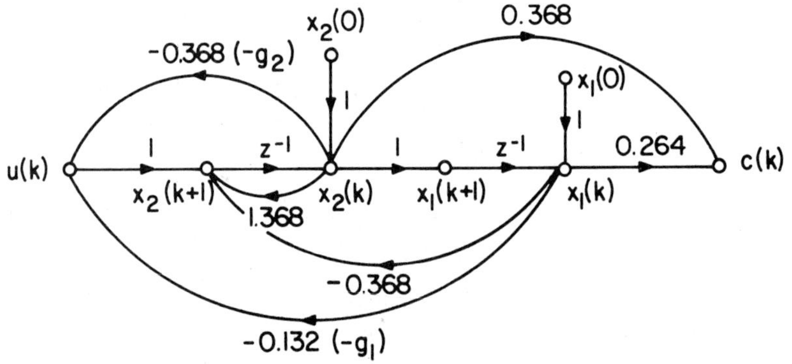

Fig. 10-63. State diagram of a digital control system with state feedback.

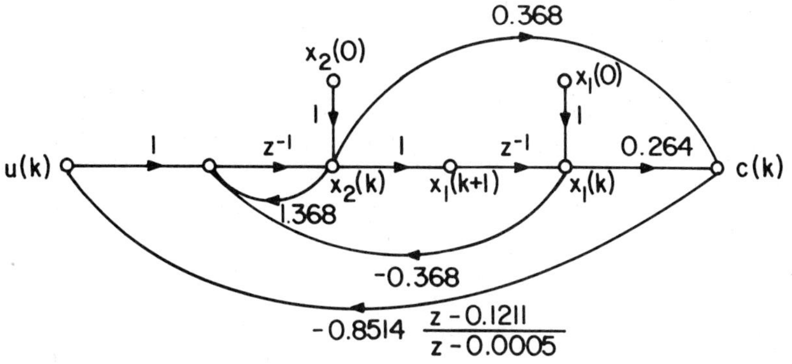

Fig. 10-64. Digital control system with dynamic controller equivalent to that in Fig. 10-63.

The roots of the last equation are $z = 0.5 + j0.5$ and $z = 0.5 - j0.5$, as specified. The characteristic equation of the system with the dynamic controller feeding back from the output is

$$z^3 - 1.05202z^2 + 0.557398z - 0.027678 = 0 \qquad (10\text{-}426)$$

The roots of the last equation are

$$z = 0.0548 \;, \quad z = 0.4986 + j0.5043 \quad \text{and} \quad z = 0.4986 - j0.5043 \;.$$

Thus we see that although the use of the dynamic controller in place of the state feedback has made the overall system into a third-order one, the dominant characteristic roots are still very close to $0.5 \pm j0.5$. The root at $z = 0.0548$, which is very close to the origin of the z-plane, should not affect the dynamics of the system significantly.

Let us consider the initial state, $x_1(0) = 1$ and $x_2(0) = 0$. From Fig. 10-63 we have

$$C(z) = \frac{0.264z^2 - 0.448z}{z^2 - z + 0.5} x_1(0) \qquad (10\text{-}427)$$

Figure 10-65 shows the response of c(k). From Fig. 10-64, the z-transform of the output of the system with the dynamic controller, when the above initial states are applied, is obtained as

$$C(z) = \frac{0.264z^3 - 0.496712z^2 + 0.000248z}{z^3 - 1.05202z^2 + 0.557398z - 0.027678} x_1(0) \qquad (10\text{-}428)$$

The time response of c(k) that corresponds to the last equation is shown in Fig. 10-65. From these responses we see that the system with the dynamic controller gives a higher overshoot, but the two responses are generally very close.

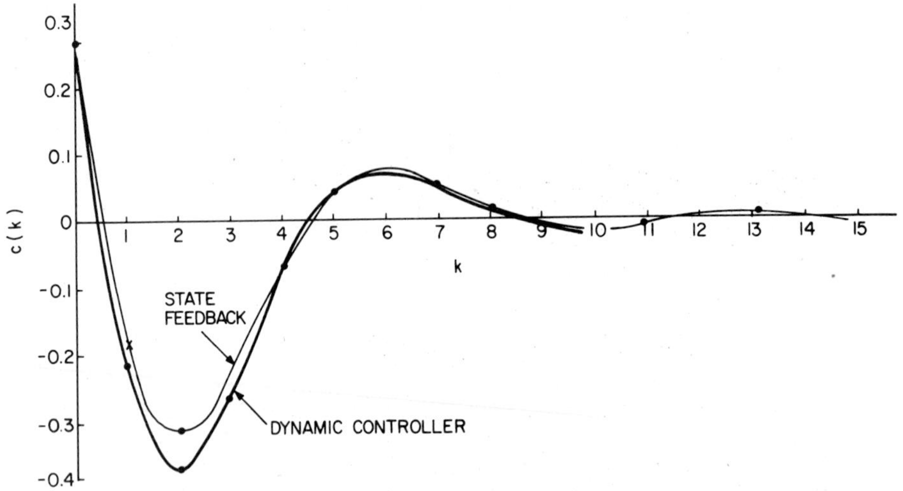

Fig. 10-65. Time responses of the systems in Figs. 10-63 and 10-64.

REFERENCES

1. Davison, E. J., "On Pole Assignment in Linear Systems with Incomplete State Feedback," *IEEE Trans. on Automatic Control,* Vol. AC-15, June 1970, pp. 348-351.
2. Wonham, W. M., "On Pole Assignment in Multi-Input Controllable Linear Systems," *IEEE Trans. on Automatic Control,* Vol. AC-12, December 1976, pp. 660-665.
3. Smith, H. W. and Davison, E. J., "Design of Industrial Regulations," *Proc. IEE* (London), Vol. 119, No. 8, August 1972, pp. 1210-1216.
4. Whitbeck, R. F. and Hofmann, L. G., "Digital Control Law Synthesis in the w′ Domain," *Journal of Guidance and Control,* Vol. 1, No. 5, September-October 1978, pp. 319-326.

PROBLEMS

10-1 The block diagram of a digital control system with a cascade digital controller is shown in Fig. 10P-1(a). The transfer function of the digital controller is given as

$$G_c(z) = \frac{(z-0.5)(z-0.2)}{z(z-0.1)}$$

The transfer function of the controlled process is

$$G_p(s) = \frac{1}{s(s+0.694)}$$

Find the transfer function of the continuous-data feedback controller of the system in Fig. 10P-1(b) so that the two systems will have the same closed-loop transfer function. H(s) must be of the form: $sH_1(s)$, where $H_1(s)$ does not have any poles or zeros at $s = 0$.

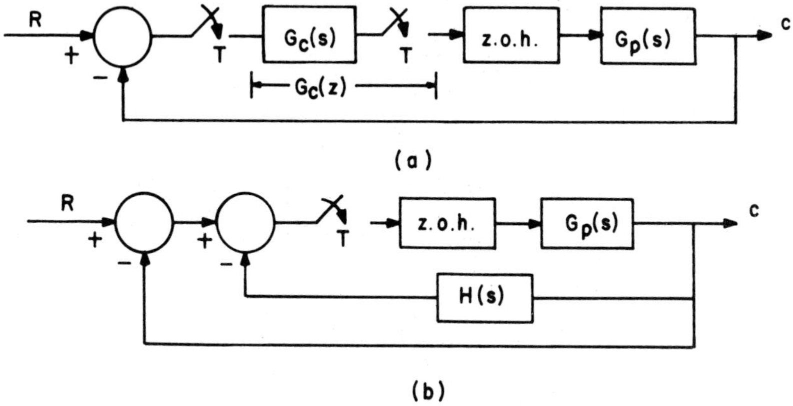

Fig. 10P-1.

10-2 Given the transfer function

$$G_c(z) = \frac{E_2(z)}{E_1(z)} = \frac{z^{-5}}{(1 + z^{-1} + z^{-2})^2}$$

(a) Draw the block diagram of a direct digital program of $G_c(z)$, using a minimum number of data storage units.
(b) Draw the block diagram of a cascade digital program of $G_c(z)$.
(c) Draw the block diagram of a parallel digital program of $G_c(z)$, using a minimum number of data storage units.

10-3 When traveling through the atmosphere, a guided missile usually encounters aerodynamic forces which tends to cause instability in the attitude of the missile. As shown in Fig. 10P-3(a), the side force F which is applied at the center of pressure P, tends to cause the missile to tumble. Let the angular acceleration of the missile about the center of gravity C be denoted by a. Normally, a is directly proportional to the angle of attack θ_c, and is given by

$$a = K\theta_c$$

where K is a constant described by

$$K = K_F d_1/J$$

where K_F is a constant which depends on such parameters as dynamic pressure, velocity of the missile, air density, and so on, and J is the moment of inertia of the missile about C. The main object of the flight control system is to provide the stabilization action to counter the effect of the side force by use of gas injection at the tail of the missile to deflect the direction of the rocket engine thrust T_s.

For small angle δ, $T_s \sin\delta \cong T_s\delta$. The torque due to T_s about the point C is $T_s\delta d_2$. Thus, the torque equation is written

$$T_s d_2 \delta + JK\theta_c = J\ddot{\theta}_c$$

The block diagram of the closed-loop flight control system is shown in Fig. 10P-3(b). The sampling period of the digital control system is 0.1 sec.

Let $T_s d_2/J = 10$ and $K = 1$. The transfer function of the digital controller is of the form:

$$G_c(z) = \frac{z + a}{z + b}$$

where a and b are real constants. Find the values of a and b so that the dominant roots of the characteristic equation correspond to critical damping (two equal positive real roots inside the unit circle in the z-plane).

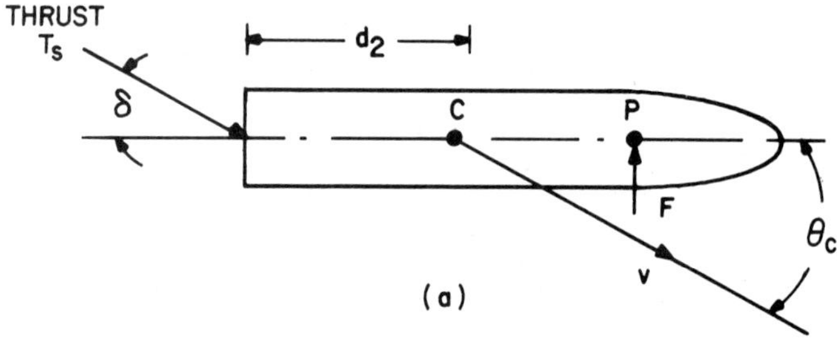

(a)

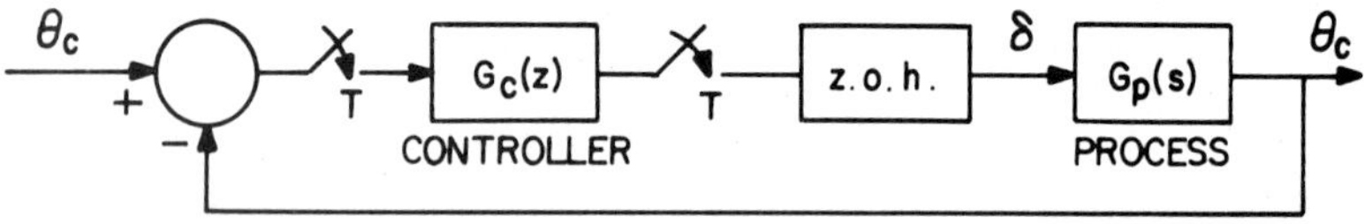

(b)

Fig. 10P-3.

10-4 Consider the level-control system described in Problem 7-17. The root locus problem of the system is investigated with N, the number of inlet, as the variable parameter. It is found in Problem 7-17 that when $N = 5$, the system without a controller is stable, but the step response is quite oscillatory, and the overshoot is large. The objective of this problem is to design a digital controller using the root locus and the z-plane design methods discussed in this chapter.

Referring to Fig. 7P-17, the open-loop transfer function of the system is obtained as $(N = 5)$

$$G(z) = \frac{H(z)}{E(z)} = \frac{0.00147635(z^2 + 3.4030929z + 0.71390672)}{(z-1)(z-0.535261429)(z-0.951229425)}$$

where E(z) denotes the input to the zero-order hold which represents the D/A. Let the digital controller have the transfer function

$$G_c(z) = K_c \frac{z - z_1}{z - p_1}$$

Consider that it is appropriate to cancel a pole and a zero of G(z) by the zero and pole of $G_c(z)$, except for the pole at $z = 1$. Determine the values of K_c, z_1 and p_1 so that the ramp-error constant K_v is unchanged, and the peak overshoot is the smallest that can be obtained with this pole-zero cancellation design. Note that several pole-zero cancellation combinations are possible.

10.5 For the control system described in Problem 7-17, sketch the Bode diagram of the open-loop transfer function $G(z) = H(z)/E(z)$ in the magnitude (db) versus ω_w coordinates, where $\omega_w = \text{Im}\, w$; w is the complex variable $w = (z-1)/(z+1)$. Consider that there is no digital controller; i.e., $G_c(z) = 1$. Use $N = 5$, and find the gain margin and the phase margin of the closed-loop system.

Apply a cascade digital controller that has the transfer function in the w-domain, of the form,

$$G_c(w) = \frac{1 + aw}{1 + bw}$$

Find the values of a and b so that the phase margin of the system is approximately 45 degrees. Find the corresponding $G_c(z)$. Find the unit-step response of h(t) at the sampling instants of the compensated system.

10-6 The block diagram of a control system with a digital controller is shown in Fig. 10P-6. The transfer function of the controlled process is

$$G_p(s) = \frac{1}{s(s + 0.5)}$$

The sampling period T is 1 second.

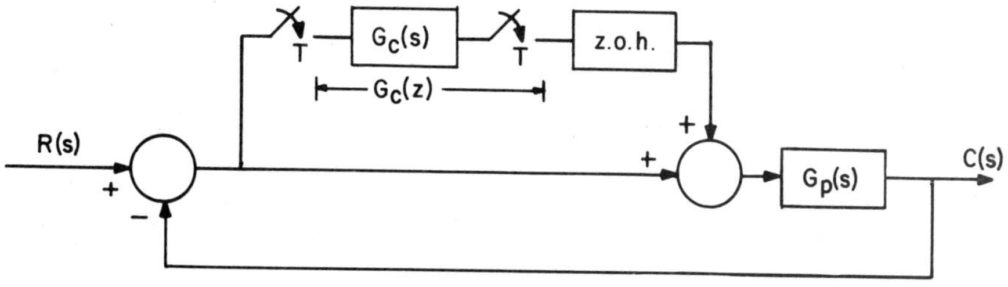

Fig. 10P-6.

(a) Find the transfer function of the digital controller $G_c(z)$ if the output sequence $c(kT)$ is to be $(0, 0.5, 1, 1, 1, \ldots)$ for a unit-step input as $r(t)$.
(b) Find the output $c(t)$ if the digital controller fails; i.e., the digital controller fails to process any signal.

10-7 The digital process of a unity-feedback control system is described by the transfer function

$$G_p(z) = \frac{K(z+0.5)}{(z-1)(z-0.5)}$$

Design a cascade digital controller with the transfer function

$$G_c(z) = K_c \frac{z - z_1}{z - p_1}$$

so that the following design specifications are satisfied:
(a) The ramp-error constant $K_v = 6$.
(b) The dominant roots of the closed-loop characteristic equation are approximately equal to $z = 0.5 + j0.707$ and $z = 0.5 - j0.707$. Find the closed-loop transfer function $C(z)/R(z)$, and give the characteristic equation roots.

10-8 Consider the stoichiometric air-fuel-ratio control system described in Problem 7-15, Fig. 7P-15(b). In that problem it is shown that when the controller transfer function $G_c(s)$ is a constant, the steady-state error due to a step input is nonzero. In order to eliminate the steady-state error, we must introduce integral control. Let

$$G_c(s) = K_p + \frac{K_i}{s}$$

(a) Show that the steady-state error

$$\lim_{k \to \infty} e(kT)$$

is zero when $r(t)$ and $d(t)$ are step functions, and when the overall system is stable.
(b) Find the characteristic equation of the system in terms of K_p and K_i.
(c) Sketch the root loci of the closed-loop system with $K_i = 1$ and K_p varies from 0 to ∞.
(d) Plot the time responses of $c(t)$ when $r(t) = u_s(t)$, $d(t) = 0$, with $K_i = 1$ and $K_p = 0.1, 0.5$ and 1.

10-9 The block diagram shown in Fig. 10P-9 represents the sampled-data automatic forage and concentrate-proportioning system described in Problem 5-15. Since the system described in Problem 5-15 is unstable, it is necessary to introduce a digital controller with the transfer function $G_c(z)$. The transfer function of the process is given as

$$G_p(z) = \frac{1.3359(z + 0.7542)}{z^2 + 0.3957z + 1.4422}$$

The sampling period is 0.01 sec. Let the digital controller be described by the transfer function of a PID controller,

$$G_c(z) = K_p + \frac{K_I T(z+1)}{2(z-1)} + \frac{K_d(z-1)}{Tz}$$

Find the values of K_p, K_I and K_d so that the zeros of $G_c(z)$ cancel the poles of $G_p(z)$, and the ramp-error constant K_v is equal to 0.5. Find the closed-loop transfer function $C(z)/V_s(z)$. Find the steady-state value of c(kT) and the maximum overshoot when the input v_s is a unit step function.

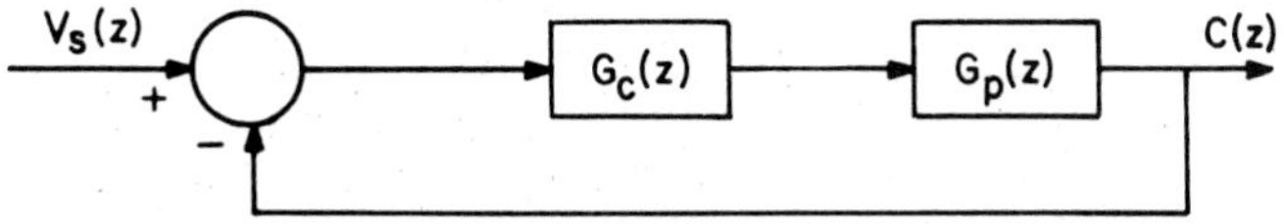

Fig. 10P-9.

10-10 Given a single-input digital control system

$$\mathbf{x}(k+1) = \mathbf{A}\mathbf{x}(k) + \mathbf{B}u(k)$$

where $\mathbf{x}(k)$ is an n-vector, and the pair $(\mathbf{A}, \mathbf{B})$ is completely controllable.

(a) Show that given the closed-loop characteristic equation

$$\Delta_c(z) = z^n + a_1 z^{n-1} + a_2 z^{n-2} + \ldots + a_{n-1} z + a_n = 0$$

the state-feedback control is $u(k) = -\mathbf{G}\mathbf{x}(k)$, where

$$\mathbf{G} = \mathbf{F}\Delta_c(\mathbf{A})$$

$$\mathbf{F} = [0 \quad 0 \quad \ldots \quad 0 \quad 1][\mathbf{B} \quad \mathbf{AB} \quad \mathbf{A}^2\mathbf{B} \quad \ldots \quad \mathbf{A}^{n-1}\mathbf{B}]^{-1}$$

(b) Show that the feedback gain matrix can be written as

$$\mathbf{G} = [1 \quad a_1 \quad a_2 \quad \ldots \quad a_n]\begin{bmatrix} \mathbf{F}\mathbf{A}^n \\ \mathbf{F}\mathbf{A}^{n-1} \\ \vdots \\ \mathbf{F}\mathbf{A} \\ \mathbf{F} \end{bmatrix}$$

Note that the results presented in this problem are alternates to the state-feedback algorithms for pole placement that are given by Eqs. (10-244) and (10-254).

(c) Let

$$\mathbf{A} = \begin{bmatrix} 0 & 1 \\ -1 & 2 \end{bmatrix} \qquad \mathbf{B} = \begin{bmatrix} 0 \\ 1 \end{bmatrix}$$

Find the state feedback matrix **G** such that the eigenvalues of **A** − **BG** are at 0 and 0.3.

10-11 (a) Given the discrete-data system

$$\mathbf{x}(k+1) = \mathbf{A}\mathbf{x}(k) + \mathbf{B}u(k)$$

where **A** is n × n, **B** is n × 1; **A** and **B** are in the phase-variable canonical form

$$\mathbf{A} = \begin{bmatrix} 0 & 1 & 0 & 0 & 0 & \cdots & 0 \\ 0 & 0 & 1 & 0 & 0 & \cdots & 0 \\ 0 & 0 & 0 & 1 & 0 & \cdots & 0 \\ \cdots & \cdots & \cdots & \cdots & \cdots & \cdots & \cdots \\ 0 & 0 & 0 & 0 & 0 & \cdots & 1 \\ -a_1 & -a_2 & -a_3 & -a_4 & -a_5 & \cdots & -a_n \end{bmatrix}$$

$$\mathbf{B} = \begin{bmatrix} 0 \\ 0 \\ 0 \\ \vdots \\ 0 \\ 1 \end{bmatrix}$$

Let

$$u(k) = -\mathbf{G}\mathbf{x}(k)$$

It is desired to place the eigenvalues of the closed-loop system $\mathbf{x}(k+1) = (\mathbf{A} - \mathbf{BG})\mathbf{x}(k)$ at $z_1, z_2, \ldots, z_n$, all distinct. Show that the feedback matrix **G** is given by

$$\mathbf{G} = -[\Delta_{01} \quad \Delta_{02} \quad \cdots \quad \Delta_{0n}]\mathbf{K}^{-1}$$

where

$$\Delta_{0i} = \Delta_0(z_i) = |z_i\mathbf{I} - \mathbf{A}|$$

and

$$K = \begin{bmatrix} 1 & 1 & \cdots & 1 \\ z_1 & z_2 & \cdots & z_n \\ z_1^2 & z_2^2 & \cdots & z_n^2 \\ \vdots & \vdots & & \vdots \\ z_1^{n-1} & z_2^{n-1} & \cdots & z_n^{n-1} \end{bmatrix}$$

(b) Let

$$A = \begin{bmatrix} 0 & 1 \\ -1 & 2 \end{bmatrix} \qquad B = \begin{bmatrix} 0 \\ 1 \end{bmatrix}$$

Find the state feedback matrix G such that the eigenvalues of A − BG are at 0 and 0.3.

10-12 An inventory control system is modelled by the following differential equations:

$$\frac{dx_1}{dt} = -x_2 + u$$

$$\frac{dx_2}{dt} = -bu$$

where

x_1 = level of inventory

x_2 = rate of sales of product

u = production rate

b = constant

Let the production rate be controlled by discrete-time control, so that

$$u(t) = u(kT) \qquad kT \leqslant t < (k+1)T$$

where T is the sampling period. Furthermore, let

$$u(kT) = r(kT) - g_1 x_1(kT) - g_2 x_2(kT)$$

where r(kT) is a reference set point used to control the inventory level; g_1 and g_2 are feedback gains.

(a) Find the values of g_1 and g_2 so that the characteristic equation roots of the closed-loop system are all at the origin of the z-plane.
(b) Find $x_1(kT)$ and $x_2(kT)$ for $k = 0,1,2,\ldots$ where $r(kT)$ is unity for all k. Assume that $x_1(0) = x_{10}$ and $x_2(0) = 0$.
(c) Find the value of x_{10} so that following from part (b) $x_1(kT)$ is constant for all k. $x_2(0) = 0$. Find the corresponding $x_2(kT)$.

10-13 Given the digital control system

$$\mathbf{x}(k+1) = \mathbf{A}\mathbf{x}(k) + \mathbf{B}\mathbf{u}(k)$$

where

$$\mathbf{A} = \begin{bmatrix} 0 & 1 \\ -1 & -2 \end{bmatrix} \qquad \mathbf{B} = \begin{bmatrix} 1 & 0 \\ 0 & 1 \end{bmatrix}$$

(a) Find the state feedback $\mathbf{u}(k) = -\mathbf{G}\mathbf{x}(k)$ such that the eigenvalues of $\mathbf{A} - \mathbf{B}\mathbf{G}$ are at $z = 0, 0$. It is also required that feedback is brought only to $u_1(k)$ and not to $u_2(k)$. Can this be achieved?
(b) Repeat part (a) by requiring that feedback be brought only to $u_2(k)$.
(c) Repeat part (a) by requiring that the feedback to u_1 is weighed twice that to u_2.

10-14 The open-loop space-shuttle control system described in Problem 7-6 is represented by the block diagram shown in Fig. 10P-14. Find the gain matrix $\mathbf{G}$ such that the state feedback $u(kT) = -\mathbf{G}\mathbf{x}(kT)$ places the closed-loop eigenvalues at $z = 0.2$ and 0.5. The control u is subject to sample-and-hold. The sampling period is 1 sec. $M = 600$, $K_s = 10$.

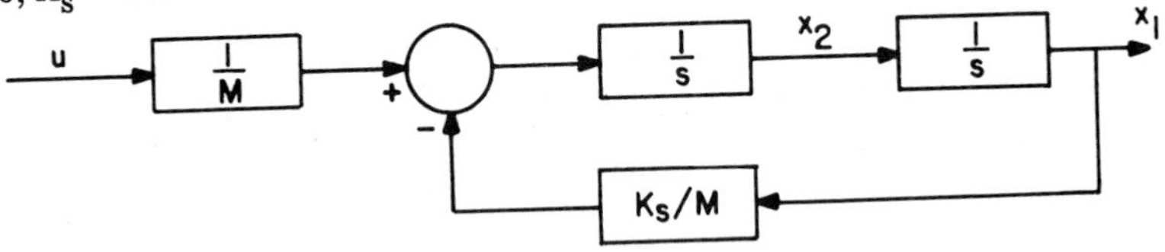

Fig. 10P-14.

10-15 Consider the inventory control system described in Problem 10-12. The discrete-time control is given by

$$u(t) = u(kT) \qquad kT \leq t < (k+1)T$$

where $u(t)$ is the output of the zero-order hold. The input of the zero-order hold is generated by a digital controller whose transfer function is designated as

$$\frac{U(z)}{E(z)} = G_c(z)$$

Let $e(kT) = r(kT) - x_1(kT)$, where $r(kT)$ is the reference set point used to control the inventory level.

(a) Given that $x_1(0) = 0$ and $x_2(0) = 0$, find the transfer function $G_c(z)$ so that the response of $x_1(kT)$ is deadbeat for $r(kT) = 1$ for all $k \geqslant 0$. Let $T = 1$ and $b = 1$. Find $x_2(kT)$ for $k \geqslant 0$.

(b) Let the transfer function of the controller be

$$G_c(z) = K_p + \frac{K_d(z+1)}{Tz}$$

Find the values of K_p and K_d so that two of the roots of the characteristic equation are at $z = 0.5$ and 0.5. Where is the other root?

10-16 Consider the dc-motor speed control system described in Problem 7-2. The block diagram of the system is shown in Fig. 7P-2(b). All the system parameters are as given in Problem 7-2.

(a) Find the transfer function of the digital controller $G_c(z)$ so that the output signal $\omega_L(t)$ is deadbeat at the sampling instants when the input $\omega_r(t)$ is a unit-step function.

(b) Sketch the sampled output response $\omega_L^*(t)$ at the sampling instants.

(c) For the system designed in (a), find the modified z-transform of the output, $\Omega_L(z,m)$ when the input is a unit-step function, and for $m = 0.5$. Plot the response $\omega_L^*(t,m)$ for $m = 0.5$, and compare that with $\omega_L^*(t)$.

(d) Find the transfer function of the digital controller $G_c(z)$ so that the output $\omega_L(t)$ is deadbeat at the sampling instants when the input $\omega_r(t)$ is a unit-ramp function. Sketch the output $\omega_L^*(t)$ at the sampling instants when the input is a unit ramp, and then when it is a unit-step function.

(e) In order to reach a compromised design for both a step input and a ramp input, use the staleness weighting factor design; that is, let the closed-loop transfer function be of the form of Eq. (10-205). Let the staleness weighting factor be 0.5. Find the transfer function of $G_c(z)$. Sketch the unit-step and the unit-ramp responses of $\omega_L^*(t)$.

10-17 The temperature $x(t)$ in the electric furnace shown in Fig. 10P-17 is governed by the differential equation

$$\dot{x}(t) = -x(t) + u(t) + w_2(t)$$

where $u(t)$ is the control and $w_2(t)$ is an unknown constant disturbance due to heat losses. It is desired that the equilibrium temperature be at a set point w_1 = constant. Design a digital control with state and dynamic feedback by sampling the temperature $x(t)$ once every 0.2 second, and $u(t) = u(kT)$, $kT \leqslant t < (k+1)T$, such that the eigenvalues of the digital closed-loop system are at $z = 0$, and $x(kT) \to w_1$ as $k \to \infty$. Assume zero initial conditions. Sketch $x(kT)$ for $k = 0,1,2,\ldots$ for positive w_1 and w_2.

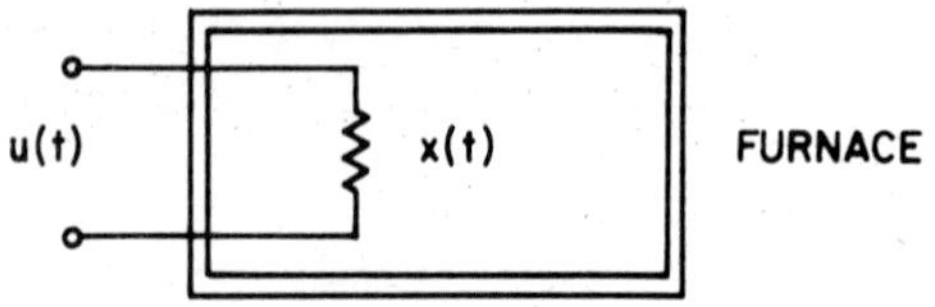

Fig. 10P-17.

10-18 Figure 10P-18(a) shows the schematic diagram of a DC motor which is used for the speed-control of a load. The characteristics of the permanent-magnet DC motor are described by the following parameters:

Armature resistance	$R = 2$ ohms
Armature inductance	$L =$ negligible
Torque constant	$K_m = 10$ oz-in/amp
Back emf constant	$K_b = 0.052$ volts/rad/sec
Load and motor inertia at motor shaft	$J = 0.01$ oz-in-sec^2
Load and motor friction	negligible

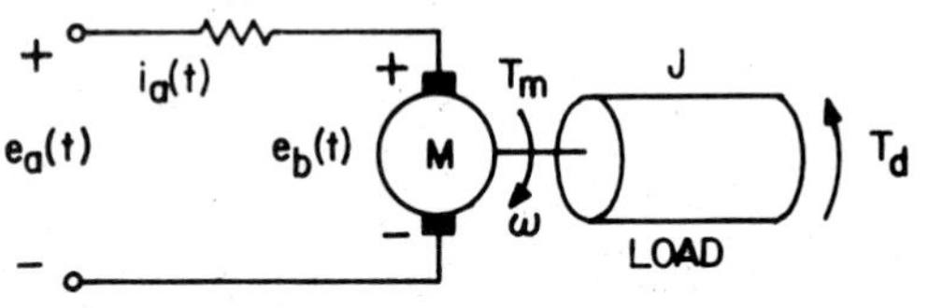

(a)

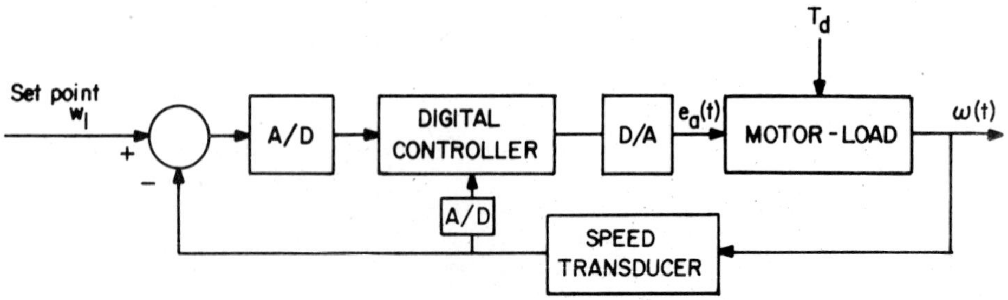

(b)

Fig. 10P-18.

The dynamics of the motor control system are described by the following equations:

$$e_a(t) = Ri_a(t) + e_b(t)$$

$$e_b(t) = K_b\omega(t)$$

$$T_m(t) = K_m i_a(t)$$

$$T_m(t) = J\dot{\omega}(t) + T_d$$

where

$e_a(t)$ = motor applied voltage

$i_a(t)$ = armature current

$e_b(t)$ = back emf

$T_m(t)$ = motor torque

T_d = constant load torque disturbance

$\omega(t)$ = motor speed

It is desired that the equilibrium speed of the motor-load shaft be at a set point w_1 = constant. The block diagram of the closed-loop digital control system is shown in Fig. 10P-18(b).

Let the sampling period be 1/26 seconds. Design the configuration of the digital controller and find the values of its parameters so that $\omega(kT) \to w_1$ as $k \to \infty$. The eigenvalues of the closed-loop system should be located at $z = 0$.

10-19 For the controlled process shown in Fig. 10P-19, design state and dynamic feedback control so that the state variable $x_1(k)$ will follow the set point w_1 = constant as $k \to \infty$. The noise signals w_2 and w_3 are unknown constants. The roots of the characteristic equation of the closed-loop system should all be at the origin of the z-plane. The control is subject to sample-and-hold; that is, $u(t) = u(kT)$ for $kT \leqslant t < (k+1)T$. The sampling period is 1 sec.

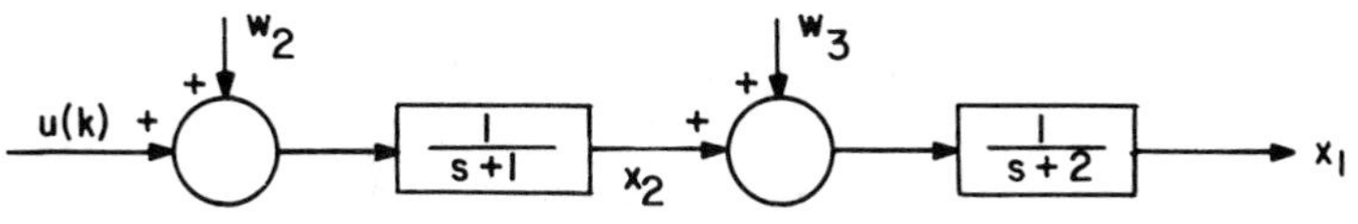

Fig. 10P-19.

10-20 Given the system

$$\mathbf{x}(k+1) = \mathbf{A}\mathbf{x}(k) + \mathbf{B}u(k)$$

$$c(k) = \mathbf{D}\mathbf{x}(k)$$

$$\mathbf{A} = \begin{bmatrix} 0 & 1 \\ -1 & -2 \end{bmatrix} \qquad \mathbf{B} = \begin{bmatrix} 0 \\ 1 \end{bmatrix} \qquad \mathbf{D} = [1 \quad 1]$$

Show that the output feedback $u(k) = -Gc(k)$ is equivalent to incomplete state feedback with only $x_2(k)$ available for feedback.

11

DESIGN BY MAXIMUM PRINCIPLE

11.1 THE DISCRETE EULER-LAGRANGE EQUATION

A large class of optimal design of digital control systems has the objective of minimizing or maximizing a performance index which is of the form

$$J = \sum_{k=0}^{N-1} F[\mathbf{x}(k),\mathbf{x}(k+1),\mathbf{u}(k),k] \tag{11-1}$$

where $F[\mathbf{x}(k),\mathbf{x}(k+1),\mathbf{u}(k),k]$ is a differentiable scalar function.

The minimization or maximization of J is subject to the equality constraint of

$$\mathbf{x}(k+1) = \mathbf{f}[\mathbf{x}(k),\mathbf{u}(k),k] \tag{11-2}$$

which is recognized as the state equation of the system, as well as other equality and inequality constraints. For Eq. (11-2), $\mathbf{x}(k)$ is n-dimensional and $\mathbf{u}(k)$ is p-dimensional.

A majority of the design techniques of optimal control rely on the principle of calculus of variation. According to the principle of variation, the problem of minimizing one function while subject to equality constraints is solved by adjoining the constraint to the function to be minimized or maximized.

Let $\lambda(k+1)$, an $n \times 1$ vector, be defined as the *Lagrange multiplier*. We adjoin the performance index J of Eq. (11-1) with the equality constraint in Eq. (11-2) to form the adjoined performance index

$$J_c = \sum_{k=0}^{N-1} F[\mathbf{x}(k),\mathbf{x}(k+1),\mathbf{u}(k),k] + <\lambda(k+1),[\mathbf{x}(k+1) - \mathbf{f}(\mathbf{x},\mathbf{u},k)]> \tag{11-3}$$

where $<\cdot>$ denotes the inner product of matrix vectors.

The calculus of variation asserts that the minimization or maximization of J subject to Eq. (11-2) is equivalent to minimizing or maximizing J_c without any constraints.

Let $\mathbf{x}(k)$, $\mathbf{x}(k+1)$, $\mathbf{u}(k)$, and $\lambda(k+1)$ take on various variations as follows:

$$\mathbf{x}(k) = \mathbf{x}^\circ(k) + \epsilon\eta(k) \tag{11-4}$$

$$\mathbf{x}(k+1) = \mathbf{x}^\circ(k+1) + \epsilon\eta(k+1) \tag{11-5}$$

$$\mathbf{u}(k) = \mathbf{u}^\circ(k) + \delta\mu(k) \tag{11-6}$$

$$\lambda(k+1) = \lambda^\circ(k+1) + \gamma\omega(k+1) \tag{11-7}$$

where $\mathbf{x}^\circ(k)$, $\mathbf{x}^\circ(k+1)$, $\mathbf{u}^\circ(k)$ and $\lambda^\circ(k+1)$ represent the vectors that correspond to the optimal trajectories; $\eta(k)$, $\mu(k)$ and $\omega(k)$ are arbitrary vector variables.

Substitution of Eqs. (11-4) through (11-7) into Eq. (11-3) yields

$$\begin{aligned} J_c = \sum_{k=0}^{N-1} \; & F[\mathbf{x}^\circ(k) + \epsilon\eta(k), \mathbf{x}^\circ(k+1) + \epsilon\eta(k+1), \mathbf{u}^\circ(k) + \delta\mu(k), k] \\ & + <\lambda^\circ(k+1) + \gamma\omega(k+1), \quad \mathbf{x}^\circ(k+1) + \epsilon\eta(k+1) \\ & - \mathbf{f}[\mathbf{x}^\circ(k) + \epsilon\eta(k), \mathbf{u}^\circ(k) + \delta\mu(k), k] > \end{aligned} \tag{11-8}$$

To simplify the notation, we express J_c as

$$J_c = \sum_{k=0}^{N-1} F_c[\mathbf{x}(k), \mathbf{x}(k+1), \lambda(k+1), \mathbf{u}(k), k] \tag{11-9}$$

Expanding F_c into a Taylor series about $\mathbf{x}^\circ(k)$, $\mathbf{x}^\circ(k+1)$, $\lambda^\circ(k+1)$, $\mathbf{u}^\circ(k)$, we get

$$\begin{aligned} F_c[\mathbf{x}(k), \mathbf{x}(k+1), \lambda(k+1), \mathbf{u}(k), k] &= F_c[\mathbf{x}^\circ(k), \mathbf{x}^\circ(k+1), \lambda^\circ(k+1), \mathbf{u}^\circ(k), k] \\ & + <\epsilon\eta(k), \frac{\partial F_c^\circ(k)}{\partial \mathbf{x}^\circ(k)}> + <\epsilon\eta(k+1), \frac{\partial F_c^\circ(k)}{\partial \mathbf{x}^\circ(k+1)}> + <\gamma\omega(k+1), \frac{\partial F_c^\circ(k)}{\partial \lambda^\circ(k+1)}> \\ & + <\delta\mu(k), \frac{\partial F_c^\circ(k)}{\partial \mathbf{u}^\circ(k)}> + \text{higher-order terms} \end{aligned} \tag{11-10}$$

where

$$F_c^\circ(k) = F_c[\mathbf{x}^\circ(k), \mathbf{x}^\circ(k+1), \lambda^\circ(k+1), \mathbf{u}^\circ(k), k] \tag{11-11}$$

The necessary condition for J_c to be a minimum is

$$\left.\frac{\partial J_c}{\partial \epsilon}\right|_{\epsilon=\delta=\gamma=0} = 0 \tag{11-12}$$

$$\left.\frac{\partial J_c}{\partial \gamma}\right|_{\epsilon=\delta=\gamma=0} = 0 \tag{11-13}$$

$$\left.\frac{\partial J_c}{\partial \delta}\right|_{\epsilon=\delta=\gamma=0} = 0 \tag{11-14}$$

Substituting the Taylor series expansion of F_c into Eq. (11-9) and applying the necessary condition of minimum J_c in Eqs. (11-12) through (11-14), we get

$$\sum_{k=0}^{N-1}\left[<\eta(k), \frac{\partial F_c^\circ(k)}{\partial \mathbf{x}^\circ(k)}> + <\eta(k+1), \frac{\partial F_c^\circ(k)}{\partial \mathbf{x}^\circ(k+1)}>\right] = 0 \tag{11-15}$$

$$\sum_{k=0}^{N-1} <\omega(k+1), \frac{\partial F_c^\circ(k)}{\partial \lambda^\circ(k+1)}> = 0 \tag{11-16}$$

$$\sum_{k=0}^{N-1} <\mu(k), \frac{\partial F_c^\circ(k)}{\partial \mathbf{u}^\circ(k)}> = 0 \tag{11-17}$$

Equation (11-15) can be written as

$$\begin{aligned}\sum_{k=0}^{N-1} <\eta(k), \frac{\partial F_c^\circ(k)}{\partial \mathbf{x}^\circ(k)}> &= -\sum_{k=1}^{N} <\eta(k), \frac{\partial F_c^\circ(k-1)}{\partial \mathbf{x}^\circ(k)}> \\ &= -\sum_{k=0}^{N-1} <\eta(k), \frac{\partial F_c^\circ(k-1)}{\partial \mathbf{x}^\circ(k)}> \\ &\quad + \left.<\eta(k), \frac{\partial F_c^\circ(k-1)}{\partial \mathbf{x}^\circ(k)}>\right|_{k=0} - \left.<\eta(k), \frac{\partial F_c^\circ(k-1)}{\partial \mathbf{x}^\circ(k)}>\right|_{k=N}\end{aligned} \tag{11-18}$$

where

$$F_c^\circ(k-1) = F_c[\mathbf{x}^\circ(k-1), \mathbf{x}^\circ(k), \lambda^\circ(k), \mathbf{u}^\circ(k-1), k-1] \tag{11-19}$$

Rearranging the terms in the last equation, we have

$$\sum_{k=0}^{N-1} <\eta(k), \frac{\partial F_c^\circ(k)}{\partial \mathbf{x}^\circ(k)} + \frac{\partial F_c^\circ(k-1)}{\partial \mathbf{x}^\circ(k)}> + <\eta(k), \frac{\partial F_c^\circ(k-1)}{\partial \mathbf{x}^\circ(k)}>\Bigg|_{k=0}^{k=N} = 0 \tag{11-20}$$

According to the fundamental lemma of the calculus of variation, Eq. (11-20) is satisfied for any $\eta(k)$ when the following equations are satisfied:

$$\frac{\partial F_c^\circ(k)}{\partial \mathbf{x}^\circ(k)} + \frac{\partial F_c^\circ(k-1)}{\partial \mathbf{x}^\circ(k)} = 0 \tag{11-21}$$

$$<\eta(k), \frac{\partial F_c^\circ(k-1)}{\partial \mathbf{x}^\circ(k)}>\Bigg|_{k=0}^{k=N} = 0 \tag{11-22}$$

The reason behind the fundamental lemma is that for *arbitrary* $\eta(k)$, the only way that Eq. (11-20) can be satisfied is that the two components of the equation be individually zero.

Equation (11-21) is called the *discrete Euler-Lagrange equation* which is the necessary condition that must be satisfied for J_c to be an extremum (maximum or minimum). The equation in Eq. (11-22) is known as the *transversality condition*, or simply the boundary condition which is needed to solve the partial differential equations in Eq. (11-21).

Now referring to the two additional conditions in Eqs. (11-16) and (11-17), we have for arbitrary $\mu(k)$ and $\omega(k+1)$,

$$\frac{\partial F_c^\circ(k)}{\partial \lambda_i^\circ(k+1)} = 0 \qquad i = 1,2,\ldots,n \tag{11-23}$$

$$\frac{\partial F_c^\circ(k)}{\partial u_j^\circ(k)} = 0 \qquad j = 1,2,\ldots,p \tag{11-24}$$

Equation (11-23) leads to

$$\mathbf{x}^\circ(k+1) = \mathbf{f}[\mathbf{x}^\circ(k), \mathbf{u}^\circ(k), k] \tag{11-25}$$

which implies that the state equation must satisfy the optimal trajectory. Equation (11-24), when applied to $F_c^\circ(k)$, gives the optimal control $\mathbf{u}^\circ(k)$ in terms of $\lambda^\circ(k+1)$.

In a majority of the design problems the initial state $\mathbf{x}(0)$ is given at the outset. Then, the perturbation to $\mathbf{x}(k)$ at $k = 0$ is zero since $\mathbf{x}(0)$ is fixed; thus, $\eta(0) = \mathbf{0}$. The transversality condition in Eq. (11-22) is reduced to

$$<\eta(k), \frac{\partial F_c^\circ(k-1)}{\partial \mathbf{x}^\circ(k)}>\bigg|_{k=N} = 0 \tag{11-26}$$

Furthermore, most optimal control design problems are classified according to the end conditions. For instance, if $\mathbf{x}(N)$ is given and fixed, the problem is defined as a *fixed-endpoint* design problem. On the other hand, if $\mathbf{x}(N)$ is free or belongs to a certain target set, we have a free-endpoint problem. The transversality condition in Eq. (11-26) should be applied according to the end conditions:

Fixed-endpoint: $\mathbf{x}(N) = \text{fixed}, \eta(N) = \mathbf{0}$

then nothing can be said about

$$\frac{\partial F_c^\circ(k-1)}{\partial \mathbf{x}^\circ(k)}\bigg|_{k=N}$$

and no transversality condition is needed to solve Eq. (11-21).

Free-endpoint: $\mathbf{x}(N) = \text{free}, \eta(N) \neq 0$

then

$$\frac{\partial F_c^\circ(k-1)}{\partial \mathbf{x}^\circ(k)}\bigg|_{k=N} = 0 \tag{11-27}$$

which is the transversality condition required to solve Eq. (11-21).

In many cases some components of $\mathbf{x}(N)$ are fixed and the others are free; then the transversality conditions discussed above can be applied accordingly.

The following example illustrates the application of the calculus of variation and the discrete Euler-Lagrange equation method to the design of digital control systems.

Example 11-1.

Find the optimal control $u^\circ(k)$, $k = 0,1,2,\ldots,10$, such that the performance index

$$J = \frac{1}{2}\sum_{k=0}^{10} [x^2(k) + 2u^2(k)] \tag{11-28}$$

is minimized, subject to the equality constraint

$$x(k+1) = x(k) + 2u(k) \tag{11-29}$$

(a) The initial state is $x(0) = 1$ and the final state is $x(11) = 0$.

(b) The initial state is $x(0) = 1$ but the final state $x(11)$ is free.

Solution: (a) We first form the adjoined performance index

$$J_c = \sum_{k=0}^{10} F_c[x(k), u(k)] \tag{11-30}$$

where

$$F_c[x(k), u(k)] = \frac{1}{2}[x^2(k) + 2u^2(k)] + \lambda(k+1)[x(k+1) - x(k) - 2u(k)]$$

$$= F_c(k) \tag{11-31}$$

Using Eq. (11-21), the discrete Euler-Lagrange equation is determined,

$$\lambda^\circ(k+1) - \lambda^\circ(k) - x^\circ(k) = 0 \tag{11-32}$$

From Eq. (11-23),

$$\frac{\partial F_c^\circ(k)}{\partial \lambda^\circ(k+1)} = x^\circ(k+1) - x^\circ(k) - 2u^\circ(k) = 0 \tag{11-33}$$

which is equality constraint or the state equation in Eq. (11-29) under the optimal condition. The optimal control is determined from Eq. (11-24),

$$\frac{\partial F_c^\circ(k)}{\partial u^\circ(k)} = 2u^\circ(k) - 2\lambda^\circ(k+1) = 0 \tag{11-34}$$

Thus,

$$u^\circ(k) = \lambda^\circ(k+1) \tag{11-35}$$

After substituting Eq. (11-35) into Eq. (11-33), Eqs. (11-32) and (11-33) form a set of two simultaneous first-order difference equations which must be solved to give $\lambda^\circ(k+1)$ for substitution in Eq. (11-35). In the present case, since $x(0)$ and $x(11)$ are fixed, these values provide the two necessary boundary conditions for solving Eqs. (11-32) and (11-33), and the transversality condition of Eq. (11-26) is not needed.

The two simultaneous difference equations to be solved are:

$$\lambda^\circ(k+1) - \lambda^\circ(k) - x^\circ(k) = 0 \tag{11-36}$$

$$x^\circ(k+1) - 2\lambda^\circ(k+1) - x^\circ(k) = 0 \tag{11-37}$$

with $x(0) = 1$ and $x(11) = 0$. These equations may be solved by the z-transform method or the state transition method, and the solutions are:

$$x^\circ(k) = 0.289[2.732 + 2\lambda^\circ(0)](3.732)^k + 0.289[0.732 - 2\lambda^\circ(0)](0.268)^k \tag{11-38}$$

$$\lambda^\circ(k) = [0.289 + 0.211\lambda^\circ(0)](3.732)^k + [-0.289 + 0.789\lambda^\circ(0)](0.268)^k \tag{11-39}$$

The initial value of $\lambda^\circ(k)$ is found by substituting $x^\circ(11) = 0$ into Eq. (11-38),

$$\lambda^\circ(0) = -1.366 \tag{11-40}$$

Then, the optimal trajectory of $x^\circ(k)$ is described by

$$x^\circ(k) = (0.268)^k \tag{11-41}$$

and the optimal control is

$$\begin{aligned} u^\circ(k) &= -2.732(0.268)^{k+1} \\ &= -0.732x^\circ(k) \end{aligned} \tag{11-42}$$

Strictly, when $k = 11$, $x^\circ(11)$ should be equal to zero; however, due to the numerical nature of the problem, an error of $(0.268)^{11}$ is generated.

(b) When $x(0) = 1$ and $x(11)$ is free, we need to use the transversality condition:

$$\left.\frac{\partial F_c^\circ(k-1)}{\partial x^\circ(k)}\right|_{k=11} = 0 \tag{11-43}$$

The last condition leads to

$$\lambda^\circ(11) = 0 \tag{11-44}$$

Now substituting Eq. (11-44) into Eq. (11-39), we get $\lambda^\circ(0) = -1.366$ and the same results as in Eqs. (11-41) and (11-42).

The reason that the same results are obtained when the endpoint $x^\circ(11)$ is fixed at 0 or free is because the performance index already includes a constraint on $x(k)$, so that the design naturally forces $x(k)$ to go to zero as quickly as possible. In general, of course, when $x^\circ(N)$ is not fixed it may reach any final value.

In summary, the calculus of variation method requires the solution of the discrete Euler-Lagrange equation. However, for an nth-order system the Euler-Lagrange equation is of the order 2n. This means that solution of the Euler-Lagrange equation is usually a tedious task.

Although controllability was not mentioned in the formulation of the design using the discrete Euler-Lagrange equation, it is apparent that state controllability is required when $\mathbf{x}(N)$ is specified and when N is a finite integer.

11.2 THE DISCRETE MAXIMUM (MINIMUM) PRINCIPLE

The maximum principle (or minimum principle) introduced by Pontryagin is a very powerful method of solving a wide class of continuous-data control systems. The design by maximum principle is also based on the calculus of variation, but the mechanics are more refined and elegant than the use of the Euler-Lagrange equation. The discrete maximum principle may be considered as an extension to the design of digital control systems. Strictly speaking, the application of the discrete maximum principle requires the investigation of the condition of convexity of the system. The presentation of the subject in this section is intended only for a working knowledge of the method so that abstract mathematics are kept to a minimum.

The design problem can be stated as:

Find the optimal control $\mathbf{u}^\circ(k)$ over $[0, N]$ such that the performance index

$$J = G[\mathbf{x}(N), N] + \sum_{k=0}^{N-1} F[\mathbf{x}(k), \mathbf{u}(k), k] \tag{11-45}$$

is minimized, subject to the equality constraint,

$$\mathbf{x}(k+1) = \mathbf{f}[\mathbf{x}(k), \mathbf{u}(k), k] \tag{11-46}$$

The term $G[\mathbf{x}(N), N]$, in Eq. (11-45), is the terminal cost of the performance index. It is required as a terminal constraint on the end condition only if $\mathbf{x}(N)$ is not fixed.

As with the Lagrange multiplier, we define a *costate* vector $\mathbf{p}(k)$ $(n \times 1)$. Then, the optimization problem is equivalent to minimizing

$$J_c = G[\mathbf{x}(N), N] + \sum_{k=0}^{N-1} \Big[F[\mathbf{x}(k), \mathbf{u}(k), k] - \langle \mathbf{p}(k+1), [\mathbf{x}(k+1) - \mathbf{f}(\mathbf{x}, \mathbf{u}, k)] \rangle \Big] \tag{11-47}$$

Let us define the scalar function $H[\mathbf{x}(k), \mathbf{u}(k), \mathbf{p}(k+1), k]$ as the Hamiltonian, such that

$$H[\mathbf{x}(k), \mathbf{u}(k), \mathbf{p}(k+1), k] = F[\mathbf{x}(k), \mathbf{u}(k), k] - \langle \mathbf{p}(k+1), \mathbf{f}[\mathbf{x}(k), \mathbf{u}(k), k] \rangle \tag{11-48}$$

When the Hamiltonian is defined as in the last equation, it forms the basis of the discrete maximum principle. For discrete minimum principle, the Hamiltonian is defined as

$$H[\mathbf{x}(k), \mathbf{u}(k), \mathbf{p}(k+1), k] = F[\mathbf{x}(k), \mathbf{u}(k), k] + \langle \mathbf{p}(k+1), \mathbf{f}[\mathbf{x}(k), \mathbf{u}(k), k] \rangle \tag{11-49}$$

As we shall see later in the discussions, maximum principle refers to the property that the Hamiltonian is a maximum along the optimal trajectory, whereas for the minimum principle the Hamiltonian is a minimum.

Substituting the definition of the Hamiltonian of Eq. (11-49) into Eq.

(11-47), we have

$$J_c = G[\mathbf{x}(N), N] + \sum_{k=0}^{N-1} \Big[H[\mathbf{x}(k), \mathbf{u}(k), \mathbf{p}(k+1), k] - <\mathbf{p}(k+1), \mathbf{x}(k+1)> \Big] \tag{11-50}$$

which is for the minimum principle.

Let $\mathbf{x}(k)$, $\mathbf{x}(k+1)$ and $\mathbf{u}(k)$ take on variations as follows:

$$\mathbf{x}(k) = \mathbf{x}^\circ(k) + \epsilon\eta(k) \qquad (n \times 1) \tag{11-51}$$

$$\mathbf{x}(k+1) = \mathbf{x}^\circ(k+1) + \epsilon\eta(k+1) \qquad (n \times 1) \tag{11-52}$$

$$\mathbf{u}(k) = \mathbf{u}^\circ(k) + \delta\mu(k) \qquad (p \times 1) \tag{11-53}$$

$$\mathbf{p}(k+1) = \mathbf{p}^\circ(k+1) + \gamma\omega(k+1) \qquad (n \times 1) \tag{11-54}$$

Equation (11-50) is now written

$$J_c = G[\mathbf{x}^\circ(N) + \epsilon\eta(N), N]$$

$$+ \sum_{k=0}^{N-1} H[\mathbf{x}^\circ(k) + \epsilon\eta(k), \mathbf{u}^\circ(k) + \delta\mu(k), \mathbf{p}^\circ(k+1) + \gamma\omega(k+1), k]$$

$$- <\mathbf{p}^\circ(k+1) + \gamma\omega(k+1), \mathbf{x}^\circ(k+1) + \epsilon\eta(k+1)> \tag{11-55}$$

Expanding $G[\mathbf{x}(N), N]$ into a Taylor series about $G[\mathbf{x}^\circ(N), N]$, we get

$$G[\mathbf{x}(N), N] = G[\mathbf{x}^\circ(N), N] + \epsilon<\eta(N), \frac{\partial G^\circ(N)}{\partial \mathbf{x}^\circ(N)}> + \ldots \tag{11-56}$$

Similarly, we expand $H[\mathbf{x}(k), \mathbf{u}(k), \mathbf{p}(k+1), k]$ into a Taylor series about $\mathbf{x}^\circ(k)$, $\mathbf{u}^\circ(k)$, $\mathbf{p}^\circ(k+1)$ and $\mathbf{x}^\circ(k+1)$,

$$H[\mathbf{x}(k), \mathbf{u}(k), \mathbf{p}(k+1), k] = H[\mathbf{x}^\circ(k), \mathbf{u}^\circ(k), \mathbf{p}^\circ(k+1), k] + \epsilon<\eta(k), \frac{\partial H^\circ(k)}{\partial \mathbf{x}^\circ(k)}>$$

$$+ \delta<\mu(k), \frac{\partial H^\circ(k)}{\partial \mathbf{u}^\circ(k)}>$$

$$+ \gamma<\omega(k+1), \frac{\partial H^\circ(k)}{\partial \mathbf{p}^\circ(k+1)}> + \ldots \tag{11-57}$$

where

$$H^\circ(k) = H[\mathbf{x}^\circ(k), \mathbf{u}^\circ(k), \mathbf{p}^\circ(k+1), k] \tag{11-58}$$

Substituting, into Eq. (11-55), the Taylor series expansions of $G[\mathbf{x}(N), N]$ and $H[\mathbf{x}(k), \mathbf{u}(k), \mathbf{p}(k+1), k]$ and carrying out the following necessary conditions for minimum J_c:

$$\left.\frac{\partial J_c}{\partial \epsilon}\right|_{\epsilon=\delta=\gamma=0} = 0 \tag{11-59}$$

$$\left.\frac{\partial J_c}{\partial \delta}\right|_{\epsilon=\delta=\gamma=0} = 0 \tag{11-60}$$

$$\left.\frac{\partial J_c}{\partial \gamma}\right|_{\epsilon=\delta=\gamma=0} = 0 \tag{11-61}$$

we have,

$$<\eta(N), \frac{\partial G^\circ(N)}{\partial \mathbf{x}^\circ(N)}> + \sum_{k=0}^{N-1} <\eta(k), \frac{\partial H^\circ(k)}{\partial \mathbf{x}^\circ(k)}> - \sum_{k=0}^{N-1} <\mathbf{p}^\circ(k+1), \eta(k+1)> = 0 \tag{11-62}$$

$$<\mu(k), \frac{\partial H^\circ(k)}{\partial \mu^\circ(k)}> = 0 \tag{11-63}$$

$$<\omega(k+1), \frac{\partial H^\circ(k)}{\partial \mathbf{p}^\circ(k+1)} - \mathbf{x}^\circ(k+1)> = 0 \tag{11-64}$$

Equation (11-64) leads to

$$\frac{\partial H^\circ(k)}{\partial \mathbf{p}^\circ(k+1)} = \mathbf{x}^\circ(k+1) \tag{11-65}$$

which is recognized as the original state equation, Eq. (11-46). Equation (11-63) gives

$$\frac{\partial H^\circ(k)}{\partial \mathbf{u}^\circ(k)} = 0 \tag{11-66}$$

which has the significance that the Hamiltonian has an extremum along the optimal trajectory with respect to the optimal control.

The last term on the left-hand side of Eq. (11-62) is written as

$$\begin{aligned}\sum_{k=0}^{N-1} <\mathbf{p}^\circ(k+1), \eta(k+1)> &= \sum_{k=1}^{N} <\mathbf{p}^\circ(k), \eta(k)> \\ &= \sum_{k=0}^{N-1} <\mathbf{p}^\circ(k), \eta(k)> + <\mathbf{p}^\circ(N), \eta(N)> \\ &\quad - <\mathbf{p}^\circ(0), \eta(0)>\end{aligned} \tag{11-67}$$

Since $\mathbf{x}(0)$ is given, $\eta(0) = 0$; the last equation becomes

$$\sum_{k=0}^{N-1} <\mathbf{p}^\circ(k+1), \eta(k+1)> = \sum_{k=0}^{N-1} <\mathbf{p}^\circ(k), \eta(k)> + <\mathbf{p}^\circ(N), \eta(N)> \qquad (11\text{-}68)$$

After substitution of the last expression into Eq. (11-62) and rearranging, we have

$$< \frac{\partial G^\circ(N)}{\partial \mathbf{x}^\circ(N)} - \mathbf{p}^\circ(N), \eta(N)> + \sum_{k=0}^{N-1} < \frac{\partial H^\circ(k)}{\partial \mathbf{x}^\circ(k)} - \mathbf{p}^\circ(k), \eta(k)> = 0 \qquad (11\text{-}69)$$

Since the variations are mutually independent, the only way to satisfy the last equation is

$$\frac{\partial G^\circ(N)}{\partial \mathbf{x}^\circ(N)} = \mathbf{p}^\circ(N) \qquad (11\text{-}70)$$

$$\frac{\partial H^\circ(k)}{\partial \mathbf{x}^\circ(k)} = \mathbf{p}^\circ(k) \qquad (11\text{-}71)$$

In summarizing the results, the necessary condition for J_c to be an extremum is that

$$\frac{\partial H^\circ(k)}{\partial \mathbf{x}^\circ(k)} = \mathbf{p}^\circ(k) \qquad (11\text{-}72)$$

$$\frac{\partial H^\circ(k)}{\partial \mathbf{p}^\circ(k+1)} = \mathbf{x}^\circ(k+1) \qquad (11\text{-}73)$$

$$\frac{\partial H^\circ(k)}{\partial \mathbf{u}^\circ(k)} = \mathbf{0} \qquad (11\text{-}74)$$

$$\frac{\partial G^\circ(N)}{\partial \mathbf{x}^\circ(N)} = \mathbf{p}^\circ(N) \qquad (11\text{-}75)$$

Equations (11-72) and (11-73) represent 2n first-order difference equations, which are defined as the *canonical state equations*. Equation (11-74) gives the relation for the optimal control $\mathbf{u}^\circ(k)$, and Eq. (11-75) gives the transversality condition, which should be used when $\mathbf{x}(N)$ is not fixed. If any component of $\mathbf{x}(N)$ is fixed, the corresponding transversality condition of $\mathbf{p}^\circ(N)$ does not apply.

The following example illustrates the application of the discrete minimum principle to the design of a digital control system.

Example 11-2.

Find the optimal control $u^\circ(k)$, $k = 0,1,2,\ldots,10$, such that the performance index

$$J = \frac{1}{2} \sum_{k=0}^{10} [x^2(k) + 2u^2(k)] \tag{11-76}$$

is minimized, subject to the equality constraint

$$x(k+1) = x(k) + 2u(k) \tag{11-77}$$

The initial state is $x(0) = 1$ and the final state is $x(11) = 0$. Notice that this is the same design problem stated in part (a) of Example 11-1.

The first step in the discrete minimum principle design is to define the Hamiltonian, according to Eq. (11-49),

$$H[\mathbf{x}(k), \mathbf{u}(k), \mathbf{p}(k+1)] = \frac{1}{2}[x^2(k) + 2u^2(k)] + p(k+1)[x(k) + 2u(k)] \tag{11-78}$$

The canonical state equations are obtained from Eqs. (11-72) and (11-73). These are:

$$p^\circ(k+1) - p^\circ(k) = -x^\circ(k) \tag{11-79}$$

$$x^\circ(k+1) - x^\circ(k) = 2u^\circ(k) \tag{11-80}$$

From Eq. (11-74), the optimal control is found from Eq. (11-74),

$$u^\circ(k) = -p^\circ(k+1) \tag{11-81}$$

Since the endpoint $x(11)$ is fixed, the transversality condition of Eq. (11-75) is not needed.

The solutions of Eqs. (11-79), (11-80) and (11-81) are

$$x^\circ(k) = 0.289[2.732 - 2p^\circ(0)](3.732)^k + 0.289[0.732 + 2p^\circ(0)](0.268)^k \tag{11-82}$$

$$p^\circ(k) = [-0.289 + 0.211p^\circ(0)](3.732)^k + [0.289 + 0.789p^\circ(0)](0.268)^k \tag{11-83}$$

These equations are very similar to the solutions in Eqs. (11-38) and (11-39) except for a few differences in signs. Substitution of $x^\circ(11) = 0$ in Eq. (11-82) yields the result $p^\circ(0) = 1.366$. Thus,

$$x^\circ(k) = (0.268)^k$$

and

$$u^\circ(k) = -2.732(0.268)^{k+1}$$

which are the same results as in Example 11-1, part (a).

Now let us make $x(11)$ a free-end condition, and add a terminal cost to the performance index of Eq. (11-76),

$$J = \frac{1}{2}x^2(11) + \frac{1}{2} \sum_{k=0}^{10} [x^2(k) + 2u^2(k)] \tag{11-84}$$

The transversality condition of Eq. (11-75) gives

$$x^\circ(11) = p^\circ(11) \tag{11-85}$$

Now equating Eqs. (11-82) and (11-83) at k = 11 again results in

$$p^\circ(0) = 1.366$$

and the same results are obtained for $x^\circ(k)$ and $u^\circ(k)$.

It should be pointed out that the use of the terminal cost of the performance index of the minimum principle design of free-endpoint problems may yield different results than the calculus of variation method.

11.3 TIME-OPTIMAL CONTROL WITH ENERGY CONSTRAINT

In general, the time-optimal control design can be described as the problem of bringing $\mathbf{x}(0)$ to $\mathbf{x}(N)$ in minimum time. The control is subject to the amplitude constraint $|u(kT)| \leqslant U$. In this section we shall present a time-optimal control design with a quadratic constraint on the control. The design is carried out by use of the discrete minimum principle. The development will also pave the way for the discussions on the linear digital regulator design which are presented in the next chapter. The design problem is stated as:

The digital system

$$\mathbf{x}(k + 1) = \mathbf{A}\mathbf{x}(k) + \mathbf{B}\mathbf{u}(k) \tag{11-86}$$

where $\mathbf{x}(k)$ is $n \times 1$, $\mathbf{u}(k)$ is $p \times 1$, and $\mathbf{A}$ is nonsingular is given. It is assumed that the pair $[\mathbf{A},\mathbf{B}]$ is completely controllable. Find the optimal control $\mathbf{u}^\circ(k)$, $k = 0,1,2,\ldots,N-1$, such that the initial state $\mathbf{x}(0)$ is driven to the final state $\mathbf{x}(N) = \mathbf{0}$, subject to the control constraint

$$J = \frac{1}{2}\sum_{k=0}^{N-1} \mathbf{u}'(k)\mathbf{R}\mathbf{u}(k) = \text{minimum} \tag{11-87}$$

where $\mathbf{R}$ is symmetric and positive definite.

The performance index in Eq. (11-87) is generally known as the quadratic form, and in this case it represents an energy constraint on the designed system. Therefore, the performance index J in Eq. (11-87) represents an alternate, although not equivalent, constraint on the control $\mathbf{u}(k)$.

The solution of the problem begins by defining the Hamiltonian function

$$H[\mathbf{x}(k), \mathbf{p}(k + 1), \mathbf{u}(k)] = \frac{1}{2}\langle\mathbf{u}(k), \mathbf{R}\mathbf{u}(k)\rangle + \langle\mathbf{p}(k + 1), \mathbf{x}(k + 1)\rangle \tag{11-88}$$

The necessary conditions for J to be a minimum are:

$$\frac{\partial H^\circ(k)}{\partial \mathbf{x}^\circ(k)} = \mathbf{p}^\circ(k) = \mathbf{A}'\mathbf{p}^\circ(k + 1) \tag{11-89}$$

$$\frac{\partial H^\circ(k)}{\partial \mathbf{p}^\circ(k + 1)} = \mathbf{x}^\circ(k + 1) = \mathbf{A}\mathbf{x}^\circ(k) + \mathbf{B}\mathbf{u}^\circ(k) \tag{11-90}$$

$$\frac{\partial H^\circ(k)}{\partial \mathbf{u}^\circ(k)} = \mathbf{R}\mathbf{u}^\circ(k) + \mathbf{B}'\mathbf{p}^\circ(k+1) = \mathbf{0} \tag{11-91}$$

where

$$H^\circ(k) = \frac{1}{2} \langle \mathbf{u}^\circ(k), \mathbf{R}\mathbf{u}^\circ(k) \rangle + \langle \mathbf{p}^\circ(k+1), \mathbf{A}\mathbf{x}^\circ(k) + \mathbf{B}\mathbf{u}^\circ(k) \rangle \tag{11-92}$$

Since $\mathbf{x}(0)$ and $\mathbf{x}(N)$ are all fixed, the transversality condition is not needed.

The optimal control is obtained from Eq. (11-91),

$$\mathbf{u}^\circ(k) = -\mathbf{R}^{-1}\mathbf{B}'\mathbf{p}^\circ(k+1) \tag{11-93}$$

where the inverse of $\mathbf{R}$ exists since it is positive definite.

In the present problem the costate equation, Eq. (11-89), is not coupled to the state variable $\mathbf{x}(k)$. Thus, Eq. (11-89) can be solved by itself to give

$$\mathbf{p}^\circ(k) = (\mathbf{A}^{-k})'\mathbf{p}^\circ(0) \tag{11-94}$$

where we have assumed that $\mathbf{A}$ has an inverse.

Substitution of Eq. (11-93) into Eq. (11-90) gives

$$\mathbf{x}^\circ(k+1) = \mathbf{A}\mathbf{x}^\circ(k) - \mathbf{B}\mathbf{R}^{-1}\mathbf{B}'\mathbf{p}^\circ(k+1) \tag{11-95}$$

The solution of the last equation is

$$\mathbf{x}^\circ(N) = \mathbf{A}^N\mathbf{x}(0) - \sum_{k=0}^{N-1} \mathbf{A}^{N-k-1}\mathbf{B}\mathbf{R}^{-1}\mathbf{B}'(\mathbf{A}^{-1})'\mathbf{p}^\circ(k) \tag{11-96}$$

Setting $\mathbf{x}^\circ(N) = \mathbf{0}$ and solving for $\mathbf{x}(0)$ in the last equation, we get

$$\mathbf{x}(0) = \sum_{k=0}^{N-1} \mathbf{A}^{-k-1}\mathbf{B}\mathbf{R}^{-1}\mathbf{B}'(\mathbf{A}^{-k-1})'\mathbf{p}^\circ(0) \tag{11-97}$$

where Eq. (11-94) has been used.

It is interesting to note that the matrix

$$\mathbf{W} = \sum_{k=0}^{N-1} \mathbf{A}^{-k-1}\mathbf{B}\mathbf{R}^{-1}\mathbf{B}'(\mathbf{A}^{-k-1})' = \sum_{k=0}^{N-1} \mathbf{S}_k\mathbf{S}_k' \tag{11-98}$$

is the controllability matrix if $\mathbf{R}$ is an identity matrix [Refer to Eq. (9-11)]. For any $\mathbf{R}$ that is not an identity matrix, but still symmetric and positive definite, we let

$$\mathbf{R}^{-1} = \mathbf{K}\mathbf{K}' \tag{11-99}$$

where $\mathbf{K}$ is a $p \times p$ matrix. Then in Eq. (11-98),

$$\mathbf{S}_k = \mathbf{A}^{-k-1}\mathbf{B}\mathbf{K} \tag{11-100}$$

Since $\mathbf{R}$ is symmetric and positive definite, $\mathbf{R}^{-1}$ has the same properties. Equation (11-99) has a unique solution $\mathbf{K}$ which is symmetric and positive definite, since the equation is a simple form of the Riccati-type nonlinear matrix equation.

$$\mathbf{K}\mathbf{\Theta}\mathbf{Q}^{-1}\mathbf{\Theta}'\mathbf{K}' - \mathbf{K}\mathbf{\Phi} - \mathbf{\Phi}'\mathbf{K} - \mathbf{R}^{-1} = 0 \tag{11-101}$$

with $\mathbf{\Theta} = \mathbf{I}$, $\mathbf{Q}^{-1} = \mathbf{I}$, and $\mathbf{\Phi} = \mathbf{0}$.

The $n \times n$ matrix $\mathbf{W}$ in Eq. (11-98) is nonsingular if the matrix

$$[\mathbf{A}^{-1}\mathbf{B}\mathbf{K} \quad \mathbf{A}^{-2}\mathbf{B}\mathbf{K} \quad \ldots \quad \mathbf{A}^{-N}\mathbf{B}\mathbf{K}] \qquad (n \times pN) \tag{11-102}$$

is of rank n. The last matrix can be written as

$$\underset{(n \times pN)}{[\mathbf{A}^{-1}\mathbf{B} \quad \mathbf{A}^{-2}\mathbf{B} \quad \ldots \quad \mathbf{A}^{-N}\mathbf{B}]} \underset{(pN \times pN)}{\begin{bmatrix} \mathbf{K} & & & \\ & \mathbf{K} & & \\ & & \ddots & \\ & & & \mathbf{K} \end{bmatrix}} \tag{11-103}$$

The rank of the matrix in Eq. (11-102) is the same as the rank of $[\mathbf{A}^{-1}\mathbf{B} \quad \mathbf{A}^{-2}\mathbf{B} \quad \ldots \quad \mathbf{A}^{-N}\mathbf{B}]$ or of $[\mathbf{B} \quad \mathbf{A}\mathbf{B} \quad \ldots \quad \mathbf{A}^{N-1}\mathbf{B}]$, since $\mathbf{K}$ is positive definite. Therefore, if the pair $[\mathbf{A},\mathbf{B}]$ is completely controllable, so is $[\mathbf{A},\mathbf{B}\mathbf{K}]$.

We have now established that if $[\mathbf{A},\mathbf{B}]$ is completely controllable, $\mathbf{W}$ is nonsingular. From Eq. (11-97) we have

$$\mathbf{p}^\circ(0) = \mathbf{W}^{-1}\mathbf{x}(0) \tag{11-104}$$

The optimal control can be expressed in terms of the initial state $\mathbf{x}(0)$; from Eq. (11-93),

$$\begin{aligned} \mathbf{u}^\circ(k) &= -\mathbf{R}^{-1}\mathbf{B}'(\mathbf{A}^{-1})'\mathbf{p}(k) \\ &= -\mathbf{R}^{-1}\mathbf{B}'(\mathbf{A}^{-1})'(\mathbf{A}^{-k})'\mathbf{p}(0) \end{aligned} \tag{11-105}$$

Thus,

$$\mathbf{u}^\circ(k) = -\mathbf{R}^{-1}\mathbf{B}'(\mathbf{A}^{-k-1})'\mathbf{W}^{-1}\mathbf{x}(0) \tag{11-106}$$

Substituting the last equation into Eq. (11-87), we get, after simplification, the optimal performance index

$$J^\circ = \frac{1}{2}\,\mathbf{x}'(0)\mathbf{W}^{-1}\mathbf{x}(0) \tag{11-107}$$

This result is significant in the sense that the optimal performance index is dependent upon the initial state $\mathbf{x}(0)$. Since $\mathbf{W}$ depends only upon $\mathbf{A}$, $\mathbf{B}$ and $\mathbf{R}$, which are all given, Eq. (11-107) implies that once a maximum bound on J° is fixed, a region of controllable states can be established for $\mathbf{x}(0)$. In other words, Eq. (11-107) defines a domain in the state space for $\mathbf{x}(0)$ which can be brought to $\mathbf{x}(N) = \mathbf{0}$ for a given N and a given J°.

Example 11-3.

Given the digital system

$$\mathbf{x}(k+1) = \mathbf{A}\mathbf{x}(k) + \mathbf{B}u(k) \tag{11-108}$$

where

$$A = \begin{bmatrix} 0 & 1 \\ -0.5 & -0.2 \end{bmatrix} \qquad B = \begin{bmatrix} 0 \\ 1 \end{bmatrix}$$

the following parts are to be carried out in this problem.

(a) Find the constant state feedback gain matrix $\mathbf{G}$ such that $u(k) = -\mathbf{G}\mathbf{x}(k)$ will bring any initial state $\mathbf{x}(0)$ to $\mathbf{x}(N) = \mathbf{0}$, for $N = 2$. Determine the optimal control $u^\circ(k)$ for $k = 0,1$, and the optimal state trajectory $\mathbf{x}^\circ(k)$ when $\mathbf{x}(0) = [1 \quad 1]'$ and $\mathbf{x}^\circ(2) = \mathbf{0}$.

(b) The control is subject to amplitude constraint $|u(k)| \leqslant 1$. Determine the region of controllable states of $\mathbf{x}(0)$ in the state plane for $N \leqslant 2$, such that $\mathbf{x}(N) = \mathbf{0}$.

(c) Find the optimal control $u^\circ(k)$ which will bring the initial state $\mathbf{x}(0) = [1 \quad 1]'$ to $\mathbf{x}(2) = \mathbf{0}$, and simultaneously satisfy

$$J = \frac{1}{2}\sum_{k=0}^{1} u^2(k) = \text{minimum} \tag{11-109}$$

Determine the optimal trajectory $\mathbf{x}^\circ(k)$ and the optimal value of J.

(d) Determine the region of controllable states of $\mathbf{x}(0)$ in the state plane for $N = 2$ such that $\mathbf{x}(N) = \mathbf{0}$, and $J \leqslant 1$, with the J given in Eq. (11-109). Repeat the problem for $J \leqslant 0.25$.

Solution: (a) Setting $\mathbf{x}(2)$ to $\mathbf{0}$ in the solution of Eq. (11-108), it can be shown that the time optimal control is

$$u^\circ(k) = -[1 \quad 0][A^{-1}B \quad A^{-2}B]^{-1}\mathbf{x}(k) \tag{11-110}$$

Thus,

$$u^\circ(k) = [0.5 \quad 0.2]\mathbf{x}(k) \tag{11-111}$$

The optimal state feedback gain matrix is

$$\mathbf{G} = [-0.5 \quad -0.2] \tag{11-112}$$

For the given initial state, the following results are obtained:

$$u^\circ(0) = 0.7 \qquad \mathbf{x}^\circ(1) = \begin{bmatrix} 1 \\ 0 \end{bmatrix}$$

$$u^\circ(1) = 0.5 \qquad \mathbf{x}^\circ(2) = \begin{bmatrix} 0 \\ 0 \end{bmatrix}$$

The optimal state trajectory is shown in Fig. 11-1.

(b) The control is now subject to the amplitude constraint, $|u(k)| \leqslant 1$. For $N = 2$, the state transition equation is written

$$\mathbf{x}(2) = \mathbf{A}^2\mathbf{x}(0) + \mathbf{ABu}(0) + \mathbf{Bu}(1) = 0 \tag{11-113}$$

Solving for $\mathbf{x}(0)$ from the last equation, we have

$$\mathbf{x}(0) = \begin{bmatrix} 2 & -0.8 \\ 0 & 2 \end{bmatrix} \begin{bmatrix} u(0) \\ u(1) \end{bmatrix} \tag{11-114}$$

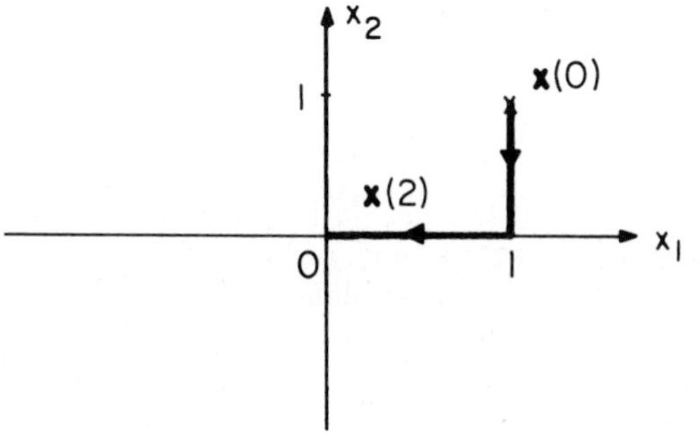

Fig. 11-1. Optimal state trajectory.

The vertices of the region of controllable states are found by substituting the four possible combinations of $u(k) = +1$ and -1 into Eq. (11-114). The results are:

$$u(0) = u(1) = 1 \qquad \mathbf{x}(0) = \begin{bmatrix} 1.2 \\ 2 \end{bmatrix} \qquad u(0) = -u(1) = 1 \qquad \mathbf{x}(0) = \begin{bmatrix} 2.8 \\ -2 \end{bmatrix}$$

$$u(0) = u(1) = -1 \qquad \mathbf{x}(0) = \begin{bmatrix} -1.2 \\ -2 \end{bmatrix} \qquad -u(0) = u(1) = 1 \qquad \mathbf{x}(0) = \begin{bmatrix} -2.8 \\ 2 \end{bmatrix}$$

The convex polygon bound by these four vertices is shown in Fig. 11-2.

(c) Referring to the performance index in Eq. (11-109), $R = 1$. The matrix $\mathbf{W}$ is determined using Eq. (11-98), for $N = 2$.

$$\mathbf{W} = \mathbf{A}^{-1}\mathbf{B}\mathbf{B}'(\mathbf{A}^{-1})' + \mathbf{A}^{-2}\mathbf{B}\mathbf{B}'(\mathbf{A}^{-2})' = \begin{bmatrix} 4.64 & -1.6 \\ -1.6 & 4 \end{bmatrix} \tag{11-115}$$

The optimal control is given by Eq. (11-106),

$$u^{\circ}(k) = -\mathbf{B}'(\mathbf{A}^{-k-1})'\mathbf{W}^{-1}\mathbf{x}(0) \tag{11-116}$$

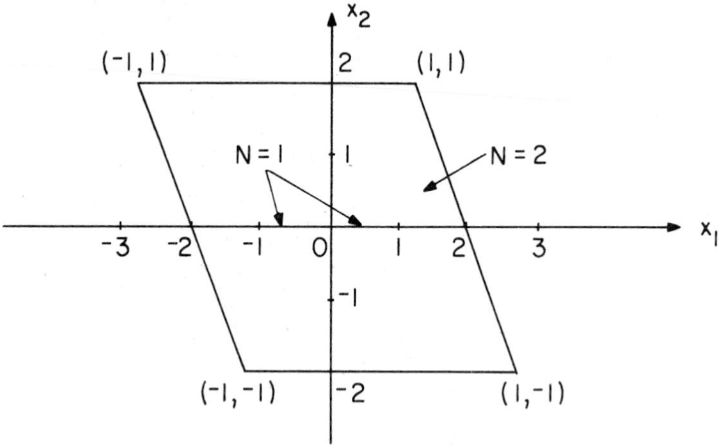

Fig. 11-2. Region of controllable states.

Thus, $u^{\circ}(0) = 0.7$ and $u^{\circ}(1) = 0.5$. Notice that these solutions are identical to those obtained in part (a). This is because as long as N is less than or equal to 2 for the second-order system, the solution for $\mathbf{x}(N) = \mathbf{0}$ is unique.

The optimal performance index is

$$J^{\circ} = \frac{1}{2}\mathbf{x}'(0)\mathbf{W}^{-1}\mathbf{x}(0) = 0.37 \tag{11-117}$$

(d) For $J \leqslant 1$, we have

$$\frac{1}{2}\mathbf{x}'(0)\mathbf{W}^{-1}\mathbf{x}(0) \leqslant 1 \tag{11-118}$$

or

$$0.125x_1^2(0) + 0.1x_1(0)x_2(0) + 0.145x_2^2(0) \leqslant 1 \tag{11-119}$$

The boundary described by the last expression is an ellipse in the state plane. The regions of controllable states for $J \leqslant 1$ and $J \leqslant 0.25$ are shown in Fig. 11-3.

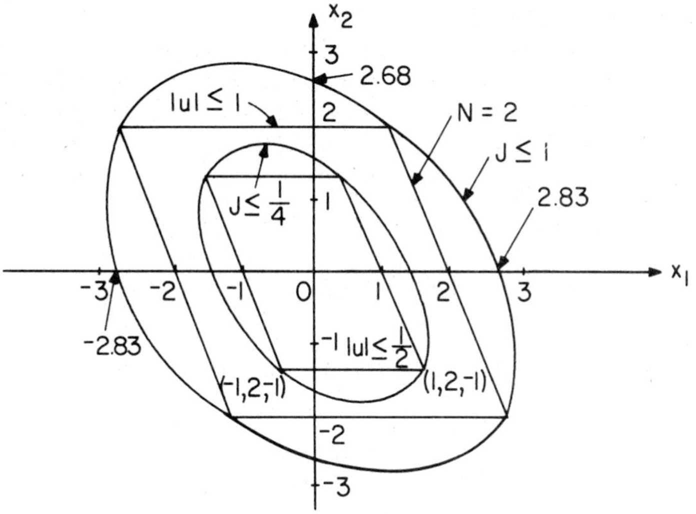

Fig. 11-3. Region of controllable states.

REFERENCES

1. Rozonoer, L. I., "The Maximum Principle of L. S. Pontryagin in Optimal System Theory," *Automation and Remote Control,* Parts I, II, III, Vol. 20, 1959.
2. Chang, S. S. L., "Digitized Maximum Principle," *Proc. IRE,* Vol. 48, December 1960, pp. 2030-2031.
3. Butkovskii, A. G., "The Necessary and Sufficient Conditions for Optimality of Discrete Control Systems," *Automation and Remote Control,* Vol. 24, August 1963, pp. 1056-1064.
4. Katz, S., "A Discrete Version of Pontryagin's Maximum Principle," *J. Electron. Control,* Vol. 13, No. 2, 1962, pp. 179-184.
5. Halkin, H., "Optimal Control for Systems Described by Difference Equations," *Advances in Control Systems,* Chapter 4, C. T. Leondes, ed., Academic Press, New York, 1964.
6. Sage, A. P., *Optimum System Control.* Prentice-Hall, Englewood Cliffs, N.J., 1968.

PROBLEMS

11-1 For a linear digital control process

$$\mathbf{x}(k+1) = \mathbf{A}\mathbf{x}(k) + \mathbf{B}\mathbf{u}(k)$$

where $\mathbf{A}$ is $n \times n$ and is nonsingular, show that the necessary conditions for minimizing the performance index

$$J = \frac{1}{2}\mathbf{x}'(N)\mathbf{S}\mathbf{x}(N) + \frac{1}{2}\sum_{k=0}^{N-1}[\mathbf{x}'(k)\mathbf{Q}\mathbf{x}(k) + \mathbf{u}'(k)\mathbf{R}\mathbf{u}(k)]$$

where $\mathbf{S}$ and $\mathbf{Q}$ are symmetric positive semidefinite matrices and $\mathbf{R}$ is a symmetric positive definite matrix, are

$$\mathbf{p}(k+1) = (\mathbf{A}')^{-1}\mathbf{p}(k) - (\mathbf{A}')^{-1}\mathbf{Q}\mathbf{x}(k)$$

$$\mathbf{x}(k+1) = \mathbf{A}\mathbf{x}(k) - \mathbf{B}\mathbf{R}^{-1}\mathbf{B}'\mathbf{p}(k+1)$$

11-2 The dynamic equations of a permanent-magnet dc motor are

$$v(t) = Ri(t) + L\frac{di(t)}{dt} + K_b\omega(t)$$

$$T_m(t) = K_i i(t) = J\frac{d\omega(t)}{dt}$$

where

R = armature resistance = 0.45 ohm

L = armature inductance = 0.05 h

K_b = back emf constant = 1.165 volts/rad/sec

K_i = torque constant = 0.858 ft-lb/amp

J = motor and load inertia = 0.05 ft-lb/rad/sec

$i(t)$ = armature current (amp)

$T_m(t)$ = motor torque (ft-lb)

$\omega(t)$ = motor angular velocity (rad/sec)

$v(t)$ = applied voltage = $v(kT)$ $\quad kT \leqslant t < (k+1)T$
(output of zero-order hold)

T = sampling period = 0.5 sec

Write the state equations in the form of

$$\mathbf{x}[(k+1)T] = \phi(T)\mathbf{x}(kT) + \theta(T)v(kT)$$

where

$$\mathbf{x}(kT) = \begin{bmatrix} i(kT) \\ \omega(kT) \end{bmatrix}$$

For $\mathbf{x}(0) = [0 \quad 1]'$ find the optimal control v(kT) for k = 0,1,...,9, such that

$$J_{10} = \frac{1}{2}\sum_{k=0}^{9} [x_1^2(kT) + x_2^2(kT) + v^2(kT)] = \min$$

11-3 Given the digital control system

$$\mathbf{x}(k+1) = \mathbf{A}\mathbf{x}(k) + \mathbf{B}u(k)$$

where

$$\mathbf{A} = \begin{bmatrix} 0 & 1 \\ -1 & 1 \end{bmatrix} \qquad \mathbf{B} = \begin{bmatrix} 0 \\ 1 \end{bmatrix} \qquad \mathbf{x}(0) = \begin{bmatrix} 1 \\ 1 \end{bmatrix}$$

Find the optimal control sequence u(0), u(1) and u(2) so that the performance index

$$J_3 = \frac{1}{2}x_1^2(3) + \frac{1}{2}\sum_{k=0}^{2} [x_1^2(k) + u^2(k)]$$

is minimized. The end condition is $x_1(3)$ = free and $x_2(3) = 0$.

11-4 The dynamic equations of a permanent-magnet step motor are written as

$$\frac{d\lambda_a(\theta)}{d\theta} = \frac{1}{2}\frac{v_a}{\omega(\theta)} - \frac{R}{2L\omega(\theta)}\lambda_a(\theta) + \frac{RK_1\cos\theta}{2L\omega(\theta)}$$

$$\frac{d\lambda_b(\theta)}{d\theta} = \frac{1}{2}\frac{v_b}{\omega(\theta)} - \frac{R}{2L\omega(\theta)}\lambda_b(\theta) + \frac{RK_1\sin\theta}{2L\omega(\theta)}$$

$$\frac{d\omega(\theta)}{d\theta} = -\frac{N_rK_T}{JL}\,\frac{\lambda_a(\theta)\sin\theta - \lambda_b(\theta)\cos\theta}{\omega(\theta)} - \frac{T_DN_r}{J\omega(\theta)} - \frac{BN_r}{J}$$

where

v_a, v_b = constant applied voltages

λ_a, λ_b = flux linkages of motor phases

L = inductance of each phase

R = resistance of each phase

N_r = number of rotor teeth

K_T = torque constant

K_1 = constant linking the flux between the permanent-magnet rotor an and the stator

ω = angular velocity of rotor shaft

θ = angular displacement of rotor shaft

J = inertia of rotor and load

B = viscous friction coefficient

T_D = Coulomb friction torque

The initial state is given as $\lambda_a(\theta_0) = \lambda_{a0}$, $\lambda_b(\theta_0) = \lambda_{b0}$ and $\omega(\theta_0) = \omega_0$. The performance index to be minimized is

$$J = \int_{\theta_0}^{\theta_f} \frac{1}{\omega(\theta)} d\theta$$

The controls v_a and v_b are subject to the constraints $-V \leqslant v_a \leqslant V$, $-V \leqslant v_b \leqslant V$.

(a) Discretize the state equations and the performance index by using

$$\frac{dx(\theta)}{d\theta} \cong \frac{1}{\Delta\theta} [x(k+1) - x(k)]$$

where $x(k)$ is a state variable evaluated at $\theta = k$.

(b) Set up the Hamiltonian, the necessary conditions for optimal control, and the transversality condition.

12

OPTIMAL LINEAR DIGITAL REGULATOR DESIGN

12.1 INTRODUCTION

One of the modern optimal control design techniques that has found general practical applications is the linear regulator design. A regulator problem is defined with reference to a system with zero reference inputs, and the design objective is to drive the states or outputs to the neighborhood of the equilibrium state. The condition of zero inputs is not a severe limitation to the design since the linear regulator design (infinite-time) assures that the resultant system is stable, and possesses certain damping characteristics, so that the performance of the system will be satisfactory in practice even if the inputs are nonzero.

In the following discussions we shall first formulate the performance index with reference to a linear continuous-data process with sampled data, and then with respect to a completely digital process.

The linear digital regulator problem may be stated as:

Given the linear system

$$\dot{\mathbf{x}}(t) = \mathbf{A}\mathbf{x}(t) + \mathbf{B}\mathbf{u}(t) \tag{12-1}$$

where $\mathbf{x}(t)$ is the $n \times 1$ state vector, $\mathbf{u}(t)$ is the $p \times 1$ control vector that is given by

$$\mathbf{u}(t) = \mathbf{u}(kT) \qquad kT \leqslant t < (k+1)T \tag{12-2}$$

Find the optimal control $\mathbf{u}^{\circ}(kT)$ for $k = 0,1,2,\ldots,N-1$, such that the quadratic performance index

$$J = \frac{1}{2}\langle \mathbf{x}(t_f), \mathbf{S}\mathbf{x}(t_f)\rangle + \frac{1}{2}\int_0^{t_f} [\langle \mathbf{x}(t), \mathbf{Q}\mathbf{x}(t)\rangle + \langle \mathbf{u}(t), \mathbf{R}\mathbf{u}(t)\rangle]\,dt \quad (12\text{-}3)$$

is minimized. In the last equation, $t_f = NT$, and

$\mathbf{S}$ = symmetric positive semidefinite matrix $(n \times n)$

$\mathbf{Q}$ = symmetric positive semidefinite matrix $(n \times n)$

$\mathbf{R}$ = symmetric positive definite matrix $(p \times p)$

We first discretize the system of Eq. (12-1) by forming the difference equation,

$$\mathbf{x}[(k+1)T] = \phi(T)\mathbf{x}(kT) + \theta(T)\mathbf{u}(kT) \quad (12\text{-}4)$$

where

$$\phi(T) = e^{\mathbf{A}T} \quad (12\text{-}5)$$

$$\theta(T) = \int_0^T \phi(T-\tau)\mathbf{B}\,d\tau \quad (12\text{-}6)$$

To discretize the performance index of Eq. (12-3), we write

$$J_N = \frac{1}{2}\langle \mathbf{x}(NT), \mathbf{S}\mathbf{x}(NT)\rangle + \frac{1}{2}\sum_{k=0}^{N-1}\int_{kT}^{(k+1)T} [\langle \mathbf{x}(t), \mathbf{Q}\mathbf{x}(t)\rangle + \langle \mathbf{u}(t), \mathbf{R}\mathbf{u}(t)\rangle]\,dt \quad (12\text{-}7)$$

The state transition equation of Eq. (12-1) is written for $t \geqslant kT$,

$$\mathbf{x}(t) = \phi(t-kT)\mathbf{x}(kT) + \theta(t-kT)\mathbf{u}(kT) \quad (12\text{-}8)$$

Then,

$$\mathbf{x}'(t) = \mathbf{x}'(kT)\phi'(t-kT) + \mathbf{u}'(kT)\theta'(t-kT) \quad (12\text{-}9)$$

Substitution of Eqs. (12-2), (12-8) and (12-9) into Eq. (12-7) yields

$$J_N = \frac{1}{2}\mathbf{x}'(NT)\mathbf{S}\mathbf{x}(NT) + \frac{1}{2}\sum_{k=0}^{N-1} [\mathbf{x}'(kT)\hat{\mathbf{Q}}(T)\mathbf{x}(kT) + 2\mathbf{x}'(kT)\mathbf{M}(T)\mathbf{u}(kT) + \mathbf{u}'(kT)\hat{\mathbf{R}}(T)\mathbf{u}(kT)] \quad (12\text{-}10)$$

where

$$\hat{\mathbf{Q}}(T) = \int_{kT}^{(k+1)T} \phi'(t - kT)\mathbf{Q}\phi(t - kT)dt \tag{12-11}$$

$$\mathbf{M}(T) = \int_{kT}^{(k+1)T} \phi'(t - kT)\mathbf{Q}\theta(t - kT)dt \tag{12-12}$$

$$\hat{\mathbf{R}}(T) = \int_{kT}^{(k+1)T} [\theta'(t - kT)\mathbf{Q}\theta(t - kT) + \mathbf{R}]\,dt \tag{12-13}$$

In view of the properties of $\mathbf{Q}$ and $\mathbf{R}$, we see that $\hat{\mathbf{Q}}(T)$ is symmetric and positive semidefinite; $\hat{\mathbf{R}}(T)$ is symmetric and positive definite. However, nothing can be said with regard to $\mathbf{M}(T)$.

The problem now is that given the digital system of Eq. (12-4), find the optimal control so that the performance index in Eq. (12-10) is minimized.

In general, the digital system may be described by the state equation of Eq. (12-4) at the outset; there is no special reason why the performance index should be as complex as that in Eq. (12-10), since it is not clear, in general, how the weighting matrix $\mathbf{M}$ should be chosen. It is more natural to define the quadratic performance index as

$$J_N = \frac{1}{2} <\mathbf{x}(N), \mathbf{S}\mathbf{x}(N)> + \frac{1}{2} \sum_{k=0}^{N-1} [<\mathbf{x}(k), \mathbf{Q}\mathbf{x}(k)> + <\mathbf{u}(k), \mathbf{R}\mathbf{u}(k)>] \tag{12-14}$$

where $\mathbf{Q}$ and $\mathbf{R}$ are weighting matrices with properties specified earlier, and the sampling period T has been dropped for convenience.

12.2 LINEAR DIGITAL REGULATOR DESIGN (FINITE-TIME PROBLEM)

The linear digital regulator design problem stated in the last section is to be solved by use of the discrete minimum principle.

The digital control process is described by

$$\mathbf{x}(k + 1) = \phi\mathbf{x}(k) + \theta\mathbf{u}(k) \tag{12-15}$$

with $\mathbf{x}(0)$ given. The design objective is to find $\mathbf{u}^\circ(k)$ so that the performance index

$$J_N = \frac{1}{2} <\mathbf{x}(N), \mathbf{S}\mathbf{x}(N)>$$

$$+ \frac{1}{2} \sum_{k=0}^{N-1} [\mathbf{x}(k), \hat{\mathbf{Q}}\mathbf{x}(k)> + <\mathbf{x}(k), 2\mathbf{M}\mathbf{u}(k)> + <\mathbf{u}(k), \hat{\mathbf{R}}\mathbf{u}(k)>] \tag{12-16}$$

is minimized.

We form the Hamiltonian,

$$H(k) = H[\mathbf{x}(k), \mathbf{p}(k+1), \mathbf{u}(k)]$$
$$= \frac{1}{2} \langle \mathbf{x}(k), \hat{\mathbf{Q}}\mathbf{x}(k) \rangle + \langle \mathbf{x}(k), \mathbf{M}\mathbf{u}(k) \rangle$$
$$+ \frac{1}{2} \langle \mathbf{u}(k), \hat{\mathbf{R}}\mathbf{u}(k) \rangle + \langle \mathbf{p}(k+1), \phi\mathbf{x}(k) + \theta\mathbf{u}(k) \rangle \tag{12-17}$$

The necessary conditions for J_N to be an extremum are:

$$\frac{\partial H^\circ(k)}{\partial \mathbf{x}^\circ(k)} = \mathbf{p}^\circ(k) = \hat{\mathbf{Q}}\mathbf{x}^\circ(k) + \phi'\mathbf{p}^\circ(k+1) + \mathbf{M}\mathbf{u}^\circ(k) \tag{12-18}$$

$$\frac{\partial H^\circ(k)}{\partial \mathbf{p}^\circ(k+1)} = \mathbf{x}^\circ(k+1) = \phi\mathbf{x}^\circ(k) + \theta\mathbf{u}^\circ(k) \tag{12-19}$$

$$\frac{\partial H^\circ(k)}{\partial \mathbf{u}^\circ(k)} = \mathbf{M}'\mathbf{x}^\circ(k) + \hat{\mathbf{R}}\mathbf{u}^\circ(k) + \theta'\mathbf{p}^\circ(k+1) = \mathbf{0} \tag{12-20}$$

In this case since $\mathbf{x}^\circ(N)$ is not specified, the transversality condition is

$$\frac{\partial G[\mathbf{x}(N), N]}{\partial \mathbf{x}(N)} = \frac{\partial}{\partial \mathbf{x}(N)} \left[\frac{1}{2} \langle \mathbf{x}(N), \mathbf{S}\mathbf{x}(N) \rangle \right] = \mathbf{S}\mathbf{x}(N) = \mathbf{p}(N) \tag{12-21}$$

The optimal control is obtained from Eq. (12-20),

$$\mathbf{u}^\circ(k) = -\hat{\mathbf{R}}^{-1} [\theta'\mathbf{p}^\circ(k+1) + \mathbf{M}'\mathbf{x}^\circ(k)] \tag{12-22}$$

Substituting the last equation into Eqs. (12-18) and (12-19), the canonical state equations are written

$$\mathbf{x}^\circ(k+1) = (\phi - \theta\hat{\mathbf{R}}^{-1}\mathbf{M}')\mathbf{x}^\circ(k) - \theta\hat{\mathbf{R}}^{-1}\theta'\mathbf{p}^\circ(k+1) \tag{12-23}$$

$$(\phi' - \mathbf{M}\hat{\mathbf{R}}^{-1}\theta')\mathbf{p}^\circ(k+1) = \mathbf{p}^\circ(k) - (\hat{\mathbf{Q}} - \mathbf{M}\hat{\mathbf{R}}^{-1}\mathbf{M}')\mathbf{x}^\circ(k) \tag{12-24}$$

These represent 2n difference equations which are to be solved with the known boundary conditions of $\mathbf{x}(0)$ and $\mathbf{p}^\circ(N) = \mathbf{S}\mathbf{x}^\circ(N)$. Notice that the two equations in Eqs. (12-23) and (12-24) are coupled in $\mathbf{x}^\circ(k)$ and $\mathbf{p}^\circ(k)$. These are more general than Eqs. (11-89) and (11-90), which are derived for a much simpler performance index. Since Eq. (11-89) is completely decoupled from $\mathbf{x}^\circ(k)$, it can be solved directly.

The coupled canonical state equations in Eqs. (12-23) and (12-24) are difficult to solve directly. However, it can be shown that the solution is of the form

$$\mathbf{p}(k) = \mathbf{K}(k)\mathbf{x}(k) \tag{12-25}$$

where $\mathbf{K}(k)$ is an $n \times n$ matrix with yet unknown properties, except that at $k = N$, from Eq. (12-21),

$$\mathbf{K}(N) = \mathbf{S} \tag{12-26}$$

Substituting Eq. (12-25) into Eq. (12-23) and rearranging terms, we get

$$\mathbf{x}^\circ(k+1) = [\mathbf{I} + \theta\hat{\mathbf{R}}^{-1}\theta'\mathbf{K}(k+1)]^{-1}(\phi - \theta\hat{\mathbf{R}}^{-1}\mathbf{M}')\mathbf{x}^\circ(k) \tag{12-27}$$

where it is assumed that the inverse of $[\mathbf{I} + \theta\hat{\mathbf{R}}^{-1}\theta'\mathbf{K}(k+1)]$ exists. Later it will be shown that $\mathbf{K}(k+1)$ is at least positive semidefinite so that the assertion is true. Similarly, substituting Eq. (12-25) into Eq. (12-24), we have

$$(\phi' - \mathbf{M}\hat{\mathbf{R}}^{-1}\theta')\mathbf{K}(k+1)\mathbf{x}^\circ(k+1) = [\mathbf{K}(k) - \hat{\mathbf{Q}} + \mathbf{M}\hat{\mathbf{R}}^{-1}\mathbf{M}']\mathbf{x}^\circ(k) \tag{12-28}$$

Now substitute Eq. (12-27) into Eq. (12-28); we get

$$(\phi' - \mathbf{M}\hat{\mathbf{R}}^{-1}\theta')\mathbf{K}(k+1)[\mathbf{I} + \theta\hat{\mathbf{R}}^{-1}\theta'\mathbf{K}(k+1)]^{-1}(\phi - \theta\hat{\mathbf{R}}^{-1}\mathbf{M}')\mathbf{x}^\circ(k)$$
$$= [\mathbf{K}(k) - \hat{\mathbf{Q}} + \mathbf{M}\hat{\mathbf{R}}^{-1}\mathbf{M}']\mathbf{x}^\circ(k) \tag{12-29}$$

For any $\mathbf{x}^\circ(k)$, the following equation must hold:

$$(\phi' - \mathbf{M}\hat{\mathbf{R}}^{-1}\theta')\mathbf{K}(k+1)[\mathbf{I} + \theta\hat{\mathbf{R}}^{-1}\theta'\mathbf{K}(k+1)]^{-1}(\phi - \theta\hat{\mathbf{R}}^{-1}\mathbf{M}')$$
$$= \mathbf{K}(k) - \mathbf{Q} + \mathbf{M}\hat{\mathbf{R}}^{-1}\mathbf{M}' \tag{12-30}$$

The last equation represents a nonlinear matrix difference equation in $\mathbf{K}(k)$ which is of the Riccati type, and is generally called the *discrete Riccati equation.* The $n \times n$ matrix $\mathbf{K}(k)$ is often referred to as the *Riccati gain.* The boundary condition for the Riccati equation is given in Eq. (12-26). In general, Eq. (12-30) contains n^2 scalar equations with the same number of unknowns in the elements of $\mathbf{K}(k)$. However, we shall show later that $\mathbf{K}(k)$ is symmetrical so that there are only $n(n+1)$ unknowns.

The optimal control is determined by substituting Eq. (12-25) into Eq. (12-22). Thus,

$$\mathbf{u}^\circ(k) = -[\mathbf{I} + \hat{\mathbf{R}}^{-1}\theta'\mathbf{K}(k+1)\theta]^{-1}\hat{\mathbf{R}}^{-1}[\theta'\mathbf{K}(k+1)\phi + \mathbf{M}']\mathbf{x}^\circ(k)$$
$$= -[\hat{\mathbf{R}} + \theta'\mathbf{K}(k+1)\theta]^{-1}[\theta'\mathbf{K}(k+1)\phi + \mathbf{M}']\mathbf{x}^\circ(k) \tag{12-31}$$

which is in the form of state feedback.

When $\mathbf{M} = \mathbf{0}$, $\hat{\mathbf{Q}} = \mathbf{Q}$ and $\hat{\mathbf{R}} = \mathbf{R}$, the Riccati equation in Eq. (12-30) becomes

$$\boldsymbol{\phi}'\mathbf{K}(k+1)[\mathbf{I}+\boldsymbol{\theta}\mathbf{R}^{-1}\boldsymbol{\theta}'\mathbf{K}(k+1)]^{-1}\boldsymbol{\phi}+\mathbf{Q}=\mathbf{K}(k) \tag{12-32}$$

The corresponding optimal control is given by

$$\mathbf{u}^{\circ}(k)=-[\mathbf{R}+\boldsymbol{\theta}'\mathbf{K}(k+1)\boldsymbol{\theta}]^{-1}\boldsymbol{\theta}'\mathbf{K}(k+1)\boldsymbol{\phi}\mathbf{x}^{\circ}(k) \tag{12-33}$$

Before embarking on the discussion of the various methods of solving the Riccati equation, let us investigate the important properties of the Riccati equation and the Riccati gain $\mathbf{K}(k)$.

The Riccati equation given in Eq. (12-30) is only one of many equivalent forms which satisfy the optimal linear discrete regulator design. In general, a particular form of the Riccati equation is best suited for a given purpose.

First we shall show that Eq. (12-30) is equivalent to the following form:

$$\mathbf{K}(k)=\boldsymbol{\phi}'\mathbf{K}(k+1)\boldsymbol{\phi}+\hat{\mathbf{Q}}$$
$$-[\boldsymbol{\theta}'\mathbf{K}(k+1)\boldsymbol{\phi}+\mathbf{M}']'[\hat{\mathbf{R}}+\boldsymbol{\theta}'\mathbf{K}(k+1)\boldsymbol{\theta}]^{-1}[\boldsymbol{\theta}'\mathbf{K}(k+1)\boldsymbol{\phi}+\mathbf{M}'] \tag{12-34}$$

The equivalence of the two Riccati equations in Eqs. (12-30) and (12-34) is not obvious. To show that these two equations are equivalent, we shall first prove the following identity:

$$\mathbf{K}(k+1)[\mathbf{I}+\boldsymbol{\theta}\hat{\mathbf{R}}^{-1}\boldsymbol{\theta}'\mathbf{K}(k+1)]^{-1}$$
$$=\mathbf{K}(k+1)-\mathbf{K}(k+1)\boldsymbol{\theta}[\hat{\mathbf{R}}+\boldsymbol{\theta}'\mathbf{K}(k+1)\boldsymbol{\theta}]^{-1}\boldsymbol{\theta}'\mathbf{K}(k+1) \tag{12-35}$$

We let

$$\mathbf{P}=\mathbf{K}(k+1)[\mathbf{I}+\boldsymbol{\theta}\hat{\mathbf{R}}^{-1}\boldsymbol{\theta}'\mathbf{K}(k+1)]^{-1} \tag{12-36}$$

Post-multiplying both sides of the last equation by $[\mathbf{I}+\boldsymbol{\theta}\hat{\mathbf{R}}^{-1}\boldsymbol{\theta}'\mathbf{K}(k+1)]\boldsymbol{\theta}$, we have

$$\mathbf{P}\boldsymbol{\theta}\hat{\mathbf{R}}^{-1}[\hat{\mathbf{R}}+\boldsymbol{\theta}'\mathbf{K}(k+1)\boldsymbol{\theta}]=\mathbf{K}(k+1)\boldsymbol{\theta} \tag{12-37}$$

We then post-multiply both sides of the last equation by $[\hat{\mathbf{R}}+\boldsymbol{\theta}'\mathbf{K}(k+1)\boldsymbol{\theta}]^{-1}\cdot\boldsymbol{\theta}'\mathbf{K}(k+1)$; the result is

$$\mathbf{P}\boldsymbol{\theta}\hat{\mathbf{R}}^{-1}\boldsymbol{\theta}'\mathbf{K}(k+1)=\mathbf{K}(k+1)\boldsymbol{\theta}[\hat{\mathbf{R}}+\boldsymbol{\theta}'\mathbf{K}(k+1)\boldsymbol{\theta}]^{-1}\boldsymbol{\theta}'\mathbf{K}(k+1) \tag{12-38}$$

Equation (12-36) indicates that the left-hand side of the last equation is $\mathbf{K}(k+1)-\mathbf{P}$. Thus, Eq. (12-38) is written

$$\mathbf{P}=\mathbf{K}(k+1)-\mathbf{K}(k+1)\boldsymbol{\theta}[\hat{\mathbf{R}}+\boldsymbol{\theta}'\mathbf{K}(k+1)\boldsymbol{\theta}]^{-1}\boldsymbol{\theta}'\mathbf{K}(k+1) \tag{12-39}$$

Comparing Eq. (12-39) with Eq. (12-36), we clearly have the identity in Eq. (12-35). Now substituting Eq. (12-35) into Eq. (12-30), we get

$$\mathbf{K}(k) - \hat{\mathbf{Q}} + \mathbf{M}\mathbf{R}^{-1}\mathbf{M}' = (\phi' - \mathbf{M}\hat{\mathbf{R}}^{-1}\theta')\Big[\mathbf{K}(k+1) - \mathbf{K}(k+1)\theta[\hat{\mathbf{R}} + \theta'\mathbf{K}(k+1)\theta]^{-1}\theta'\mathbf{K}(k+1)\Big](\phi - \theta\hat{\mathbf{R}}^{-1}\mathbf{M}') \tag{12-40}$$

The last expression may be regarded as still another form of the discrete Riccati equation. Equation (12-40) is further manipulated to show that it is equivalent to Eq. (12-34).

Multiplying out the terms in Eq. (12-40) gives

$$\mathbf{K}(k) - \hat{\mathbf{Q}} + \mathbf{M}\mathbf{R}^{-1}\mathbf{M}' = (\phi' - \mathbf{M}\hat{\mathbf{R}}^{-1}\theta')[\mathbf{K}(k+1)\phi - \mathbf{K}(k+1)\theta\hat{\mathbf{R}}^{-1}\mathbf{M}'] - (\phi' - \mathbf{M}\hat{\mathbf{R}}^{-1}\theta')\mathbf{K}(k+1)\theta[\hat{\mathbf{R}} + \theta'\mathbf{K}(k+1)\theta]^{-1}\theta'\mathbf{K}(k+1)(\phi - \theta\hat{\mathbf{R}}^{-1}\mathbf{M}') \tag{12-41}$$

$$\mathbf{K}(k) - \hat{\mathbf{Q}} + \mathbf{M}\hat{\mathbf{R}}^{-1}\mathbf{M}' = \phi'\mathbf{K}(k+1)\phi - \phi'\mathbf{K}(k+1)\theta\hat{\mathbf{R}}^{-1}\mathbf{M}' - \mathbf{M}\hat{\mathbf{R}}^{-1}\theta'\mathbf{K}(k+1)\phi + \mathbf{M}\hat{\mathbf{R}}^{-1}\theta'\mathbf{K}(k+1)\theta\hat{\mathbf{R}}^{-1}\mathbf{M}' - [\phi'\mathbf{K}(k+1)\theta - \mathbf{M}\hat{\mathbf{R}}^{-1}\theta'\mathbf{K}(k+1)\theta] \cdot [\hat{\mathbf{R}} + \theta'\mathbf{K}(k+1)\theta]^{-1}[\theta'\mathbf{K}(k+1)\phi - \theta'\mathbf{K}(k+1)\theta\hat{\mathbf{R}}^{-1}\mathbf{M}'] \tag{12-42}$$

The last term inside the bracket in Eq. (12-42) is written as

$$\theta'\mathbf{K}(k+1)\phi - \theta'\mathbf{K}(k+1)\theta\hat{\mathbf{R}}^{-1}\mathbf{M}' = \theta'\mathbf{K}(k+1)\phi - [\hat{\mathbf{R}} + \theta'\mathbf{K}(k+1)\theta]\hat{\mathbf{R}}^{-1}\mathbf{M}' + \mathbf{M}' \tag{12-43}$$

Substitution of the right-hand side of the last equation into Eq. (12-42), we get

$$\mathbf{K}(k) - \hat{\mathbf{Q}} + \mathbf{M}\hat{\mathbf{R}}^{-1}\mathbf{M}' = \phi'\mathbf{K}(k+1)\phi - \phi'\mathbf{K}(k+1)\theta\hat{\mathbf{R}}^{-1}\mathbf{M}' - \mathbf{M}\hat{\mathbf{R}}^{-1}\theta'\mathbf{K}(k+1)\phi + \mathbf{M}\hat{\mathbf{R}}^{-1}\theta'\mathbf{K}(k+1)\theta\hat{\mathbf{R}}^{-1}\mathbf{M}' - \phi'\mathbf{K}(k+1)\theta[\hat{\mathbf{R}} + \theta'\mathbf{K}(k+1)\theta]^{-1} \cdot [\mathbf{M}' + \theta'\mathbf{K}(k+1)\phi] + \phi'\mathbf{K}(k+1)\theta\hat{\mathbf{R}}^{-1}\mathbf{M}' - \mathbf{M}\hat{\mathbf{R}}^{-1}\theta'\mathbf{K}(k+1)\theta\hat{\mathbf{R}}^{-1}\mathbf{M}' + \mathbf{M}\hat{\mathbf{R}}^{-1}\theta'\mathbf{K}(k+1)\theta[\hat{\mathbf{R}} + \theta'\mathbf{K}(k+1)\theta]^{-1}[\mathbf{M}' + \theta'\mathbf{K}(k+1)\phi] \tag{12-44}$$

After cancellation of equal and opposite terms, and conditioning the last term, the last equation becomes

$$\mathbf{K}(k) - \hat{\mathbf{Q}} + \mathbf{M}\hat{\mathbf{R}}^{-1}\mathbf{M}' = \phi'\mathbf{K}(k+1)\phi - \mathbf{M}\hat{\mathbf{R}}^{-1}\theta'\mathbf{K}(k+1)\phi - \phi'\mathbf{K}(k+1)\theta[\hat{\mathbf{R}} + \theta'\mathbf{K}(k+1)\theta]^{-1}[\mathbf{M}' + \theta'\mathbf{K}(k+1)\phi] + \mathbf{M}\hat{\mathbf{R}}^{-1}[\theta'\mathbf{K}(k+1)\theta + \hat{\mathbf{R}} - \hat{\mathbf{R}}][\hat{\mathbf{R}} + \theta'\mathbf{K}(k+1)\theta]^{-1}[\mathbf{M}' + \theta'\mathbf{K}(k+1)\phi] \tag{12-45}$$

The last equation is finally simplified to

$$\mathbf{K}(k) = \phi'\mathbf{K}(k+1)\phi + \hat{\mathbf{Q}} - [\theta'\mathbf{K}'(k+1)\phi + \mathbf{M}']'[\hat{\mathbf{R}} + \theta'\mathbf{K}(k+1)\theta]^{-1} \cdot [\theta'\mathbf{K}(k+1)\phi + \mathbf{M}'] \tag{12-46}$$

which is identical to Eq. (12-34).

The following theorems concern the properties of the Riccati gain $\mathbf{K}(k)$.

Theorem 12-1. $\mathbf{K}(k)$ is a symmetric matrix.

Proof: Taking the transpose on both sides of the Riccati equation in Eq. (12-46), we have

$$\begin{aligned}\mathbf{K}'(k) &= \phi'\mathbf{K}'(k+1)\phi + \hat{\mathbf{Q}} \\ &\quad - [\theta'\mathbf{K}(k+1)\phi + \mathbf{M}']'[\hat{\mathbf{R}} + \theta'\mathbf{K}'(k+1)\theta]^{-1}[\phi'\mathbf{K}(k+1)\theta + \mathbf{M}]' \\ &= \phi'\mathbf{K}'(k+1)\phi + \hat{\mathbf{Q}} \\ &\quad - [\theta'\mathbf{K}(k+1)\phi + \mathbf{M}']'[\hat{\mathbf{R}} + \theta'\mathbf{K}'(k+1)\theta]^{-1}[\theta'\mathbf{K}'(k+1)\phi + \mathbf{M}']\end{aligned} \tag{12-47}$$

Since Eq. (12-47) is identical to Eq. (12-46) except that $\mathbf{K}(k)$ and $\mathbf{K}(k+1)$ are replaced by $\mathbf{K}'(k)$ and $\mathbf{K}'(k+1)$, respectively, these equations should have the same solution. Therefore,

$$\mathbf{K}(k) = \mathbf{K}'(k) \tag{12-48}$$

Theorem 12-2. The performance index J_N in Eq. (12-16) is for the time interval $[0, N]$, or a total of N stages if time is not involved as the independent variable. We define $J_{N-i}[\mathbf{x}(i)]$ as the performance index over the interval from i to N, or the last $N - i$ stages; i.e.,

$$J_{N-i}[\mathbf{x}(i)] = \frac{1}{2}<\mathbf{x}(N), \mathbf{S}\mathbf{x}(N)> + \frac{1}{2}\sum_{k=i}^{N-1}[<\mathbf{x}(k), \hat{\mathbf{Q}}\mathbf{x}(k)> + <\mathbf{x}(k), 2\mathbf{M}\mathbf{u}(k)> + <\mathbf{u}(k), \hat{\mathbf{R}}\mathbf{u}(k)>] \tag{12-49}$$

When $i = 0$,

$$J_N[\mathbf{x}(0)] = J_N \tag{12-50}$$

When $i = N$,

$$J_0[\mathbf{x}(N)] = \frac{1}{2}<\mathbf{x}(N), \mathbf{S}\mathbf{x}(N)> \tag{12-51}$$

which is the performance index over the last stage only.

The theorem states that

$$\underset{\mathbf{u}(i)}{\text{Min}}\ J_{N-i}[\mathbf{x}(i)] = \frac{1}{2}\mathbf{x}'(i)\mathbf{K}(i)\mathbf{x}(i) \tag{12-52}$$

where $\mathbf{K}(i)$ is the Riccati gain matrix.

The proof of this theorem is given in the process of derivation of the Riccati equation using the principle of optimality which is found in Sec. 12.4.

Theorem 12-3. The Riccati gain $\mathbf{K}(k)$ is at least positive semidefinite for $k = 0,1,2,\ldots,N$.

Proof: Since the performance index J_N is in a quadratic form, it is nonnegative. In view of Eq. (12-52), $\mathbf{K}(k)$ must be nonnegative.

Remarks on Controllability, Observability and Stability

For the finite-time (N = finite) linear digital regulator problem we do not require that the process be controllable, observable, or even stable. The performance index J_N can be finite for finite N even if an uncontrollable state is unstable. The design objective using the quadratic performance index is to drive any initial state $\mathbf{x}(0)$ to the equilibrium state $\mathbf{0}$ as closely as possible, but $\mathbf{x}(N)$ is not specified. Therefore, for finite N, stability is not a problem.

12.3 LINEAR DIGITAL REGULATOR DESIGN (INFINITE-TIME PROBLEM)

For infinite time or infinite number of stages, $N = \infty$, the performance index in Eq. (12-16) is written

$$J = \frac{1}{2}\sum_{k=0}^{\infty}[\langle\mathbf{x}(k), \hat{\mathbf{Q}}\mathbf{x}(k)\rangle + \langle\mathbf{x}(k), 2\mathbf{M}\mathbf{u}(k)\rangle + \langle\mathbf{u}(k), \hat{\mathbf{R}}\mathbf{u}(k)\rangle] \tag{12-53}$$

In this case the terminal cost is eliminated, since as N approaches infinity, the final state $\mathbf{x}(N)$ should approach the equilibrium state $\mathbf{0}$, so that the terminal constraint is no longer necessary.

One important requirement of the infinite-time linear regulator design is that the closed-loop system should be asymptotically stable. The following conditions are required with regard to the system modeled by Eq. (12-15):

1. The pair $[\phi, \theta]$ must be either completely controllable or stabilizable by state feedback (necessary condition).
2. The pair $[\phi, \mathbf{D}]$ must be completely observable, where $\mathbf{D}$ is any $n \times n$ matrix such that $\mathbf{D}\mathbf{D}' = \mathbf{Q}$ (sufficient condition).

Then, the solution to the infinite-time linear digital regulator problem can be obtained by setting $k \to -\infty$. As $N \to \infty$, the Riccati gain matrix $\mathbf{K}(k)$ becomes a constant matrix, i.e.,

$$\lim_{k \to -\infty} \mathbf{K}(k) = \mathbf{K} \tag{12-54}$$

Replacing $\mathbf{K}(k+1)$ and $\mathbf{K}(k)$ by $\mathbf{K}$ in Eq. (12-34), we have the steady-state or infinite-time Riccati equation

$$\mathbf{K} = \boldsymbol{\phi}'\mathbf{K}\boldsymbol{\phi} + \hat{\mathbf{Q}} - (\boldsymbol{\phi}'\mathbf{K}\boldsymbol{\theta} + \mathbf{M})(\hat{\mathbf{R}} + \boldsymbol{\theta}'\mathbf{K}\boldsymbol{\theta})^{-1}(\boldsymbol{\theta}'\mathbf{K}\boldsymbol{\phi} + \mathbf{M}') \tag{12-55}$$

This equation is often referred to as the *algebraic Riccati equation.*

The optimal control is, from Eq. (12-31),

$$\mathbf{u}^{\circ}(k) = -(\hat{\mathbf{R}} + \boldsymbol{\theta}'\mathbf{K}\boldsymbol{\theta})^{-1}(\boldsymbol{\theta}'\mathbf{K}\boldsymbol{\phi} + \mathbf{M}')\mathbf{x}^{\circ}(k) \tag{12-56}$$

In this case the feedback matrix

$$\mathbf{G} = (\hat{\mathbf{R}} + \boldsymbol{\theta}'\mathbf{K}\boldsymbol{\theta})^{-1}(\boldsymbol{\theta}'\mathbf{K}\boldsymbol{\phi} + \mathbf{M}') \tag{12-57}$$

is a constant matrix.

The optimal performance index for $N = \infty$ follows directly from Eq. (12-52),

$$J_{\infty}^{\circ} = \frac{1}{2}\mathbf{x}'(0)\mathbf{K}\mathbf{x}(0) \tag{12-58}$$

The controllability, stabilizability, and observability conditions given earlier in (1) and (2) require further discussion.

For the infinite-time regulator it is necessary that the process in Eq. (12-15) be either controllable or stabilizable by state feedback. Controllability is a stronger requirement than stabilizability, since an uncontrollable system can still be stabilized if the uncontrollable states are stable. However, just because the system in Eq. (12-15) is controllable or stabilizable does not mean that the closed-loop system designed by the optimal linear regulator theory is asymptotically stable. As it turns out, the observability condition stated in (2) should also be satisfied. Therefore, controllability and stabilizability are necessary conditions, whereas observability of $[\boldsymbol{\phi}, \mathbf{D}]$ is a sufficient condition. The following example and theorems will further amplify the ideas behind the requirements on controllability, observability, and stabilizability.

Example 12-1.

Consider the first-order digital process

$$x(k+1) = x(k) + u(k) \tag{12-59}$$

It is apparent that the process is controllable, unstable, but stabilizable. With the constant state feedback $u(k) = -Gx(k)$, the closed-loop system is asymptotically stable for $|1 - G| < 1$.

For the infinite-time linear regulator design, let us choose the performance index to be

$$J = \frac{1}{2} \sum_{k=0}^{\infty} u^2(k) \tag{12-60}$$

Then, $\phi = 1$, $\theta = 1$, $\hat{Q} = 0$, $M = 0$, and $\hat{R} = 1$. The optimal control which minimizes J is given by Eq. (12-56), reduced to

$$u^\circ(k) = -\frac{K}{1+K} x^\circ(k) \tag{12-61}$$

where K is the scalar constant Riccati gain which is the solution of the Riccati equation [Eq. (12-55)]:

$$K = K - \frac{K^2}{1+K} \tag{12-62}$$

The solution of the last equation is $K = 0$. Therefore, $u^\circ(k) = 0$ for all k, and the closed-loop system is not asymptotically stable.

The reason that the optimal linear regulator design does not provide an asymptotically stable system in this case is because $\hat{Q} = 0$, so that the state variable is not observed by the performance index. More important, it is the unstable state that is unobservable. To insure that all the states are observed by J, it is *sufficient* that $\hat{Q}$ is positive definite. However, in general, we may require the pair $[\phi, \mathbf{D}]$ to be completely observable, where $\mathbf{D}$ is any matrix such that $\mathbf{DD}' = \hat{\mathbf{Q}}$. Of course, if $\hat{\mathbf{Q}}$ is positive definite, we can always find a $\mathbf{D}$ which is a square matrix, then $[\phi, \mathbf{D}]$ is always observable or of rank n for any ϕ.

To see why the observability of $[\phi, \mathbf{D}]$ is important for the states to be reflected in J, let us consider only the homogeneous part of the state transition equation,

$$\mathbf{x}(k) = \phi(k)\mathbf{x}(0) \tag{12-63}$$

and let the performance index be

$$J = \sum_{k=0}^{\infty} \mathbf{x}'(k)\hat{\mathbf{Q}}\mathbf{x}(k) \tag{12-64}$$

Substitution of Eq. (12-63) into Eq. (12-64) yields

$$J = \sum_{k=0}^{\infty} \mathbf{x}'(0)\phi'(k)\hat{\mathbf{Q}}\phi(k)\mathbf{x}(0) \tag{12-65}$$

From Chapter 9 we learned that $\mathbf{x}(0)$ will be observable by J if the matrix

$$\sum_{k=0}^{\infty} \phi'(k)\hat{Q}\phi(k)$$

is nonsingular or positive definite. Therefore, letting $\mathbf{DD}' = \hat{\mathbf{Q}}$, we have the condition that $[\phi, \mathbf{D}]$ must be observable.

Theorem 12-4. For the performance index

$$J = \frac{1}{2}\sum_{k=0}^{\infty} [\mathbf{x}'(k)\mathbf{Q}\mathbf{x}(k) + \mathbf{u}'(k)\mathbf{R}\mathbf{u}(k)] \tag{12-66}$$

if $\mathbf{Q}$ and $\mathbf{R}$ are both positive definite, $\mathbf{K}$ is positive definite.

Proof: Since J is of the quadratic form, for positive definite $\mathbf{Q}$ and $\mathbf{R}$, J is positive. Since J and $\mathbf{K}$ are related through Eq. (12-58) it follows that $\mathbf{K}$ is positive definite.

Theorem 12-5. For the digital system in Eq. (12-15), if the performance index is

$$J = \frac{1}{2}\sum_{k=0}^{\infty} [\mathbf{x}'(k)\mathbf{Q}\mathbf{x}(k) + \mathbf{u}'(k)\mathbf{R}\mathbf{u}(k)] \tag{12-67}$$

where $\mathbf{Q}$ and $\mathbf{R}$ are both positive definite, then the optimal control which minimizes J,

$$\mathbf{u}^{\circ}(k) = -(\mathbf{R} + \theta'\mathbf{K}\theta)^{-1}\theta'\mathbf{K}\phi\mathbf{x}^{\circ}(k) \tag{12-68}$$

provides an asymptotically stable closed-loop system

$$\mathbf{x}^{\circ}(k+1) = [\phi - \theta(\mathbf{R} + \theta'\mathbf{K}\theta)^{-1}\theta'\mathbf{K}\phi]\mathbf{x}^{\circ}(k) \tag{12-69}$$

Proof: From Theorem 12-4, since $\mathbf{Q}$ and $\mathbf{R}$ are both positive definite, $\mathbf{K}$ is also positive definite. Let the Liapunov function be defined as

$$V[\mathbf{x}(k)] = \frac{1}{2}\mathbf{x}'(k)\mathbf{K}\mathbf{x}(k) \tag{12-70}$$

which is positive definite. Then,

$$\begin{aligned}\Delta V[\mathbf{x}(k)] &= V[\mathbf{x}(k+1)] - V[\mathbf{x}(k)] \\ &= \frac{1}{2}\mathbf{x}'(k+1)\mathbf{K}\mathbf{x}(k+1) - \frac{1}{2}\mathbf{x}'(k)\mathbf{K}\mathbf{x}(k)\end{aligned} \tag{12-71}$$

Substituting Eq. (12-69) into Eq. (12-71), we have

$$\Delta V[\mathbf{x}(k)] = \frac{1}{2}\mathbf{x}'(k)[\phi'\mathbf{K}\phi - \phi'\mathbf{K}\theta(\mathbf{R} + \theta'\mathbf{K}\theta)^{-1}\theta'\mathbf{K}\phi$$
$$+ \phi'\mathbf{K}\theta(\mathbf{R} + \theta'\mathbf{K}\theta)^{-1}\theta'\mathbf{K}\theta(\mathbf{R} + \theta'\mathbf{K}\theta)^{-1}\theta'\mathbf{K}\phi$$
$$- \phi'\mathbf{K}\theta(\mathbf{R} + \theta'\mathbf{K}\theta)^{-1}\theta'\mathbf{K}\phi - \mathbf{K}]\mathbf{x}(k) \quad (12\text{-}72)$$

The Riccati equation for this case is

$$\mathbf{K} = \mathbf{Q} + \phi'\mathbf{K}\phi - \phi'\mathbf{K}\theta(\mathbf{R} + \theta'\mathbf{K}\theta)^{-1}\theta'\mathbf{K}\phi \quad (12\text{-}73)$$

Using the last equation, Eq. (12-72) is simplified to

$$\Delta V[\mathbf{x}(k)] = \frac{1}{2}\mathbf{x}'(k)\left[-\mathbf{Q} - \phi'\mathbf{K}\theta(\mathbf{R} + \theta'\mathbf{K}\theta)^{-1}\theta'\mathbf{K}[\phi - \theta(\mathbf{R} + \theta'\mathbf{K}\theta)^{-1}\theta'\mathbf{K}\phi]\right]\mathbf{x}(k) \quad (12\text{-}74)$$

Since **Q** and **K** are both positive definite, the quantity inside the bracket on the right-hand side of Eq. (12-74) is negative definite. Therefore, $\Delta V[\mathbf{x}(k)]$ is negative, and according to the Liapunov stability theorem, the system in Eq. (12-69) is asymptotically stable.

12.4 PRINCIPLE OF OPTIMALITY AND DYNAMIC PROGRAMMING

The optimal linear digital regulator design carried out in the preceding sections can be affected by use of the principle of optimality. The design using the principle of optimality is also known as the method of dynamic programming. Let us first state the principle of optimality.

Principle of Optimality

An optimal control strategy has the property that whatever the initial state and the control of the initial stages, the remaining control must form an optimal control with respect to the state resulting from the control of the initial stages. Stated in another way, any control strategy that is optimal over the interval $[i, N]$ is necessarily optimal over $[i + 1, N]$ for $i = 0, 1, 2, \ldots, N - 1$.

The finite-time optimal linear digital regulator design problem is repeated as:

Find the optimal control $\mathbf{u}^\circ(k)$, $k = 0, 1, 2, \ldots, N - 1$, so that

$$J_N = G[\mathbf{x}(N)] + \sum_{k=0}^{N-1} F_k[\mathbf{x}(k), \mathbf{u}(k)] = \text{minimum} \quad (12\text{-}75)$$

where

$$G[\mathbf{x}(N)] = \frac{1}{2}\mathbf{x}'(N)\mathbf{S}\mathbf{x}(N) \quad (12\text{-}76)$$

$$F_k[\mathbf{x}(k), \mathbf{u}(k)] = \frac{1}{2}\mathbf{x}'(k)\hat{\mathbf{Q}}\mathbf{x}(k) + \mathbf{x}'(k)\mathbf{M}\mathbf{u}(k) + \frac{1}{2}\mathbf{u}'(k)\hat{\mathbf{R}}\mathbf{u}(k) \tag{12-77}$$

subject to the constraint

$$\mathbf{x}(k+1) = \phi\mathbf{x}(k) + \theta\mathbf{u}(k)$$

with $\mathbf{x}(0)$ given.

Let $J_{N-i}[\mathbf{x}(i)]$ be the performance index over the interval $[i, N]$; that is, over the last $N - i$ intervals or stages. Then,

$$J_{N-i}[\mathbf{x}(i)] = G[\mathbf{x}(N)] + \sum_{k=i}^{N-1} F_k[\mathbf{x}(k), \mathbf{u}(k)] \qquad i = 0,1,2,\ldots,N \tag{12-78}$$

Let the minimum value of $J_{N-i}[\mathbf{x}(i)]$ be represented by

$$f_{N-i}[\mathbf{x}(i)] = \min_{\mathbf{u}(i)} J_{N-i}[\mathbf{x}(i)] \tag{12-79}$$

For $i = N$, the last equation represents the performance index or return over the last 0 stage which is the terminal cost. Therefore,

$$f_0[\mathbf{x}(N)] = G[\mathbf{x}(N)] = \frac{1}{2}\mathbf{x}'(N)\mathbf{S}\mathbf{x}(N) \tag{12-80}$$

For $i = N - 1$, we have a one-stage or one-interval process which is the last stage. Then, the optimal performance index is

$$\begin{aligned} f_1[\mathbf{x}(N-1)] &= \min_{\mathbf{u}(N-1)} J_1[\mathbf{x}(N-1)] \\ &= \min_{\mathbf{u}(N-1)} \Big[G[\mathbf{x}(N)] + F_{N-1}[\mathbf{x}(N-1), \mathbf{u}(N-1)]\Big] \end{aligned} \tag{12-81}$$

Substituting Eq. (12-77) and

$$G[\mathbf{x}(N)] = \frac{1}{2}[\phi\mathbf{x}(N-1) + \theta\mathbf{u}(N-1)]'\mathbf{S}[\phi\mathbf{x}(N-1) + \theta\mathbf{u}(N-1)] \tag{12-82}$$

into Eq. (12-81) and rearranging, we have

$$\begin{aligned} f_1[\mathbf{x}(N-1)] = \min_{\mathbf{u}(N-1)} \Bigg[&\frac{1}{2}\mathbf{x}'(N-1)(\hat{\mathbf{Q}} + \phi'\mathbf{S}\phi)\mathbf{x}(N-1) \\ &+ \mathbf{x}'(N-1)(\mathbf{M} + \frac{1}{2}\phi'\mathbf{S}\theta)\mathbf{u}(N-1) \\ &+ \frac{1}{2}\mathbf{u}'(N-1)\theta'\mathbf{S}\phi\mathbf{x}(N-1) \end{aligned}$$

$$+ \frac{1}{2} \mathbf{u}'(N-1)(\hat{\mathbf{R}} + \theta'\mathbf{S}\theta)\mathbf{u}(N-1)\Big]$$

$$= \min_{\mathbf{u}(N-1)} J_1[\mathbf{x}(N-1)] \tag{12-83}$$

For minimum $J_1[\mathbf{x}(N-1)]$ we set

$$\frac{\partial J_1[\mathbf{x}(N-1)]}{\partial \mathbf{u}(N-1)} = \mathbf{0} \tag{12-84}$$

The result is

$$[(\mathbf{M} + \frac{1}{2}\phi'\mathbf{S}\theta)' + \frac{1}{2}\theta'\mathbf{S}\phi]\mathbf{x}^\circ(N-1) + (\hat{\mathbf{R}} + \theta'\mathbf{S}\theta)\mathbf{u}^\circ(N-1) = 0 \tag{12-85}$$

Thus, the optimal control is

$$\mathbf{u}^\circ(N-1) = -(\hat{\mathbf{R}} + \theta'\mathbf{S}\theta)^{-1}(\mathbf{M}' + \theta'\mathbf{S}\phi)\mathbf{x}^\circ(N-1) \tag{12-86}$$

Notice that this is the same result as in Eq. (12-31) by setting $k = N - 1$. Now substituting Eq. (12-86) into Eq. (12-83) for $\mathbf{x}(N-1)$, we get after simplification,

$$f_1[\mathbf{x}(N-1)] = \frac{1}{2}\mathbf{x}'(N-1)[\hat{\mathbf{Q}} + \phi'\mathbf{S}\phi - (\mathbf{M}' + \theta'\mathbf{S}\phi)'(\hat{\mathbf{R}} + \theta'\mathbf{S}\theta)^{-1}(\mathbf{M}' + \theta'\mathbf{S}\phi)]\mathbf{x}(N-1) \tag{12-87}$$

We make the following definitions:

$$\mathbf{K}(N) = \mathbf{S} \tag{12-88}$$

$$\mathbf{K}(N-1) = \hat{\mathbf{Q}} + \phi'\mathbf{S}\phi - (\mathbf{M}' + \theta'\mathbf{S}\phi)'(\hat{\mathbf{R}} + \theta'\mathbf{S}\theta)^{-1}(\mathbf{M}' + \theta'\mathbf{S}\phi) \tag{12-89}$$

We see that Eq. (12-89) is the same as the Riccati equation of Eq. (12-34) with $k = N - 1$.

The optimal returns in Eqs. (12-80) and (12-81) are, respectively,

$$f_0[\mathbf{x}(N)] = \frac{1}{2}\mathbf{x}'(N)\mathbf{K}(N)\mathbf{x}(N) \tag{12-90}$$

$$f_1[\mathbf{x}(N-1)] = \frac{1}{2}\mathbf{x}'(N-1)\mathbf{K}(N-1)\mathbf{x}(N-1) \tag{12-91}$$

Continuing the process, we let $i = N - 2$; that is, an optimization problem which consists of two (last) stages. The optimal performance index for the two-stage process is written

$$f_2[\mathbf{x}(N-2)] = \min_{\substack{\mathbf{u}(N-2)\\ \mathbf{u}(N-1)}} J_2[\mathbf{x}(N-2)]$$

$$= \min_{\substack{\mathbf{u}(N-2)\\ \mathbf{u}(N-1)}} \Big[F_{N-2}[\mathbf{x}(N-2), \mathbf{u}(N-2)] + F_{N-1}[\mathbf{x}(N-1), \mathbf{u}(N-1)]$$

$$+ G[\mathbf{x}(N)]\Big] \tag{12-92}$$

Applying the principle of optimality which asserts that for the two-stage process to be optimal, regardless of what the control strategy for the first stage is, the last stage must be optimal by itself. Therefore, Eq. (12-92) is written

$$f_2[\mathbf{x}(N-2)] = \min_{\mathbf{u}(N-2)} \Big[F_{N-2}[\mathbf{x}(N-2), \mathbf{u}(N-2)] + f_1[\mathbf{x}(N-1)]\Big] \tag{12-93}$$

where $f_1[\mathbf{x}(N-1)]$ is the optimal return from the last stage and is given by Eq. (12-91). Substituting

$$F_{N-2}[\mathbf{x}(N-2), \mathbf{u}(N-2)] = \frac{1}{2}\mathbf{x}'(N-2)\hat{\mathbf{Q}}\mathbf{x}(N-2) + \mathbf{x}'(N-2)\mathbf{M}\mathbf{u}(N-2)$$

$$+ \frac{1}{2}\mathbf{u}'(N-2)\hat{\mathbf{R}}\mathbf{u}(N-2) \tag{12-94}$$

and

$$f_1[\mathbf{x}(N-1)] = \frac{1}{2}[\phi\mathbf{x}(N-2) + \theta\mathbf{u}(N-2)]'\mathbf{K}(N-1)[\phi\mathbf{x}(N-2) + \theta\mathbf{u}(N-2)] \tag{12-95}$$

into Eq. (12-93), rearranging terms, and setting

$$\frac{\partial J_2[\mathbf{x}(N-2)]}{\partial \mathbf{u}(N-2)} = 0 \tag{12-96}$$

we can show that the optimal control is

$$\mathbf{u}^\circ(N-2) = -[\hat{\mathbf{R}} + \theta'\mathbf{K}(N-1)\theta]^{-1}[\mathbf{M}' + \theta'\mathbf{K}(N-1)\phi]\mathbf{x}^\circ(N-2) \tag{12-97}$$

and

$$f_2[\mathbf{x}(N-2)] = \frac{1}{2}\mathbf{x}'(N-2)\Big[\hat{\mathbf{Q}} + \phi'\mathbf{K}(N-1)\phi$$

$$- [\mathbf{M}' + \theta'\mathbf{K}(N-1)\phi]'[\hat{\mathbf{R}} + \theta'\mathbf{K}(N-1)\theta]^{-1}[\mathbf{M}'$$

$$+ \theta'\mathbf{K}(N-1)\phi]\Big]\mathbf{x}(N-2) \tag{12-98}$$

Letting

$$\mathbf{K}(N-2) = \hat{\mathbf{Q}} + \phi'\mathbf{K}(N-1)\phi$$

$$- [\mathbf{M}' + \theta'\mathbf{K}(N-1)\phi]'[\hat{\mathbf{R}} + \theta'\mathbf{K}(N-1)\theta]^{-1}[\mathbf{M}' + \theta'\mathbf{K}(N-1)\phi] \tag{12-99}$$

Eq. (12-98) is simplified to

$$f_2[\mathbf{x}(N-2)] = \frac{1}{2}\mathbf{x}'(N-2)\mathbf{K}(N-2)\mathbf{x}(N-2) \tag{12-100}$$

Continuing the induction process, we can show that, in general,

$$f_{N-i}[\mathbf{x}(i)] = \frac{1}{2}\mathbf{x}'(i)\mathbf{K}(i)\mathbf{x}(i) \tag{12-101}$$

where

$$\begin{aligned}\mathbf{K}(i) = {} & \hat{\mathbf{Q}} + \boldsymbol{\phi}'\mathbf{K}(i+1)\boldsymbol{\phi} \\ & - [\mathbf{M}' + \boldsymbol{\theta}'\mathbf{K}(i+1)\boldsymbol{\phi}]'[\hat{\mathbf{R}} + \boldsymbol{\theta}'\mathbf{K}(i+1)\boldsymbol{\theta}]^{-1}[\mathbf{M}' + \boldsymbol{\theta}'\mathbf{K}(i+1)\boldsymbol{\phi}]\end{aligned} \tag{12-102}$$

The optimal control is

$$\mathbf{u}^\circ(i) = -[\hat{\mathbf{R}} + \boldsymbol{\theta}'\mathbf{K}(i+1)\boldsymbol{\theta}]^{-1}[\mathbf{M}' + \boldsymbol{\theta}'\mathbf{K}(i+1)\boldsymbol{\phi}]\mathbf{x}^\circ(i) \tag{12-103}$$

Thus, we have derived the Riccati equation using the principle of optimality. The method of solution is also known as dynamic programming.

It is interesting to note that the dynamic programming method does not require that $\hat{\mathbf{R}}$ be positive definite, since $\hat{\mathbf{R}}^{-1}$ was not encountered. Furthermore, $\hat{\mathbf{R}}$ can be a null matrix. However, the matrix $\hat{\mathbf{R}} + \boldsymbol{\theta}'\mathbf{K}(i+1)\boldsymbol{\theta}$ must have an inverse.

For infinite time or infinite number of stages, $N = \infty$, similar to the results obtained in the preceding sections, $\mathbf{K}(i)$ becomes $\mathbf{K}$, and Eqs. (12-101), (12-102) and (12-103) are reduced to Eqs. (12-58), (12-55) and (12-56), respectively.

12.5 SOLUTION OF THE DISCRETE RICCATI EQUATION

The amount of work that has been published on the solution and properties of the Riccati equation can probably fill a book by itself. In general, the algebraic Riccati equation of Eq. (12-55) is more difficult to solve than the difference equation of Eq. (12-102).

The difference Riccati equation is generally solved by one of the following methods:

1. Numerical Computation Methods
2. Recursive Method
3. Eigenvalue-Eigenvector Method.

The numerical computation method involves iterative solution of the nonlinear difference Riccati equation. The recursive and the eigenvalue-eigenvector methods are discussed in the following.

Recursive Method of Solving the Riccati Equation

The dynamic programming method presented in the last section is generally known as a recursive solution of optimal control problems.

Let

$$\mathbf{G}(i) = [\hat{\mathbf{R}} + \theta'\mathbf{K}(i+1)\theta]^{-1}[\mathbf{M}' + \theta'\mathbf{K}(i+1)\phi] \tag{12-104}$$

which is the feedback gain matrix of the optimal linear digital regulator; the optimal control in Eq. (12-103) is written

$$\mathbf{u}^{\circ}(i) = -\mathbf{G}(i)\mathbf{x}^{\circ}(i) \tag{12-105}$$

Similarly, the Riccati equation in Eq. (12-102) is simplified to

$$\mathbf{K}(i) = \hat{\mathbf{Q}} + \phi'\mathbf{K}(i+1)\phi - [\mathbf{M}' + \theta'\mathbf{K}(i+1)\phi]'\mathbf{G}(i) \tag{12-106}$$

Starting with the boundary condition $\mathbf{K}(N) = \mathbf{S}$, Eqs. (12-104) and (12-106) are solved by recursion in the backward direction.

As an alternative, we can use the Riccati equation in Eq. (12-30). Letting

$$\mathbf{H}(k+1) = \mathbf{K}(k+1)[\mathbf{I} + \theta\hat{\mathbf{R}}^{-1}\theta'\mathbf{K}(k+1)]^{-1} \tag{12-107}$$

then Eq. (12-30) becomes

$$\mathbf{K}(k) = (\phi' - \mathbf{M}\hat{\mathbf{R}}^{-1}\theta')\mathbf{H}(k+1)(\phi - \theta\hat{\mathbf{R}}^{-1}\mathbf{M}') + \hat{\mathbf{Q}} - \mathbf{M}\hat{\mathbf{R}}^{-1}\mathbf{M}' \tag{12-108}$$

Again, these last two equations can be solved recursively backward in time starting from the boundary condition $\mathbf{K}(N) = \mathbf{S}$.

Equations (12-107) and (12-108) have the nice property that $\mathbf{K}(k+1)$ appears only in $\mathbf{H}(k+1)$. However, the recursive method using Eqs. (12-104) and (12-106) has the advantage that the feedback matrix is given directly by Eq. (12-104).

Example 12-2.

Consider that a first-order digital process is described by

$$x(k+1) = x(k) + u(k) \tag{12-109}$$

with $x(0) = x_0$. The design objective is to find the optimal control $u^{\circ}(k)$, $k = 0,1,2,\ldots,9$, such that the following performance index is minimized:

$$J_{10} = \frac{1}{2}[10x^2(10)] + \frac{1}{2}\sum_{k=0}^{9}[x^2(k) + u^2(k)] \tag{12-110}$$

For the problem we can identify that $S = 10$, $\hat{Q} = 1$, $\hat{R} = 1$, $M = 0$, and $\phi = \theta = 1$. The Riccati equation is obtained by substituting these parameters into Eq. (12-106),

$$K(i) = 1 + K(i+1) - K(i+1)G(i) \tag{12-111}$$

where

$$G(i) = \frac{K(i+1)}{1+K(i+1)} \tag{12-112}$$

The optimal control is

$$u^{\circ}(i) = -G(i)x^{\circ}(i) \tag{12-113}$$

Equations (12-111) and (12-112) are solved recursively starting with the boundary condition K(10) = S = 10. The results are tabulated in the following.

i	K(i)	G(i)
0	1.6180	0.6180
1	1.6180	0.6180
2	1.6180	0.6180
3	1.6180	0.6180
4	1.6180	0.6180
5	1.6182	0.6182
6	1.6188	0.6188
7	1.6236	0.6236
8	1.6562	0.6562
9	1.9091	0.9091
10	10	0

Substitution of the optimal control into Eq. (12-109) we can show that the optimal trajectory of x(k) is given by

$$x^{\circ}(k) = \prod_{i=0}^{k-1} [1 - G(i)]x_0 \tag{12-114}$$

for $k = 0,1,\ldots,10$. For any nonzero x_0, we see that $x^{\circ}(k)$ approaches zero rapidly as k increases.

Example 12-3.

A second-order digital process is described by the state equation,

$$\mathbf{x}(k+1) = \phi\mathbf{x}(k) + \theta u(k) \tag{12-115}$$

where

$$\phi = \begin{bmatrix} 0 & 1 \\ -1 & 1 \end{bmatrix} \qquad \theta = \begin{bmatrix} 0 \\ 1 \end{bmatrix} \tag{12-116}$$

Given that $\mathbf{x}(0) = [1 \quad 1]'$, find the optimal control u(k), $k = 0,1,2,\ldots,7$, such that the performance index

$$J_8 = \sum_{k=0}^{7} [x_1^2(k) + u^2(k) \tag{12-117}$$

is minimized.

For this problem we identify that $\mathbf{M} = 0$, $\hat{R} = 2$,

$$\hat{\mathbf{Q}} = \begin{bmatrix} 2 & 0 \\ 0 & 0 \end{bmatrix} \quad \text{and} \quad \mathbf{S} = \begin{bmatrix} 0 & 0 \\ 0 & 0 \end{bmatrix} \tag{12-118}$$

The Riccati equation is obtained by substituting these parameters to Eq. (12-106); we have

$$\mathbf{K}(i) = \begin{bmatrix} 2 & 0 \\ 0 & 0 \end{bmatrix} + \begin{bmatrix} 0 & -1 \\ 1 & 1 \end{bmatrix} \mathbf{K}(i+1) \left[\begin{bmatrix} 0 & 1 \\ -1 & 1 \end{bmatrix} - \begin{bmatrix} 0 \\ 1 \end{bmatrix} \mathbf{G}(i) \right] \tag{12-119}$$

The optimal control is

$$u^\circ(i) = -\mathbf{G}(i)\mathbf{x}^\circ(i) \tag{12-120}$$

where from Eq. (12-104),

$$\mathbf{G}(i) = \left[2 + [0 \quad 1]\mathbf{K}(i+1) \begin{bmatrix} 0 \\ 1 \end{bmatrix} \right]^{-1} [0 \quad 1]\mathbf{K}(i+1) \begin{bmatrix} 0 & 1 \\ -1 & 1 \end{bmatrix} \tag{12-121}$$

Starting with the boundary condition,

$$\mathbf{K}(8) = \mathbf{S} = \begin{bmatrix} 0 & 0 \\ 0 & 0 \end{bmatrix} \tag{12-122}$$

Eqs. (12-119) and (12-121) are solved recursively to give

$$\mathbf{K}(7) = \begin{bmatrix} 2 & 0 \\ 0 & 0 \end{bmatrix} \qquad \mathbf{G}(7) = [0 \quad 0]$$

$$\mathbf{K}(6) = \begin{bmatrix} 2 & 0 \\ 0 & 2 \end{bmatrix} \qquad \mathbf{G}(6) = [0 \quad 0]$$

$$\mathbf{K}(5) = \begin{bmatrix} 3 & -1 \\ -1 & 3 \end{bmatrix} \qquad \mathbf{G}(5) = [-0.5 \quad 0.5]$$

$$\mathbf{K}(4) = \begin{bmatrix} 3.2 & -0.8 \\ -0.8 & 3.2 \end{bmatrix} \qquad \mathbf{G}(4) = [-0.6 \quad 0.4]$$

$$K(3) = \begin{bmatrix} 3.23 & -0.922 \\ -0.922 & 3.69 \end{bmatrix} \qquad G(3) = [-0.615 \quad 0.462]$$

$$K(2) = \begin{bmatrix} 3.297 & -0.973 \\ -0.973 & 3.729 \end{bmatrix} \qquad G(2) = [-0.651 \quad 0.481]$$

$$K(1) = \begin{bmatrix} 3.301 & -0.962 \\ -0.962 & 3.75 \end{bmatrix} \qquad G(1) = [-0.652 \quad 0.485]$$

$$K(0) = \begin{bmatrix} 3.305 & -0.97 \\ -0.97 & 3.777 \end{bmatrix} \qquad G(0) = [-0.6538 \quad 0.486]$$

The optimal control and the optimal trajectories are computed by substituting the feedback gains into Eqs. (12-120) and (12-115). The results are tabulated as follows:

i	$u^{\circ}(i)$	$x_1^{\circ}(i)$	$x_2^{\circ}(i)$
0	0.1678	1	1
1	0.5708	1	0.1678
2	0.235	0.1678	−0.2614
3	−0.071	−0.2614	−0.1942
4	−0.115	−0.1942	−0.0038
5	0.0358	−0.0038	0.0754
6	0	0.0754	0.115
7	0	0.115	0.0396
8	0	0.0396	−0.0754

For large values of N we can show that the Riccati gain approaches to the steady-state solution of

$$K = \begin{bmatrix} 3.308 & -0.972 \\ -0.972 & 3.780 \end{bmatrix} \tag{12-123}$$

and the constant optimal control is

$$G = [-0.654 \quad 0.486]$$

For N = 8, the finite-time problem has solutions already approaching these values rapidly. In general, however, the recursive method could only lead to the steady-state solutions by using as large a value of N as necessary. In this case, since the pair $[\phi, \theta]$ is completely controllable, and we can find a 2 × 2 matrix **D** such that $\mathbf{DD}' = \mathbf{Q}$ and $[\phi, \mathbf{D}]$ is observable, the closed-loop system will be asymptotically stable for $N = \infty$.

The Eigenvalue-Eigenvector Method

The nonlinear difference Riccati equation given in Eq. (12-102) can be solved by the eigenvalue-eigenvector method. The result of the method is a closed-form solution for the Riccati gain, and the solution of the algebraic Riccati equation can be obtained by applying a limiting process.

The canonical state equations in Eqs. (12-23) and (12-24) are rewritten as

$$\mathbf{x}^\circ(k+1) = \Omega\mathbf{x}^\circ(k) - \theta\hat{\mathbf{R}}^{-1}\theta'\mathbf{p}^\circ(k+1) \tag{12-124}$$

$$\mathbf{p}^\circ(k) = \Gamma\mathbf{x}^\circ(k) + \Omega'\mathbf{p}^\circ(k+1) \tag{12-125}$$

where

$$\Omega = \phi - \theta\hat{\mathbf{R}}^{-1}\mathbf{M}' \tag{12-126}$$

$$\Gamma = \hat{\mathbf{Q}} - \mathbf{M}\hat{\mathbf{R}}^{-1}\mathbf{M}' \tag{12-127}$$

Solving for $\mathbf{x}^\circ(k)$ from Eq. (12-124), and writing the canonical state equations in vector-matrix form, we have

$$\begin{bmatrix} \mathbf{x}^\circ(k) \\ \mathbf{p}^\circ(k) \end{bmatrix} = \mathbf{V} \begin{bmatrix} \mathbf{x}^\circ(k+1) \\ \mathbf{p}^\circ(k+1) \end{bmatrix} \tag{12-128}$$

where

$$\mathbf{V} = \begin{bmatrix} \Omega^{-1} & \Omega^{-1}\theta\hat{\mathbf{R}}^{-1}\theta' \\ \Gamma\Omega^{-1} & \Omega' + \Gamma\Omega^{-1}\theta\hat{\mathbf{R}}^{-1}\theta' \end{bmatrix} \tag{12-129}$$

Equation (12-128) represents 2n difference equations in backward time with boundary conditions, $\mathbf{x}(0) = \mathbf{x}_0$ and $\mathbf{p}(N) = \mathbf{S}\mathbf{x}(N)$.

An important property of the matrix $\mathbf{V}$ is that the reciprocal of every eigenvalue is also an eigenvalue. We can show this by letting λ be an eigenvalue of $\mathbf{V}$ and $\mathbf{h}$ be the corresponding eigenvector. Then, by definition of eigenvector,

$$\mathbf{V}\mathbf{h} = \lambda\mathbf{h} \tag{12-130}$$

Let us partition $\mathbf{h}$ so that the last equation is written

$$\begin{bmatrix} \Omega^{-1} & \Omega^{-1}\theta\hat{\mathbf{R}}^{-1}\theta' \\ \Gamma\Omega^{-1} & \Omega' + \Gamma\Omega^{-1}\theta\hat{\mathbf{R}}^{-1}\theta' \end{bmatrix} \begin{bmatrix} \mathbf{f} \\ \hline \mathbf{g} \end{bmatrix} = \lambda \begin{bmatrix} \mathbf{f} \\ \hline \mathbf{g} \end{bmatrix} \tag{12-131}$$

The determinant of $\mathbf{V}$ is written as

$$\Delta = |\Omega^{-1}(\Omega' + \Gamma\Omega^{-1}\theta\hat{\mathbf{R}}^{-1}\theta') - \Omega^{-1}\Gamma\Omega^{-1}\Omega\Omega^{-1}\theta\hat{\mathbf{R}}^{-1}\theta'|$$

$$= |\Omega^{-1}\Omega' + \Omega^{-1}\Gamma\Omega^{-1}\theta\hat{\mathbf{R}}^{-1}\theta' - \Omega^{-1}\Gamma\Omega^{-1}\theta\hat{\mathbf{R}}^{-1}\theta'|$$

$$= |\Omega^{-1}\Omega'| = |\Omega^{-1}||\Omega'| = 1 \tag{12-132}$$

Therefore, the determinant of $\mathbf{V}$ is unity.

Now taking the inverse of $\mathbf{V}$ and then the transpose, the following identity results:

$$(\mathbf{V}^{-1})' \begin{bmatrix} \mathbf{g} \\ --- \\ -\mathbf{f} \end{bmatrix} = \lambda \begin{bmatrix} \mathbf{g} \\ --- \\ -\mathbf{f} \end{bmatrix} \tag{12-133}$$

The significance of the last equation is that λ is an eigenvalue of $(\mathbf{V}^{-1})'$ and also of $\mathbf{V}^{-1}$. Thus, $1/\lambda$ is an eigenvalue of $\mathbf{V}$. This also means that n of the eigenvalues of $\mathbf{V}$ are inside the unit circle and n are outside. Let us introduce a nonsingular transformation

$$\begin{bmatrix} \mathbf{x}^\circ(k) \\ \mathbf{p}^\circ(k) \end{bmatrix} = \mathbf{W} \begin{bmatrix} \mathbf{q}(k) \\ \mathbf{r}(k) \end{bmatrix} \tag{12-134}$$

where $\mathbf{W}$ is of the form,

$$\mathbf{W} = \begin{bmatrix} \mathbf{W}_{11} & \mathbf{W}_{12} \\ \mathbf{W}_{21} & \mathbf{W}_{22} \end{bmatrix} \tag{12-135}$$

and $\mathbf{W}$ has the property that

$$\mathbf{W}^{-1}\mathbf{V}\mathbf{W} = \begin{bmatrix} \Lambda & 0 \\ 0 & \Lambda^{-1} \end{bmatrix} \tag{12-136}$$

For distinct eigenvalues, the elements of Λ form either a diagonal matrix with λ_i on the main diagonal, where λ_i represents the eigenvalues of $\mathbf{V}$, that are outside the unit circle, or a modal form matrix for complex conjugate eigenvalues. For example, let λ_1 and λ_2 be the real eigenvalues of $\mathbf{V}$ that are outside the unit circle, then Eq. (12-136) is written

$$
W^{-1}VW = \left[\begin{array}{cc|cc} \lambda_1 & 0 & 0 & 0 \\ 0 & \lambda_2 & 0 & 0 \\ \hline 0 & 0 & \dfrac{1}{\lambda_1} & 0 \\ 0 & 0 & 0 & \dfrac{0}{\lambda_2} \end{array}\right] = \left[\begin{array}{c|c} \Lambda & 0 \\ \hline 0 & \Lambda^{-1} \end{array}\right] \tag{12-137}
$$

If $\mathbf{V}$ has complex conjugate eigenvalues, $\sigma_1 + j\omega_1$ and $\sigma_1 - j\omega_1$, all outside the unit circle, then Eq. (12-136) becomes

$$
W^{-1}VW = \left[\begin{array}{cc|cc} \sigma_1 & \omega_1 & 0 & 0 \\ -\omega_1 & \sigma_1 & 0 & 0 \\ \hline 0 & 0 & \sigma_2 & -\omega_2 \\ 0 & 0 & \omega_2 & \sigma_2 \end{array}\right] = \left[\begin{array}{c|c} \Lambda & 0 \\ \hline 0 & \Lambda^{-1} \end{array}\right] \tag{12-138}
$$

where

$$
\sigma_2 - j\omega_2 = \frac{1}{\sigma_1 + j\omega_1} \tag{12-139}
$$

Of course, when $\mathbf{V}$ has both real and complex eigenvalues, a combination of Eqs. (12-137) and (12-138) should be used.

Now writing Eq. (12-134) as

$$
\begin{bmatrix} \mathbf{q}(k) \\ \mathbf{r}(k) \end{bmatrix} = W^{-1} \begin{bmatrix} \mathbf{x}^{\circ}(k) \\ \mathbf{p}^{\circ}(k) \end{bmatrix}
$$

and using Eq. (12-128), we have

$$
\begin{bmatrix} \mathbf{q}(k) \\ \mathbf{r}(k) \end{bmatrix} = W^{-1}VW \begin{bmatrix} \mathbf{q}(k+1) \\ \mathbf{r}(k+1) \end{bmatrix} = \begin{bmatrix} \Lambda & 0 \\ 0 & \Lambda^{-1} \end{bmatrix} \begin{bmatrix} \mathbf{q}(k+1) \\ \mathbf{r}(k+1) \end{bmatrix} \tag{12-140}
$$

The last equation is solved by recursion in the backward direction with the boundary condition $[\mathbf{q}(N) \quad \mathbf{r}(N)]'$ to give

$$
\begin{bmatrix} \mathbf{q}(N-k) \\ \mathbf{r}(N-k) \end{bmatrix} = \begin{bmatrix} \Lambda^{k} & 0 \\ 0 & \Lambda^{-k} \end{bmatrix} \begin{bmatrix} \mathbf{q}(N) \\ \mathbf{r}(N) \end{bmatrix} \tag{12-141}
$$

which is rearranged to the form:

$$\begin{bmatrix} \mathbf{q}(N) \\ \mathbf{r}(N-k) \end{bmatrix} = \begin{bmatrix} \Lambda^{-k} & 0 \\ 0 & \Lambda^{-k} \end{bmatrix} \begin{bmatrix} \mathbf{q}(N-k) \\ \mathbf{r}(N) \end{bmatrix} \tag{12-142}$$

Let $\mathbf{r}(N)$ and $\mathbf{q}(N)$ be related through

$$\mathbf{r}(N) = \mathbf{U}\mathbf{q}(N) \tag{12-143}$$

where $\mathbf{U}$ is an $n \times n$ matrix. Substitution of Eq. (12-143) into Eqs. (12-142) and (12-134) for $k = N$, we have

$$\mathbf{U} = -(\mathbf{W}_{22} - \mathbf{S}\mathbf{W}_{12})^{-1}(\mathbf{W}_{21} - \mathbf{S}\mathbf{W}_{11}) \tag{12-144}$$

and

$$\mathbf{r}(N-k) = \Lambda^{-k}\mathbf{U}\Lambda^{-k}\mathbf{q}(N-k) \tag{12-145}$$

Letting

$$\mathbf{H}(k) = \Lambda^{-k}\mathbf{U}\Lambda^{-k} \tag{12-146}$$

Eq. (12-145) is written

$$\mathbf{r}(N-k) = \mathbf{H}(k)\mathbf{q}(N-k) \tag{12-147}$$

Now substituting the last equation into Eq. (12-134) we can show that the following relations are true:

$$\mathbf{x}(N-k) = [\mathbf{W}_{11} + \mathbf{W}_{12}\mathbf{H}(k)]\,\mathbf{q}(N-k) \tag{12-148}$$

$$\mathbf{p}(N-k) = \mathbf{K}(N-k)\mathbf{x}(N-k) = [\mathbf{W}_{21} + \mathbf{W}_{22}\mathbf{H}(k)]\,\mathbf{q}(N-k) \tag{12-149}$$

Comparing the last two equations, the Riccati gain is written as

$$\mathbf{K}(N-k) = [\mathbf{W}_{21} + \mathbf{W}_{22}\mathbf{H}(k)]\,[\mathbf{W}_{11} + \mathbf{W}_{12}\mathbf{H}(k)]^{-1} \tag{12-150}$$

Although the last equation gives a closed-form solution of the difference Riccati equation, in general, because of the computations involved in finding the matrix $\mathbf{W}$ and the eigenvalues of $\mathbf{V}$ the eigenvalue-eigenvector method does not have a clear-cut advantage over the recursive method. However, the eigenvalue-eigenvector method leads directly to the solution of the algebraic Riccati equation for infinite-time problems.

Letting $i = N - k$, $\mathbf{K}(N-k)$ becomes $\mathbf{K}(i)$ which is the Riccati gain used in the preceding sections, $i = 0,1,2,\ldots,N-1$. The constant Riccati gain is written

$$\mathbf{K} = \lim_{i=-\infty} \mathbf{K}(i) = \lim_{k\to\infty} \mathbf{K}(N-k) \tag{12-151}$$

Since

$$\lim_{k\to\infty} \mathbf{H}(k) = \lim_{k\to\infty} \Lambda^{-k}\mathbf{U}\Lambda^{-k} = 0 \tag{12-152}$$

Eq. (12-150) leads to

$$\mathbf{K} = \lim_{k\to\infty} \mathbf{K}(N-k) = \mathbf{W}_{21}\mathbf{W}_{11}^{-1} \tag{12-153}$$

Example 12-4.

The first-order system treated in Example 12-2 is considered here again. The finite-time problem is to be solved by the eigenvalue-eigenvector method, and in addition, the problem is to be solved for $N = \infty$.

Substitution of the system parameters into Eq. (12-129), we have

$$V = \begin{bmatrix} 1 & 1 \\ 1 & 2 \end{bmatrix} \tag{12-154}$$

The eigenvalues of V are $\lambda_1 = 2.618$ and $\lambda_2 = 1/\lambda_1 = 0.382$. Therefore, we form the matrix

$$W^{-1}VW = \begin{bmatrix} \Lambda & 0 \\ 0 & \Lambda^{-1} \end{bmatrix} = \begin{bmatrix} 2.618 & 0 \\ 0 & 0.382 \end{bmatrix} \tag{12-155}$$

Since **W** is a similarity transformation, the columns of **W** are the eigenvectors of V. Therefore,

$$W = \begin{bmatrix} W_{11} & W_{12} \\ W_{21} & W_{22} \end{bmatrix} = \begin{bmatrix} 1 & 1 \\ 1.618 & -0.618 \end{bmatrix} \tag{12-156}$$

From Eq. (12-144) we have

$$U = -(-0.618 - 10)^{-1}(1.618 - 10) = -0.789 \tag{12-157}$$

and Eq. (12-146) gives

$$H(k) = \Lambda^{-k}U\Lambda^{-k} = -0.789(0.382)^{2k} \tag{12-158}$$

The Riccati gain of the finite-time problem is

$$K(N-k) = [1.618 - 0.618(-0.789)(0.382)^{2k}][1 - 0.789(0.382)^{2k}]^{-1} \tag{12-159}$$

or

$$K(i) = [1.618 + 0.488(0.382)^{2(N-i)}][1 + 0.789(0.382)^{2(N-i)}]^{-1} \tag{12-160}$$

It is simple to show that this result will give the same results as those obtained in Example 12-2 for $N = 10$.

For N = ∞, the constant Riccati gain is given by Eq. (12-153); we have

$$K = W_{21}W_{11}^{-1} = 1.618 \tag{12-161}$$

Example 12-5.

It is informative to apply the eigenvalue-eigenvector method to a system higher than the first order. The second-order system in Example 12-3 is considered.

The matrix V is written

$$V = \left[\begin{array}{cc|cc} 1 & -1 & 0 & -0.5 \\ 1 & 0 & 0 & 0 \\ \hline 2 & -2 & 0 & -2 \\ 0 & 0 & 1 & 1 \end{array}\right] \tag{12-162}$$

The eigenvalues of V are $\lambda_1 = \sigma_1 \pm j\omega_1 = 0.743 \pm j1.529$ and $\lambda_2 = 1/\lambda_1 = \sigma_2 \pm j\omega_2 = 0.257 \pm j0.529$. These eigenvalues are complex conjugate pairs with two outside and two inside the unit circle. Then,

$$W^{-1}VW = \left|\begin{array}{c|c} \Lambda & 0 \\ \hline 0 & \Lambda^{-1} \end{array}\right| = \left[\begin{array}{cc|cc} \sigma_1 & \omega_1 & 0 & 0 \\ -\omega_1 & \sigma_1 & 0 & 0 \\ \hline 0 & 0 & \sigma_2 & -\omega_2 \\ 0 & 0 & \omega_2 & \sigma_2 \end{array}\right]$$

$$= \left[\begin{array}{cc|cc} 0.743 & 1.529 & 0 & 0 \\ -1.529 & 0.743 & 0 & 0 \\ \hline 0 & 0 & 0.257 & -0.529 \\ 0 & 0 & 0.529 & 0.257 \end{array}\right] \tag{12-163}$$

It is known that given the coefficient matrix

$$A = \begin{bmatrix} \sigma_1 + j\omega_1 & 0 \\ 0 & \sigma_1 - j\omega_1 \end{bmatrix} \tag{12-164}$$

with its eigenvectors designated as $\alpha_1 + j\beta_1$ and $\alpha_1 - j\beta_1$, then the eigenvectors of the modal form

$$\Lambda = \begin{bmatrix} \sigma_1 & \omega_1 \\ -\omega_1 & \sigma_1 \end{bmatrix} \tag{12-165}$$

are α_1 and β_1. For the modal form matrix in Eq. (12-138) the matrix **W** is found by using the real and imaginary parts of the eigenvectors of the eigenvalues of $\mathbf{W}^{-1}\mathbf{V}\mathbf{W}$. Thus,

$$W = [\alpha_1 \mid \beta_1 \mid \alpha_2 \mid \beta_2] = \left[\begin{array}{cc|cc} 0.318 & -0.053 & 0 & -0.5 \\ 0.053 & -0.182 & -0.765 & -0.371 \\ \hline 1 & 0 & -0.743 & 0.529 \\ -0.107 & -0.636 & 1 & 0 \end{array}\right] \tag{12-166}$$

where $\alpha_1 + j\beta_1$ is the eigenvector of $\sigma_1 + j\omega_1$ and $\alpha_2 + j\beta_2$ is the eigenvector of $\sigma_2 + j\omega_2$.

From Eq. (12-144), we get

$$U = \begin{bmatrix} 0.108 & 0.635 \\ -0.174 & 0.894 \end{bmatrix} \tag{12-167}$$

and H(k) is given by Eq. (12-146). Then, the time-varying Riccati gain, K(N − k), is determined by using Eq. (12-150). It is apparent that the solution process requires lengthy matrix multiplications, and even for a second-order system a digital computer is needed.

The infinite-time solution is obtained easily from Eq. (12-153). Therefore, for $N = \infty$,

$$K = W_{21}W_{11}^{-1} = \begin{bmatrix} 1 & 0 \\ -0.107 & -0.636 \end{bmatrix} \begin{bmatrix} 0.318 & -0.053 \\ 0.053 & -0.182 \end{bmatrix}^{-1}$$

$$= \begin{bmatrix} 3.308 & -0.972 \\ -0.972 & 3.780 \end{bmatrix} \tag{12-168}$$

which is the same result predicted by the recursive method in Eq. (12-123). It has been pointed out that the closed-loop system with $N = \infty$ will be asymptotically stable.

12.6 SAMPLING PERIOD SENSITIVITY

The unique properties of the discrete Riccati equation lead to a sensitivity study of the optimal digital regulator system with respect to the sampling period T, when the system model is described by Eqs. (12-1) and (12-2). The sampling period sensitivity also provides a method of approximating the constant Riccati gain **K** for the infinite-time linear regulator problem, based on the knowledge of the Riccati gain of the continuous-data system, that is, when T = 0. In the end it is possible to approximate the feedback gain matrix **G** of the optimal linear digital regulator in terms of the feedback gain matrix of the optimal continuous-data linear regulator; i.e., when T = 0. It is interesting to note that the development on the sampling period sensitivity and its application to the optimal linear digital regulator design originally motivated the solution to the digital redesign problem presented in Chapter 8. Therefore, in this section we shall look at the digital redesign problem again, but from a different viewpoint.

We begin by referring to the linear digital regulator design problem stated in Eqs. (12-1), (12-2) and (12-3). It has been shown in Sec. 12.2 that the optimal digital control is [Eq. (12-31)]

$$\begin{aligned}\mathbf{u}^\circ(k) &= -[\hat{\mathbf{R}} + \theta'\mathbf{K}(k+1)\theta]^{-1}[\theta'\mathbf{K}(k+1)\phi + \mathbf{M}']\mathbf{x}^\circ(k)\\ &= -\mathbf{G}(k)\mathbf{x}^\circ(k) \qquad (12\text{-}169)\end{aligned}$$

where $\mathbf{K}(k)$ is the positive semidefinite solution of the Riccati equation [Eq. (12-34)]

$$\begin{aligned}\mathbf{K}(k) &= \phi'\mathbf{K}(k+1)\phi + \hat{\mathbf{Q}}\\ &\quad - [\theta'\mathbf{K}(k+1)\phi + \mathbf{M}']'[\hat{\mathbf{R}} + \theta'\mathbf{K}(k+1)\theta]^{-1}[\theta'\mathbf{K}(k+1)\phi + \mathbf{M}'] \qquad (12\text{-}170)\end{aligned}$$

with the boundary condition $\mathbf{K}(t_f) = \mathbf{S}$.

The last equation is conditioned by rearranging the terms, adding and subtracting $\mathbf{K}(k+1)$ on the same side, and dividing both sides by T; we have

$$\frac{\mathbf{K}(k+1) - \mathbf{K}(k)}{T} + \frac{\phi'\mathbf{K}(k+1)\phi - \mathbf{K}(k+1) + \hat{\mathbf{Q}}}{T}$$

$$= \left[\frac{\mathbf{M} + \phi'\mathbf{K}(k+1)\theta}{T}\right]\left[\frac{\hat{\mathbf{R}} + \theta'\mathbf{K}(k+1)\theta}{T}\right]^{-1}\left[\frac{\theta'\mathbf{K}(k+1)\phi + \mathbf{M}'}{T}\right] \qquad (12\text{-}171)$$

It can be shown that as the sampling period T approaches zero, the last equation becomes

$$\dot{\mathbf{K}}(t) = \mathbf{A}'\mathbf{K}(t) + \mathbf{K}(t)\mathbf{A} + \mathbf{Q} = \mathbf{K}(t)\mathbf{B}\mathbf{R}^{-1}\mathbf{B}'\mathbf{K}(t) \qquad (12\text{-}172)$$

which is known as the differential Riccati equation of the optimal continuous-data linear regulator, and $\mathbf{K}(t)$ is the Riccati gain, that is, when the system in Eq. (12-1) is subject to continuous-data control $\mathbf{u}(t)$.

For infinite time, $t_f = \infty$,

$$\mathbf{K}(k) = \mathbf{K}(k+1) = \mathbf{K}(T) \qquad (12\text{-}173)$$

where instead of using the notation $\mathbf{K}$ as the steady-state discrete Riccati gain we have adopted the notation $\mathbf{K}(T)$ to indicate the dependence on T. For infinite time, Eq. (12-171) becomes the algebraic Riccati equation,

$$\frac{\phi'\mathbf{K}(T)\phi - \mathbf{K}(T) + \hat{\mathbf{Q}}}{T} = \left[\frac{\mathbf{M} + \phi'\mathbf{K}(T)\theta}{T}\right]\left[\frac{\hat{\mathbf{R}} + \theta'\mathbf{K}(T)\theta}{T}\right]^{-1}\left[\frac{\theta'\mathbf{K}(T)\phi + \mathbf{M}'}{T}\right] \qquad (12\text{-}174)$$

As T approaches zero, the last equation becomes

$$\mathbf{K}(0)\mathbf{A} + \mathbf{A}'\mathbf{K}(0) + \mathbf{Q} = \mathbf{K}(0)\mathbf{B}\mathbf{R}^{-1}\mathbf{B}'\mathbf{K}(0) \tag{12-175}$$

where

$$\mathbf{K}(0) = \lim_{T\to 0} \mathbf{K}(T) \tag{12-176}$$

or, $\mathbf{K}(0)$ is also the steady-state solution of the differential Riccati equation in Eq. (12-172).

It follows that the optimal feedback gain matrix of the infinite-time digital regulator can be written as

$$\mathbf{G}(T) = \left[\frac{\hat{\mathbf{R}} + \theta'\mathbf{K}(T)\theta}{T}\right]^{-1}\left[\frac{\theta'\mathbf{K}(T)\phi + \mathbf{M}'}{T}\right] \tag{12-177}$$

For $T = 0$, the optimal feedback gain matrix of the infinite-time continuous-data regulator is

$$\mathbf{G}(0) = \lim_{T\to 0} \mathbf{G}(T) = \mathbf{R}^{-1}\mathbf{B}'\mathbf{K}(0) \tag{12-178}$$

Sampling Period Sensitivity Study of the Riccati Gain Matrix

Let us expand the steady-state discrete Riccati gain $\mathbf{K}(T)$ into a Taylor series about $T = 0$:

$$\mathbf{K}(T) = \mathbf{K}(0) + \sum_{i=0}^{\infty} \mathbf{S}_i^K \frac{T^i}{i!} \tag{12-179}$$

where

$$\mathbf{S}_i^K = \left.\frac{\partial^i \mathbf{K}(T)}{\partial T^i}\right|_{T=0} \tag{12-180}$$

is defined as the *ith-order sampling period sensitivity* of $\mathbf{K}(T)$. It is assumed that the infinite series in Eq. (12-179) converges over the sampling period $[0, T_c]$. For most practical problems it is adequate to use only a few terms of the series to approximate $\mathbf{K}(T)$. In the following, the first-order and the second-order sensitivities of $\mathbf{K}(T)$ are derived. For simplicity of notation, it is understood that all the partial derivatives are evaluated at $T = 0$.

For convenience, we define

$$\mathbf{X}(T) = \frac{\phi'\mathbf{K}(T)\phi - \mathbf{K}(T) + \hat{\mathbf{Q}}}{T} \tag{12-181}$$

$$\mathbf{X}(0) = \lim_{T\to 0} \mathbf{X}(T) = \mathbf{K}(0)\mathbf{A} + \mathbf{A}'\mathbf{K}(0) + \mathbf{Q} \tag{12-182}$$

$$\mathbf{Y}(T) = \frac{\mathbf{M} + \boldsymbol{\phi}'\mathbf{K}(T)\boldsymbol{\theta}}{T} \tag{12-183}$$

$$\mathbf{Y}(0) = \lim_{T \to 0} \mathbf{Y}(T) = \mathbf{K}(0)\mathbf{B} \tag{12-184}$$

$$\mathbf{Z}(T) = \frac{\hat{\mathbf{R}} + \boldsymbol{\theta}'\mathbf{K}(T)\boldsymbol{\theta}}{T} \tag{12-185}$$

$$\mathbf{Z}(0) = \mathbf{R} \tag{12-186}$$

Then, the optimal feedback gain in Eq. (12-177) is simply

$$\mathbf{G}(T) = \mathbf{Z}^{-1}(T)\mathbf{Y}'(T) \tag{12-187}$$

and the discrete Riccati equation in Eq. (12-174) becomes

$$\begin{aligned}\mathbf{X}(T) &= \mathbf{Y}(T)\mathbf{Z}^{-1}(T)\mathbf{Y}'(T) \\ &= \mathbf{Y}(T)\mathbf{G}(T)\end{aligned} \tag{12-188}$$

For the first-order sensitivities, Eqs. (12-187) and (12-188) are differentiated with respect to T and then evaluated at T = 0. We have

$$\frac{\partial \mathbf{G}(T)}{\partial T} = \mathbf{R}^{-1}\frac{\partial \mathbf{Y}'(T)}{\partial T} - \mathbf{R}^{-1}\frac{\partial \mathbf{Z}(T)}{\partial T}\mathbf{R}^{-1}\mathbf{B}'\mathbf{K}(0) \tag{12-189}$$

$$\frac{\partial \mathbf{X}(T)}{\partial T} = \mathbf{K}(0)\mathbf{B}\frac{\partial \mathbf{G}(T)}{\partial T} + \frac{\partial \mathbf{Y}(T)}{\partial T}\mathbf{R}^{-1}\mathbf{B}'\mathbf{K}(0) \tag{12-190}$$

where $\partial^i \mathbf{G}(T)/\partial T^i$ denotes the ith-order derivative of $\mathbf{G}(T)$ with respect to T, evaluated at T = 0, and is called the ith-order sampling period sensitivity of $\mathbf{G}(T)$.

The partial derivatives of $\mathbf{X}(T)$, $\mathbf{Y}(T)$ and $\mathbf{Z}(T)$ with respect to T can be found from Eqs. (12-181), (12-183) and (12-185), respectively, by using the infinite series expansion of $\phi = e^{\mathbf{A}T}$ in Eqs. (12-6), (12-11), (12-12) and (12-13), retaining terms only up to the third order in T. The following results are obtained:

$$\frac{\partial \mathbf{X}(T)}{\partial T} = \mathbf{A}'\frac{\partial \mathbf{K}(T)}{\partial T} + \frac{\partial \mathbf{K}(T)}{\partial T}\mathbf{A} + \frac{1}{2}\mathbf{A}'\mathbf{K}(0)\mathbf{B}\mathbf{R}^{-1}\mathbf{B}'\mathbf{K}(0) + \frac{1}{2}\mathbf{K}(0)\mathbf{B}\mathbf{R}^{-1}\mathbf{B}'\mathbf{K}(0)\mathbf{A} \tag{12-191}$$

$$\frac{\partial \mathbf{Y}(T)}{\partial T} = \frac{\partial \mathbf{K}(T)}{\partial T}\mathbf{B} + \frac{1}{2}[\mathbf{A}'\mathbf{K}(0) + \mathbf{K}(0)\mathbf{B}\mathbf{R}^{-1}\mathbf{B}'\mathbf{K}(0)]\mathbf{B} \tag{12-192}$$

$$\frac{\partial \mathbf{Z}(T)}{\partial T} = \mathbf{B}'\mathbf{K}(0)\mathbf{B} \tag{12-193}$$

Substitution of Eqs. (12-192) and (12-193) into Eq. (12-189), we have after simplification,

$$\frac{\partial \mathbf{G}(T)}{\partial T} = \mathbf{R}^{-1}\mathbf{B}' \frac{\partial \mathbf{K}(T)}{\partial T} + \frac{1}{2}\mathbf{R}^{-1}\mathbf{B}'\mathbf{K}(0)[\mathbf{A} - \mathbf{B}\mathbf{R}^{-1}\mathbf{B}'\mathbf{K}(0)] \tag{12-194}$$

Substitution of Eqs. (12-191), (12-192) and (12-194) into Eq. (12-190), we get after simplification,

$$[\mathbf{A}' - \mathbf{K}(0)\mathbf{B}\mathbf{R}^{-1}\mathbf{B}'\mathbf{K}(0)] \frac{\partial \mathbf{K}(T)}{\partial T} + \frac{\partial \mathbf{K}(T)}{\partial T} [\mathbf{A} - \mathbf{B}\mathbf{R}^{-1}\mathbf{B}'\mathbf{K}(0)] = \mathbf{0} \tag{12-195}$$

which is known as the Liapunov equation. Since $\mathbf{A} - \mathbf{B}\mathbf{R}^{-1}\mathbf{B}'\mathbf{K}(0)$ is the coefficient matrix of the optimal closed-loop continuous-data linear regulator system which is asymptotically stable, it follows that the only possible solution to Eq. (12-195) is

$$\frac{\partial \mathbf{K}(T)}{\partial T} = \mathbf{0} \tag{12-196}$$

Therefore, Eq. (12-194) gives

$$\frac{\partial \mathbf{G}(T)}{\partial T} = \frac{1}{2}\mathbf{R}^{-1}\mathbf{B}'\mathbf{K}(0)[\mathbf{A} - \mathbf{B}\mathbf{R}^{-1}\mathbf{B}'\mathbf{K}(0)] \tag{12-197}$$

Equation (12-196) also leads to the interesting result on the sensitivity of the performance index with respect to T for small T. Equation (12-58) gives the optimal performance index for the infinite-time digital regulator and is rewritten as follows.

$$J_\infty^\circ = \frac{1}{2}\mathbf{x}'(0)\mathbf{K}(T)\mathbf{x}(0) \tag{12-198}$$

Therefore,

$$\left.\frac{\partial J_\infty^\circ}{\partial T}\right|_{T=0} = \frac{1}{2}\mathbf{x}'(0) \frac{\partial \mathbf{K}(T)}{\partial T} \mathbf{x}(0) = 0 \tag{12-199}$$

which means that the optimal performance index J_∞° is insensitive to variations in T for small T.

For second-order sensitivities we must evalute the second-order derivatives of $\mathbf{G}(T)$, $\mathbf{X}(T)$, $\mathbf{Y}(T)$ and $\mathbf{Z}(T)$. These are given as follows:

$$\frac{\partial^2 \mathbf{X}(T)}{\partial T^2} = \mathbf{A}' \frac{\partial^2 \mathbf{K}(T)}{\partial T^2} + \frac{\partial^2 \mathbf{K}(T)}{\partial T^2}\mathbf{A} + \frac{1}{3}(\mathbf{A}')^2[\mathbf{K}(0)\mathbf{B}\mathbf{R}^{-1}\mathbf{B}'\mathbf{K}(0)$$
$$+ \frac{1}{3}\mathbf{K}(0)\mathbf{B}\mathbf{R}^{-1}\mathbf{B}'\mathbf{K}(0)]\mathbf{A}^2 + \frac{2}{3}\mathbf{A}'\mathbf{K}(0)\mathbf{B}\mathbf{R}^{-1}\mathbf{B}'\mathbf{K}(0)\mathbf{A} \tag{12-200}$$

$$\frac{\partial^2 \mathbf{Y}(T)}{\partial T^2} = \frac{\partial^2 \mathbf{K}(T)}{\partial T^2}\mathbf{B} + \frac{1}{3}(\mathbf{A}')^2\mathbf{K}(0)\mathbf{B} + \frac{2}{3}\mathbf{A}'\mathbf{K}(0)\mathbf{B}\mathbf{R}^{-1}\mathbf{B}'\mathbf{K}(0)\mathbf{B}$$

$$+ \frac{1}{3}\mathbf{K}(0)\mathbf{B}\mathbf{R}^{-1}\mathbf{B}'\mathbf{K}(0)\mathbf{A}\mathbf{B} \qquad (12\text{-}201)$$

$$\frac{\partial^2 \mathbf{Z}(T)}{\partial T^2} = \frac{1}{3}\mathbf{B}'\mathbf{K}(0)\mathbf{A}\mathbf{B} + \frac{1}{3}\mathbf{B}'\mathbf{A}'\mathbf{K}(0)\mathbf{B} + \frac{2}{3}\mathbf{B}'\mathbf{K}(0)\mathbf{B}\mathbf{R}^{-1}\mathbf{B}'\mathbf{K}(0)\mathbf{B} \qquad (12\text{-}202)$$

Differentiating Eqs. (12-189) and (12-190) once with respect to T gives

$$\frac{\partial^2 \mathbf{G}(T)}{\partial T^2} = \mathbf{R}^{-1}\left[\frac{\partial^2 \mathbf{Y}'(T)}{\partial T^2} - 2\frac{\partial \mathbf{Z}(T)}{\partial T}\frac{\partial \mathbf{G}(T)}{\partial T} - \frac{\partial^2 \mathbf{Z}(T)}{\partial T^2}\mathbf{R}^{-1}\mathbf{B}'\mathbf{K}(0)\right] \qquad (12\text{-}203)$$

$$\frac{\partial^2 \mathbf{X}(T)}{\partial T^2} = 2\frac{\partial \mathbf{Y}(T)}{\partial T}\frac{\partial \mathbf{G}(T)}{\partial T} + \mathbf{K}(0)\mathbf{B}\frac{\partial^2 \mathbf{G}(T)}{\partial T^2} + \frac{\partial^2 \mathbf{Y}(T)}{\partial T^2}\mathbf{R}^{-1}\mathbf{B}'\mathbf{K}(0) \qquad (12\text{-}204)$$

It should be reminded that all the derivatives are evaluated at $T = 0$.

Substitution of Eqs. (12-193), (12-197) and (12-202) into Eq. (12-203) and after simplification, we have

$$\frac{\partial^2 \mathbf{G}(T)}{\partial T^2} = \mathbf{R}^{-1}\mathbf{B}'\frac{\partial^2 \mathbf{K}(T)}{\partial T^2} + \frac{1}{3}\mathbf{R}^{-1}\mathbf{B}'\mathbf{K}(0)[\mathbf{A} - \mathbf{B}\mathbf{R}^{-1}\mathbf{B}'\mathbf{K}(0)]^2 \qquad (12\text{-}205)$$

Similarly, it can be shown that $\partial^2\mathbf{K}(T)/\partial T^2$ is the solution of the Liapunov equation

$$[\mathbf{A}' - \mathbf{K}(0)\mathbf{B}\mathbf{R}^{-1}\mathbf{B}']\frac{\partial^2 \mathbf{K}(T)}{\partial T^2} + \frac{\partial^2 \mathbf{K}(T)}{\partial T^2}[\mathbf{A} - \mathbf{B}\mathbf{R}^{-1}\mathbf{B}'\mathbf{K}(0)]$$

$$+ \frac{1}{6}[\mathbf{A}' - \mathbf{K}(0)\mathbf{B}\mathbf{R}^{-1}\mathbf{B}'][\mathbf{K}(0)\mathbf{B}\mathbf{R}^{-1}\mathbf{B}'\mathbf{K}(0)][\mathbf{A} - \mathbf{B}\mathbf{R}^{-1}\mathbf{B}'\mathbf{K}(0)] = \mathbf{0}$$

$$(12\text{-}206)$$

Since the third term of the last equation is always positive semidefinite, it follows that $\partial^2\mathbf{K}(T)/\partial T^2$ is unique and positive semidefinite.

The amount of work involved in generating the third and higher-order sensitivities is apparently excessive, and the results are not available. However, in general, the second-order approximation usually provides adequate accuracy.

Equation (12-206) can be solved using other equivalent forms. Assume that $\partial^2\mathbf{K}(T)/\partial T^2$ is of the following form:

$$\frac{\partial^2 \mathbf{K}(T)}{\partial T^2} = [\mathbf{A}' - \mathbf{K}(0)\mathbf{B}\mathbf{R}^{-1}\mathbf{B}']\frac{\Pi}{6}[\mathbf{A} - \mathbf{B}\mathbf{R}^{-1}\mathbf{B}'\mathbf{K}(0)] \qquad (12\text{-}207)$$

Then Π must be the unique positive semidefinite solution to

$$[\mathbf{A}' - \mathbf{K}(0)\mathbf{B}\mathbf{R}^{-1}\mathbf{B}']\Pi + \Pi[\mathbf{A} - \mathbf{B}\mathbf{R}^{-1}\mathbf{B}'\mathbf{K}(0)] + \mathbf{K}(0)\mathbf{B}\mathbf{R}^{-1}\mathbf{B}'\mathbf{K}(0) = \mathbf{0} \quad (12\text{-}208)$$

Equation (12-175) is reconditioned to read

$$[\mathbf{A}' - \mathbf{K}(0)\mathbf{B}\mathbf{R}^{-1}\mathbf{B}']\mathbf{K}(0) + \mathbf{K}(0)[\mathbf{A} - \mathbf{B}\mathbf{R}^{-1}\mathbf{B}'\mathbf{K}(0)] + \mathbf{K}(0)\mathbf{B}\mathbf{R}^{-1}\mathbf{B}'\mathbf{K}(0) + \mathbf{Q} = \mathbf{0} \quad (12\text{-}209)$$

Subtraction of Eq. (12-208) from Eq. (12-209) yields

$$[\mathbf{A}' - \mathbf{K}(0)\mathbf{B}\mathbf{R}^{-1}\mathbf{B}'][\mathbf{K}(0) - \Pi] + [\mathbf{K}(0) - \Pi][\mathbf{A} - \mathbf{B}\mathbf{R}^{-1}\mathbf{B}'\mathbf{K}(0)] + \mathbf{Q} = \mathbf{0} \quad (12\text{-}210)$$

The unique positive semidefinite solution of Eq. (12-210), $\mathbf{K}(0) - \Pi$, is enough to determine $\partial^2\mathbf{K}(T)/\partial T^2$ from Eq. (12-207).

Sampling Period Sensitivity Study of the Feedback Gain Matrix

The best approximation of $\mathbf{K}(T)$ given by a finite number (two) of terms of Eq. (12-179) can be substituted in Eq. (12-177) to give an approximation to the optimal feedback matrix $\mathbf{G}(T)$. For the second-order approximation, $\mathbf{G}(T)$ is given by

$$\mathbf{G}(T) = \left[\frac{\hat{\mathbf{R}} + \theta'\left[\mathbf{K}(0) + \dfrac{\partial^2\mathbf{K}(T)}{\partial T^2}\dfrac{T^2}{2}\right]\theta}{T}\right]^{-1}\left[\frac{\mathbf{M}' + \theta'\left[\mathbf{K}(0) + \dfrac{\partial^2\mathbf{K}(T)}{\partial T^2}\dfrac{T^2}{2}\right]\phi}{T}\right] \quad (12\text{-}211)$$

As an alternative, we can expand $\mathbf{G}(T)$ into a Taylor series about $T = 0$; that is

$$\mathbf{G}(T) = \mathbf{G}(0) + \sum_{i=1}^{\infty} \frac{\partial^i\mathbf{G}(T)}{\partial T^i}\frac{T^i}{i!} \quad (12\text{-}212)$$

where $\mathbf{G}(0)$ is given by Eq. (12-178), and the first and second derivatives of $\mathbf{G}(T)$ are given by Eqs.(12-197) and (12-205), respectively.

Example 12-6.

Consider that the state equations of a second-order process are given by

$$\dot{\mathbf{x}}(t) = \mathbf{A}\mathbf{x}(t) + \mathbf{B}\mathbf{u}(t) \quad (12\text{-}213)$$

where

$$\mathbf{A} = \begin{bmatrix} 0 & 0 \\ 0 & 0 \end{bmatrix} \qquad \mathbf{B} = \begin{bmatrix} 1 & 0 \\ 0 & 1 \end{bmatrix} \tag{12-214}$$

The problem is to find the optimal sampled-data control $\mathbf{u}(t) = \mathbf{u}(kT)$, $k = 0,1,2,\ldots$, such that the performance index

$$J = \frac{1}{2}\int_0^\infty \left[[x_1(t) - x_2(t)]^2 + x_1^2(t) + x_2^2(t) + u_1^2(t) + u_2^2(t) \right] dt \tag{12-215}$$

is minimized.

According to the notation used in Eq. (12-3), the following weighting matrices are defined:

$$\mathbf{Q} = \begin{bmatrix} 2 & -1 \\ -1 & 2 \end{bmatrix} \quad , \quad \mathbf{R} = \begin{bmatrix} 1 & 0 \\ 0 & 1 \end{bmatrix} \tag{12-216}$$

The sampling period of the system is not specified at the outset since its effect on the optimal linear digital regulator is to be investigated.

Since the process in Eq. (12-213) is controllable (as well as stabilizable), and the pair $[\mathbf{A}, \mathbf{D}]$ is observable, where $\mathbf{DD}' = \mathbf{Q}$, the linear continuous-data regulator problem has an asymptotically stable solution; that is, for $T = 0$. The coefficient matrix of the sampled-data process is $\phi(T) = e^{\mathbf{A}T} = \mathbf{I}$, and

$$\theta(T) = \int_0^T \phi(\lambda)\mathbf{B}d\lambda = \begin{bmatrix} T & 0 \\ 0 & T \end{bmatrix} \tag{12-217}$$

Therefore, the sampled-data system

$$\mathbf{x}[(k+1)T] = \phi(T)\mathbf{x}(kT) + \theta(T)\mathbf{u}(kT) \tag{12-218}$$

is completely controllable, and since $\hat{\mathbf{R}}$ as given by Eq. (12-13) is nonsingular, the pair $[\phi, \mathbf{D}]$ is observable for any $\mathbf{DD}' = \mathbf{Q}$. Thus, the system in Eq. (12-218) has an asymptotically stable solution for the optimal linear digital regulator problem.

Ordinarily, the optimal linear digital regulator problem can be solved by formulating the discrete Riccati equation of Eq. (12-30) or Eq. (12-34), solve for the Riccati gain $\mathbf{K}(T)$, and then the optimal control is given by Eq. (12-56). In this case since the sampling period T is not specified, it would be difficult to solve the Riccati equation whether the recursive method or the eigenvalue-eigenvector method is used.

Let us solve the problem by use of the sampling sensitivity method. Since the matrices $\mathbf{K}(0)$ and $\mathbf{G}(0)$ are required, we must first solve the optimal linear continuous-data regulator problem. The Riccati equation for the continuous-data regulator is given in Eq. (12-175), and upon substitution of the system parameters, it is simplified to

$$\mathbf{K}^2(0) = \mathbf{Q} = \begin{bmatrix} 2 & -1 \\ -1 & 2 \end{bmatrix} \tag{12-219}$$

Thus,

$$\mathbf{K}(0) = \frac{1}{2}\begin{bmatrix} 1+\sqrt{3} & 1-\sqrt{3} \\ 1-\sqrt{3} & 1+\sqrt{3} \end{bmatrix} \tag{12-220}$$

The optimal feedback gain is determined by using Eq. (12-178), which leads to

$$\mathbf{G}(0) = \mathbf{R}^{-1}\mathbf{B}'\mathbf{K}(0) = \mathbf{K}(0) \tag{12-221}$$

The first-order sampling period sensitivity of $\mathbf{K}(T)$ is always zero. The second-order sensitivity $\partial^2\mathbf{K}(T)/\partial T^2$, is obtained by first solving for $\mathbf{\Pi}$ from Eq. (12-210). Equation (12-210) becomes

$$-2\mathbf{K}^2(0) + 2\mathbf{K}(0)\mathbf{\Pi} + \mathbf{Q} = \mathbf{0} \tag{12-222}$$

from which we have

$$\mathbf{\Pi} = \frac{1}{4}\begin{bmatrix} 1+\sqrt{3} & 1-\sqrt{3} \\ 1-\sqrt{3} & 1+\sqrt{3} \end{bmatrix} \tag{12-223}$$

Substitution of $\mathbf{\Pi}$ into Eq. (12-207) gives

$$\frac{\partial^2\mathbf{K}(T)}{\partial T^2} = \frac{1}{6}\mathbf{K}(0)\mathbf{\Pi}\mathbf{K}(0) = \frac{1}{24}\begin{bmatrix} 1+3\sqrt{3} & 1-3\sqrt{3} \\ 1-3\sqrt{3} & 1+3\sqrt{3} \end{bmatrix} \tag{12-224}$$

Thus, the second-order approximation of $\mathbf{K}(T)$ is

$$\mathbf{K}(T) \cong \mathbf{K}(0) + \frac{T^2}{2}\frac{\partial^2\mathbf{K}(T)}{\partial T^2} = \begin{bmatrix} e_1+e_2 & e_1-e_2 \\ e_1-e_2 & e_1+e_2 \end{bmatrix} \tag{12-225}$$

where

$$e_1 = \frac{1}{2} + \frac{T^2}{48} \tag{12-226}$$

$$e_2 = \frac{\sqrt{3}}{2} + \frac{\sqrt{3}\,T^2}{16} \tag{12-227}$$

We can show that by solving the Riccati equation in Eq. (12-174) the exact Riccati gain is found to be

$$\mathbf{K}(T) = \begin{bmatrix} d_1+d_2 & d_1-d_2 \\ d_1-d_2 & d_1+d_2 \end{bmatrix} \tag{12-228}$$

where

$$d_1 = \frac{1}{2}\left[1 + \frac{T^2}{12}\right]^{1/2} \tag{12-229}$$

$$d_2 = \frac{\sqrt{3}}{2} + \frac{\sqrt{3}\,T^2}{8} \tag{12-230}$$

Substitution of the approximation for K(T) in Eq. (12-225) into Eq. (12-177), G(T) is approximated by (after lengthy matrix manipulation)

$$G(T) \cong \frac{1}{2}\begin{bmatrix} f_1 + f_2 & f_1 - f_2 \\ f_1 - f_2 & f_1 + f_2 \end{bmatrix} \tag{12-231}$$

where

$$f_1 = \frac{24 + 12T + T^2}{24 + 24T + 8T^2 + T^3} \tag{12-232}$$

$$f_2 = \frac{8\sqrt{3} + 12T + \sqrt{3}\,T^2}{8 + 8\sqrt{3}\,T + 8T^2 + \sqrt{3}\,T^3} \tag{12-233}$$

As an alternative we can approximate G(T) by the first three terms of the series expansion in Eq. (12-212). From Eq. (12-221), we have found that G(0) = K(0). The first-order sensitivity of G(T) is given by Eq. (12-194); we have

$$\frac{\partial G(T)}{\partial T} = \frac{1}{2}\begin{bmatrix} -2 & 1 \\ 1 & -2 \end{bmatrix} \tag{12-234}$$

The second-order sensitivity of G(T) is determined from Eq. (12-205),

$$\frac{\partial^2 G(T)}{\partial T^2} = \frac{5}{24}\begin{bmatrix} 1 + 3\sqrt{3} & 1 - 3\sqrt{3} \\ 1 - 3\sqrt{3} & 1 + 3\sqrt{3} \end{bmatrix} \tag{12-235}$$

Thus, the series approximation of G(T) gives

$$G(T) \cong G(0) + T\frac{\partial G(T)}{\partial T} + \frac{T^2}{2!}\frac{\partial^2 G(T)}{\partial T^2}$$

$$= \begin{bmatrix} \frac{1+\sqrt{3}}{2} + T + \frac{5T^2}{48}(1 + 3\sqrt{3}) & \frac{1-\sqrt{3}}{2} + \frac{T}{2} + \frac{5T^2}{48}(1 - 3\sqrt{3}) \\ \frac{1-\sqrt{3}}{2} + \frac{T}{2} + \frac{5T^2}{48}(1 - 3\sqrt{3}) & \frac{1+\sqrt{3}}{2} + T + \frac{5T^2}{48}(1 + 3\sqrt{3}) \end{bmatrix} \tag{12-236}$$

The exact feedback gain matrix is obtained by substituting the exact expression of

K(T) in Eq. (12-228) into Eq. (12-177), and the result is

$$\mathbf{G}(T) = \frac{1}{2}\begin{bmatrix} g_1 + g_2 & g_1 - g_2 \\ g_1 - g_2 & g_1 + g_2 \end{bmatrix} \tag{12-237}$$

where

$$g_1 = \frac{\left[1 + \frac{T^2}{12}\right]^{1/2} + \frac{T}{2}}{1 + T\left[1 + \frac{T^2}{12}\right]^{1/2} + \frac{T^2}{3}} \tag{12-238}$$

$$g_2 = \frac{\sqrt{3}\left[1 + \frac{T^2}{4}\right]^{1/2} + \frac{3T}{2}}{1 + \sqrt{3}\,T\left[1 + \frac{T^2}{4}\right]^{1/2} + T^2} \tag{12-239}$$

Sampling Period Sensitivity And Digital Redesign

The digital redesign problem studied in Chapter 6 was motivated by the development of the sampling period sensitivity of the Riccati gain and the optimal feedback gain. In Chapter 6 the digital redesign problem is solved by means of a point-by-point comparison of the state responses of the continuous-data system and the digital system at the sampling instants. The development given in the preceding section leads directly to another solution to the digital redesign problem using the sampling period sensitivity.

With reference to the notation used in this chapter, the digital redesign problem may be stated as: given the continuous-data system of Eq. (12-1) which has the optimal control

$$\mathbf{u}(t) = -\mathbf{G}(0)\mathbf{x}(t) \tag{12-240}$$

that minimizes the performance index in Eq. (12-3) for $t_f = \infty$. The optimal feedback gain $\mathbf{G}(0)$ is given by Eq. (12-178); that is,

$$\mathbf{G}(0) = \mathbf{R}^{-1}\mathbf{B}'\mathbf{K}(0) \tag{12-241}$$

where $\mathbf{K}(0)$ is the positive semidefinite solution of the Riccati equation in Eq. (12-175).

Let the continuous-data system in Eq. (12-1) be approximated by a digital system in such a way that the control $\mathbf{u}(t)$ is derived through sample-and-hold units. Thus,

$$\mathbf{u}(t) = \mathbf{u}(kT) \qquad kT \leqslant t < (k+1)T \tag{12-242}$$

The problem is to find the optimal state-feedback control

$$\mathbf{u}(kT) = -\mathbf{G}(T)\mathbf{x}(kT) \tag{12-243}$$

such that the discretized performance index of Eq. (12-7) is minimized for $N = \infty$.

Summarizing the results obtained in the preceding section, the feedback gain of the digital system is expanded into a Taylor series about $T = 0$, and by taking only the first three terms, we have

$$\mathbf{G}(T) = \mathbf{G}(0) + T\,\frac{\partial \mathbf{G}(T)}{\partial T} + \frac{T^2}{2!}\,\frac{\partial^2 \mathbf{G}(T)}{\partial T^2} \tag{12-244}$$

where $\mathbf{G}(0)$ is given by Eq. (12-241) and from Eq. (12-194),

$$\frac{\partial \mathbf{G}(T)}{\partial T} = \frac{1}{2}\,\mathbf{G}(0)[\mathbf{A} - \mathbf{B}\mathbf{G}(0)] \tag{12-245}$$

Equation (12-205) gives

$$\frac{\partial^2 \mathbf{G}(T)}{\partial T^2} = \mathbf{R}^{-1}\mathbf{B}'\,\frac{\partial^2 \mathbf{K}(T)}{\partial T^2} + \frac{1}{3}\,\mathbf{G}(0)[\mathbf{A} - \mathbf{B}\mathbf{G}(0)]^2 \tag{12-246}$$

where $\partial^2\mathbf{K}(T)/\partial T^2$ is solved from Eq. (12-206).

Comparing these results with those in Chapter 6 it is interesting to note that the first two terms of the Taylor series expansion of $\mathbf{G}(T)$ by the two methods are identical.

REFERENCES

1. Dorato, P., and Levis, A. H., "Optimal Linear Regulator: The Discrete-Time Case," *IEEE Trans. on Automatic Control,* Vol. AC-16, December 1971, pp. 613-620.
2. Kleinman, D. L., "Stabilizing A Discrete, Constant, Linear System With Application to Iterative Methods for Solving the Riccati Equation," *IEEE Trans. on Automatic Control,* Vol. AC-19, June 1974, pp. 252-254.
3. Vaughan, D. R., "A Nonrecursive Algebraic Solution for the Discrete Riccati Equation," *IEEE Trans. on Automatic Control,* Vol. AC-15, October 1970, pp. 597-599.
4. Howerton, R. D., "A New Solution of the Discrete Algebraic Riccati Equation," *IEEE Trans. on Automatic Control,* Vol. AC-19, February 1974, pp. 90-92.
5. Lainiotis, D. G., "Discrete Riccati Equation Solutions: Partitioned Algorithms," *IEEE Trans. on Automatic Control,* Vol. AC-20, August 1975, pp. 555-556.
6. Molinari, B. P., "The Stabilizing Solution of the Discrete Algebraic Riccati Equation," *IEEE Trans. on Automatic Control,* Vol. AC-20, June 1975, pp. 396-399.

7. Molinari, B. P., "The Stabilizing Solution of the Algebraic Riccati Equation," *SIAM Journal on Control,* Vol. 11, May 1973, pp. 262-271.
8. Payne, H. J., and Silverman, L. M., "On the Discrete Time Algebraic Riccati Equation," *IEEE Trans. on Automatic Control,* Vol. AC-18, June 1973, pp. 226-234.
9. Caines, P. E., and Mayne, D. Q., "On the Discrete-Time Matrix Riccati Equation of Optimal Control," *International J. on Control,* Vol. 12, November 1970, pp. 785-794.
10. Hewer, G. A., "An Iterative Technique for the Computation of the Steady-State Gains For the Discrete Optimal Regulator," *IEEE Trans. on Automatic Control,* Vol. AC-16, August 1971, pp. 382-384.
11. Hewer, G. A., "Analysis of a Discrete Matrix Riccati Equation of Linear Control and Kalman Filtering," *J. Math. Analysis and Applications,* Vol. 42, April 1973, pp. 226-236.
12. Cadzow, J. A., "Nilpotency Property of the Discrete Regulator," *IEEE Trans. on Automatic Control,* Vol. AC-13, December 1968, pp. 734-735.

PROBLEMS

12-1 Consider the optimal capital allocation problem:

c = initial amount of capital to be invested in investment A or investment B over a total of N periods

x = amount of capital to be invested in the first period in investment choice A

$g(x) = 0.4\sqrt{x}$ = return from investment A for one period

$h(c-x) = 0.2\sqrt{c-x}$ = return from investment B for one period

ax = capital left at the end of the first stage from investment A

$b(c-x)$ = capital left at the end of the first stage from investment B

$a = 0.6$ and $b = 0.8$.

(a) Set up the capital allocation problem as an optimal control problem. Find the maximum return for a three-stage process using the principle of optimality.

(b) Determine the optimum allocation of capital at the beginning of each period.

12-2 A linear digital system is characterized by the state equation

$$\frac{d\mathbf{x}(t)}{dt} = \mathbf{A}\mathbf{x}(t) + \mathbf{B}u(t)$$

where

$$\mathbf{A} = \begin{bmatrix} 0 & 1 \\ -4 & -5 \end{bmatrix} \qquad \mathbf{B} = \begin{bmatrix} 0 \\ 1 \end{bmatrix}$$

The control is described by $u(t) = u(kT)$ for $kT \leqslant t < (k+1)T$, where $T = 0.5$ sec. Determine the optimal control $u(kT)$ as a function of $\mathbf{x}(kT)$ for $k = 0,1,2,\ldots,4$, which minimizes the performance index

$$J = \sum_{k=0}^{4} x_1^2[(k+1)T]$$

Determine the optimal trajectories for $x_1(kT)$ and $x_2(kT)$ as functions of $\mathbf{x}(0)$. Repeat the problem for $N = \infty$.

12-3 A linear digital process is represented by the state equation

$$\mathbf{x}(k+1) = \mathbf{A}\mathbf{x}(k) + \mathbf{B}u(k)$$

where

$$\mathbf{A} = \begin{bmatrix} 0 & 1 \\ -1 & -2 \end{bmatrix} \qquad \mathbf{B} = \begin{bmatrix} 0 \\ 1 \end{bmatrix}$$

(a) Determine the optimal control u(0), u(1), ..., u(4), in the form of state feedback so that the performance index

$$J = \frac{1}{2} \sum_{k=0}^{5} [x_1^2(k) + x_2^2(k)]$$

is minimized.

(b) Repeat part (a) when $N = \infty$, infinite time.

12-4 Repeat Problem 12-3 when the performance index is

$$J = \frac{1}{2} \sum_{k=0}^{5} [x_1^2(k) + x_2^2(k) + u^2(k)]$$

12-5 (a) Find the optimal control sequence u(0), u(1), u(2), such that the performance index

$$J = \frac{1}{2} \sum_{k=0}^{2} [x_1^2(k) + u^2(k)]$$

is minimized, subject to the constraint

$$\mathbf{x}(k+1) = \mathbf{A}\mathbf{x}(k) + \mathbf{B}u(k)$$

$$\mathbf{A} = \begin{bmatrix} 0 & 1 \\ -1 & 1 \end{bmatrix} \qquad \mathbf{B} = \begin{bmatrix} 0 \\ 1 \end{bmatrix}$$

$$\mathbf{x}(0) = \begin{bmatrix} 1 \\ 1 \end{bmatrix} \qquad \mathbf{x}(3) = \text{free}$$

Use the Riccati equation method, and solve for the Riccati gain matrices $\mathbf{K}(0)$, $\mathbf{K}(1)$, $\mathbf{K}(2)$, and $\mathbf{K}(3)$.

(b) Repeat part (a) for the performance index

$$J = \frac{1}{2}\mathbf{x}'(3)\mathbf{S}\mathbf{x}(3) + \sum_{k=0}^{2} [x_1^2(k) + u^2(k)]$$

where $\mathbf{S} = \mathbf{I}$ (identity matrix).

(c) Repeat part (a) for the infinite-time problem.

12-6 A linear digital process with input delays is described by the state equation

$$\mathbf{x}[(k+1)T] = \mathbf{A}\mathbf{x}(kT) + \mathbf{B}u[(k-p)T]$$

where $\mathbf{x}(kT)$ is an $n \times 1$ state vector and $u[(k-p)T]$ is the delayed scalar control. The delay time pT is assumed to be an integral multiple of T. Given the initial state $\mathbf{x}(0)$ and the past inputs, $u(-pT)$, $u[(-p+1)T], \ldots, u(-T)$, determine the optimal control $u(0), u(T), \ldots, u[(N-1)T]$ in terms of state feedback and the past inputs so that the performance index

$$J = \frac{1}{2}\sum_{k=0}^{N+p-1} \left[\mathbf{x}'(kT)\mathbf{Q}(kT)\mathbf{x}(kT) + u^2[(k-p-1)T]\right]$$

is minimized for $N > p$.

12-7 Consider that a linear digital process with input delay is described by the first-order difference equation

$$x(k+1) = Ax(k) + Bu(k-1)$$

where $A = 0.368$ and $B = 0.632$. Determine the optimal control $u(0)$ through $u(3)$ in terms of state feedback so that the performance index

$$J = \frac{1}{2}\sum_{k=0}^{4} [x^2(k) + u^2(k-1)]$$

is minimized. For $x(0) = -1$ and $u(-1) = 0.178$, determine the optimal trajectory for $x(k)$, $k = 0,1,2,3,4$.

12-8 A first-order digital control process with state delay is described by the state equation

$$x(k+1) = x(k) + 2x(k-1) + u(k)$$

The performance index is

$$J = \frac{1}{2}\sum_{k=0}^{N-1} [x^2(k) + u^2(k)]$$

Find the optimal control u(0), u(1) so that J is minimized for N = 2. The initial states are given as x(0) and x(−1).

12-9 A tracking problem of the linear regulator design is described as one which attempts to match the system output **c**(k) with a desired output $\mathbf{c}_d(k)$. The tracking problem can be stated as: given the linear process

$$\mathbf{x}(k+1) = \mathbf{A}\mathbf{x}(k) + \mathbf{B}\mathbf{u}(k)$$

$$\mathbf{c}(k) = \mathbf{D}\mathbf{x}(k)$$

where **x**(k) is the n × 1 state vector, **u**(k) is the r × 1 input vector, and **c**(k) is the p × 1 output vector. It is assumed that **A** is nonsingular. Find the optimal control **u**(k), k = 0,1,...,N − 1, so that the performance index

$$J = \frac{1}{2}\mathbf{e}'(N)\mathbf{S}\mathbf{e}(N) + \frac{1}{2}\sum_{k=0}^{N-1}[\mathbf{e}'(k)\mathbf{Q}\mathbf{e}(k) + \mathbf{u}'(k)\mathbf{R}\mathbf{u}(k)]$$

is a minimum. **S** and **Q** are positive semidefinite, and **R** is positive definite, and all are symmetric matrices.

$$\mathbf{e}(k) = \mathbf{c}_d(k) - \mathbf{c}(k)$$

where $\mathbf{c}_d(k)$ is a constant vector for a given k.

12-10 Given the digital control system

$$\mathbf{x}(k+1) = \mathbf{A}\mathbf{x}(k) + \mathbf{B}u(k)$$

where

$$\mathbf{A} = \begin{bmatrix} 0 & 1 \\ -0.5 & -1 \end{bmatrix} \qquad \mathbf{B} = \begin{bmatrix} 0 \\ 1 \end{bmatrix}$$

The control is

$$u(k) = -\mathbf{G}\mathbf{x}(k)$$

where

$$\mathbf{G} = [-0.4 \quad -0.2]$$

Determine for what **Q** is the system optimal in the sense that

$$J = \frac{1}{2}\sum_{k=0}^{\infty}[\mathbf{x}'(k)\mathbf{Q}\mathbf{x}(k) + u^2(k)]$$

is a minimum.

12-11 Given the digital process

$$\mathbf{x}(k+1) = \mathbf{A}\mathbf{x}(k) + \mathbf{B}u(k)$$

where

$$\mathbf{A} = \begin{bmatrix} 0 & 1 \\ -0.4 & -0.6 \end{bmatrix} \qquad \mathbf{B} = \begin{bmatrix} 0 \\ 1 \end{bmatrix}$$

The control u(k) is obtained through partial state feedback,

$$u(k) = -\mathbf{G}\mathbf{x}(k)$$

where

$$\mathbf{G} = [-0.3 \quad 0]$$

Determine the admissible **Q** so that the system is optimal in the sense that

$$J = \frac{1}{2} \sum_{k=0}^{\infty} [\mathbf{x}'(k)\mathbf{Q}\mathbf{x}(k) + u^2(k)]$$

is a minimum.

12-12 Given the digital control process

$$\mathbf{x}(k+1) = \mathbf{A}\mathbf{x}(k) + \mathbf{B}u(k)$$

where

$$\mathbf{A} = \begin{bmatrix} 0 & 1 \\ -0.4 & -0.6 \end{bmatrix} \qquad \mathbf{B} = \begin{bmatrix} 0 \\ 1 \end{bmatrix}$$

$$u(k) = -\mathbf{G}\mathbf{x}(k)$$

The characteristic equation of the closed-loop system is specified as

$$|\lambda\mathbf{I} - \mathbf{A} + \mathbf{B}\mathbf{G}| = \lambda^2 + 6\lambda = 0$$

Find the admissible **Q** so that the system is optimal in the sense that

$$J = \frac{1}{2} \sum_{k=0}^{\infty} [\mathbf{x}'(k)\mathbf{Q}\mathbf{x}(k) + u^2(k)] = \text{minimum}$$

12-13 Consider that a linear process is described by

$$\dot{\mathbf{x}}(t) = \mathbf{A}\mathbf{x}(t) + \mathbf{B}u(t)$$

where u(t) is a scalar input. The input is derived from a pulsewidth modulator so that u(t) is described by

$$u(t) = \epsilon_k U[u(kT) - u(kT - \rho_k)] \qquad kT \leqslant t < (k+1)T$$

$$0 \leqslant \rho_k \leqslant T \qquad \epsilon_k = \pm 1$$

$$U = \text{constant}$$

Devise a way to find the optimal control u(t) so that the performance index

$$J = \sum_{k=0}^{N-1} \left[\left[\mathbf{x}_d[(k+1)T] - \mathbf{x}[(k+1)T] \right]' \mathbf{Q} \left[\mathbf{x}_d[(k+1)T] - \mathbf{x}[(k+1)T] \right] \right]$$

is minimized. $\mathbf{x}_d[(k+1)T]$, $k = 0,1,2,3,\ldots,N-1$, is the desired state trajectory.

12-14 Consider that the state equation of a first-order linear process is given by

$$\dot{x}(t) = -2x(t) + u(t)$$

with

$$u(t) = u(kT) \qquad kT \leqslant t < (k+1)T$$

with the initial state given as $x(t_0)$. It is desired to find the optimal control u(kT) such that the performance index

$$J = \frac{1}{2} \sum_{0}^{\infty} [x^2(t) + u^2(t)]\,dt = \text{minimum}$$

Find the optimal control $u(kT) = -G(T)x(kT)$ by approximating the Riccati gain K(T) and the feedback matrix G(T) by the first three terms of their power series expansions. Express K(T) and G(T) as functions of T. Evaluate G(T) for $T = 0.1$, 0.5, and 1.0. Determine for what values of T the closed-loop system will be unstable.

13

DIGITAL STATE OBSERVER

13.1 INTRODUCTION

A significant portion of the optimal control theory of digital control systems rely on the feeding back of the state variables to form the control. Unfortunately, in practice, not all the state variables are accessible, and, in general, only the outputs of the system are measurable. Therefore, when feedback from all the state variables is required in a given design, and not all the state variables are accessible, it is necessary to "observe" the states from information contained in the output as well as the input variables. The subsystem that performs the observation of the state variables based on information received from the measurements of the input and the output is called a *state observer*, or simply an *observer*. Figure 13-1 shows the block diagram of a digital control system which has a state observer. The observed state vector $\mathbf{x}_e(k)$ is fed to the feedback gain $\mathbf{G}$ which in turn generates the control signal $\mathbf{u}(k)$. From Fig. 13-1 the

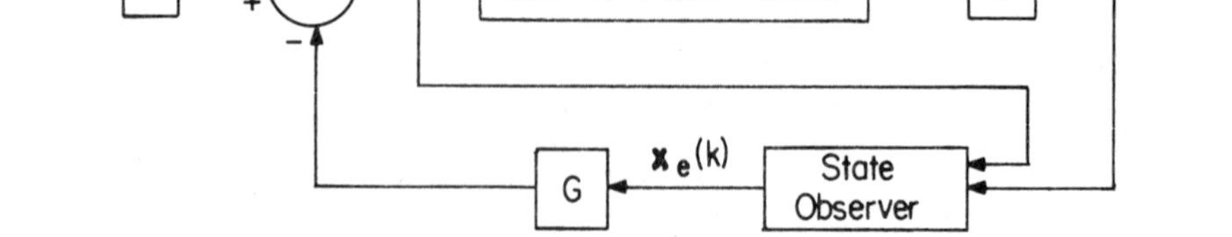

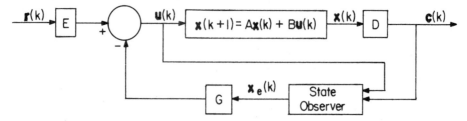

Fig. 13-1. Linear digital system with state observer and state feedback.

control signal is expressed as

$$\mathbf{u}(k) = \mathbf{E}\mathbf{r}(k) - \mathbf{G}\mathbf{x}_e(k) \tag{13-1}$$

We must first establish the condition under which an observer exists. The following theorem indicates that the design of the digital state observer is closely related to the observability criterion.

Theorem 13-1. Given a linear digital system that is described by the dynamic equations,

$$\mathbf{x}(k+1) = \mathbf{A}\mathbf{x}(k) + \mathbf{B}\mathbf{u}(k) \tag{13-2}$$

$$\mathbf{c}(k) = \mathbf{D}\mathbf{x}(k) \tag{13-3}$$

where $\mathbf{x}(k)$ is an n-vector, $\mathbf{u}(k)$ is a p-vector and $\mathbf{c}(k)$ is a q-vector. It is assumed that the matrix $\mathbf{A}$ is nonsingular. The state vector $\mathbf{x}(k)$ may be constructed from linear combinations of the output, $\mathbf{c}(k)$, input $\mathbf{u}(k)$, and the past values of these variables, if the digital system is completely observable.

Proof: Equation (13-2) leads to

$$\mathbf{x}(k-1) = \mathbf{A}^{-1}\mathbf{x}(k) - \mathbf{A}^{-1}\mathbf{B}\mathbf{u}(k-1) \qquad (k \geqslant 1) \tag{13-4}$$

In general,

$$\mathbf{x}(k-n) = \mathbf{A}^{-1}\mathbf{x}(k-n+1) - \mathbf{A}^{-1}\mathbf{B}\mathbf{u}(k-n) \qquad (k \geqslant n) \tag{13-5}$$

From Eq. (13-3) we write

$$\mathbf{c}(k-1) = \mathbf{D}\mathbf{x}(k-1) \qquad (k \geqslant 1) \tag{13-6}$$

Substitution of Eq. (13-4) into Eq. (13-6) yields

$$\mathbf{c}(k-1) = \mathbf{D}\mathbf{A}^{-1}\mathbf{x}(k) - \mathbf{D}\mathbf{A}^{-1}\mathbf{B}\mathbf{u}(k-1) \qquad (k \geqslant 1) \tag{13-7}$$

Similarly,

$$\begin{aligned}\mathbf{c}(k-2) &= \mathbf{D}\mathbf{x}(k-2)\\ &= \mathbf{D}\mathbf{A}^{-2}\mathbf{x}(k) - \mathbf{D}\mathbf{A}^{-2}\mathbf{B}\mathbf{u}(k-1) - \mathbf{D}\mathbf{A}^{-1}\mathbf{B}\mathbf{u}(k-2)\end{aligned} \tag{13-8}$$

Continuing the process, we have

$$\mathbf{c}(k-N) = \mathbf{D}\mathbf{A}^{-N}\mathbf{x}(k) - \sum_{i=1}^{N} \mathbf{D}\mathbf{A}^{-N+i-1}\mathbf{B}\mathbf{u}(k-i) \qquad (k \geqslant N) \tag{13-9}$$

The preceding equations are written in matrix form as follows:

$$\begin{bmatrix} \mathbf{c}(k-1) \\ \mathbf{c}(k-2) \\ \cdot \\ \cdot \\ \cdot \\ \mathbf{c}(k-N) \end{bmatrix} = \begin{bmatrix} \mathbf{DA}^{-1} \\ \mathbf{DA}^{-2} \\ \cdot \\ \cdot \\ \cdot \\ \mathbf{DA}^{-N} \end{bmatrix} \mathbf{x}(k) - \begin{bmatrix} \mathbf{DA}^{-1}\mathbf{B} & \mathbf{0} & \mathbf{0} & \cdots & \mathbf{0} \\ \mathbf{DA}^{-2}\mathbf{B} & \mathbf{DA}^{-1}\mathbf{B} & \mathbf{0} & \cdots & \mathbf{0} \\ \mathbf{DA}^{-3}\mathbf{B} & \mathbf{DA}^{-2}\mathbf{B} & \mathbf{DA}^{-1}\mathbf{B} & \cdots & \mathbf{0} \\ \vdots & \vdots & \vdots & & \vdots \\ \mathbf{DA}^{-N}\mathbf{B} & \mathbf{DA}^{-N+1} & \mathbf{DA}^{-N+2} & \cdots & \mathbf{DA}^{-1}\mathbf{B} \end{bmatrix} \begin{bmatrix} \mathbf{u}(k-1) \\ \mathbf{u}(k-2) \\ \mathbf{u}(k-3) \\ \vdots \\ \mathbf{u}(k-N) \end{bmatrix}$$

(13-10)

This matrix equation represents Nq equations with n unknowns in the state vector $\mathbf{x}(k)$. With $Nq \geqslant n$, given the control inputs and the outpts in Eq. (13-10), $\mathbf{x}(k)$ can be determined if the matrix

$$[\mathbf{DA}^{-1} \quad \mathbf{DA}^{-2} \quad \ldots \quad \mathbf{DA}^{-N}]' \qquad (Nq \times n) \tag{13-11}$$

is of rank n. It is apparent that the rank condition on the matrix in Eq. (13-11) is the observability criterion on the pair $[\mathbf{A},\mathbf{D}]$ for $N = n$.

Under certain conditions the state vector $\mathbf{x}(k)$ can be uniquely solved from Eq. (13-11). If $Nq = n$, the matrix in Eq. (13-11) is square, and if it is nonsingular, $\mathbf{x}(k)$ can be determined for $k \geqslant N$ by obtaining N past output measurements and N past input measurements.

13.2 DESIGN OF THE FULL-ORDER STATE OBSERVER

In general, the state observer shown in Fig. 13-1 is to be designed so that the observed state $\mathbf{x}_e(k)$ will be as close as possible to the actual state $\mathbf{x}(k)$. There are many ways of designing a digital state observer, and generally there is more than one way of judging the closeness of $\mathbf{x}_e(k)$ to $\mathbf{x}(k)$. Intuitively, the state observer should have the same state equations as the original system. However, the observer should have a configuration with $\mathbf{u}(k)$ and $\mathbf{c}(k)$ as inputs and should have the capability of minimizing the error between $\mathbf{x}(k)$ and $\mathbf{x}_e(k)$ automatically.

The digital state observer design presented in the following is a parallel of the observer design for continuous-data systems.

Since $\mathbf{x}(k)$ cannot be measured directly, we cannot compare $\mathbf{x}_e(k)$ with $\mathbf{x}(k)$. As an alternative, $\mathbf{c}_e(k)$ is compared with $\mathbf{c}(k)$, where

$$\mathbf{c}_e(k) = \mathbf{D}\mathbf{x}_e(k) \tag{13-12}$$

Based on the above discussion, a logical configuration for a digital state observer is shown in Fig. 13-2. The observer is formulated as a feedback control system with $\mathbf{G}_e$ as the feedback gain matrix. The design objective is to select $\mathbf{G}_e$

such that $\mathbf{c}_e(k)$ will approach $\mathbf{c}(k)$ as fast as possible.

The state equation of the closed-loop observer is

$$\mathbf{x}_e(k+1) = (\mathbf{A} - \mathbf{G}_e\mathbf{D})\mathbf{x}_e(k) + \mathbf{B}\mathbf{u}(k) + \mathbf{G}_e\mathbf{c}(k) \tag{13-13}$$

where the matrices **A**, **B** and **D** are the same as those in Eqs. (13-2) and (13-3), and $\mathbf{G}_e$ is an $n \times q$ feedback matrix. When $\mathbf{c}_e(k)$ is equal to $\mathbf{c}(k)$, Eq. (13-13) becomes

$$\mathbf{x}_e(k+1) = \mathbf{A}\mathbf{x}_e(k) + \mathbf{B}\mathbf{u}(k) \tag{13-14}$$

which is identical to the state equation of the original system.

Substituting the block diagram in Fig. 13-2 in the state observer block of Fig. 13-1, we have the combined system shown in Fig. 13-3.

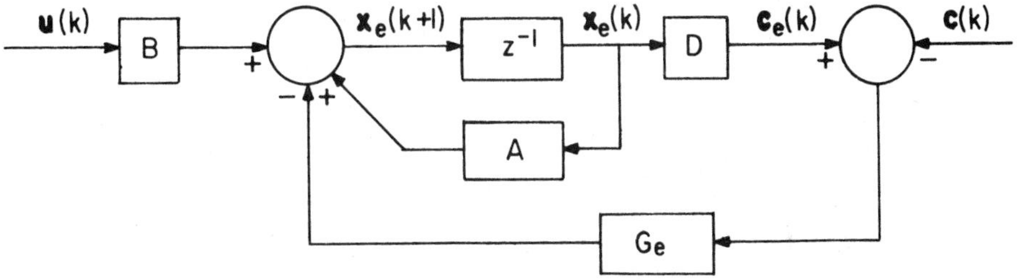

Fig. 13-2. A digital state observer with feedback.

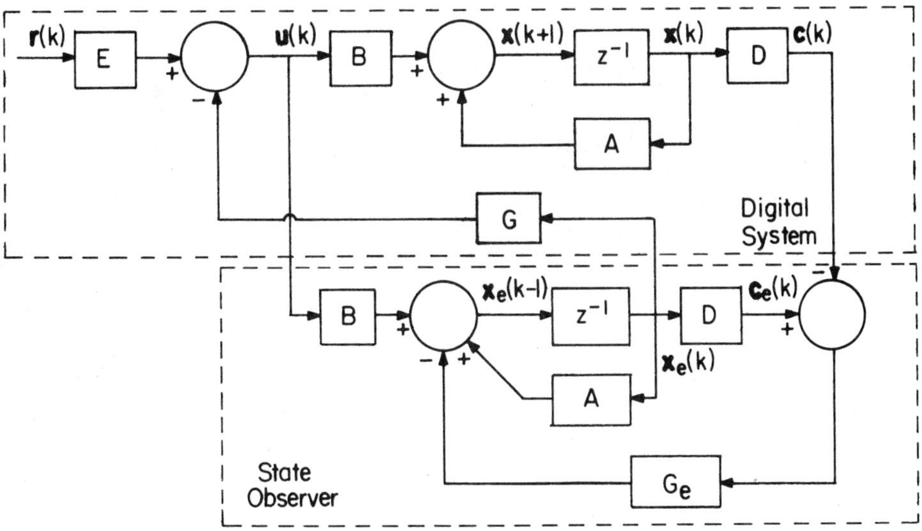

Fig. 13-3. Digital control system with state observer.

Since $\mathbf{c}(k)$ and $\mathbf{x}(k)$ are related through Eq. (13-3), Eq. (13-13) is written

$$\mathbf{x}_e(k+1) = \mathbf{A}\mathbf{x}_e(k) + \mathbf{B}\mathbf{u}(k) + \mathbf{G}_e\mathbf{D}[\mathbf{x}(k) - \mathbf{x}_e(k)] \tag{13-15}$$

The significance of this equation is that if the initial states $\mathbf{x}(0)$ and $\mathbf{x}_e(0)$ are identical, the equation is identical to Eq. (13-14), and the response of the observer will be identical to that of the original system. Therefore, the design of the observer is significant only if the initial conditions of $\mathbf{x}(k)$ and $\mathbf{x}_e(k)$ are different.

Subtracting Eq. (13-15) from Eq. (13-2) leads to the following equation:

$$\mathbf{x}(k+1) - \mathbf{x}_e(k+1) = (\mathbf{A} - \mathbf{G}_e\mathbf{D})[\mathbf{x}(k) - \mathbf{x}_e(k)] \tag{13-16}$$

which may be regarded as the homogeneous difference state equation of a linear digital system with coefficient matrix $\mathbf{A} - \mathbf{G}_e\mathbf{D}$. One way of achieving a rapid convergence of $\mathbf{x}_e(k)$ to $\mathbf{x}(k)$ would be to design $\mathbf{G}_e$ in such a way that the eigenvalues of $\mathbf{A} - \mathbf{G}_e\mathbf{D}$ are appropriately placed in the z-plane. Using the eigenvalue-assignment technique discussed in Chapter 4, the elements of $\mathbf{G}_e$ can be so selected that the natural response of the system in Eq. (13-16) decays to zero as quickly as possible.

Since the eigenvalues of $\mathbf{A} - \mathbf{G}_e\mathbf{D}$ and $(\mathbf{A} - \mathbf{G}_e\mathbf{D})' = \mathbf{A}' - \mathbf{D}'\mathbf{G}_e'$ are identical, it follows from the discussions in Sec. 4.13 that the requirement for arbitrary assignment of the eigenvalues of $\mathbf{A} - \mathbf{G}_e\mathbf{D}$ is that the pair $[\mathbf{A}', \mathbf{D}']$ be completely controllable. Since the controllability of $[\mathbf{A}', \mathbf{D}']$ is equivalent to the complete observability of $[\mathbf{A}, \mathbf{D}]$, satisfying this requirement would assure not only the existence of a state observer but also the solution of arbitrary eigenvalue assignment.

The following example will illustrate the design of a state observer by eigenvalue assignment.

Example 13-1.

Consider the digital process in Example 12-3, with the state equations described by

$$\mathbf{x}(k+1) = \mathbf{A}\mathbf{x}(k) + \mathbf{B}\mathbf{u}(k) \tag{13-17}$$

where

$$\mathbf{A} = \begin{bmatrix} 0 & 1 \\ -1 & 1 \end{bmatrix} \qquad \mathbf{B} = \begin{bmatrix} 0 \\ 1 \end{bmatrix}$$

Let the output equation be

$$\mathbf{c}(k) = \mathbf{D}\mathbf{x}(k) \tag{13-18}$$

where

$$\mathbf{D} = [2 \quad 0] \tag{13-19}$$

A digital observer is to be designed which will observe the states $x_1(k)$ and $x_2(k)$ from the output $c(k)$.

The digital observer has the block diagram shown in Fig. 13-2. The characteristic equation of the observer is

$$|\lambda\mathbf{I} - \mathbf{A} + \mathbf{G}_e\mathbf{D}| = 0 \tag{13-20}$$

or

$$\lambda^2 + (2g_{e1} - 1)\lambda + 1 + 2g_{e2} - 2g_{e1} = 0 \tag{13-21}$$

where g_{e1} and g_{e2} are the elements of the 2×1 feedback matrix $\mathbf{G}_e$. We can attempt to design the observer to have a deadbeat response so that $\mathbf{x}_e(k)$ will reach $\mathbf{x}(k)$ in one sampling period. For a deadbeat response the characteristic equation should be of the form:

$$\lambda^2 = 0 \tag{13-22}$$

Thus, from Eq. (13-21), $g_{e1} = 0.5$ and $g_{e2} = 0$. The corresponding coefficient matrix for the closed-loop observer is now

$$\mathbf{A} - \mathbf{G}_e\mathbf{D} = \begin{bmatrix} -1 & 1 \\ -1 & 1 \end{bmatrix} \tag{13-23}$$

For $u(k) = 0$, the state equations of the observer are written

$$\mathbf{x}_e(k+1) = \begin{bmatrix} -1 & 1 \\ -1 & 1 \end{bmatrix} \mathbf{x}_e(k) + \begin{bmatrix} 1 & 0 \\ 0 & 0 \end{bmatrix} \mathbf{x}(k) \tag{13-24}$$

Let us arbitrarily assume that the initial states of the system and the observer are

$$\mathbf{x}(0) = \begin{bmatrix} 1 \\ 0 \end{bmatrix} \quad \text{and} \quad \mathbf{x}_e(0) = \begin{bmatrix} 0.5 \\ 0 \end{bmatrix}$$

respectively. Equations (13-17) and (13-24) are now solved, starting with the given initial states. The true states, $x_1(k)$, $x_2(k)$, and the observed states $x_{e1}(k)$ and $x_{e2}(k)$ are sketched as shown in Fig. 13-4. The observed states reach the actual states in at most two sampling periods.

Closed-Loop Control With Observer

The digital observer design carried out thus far is for an open-loop control system; that is, $\mathbf{G} = \mathbf{0}$ in Fig. 13-3. For the general case shown in Fig. 13-3, the state equations are written as follows:

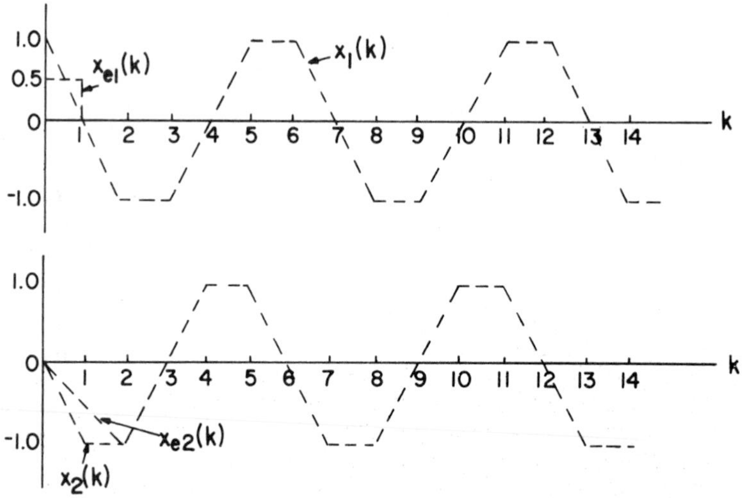

Fig. 13-4. State responses of digital system and observer.

$$\mathbf{x}(k+1) = \mathbf{A}\mathbf{x}(k) - \mathbf{B}\mathbf{G}\mathbf{x}_e(k) + \mathbf{B}\mathbf{E}\mathbf{r}(k) \tag{13-25}$$

$$\mathbf{x}_e(k+1) = (\mathbf{A} - \mathbf{B}\mathbf{G} - \mathbf{G}_e\mathbf{D})\mathbf{x}_e(k) + \mathbf{G}_e\mathbf{D}\mathbf{x}(k) + \mathbf{B}\mathbf{E}\mathbf{r}(k) \tag{13-26}$$

We can investigate the effects of the initial states of the system, the observer, and the mutual effects of these two systems by taking the z-transforms on both sides of Eqs. (13-25) and (13-26). After rearranging the terms, the z-transform equations are,

$$(z\mathbf{I} - \mathbf{A})\mathbf{X}(z) = z\mathbf{x}(0) + \mathbf{B}\mathbf{E}\mathbf{R}(z) - \mathbf{B}\mathbf{G}\mathbf{X}_e(z) \tag{13-27}$$

$$(z\mathbf{I} - \mathbf{A} + \mathbf{B}\mathbf{G} + \mathbf{G}_e\mathbf{D})\mathbf{X}_e(z) = z\mathbf{x}_e(0) + \mathbf{G}_e\mathbf{D}\mathbf{X}(z) + \mathbf{B}\mathbf{E}\mathbf{R}(z) \tag{13-28}$$

When $\mathbf{x}(0) = \mathbf{x}_e(0)$, subtracting Eq. (13-28) from Eq. (13-27), we get

$$(z\mathbf{I} - \mathbf{A} + \mathbf{G}_e\mathbf{D})\mathbf{X}(z) = (z\mathbf{I} - \mathbf{A} + \mathbf{G}_e\mathbf{D})\mathbf{X}_e(z) \tag{13-29}$$

Thus,

$$\mathbf{x}(k) = \mathbf{x}_e(k) \tag{13-30}$$

and Eq. (13-27) becomes

$$(z\mathbf{I} - \mathbf{A} + \mathbf{B}\mathbf{G})\mathbf{X}(z) = z\mathbf{x}(0) + \mathbf{B}\mathbf{E}\mathbf{R}(z) \tag{13-31}$$

This result shows that when $\mathbf{x}(0) = \mathbf{x}_e(0)$, the dynamics of the original system are independent of that of the observer. However, in general, the transient response of the original system will be effected by the observer when $\mathbf{x}(0) \neq \mathbf{x}_e(0)$.

Subtracting Eq. (13-26) from Eq. (13-25), we have

$$\mathbf{x}(k+1) - \mathbf{x}_e(k+1) = (\mathbf{A} - \mathbf{G}_e\mathbf{D})[\mathbf{x}(k) - \mathbf{x}_e(k)] \tag{13-32}$$

which shows that given **A** and **D**, the natural response of $\mathbf{x}(k) - \mathbf{x}_e(k)$ depends only upon $\mathbf{G}_e$. Therefore, the design of the observer for a system with a state feedback **G** can still be accomplished by placing the eigenvalues of $\mathbf{A} - \mathbf{G}_e\mathbf{D}$. The only difference is that, in general, when $\mathbf{x}(0) \neq \mathbf{x}_e(0)$, the response of the system, $\mathbf{x}(k)$, will be affected by the dynamics of the observer, due to the feedback of $\mathbf{x}_e(k)$ through **G**. The following example will illustrate the performance of the system and its observer in Example 13-1 when there is a state feedback matrix **G**.

Example 13-2.

For the digital process in Example 13-1, let us adopt the optimal feedback gain designed in Example 12-3,

$$\mathbf{G} = [-0.654 \quad 0.486] \tag{13-33}$$

As designed in Example 13-1, the feedback gain matrix of the digital observer for the response $\mathbf{x}(k) - \mathbf{x}_e(k)$ to be deadbeat is

$$\mathbf{G}_e = \begin{bmatrix} 0.5 \\ 0 \end{bmatrix} \tag{13-34}$$

Thus, for $\mathbf{r}(k) = \mathbf{0}$, the state equations of the digital system with feedback and the observer in Eqs. (13-25) and (13-26) are written

$$\begin{bmatrix} \mathbf{x}(k+1) \\ \hline \mathbf{x}_e(k+1) \end{bmatrix} = \left[\begin{array}{c|c} \mathbf{A} & -\mathbf{BG} \\ \hline \mathbf{G}_e\mathbf{D} & \mathbf{A} - \mathbf{BG} - \mathbf{G}_e\mathbf{D} \end{array}\right] \begin{bmatrix} \mathbf{x}(k) \\ \hline \mathbf{x}_e(k) \end{bmatrix}$$

$$= \left[\begin{array}{cc|cc} 0 & 1 & 0 & 0 \\ -1 & 1 & 0.654 & -0.486 \\ \hline 1 & 0 & -1 & 1 \\ 0 & 0 & -0.346 & 0.514 \end{array}\right] \begin{bmatrix} \mathbf{x}(k) \\ \hline \mathbf{x}_e(k) \end{bmatrix} \tag{13-35}$$

Figure 13-5 shows the response of $x_1(k)$ and $x_2(k)$ when $\mathbf{x}(0) = \mathbf{x}_e(0)$, which is equivalent to the condition that the observer is absent. The responses of the original system with the observer are also shown in Fig. 13-5. As explained earlier, the observer dynamics *do* affect the transient responses of the system. For the initial states chosen, $\mathbf{x}_e(k)$ reaches $\mathbf{x}(k)$ after two sampling periods.

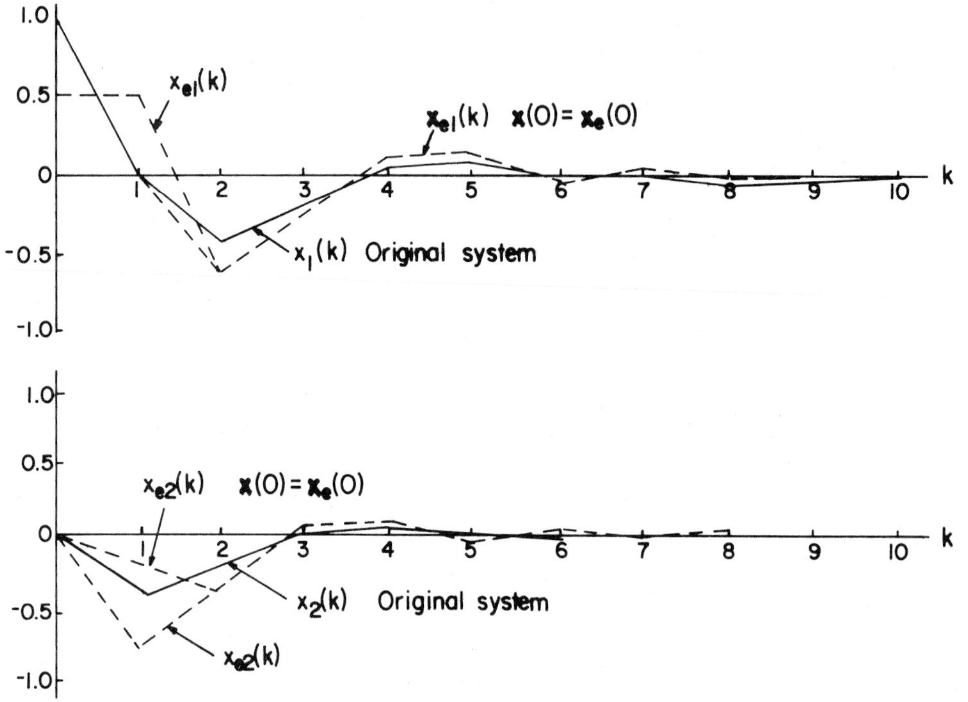

Fig. 13-5. State responses of digital feedback system with state observer.

Design By Adjoint Phase-Variable Canonical Form

Since the observer design presented in the preceding section is based on the eigenvalue assignment of the closed-loop observer, the phase-variable canonical form transformation discussed in Sec. 4.13 may be utilized. We shall consider only systems with single input and output.

Let us restate the observer design problem. The states to be observed are described by the state equation

$$\mathbf{x}(k+1) = \mathbf{A}\mathbf{x}(k) + \mathbf{B}u(k) \tag{13-36}$$

where $\mathbf{x}(k)$ is an n-vector and $u(k)$ is the scalar input. The control is obtained through state feedback, $u(k) = -\mathbf{G}\mathbf{x}(k)$, and the output equation is

$$c(k) = \mathbf{D}\mathbf{x}(k) \tag{13-37}$$

where c(k) is the scalar output, and **D** is an $1 \times n$ matrix. The observer dynamics are described by [Eq. (13-13)]

$$\mathbf{x}_e(k+1) = (\mathbf{A} - \mathbf{G}_e\mathbf{D})\mathbf{x}_e(k) + \mathbf{B}u(k) + \mathbf{G}_e c(k) \tag{13-38}$$

The original closed-loop system is represented by

$$\mathbf{x}(k+1) = (\mathbf{A} - \mathbf{B}\mathbf{G})\mathbf{x}(k) \tag{13-39}$$

It is known that if the pair [**A**,**B**] is completely controllable, then the eigenvalues of $\mathbf{A} - \mathbf{B}\mathbf{G}$ can be arbitrarily assigned by properly choosing the elements of the feedback matrix **G**. In Sec. 4.16 we have shown that if [**A**,**B**] is controllable, **A** and **B** can be transformed into the phase-variable canonical form of Eqs. (4-295) and (4-296). Then each of the feedback gain elements is associated only with one of the coefficients of the characteristic equation. In view of Eq. (13-38), the design of the observer by eigenvalue assignment creates a similar problem. In this case the eigenvalues of $\mathbf{A} - \mathbf{G}_e\mathbf{D}$ are to be located specifically by proper selection of the elements of the feedback matrix $\mathbf{G}_e$. Since the eigenvalues of $\mathbf{A} - \mathbf{G}_e\mathbf{D}$ are the same as those of $(\mathbf{A} - \mathbf{G}_e\mathbf{D})' = \mathbf{A}' - \mathbf{D}'\mathbf{G}_e'$, we can identify a parallel between the **A**, **B**, **G**, system and the $\mathbf{A}'$, $\mathbf{D}'$, $\mathbf{G}_e'$, system. Notice that the prerequisite of arbitrary eigenvalue assignment for the observer is the complete controllability of the pair $[\mathbf{A}',\mathbf{D}']$ which is equivalent to the observability of [**A**,**D**], a requirement that has been established in Sec. 13.1.

We must first transform the observer system into a so-called *adjoint phase-variable canonical form*. The reason for this modification is because the coefficient matrix in the form of $\mathbf{A}' - \mathbf{D}'\mathbf{G}_e'$ is required for isolation of the feedback matrix elements in the coefficients of the characteristic equation, and the matrix corresponds to a system with the state equation

$$\mathbf{y}(k+1) = \mathbf{A}'\mathbf{y}(k) + \mathbf{D}'v(k) \tag{13-40}$$

where

$$v(k) = -\mathbf{G}_e'\mathbf{y}(k) \tag{13-41}$$

$$\mathbf{y}(k+1) = (\mathbf{A}' - \mathbf{D}'\mathbf{G}_e')\mathbf{y}(k) \tag{13-42}$$

We need to transform the observer system in Eq. (13-38) into the following form:

$$\mathbf{y}_e(k+1) = (\mathbf{A}_1 - \mathbf{K}_e\mathbf{D}_1)\mathbf{y}_e(k) + \mathbf{B}_1u(k) + \mathbf{K}_e c(k) \tag{13-43}$$

where $\mathbf{A}_1$ is $n \times n$, $\mathbf{K}_e$ is $n \times 1$, $\mathbf{D}_1$ is $1 \times n$, and $\mathbf{B}_1$ is $n \times 1$. In general, we do not have to restrict u(k) to be a scalar input as $\mathbf{B}_1$ will not enter the design equations. The transformation is

$$\mathbf{y}_e(k) = \mathbf{P}\mathbf{x}_e(k) \tag{13-44}$$

where $\mathbf{P}$ is nonsingular. Then,

$$\mathbf{K}_e = \mathbf{P}\mathbf{G}_e = [k_{e1} \quad k_{e2} \quad \cdots \quad k_{en}]' \tag{13-45}$$

$$\mathbf{D}_1 = \mathbf{D}\mathbf{P}^{-1} = [0 \quad 0 \quad \ldots \quad 1] \tag{13-46}$$

$$\mathbf{B}_1 = \mathbf{P}\mathbf{B} \tag{13-47}$$

and

$$\mathbf{A}_1 = \mathbf{P}\mathbf{A}\mathbf{P}^{-1}$$

$$= \begin{bmatrix} 0 & 0 & \ldots & 0 & -a_n \\ 1 & 0 & \ldots & 0 & -a_{n-1} \\ 0 & 1 & \ldots & 0 & -a_{n-2} \\ \cdots & \cdots & \cdots & \cdots & \cdots \\ 0 & 0 & \ldots & 1 & -a_1 \end{bmatrix} \tag{13-48}$$

Thus, the system with $\mathbf{A}_1$ and $\mathbf{B}_1$ is known as the adjoint phase-variable canonical form since these matrices are the transposes of the phase-variable canonical form matrices $\mathbf{A}$ and $\mathbf{B}$, respectively. Then,

$$\mathbf{A}_1 - \mathbf{K}_e\mathbf{D}_1 = \begin{bmatrix} 0 & 0 & \ldots & 0 & -a_n - k_{e1} \\ 1 & 0 & \ldots & 0 & -a_{n-1} - k_{e2} \\ 0 & 1 & \ldots & 0 & -a_{n-2} - k_{e3} \\ \cdots & \cdots & \cdots & \cdots & \cdots \\ 0 & 0 & \ldots & 1 & -a_1 - k_{en} \end{bmatrix} \tag{13-49}$$

The characteristic equation of the transformed observer in Eq. (13-43) is

$$|\lambda\mathbf{I} - \mathbf{A}_1 + \mathbf{K}_e\mathbf{D}_1| = \lambda^n + (a_1 + k_{en})\lambda^{n-1} + \ldots + (a_{n-1} + k_{e2})\lambda + (a_n + k_{e1}) = 0 \tag{13-50}$$

Therefore, the coefficients of the feedback matrix $\mathbf{K}_e$ are all isolated in the coefficients of the characteristic equation.

To determine the matrix $\mathbf{P}$, we let

$$\mathbf{P} = \begin{bmatrix} \mathbf{P}_1 \\ \mathbf{P}_2 \\ \cdot \\ \cdot \\ \cdot \\ \mathbf{P}_n \end{bmatrix} \tag{13-51}$$

where $\mathbf{P}_i$ is an $1 \times n$ row matrix, $i = 1,2,\ldots,n$. Substitution of Eq. (13-51) into Eq. (13-48), we have

$$\mathbf{A}_1\mathbf{P} = \mathbf{PA} \begin{bmatrix} -a_n\mathbf{P}_n \\ \mathbf{P}_1 - a_{n-1}\mathbf{P}_n \\ \mathbf{P}_2 - a_{n-2}\mathbf{P}_n \\ \cdot \\ \cdot \\ \cdot \\ \mathbf{P}_{n-1} - a_1\mathbf{P}_n \end{bmatrix} \tag{13-52}$$

From Eq. (13-46),

$$\mathbf{D}_1\mathbf{P} = \mathbf{D} = [0 \quad 0 \quad \ldots \quad 1]\mathbf{P}$$

$$= \mathbf{P}_n \tag{13-53}$$

Equation (13-52) becomes

$$\begin{bmatrix} -a_n\mathbf{D} \\ \mathbf{P}_1 - a_{n-1}\mathbf{D} \\ \mathbf{P}_2 - a_{n-2}\mathbf{D} \\ \cdot \\ \cdot \\ \cdot \\ \mathbf{P}_{n-1} - a_1\mathbf{D} \end{bmatrix} = \begin{bmatrix} \mathbf{P}_1\mathbf{A} \\ \mathbf{P}_2\mathbf{A} \\ \cdot \\ \cdot \\ \cdot \\ \mathbf{P}_n\mathbf{A} \end{bmatrix} \tag{13-54}$$

The last equation leads to

$$\mathbf{P}_{n-1} = a_1\mathbf{D} + \mathbf{DA} \tag{13-55}$$

$$\mathbf{P}_{n-2} = a_2\mathbf{D} + \mathbf{P}_{n-1}\mathbf{A}$$

$$= a_2\mathbf{D} + a_1\mathbf{DA} + \mathbf{DA}^2 \tag{13-56}$$

$$\vdots$$

$$\mathbf{P}_2 = a_{n-2}\mathbf{D} + a_{n-3}\mathbf{DA} + \dots + a_1\mathbf{DA}^{n-3} + \mathbf{DA}^{n-2} \tag{13-57}$$

$$\mathbf{P}_1 = a_{n-1}\mathbf{D} + a_{n-2}\mathbf{DA} + \dots + a_1\mathbf{DA}^{n-2} + \mathbf{DA}^{n-1} \tag{13-58}$$

Combining Eqs. (13-53), (13-55) through (13-58), the matrix **P** is given by

$$\mathbf{P} = \begin{bmatrix} a_{n-1} & a_{n-2} & \cdots & a_1 & 1 \\ a_{n-2} & a_{n-3} & \cdots & 1 & 0 \\ a_{n-3} & a_{n-4} & \cdots & 0 & 0 \\ \cdots & \cdots & \cdots & \cdots & \cdots \\ a_1 & 1 & & 0 & 0 \\ 1 & 0 & & 0 & 0 \end{bmatrix} \begin{bmatrix} \mathbf{D} \\ \mathbf{DA} \\ \mathbf{DA}^2 \\ \vdots \\ \mathbf{DA}^{n-2} \\ \mathbf{DA}^{n-1} \end{bmatrix} \tag{13-59}$$

where it should be emphasized that the coefficients $a_1, a_2, \dots, a_{n-1}$, are identified from the last column of $\mathbf{A}_1$ in Eq. (13-48), but more important, these coefficients are found from the characteristic equation of **A** which has the form:

$$\lambda^n + a_1\lambda^{n-1} + a_2\lambda^{n-2} + \dots + a_{n-1}\lambda + a_n = 0 \tag{13-60}$$

The adjoint phase-variable canonical form method just described can be applied to systems with multiple outputs. Consider that in Eq. (13-37), c(k) is now a q-vector, and **D** is $q \times n$. Since changing the state variables of an observable system does not destroy the observability of the system, we may transform the state equations to the following form:

$$\mathbf{y}(k+1) = \begin{bmatrix} \mathbf{A}_1 & & & 0 \\ & \mathbf{A}_2 & & \\ & & \ddots & \\ 0 & & & \mathbf{A}_q \end{bmatrix} \mathbf{y}(k) + \begin{bmatrix} \mathbf{B}_1 \\ \mathbf{B}_2 \\ \vdots \\ \mathbf{B}_q \end{bmatrix} \mathbf{u}(k) \tag{13-61}$$

where $\mathbf{A}_i$, $i = 1,2,\ldots,q$, are all in the adjoint phase-variable canonical form of Eq. (13-48). The output equation is transformed to

$$\mathbf{c}(k) = \begin{bmatrix} \mathbf{D}_1 & & & 0 \\ & \mathbf{D}_2 & & \\ & & \ddots & \\ 0 & & & \mathbf{D}_q \end{bmatrix} \mathbf{y}(k) \tag{13-62}$$

where $\mathbf{D}_i$, $i = 1,2,\ldots,q$, are of the form of the row matrix in Eq. (13-46).

We shall use the same observer design problem in Example 13-2 to illustrate the adjoint phase-variable canonical form method.

Example 13-3.

For the digital process described in Examples 13-1 and 13-2, the characteristic equation of **A** is

$$|\lambda\mathbf{I} - \mathbf{A}| = \begin{vmatrix} \lambda & -1 \\ 1 & \lambda - 1 \end{vmatrix} = \lambda^2 - \lambda + 1 = 0 \tag{13-63}$$

Comparing Eqs. (13-60) and (13-63) we see that $a_1 = -1$ and $a_2 = 1$. The adjoint phase-variable canonical form of **A** is

$$\mathbf{A}_1 = \begin{bmatrix} 0 & -a_2 \\ 1 & -a_1 \end{bmatrix} = \begin{bmatrix} 0 & -1 \\ 1 & 1 \end{bmatrix} \tag{13-64}$$

For a deadbeat response for $\mathbf{x}_e(k) - \mathbf{x}(k)$, Eq. (13-50) gives

$$k_{e2} = -a_1 = 1$$

$$k_{e1} = -a_2 = -1$$

Thus,

$$\mathbf{K}_e = \begin{bmatrix} -1 \\ 1 \end{bmatrix} \tag{13-65}$$

The transformation matrix **P** is determined from Eq. (13-59):

$$\mathbf{P} = \begin{bmatrix} a_1 & 1 \\ 1 & 0 \end{bmatrix} \begin{bmatrix} \mathbf{D} \\ \mathbf{DA} \end{bmatrix} = \begin{bmatrix} -1 & 1 \\ 1 & 0 \end{bmatrix} \begin{bmatrix} 2 & 0 \\ 0 & 2 \end{bmatrix}$$

$$= \begin{bmatrix} -2 & 2 \\ 2 & 0 \end{bmatrix} \tag{13-66}$$

The feedback matrix of the observer is found by use of Eq. (13-45). We have

$$\mathbf{G}_e = \mathbf{P}^{-1}\mathbf{K}_e = \begin{bmatrix} \frac{1}{2} \\ 0 \end{bmatrix} \tag{13-67}$$

which is the same result as in Eq. (13-34).

13.3 DESIGN OF THE REDUCED-ORDER STATE OBSERVER

The observer design discussed in the preceding sections is for the full-order observer. In other words, the order of the observer is the same as that of the system. In general, since the q outputs are linear combinations of the n state variables, only a maximum of n − q states need be observed. This results in a reduced-order observer. In fact, in the system described in Example 13-1, since $c(k) = 2x_1(k)$, we can get $x_1(k)$ directly from c(k) without any observer dynamics. In this case only a first-order observer is necessary. However, it should be pointed out that while q of the states are obtained directly from the output, we may have less flexibility in the design of the dynamics of the reduced-order observer.

The principle of the reduced-order observer is illustrated by the block diagram of Fig. 13-6. The reduced-order observer is of the (n − q)th order. The n observed states, $\mathbf{x}_e(k)$, are obtained from the (n − q) observed states $\bar{\mathbf{w}}_e(k)$ and the q × 1 output vector **c**(k).

The reduced-order observer can be designed using the principle of the adjoint phase-variable canonical form transformation. We start with the system with single output and input,

$$\mathbf{x}(k+1) = \mathbf{A}\mathbf{x}(k) + \mathbf{B}u(k) \tag{13-68}$$

$$c(k) = \mathbf{D}\mathbf{x}(k) \tag{13-69}$$

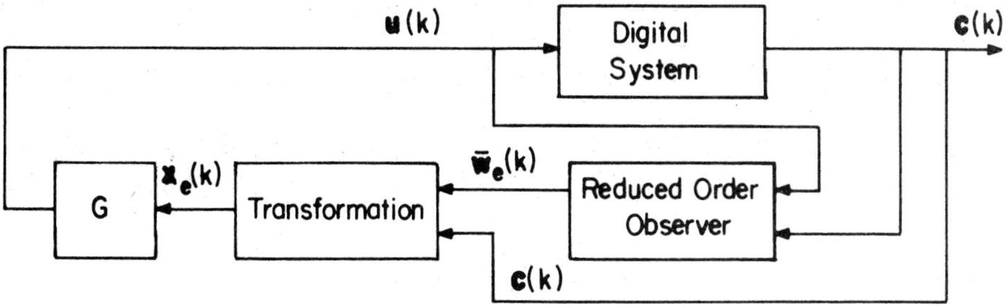

Fig. 13-6. Digital system with a reduced-order observer.

which is controllable and observable. The system is transformed to the adjoint phase-variable canonical form

$$\mathbf{y}(k+1) = \mathbf{A}_1\mathbf{y}(k) + \mathbf{B}_1 u(k) \tag{13-70}$$

where $\mathbf{A}_1 = \mathbf{PAP}^{-1}$ is given by Eq. (13-48), and $\mathbf{B}_1 = \mathbf{PB}$. The output equation is transformed to

$$c(k) = \mathbf{D}_1\mathbf{y}(k) \tag{13-71}$$

where

$$\mathbf{y}(k) = \mathbf{Px}(k)$$

$$\mathbf{D}_1 = \mathbf{DP}^{-1} = [0 \quad 0 \quad \ldots \quad 1] \qquad (1 \times n) \tag{13-72}$$

The full-order observer of the system described by Eqs. (13-68) and (13-69) is

$$\mathbf{x}_e(k+1) = (\mathbf{A} - \mathbf{G}_e\mathbf{D})\mathbf{x}_e(k) + \mathbf{B}u(k) + \mathbf{G}_e c(k) \tag{13-73}$$

where $\mathbf{G}_e$ is the $n \times 1$ feedback gain matrix of the observer. The full-order observer for the system described by the adjoint phase-variable canonical form system in Eqs. (13-70) and (13-71) is represented by

$$\mathbf{y}_e(k+1) = (\mathbf{A}_1 - \mathbf{K}_e\mathbf{D}_1)\mathbf{y}_e(k) + \mathbf{B}_1 u(k) + \mathbf{K}_e c(k) \tag{13-74}$$

where $\mathbf{K}_e$ is the $n \times 1$ feedback gain matrix of the observer. Since $\mathbf{D}_1$ is of the form given in Eq. (13-72), it implies that $c(k) = y_n(k)$, so that $y_n(k)$ does not need to be observed. To observe the remaining $n - 1$ states of $\mathbf{y}(k)$, and be able to arbitrarily place the eigenvalues, we let

$$\mathbf{Q} = \begin{bmatrix} 1 & 0 & \dots & 0 & -a_{n-1} \\ 0 & 1 & \dots & 0 & -a_{n-2} \\ \cdots & \cdots & \cdots & \cdots & \cdots \\ 0 & 0 & \dots & 1 & -a_1 \\ 0 & 0 & \dots & 0 & 1 \end{bmatrix} \quad (n \times n) \tag{13-75}$$

Then

$$\mathbf{Q}^{-1} = \begin{bmatrix} 1 & 0 & \dots & 0 & a_{n-1} \\ 0 & 1 & \dots & 0 & a_{n-2} \\ \cdots & \cdots & \cdots & \cdots & \cdots \\ 0 & 0 & \dots & 1 & a_1 \\ 0 & 0 & \dots & 0 & 1 \end{bmatrix} \tag{13-76}$$

Note that $\mathbf{QQ}^{-1} = \mathbf{I}$ (identity matrix).

Let

$$\mathbf{A}_2 = \mathbf{QA}_1\mathbf{Q}^{-1} = \begin{bmatrix} 0 & 0 & \dots & -a_{n-1} & -a_1 a_{n-1} - a_n + a_1 a_{n-1} \\ 1 & 0 & \dots & -a_{n-2} & -a_1 a_{n-2} - a_{n-1} + a_1 a_{n-2} \\ 0 & 1 & \dots & -a_{n-3} & -a_1 a_{n-3} - a_{n-2} + a_1 a_{n-3} \\ \cdots & \cdots & \cdots & \cdots & \cdots \\ 0 & 0 & \dots & -a_1 & -a_1^2 - a_2 + a_1 a_1 \\ 0 & 0 & \dots & 1 & -a_1 + a_1 \end{bmatrix} \tag{13-77}$$

Now we transform the system in $\mathbf{y}(k)$ to the following system:

$$\mathbf{w}(k+1) = \mathbf{A}_2\mathbf{w}(k) + \mathbf{B}_2 u(k) \tag{13-78}$$

$$c(k) = \mathbf{D}_2\mathbf{w}(k) \tag{13-79}$$

where

$$\mathbf{w}(k) = \mathbf{Qy}(k)$$

$\mathbf{A}_2$ is defined in Eq. (13-77), and $\mathbf{B}_2 = \mathbf{QB}_1$,

$$\mathbf{D}_2 = \mathbf{D}_1\mathbf{Q}^{-1} = [0 \quad 0 \quad \dots \quad 1] \tag{13-80}$$

The observer which observes $\mathbf{w}(k)$ is of the form,

$$\mathbf{w}_e(k+1) = \mathbf{A}_2\mathbf{w}_e(k) + \mathbf{B}_2u(k) + \mathbf{L}_e\mathbf{D}_2[\mathbf{w}(k) - \mathbf{w}_e(k)] \tag{13-81}$$

In view of the form of $\mathbf{D}_2$ in Eq. (13-80), we have $c(k) = w_n(k)$; that is, the nth state of $\mathbf{w}(k)$ is the output so that it does not have to be observed. Therefore, $w_{en}(k) = w_n(k)$, and

$$\mathbf{L}_e\mathbf{D}_2[\mathbf{w}(k) - \mathbf{w}_e(k)] = \mathbf{0} \tag{13-82}$$

Equation (13-81) becomes

$$\mathbf{w}_e(k+1) = \mathbf{A}_2\mathbf{w}_e(k) + \mathbf{B}_2u(k) \tag{13-83}$$

The main objective now is to reduce the observer in $\mathbf{w}_e(k)$ to an $(n-1)$th order one and at the same time to be able to place its eigenvalues. Equation (13-83) is written

$$\begin{bmatrix} \bar{\mathbf{w}}_e(k+1) \\ \hline w_{en}(k+1) \end{bmatrix} = \left[\begin{array}{c|c} \bar{\mathbf{A}}_2 & \mathbf{E}_2 \\ \hline 0 \quad 0 \quad \dots \quad 1 & -a_1 + \alpha_1 \end{array}\right] \begin{bmatrix} \bar{\mathbf{w}}_e(k) \\ \hline c(k) \end{bmatrix} + \begin{bmatrix} \bar{\mathbf{B}}_2 \\ \hline b_2 \end{bmatrix} u(k) \tag{13-84}$$

Thus, the $(n-1)$th-order observer is described by

$$\bar{\mathbf{w}}_e(k+1) = \bar{\mathbf{A}}_2\bar{\mathbf{w}}_e(k) + \mathbf{E}_2c(k) + \bar{\mathbf{B}}_2u(k) \tag{13-85}$$

where

$$\bar{\mathbf{A}}_2 = \begin{bmatrix} 0 & 0 & \dots & 0 & -\alpha_{n-1} \\ 1 & 0 & \dots & 0 & -\alpha_{n-2} \\ 0 & 1 & \dots & 0 & -\alpha_{n-3} \\ \cdots & \cdots & \cdots & \cdots & \cdots \\ 0 & 0 & \dots & 1 & -\alpha_1 \end{bmatrix} \tag{13-86}$$

$$\mathbf{E}_2 = \begin{bmatrix} -\alpha_1\alpha_{n-1} - a_n + a_1\alpha_{n-1} \\ -\alpha_1\alpha_{n-2} - a_{n-1} + a_1\alpha_{n-2} + \alpha_{n-1} \\ \cdots\cdots\cdots \\ -\alpha_1^2 - a_2 + a_1\alpha_1 \end{bmatrix} \tag{13-87}$$

and $\bar{\mathbf{B}}_2$ is an $(n-1) \times 1$ matrix.

Since $\bar{\mathbf{A}}_2$ is in the adjoint phase-variable canonical form the last column of the matrix contains the coefficients of the characteristic equation

$$|\lambda \mathbf{I} - \bar{\mathbf{A}}_2| = \lambda^{n-1} + a_1\lambda^{n-2} + a_2\lambda^{n-3} + \ldots + a_{n-2}\lambda + a_{n-1} = 0 \tag{13-88}$$

Once the reduced-order observed state vector $\bar{\mathbf{w}}_e(k)$ is determined, $\mathbf{w}_e(k)$ is given by

$$\mathbf{w}_e(k) = \begin{bmatrix} \bar{\mathbf{w}}_e(k) \\ c(k) \end{bmatrix} \tag{13-89}$$

Then,

$$\mathbf{x}_e(k) = (\mathbf{QP})^{-1}\mathbf{w}_e(k) \tag{13-90}$$

which is the transformation indicated in Fig. 13-6.

The effectiveness of the reduced-order observer is studied by first comparing the observed state $\mathbf{w}_e(k)$ with the state $\mathbf{w}(k)$. Subtracting Eq. (13-83) from Eq. (13-78), we get

$$\mathbf{w}(k+1) - \mathbf{w}_e(k+1) = \mathbf{A}_2[\mathbf{w}(k) - \mathbf{w}_e(k)] \tag{13-91}$$

Since $w_n(k) = c(k) = w_{en}(k)$, the last equation leads to

$$\bar{\mathbf{w}}(k+1) - \bar{\mathbf{w}}_e(k+1) = \bar{\mathbf{A}}_2[\bar{\mathbf{w}}(k) - \bar{\mathbf{w}}_e(k)] \tag{13-92}$$

Thus, $\bar{\mathbf{w}}_e(k)$ approaches $\bar{\mathbf{w}}(k)$ according to the dynamics determined by the eigenvalues of $\bar{\mathbf{A}}_2$. Since the transformation in Eq. (13-90) is nondynamic, and $w_{en}(k) = c(k)$, the $(n-1)$ states of $\mathbf{x}(k)$ that have to be observed are also controlled by the eigenvalues of $\bar{\mathbf{A}}_2$.

Example 13-4.

The digital process described in Example 13-1 is again considered in this example. We shall first consider that the system is open loop. The state equation is

$$\mathbf{x}(k+1) = \mathbf{A}\mathbf{x}(k) + \mathbf{B}u(k) \tag{13-93}$$

where

$$\mathbf{A} = \begin{bmatrix} 0 & 1 \\ -1 & 1 \end{bmatrix} \qquad \mathbf{B} = \begin{bmatrix} 0 \\ 1 \end{bmatrix}$$

The output equation is

$$c(k) = \mathbf{D}\mathbf{x}(k) = [2 \quad 0]\mathbf{x}(k) \tag{13-94}$$

The objective is to design a first-order observer for the system. The matrix **P** for the transformation of **A** into the adjoint phase-variable canonical form is given in Eq. (13-66). The matrix $\mathbf{A}_2$ in Eq. (13-77) is written for the second-order case as

$$\mathbf{A}_2 = \begin{bmatrix} -\alpha_1 & -\alpha_1^2 - a_2 + a_1\alpha_1 \\ 1 & -a_1 + \alpha_1 \end{bmatrix} \tag{13-95}$$

where a_1 and a_2 are the coefficients of the characteristic equation of **A**, i.e.,

$$|\lambda\mathbf{I} - \mathbf{A}| = \lambda^2 - \lambda + 1 = 0 \tag{13-96}$$

Thus, according to Eq. (13-60), $a_1 = -1$ and $a_2 = 1$. The coefficient α_1 is identified with the characteristic equation of $\bar{\mathbf{A}}_2$ which is

$$\lambda + \alpha_1 = 0 \tag{13-97}$$

The reduced-order observer in the transformed domain is described by

$$\bar{w}_e(k+1) = \bar{\mathbf{A}}_2\bar{w}_e(k) + \mathbf{E}_2 c(k) + \bar{\mathbf{B}}_2 u(k) \tag{13-98}$$

and

$$\mathbf{w}_e(k) = [\bar{w}_e(k) \quad c(k)]' \tag{13-99}$$

$$\bar{\mathbf{A}}_2 = -\alpha_1$$

$$\mathbf{E}_2 = -\alpha_1^2 - a_2 + a_1\alpha_1 = -\alpha_1^2 - \alpha_1 - 1$$

$$\bar{\mathbf{B}}_2 = 2$$

For a deadbeat response for the error of the observer we set α_1 to zero. Equation (13-98) becomes

$$\bar{w}_e(k+1) = -c(k) + 2u(k) \tag{13-100}$$

A state diagram for the system and the first-order observer is shown in Fig. 13-7. For arbitrary initial states $\mathbf{x}(0)$ and $\bar{w}_e(0)$ we can show that $\mathbf{x}_e(k)$ reaches $\mathbf{x}(k)$ for $k \geqslant 1$. For $\mathbf{x}(0) = [1 \quad 0]'$, $\bar{w}_e(0) = 0.5$, and $u(k) = 0$ for all k, the responses of the system are tabulated below for $k \leqslant 5$.

k	0	1	2	3	4	5
$x_1(k)$	1	0	−1	−1	0	1
$x_{e1}(k)$	1	0	−1	−1	0	1
$x_2(k)$	0	−1	−1	0	1	1
$x_{e2}(k)$	1.25	−1	−1	0	1	1
$\bar{w}_e(k)$	0.50	−2	0	2	2	0
$c(k)$	2	0	−2	−2	0	2

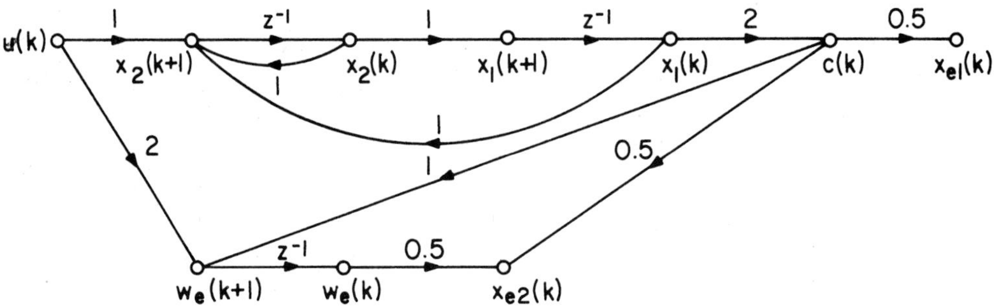

Fig. 13-7. Reduced-order observer and open-loop digital system in Example 13-4.

For the closed-loop system $u(k) = -\mathbf{G}\mathbf{x}_e(k)$ where the feedback gain matrix $\mathbf{G}$ is given in Eq. (13-33). Substitution of Eqs. (13-99), (13-90) and the control into Eq. (13-100), we have

$$\bar{w}_e(k+1) = -c(k) - 2\mathbf{G}\mathbf{x}_e(k)$$

$$= -2x_1(k) + [1.308 \quad -0.972]\begin{bmatrix} x_1(k) \\ 0.5\bar{w}_e(k) + x_1(k) \end{bmatrix} \tag{13-101}$$

or

$$\bar{w}_e(k+1) = -1.664x_1(k) - 0.486\bar{w}_e(k) \tag{13-102}$$

The state equations of the digital system in Eq. (13-93) are conditioned with the state feedback, and the resulting state equations are,

$$x_1(k+1) = x_2(k) \tag{13-103}$$

$$x_2(k+1) = -0.832x_1(k) + x_2(k) - 0.243\bar{w}_e(k) \tag{13-104}$$

Figure 13-8 shows the state diagram of the feedback system and its first-order observer.

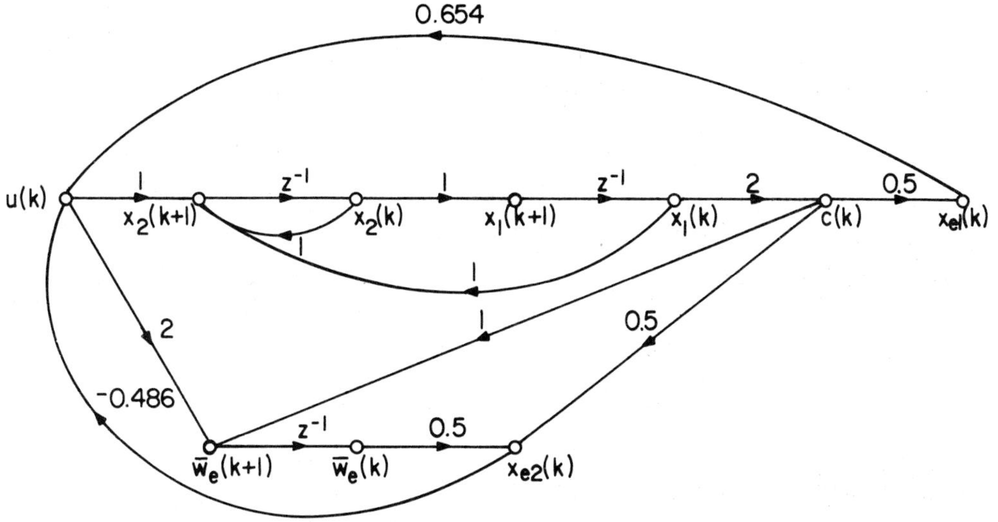

Fig. 13-8. Reduced-order observer and closed-loop digital system in Example 13-4.

For the initial states $\mathbf{x}(0) = [1 \quad 0]'$ and $\bar{w}_e(0) = 0.5$, the last three equations are solved to give the responses of $\mathbf{x}(k)$ and $\bar{w}_e(k)$, $k = 1,2,\ldots$. The observed states $\mathbf{x}_e(k)$ are determined from Eq. (13-90); that is,

$$\mathbf{x}_e(k) = (\mathbf{QP})^{-1}\mathbf{w}_e(k)$$

$$= \begin{bmatrix} x_1(k) \\ 0.5\bar{w}_e(k) + x_1(k) \end{bmatrix} \tag{13-105}$$

Thus, in this system, $x_{e1}(k)$ is directly $x_1(k)$, and $x_2(k)$ is the only state that is observed through the first-order observer.

Figure 13-9 shows the time responses of $\mathbf{x}(k)$ and $\mathbf{x}_e(k)$ for the closed-loop system. It is interesting to note that since the reduced-order observer is of the first order, $\mathbf{x}_e(k)$ reaches $\mathbf{x}(k)$ in only one sampling period for the deadbeat observer design. However, comparing the responses in Fig. 13-9 with those of the system with full observer (Fig. 13-5), we see that the reduced-order observer causes the closed-loop system to have a higher overshoot.

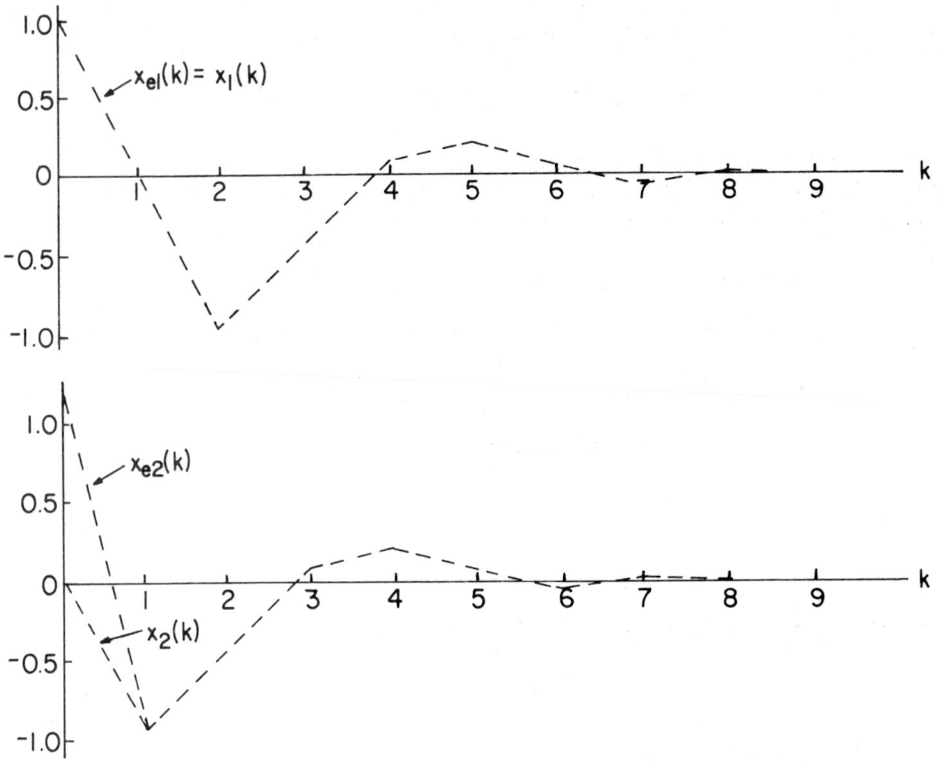

Fig. 13-9. State responses of digital feedback system with reduced-order state observer.

REFERENCES

1. Tse, E., and Athans, M., "Optimal Minimal-Order Observer-Estimators For Discrete Linear Time-Varying Systems," *IEEE Trans. on Automatic Control*, Vol. AC-15, August 1970, pp. 416-426.
2. Luenberger, D. G., "An Introduction to Observers," *IEEE Trans. on Automatic Control*, Vol. AC-16, December 1971, pp. 596-602.
3. Leondes, C. T., and Novak, L. M., "Reduced-Order Observers for Linear Discrete-Time Systems," *IEEE Trans. on Automatic Control*, Vol. AC-19, February 1974, pp. 42-46.

PROBLEMS

13-1 The dynamic equations of a digital control system are described by

$$\mathbf{x}(k+1) = \mathbf{A}\mathbf{x}(k) + \mathbf{B}u(k)$$

$$c(k) = \mathbf{D}\mathbf{x}(k)$$

where

$$\mathbf{A} = \begin{bmatrix} 0 & 0 \\ 0.5 & 0.5 \end{bmatrix} \qquad \mathbf{B} = \begin{bmatrix} 0 \\ 1 \end{bmatrix}$$

$$\mathbf{D} = [1 \quad 1]$$

The state vector $\mathbf{x}(k)$ at $k = N$ is to be determined from the measurements of $u(k)$ and $c(k)$ at instants previous to $k = N$. Determine how many samples of $c(k)$ and $u(k)$ are needed for exact measurements for $\mathbf{x}(k)$. Express $\mathbf{x}(k)$ in terms of these measurements of $c(k)$ and $u(k)$.

13-2 The dynamic equations of a digital process are given as

$$\mathbf{x}(k+1) = \mathbf{A}\mathbf{x}(k) + \mathbf{B}u(k)$$

$$c(k) = \mathbf{D}\mathbf{x}(k)$$

where

$$\mathbf{A} = \begin{bmatrix} 0 & 1 \\ 0 & 0 \end{bmatrix} \qquad \mathbf{B} = \begin{bmatrix} 0 \\ 1 \end{bmatrix} \qquad \mathbf{D} = [1 \quad 1]$$

The state feedback control is $u(k) = -\mathbf{G}\mathbf{x}(k)$, where

$$\mathbf{G} = [g_1 \quad g_2]$$

(a) Assuming that the state variables are unaccessible, design a full-order state observer so that $\mathbf{x}(k)$ is observed from $c(k)$. Find the elements of $\mathbf{G}_e$ in terms of g_1 and g_2 so that the dynamics of the observer are the same as that of the closed-loop digital process.

(b) Design a first-order observer for the system with the eigenvalue at 0.5. Write the state equations of the system with feedback realized from the first-order observer.

13-3 For the stick-balancing control system described in Problem 9-6, design a full-order observer which will observe the states $x_1(kT)$, $x_2(kT)$, $x_3(kT)$ and $x_4(kT)$ from the output $c(k) = [0 \quad 0 \quad 1 \quad 0]\mathbf{x}(kT)$. The eigenvalues of the closed-loop observer should be zero. Solve the problem first by working with the characteristic equation and then by means of the adjoint phase-variable canonical form.

13-4 Design a reduced-order observer for the system described in Problem 13-3.

14

MICROPROCESSOR CONTROL

14.1 INTRODUCTION

In the preceding chapters we have presented the analysis and design of digital control systems mainly from an analytical standpoint. We have covered the z-transform method and the state-variable method of analysis and design. In all instances, the sampling periods were chosen either based on the sampling theorem, or on the stability and overall performance of the digital control system. The digital controller parameters and other system parameters were selected purely from the standpoint of performance and physical realizability in the analytical and theoretical sense. However, we must realize that in digital control systems there are physical limitations due to the inherent discrete nature of the system components which are not found in analog or continuous-data control systems. For example, the sampling period of a digital control system is governed by the clock rate and how fast the numerical operations and instructions are executed by the digital processor. In the case of the digital processor being a microprocessor, the speed of execution can be relatively slow. Thus, there is an inherent upper limit on the sampling rate based on the hardware used in a digital control system.

Another limitation of the digital control system design is the finite-word-length characteristic of a digital computer. This means that not all numbers can be represented by a digital processor. Many microprocessors have a wordlength of only eight bits. From Chapter 2 we learned that an eight-bit word would allow us only $2^8 = 256$ levels of resolution. This is also known as quantization. In the preceding sections, the design problems often yield results for controller parameters such as 0.995 or 1.316, etc. It is apparent that we cannot represent these numbers accurately on an eight-bit microprocessor; so we must study the effects of quantization. It is important to be aware of these limitations, and consider them as the design is being carried out. This also points to the fact that while the pencil-and-paper design of digital control systems could easily yield us a deadbeat response, in practice, we have to face the reality of what the hardware can provide.

In this chapter we shall first give introductory material on microprocessors and microcomputers. Since space is limited, it is impossible to cover the subject in a comprehensive way, especially since there are numerous books now available that are devoted to the subject. We will then give examples on how a digital controller is programmed on a microprocessor, and demonstrate all the limitations.

The remaining portion of this chapter will be devoted to the discussions on the effects of finite wordlength, quantization, and time delay due to finite-time computation in the processor.

14.2 BASIC COMPUTER ARCHITECTURE

The main purpose of this section is to familiarize the reader with the basic components of a digital computer. Although each machine may have its own special features and unique hardware, certain pieces of hardware are common to most existing computers.

A basic digital computer consists of the following components:

Central Processing Unit (CPU)
Memory
Input and Output Devices (I/O)

A microprocessor is essentially the central processing unit (CPU) of a digital computer. When a microprocessor is equipped with memory and input-output (I/O) devices, the microcomputer is complete. Figure 14-1 illustrates the interconnections of the basic building blocks of a microcomputer in block diagram form.

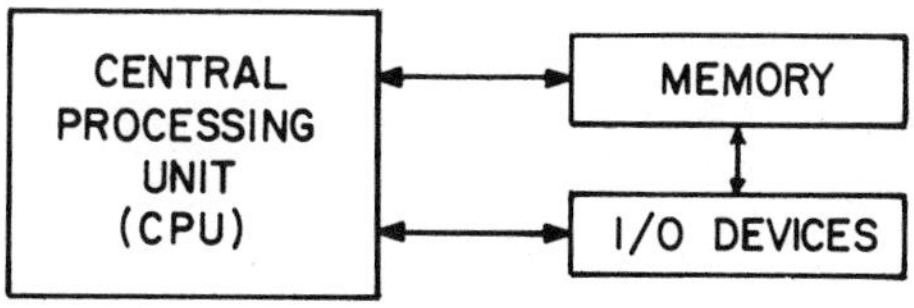

Fig. 14-1. Basic components of a microcomputer.

In general, the CPU also contains the arithmetic logic unit (ALU), registers, instruction register and decoder (IRD), timing and control circuits. These basic elements of the microcomputers are described separately in the following.

Input-Output Devices

Information is transferred in and out of the computer by means of input/output (I/O) devices. The computer is generally provided with one or more INPUT PORTS. The CPU can address these input ports and issue appropriate control signals to receive data at these ports. The addition of input ports enables the computer to receive information from external input devices such as manual keyboard, card reader, punched paper-tape reader, switches, thumbwheels, or floppy-disk units.

In order for the CPU to communicate the result of its processing to the outside world, a computer also needs one or more OUTPUT PORTS. The CPU can then address these output ports and send the data to the output devices. Typical output devices are: teletypewriter, paper-tape punch, card punch, line printer, plotter or CRT display. The outputs of the computer may also constitute process control signals that control the operation of another part of a system, or another system such as an automated assembly line.

Memory

One device which all digital computers have is the memory. The memory serves as a place to store INSTRUCTIONS, the operation codes that direct the activities of the CPU, and DATA, the binary-coded information that are processed by the CPU.

The memory of a computer contains a large number of memory cells or locations. Each location can be used to store n-bits of binary information. The smallest unit of the memory is a BIT which can be put into one of two states, 0 or 1.

A basic memory cell which is identifiable by a distinct memory location is called a WORD. In general, the words of a computer memory may be of almost any bit length, from 4 bits in the simple hand-held calculators to 128 bits or more for large-scale scientific computers. Microprocessors generally have a wordlength of 4 to 16 bits. For instance, the wordlength of the INTEL 8080 and the MOTOROLA M6800 microprocessors is 8 bits.

Each of the words in the memory has a unique and permanent identifying number called the ADDRESS. The individual numbering of the memory locations is called ADDRESSING. Access to the particular word or location is accomplished by specifying the address of the word location.

It is important to distinguish the difference between the address and the numerical content of that address in the memory. The memory address is normally occupied by one word, but the numerical address itself may occupy more than one word. For instance, on the INTEL 8080 two consecutive words are used for addressing. With the word in this case being 8 bits long, each address is 16 bits. Thus, the largest number the 8080 can address is 2^{16}, or 65,536. In computer jargon we say that a maximum of 64K bytes of memory can be attached to the INTEL 8080, whereas in reality it is 65,536 bytes. Actually, the address number runs from 0 through 65,535. Similarly, when a memory capacity of 1K bytes is referred to, it is actually $2^{10} = 1024$ bytes, and 2K bytes is $2^{11} = 2048$ bytes, etc.

Figure 14-2 shows the two characteristics of a typical stack of memory locations, the address, and the contents of the address. For instance, memory location number 0 contains the 8-bit binary word 00010000 which represents the decimal number 16. Similarly, each of the other illustrated memory addresses contains an 8-bit word. As mentioned earlier, each of these words may represent a piece of data, an operation code (OP code) or a part of an operation code. The CPU when executing these memory addresses should know how to interpret the contents of these locations.

ADDRESS	CONTENTS
0	0 0 0 1 0 0 0 0
1	0 1 0 0 0 0 0 1
2	0 1 0 1 0 1 0 0
3	0 0 0 0 1 0 0 0
4	0 0 0 0 1 0 0 1

Fig. 14-2. Memory addresses and contents.

Random Access Memory (RAM)

The memory of a microcomputer is available in several forms. The random-access memory (RAM) can be accessed by the computer at any time and at any location in a random fashion. The RAM can be used for both read and write operations. The important feature of the RAM is that when the power to the computer is turned off the contents of the memory are entirely lost. Thus, the RAM can be used for data storage, temporary instructions and OP CODES.

Read-Only Memory (ROM)

The read-only memory (ROM) is used for the storage of programs that are executed repeatedly and must be stored permanently. Typically, the contents of the ROM are written initially and then stored permanently. The microcomputer does not write any new information into the program, but merely reads the contents of the ROM and executes them.

Certain types of ROMs can be reprogrammed by a special process and are known as programmable read-only memories (PROM).

The Central Processing Unit (CPU)

The central processing unit (CPU) of a microcomputer is the heart of the machine. It is a piece of hardware, in this case, the microprocessor, in which data processing and control take place. For example, inside the CPU two operands may be added together, data may be transferred into the CPU from some input device through an input port, or the CPU may transmit information to the output through an output port.

The CPU is responsible for centrally controlling almost the entire computer. It generates most of the key timing pulses and control signals which the external devices will need to interact properly with the computer. Most of the computer control functions are derived from a finite-state machine (FSM) which is a part of the CPU. The FSM will direct the processor to follow the sequential processing of the stored program. When the computer is first turned on, the FSM knows where to retrieve the first instruction from the memory. It retrieves this instruction and places it in some decoder so that the CPU can understand what to do with the instruction.

A typical CPU consists of the following interconnected functional blocks:

Registers
Arithmetic-logic unit (ALU)
Control circuitry

These units and the other essential elements and operations of a microcomputer are described in the following.

Arithmetic-Logic Unit (ALU)

The ALU is a device which performs all the arithmetic computation and logical operations of the computer. The ALU can receive two numbers and perform arithmetic or logic operations such as addition, subtraction, AND, OR, NOT, and EXCLUSIVE-OR. Most microprocessor ALUs are fairly simple so that they cannot perform hardware operations such as multiplication and division. However, through the use of software, operations such as multiplication can be performed (by repeated addition, for example).

The ALU has some registers for storing sequences of digits. The main register is called the ACCUMULATOR. At the beginning of an operation data are sent to the ALU by the CPU control circuitry. The ALU performs the arithmetic or logic operation requested by the CPU controller and then stores the answer in some register such as the ACCUMULATOR in the CPU.

One final function performed by the ALU is to report interpretive information on the results. For example, if the ALU subtracts two numbers, often it is necessary for the ALU to report if the result is greater than zero, equal to zero, or less than zero. Similarly, it may be necessary to have the ALU report if the sum of two numbers or the result of any operation overflows the 8 bits of a byte and is carried over to a ninth bit. Therefore, one of the functions of the ALU is to wave a FLAG when the result of an operation satisfies a certain condition. The operation of FLAGS of the microprocessor will be discussed in more detail later.

Registers

Registers are special memory locations located inside the CPU. For example, on the INTEL 8080 there are several registers, each serves a specific function of storage. Some of these registers are 8 bits long so that they can hold a single byte of data. Other registers are used to store addresses and so must be 16 bits in length.

Because registers are physically a part of the CPU, they are directly wired to the control logic, the ALU and other internal CPU devices. On some machines data must be first transferred from addressable memory locations to a register before they can be operated on. Even on machines which can operate directly on data that are stored in the core memory, it is always faster to execute an instruction in which all the operands are already in the registers. This is simply because in order to extract an operand out of the core memory the CPU must access the memory by first specifying which address the operand is in, wait for the memory to decode the address, and finally have the correct memory location supply the operand to the CPU. On the other hand, an operand in a register is immediately available to the microprocessor. Thus, instructions executed through registers are done faster. For this reason, and, more importantly, for the sake of simplicity, we assume that initially each operand must be in a register before it can be further processed. For example, to add the contents of two given core memory locations, we first execute instructions to bring the contents of these locations to two registers. We then execute the addition of the contents of these registers and store the result in either one of the two registers or some other register for future use. Of course, we can also store the result in some core memory location.

As mentioned earlier, many machines have a special register called the ACCUMULATOR (AC), or the "result" register. The AC is special in two respects. On some computers we cannot process the contents of any two registers directly. On these machines, one of the operands to be processed in an operation must be in the AC. Furthermore, the result of the operation always stays in the AC. For example, if we add the content of the AC to that of register B, the result will be placed in the AC; the original number that was in the AC was automatically replaced by the sum. Therefore, in general, we can regard the ACCUMULATOR as both a source (operand) and a destination (result) register.

Another important feature of the AC is that in order to transfer data between the core memory and the CPU, the data must first be transferred to the AC. Thus, we must have instructions to load the AC with the contents of any memory location, as well as instructions to store the number in the AC into any memory location. We assume that no other register is wired to have this transfer characteristic with the memory. Once the information is in the AC, it can be freely moved to any other register if desired. Information can always be moved between any two registers.

In addition to the AC, the INTEL 8080, for example, has six other general-purpose registers. These are denoted as registers B, C, D, E, H and L; all are 8 bits in length. These six general-purpose registers can be used either as single 8-bit registers or as register pairs (16 bits).

On the INTEL 8080 there are also three special registers: the *program counter* (PC), the *stack pointer* (SP) and the *flag*. The PC and the SP are 16-bit registers since they have to store the addresses of the memory locations. The FLAG register is only 5 bits. The operations of the general-purpose and the special-purpose registers will be discussed in later sections.

Busses

In the preceding sections we have described the microcomputer as a machine which consists of a memory, input/output devices, and a CPU. One important aspect on the operation of the microcomputer is that rapid and accurate information link between these hardware components must be established and maintained. Thus, these blocks are interconnected by various busses. A bus is a set of wires, grouped together because of the similarity of their functions, which run between all the functional blocks of a microcomputer system. Figure 14-3 shows the simplified block diagram of a microcomputer system with interconnecting busses.

For the INTEL 8080 microcomputer system, for example, there are three busses that interconnect the memory with the CPU and the I/O devices. These three busses are briefly described as follows:

DATA BUS: This is a set of wires in which data can flow (in either direction) between the CPU, the memory and the I/O. The data bus is 8 bits in length.

ADDRESS BUS: This is a set of wires that connects the memory and the I/O devices to the CPU. The bus is used to identify a particular memory location or I/O device. Information flows only in the direction from the CPU to the memory or the I/O. The address bus on the 8080 is 16 bits in length.

CONTROL BUS: This is a set of wires that connects the memory and the

I/O devices to the CPU, in order to indicate the type of activity in the current process. Information flows only in the direction from the CPU to the memory or the I/O. The activities may be a MEMORY READ, MEMORY WRITE, I/O READ, I/O WRITE, or INTERRUPT ACKNOWLEDGE. On the 8080, the control bus is 6 bits; that is, it consists of 6 wires.

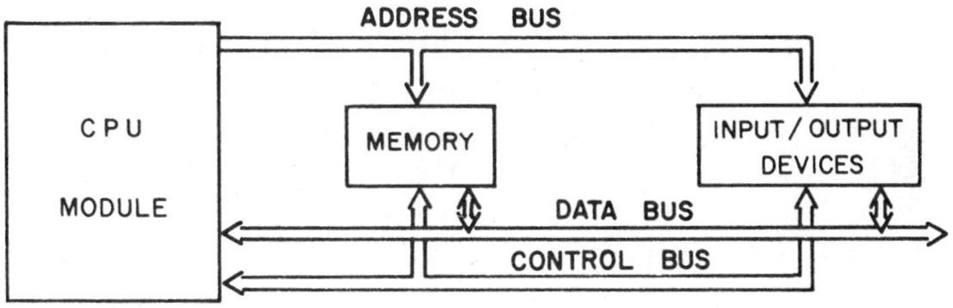

Fig. 14-3. Block diagram of a microcomputer system with busses.

The basic advantage of using the bus structure is that it simplifies the communication between the main hardware components of a microcomputer. Instead of having separate wires connecting each pair of devices, we have one set of wires running throughout the system. The bus effectively connects all the devices together.

The only disadvantage with the bus structure is that the system designer must make sure that two different devices do not try to put data on a bus at the same time.

Instruction Register And Decoder (IRD)

One-Byte Instructions And The IRD

Each operation that the CPU of a microcomputer can perform is identified by a unique byte of data called an INSTRUCTION CODE or OPERATION CODE (OP CODE). For example, the OPERATION CODE for the operation instruction, "load the AC with the content of memory location 50," used in the last section is, "LDA 50,". The operation code, "STA 100" stands for, "store the content of the AC into memory location 100." A given microprocessor may have several hundred different operation codes. The INTEL 8080 has 256 operation codes, and these are listed at the end of this chapter.

Instructions on some microprocessors, such as the INTEL 8080, are of variable bit lengths; that is, some instructions are of 8 bits or 1 byte, and some are up to 3 bytes in length. In other words, a 3-byte instruction would require three consecutive memory locations to store.

The purpose of the INSTRUCTION REGISTER AND DECODER is to

inform the CPU which instruction the programmer has requested, given the binary coded version of the instruction. Suppose the CPU knows that the next instruction is stored in memory location 15. The CPU will issue signals to the memory on the CONTROL BUS and place the binary number "0000 0000 0000 1111" (decimal 15) on the DATA BUS. The memory will respond by placing the contents of location 15 on the Data Bus. Let us assume that the instruction code is represented by the binary number "01001000", and this has been placed on the Data Bus from memory location 15. Since the CPU knows that the number 01001000 in this case is not a piece of data but rather a binary-coded instruction, it will *not* latch the number into the ACCUMULATOR. Instead, the CPU will latch the number on the Data Bus into the Instruction Register and Decoder (IRD). The register part of the IRD is used for storage of the data, whereas the decoder part consists of a large amount of logic which is wired to decode every possible instruction code available on the microprocessor. The outputs of the IRD signify which instruction code is being requested and what steps to be taken next. In the present example the binary number 01001000 actually represents the instruction code "MOV C,B", or "move the content of register B to register C".

Thus, we have illustrated that the IRD is a vital part of the stored program implementation.

Variable-Length Instructions And The IRD

Some microprocessors such as the INTEL 8080 have variable-length instruction codes. In other words, some instructions, such as "MOV C,B", can be completely specified in 8 bits (1 byte), but other instructions may take more than 1 byte. For instance, if we desire to execute the instruction, "load the ACCUMULATOR with the content of memory location 359", this is coded as "LDA 359". However, on the 8080 addresses are 16 bits long so that just to store the binary version of "359" would take 16 bits or two full bytes. The operation code "OP CODE" part of the instruction, LDA, which stands for "load the ACCUMULATOR" will require another byte to store. Thus, the instruction, "LDA 359" is 3 bytes in length. Typically, the first byte will contain the OP CODE LDA, then followed by the two bytes which contain the 16-bit address of the memory whose contents are to be loaded. The upper 8 bits of the address (most-significant part) are known as the high-order address byte, whereas the last 8 bits (least-significant part) are known as the low-order address byte. For example, Fig. 14-4 illustrates the number 359 expressed as a 16-bit, 2-byte binary number. The high-order byte contains the number 1 while the low-order byte contains the number 103. This combination of the two bytes gives $1 \times 256 + 103 = 359$. In general, the number N stored in the high-order byte is multiplied by 256 (2^8) since it represents the upper 8 bits of a 16-bit binary number.

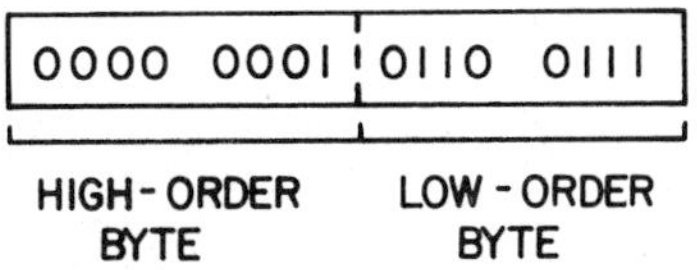

Fig. 14-4. The number 359 expressed as a 16-bit, 2-byte, binary number.

For storage in the memory a particular microprocessor may have a special way of storing multiple-byte data. For instance, on the INTEL 8080 the low-order byte is stored first and then followed by the high-order byte. Figure 14-5 shows the instruction "LDA 359" being stored in three consecutive memory locations.

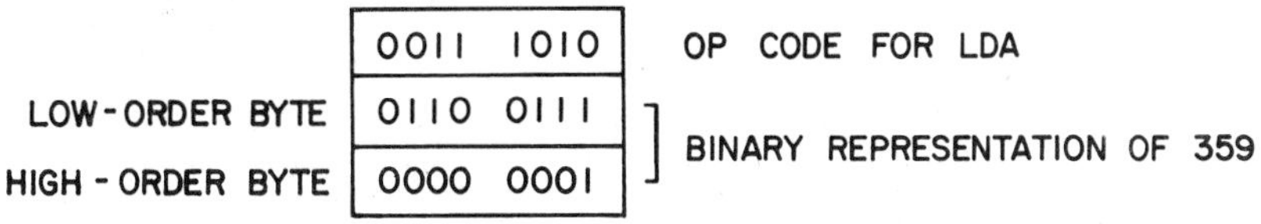

Fig. 14-5. Storing the instruction "LDA 359" on INTEL 8080 in three successive memory locations.

In addition to the representation and the storing of a variable-length (multiple-byte) instruction, there is one more problem the CPU must encounter when reading an instruction. Once the CPU reads the first byte of an instruction, which may be one or more bytes in length, it must determine and know the length of the instruction. If the instruction is only one byte long, it should be executed before reading the next byte. Since the OP-CODE of an instruction is always found in the first byte, the IRD always knows the length of each instruction. For instance, the IRD knows that "MOV C,B" is a one-byte instruction, and when the IRD decodes the LDA OP-CODE, it signifies the CPU to access the memory for the two additional bytes which contain the low-order and the high-order bytes of the address to be fetched.

The Program Counter

The Program Counter (PC) is a special register which holds the address of the next instruction to be executed. Since data and OP-CODES are all represented by binary numbers on a microprocessor, we need to make a distinction between these two categories. For example, on the INTEL 8080 the bit pattern "01001000" means "MOV C,B" (move the content of register B to register C), but it is also the binary representation of the decimal number 72. If the bit pattern "01001000" is stored in a given memory location how does the CPU know whether it is meant to be "MOV C,B" or the decimal number 72?

In order to distinguish the difference between data and instruction we use the PC to point to the next instruction. Therefore, the PC contains the address of the next instruction or data to be executed. (The PC on the INTEL 8080 is a special register which is 16 bits long, since it must be able to hold an address.)

After an instruction has been executed, the CPU will increment the PC automatically so that the PC will point to the next instruction. Therefore, the main function of the PC is to keep track of where the computer is in the execution of the program, and enables the CPU to distinguish program from data.

Up until now we have described the basic architecture and components of a microprocessor. We have described the functions and the operations of the memory, CPU, registers, ALU, Accumulator, the IRD and the Program Counter. The interactions between the vital elements of a microcomputer (INTEL 8080) are now summarized by the block diagram of Fig. 14-6.

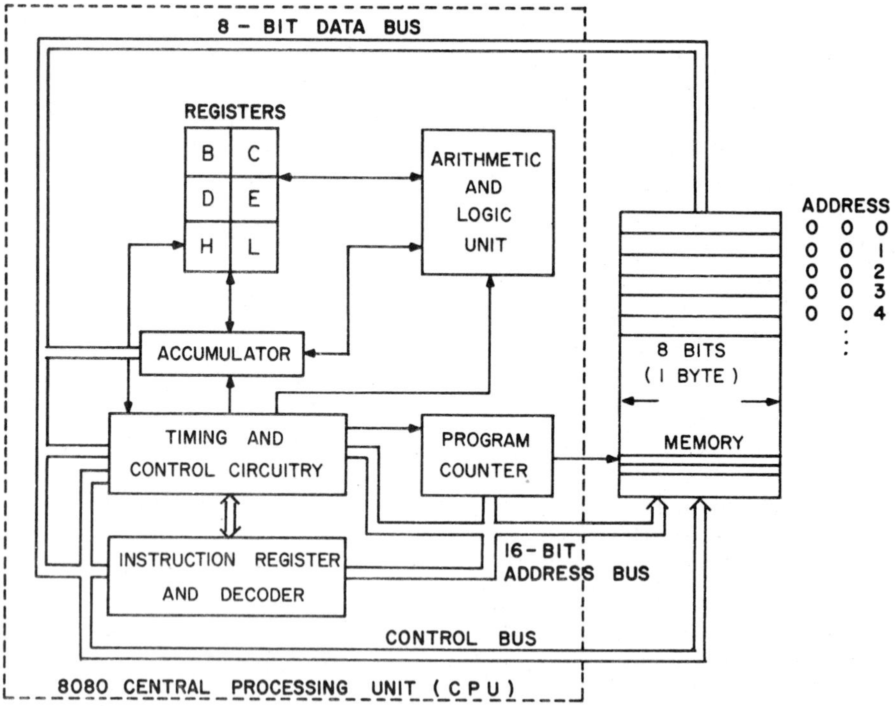

Fig. 14-6. Block diagram of a typical (8080) CPU and bus structure.

14.3 MACHINE-LANGUAGE PROGRAMMING

The objective of this section is to introduce to the reader the basic elements of machine-language programming, which is necessary to program a microprocessor. Due to the limitation of space and objectives, the coverage on the subject in this section is by no means exhaustive. The material presented here is purely intended to familiarize the reader with the basics of how to program a microprocessor, and its ramifications, so that practical considerations may be incorporated when designing digital control systems.

Let us begin the subject of machine-language programming by following through the structure of a simple computer program. Consider that we wish to add the numbers 3 and 4 and then store the result in some memory location. Let us assume that the number 3 is stored in memory location 12 and the number 4 is stored in location 13, and the result of the addition is to be stored in location 14.

The FORTRAN program for such an arithmetic operation would simply be

X = 3 + 4
STOP

and we assume that the reader is familiar with the FORTRAN language. In terms of machine language, the simple operation described by the last two FORTRAN instructions consists of the sequence of steps described by the flow chart shown on the left side of Fig. 14-7. The right-hand columns of Fig. 14-7 illustrate the 8080 mnemonic coding of instructions and the binary equivalent of the codes and data. The reader should refer to the end of the chapter for the explanation of the 8080 OP codes.

We assume that the first instruction of the program starts at location 0 in the memory. The first instruction "LDA 12" loads the ACCUMULATOR with the number 3 from location 12. Since LDA 12 is a 3-byte instruction, it requires three memory locations to store; 000, 001 and 002. Starting from location 003 we have the single-byte instruction "MOV E,A" which represents "move the content of the AC to register E". Since data cannot be transferred directly from the memory to register E, they must go through the AC. Now the number 3 is in register E for temporary storage. Next, we place the instruction "LDA 13" in locations 004, 005 and 006. This instruction loads the number in memory location 13, which is 4, in the AC. The memory location 007 contains the single-byte instruction "ADD E" which means "add the content of register E to the content of the AC, and place the sum back in the AC". The next instruction, "STA 14" stores the content of the AC in memory location 14. Thus, we see from Fig. 14-7 that the machine-language program of X = 3 + 4, STOP, consists of six instructions. In addition, it requires three data bytes to store the numbers 3, 4 and the sum.

The right-hand most column of Fig. 14-7 gives the 8080 binary-coded version of the program. Since several instructions take up more than one byte each, there are a total of 15 bytes in the binary-coded program for as simple as

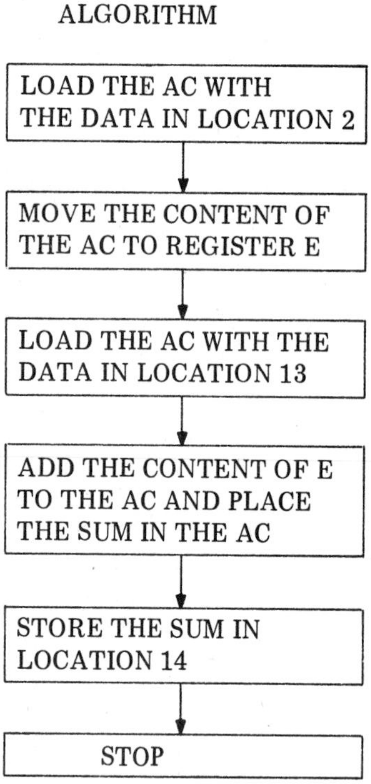
ALGORITHM
LOAD THE AC WITH THE DATA IN LOCATION 2
MOVE THE CONTENT OF THE AC TO REGISTER E
LOAD THE AC WITH THE DATA IN LOCATION 13
ADD THE CONTENT OF E TO THE AC AND PLACE THE SUM IN THE AC
STORE THE SUM IN LOCATION 14
STOP

ADDRESS	CONTENTS OF ADDRESS MNEMONIC	BINARY
000	LDA	0011 1010
001	12	0000 1100
002	0	0000 0000
003	MOV E,A	0101 1111
004	LDA	0011 1010
005	13	0000 1101
006	0	0000 0000
007	ADD E	1000 0011
008	STA	0011 0010
009	14	0000 1110
010	0	0000 0000
011	HLT	0111 0110
012	3	0000 0011
013	4	0000 0100
014	0	0000 0000

Fig. 14-7. Flow chart, mnemonic and binary codes of the program, X = 3 + 4, STOP.

the program just illustrated. This is precisely one of the difficulties that the user of a microprocessor will find that the processor can easily run out of storage and execution capacity when the program gets more complicated.

Branching

The sample program just discussed provides an illustration of a sequentially-executed program. The instructions are performed one after the other in a sequential manner as controlled by the Program Counter. However, it is possible to branch off programs so that instructions may be executed in any arbitrary order. For instance, FORTRAN statements such as "GO TO 25" and "IF (X+4) 10,20,20" force a transfer or jump of control to a different part of the program. In addition, in FORTRAN language a do-loop forces a block of code to be executed repeatedly any specified number of times. Since jumping and looping are such an important part of programming, we must investigate how a microcomputer can be made to perform "branching" or nonsequential processing.

Unconditional Branching

The unconditional branching instruction causes the computer to begin executing instructions located anywhere in the memory, as specified by the programmer. The next instruction to be executed need not be physically located in the byte following the unconditional branch instruction. Once the branching or jump is made, the computer again resumes the normal sequential processing.

In FORTRAN, the unconditional jump instruction is "GO TO X". On the INTEL 8080, the instruction is "jump to location X," abbreviated, "JMP X," where X is some integer that lies between 0 and 65,535.

Conditional Branching (Flags)

A conditional branching instruction causes a jump to occur only if some specific condition is met. If the condition is not met, the jump is not initiated, and the next instruction to be executed will be the one sequentially following the conditional jump instruction; that is, if the condition is not met, the jump instruction is not executed and has no effect on the sequential flow of the program. In FORTRAN language, for instance, a conditional jump statement could be, "IF(J.LT.0) GO TO 20," which means that if J is less than zero jump to location 20 (otherwise, go to the next line of instruction).

The special hardware needed by the CPU to implement conditional branching are a series of *single-bit* flip-flops called FLAGS. Each flag represents some condition that has either occured or not (yes or no, thus each flag needs only be one bit). The flags are set to 1 or 0 by certain instructions and tested by other instructions. To understand how flags work, we begin by discussing specific flags and how they get set.

Among all the instructions of a microprocessor, certain instructions are flag-setting instructions*; that is, the flags are automatically set by the ALU at the completion of any one of these flag-setting instructions, before the PC is incremented to the next instruction location.

On the INTEL 8080 there are a total of five flags; but the ones that are most frequently used are the ZERO FLAG, the CARRY FLAG, and the SIGN FLAG. The functions of these flags are described as follows:

ZERO FLAG: If the result of the instruction currently being executed is exactly zero, and *if the instruction is a flag-setting instruction*, the CPU will set the ZERO FLAG to 1, and otherwise the flag is set to 0. For example, suppose the ACCUMULATOR contains the number 4 and the instruction being executed by the CPU is SUB 11, that is, subtract the number in memory location

---

*The last column of the INTEL 8080 Instruction Set given at the end of the chapter specifies the flag-setting conditions of the instructions. Some instructions set only certain flags.

11 from the number in the AC. If the number stored in location 11 is 4, the result in the ACCUMULATOR after the execution of SUB 11 will be zero. Since "SUB 11" is a flag-setting instruction, after the CPU has found that the content of the AC is zero, it will set the ZERO FLAG to 1*. If the result in the AC were not zero, the CPU would set the ZERO-FLAG bit to 0. On the other hand, if the instruction just executed is not a flag-setting instruction (e.g., HLT), then the CPU will not change the value of any of the flags; that is, the flags will remain at their values resulting from the *last* flag-setting instruction.

CARRY FLAG: The CARRY FLAG is used to indicate an overflow out of the 8th bit due to addition or a borrow caused by subtraction. Therefore, whenever an overflow or a borrow occurs, the CARRY FLAG is set to 1, otherwise it is set to 0. For example, if register E contains the number 250 and the AC contains the number 9, and the contents of the two registers are added together, by the flag-setting instruction "ADD E", an overflow will occur, since the sum is 259 which is greater than 2^8 or 256. As a result, the number 3 is stored as the result of 250 + 9 in the AC, but the CARRY FLAG is set to 1 to indicate that if there were a 9th bit, the bit pattern in the AC would be 100000011 which is 259 decimal.

Similarly, if we subtract 250 from 9, using the instruction "SUB E", the subtraction can only be performed by borrowing from a "9th bit". Thus, the CARRY FLAG is again set to 1 at the completion of executing "SUB E", to indicate that a borrow has taken place.

SIGN FLAG: The SIGN FLAG on the INTEL 8080 is actually an "8th-bit tester"; that is, if the 8th bit of the result of a flag-setting instruction is 1, the SIGN FLAG is set to 1, otherwise it is set to zero. The reason for referring to the SIGN FLAG is that the 8th bit is often thought of as a sign bit, determining whether a number is positive or negative. When using the two's complement [18] mode to represent a negative number in binary code, the 8th bit is set to 1. However, to represent a large positive number such as 255, the 8th bit is also set to 1. There is no way the computer can distinguish between negative and large positive numbers. It is up to the programmer to interpret the "8th bit on" condition as indicating a negative number in two's complement or a large positive number.

The FLAGS play a key role in implementing conditional branching. Once the FLAGS are set it is necessary to have instructions to test these FLAGS for conditional branching. For instance, on the INTEL 8080 the instruction "JZ X" stands for "jump to location X if the ZERO FLAG is 1", and "JNZ X" stands for "jump to location X if the ZERO FLAG is 0, or jump on not zero".

*The other flags will also be set simultaneously according to their conditions.

For testing the CARRY FLAG, the instruction "JC" followed by a two-type address means "jump if the CARRY FLAG is 1". Similarly, "JNC" stands for "jump if the CARRY FLAG is 0". To test the SIGN FLAG we use "JM", "jump on minus", or "JP", "jump on positive".

The simple program shown in Fig. 14-8 illustrates the setting and testing of FLAGS.

	FORTRAN	ADDRESS	INTEL 8080 CODE
	J=4	50	MVI A,4
52	J=J–2	52	SUI 2
	IF(J.LT.0)GO TO 150	54	JM 150
	IF(J.EQ.0)GO TO 52	57	JZ 52
	GO TO 52	60	JMP 52
	⋮	⋮	⋮
150	STOP	150	HLT

Fig. 14-8. A sample program on using conditional branching.

The first instruction found in location 050 is used to initialize the content of the ACCUMULATOR to 4. The OP-CODE "MVI" stands for "move immediate to the AC". The data to be moved are contained in the next byte of the instruction. Thus, the instruction, "MVI A,4" instructs the CPU to move the number 4 into the AC (*not* the content of location 4). Since "MVI" is a data-transfer instruction, it does not set any one of the FLAGS. The PC is then incremented to 52. At locations 052 and 053 the instruction is "SUI 2", which is for "subtract immediately the number 2 from the AC". At this time the content of the AC is decremented from 4 to 2. Since "SUI" is a flag-setting OP-CODE, after its execution and before the CPU goes on to the next instruction it sets all the FLAGS according to the outcome or result of the "SUI 2" instruction. Since no borrowing occurred in performing the subtraction, the CARRY FLAG is set to 0, and since the 8th bit of the AC is 0, the SIGN FLAG is set to 0. Since the content of the AC is 2, not 0, the ZERO FLAG is also set to 0. Only after all the FLAGS have been properly set does the CPU increment the PC to the next location at 54.

At location 54 the first flag-testing instruction is encountered, "JM 150", which stands for "jump to location 150 if the SIGN FLAG is 1". In reality if the SIGN FLAG were equal to 1, the PC would be set to 150 (instead of the next location in the sequence). However, since the SIGN FLAG is 0, no jump takes place, and the PC is incremented to 57. Note that the conditional jump instructions are not flag-setting instructions so that the FLAGS are not disturbed by their executions.

At location 57 we find the conditional branching instruction, "JZ 52", which stands for "jump to location 52 if the ZERO FLAG is 1". Since the ZERO FLAG is currently set to 0, no jump occurs, and the PC is incremented to 60. Location 60 contains the unconditional jump instruction "JMP 52", thus the PC is next set to 52 and the instruction "SUI 2" is again executed. Now when 2 is subtracted from the content of the AC which is also 2, a zero results and is stored in the AC. Since "SUI 2" is a flag-setting instruction, after its execution all the FLAGS are reset according to the current content of the AC. Thus, SIGN FLAG = 0, CARRY FLAG = 0, but ZERO FLAG = 1. Now when we go through the flag testings the second round, no action is taken on "JM 150" since the SIGN FLAG is still 0. However, when we come to the "JZ 52" instruction in location 057, since the ZERO FLAG is indeed 1, the PC is incremented to 52. After this conditional jump the instruction JMP 52 in location 060 is never encountered.

Now after executing "SUI 2" in location 052 after the conditional jump, the AC now contains the number −2, stored in the two's complement notation. Thus, the 8th bit of the AC is occupied by a 1, and thus the SIGN FLAG set to 1. The CARRY FLAG is also set to 1, since it is necessary to "borrow" to perform the subtraction. The ZERO FLAG is set to zero, since the content of the AC is not zero.

When the PC is incremented to 54, the instruction "JM 150" is encountered. This conditional jump will be executed since the SIGN FLAG is now equal to 1. At location 150, the HLT instruction causes the CPU to stop the processing, and the execution of the entire program is now complete.

Subroutine Calls

A subroutine is a block of operation code which is separate from the main program, and which performs a specific task that is frequently needed by the main program. The subroutine is entered or called by a "calling" instruction that is executed in the main program. When a subroutine is called, data from the main program are passed to the subroutine for processing. When the execution of the subroutine is completed, control is passed back to the place in the main program where the subroutine was called. This is known as "returning" to the main program. In the FORTRAN language, subroutine X is called simply by executing the instruction, "CALL X(A,B,C,D)", where A, B, C, D represent variables that are to be passed to the subroutine. When the execution of subroutine X is completed, we use the instruction, "RETURN", to transfer control back to the instruction immediately following the CALL instruction in the main program.

In order to implement subroutine calls, the computer must be capable of performing the following operations when a "CALL" instruction is encountered:

1. Store the address of the instruction immediately following the "CALL" instruction. This is also known as the return address.
2. Pass all necessary data to the subroutine.
3. Enter the subroutine by branching to the first instruction of the subroutine.
4. Execute the subroutine.

When a "RETURN" instruction is encountered, the computer must perform the following operations:

1. Pass all the results back to the calling program.
2. Retrieve the previously stored return address, and load this address into the program counter. This action will cause the calling program to reenter at the correct point so that normal sequential processing can be resumed.

There are several ways of implementing subroutine calls on a microcomputer. Some methods involve no more than adding a few new instructions to the instruction set of the machine, and setting up rules that the user must follow. However, many microcomputers implement subroutine calls by adding additional hardware devices and some new instructions. The hardware devices used by the INTEL 8080 are the STACK and the STACK POINTER. The method of subroutine call using the STACK and the STACK POINTER is discussed below.

The Stack And The Stack Pointer

Before discussing what a STACK is and how it works, it is helpful to discuss the problems that arise in subroutine handling. A bookkeeping problem may arise when a string of subroutines are to be executed. Every time one subroutine calls another, a new return address must be stored. For example, suppose a main program calls subroutine X, and subroutine X then calls subroutine Y. When the execution of subroutine Y is completed, control must be returned to its calling program, which is subroutine X. Thus, one return address of subroutine X must be stored. When subroutine X has been executed, control must be returned to the main program, and a second return address has to be stored somewhere. Each time a subroutine calls some other subroutine, a return address must be stored. Then when a subroutine returns, the correct return address must be retrieved and placed in the PC. This inevitably will involve a great deal of bookkeeping, and a systematic way of keeping track of all the return addresses in the proper sequence is needed.

Another problem that arises involves the use of registers by a subroutine. In the course of writing a long program, the programmer may place the intermediate results in various registers. Since the number of registers available is limited, there are times when all or almost all of the registers are holding intermediate results of the calling program. When the subroutine is entered, it would be very helpful to be able to save the contents of all the registers in executing the calling program somewhere and then restore the registers when the subroutine has been executed. By saving the contents of the registers in certain memory locations, the subroutine can freely use the registers for its own computations. Thus, a bookkeeping system is needed to keep track of where the contents of the registers are stored by all the various called subroutines.

The key to solving the bookkeeping problems of subroutine calls is to recognize that individual layers of storage are needed by each subroutine to store its return address and save the contents of the registers associated with the calling program. These layers of storage are needed one at a time as the subroutines are called. As the subroutines return, the most recent layer which contains the latest return address is no longer needed and can be erased. When the last called subroutine finally returns, the last layer of storage is retrieved, and the bookkeeping is over. It is important to notice that the first layer of storage, which stores the return address and registers of the main program itself, will be the last layer of storage to be retrieved.

The concept of forming and erasing layers of storage is formally called the concept of STACK STRUCTURE. A STACK is a block of memory locations used to implement the STACK STRUCTURE concept. The STACK is empty at the start and is filled up with layers of return addresses and registers of each called subroutine. Each time a program calls another program, a layer of information is added to the STACK. This is known as "pushing" data onto the STACK. As the subroutines begin to return, the calling information is retrieved from the STACK, and the operation is known as "popping" data off the STACK. Since the first layer of data stored is the last layer to be retrieved, this type of STACK operation is known as a first-in-last-out (FILO) STACK. Note that layers of information are retrieved only from the top of the STACK. Since the subroutines always return in the opposite order of which they are called, the next layer to be retrieved when a subroutine returns must be the top layer, not some intermediate layer in the STACK.

To keep track of the current top of the STACK, a STACK POINTER is devised. The STACK POINTER (SP) is simply a special register which contains the address of the current top of the STACK. If new data are to be added to the STACK, the CPU will refer to the SP to find out the address of the top of the STACK. The CPU will then store the data to be added to the STACK, and change the value of the SP so that the SP once again points to the current top of the STACK. If data are to be retrieved from the STACK, the process is reversed. The top layer of the STACK, as indicated by the SP, is retrieved, and the SP is changed to point to the new top of the STACK. The function of the SP in STACK processing is analogous to the function of the PROGRAM COUNTER in sequential processing.

Some computers reserve a special block of memory for the STACK. On these machines the STACK is physically distinct from the core memory where programs and data are stored. On other computers such as the INTEL 8080, there is no memory block specially set aside as the STACK. Instead, any part of the memory may be designated as the STACK. At the beginning of a program the programmer simply decides and specifies which memory location, say location X, is the beginning of the STACK. An instruction is executed which loads the address "X" into the STACK POINTER. Then, as the program and subroutines are executed, the STACK will grow or shrink in size as the subroutines are called and executed. When the final RETURN instruction is executed to get back to the main program, the SP will once again contain address "X", and the STACK is empty.

Several special instructions are devised to facilitate the use of the STACK. To store the contents of registers on the STACK, a computer will often have "PUSH" and "POP" instructions. The format for the PUSH instruction on the INTEL 8080 is "PUSH REGISTER-PAIR," where the "REGISTER-PAIR" means register pair B-C, D-E, or H-L. The PUSH instruction of the 8080 stores both halves of a register pair in the STACK; that is, one PUSH instruction stores two registers. The OP-CODES of the PUSH and POP instructions generally specify only the first register of a register pair. For instance, to specify that the contents of the register pair B-C should be stored in the STACK, the mnemonic code for the 8080 is "PUSH B," not "PUSH BC." To retrieve the upper two bytes of the STACK and place them in register pair B-C, the OP-CODE IS "POP B." There is also a pair of instructions for storing and retrieving the contents of the ACCUMULATOR and the states of all the FLAGS. The instruction codes are "PUSH PSW" and "POP PSW," where PSW stands for "processor status word." The name comes from the observation that the information in the AC and the states of the FLAGS represent, in some sense, the "status" of the computer at the time the "PUSH PSW" instruction is executed.

14.4 MICROPROCESSOR CONTROL OF CONTROL SYSTEMS

In this section we shall describe some of the programming aspects of controlling a physical system by a microprocessor. One way of using a microprocessor as a controller is shown in Fig. 14-9. In this case the controlled process consists of a dc motor, load, and the power amplifier. The analog process and the microprocessor are interfaced through A/D and D/A converters. Thus, the overall system is analytically considered to be a digital control system, with

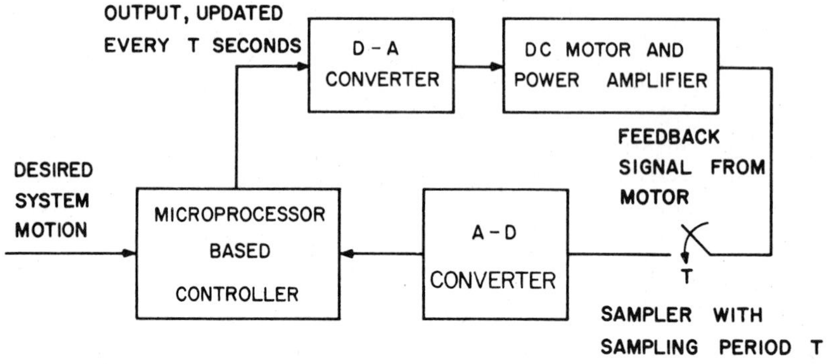

Fig. 14-9. Block diagram of a microprocessor-controlled dc motor system.

a sampling period of T seconds.

Let us consider that the purpose of the dc-motor control system is to drive the load speed $\omega(t)$ to follow the constant command speed ω_d. The error between the command speed and the load speed is

$$e(t) = \omega_d - \omega(t) \tag{14-1}$$

Thus, the input to the microprocessor is the digitized error signal e(kT), k = 0,1,2,..., and we let the output of the microprocessor be u(kT).

Consider that the microprocessor is to perform the digital computation to implement a PI (proportional-integral) controller, so that the continuous-data form,

$$u(t) = K_p e(t) + K_I \int e(t)dt \tag{14-2}$$

The integral in the last equation is written as

$$x(t) = \int_{t_0}^{t} [\omega_d - \omega(\tau)]\, d\tau + x(t_0) \tag{14-3}$$

where t_0 is the initial time, and $x(t_0)$ is the initial value of x(t). To approximate the integral by a digital model several schemes may be used. Let us use the trapezoidal integration rule here, and let t = kT, $t_0 = (k - 1)T$. Then, the definite integral in Eq. (14-3) is approximated as

$$\int_{(k-1)T}^{kT} [\omega_d - \omega(t)]\, dt \cong \omega_d T - \frac{T}{2} [\omega(kT) + \omega(k - 1)T] \tag{14-4}$$

for k = 1,2,... . Therefore, the value of the integral in the last equation at t = kT can be computed based on the given input ω_d and the data for $\omega(kT)$

and $\omega[(k-1)T]$. However, in reality it takes the microprocessor a finite amount of time to compute the integral in Eq. (14-4), so that given the input data $\omega(k-1)T$ and $\omega(kT)$, the result of the integral computation is not available at $t = kT$. In general, one has to add up all the time intervals required to execute the OP-CODES of the integration subroutine on the microprocessor to find out what this time delay is. For convenience, we shall assume that this computation time delay is equal to one sampling period T. This means that the right-hand side of Eq. (14-4) gives the computational result of the integral at $t = (k+1)T$. Thus, Eq. (14-3) is discretized to

$$x[(k+1)T] = \omega_d T - \frac{T}{2} \quad \omega(kT) + \omega[(k-1)T] \quad + x(kT) \tag{14-5}$$

Notice that we are using $x(kT)$ rather than $x[(k-1)T]$ as the initial state of $x(t)$. Substituting $x[(k+1)T]$ for the integral in Eq. (14-2), the discretized version of $u(t)$ is written as

$$u[(k+1)T] = K_p[\omega_d - \omega(kT)] + K_I x[(k+1)T] \tag{14-6}$$

This control is applied to the dc motor system at $t = (k+1)T$, $k = 0,1,2,\ldots$. The control is updated every T seconds, and is held constant between the sampling instants.

The block diagram of a typical microprocessor system used to implement the digital PI controller is shown in Fig. 14-10. The system shown in this figure

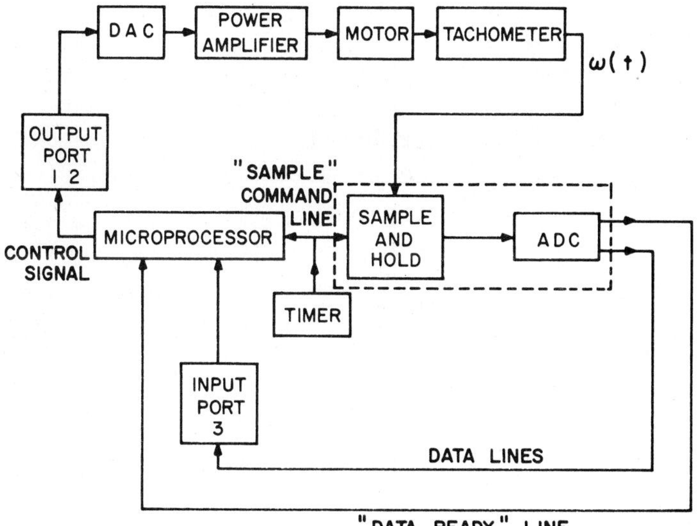

Fig. 14-10. Block diagram of implementation of discretized PI controller.

uses an analog timer to determine the start of the next sampling period. Alternatively, the microprocessor can use a software timing loop to keep track of when T seconds have elapsed. The time outputs a pulse every T seconds. This pulse is used in two ways. Firstly, the pulse is applied to the interrupt line of the microprocessor. This will cause the processor to stop what it is doing and execute the interrupt routine, which in this case would be the routine to output the next value of the control, $u[(k+1)T]$. This control is sent to the D/A whose output in turn controls the power amplifier. The timing pulse from the timer is also sent to the "sample" command line of the A/C. A pulse on the "sample" line triggers the sample-and-hold circuitry within the A/C, at which instant the motor velocity, $\omega(t)$, is sampled and held constant for one sampling period. The value of $\omega(kT)$ is then converted to an N-bit binary number by the rest of the A/D circuitry. As described in Chapter 2 there is a finite conversion time associated with the conversion process. Therefore, the A/C must signal the microprocessor via a "data ready" line that the sampled data have been converted. The "data ready" line may also be attached to the interrupt line of the microprocessor. This second interrupt will cause the processor to read in the value of $\omega(kT)$ and then compute the next value of control, $u[(k+1)T]$. After the control has been calculated, the microprocessor waits for another interrupt from the timer before it outputs the control at $t = (k+1)T$.

The main program in the 8080 code for the implementation of the PI controller is shown in Fig. 14-11. For simplicity, we have set $K_p = 1/2$ and $K_I = 4$, both powers of 2. The sampling period is chosen to be 1 second. As shown in Fig. 14-10, data from the A/D represent the sampled values of $\omega(t)$ at $t = kT$, $\omega(kT)$.

IN DATA	Read in $\omega(kT)$ from input port 3
MOV D,A	Store $\omega(kT)$ in register D
LDA WD	load the value of ω_d in AC
SUB D	compute $\omega_d - \omega(kT)$
CALL DIVIDE2	call subroutine to form $K_p[\omega_d - \omega(kT)]$, $K_p = 1/2$
MOV C,A	store the result in register C
LDA WOLD	load the AC with the "old" value of $\omega(kT)$, $\omega[(k-1)T]$
MOV B,A	store in register B
MOV A,D	move the latest value of $\omega(kT)$ in the AC
STA WOLD	Replace stored value of $\omega(kT)$ by latest value
ADD B	form $\omega(kT) + \omega[(k-1)T]$
CALL DIVIDE2	Call subroutine to form $T[\omega(kT) + \omega(k-1)T]/2$, $T = 1$

```
             MOV B,A            store result in register B
             LDA WD             load the AC with ωd
             SUB B              form ωdT − T[ω(kT) + ω(k − 1)T]/2
             MOV B,A            store result in register B
             LDA XOLD           load x(kT) in AC
             ADD B              form ωdT − T[ω(kT) + ω(k − 1)T]/2 + x(kT) =
                                x[(k + 1)T]
             STA XOLD           Replace stored value of x(kT) by x(k + 1)T
             CALL MULTIPLY4     form KIx[(k + 1)T], KI = 4
             ADD C              form u[(k + 1)T] = Kp[ωd − ω(kT)] + KIx[(k + 1)T]

DIVIDE2:     JM NEGATIVE
             ORA A
             RAR
             RET
NEGATIVE:    STC
             RAR
             RET
MULTIPLY4:   ORA A
             RAL
             ORA A
             RAL
             RET
WOLD:        DB 0
WD:          DB 25
XOLD:        DB 0
```

Fig. 14-11. An 8080 assembly-language program implementing a PI controller.

Although a proportional-plus-integral (PI) controller was illustrated in the previous example, in general, it would be simple to include the derivative operation to implement the PID controller. One common way of approximating the derivative of a signal is to use the "backward difference" equation. Let $\dot{x}(t)$ represent the time derivative of x(t), then $\dot{x}(t)$ can be approximated digitally as

$$\dot{x}(t) \cong \frac{x(kT) - x[(k-1)T]}{T} \tag{14-7}$$

where T is the sampling period in seconds.

The microprocessor programming discussions given in this section are designed to give the reader an overview of some of the techniques available in designing a direct microprocessor control system. The example given is that of the control of a dc motor with a PI controller discretized for microprocessor programming. The starting point of the problem is that the PI controller is described by a differential equation. The latter is discretized at the sampling instants by one of the numerical approximation methods, and then is programmed in the microprocessor machine language. As an alternative, and this is quite frequently the case in control system practice, the digital controller is described by a transfer function in the z domain, D(z). In order to prepare a program for the microprocessor it is necessary to first decompose D(z) by one of the decomposition methods discussed in Chapter 4. Then the microprocessor program can be generated from the state diagram or the state equations which describe the digital controller.

14.5 GENERAL DISCUSSIONS ON THE EFFECTS OF SOME LIMITATIONS OF MICROPROCESSOR-BASED CONTROL SYSTEMS

The analytical and theoretical development on digital control systems carried out thus far in this text assume that the computer implementation is generally without physical limitations. We are all aware of practical limitations in physical systems and components in the form of saturation, deadzone, hysteresis, etc., which have important effect on the performance of control systems. In the case of a microprocessor, the finite wordlength and the time delays encountered in the execution of the operational instructions of the computer all must be taken into account by the system designer. The questions on how the finite wordlength of a microprocessor affects the performance of a control system, how the quantizer nonlinearity and the time delays affect the implementation of the digital control law will be investigated in the following sections. The material presented will give the reader a feel for the kind of problems to expect when implementing a microprocessor-based control system, and to provide some solutions as to what can be done about these unavoidable problems.

14.6 EFFECTS OF FINITE WORDLENGTH ON CONTROLLABILITY AND CLOSED-LOOP POLE PLACEMENT

Because microprocessors have limited wordlength, in general, the signals in and out of the microprocessor will be truncated or quantized, and the parameters used in the control law will be truncated when the latter are being realized by the OP CODES of the processor. Therefore, it is necessary to investigate the effects on the system performance when the system parameters can be realized only by a finite set of numbers.

In Chapter 10 we discussed the design of digital control systems through placement of the closed-loop poles. The design parameters are the state feedback gains or the output feedback gains. It was pointed out in the pole-placement design that if the system is completely controllable, then the poles of the closed-loop system can be arbitrarily placed. However, if the feedback gains cannot assume an arbitrary set of numbers, this means that the closed-loop poles do not have a domain with an infinite resolution. Putting it another way, if the feedback gains can be realized by only a finite set of numbers, any initial state can be driven to only a finite number of final states in finite time. Therefore, it seems appropriate to state that in the rigorous sense of the definition of controllability, when a digital control system is subject to finite wordlength and quantization, it cannot be a controllable system. However, from a practical standpoint, if the quantization levels are small we should not be overly concerned with the controllability problems. Rather, the limitation on the resolution of the system parameters is translated to system errors and stability problems which are direct results of amplitude quantization.

The effect of quantization on the control input u(k) can be investigated by referring to the system

$$\mathbf{x}(k+1) = A\mathbf{x}(k) + Bu(k) \tag{14-8}$$

where u(k) is a scalar input. Assuming that u(k) is restricted to a quantized set of values, i.e.,

$$u(k) = n_k q \tag{14-9}$$

where $n_k = 0, \pm 1, \pm 2, \ldots$, and q is the quantization level.

The solution of Eq. (14-8) is

$$\mathbf{x}(N) = A^N \mathbf{x}(0) + \sum_{k=0}^{N-1} A^{N-k-1} Bu(k) \tag{14-10}$$

Substitution of Eq. (14-9) into Eq. (14-10), we get

$$\mathbf{x}(N) = A^N \mathbf{x}(0) + q \sum_{k=0}^{N-1} A^{N-k-1} Bn_k \tag{14-11}$$

Since u(k) is constrained to distinct levels of $n_k q$, the final state $\mathbf{x}(N)$ is parametrized by the set of integers n_k, $k = 0, 1, 2, \ldots, N-1$. If the pair [A,B] is controllable, then $\mathbf{x}(0)$ can be driven to any point $\mathbf{x}(N)$ in the state space only if u(k) can be selected from a continuum of values. However, since u(k) is limited to quantized values, $\mathbf{x}(N)$ is also restricted. The following example illustrates the controllability problem due to quantization in digital control systems.

Example 14-1.

Consider the digital control system that is described by the state equations,

$$\mathbf{x}(k+1) = \mathbf{A}\mathbf{x}(k) + \mathbf{B}u(k) \tag{14-12}$$

where

$$A = \begin{bmatrix} 0 & 1 \\ -1 & -2 \end{bmatrix} \qquad B = \begin{bmatrix} 0 \\ 1 \end{bmatrix}$$

and u(k) is subject to amplitude quantization with quantization level q. The state transition equation of Eq. (14-12) is

$$\mathbf{x}(N) = A^N \mathbf{x}(0) + \sum_{k=0}^{N-1} A^{N-k-1} Bu(k) \tag{14-13}$$

where

$$A^N = \begin{bmatrix} (-1)^{N+1}(N-1) & (-1)^{N+1}N \\ (-1)^N N & (-1)^N(N+1) \end{bmatrix} \tag{14-14}$$

Since the pair [B AB] is completely controllable, the vector $A^{N-k-1}B$ can be expressed as a linear combination of the two vectors B and AB. In the present case we can show that

$$A^{N-k-1}B = (-1)^{N-k-2}[(N-k-2)B + (N-k-1)AB] \tag{14-15}$$

Since $A^{N-k-1}B$ is linearly and integrally dependent on B and AB, we can write Eq. (14-13) as

$$\begin{aligned} \mathbf{x}(N) &= A^N \mathbf{x}(0) + n_1 qB + n_2 qAB \\ &= A^N \mathbf{x}(0) + n_1 q \begin{bmatrix} 0 \\ 1 \end{bmatrix} + n_2 q \begin{bmatrix} 1 \\ -2 \end{bmatrix} \end{aligned} \tag{14-16}$$

where n_1 and n_2 are integers, and q is the quantization level.

Equation (14-16) shows that at any particular sampling instant, $k = N$, $x_1(N)$ and $x_2(N)$ can assume a discrete set of values only. For any given initial state $\mathbf{x}(0)$, the set of reachable states for $x_1(k)$ and $x_2(k)$ are spaced at intervals of q in the x_1-versus-x_2 plane, as shown in Fig. 14-12.

When the feedback gains of state-feedback or output-feedback are implemented by a microprocess, these gains are again subject to amplitude quantization. This simply means that if the design is for the objective of pole placement, the poles of the closed-loop system cannot be placed with arbitrarily fine resolution in the z-plane. The following numerical example illustrates the effect of quantization on the pole-placement design.

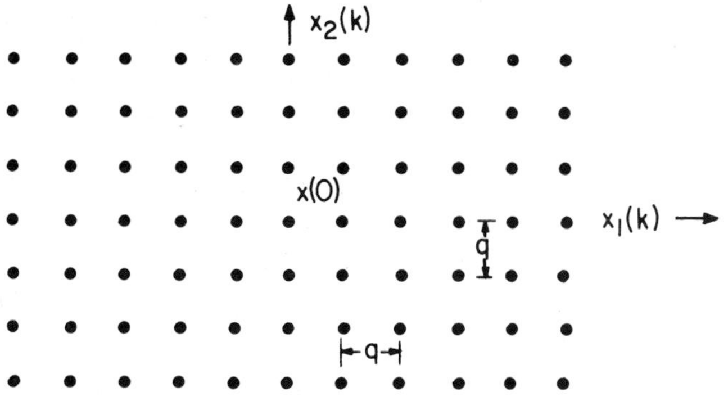

Fig. 14-12. Realizable states of $x_1(k)$ and $x_2(k)$ due to amplitude quantization in states and controls.

Example 14-2.

Consider the digital system

$$\mathbf{x}(k+1) = \mathbf{A}\mathbf{x}(k) + \mathbf{B}u(k) \tag{14-17}$$

where

$$A = \begin{bmatrix} 0 & 1 \\ 0 & 0 \end{bmatrix} \qquad B = \begin{bmatrix} 0 \\ 1 \end{bmatrix}$$

$$u(k) = -G\mathbf{x}(k) \tag{14-18}$$

$$G = [g_1 \quad g_2] \tag{14-19}$$

The values of g_1 and g_2 are quantized with quantization level q. The characteristic equation of the closed-loop system is written

$$|zI - A + BG| = z^2 + g_2 z + g_1$$

$$= (z - r_1)(z - r_2) = 0 \tag{14-20}$$

where r_1 and r_2 are the two closed-loop eigenvalues. Then

$$G = [g_1 \quad g_2] = [r_1 r_2 \quad -(r_1 + r_2)] \tag{14-21}$$

Figure 14-13 shows the region of stable operations in the $g_1 - g_2$ parameter plane when the feedback gains g_1 and g_2 are subject to fixed-point data representation with a 3-bit wordlength. In this case, the quantization level is

$$q = 2^{-3} = 1/8 \tag{14-22}$$

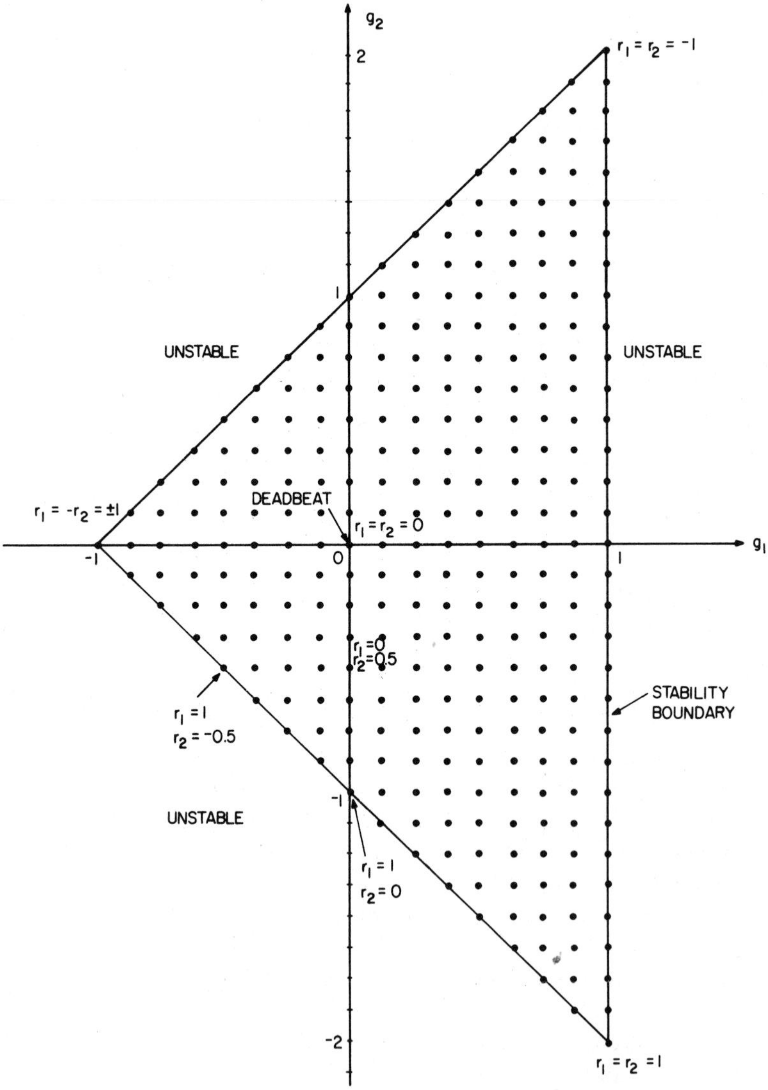

Fig. 14-13. Quantization of feedback gains g_1 and g_2 in the parameter plane.

The realizable values of g_1 and g_2 are indicated by dots in the parameter plane, and only the values that correspond to a stable closed-loop system are shown.

It is useful to investigate the realizable closed-loop poles due to the quantization of g_1 and g_2. Let the roots of Eq. (14-20) be represented in polar form,

$$z_1 = re^{j\theta} \tag{14-23}$$

and

$$z_2 = re^{-j\theta} \tag{14-24}$$

The characteristic equation of Eq. (14-20) is written

$$z^2 - 2r\cos\theta\, z + r^2 = 0 \tag{14-25}$$

Then,

$$r = \pm\sqrt{g_1} \tag{14-26}$$

Thus, quantizing g_1 with a quantization level q forces the roots to lie on concentric circles in the z-plane. These circles are all centered at the origin, with radii equal to $0, \sqrt{q}, \sqrt{2q}, \ldots$. These circles are shown in Fig. 14-14 for the present case with $q = 1/8$.

Since the real parts of the roots are $r\cos\theta$, quantizing g_2 $(= -2r\cos\theta)$ is equivalent to limiting the real parts of the roots to a finite set of numbers. Let the real part of the roots be denoted by σ, then

$$\sigma = r\cos\theta = -g_2/2 \tag{14-27}$$

The intersections of the circles and the vertical lines represent the realizable poles due to the prescribed quantization of g_1 and g_2.

14.7 TIME DELAYS IN MICROPROCESSOR-BASED CONTROL SYSTEMS

Besides having a finite wordlength, microprocessors are relatively slow digital computing machines. In many digital computing applications, which may not be concerned with real-time handling of data, the slow computing speed may not be important. However, in control systems applications, real-time computation is often necessary, and the time delays encountered in executing the data handling may have a significant effect on the system response. In general, it is important to know how large these delays are in order to deal with the problem analytically. Two immediate problems may attribute to time delays in control systems. One is that if the time delay is too large, there would not be ample time to carry out all the necessary computations required to execute the control algorithms, and the other is the adverse effect that time delay has on the stability of closed-loop control systems.

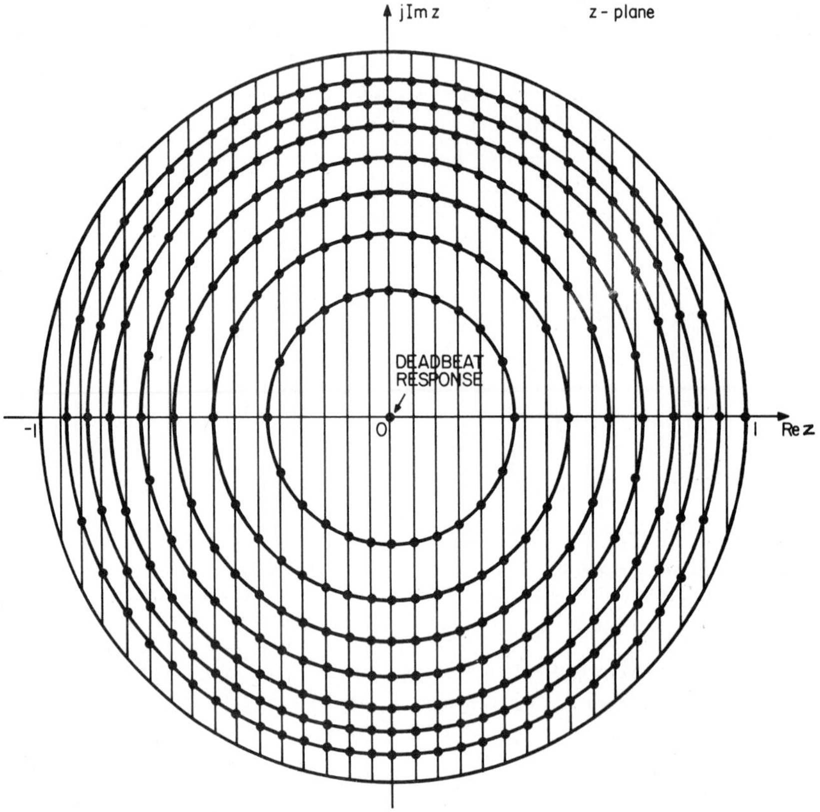

Fig. 14-14. Trajectories in the z-plane showing the realizable poles due to the quantization of feedback gains g_1 and g_2.

The time delays due to microprocessor computation may be identified by analyzing the program used for the control law along with the subroutines that may be called from any available utility package. Each program is made up of a set of instructions, and each instruction requires a particular number of machine cycles to execute. Each machine cycle in turn requires a certain number of machine states. The time required for the microprocessor to execute a particular instruction is directly proportional to the total number of machine states that the microprocessor must go through in order to complete that particular instruction. For the INTEL 8080 microprocessor, for example, each state requires 500 nanoseconds. This allows the machine to go through two million machine states per second. This may seem to be extremely fast, however, even a simple program may require thousands of machine states to execute.

In general, the information on the number of machine states necessary for a particular instruction is found in the User's Manual of the microprocessor. With this information, it is possible to go through any program, instruction by instruction, adding up the number of machine states required, to come up with both the total time necessary to complete the program, and the time required to reach a particular point in the computations.

As an illustration, the time intervals necessary to complete the necessary computations for a variety of tasks with the INTEL 8080 are given in Table 14-1.

TABLE 14-1. Some Typical Computation Times for the INTEL 8080 Assuming 500 nsec/state.

Task	Time Required
Floating-point addition	18.5 μsec — 202.5 μsec
Floating-point multiplication	63.5 μsec — 446 μsec
Floating-point to fixed-point conversion	25 μsec — 109 μsec

From this table we see that additions and multiplications in floating-point computations are time consuming.

Figure 14-15 shows the instruction codes of an implementation of the PID control. The calculation time required to execute each instruction is indicated, so that we may estimate the total time required to execute the program on the INTEL 8080. In this case the approximate time delay is 2 milliseconds. It is important to point out that, in general, this time delay will not be deterministic. One set of inputs to a program may require the microprocessor to execute a certain set of instructions, looping through these instructions in one particular way, while a different set of inputs may result in a different set of instructions being executed or possibly different looping. Thus, the number of states required to perform a calculation may vary with the inputs to the program or subroutine, and as a result, there will be a maximum time necessary for the execution of the program and a significantly different minimum time. In general, we will be able to specify a maximum delay time and a minimum delay time.

To amplify the point on the variable time delay further, consider the case of adding two floating-point numbers. First we must determine which of the two numbers has the larger exponent. Then the second number must be adjusted to have this same exponent. This adjustment may be large or small. This adjustment may be large or small, and then after all this, the numbers may be

```
                      DACA  EQU 0CH   ; DEFINE      D/A PORT
                      CNTRA EQU 07H   ; I/C PORT    A/D PORT X1
                      CNTRB EQU 0FH   ; LOCATIONS   A/D PORT X2
                      CMD   EQU 88H   ;
                      SPWT  EQU 07E6H ; DEFINE      TIMING
                      CVRT  EQU 077EH ; SUBROUTINE  A/D CONVERSIONS
                      CNRT  EQU 0759H ; LOCATIONS   FLOAT TO FIXED
                      FADD  EQU 0561H ;             FL PT ADDITION
                      FMULT EQU 05E4H ;             FL PT MULTIPLICATION
                                      ;
                      ORG   2000H     ; SET UP
                  KP: DW    2         ; MEMORY      P,I,D CONST
                  KI: DW    2         ; LOCATIONS
                  KD: DW    2         ; FOR CONST
                  X1: DW    2         ;             X1(k)
                  X2: DW    2         ;             X2(k)=dX1/dT
                  Z:  DW    2         ;             INTEGRAL STATE
                  T:  DW    2         ;             SCRATCH VARIABLE
                  U:  DW    2         ;             OUTPUT
                  S:  DW    2         ;             SET POINT
                                      ;
                      ORG   2012H     ; MAIN PROGRAM
                      MVI   A,CMD     ;
                      OUT   CNTRA     ;
                      OUT   CNTRB     ;
                                      ;
                  LP: CALL SPWT       ;  -WAIT FOR NEXT SAMPLE
Time Required         CALL CVRT       ;  -DO THE A/D CONVERSIONS
       2.5 μs         MOV H,B         ;
       3.5 μs         MVI L,OH        ;
       8   μs         SHLD X1         ;  -STORE X1 IN MEMORY
       2.5 μs         MOV H,C         ;
       8   μs         SHLD X2         ;  -STORE X2 IN MEMORY
       8   μs         LDHL KP         ;
       2.5 μs         MOV B,H         ;
       2.5 μs         MOV C,L         ;
```

```
8     µs        LDHL X1        ;
2     µs        XCHG           ;
446 + 10  µs    CALL FMULT     ;
2.5 µs          MOV H,B        ;
2.5 µs          MOV L,C        ; -MULTIPLY KP*X1 AND
8     µs        SHLD T         ;  STORE THE RESULT IN T
8     µs        LDHL Z         ;
2.5 µs          MOV B,H        ;
2.5 µs          MOV C,L        ;
8     µs        LHLD KI        ;
2     µs        SCHG           ;
446 + 10  µs    CALL FMULT     ; -NOW ADD KI*Z TO
8     µs        LHLD T         ;    T = KP*X1 + KI*Z
2     µs        XCHG           ;
202.5 + 10  µs  CALL FADD      ;
2.5 µs          MOV H,B        ;
2.5 µs          MOV L,C        ;
8     µs        SHLD T         ;
8     µs        LHLD X2        ;
2.5 µs          MOV B,H        ;
2.5 µs          MOV C,L        ;
8     µs        LHLD KD        ;
2     µs        XCHG           ;
446 + 10  µs    CALL FMULT     ;
8     µs        LHLD T         ;
2     µs        XCHG           ;
202.5 + 10  µs  CALL FADD      ;
2.5 µs          MOV H,B        ;
2.5 µs          MOV L,C        ;
8     µs        SHLD U         ; -ADD KD*X2 TO T AND STORE THE
109 + 10  µs    CALL CNRT      ;  RESULT IN U
2.5 µs          MOV A,B        ;
5     µs        OUT DACA       ; -CONVERT U FROM A FLOATING POINT
-------------                  ;  NUMBER TO A FIXED POINT NUMBER
2.072 milliseconds             ;  THEN OUTPUT IT ON DACA
                               ;
```

```
LHLD S          ; -NOW WE WANT TO UPDATE THE
MOV B,H         ;  INTEGRAL STATE FOR THE NEXT
MOV C,L         ;  SAMPLE
MVI A,OH        ;
SUB B           ;
MOV B,A         ;
LHLD X1         ;
XCHG            ;
CALL FADD       ;
LHLD Z          ;
SCHG            ;
CALL FADD       ;   Z(k + 1) = Z(k) + (X1(k) - S)
MOV H,B         ;
MOV L,C         ;
SHLD Z          ;
                ;
JMP LP          ; -NOW GO BACK TO THE BEGINNING OF
                ;  THE PROGRAM TO DO THE CALCULA-
END             ;  TIONS FOR THE NEXT SAMPLE
```

Fig. 14-15. A simple PID program for an 8080 showing the times necessary to execute each instruction assuming a 500 ns clock and the worst case times for each subroutine.

added. Thus we can see that two numbers with the same exponent can be added more quickly than two numbers for which a slight adjustment is necessary. Furthermore, two numbers with a large difference in exponents may require even a longer time to add. There may also be special cases, such as one of the numbers is zero, then there may be an early exit from the program, and the addition is executed in a shorter time interval than any of the previously mentioned cases.

In general, fixed-point arithmetics take shorter time to execute on a microprocessor. This is why in many cases fixed-point arithmetic calculations are preferred in industrial applications. For instance, typical time of fixed-point addition on an 8-bit microprocess is only 2 msec.

In summary, the main objective of this section is to point out the time delays encountered in the execution of a digital program. Since most microcomputers are slow digital machines, the time delays cannot always be neglected in the modeling of the digital controller. Since it is well-known that time delays usually have an adverse effect on the performance of closed-loop systems, digital or analog, it is essential that these imperfect conditions be considered when designing a digital control system.

14.8 EFFECTS OF QUANTIZATION — LEAST-UPPER BOUND ON QUANTIZATION ERROR

In the preceding sections of this chapter the effect of magnitude quantization and finite wordlength is demonstrated in terms of pole-placement design and controllability. In general, the direct effects of quantization on digital control systems can be described by two words: *accuracy* and *stability*. In this section we shall investigate the effect of quantization on the steady-state error of the system.

The block diagram of an amplitude quantizer, together with its input-output characteristics, are shown in Fig. 14-16. The dotted line represents the desired transfer characteristic, and the staircase function is the actual characteristic of the quantizer. Notice that the input to the quantizer, $r(t)$, could be of any form or amplitude, but the output $y(t)$ can take on only those discrete quantized levels which are nearest to the value of $r(t)$. The transfer characteristic shown in Fig. 14-16 has uniformly-spaced quantized levels, so that when the input lies between $-q/2$ and $q/2$ the output is zero; when the amplitude of the input lies between $q/2$ and $3q/2$, the output amplitude is q, and so on. As shown in Chapter 2, the relation between the quantization level q and wordlength is

$$q = 2^{-N}\,FS \tag{14-28}$$

where N is the number of significant binary bits or wordlength, and FS represents full scale. The quantization error is $q/2$, and is given by

$$\frac{q}{2} = 2^{-N-1}\,FS \tag{14-29}$$

In view of the transfer characteristic of the quantizer shown in Fig. 14-16, it is apparent that the device is nonlinear. Therefore, the analysis of the effects of quantization in a digital control system can be complex analytically.

Two most common effects of amplitude quantization in closed-loop systems are steady-state error and sustained oscillations. We shall illustrate these effects by means of a simple example.

Let us refer to the digital closed-loop systems shown in Fig. 14-17. The quantization effects have been neglected in these systems. The only difference between the two systems in Figs. 14-17(a) and (b) is that one has negative feedback and the other has positive feedback. It is simple to show that both systems are asymptotically stable. For $r(k) = 0$, the z-transform of the output of the system in Fig. 14-17(a) is written

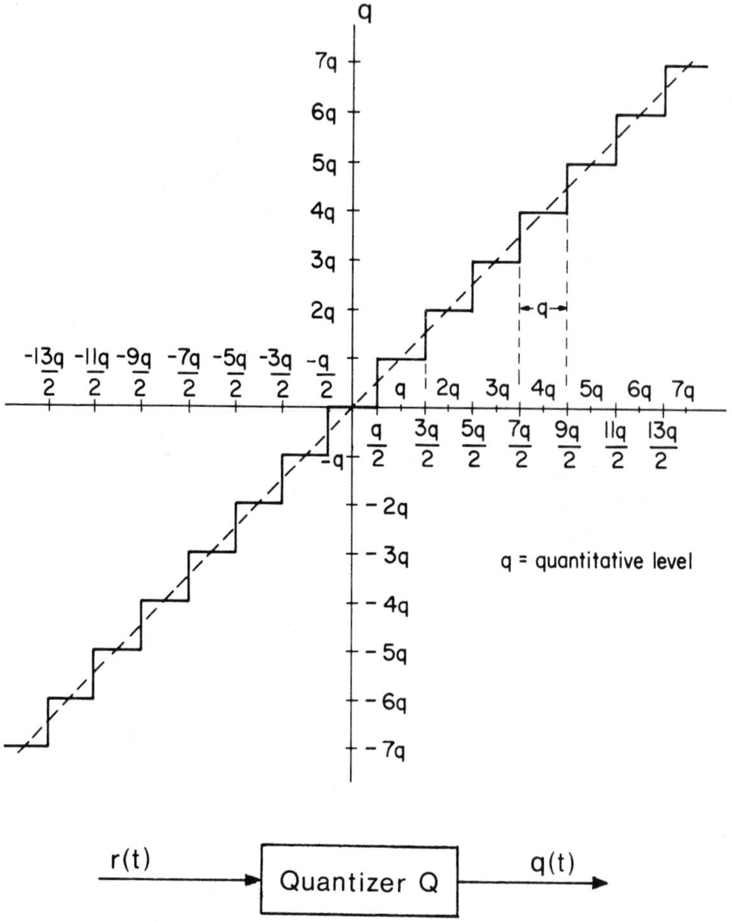

Fig. 14-16. Transfer characteristic and block diagram representation of a quantizer.

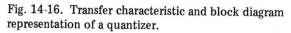

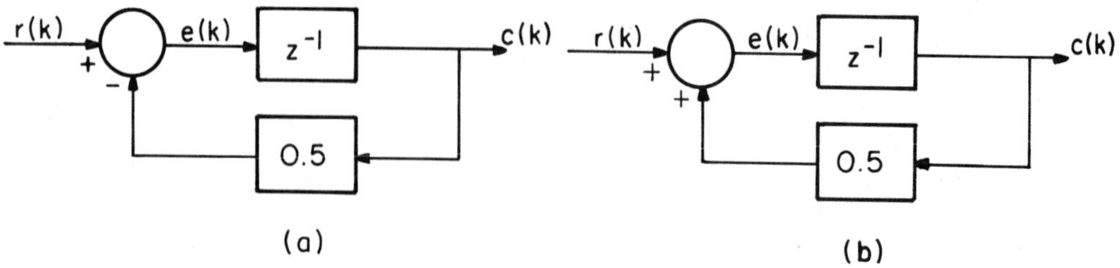

Fig. 14-17. Two digital control systems.

$$C(z) = \frac{z}{z + 0.5} c(0) \tag{14-30}$$

and that of the system in Fig. 14-15(b) is

$$C(z) = \frac{z}{z - 0.5} c(0) \tag{14-31}$$

where c(0) is the initial value of c(k).

The inverse z-transform of Eq. (14-30) is

$$c(k) = (0.5)^k \cos k\pi \cdot c(0) \tag{14-32}$$

and the inverse z transform of Eq. (14-31) is

$$c(k) = (0.5)^k c(0) \tag{14-33}$$

Thus, in both cases the output response due to any arbitrary initial condition c(0) goes to zero as k goes to infinity.

Figure 14-18 illustrates the same two digital control systems but with an amplitude quantizer in each loop. Assuming that a 4-bit wordlength is used in the digital process, so that the quantization level is $q = 2^{-4} = 0.0625$. The characteristics of the quantizer are defined as

$$e_q(k) = 0 \qquad -q/2 < e(k) < q/2$$

$$e_q(k) = nq \qquad nq - q/2 \leqslant e(k) < nq + q/2 \qquad (n \geqslant 1)$$

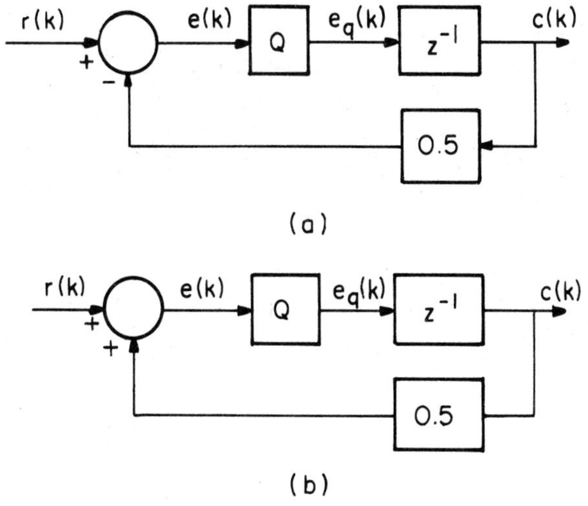

Fig. 14-18. Two digital control systems with quantizers.

$$e_q(k) = -nq \qquad -nq - q/2 < e(k) \leqslant -nq + q/2 \qquad (n \leqslant -1)$$

For the system with negative feedback, shown in Fig. 14-17(a), the response of c(k) with r(k) = 0 and c(0) = 0.58 is tabulated in Table 14-2. In this case, the output c(k) exhibits a sustained oscillation with an amplitude of ±q, and a period of 2. For the system with positive feedback shown in Fig. 14-17(b), the response c(k) for the same input and initial condition is given in Table 14-3. Note that in this case, the system has a constant steady-state error of q in c(k) as k becomes large.

TABLE 14-2.

k	c(k)	e(k)	$e_q(k)$
0	0.58000	−0.29000	−0.31250
1	−0.31250	0.15625	0.18750
2	0.18750	−0.09375	−0.12500
3	−0.12500	0.06250	0.06250
4	0.06250	−0.03125	−0.06250
5	−0.06250	0.03125	0.06250
6	0.06250	−0.03125	−0.06250
7	−0.06250	0.03125	0.06250

TABLE 14-3.

k	c(k)	e(k)	$e_q(k)$
0	0.58000	0.29000	0.31250
1	0.31250	0.15625	0.18750
2	0.18750	0.09375	0.12500
3	0.12500	0.06250	0.06250
4	0.06250	0.03125	0.06250
5	0.06250	0.03125	0.06250
6	0.06250	0.03125	0.06250
7	0.06250	0.03125	0.06250

Although the two systems with quantizers shown in Fig. 14-17 all show imperfectness in the responses that are directly dependent upon the quantization level q, the steady-state error and the sustained oscillation phenomena are grossly different in characteristics. In practice, both the steady-state error and the sustained oscillations are undesirable in control systems, and their amplitudes should be kept as small as possible.

The least-upper bound on the steady-state error due to quantization can be estimated by replacing the quantizers by equivalent noise sources. The prediction of sustained oscillations due to quantization is a nonlinear problem, and the analysis may be carried out by such techniques as the discrete describing function [19].

Since the error of the quantized signal has a least-upper bound of $\pm q/2$, the "worst" error due to quantization in a digital control system can be studied by replacing the quantizer by an external noise source with a signal magnitude of $\pm q/2$. As an illustration, the block diagram of the digital control system with quantization shown in Fig. 14-18(b) is represented by the equivalent system in Fig. 14-19, where it is understood that the latter system model is used solely for the purpose of analyzing the least-upper bound on the steady-state errors due to quantization.

From Fig. 14-19, with $r(k) = 0$, the steady-state value of $c(k)$ is obtained as

$$\lim_{k\to\infty} c(k) = \lim_{z\to 1} (1 - z^{-1})C(z) = \lim_{z\to 1} (1 - z^{-1}) \frac{z^{-1}}{1 - 0.5z^{-1}} \frac{(\pm q/2)z}{z - 1}$$

$$= \pm q \tag{14-34}$$

Therefore, in this case, the least-upper bound predicted on the quantization error is identical to the value actually calculated. However, when we apply the same technique to the system in Fig. 14-18(a), the predicted least-upper bound on the quantization error in $c(k)$ is $\pm q/3$. However, we have shown that the system actually has a sustained oscillation with an amplitude that varies between

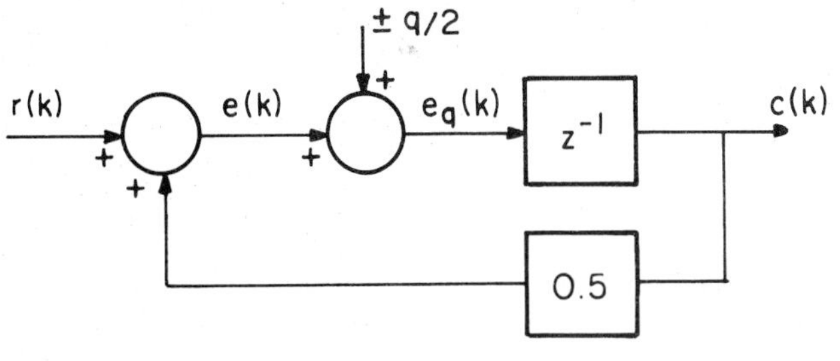

Fig. 14-19. Block diagram of digital control system with quantizer for least upper bound error analysis.

$-q$ and $+q$. Therefore, we can only say that the least-upper-bound error analysis using the equivalent noise sources does not predict the sustained-oscillation behavior of quantized systems. In general, we have to investigate both the steady-state error and the sustained-oscillation characteristics of a given digital system with quantization.

State-Variable Analysis

The least-upper-bound error analysis of quantized systems may be analyzed by the state-variable method or the z-transform method. Let the dynamic equations of a digital control system without quantization be

$$\mathbf{x}(k+1) = A\mathbf{x}(k) + B\mathbf{u}(k) \tag{14-35}$$

$$\mathbf{c}(k) = D\mathbf{x}(k) + E\mathbf{u}(k) \tag{14-36}$$

where $\mathbf{x}(k)$ is an n-vector, $\mathbf{u}(k)$ is an r-vector, and $\mathbf{c}(k)$ is a p-vector. Now let us consider that the same system described above has m quantizers which quantize the signals of the system. The quantization levels of these m quantizers are denoted by q_i, $i = 1,2,\ldots,m$. As described earlier, we may represent the m quantizers by inputs with signals of magnitudes $\pm q_i/2$, $i = 1,2,\ldots,m$. The digital control system with quantizers is now described by the following dynamic equations:

$$\mathbf{x}_q(k+1) = A\mathbf{x}_q(k) + B\mathbf{u}(k) + F\mathbf{q} \tag{14-37}$$

$$\mathbf{c}_q(k) = D\mathbf{x}_q(k) + E\mathbf{u}(k) + G\mathbf{q} \tag{14-38}$$

where $\mathbf{x}_q(k)$ is the $n \times 1$ state vector of the quantized system, and $\mathbf{c}_q(k)$ is the $p \times 1$ output vector; F is an $n \times m$ matrix which denotes the dependence of $\mathbf{x}_q(k+1)$ upon the equivalent inputs due to the quantizers, and G is a $p \times m$ matrix representing the dependence of $\mathbf{c}_q(k)$ on $\mathbf{q}$, where $\mathbf{q}$ is the vector

$$\mathbf{q} = \begin{bmatrix} \pm q_1/2 \\ \pm q_2/2 \\ \vdots \\ \pm q_m/2 \end{bmatrix} \tag{14-39}$$

Let the quantization error in the state vector at the kth sampling instant due to quantization be represented by $\mathbf{e}_x(k)$, then

$$\mathbf{e}_x(k) = \mathbf{x}(k) - \mathbf{x}_q(k) \tag{14-40}$$

Subtracting Eq. (14-37) from Eq. (14-35), we have

$$\mathbf{e_x}(k+1) = A\mathbf{e_x}(k) - \mathbf{Fq} \tag{14-41}$$

Similarly, the difference between Eqs. (14-36) and (14-38) is

$$\mathbf{e_c}(k) = \mathbf{c}(k) - \mathbf{c_q}(k) = D\mathbf{e_x}(k) - G\mathbf{q} \tag{14-42}$$

where $\mathbf{e_c}(k)$ is the quantization error in the output $\mathbf{c_q}(k)$ at the kth sampling instant.

The solution of Eq. (14-41) at k = N is

$$\mathbf{e_x}(N) = A^N \mathbf{e_x}(N) - \sum_{k=0}^{N-1} A^{N-k-1} \mathbf{Fq} \tag{14-43}$$

The ith element of $\mathbf{e}(N)$ is written

$$e_{xi}(N) = P_i A^N \mathbf{e_x}(0) - \sum_{j=1}^{m} \sum_{k=0}^{N-1} P_i A^{N-k-1} F_j \left[\pm \frac{q_j}{2} \right] \tag{14-44}$$

i = 1,2,...,n, where F_j represents the jth column of F.

For an asymptotically stable system,

$$\lim_{N\to\infty} A^N = \mathbf{0} \tag{14-45}$$

The least-upper bound of the steady-state quantization error of the ith state is

$$\left| \lim_{N\to\infty} e_{xi}(N) \right| = \left| \lim_{N\to\infty} \sum_{j=1}^{m} \sum_{k=0}^{N-1} P_i A^{N-k-1} F_j \frac{q_j}{2} \right| \qquad i = 1,2,\ldots,n \tag{14-46}$$

In a similar manner, the least-upper bound of the steady-state quantization error of the ith output is obtained by using Eq. (14-42),

$$\left| \lim_{N\to\infty} e_{ci}(N) \right| = \left| \lim_{N\to\infty} \sum_{j=1}^{m} \left[\sum_{k=0}^{N-1} D_i A^{N-k-1} F_j - g_{ij} \right] \frac{q_j}{2} \right| \tag{14-47}$$

where D_i is a $1 \times n$ matrix which is formed by the ith row of D, and g_{ij} is the ijth element of G, i = 1,2,...,p, and j = 1,2,...,m.

z-Transform Analysis

The z-transform analysis of the least-upper-bound error analysis is conceptually simpler than the state-variable method described above. Taking the z-transform on both sides of Eq. (14-41) and solving for $\mathbf{E_x}(z)$, we get

$$\mathbf{E_x}(z) = (zI - A)^{-1}\mathbf{e_x}(0) - (zI - A)^{-1} F\mathbf{q} \frac{z}{z-1} \tag{14-48}$$

The ith element of $\mathbf{E}_x(z)$ is written

$$E_{xi}(z) = P_i(zI - A)^{-1}e_x(0) - \sum_{j=1}^{m} P_i(zI - A)^{-1}F_j \frac{\pm q_j}{2} \frac{z}{z-1} \tag{14-49}$$

The least-upper bound of the quantization error in the ith state variable is

$$\left| \lim_{N\to\infty} e_{xi}(N) \right| = \left| \lim_{z\to 1} (1 - z^{-1})E_{xi}(z) \right| = \left| \lim_{z\to 1} \sum_{j=1}^{m} P_i(zI - A)^{-1}F_j \frac{q_i}{2} \right| \tag{14-50}$$

Similarly, for the output,

$$\left| \lim_{N\to\infty} e_{ci}(N) \right| = \left| \lim_{z\to 1} \sum_{j=1}^{m} \left[D_i(zI - A)^{-1}F_j - g_{ij} \right] \frac{q_j}{2} \right| \tag{14-51}$$

The following examples serve to illustrate the least-upper-bound quantization error analysis.

Example 14-3.

A digital controller in a control system is usually implemented digitally so that the round-off of digital data should be modeled by amplitude quantization. Let us consider a typical first-order digital controller with the transfer function

$$D(z) = \frac{C(z)}{U(z)} = \frac{1 + bz^{-1}}{1 + az^{-1}} \qquad (a < 1) \tag{14-52}$$

A state diagram of the controller is shown in Fig. 14-20(a), and a model with quantizers positioned at appropriate locations is shown in Fig. 14-20(b). The state diagram with the quantizers replaced by a branch with unity gain and an external source with signal magnitude of $\pm q/2$ is shown in Fig. 14-20(c). In the present case, it is assumed that all the four quantizers have the same quantization levels.

Without quantization, the dynamic equations of the controller are written directly from Fig. 14-20(a).

$$x_1(k+1) = -ax_1(k) + u(k) \tag{14-53}$$

$$c(k) = (b - a)x_1(k) + u(k) \tag{14-54}$$

For the system with quantizers, the dynamic equations are

$$x_{q1}(k+1) = -ax_{q1}(k) + u(k) \pm q \tag{14-55}$$

$$c_q(k) = (b - a)x_{q1}(k) + u(k) \pm 2q \tag{14-56}$$

The least-upper bound of the quantization error in the state variable $x_1(k)$ is defined as

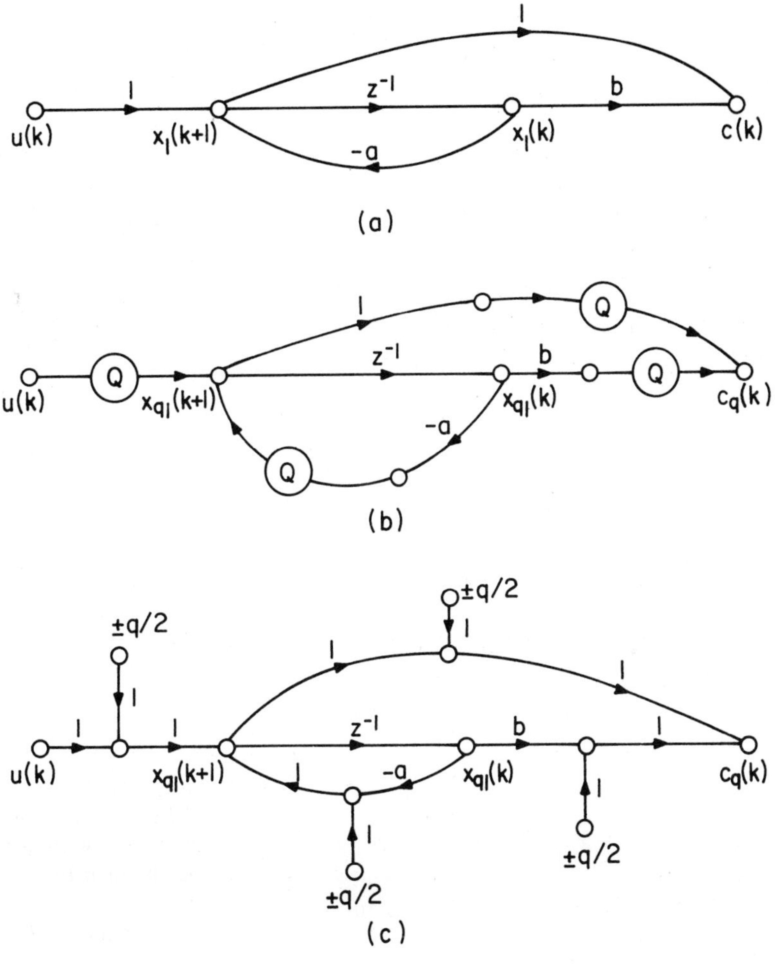

Fig. 14-20. State diagrams of the digital control system in Example 14-3.

$$e_x(k) = x_1(k) - x_{q1}(k) \tag{14-57}$$

From Eqs. (14-53) and (14-55), we have

$$e_x(k+1) = -ae_x(k) \pm q \tag{14-58}$$

The solution of the last equation is

$$e_x(N) = (-a)^N e_x(0) + \sum_{k=0}^{N-1} (-a)^{N-k-1}(\pm q) \tag{14-59}$$

The magnitude of the steady-state value of $e_x(N)$ is

$$\left| \lim_{N\to\infty} e_x(N) \right| = \left| \lim_{N\to\infty} \sum_{k=0}^{N-1} (-a)^{N-k-1}(\pm q) \right| = \frac{q}{1+a} \tag{14-60}$$

The least-upper bound of the quantization error in the output is obtained from Eqs. (14-54) and (14-56).

$$e_c(N) = c(N) - c_q(N) = (b-a)e_x(N) \pm 2q \tag{14-61}$$

Thus, the magnitude of the steady-state error bound is

$$\left| \lim_{N\to\infty} e_c(N) \right| = \frac{(b-a)q}{1+a} + 2q = \frac{2+a+b}{1+a} q \tag{14-62}$$

The z-transform analysis is carried out by evaluating the z-transforms of $x_1(k)$ and $x_{q1}(k)$ as functions of the equivalent signal sources, from Fig. 14-20. Thus,

$$E_x(z) = X_1(z) - X_{q1}(z) = \frac{-z^{-1}}{1+az^{-1}} \frac{\pm qz}{z-1} \tag{14-63}$$

where the initial states have been neglected, since they do not affect the steady-state error. Then,

$$\left| \lim_{N\to\infty} e_x(N) \right| = \left| \lim_{z\to 1} (1-z^{-1})E_x(z) \right| = \frac{q}{1+a} \tag{14-64}$$

which agrees with the answer obtained in Eq. (14-60).

Similarly, the least-upper bound of the quantization error in the output may be obtained by evaluating the z-transform of the output.

$$E_c(z) = C(z) - C_q(z) = \left[\frac{1+bz^{-1}}{1+az^{-1}} + 1 \right] \frac{\pm qz}{z-1} \tag{14-65}$$

Therefore,

$$\left| \lim_{N\to\infty} e_c(N) \right| = \left| \lim_{z\to 1} (1-z^{-1})E_c(z) \right| = \left| \lim_{z\to 1} \left[\frac{1+bz^{-1}}{1+az^{-1}} + 1 \right] q \right|$$

$$= \frac{2+a+b}{1+a} q \tag{14-66}$$

REFERENCES

1. Tou, J. T., *Digital And Sampled-Data Control Systems.* McGraw-Hill, New York, 1959.
2. Johnson, G. W., "Upper Bound on Dynamic Quantization Error in Digital Control Systems via the Direct Methods of Liapunov," *IEEE Trans. on Automatic Control,* Vol. AC-10, October 1965, pp. 439-448.
3. Sage, A. P., and Burt, R. W., "Optimum Design and Error Analysis of Digital Integrators for Discrete System Simulation," *AFIPS Proc.,* Vol. 27, Part 1, 1965, pp. 903-914.
4. Smith, F. W., "System Laplace-Transform Estimation From Sampled-Data," *IEEE Trans. on Automatic Control,* Vol. AC-13, February 1968, pp. 37-44.
5. Katzenelson, J., "On Errors Introduced by Combined Sampling and Quantization," *IRE Trans. on Automatic Control,* Vol. AC-7, April 1962, pp. 58-68.
6. Kramer, R., "Effect of Quantization on Feedback Systems With Stochastic Inputs," *Report No. 7849-R-9,* M.I.T. Electronic Systems Lab., June 1959.
7. Kramer, R., "Effects of Quantization on Feedback Systems With Stochastic Inputs," *IRE Trans. on Automatic Control,* Vol. AC-6, September 1961, pp. 292-305.
8. Greaves, C. J., and Cadzow, J. A., "The Optimal Discrete Filter Corresponding to a Given Analog Filter," *IEEE Trans. on Automatic Control,* Vol. AC-12, June 1967, pp. 304-307.
9. Brule, J. D., "Polynomial Extrapolation of Sampled Data With an Analog Computer," *IRE Trans. on Automatic Control,* Vol. AC-7, January 1962, pp. 76-77.
10. Monroe, A. J., *Digital Processes For Sampled Data Systems.* John Wiley & Sons, New York, 1962.
11. Knowles, J. B., and Edwards, R., "Finite Word-length Effects in Multirate Direct Digital Control Systems," *Proc. IEE,* Vol. 112, December 1965, pp. 76-84.
12. Phillips, C. L., "Instabilities Caused by Floating-Point Arithmetic Quantization," *IEEE Trans. on Automatic Control,* Vol. AC-17, April 1972, pp. 242-243.
13. Divieti, L. D., Rossi, C. M., Schmid, R. M., and Verschkin, A. E., "A Note on Computing Quantization Errors in Digital Control Systems," *IEEE Trans. on Automatic Control,* Vol. AC-12, October 1967, pp. 622-623.
14. Slaughter, J. B., "Quantization Errors in Digital Control Systems," *IEEE Trans. on Automatic Control,* Vol. AC-9, January 1964, pp. 70-74.
15. Bertram, J. E., "The Effects of Quantization in Sampled-Feedback Systems," *Trans. AIEE (Applications and Industry),* Vol. 77, September 1958, pp. 177-181.
16. Johnson, G. W., "Upper Bound on Dynamic Quantization Error in Digital Control Systems Via The Direct Method of Liapunov," *IEEE Trans. on Automatic Control,* Vol. AC-10, October 1965, pp. 439-448.
17. Curry, E. E., "The Analysis of Round-Off and Truncation Errors in a Hybrid Control System," *IEEE Trans. on Automatic Control,* Vol. AC-12, October 1967, pp. 601-604.

18. Kuo, B. C., and Tal, J., ed., *Incremental Motion Control, Vol. I, DC Motors And Control Systems*, SRL Publishing Company, Champaign, Ill., 1978.

19. Kuo, B. C., "The z-Transform Describing Function for Nonlinear Sampled-Data Control Systems," *Proc. of I. R. E.*, 43, No. 5, May 1960, pp. 941-942.

20. *Intel 8080 Microcomputer Systems User's Manual*, Intel Corporation, Santa Clara, Calif., September 1975.

PROBLEMS

14-1 The state diagram of a digital control system is shown in Fig. 14P-1. The digital controller is represented by the dynamics between e(k) and u(k), and the controlled process is modeled between the nodes u(k) and c(k). Consider that amplitude quantization exists in the digital controller in the branches with gains 2.72, −1 and −0.72. Also, quantization appears in the overall feedback path due to the use of a digital encoder, and at the input to the digital controller. Insert the five quantizers in Fig. 14P-1 and conduct a least-upper-bound error analysis. Determine the magnitudes of the least-upper-bound errors of the three state variables and the output. The quantization level is q.

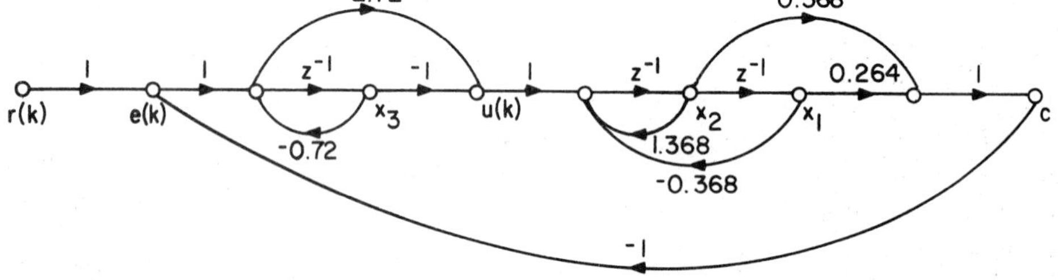

Fig. 14P-1.

14-2 The block diagram of a digital positioning control system with quantization is shown in Fig. 14P-2. Determine the maximum value of the quantization level q so that the error in the output due to quantization either in the form of steady-state error or sustained oscillation will not exceed 0.01. The input is a unit-step function, and the initial conditions are zero.

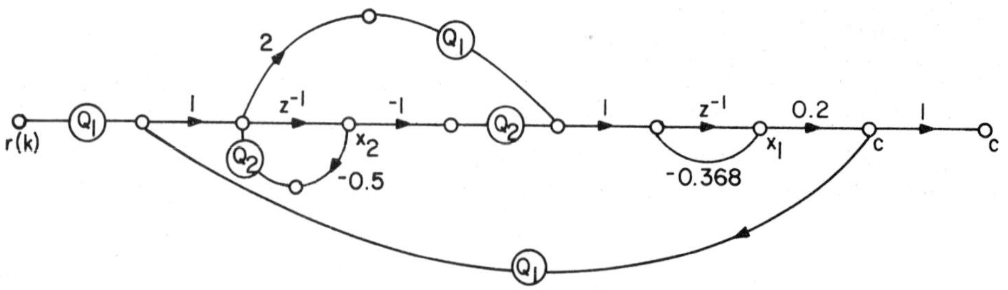

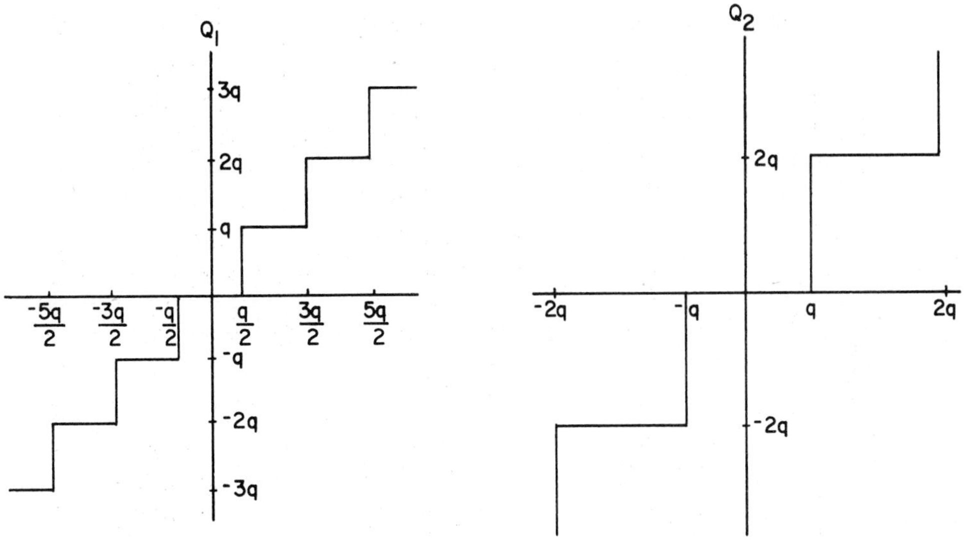

Fig. 14P-2.

14-3 The state diagram of a digital control system is shown in Fig. 14P-3. The characteristics of the quantizers that are located at the various positions in the system are given as shown. Determine the magnitudes of the least-upper-bound errors in the variables $x_1(k)$, $x_2(k)$ and $c(k)$.

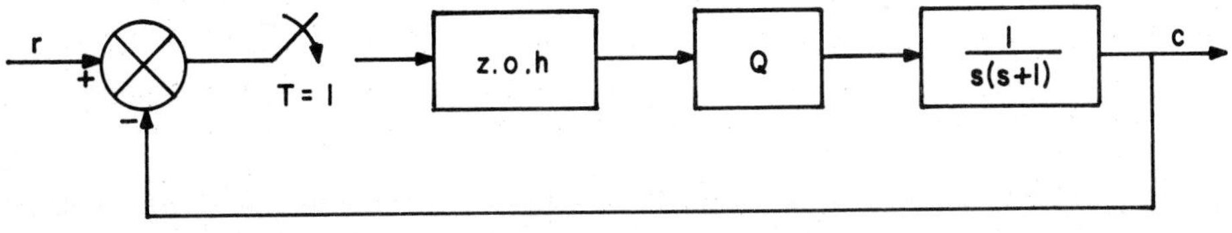

Fig. 14P-3.

14-4 Figure 14P-4 shows the block diagram of the LST system with amplitude quantization. The transfer functions of $G_a(z)$, $G_b(z)$ and $G_c(z)$ are given as

$$G_a(z) = \frac{K_p z + K_i T - K_p}{z - 1}$$

$$G_b(z) = \frac{T^2(z+1)}{2J_v(z-1)^2}$$

$$G_c(z) = \frac{K_r T}{J_v(z-1)}$$

The quantization levels of Q_d, Q_t and Q_r are q_d, q_t and q_r respectively. Find the least-upper-bound error in c(k) due to the quantization. Discuss the effect of the integral control constant K_i on the quantization error.

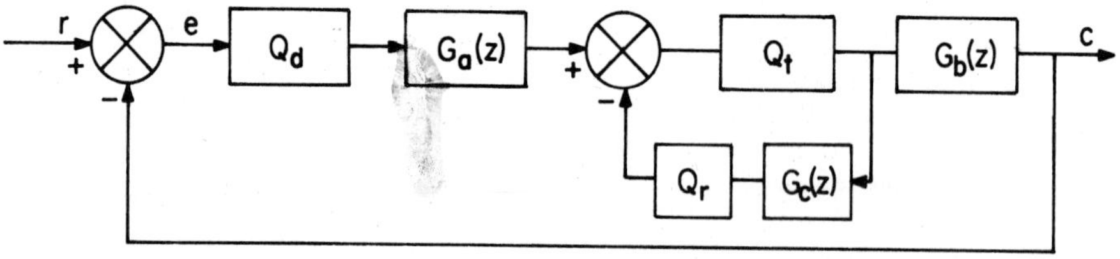

Fig. 14P-4.

8080 INSTRUCTION SET

Summary of Processor Instructions

Mnemonic	Description	D_7	D_6	D_5	D_4	D_3	D_2	D_1	D_0	Clock (2) Cycles	FLAGS Set (3)
		Instruction Code (1)									
$MOV_{r1,r2}$	Move register to register	0	1	D	D	D	S	S	S	5	None
MOV M,r	Move register to memory	0	1	1	1	0	S	S	S	7	None
MOV r,M	Move memory to register	0	1	D	D	D	1	1	0	7	None
HLT	Halt	0	1	1	1	0	1	1	0	7	None
MVI r	Move immediate register	0	0	D	D	D	1	1	0	7	None
MVI M	Move immediate memory	0	0	1	1	0	1	1	0	10	None
INR r	Increment register	0	0	D	D	D	1	0	0	5	PSZ
DCR r	Decrement register	0	0	D	D	D	1	0	1	5	PSZ
INR M	Increment memory	0	0	1	1	0	1	0	0	10	PSZ
DCR M	Decrement memory	0	0	1	1	0	1	0	1	10	PSZ
ADD r	Add register to A	1	0	0	0	0	S	S	S	4	CPSZ
ADC r	Add register to A with carry	1	0	0	0	1	S	S	S	4	CPSZ
SUB r	Subtract register from A	1	0	0	1	0	S	S	S	4	CPSZ
SBB r	Subtract register from A with borrow	1	0	0	1	1	S	S	S	4	CPSZ
ANA r	And register with A	1	0	1	0	0	S	S	S	4	LPSZ
XRA r	Exclusive Or register with A	1	0	1	0	1	S	S	S	4	LPSZ
ORA r	Or register with A	1	0	1	1	0	S	S	S	4	LPSZ
CMP r	Compare register with A	1	0	1	1	1	S	S	S	4	CPSZ
ADD M	Add memory to A	1	0	0	0	0	1	1	0	7	CPSZ
ADC M	Add memory to A with carry	1	0	0	0	1	1	1	0	7	CPSZ
SUB M	Subtract memory from A	1	0	0	1	0	1	1	0	7	CPSZ
SBB M	Subtract memory from A with borrow	1	0	0	1	1	1	1	0	7	CPSZ
ANA M	And memory with A	1	0	1	0	0	1	1	0	7	LPSZ
XRA M	Exclusive Or memory with A	1	0	1	0	1	1	1	0	7	LPSZ
ORA M	Or memory with A	1	0	1	1	0	1	1	0	7	LPSZ
CMP M	Compare memory with A	1	0	1	1	1	1	1	0	7	CPSZ
ADI	Add immediate to A	1	1	0	0	0	1	1	0	7	CPSZ
ACI	Add immediate to A with carry	1	1	0	0	1	1	1	0	7	CPSZ
SUI	Subtract immediate from A	1	1	0	1	0	1	1	0	7	CPSZ
SBI	Subtract immediate from A with borrow	1	1	0	1	1	1	1	0	7	CPSZ
ANI	And immediate with A	1	1	1	0	0	1	1	0	7	LPSZ
XRI	Exclusive Or immediate with A	1	1	1	0	1	1	1	0	7	LPSZ
ORI	Or immediate with A	1	1	1	1	0	1	1	0	7	LPSZ
CPI	Compare immediate with A	1	1	1	1	1	1	1	0	7	CPSZ
RLC	Rotate A left	0	0	0	0	0	1	1	1	4	C
RRC	Rotate A right	0	0	0	0	1	1	1	1	4	C
RAL	Rotate A left through carry	0	0	0	1	0	1	1	1	4	C
RAR	Rotate A right through carry	0	0	0	1	1	1	1	1	4	C

Mnemonic	Description	Instruction Code (1) D_7	D_6	D_5	D_4	D_3	D_2	D_1	D_0	Clock (2) Cycles	FLAGS Set (3)
JMP	Jump unconditional	1	1	0	0	0	0	1	1	10	None
JC	Jump on carry	1	1	0	1	1	0	1	0	10	None
JNC	Jump on no carry	1	1	0	1	0	0	1	0	10	None
JZ	Jump on zero	1	1	0	0	1	0	1	0	10	None
JNZ	Jump on no zero	1	1	0	0	0	0	1	0	10	None
JP	Jump on positive	1	1	1	1	0	0	1	0	10	None
JM	Jump on minus	1	1	1	1	1	0	1	0	10	None
JPE	Jump on parity even	1	1	1	0	1	0	1	0	10	None
JPO	Jump on parity odd	1	1	1	0	0	0	1	0	10	None
CALL	Call unconditional	1	1	0	0	1	1	0	1	17	None
CC	Call on carry	1	1	0	1	1	1	0	0	11/17	None
CNC	Call on no carry	1	1	0	1	0	1	0	0	11/17	None
CZ	Call on zero	1	1	0	0	1	1	0	0	11/17	None
CNZ	Call on no zero	1	1	0	0	0	1	0	0	11/17	None
CP	Call on positive	1	1	1	1	0	1	0	0	11/17	None
CM	Call on minus	1	1	1	1	1	1	0	0	11/17	None
CPE	Call on parity even	1	1	1	0	1	1	0	0	11/17	None
CPO	Call on parity odd	1	1	1	0	0	1	0	0	11/17	None
RET	Return	1	1	0	0	1	0	0	1	10	None
RC	Return on carry	1	1	0	1	1	0	0	0	5/11	None
RNC	Return on no carry	1	1	0	1	0	0	0	0	5/11	None
RZ	Return on zero	1	1	0	0	1	0	0	0	5/11	None
RNZ	Return on no zero	1	1	0	0	0	0	0	0	5/11	None
RP	Return on positive	1	1	1	1	0	0	0	0	5/11	None
RM	Return on minus	1	1	1	1	1	0	0	0	5/11	None
RPE	Return on parity even	1	1	1	0	1	0	0	0	5/11	None
RPO	Return on parity odd	1	1	1	0	0	0	0	0	5/11	None
RST	Restart	1	1	A	A	A	1	1	1	11	None
IN	Input	1	1	0	1	1	0	1	1	10	None
OUT	Output	1	1	0	1	0	0	1	1	10	None
LXI B	Load immediate register Pair B & C	0	0	0	0	0	0	0	1	10	None
LXI D	Load immediate register Pair D & E	0	0	0	1	0	0	0	1	10	None
LXI H	Load immediate register Pair H & L	0	0	1	0	0	0	0	1	10	None
LXI SP	Load immediate stack pointer	0	0	1	1	0	0	0	1	10	None
PUSH B	Push register Pair B & C on stack	1	1	0	0	0	1	0	1	11	None
PUSH D	Push register Pair D & E on stack	1	1	0	1	0	1	0	1	11	None
PUSH H	Push register Pair H & L on stack	1	1	1	0	0	1	0	1	11	None
PUSH PSW	Push A and Flags on stack	1	1	1	1	0	1	0	1	11	None
POP B	Pop register pair B & C off stack	1	1	0	0	0	0	0	1	10	None
POP D	Pop register pair D & E off stack	1	1	0	1	0	0	0	1	10	None

Mnemonic	Description	Instruction Code (1) D_7	D_6	D_5	D_4	D_3	D_2	D_1	D_0	Clock (2) Cycles	FLAGS Set (3)
POP H	Pop register pair H & L off stack	1	1	1	0	0	0	0	1	10	None
POP PSW	Pop A and Flags off stack	1	1	1	1	0	0	0	1	10	CPSZ
STA	Store A direct	0	0	1	1	0	0	1	0	13	None
LDA	Load A direct	0	0	1	1	1	0	1	0	13	None
XCHG	Exchange D & E, H & L Registers	1	1	1	0	1	0	1	1	4	None
XTHL	Exchange top of stack, H & L	1	1	1	0	0	0	1	1	18	None
SPHL	H & L to stack pointer	1	1	1	1	1	0	0	1	5	None
PCHL	H & L to program counter	1	1	1	0	1	0	0	1	5	None
DAD B	Add B & C to H & L	0	0	0	0	1	0	0	1	10	D
DAD D	Add D & E to H & L	0	0	0	1	1	0	0	1	10	D
DAD H	Add H & L to H & L	0	0	1	0	1	0	0	1	10	D
DAD SP	Add stack pointer to H & L	0	0	1	1	1	0	0	1	10	D
STAX B	Store A indirect	0	0	0	0	0	0	1	0	7	None
STAX D	Store A indirect	0	0	0	1	0	0	1	0	7	None
LDAX B	Load A indirect	0	0	0	0	1	0	1	0	7	None
LDAX D	Load A indirect	0	0	0	1	1	0	1	0	7	None
INX B	Increment B & C registers	0	0	0	0	0	0	1	1	5	None
INX D	Increment D & E registers	0	0	0	1	0	0	1	1	5	None
INX H	Increment H & L registers	0	0	1	0	0	0	1	1	5	None
INX SP	Increment stack pointer	0	0	1	1	0	0	1	1	5	None
DCX B	Decrement B & C	0	0	0	0	1	0	1	1	5	None
DCX D	Decrement D & E	0	0	0	1	1	0	1	1	5	None
DCX H	Decrement H & L	0	0	1	0	1	0	1	1	5	None
DCX SP	Decrement stack pointer	0	0	1	1	1	0	1	1	5	None
CMA	Complement A	0	0	1	0	1	1	1	1	4	None
STC	Set carry	0	0	1	1	0	1	1	1	4	C
CMC	Complement carry	0	0	1	1	1	1	1	1	4	C
DAA	Decimal adjust A	0	0	1	0	0	1	1	1	4	CPSZ
SHLD	Store H & L direct	0	0	1	0	0	0	1	0	16	None
LHLD	Load H & L direct	0	0	1	0	1	0	1	0	16	None
EI	Enable Interrupts	1	1	1	1	1	0	1	1	4	None
DI	Disable Interrupts	1	1	1	1	0	0	1	1	4	None
NOP	No operation	0	0	0	0	0	0	0	0	4	None

NOTES:
1. DDD or SSS — 000 B — 001 C — 010 D — 011 E — 100 H — 101 L — 110 Memory — 111 A.
2. Two possible cycle times, (5/11) indicate instruction cycles dependent on condition flags.
3. FLAG Codes: C = CARRY, P = PARITY (this FLAG is set to 1 on even parity; i.e., the number of 1's in the result of an operation is an even number), S = SIGN, Z = ZERO, L = CARRY FLAG is cleared to zero at the end of the operation, D = CARRY FLAG is set to 1 if a carry occurs out of the double precision addition.

APPENDIX A. TABLE OF LAPLACE TRANSFORMS, z-TRANSFORMS AND MODIFIED z-TRANSFORMS

Laplace Transform F(s)	Time Function f(t) t > 0	z-Transform F(z)	Modified z-Transform F(z,m)
1	$\delta(t)$	1	0
e^{-kTs}	$\delta(t-kT)$	z^{-k}	z^{-k-1+m}
$\frac{1}{s}$	$u_s(t)$	$\frac{z}{z-1}$	$\frac{1}{z-1}$
$\frac{1}{s^2}$	t	$\frac{Tz}{(z-1)^2}$	$\frac{mT}{z-1} + \frac{T}{(z-1)^2}$
$\frac{2}{s^3}$	t^2	$\frac{T^2z(z+1)}{(z-1)^3}$	$T^2\frac{m^2z^2 + (2m-2m^2+1)z + (m-1)^2}{(z-1)^3}$
$\frac{(k-1)!}{s^k}$	t^{k-1}	$\lim_{a\to o} (-1)^{k-1} \frac{\partial^{k-1}}{\partial a^{k-1}} \left[\frac{z}{z\ \ e^{-aT}}\right]$	$\lim_{a\to o} (-1)^{k-1} \frac{\partial^{k-1}}{\partial a^{k-1}} \left[\frac{e^{-amT}}{z-e^{-aT}}\right]$
$\frac{1}{s+a}$	e^{-at}	$\frac{z}{z-e^{-aT}}$	$\frac{e^{-amT}}{z-e^{-aT}}$
$\frac{1}{(s+a)^2}$	te^{-at}	$\frac{Tze^{-aT}}{(z-e^{-aT})^2}$	$\frac{Te^{-amT}[e^{-aT} + m(z-e^{-aT})]}{(z-e^{-aT})^2}$
$\frac{(k-1)!}{(s+a)^k}$	t^ke^{-at}	$(-1)^k \frac{\partial^k}{\partial a^k} \frac{z}{z-e^{-aT}}$	$(-1)^k \frac{\partial^k}{\partial a^k} \left[\frac{e^{-amT}}{z-e^{-aT}}\right]$
$\frac{a}{s(s+a)}$	$1-e^{-at}$	$\frac{z(1-e^{-aT})}{(z-1)(z-e^{-aT})}$	$\frac{(1-e^{-amT})z + (e^{-amT}-e^{-aT})}{(z-1)(z-e^{-aT})}$

Laplace Transform F(s)	Time Function f(t) t > 0	z-Transform F(z)	Modified z-Transform F(z,m)
$\frac{1}{(s+a)(s+b)}$	$\frac{1}{(b-a)}(e^{-at}-e^{-bt})$	$\frac{1}{(b-a)}\left[\frac{z}{z-e^{-aT}}-\frac{z}{z-e^{-bT}}\right]$	$\frac{1}{(b-a)}\left[\frac{e^{-amT}}{z-e^{-aT}}-\frac{e^{-bmT}}{z-e^{-bT}}\right]$
$\frac{a}{s^2(s+a)}$	$t-\frac{1}{a}(1-e^{-at})$	$\frac{Tz}{(z-1)^2}-\frac{(1-e^{-aT})z}{a(z-1)(z-e^{-aT})}$	$\frac{T}{(z-1)^2}+\frac{amT-1}{a(z-1)}+\frac{e^{-amT}}{a(z-e^{-aT})}$
$\frac{1}{(s+a)^2}$	te^{-at}	$\frac{Tze^{-aT}}{(z-e^{-aT})^2}$	$\frac{Te^{-amT}[e^{-aT}+m(z-e^{-aT})]}{(z-e^{-aT})^2}$
$\frac{a}{s^3(s+a)}$	$\frac{1}{2}(t^2-\frac{2}{a}t+\frac{2}{a^2}u_s(t)$ $-\frac{2}{a^2}e^{-at})$	$\frac{T^2z}{(z-1)^3}+\frac{(aT-2)Tz}{2a(z-1)^2}$ $+\frac{z}{a^2(z-1)}-\frac{z}{a^2(z-e^{-aT})}$	$\frac{T^2}{(z-1)^3}+\frac{T^2(m+½)a-T}{a(z-1)^2}$ $+\frac{(amT)^2/2-amT+1}{a^2(z-1)}-\frac{e^{-amT}}{a^2(z-e^{-aT})}$
$\frac{a^2}{s(s+a)^2}$	$u_s(t)-(1+at)e^{-at}$	$\frac{z}{z-1}-\frac{z}{z-e^{-aT}}-\frac{aTe^{-aT}z}{(z-e^{-aT})^2}$	$\frac{1}{z-1}-\left[\frac{1+amT}{z-e^{-aT}}+\frac{aTe^{-aT}}{(z-e^{-aT})^2}\right]e^{-amT}$
$\frac{a^2}{s^2(s+a)^2}$	$t-\frac{2}{a}u_s(t)+(t+\frac{2}{a})e^{-at}$	$\frac{1}{a}\left[\frac{(aT+2)z}{(z-1)^2}-\frac{2z^2}{}+\frac{2z}{z-e^{-aT}}\right.$ $\left.+\frac{aTe^{-aT}z}{(z-e^{-aT})^2}\right]$	$\frac{1}{a}\left\{\frac{aT}{(z-1)^2}+\frac{amT-2}{z-1}+\left[\frac{aTe^{-aT}}{(z-e^{-aT})^2}\right.\right.$ $\left.\left.-\frac{amT-2}{z-e^{-aT}}\right]e^{-amT}\right\}$
$\frac{\omega}{s^2+\omega^2}$	$\sin\omega t$	$z\sin\omega T$	$\frac{\sin m\omega T+\sin(1-m)\omega T}{z^2-2z\cos\omega T+1}$

Laplace Transform F(s)	Time Function f(t) t >0	z-Transform F(z)	Modified z-Transform F(z,m)
$\frac{s}{s^2 + \omega^2}$	$\cos\omega t$	$\frac{z(z - \cos\omega T)}{z^2 - 2z\cos\omega T + 1}$	$\frac{\cos m\omega T - \cos(1-m)\omega T}{z^2 - 2z\cos\omega T + 1}$
$\frac{\omega}{s^2 - \omega^2}$	$\sinh\omega t$	$\frac{z\sinh\omega T}{z^2 - 2z\cosh\omega T + 1}$	$\frac{\sinh m\omega T + \sinh(1-m)\omega T}{z^2 - 2z\cosh\omega T + 1}$
$\frac{s}{s^2 - \omega^2}$	$\cosh\omega t$	$\frac{z(z - \cosh\omega T)}{z^2 - 2z\cosh\omega T + 1}$	$\frac{\cosh m\omega T z - \cosh(1-m)\omega T}{z^2 - 2z\cosh\omega T + 1}$
$\frac{\omega}{(s+a)^2 + \omega^2}$	$e^{-at}\sin\omega t$	$\frac{ze^{-aT}\sin\omega T}{z^2 - 2ze^{-aT}\cos\omega T + e^{-2aT}}$	$\frac{e^{-amT}[z\sin m\omega T + e^{-aT}\sin(1-m)\omega T]}{z^2 - 2ze^{-aT}\cos\omega T + e^{-2aT}}$
$\frac{a^2 + \omega^2}{s[(s+a)^2 + \omega^2]}$	$1 - e^{-at}\sec\phi\cos(\omega t + \phi)$ $\phi = \tan^{-1}(-a/\omega)$	$\frac{z}{z-1} - \frac{z^2 - ze^{-aT}\sec\phi\cos(\omega T - \phi)}{z^2 - 2ze^{-aT}\cos\omega T + e^{-2aT}}$	$\frac{1}{z-1} - \frac{e^{-maT}\sec\phi(Az - B)}{z^2 - 2ze^{-aT}\cos\omega T + e^{-2aT}}$ $A = \cos(m\omega T + \phi)$ $B = e^{-aT}\cos[(1-m)\omega T - \phi]$
$\frac{s + a}{(s+a)^2 + \omega^2}$	$e^{-at}\cos\omega t$	$\frac{z^2 - ze^{-aT}\cos\omega T}{z^2 - 2ze^{-aT}\cos\omega T + e^{-2aT}}$	$\frac{e^{-maT}[z\cos m\omega T + e^{-aT}\sin(1-m)\omega T]}{z^2 - 2ze^{-aT}\cos\omega T + e^{-2aT}}$

INDEX